FLOW MEASUREMENT HANDBOOK

Flow Measurement Handbook is a reference for engineers on flow measurement techniques and instruments. It strikes a balance between laboratory ideas and realities of field experience and provides practical advice on design, operation and performance of flowmeters.

It begins with a review of essentials: accuracy, flow, selection and calibration methods. Each chapter is then devoted to a flowmeter class and includes information on design, application, installation, calibration and operation.

Among the flowmeters discussed are differential pressure devices such as orifice and Venturi; volumetric flowmeters such as positive displacement, turbine, vortex, electromagnetic, magnetic resonance, ultrasonic and acoustic; multiphase flowmeters; and mass meters such as thermal and Coriolis. There are also chapters on probes, verification and remote data access.

Roger C. Baker has worked for many years in industrial flow measurement. He studied at Cambridge and Harvard Universities and has held posts at Cambridge University, Imperial College and Cranfield, where he set up the Department of Fluid Engineering and Instrumentation. He has held visiting professorships at Cranfield and Warwick University.

Flow Measurement Handbook

INDUSTRIAL DESIGNS, OPERATING PRINCIPLES, PERFORMANCE, AND APPLICATIONS

Second Edition

Roger C. Baker

CAMBRIDGE
UNIVERSITY PRESS

One Liberty Plaza, 20th Floor, New York, NY 10006, USA

Cambridge University Press is part of the University of Cambridge.

It furthers the University's mission by disseminating knowledge in the pursuit of
education, learning, and research at the highest international levels of excellence.

www.cambridge.org
Information on this title: www.cambridge.org/9781107045866

First edition 2000

Second edition published 2016

A catalogue record for this publication is available from the British Library.

Library of Congress Cataloguing in Publication Data
Baker, R. C., author.
Flow measurement handbook : industrial designs, operating principles,
performance, and applications / Roger C. Baker – 2nd edition.
pages cm
Includes bibliographical references and index.
ISBN 978-1-107-04586-6 – ISBN 1-107-04586-6
1. Flow meters – Handbooks, manuals, etc. 2. Flow meters – Design and
construction – Handbooks, manuals, etc. 3. Fluid dynamic measurements –
Handbooks, manuals, etc. I. Title.
TA357.5.M43.B35 2016
681′.28–dc23 2015030767

ISBN 978-1-107-04586-6 Hardback

DISCLAIMER

In preparing the second edition of this book every effort has been made to provide up-to-date and accurate data and information that is in accord with accepted standards and practice at the time of publication and has been included in good faith. Nevertheless, the author, editors, and publisher can make no warranties that the data and information contained herein is totally free from error, not least because industrial design and performance is constantly changing through research, development, and regulation. Data, discussion, and conclusions developed by the author are for information only and are not intended for use without independent substantiating investigation on the part of the potential users. The author, editors, and publisher therefore disclaim all liability or responsibility for direct or consequential damages resulting from the use of data, designs, or constructions based on any of the information supplied or materials described in this book. Readers are strongly advised to pay careful attention to information provided by the manufacturer of any equipment that they plan to use and should refer to the most recent standards documents relating to their application. The author, editors, and publisher wish to point out that the inclusion or omission of a particular device, design, application, or other material in no way implies anything about its performance with respect to other devices, etc.

To Liz

and all the family

Contents

Preface *page* xxiii
Acknowledgements xxv
Nomenclature xxvii

1 Introduction 1
1.1 Initial Considerations 1
1.2 Do We Need a Flowmeter? 2
1.3 How Accurate? 4
1.4 A Brief Review of the Evaluation of Standard Uncertainty 8
1.5 Note on Monte Carlo Methods 10
1.6 Sensitivity Coefficients 10
1.7 What Is a Flowmeter? 11
1.8 Chapter Conclusions (for those who Plan to Skip the Mathematics!) 14
1.9 Mathematical Postscript 15
1.A Statistics of Flow Measurement 17
1.A.1 Introduction 17
1.A.2 The Normal Distribution 17
1.A.3 The Student *t* Distribution 20
1.A.4 Practical Application of Confidence Level 21
1.A.5 Types of Error 22
1.A.6 Combination of Uncertainties 23
1.A.7 Uncertainty Range Bars, Transfer Standards and Youden Analysis 24

2 Fluid Mechanics Essentials 27
2.1 Introduction 27
2.2 Essential Property Values 27
2.3 Flow in a Circular Cross-Section Pipe 27
2.4 Flow Straighteners and Conditioners 31
2.5 Essential Equations 34

2.6 Unsteady Flow and Pulsation 36
2.7 Compressible Flow 38
2.8 Multiphase Flow 40
2.9 Cavitation, Humidity, Droplets and Particles 42
2.10 Gas Entrapment 43
2.11 Steam 45
2.12 Chapter Conclusions 45
2.A Further Aspects of Flow Behaviour, Flow Conditioning and Flow Modelling 46
2.A.1 Further Flow Profile Equations 46
2.A.2 Non-Newtonian Flows 47
2.A.3 Flow Conditioning 47
2.A.4 Other Installation Considerations 50
2.A.5 Computational Fluid Dynamics (CFD) 50

3 Specification, Selection and Audit 52
3.1 Introduction 52
3.2 Specifying the Application 52
3.3 Notes on the Specification Form 53
3.4 Flowmeter Selection Summary Table 56
3.5 Draft Questionnaire for Flowmeter Audit 62
3.6 Final Comments 62
3.A Specification and Audit Questionnaires 63
3.A.1 Specification Questionnaire 63
3.A.2 Supplementary Audit Questionnaire 65

4 Calibration 67
4.1 Introduction 67
4.1.1 Calibration Considerations 67
4.1.2 Typical Calibration Laboratory Facilities 70
4.1.3 Calibration from the Manufacturer's Viewpoint 71
4.2 Approaches to Calibration 72
4.3 Liquid Calibration Facilities 75
4.3.1 Flying Start and Stop 75
4.3.2 Standing Start and Stop 77
4.3.3 Large Pipe Provers 80
4.3.4 Compact Provers 80
4.4 Gas Calibration Facilities 85
4.4.1 Volumetric Measurement 85
4.4.2 Mass Measurement 86
4.4.3 Gas/Liquid Displacement 86
4.4.4 *pvT* Method 87
4.4.5 Critical Nozzles 87
4.4.6 Soap Film Burette Method 88

4.5 Transfer Standards and Master Meters 88
4.6 In Situ Calibration 91
4.6.1 Provers 92
4.7 Calibration Uncertainty 98
4.8 Traceability and Accuracy of Calibration Facilities 100
4.9 Chapter Conclusions 101
4.A Calibration and Flow Measurement Facilities 101
4.A.1 Introduction 101
4.A.2 Flow Metrology Developments 102
4.A.3 Multiphase Calibration Facilities 105
4.A.4 Gas Calibration Facilities 105
4.A.5 Gas Properties 108
4.A.6 Case Study of a Water Flow Calibration Facility Which Might Be Used in a Manufacturing Plant or a Research Laboratory from the Author's Experience 108
4.A.7 Example of a Recent Large Water Calibration Facility 113

5 Orifice Plate Meters 116
5.1 Introduction 116
5.2 Essential Background Equations 118
5.3 Design Details 121
5.4 Installation Constraints 124
5.5 Other Orifice Plates 128
5.6 Deflection of Orifice Plate at High Pressure 129
5.7 Effect of Pulsation 131
5.8 Effects of More than One Flow Component 136
5.9 Accuracy Under Normal Operation 139
5.10 Industrially Constructed Designs 141
5.11 Pressure Connections 142
5.12 Pressure Measurement 144
5.13 Temperature and Density Measurement 147
5.14 Flow Computers 148
5.15 Detailed Studies of Flow through the Orifice Plate, Both Experimental and Computational 148
5.16 Application, Advantages and Disadvantages 150
5.17 Chapter Conclusions 151
5.A Orifice Discharge Coefficient Equation 152
5.A.1 Stolz Orifice Discharge Coefficient Equation as Given in ISO 5167: 1981 152
5.A.2 Orifice Discharge Coefficient Equation as set out by Gallagher (1990) 153
5.A.3 Orifice Discharge Coefficient Equation as Given in ISO 5167–2: 2003 154

5.B Review of Recent Published Research on Orifice Plates 156
5.B.1 Installation Effects on Orifice Plates 156
5.B.2 Pulsation 157
5.B.3 Contamination 157
5.B.4 Drain Holes 158
5.B.5 Flow Conditioning for Orifice Meters 158
5.B.6 Plate Thickness for Small-Diameter Orifice Plates 160
5.B.7 Variants on the Orifice Plate 160
5.B.8 Impulse Lines 160
5.B.9 Lagging Pipes 160
5.B.10 Gas Conditions 160
5.B.11 Emissions Testing Uncertainty 161
5.B.12 CFD Related to Orifice Plates 161

6 Venturi Meter and Standard Nozzles 163
6.1 Introduction 163
6.2 Essential Background Equations 165
6.3 Design Details 167
6.4 Commercially Available Devices 168
6.5 Installation Effects 168
6.6 Applications, Advantages and Disadvantages 170
6.7 Chapter Conclusions 171
6.A Research Update 172
6.A.1 Design and Installation 172
6.A.2 Meters in Nuclear Core Flows 173
6.A.3 Special Conditions 173

7 Critical Flow Venturi Nozzle 177
7.1 Introduction 177
7.2 Design Details of a Practical Flowmeter Installation 178
7.3 Practical Equations 181
7.4 Discharge Coefficient C 183
7.5 Critical Flow Function C_* 185
7.6 Design Considerations 185
7.7 Measurement Uncertainty 187
7.8 Notes on the Calculation Procedure 188
7.9 Industrial and Other Experience 189
7.10 Advantages, Disadvantages and Applications 190
7.11 Chapter Conclusions 190
7.A Critical Flow Venturi Nozzle – Recent Published Work 190

8 Other Momentum-Sensing Meters 195
8.1 Introduction 195
8.2 Variable Area Meter 196

8.2.1 Operating Principle and Background 196
8.2.2 Design Variations 196
8.2.3 Remote Readout Methods 198
8.2.4 Design Features 199
8.2.5 Calibration and Sources of Error 200
8.2.6 Installation 201
8.2.7 Unsteady and Pulsating Flows 201
8.2.8 Industrial Types, Ranges and Performance 201
8.2.9 Manufacturing Variation 202
8.2.10 Computational Analysis of the Variable Area Flowmeter 203
8.2.11 Applications 203
8.3 Spring-Loaded Diaphragm (Variable Area) Meters 204
8.4 Target (Drag Plate) Meter 208
8.5 Integral Orifice Meters 209
8.6 Dall Tubes and Devices that Approximate to Venturis and Nozzles 209
8.7 Wedge Meter 212
8.8 V-Cone Meter (Cone Meter) 213
8.9 Differential Devices with a Flow Measurement Mechanism in the Bypass 216
8.10 Slotted Orifice Plate 216
8.11 Pipework Features – Inlets and Pipe Lengths 217
8.12 Pipework Features – Bend or Elbow used as a Meter 218
8.13 Averaging Pitot 220
8.14 Laminar or Viscous Flowmeters 223
8.15 Chapter Conclusions 227
8.A History, Equations and Maximum Permissible Error Limits for the VA Meter 228
8.A.1 Some History 228
8.A.2 Equations 229
8.A.3 Maximum Permissible Error Limits 232

9 Positive Displacement Flowmeters 234
9.1 Introduction 234
9.1.1 Background 234
9.1.2 Qualitative Description of Operation 235
9.2 Principal Designs of Liquid Meters 236
9.2.1 Nutating Disc Meter 236
9.2.2 Oscillating Circular Piston (Also Known as Rotary Piston) Meter 237
9.2.3 Multirotor Meters 237
9.2.4 Oval Gear Meter 238
9.2.5 Sliding Vane Meters 240
9.2.6 Helical Rotor Meter 242
9.2.7 Reciprocating Piston Meters 243

9.2.8 Precision Gear (Spur Gear) Flowmeters 244
9.3 Calibration, Environmental Compensation and Other Factors Relating to the Accuracy of Liquid Flowmeters 245
9.3.1 Calibration Systems 246
9.3.2 Clearances 249
9.3.3 Leakage Through the Clearance Gap Between Vane and Wall 249
9.3.4 Slippage Tests 251
9.3.5 The Effects of Temperature and Pressure Changes 252
9.3.6 The Effects of Gas in Solution 252
9.4 Accuracy and Calibration 253
9.5 Principal Designs of Gas Meters 254
9.5.1 Wet Gas Meter 255
9.5.2 Diaphragm Meter 256
9.5.3 Rotary Positive Displacement Gas Meter 257
9.6 Positive Displacement Meters for Multiphase Flows 258
9.7 Meter Using Liquid Plugs to Measure Low Flows 261
9.8 Applications, Advantages and Disadvantages 261
9.9 Chapter Conclusions 262
9.A Basic Analysis and Recent Research 263
9.A.1 Theory for a Sliding Vane Meter 263
9.A.1.1 Flowmeter Equation 264
9.A.1.2 Expansion of the Flowmeter Due to Temperature 265
9.A.1.3 Pressure Effects 266
9.A.1.4 Meter Orientation 267
9.A.1.5 Analysis of Calibrators 268
9.A.1.6 Application of Equations to a Typical Meter 270
9.A.2 Recent Theoretical and Experimental Research 271

10 Turbine and Related Flowmeters 279
10.1 Introduction 279
10.1.1 Background 279
10.1.2 Qualitative Description of Operation 279
10.1.3 Basic Theory 280
10.2 Precision Liquid Meters 287
10.2.1 Principal Design Components 287
10.2.2 Dual Rotor Meters 288
10.2.3 Bearing Design Materials 288
10.2.4 Strainers 290
10.2.5 Materials 290
10.2.6 Size Ranges 290
10.2.7 Other Mechanical Design Features 291
10.2.8 Cavitation 291
10.2.9 Sensor Design and Performance 292

10.2.10 Characteristics 293
10.2.11 Accuracy 294
10.2.12 Installation 295
10.2.13 Maintenance 297
10.2.14 Viscosity, Temperature and Pressure 298
10.2.15 Unsteady Flow 299
10.2.16 Multiphase Flow 300
10.2.17 Signal Processing 301
10.2.18 Applications 301
10.2.19 Advantages and Disadvantages 302
10.3 Precision Gas Meters 303
10.3.1 Principal Design Components 303
10.3.2 Bearing Design 303
10.3.3 Materials 303
10.3.4 Size Range 303
10.3.5 Accuracy 304
10.3.6 Installation 306
10.3.7 Sensing and Monitoring 308
10.3.8 Unsteady Flow 308
10.3.9 Applications 310
10.3.10 Advantages and Disadvantages 311
10.4 Water Meters 311
10.4.1 Principal Design Components 311
10.4.2 Bearing Design 312
10.4.3 Materials 312
10.4.4 Size Range 313
10.4.5 Sensing 313
10.4.6 Characteristics and Accuracy 313
10.4.7 Installation 313
10.4.8 Special Designs 314
10.5 Other Propeller and Turbine Meters 314
10.5.1 Quantum Dynamics Flowmeter 314
10.5.2 Pelton Wheel Flowmeters 314
10.5.3 Bearingless Flowmeter 314
10.5.4 Vane Type Flowmeters 315
10.6 Chapter Conclusions 316
10.A Turbine Flowmeter Theoretical and Experimental Research 317
10.A.1 Derivation of Turbine Flowmeter Torque Equations 317
10.A.2 Transient Analysis of Gas Turbine Flowmeter 322
10.A.3 Recent Developments 324

11 Vortex Shedding, Swirl and Fluidic Flowmeters 327
11.1 Introduction 327
11.2 Vortex Shedding 327

11.3 Industrial Developments of Vortex-Shedding Flowmeters 329
11.3.1 Experimental Evidence of Performance 329
11.3.2 Bluff Body Shape 331
11.3.3 Standardisation of Bluff Body Shape 334
11.3.4 Sensing Options 334
11.3.5 Cross-Correlation and Signal Interrogation Methods 339
11.3.6 Other Aspects Relating to Design and Manufacture 339
11.3.7 Accuracy 340
11.3.8 Installation Effects 341
11.3.9 Effect of Pulsation and Pipeline Vibration 344
11.3.10 Two-Phase Flows 345
11.3.11 Size and Performance Ranges and Materials in Industrial Designs 347
11.3.12 Computation of Flow Around Bluff Bodies 348
11.3.13 Applications, Advantages, and Disadvantages 349
11.3.14 Future Developments 350
11.4 Swirl Meter – Industrial Design 351
11.4.1 Design and Operation 351
11.4.2 Accuracy and Ranges 351
11.4.3 Installation Effects 352
11.4.4 Applications, Advantages and Disadvantages 352
11.5 Fluidic Flowmeter 352
11.5.1 Design 353
11.5.2 Accuracy 355
11.5.3 Installation Effects 355
11.5.4 Applications, Advantages and Disadvantages 355
11.6 Other Proposed Designs 355
11.7 Chapter Conclusions 356
11.A Vortex Shedding Frequency 358
11.A.1 Vortex Shedding from Cylinders 358
11.A.2 Order of Magnitude Calculation of Shedding Frequency 358

12 Electromagnetic Flowmeters 362
12.1 Introduction 362
12.2 Operating Principle 362
12.3 Limitations of the Theory 364
12.4 Design Details 366
12.4.1 Sensor or Primary Element 366
12.4.2 Transmitter or Secondary Element 370
12.5 Calibration and Operation 373
12.6 Industrial and Other Designs 374
12.7 Installation Constraints – Environmental 377
12.7.1 Surrounding Pipe 377
12.7.2 Temperature and Pressure 378

12.8 Installation Constraints – Flow Profile Caused by Upstream Pipework 379
12.8.1 Introduction 379
12.8.2 Theoretical Comparison of Meter Performance Due to Upstream Flow Distortion 379
12.8.3 Experimental Comparison of Meter Performance Due to Upstream Flow Distortion 380
12.8.4 Conclusions on Installation Requirements 381
12.9 Installation Constraints – Fluid Effects 382
12.9.1 Slurries 382
12.9.2 Change of Fluid 383
12.9.3 Non-Uniform Conductivity 383
12.10 Multiphase Flow 383
12.11 Accuracy Under Normal Operation 384
12.12 New Industrial Developments 385
12.13 Applications, Advantages and Disadvantages 387
12.13.1 Applications 387
12.13.2 Advantages 388
12.13.3 Disadvantages 389
12.14 Chapter Conclusions 389
12.A Brief Review of Theory, Other Applications and Recent Research 390
12.A.1 Introduction 390
12.A.2 Electric Potential Theory 392
12.A.3 Development of the Weight Vector Theory 392
12.A.4 Rectilinear Weight Function 393
12.A.5 Axisymmetric Weight Function 394
12.A.6 Performance Prediction 395
12.A.7 Further Research 396
12.A.8 Verification 398
12.A.9 Application to Non-Conducting Dielectric Fluids 400
12.A.10 Electromagnetic Flowmeters Applied to Liquid Metals 403

13 Magnetic Resonance Flowmeters 408
13.1 Introduction and Some Early References 408
13.2 Developments in the Oil and Gas Industry 409
13.3 A Brief Introduction to the Physics 409
13.4 Outline of a Flowmeter Design 414
13.5 Chapter Conclusions 417

14 Ultrasonic Flowmeters 419
14.1 Introduction 419
14.2 Essential Background to Ultrasonics 420
14.3 Transit-Time Flowmeters 423
14.3.1 Transit-Time Flowmeters – Flowmeter Equation and the Measurement of Sound Speed 423

14.3.2 Effect of Flow Profile and Use of Multiple Paths 427
14.3.3 Transducers 432
14.3.4 Size Ranges and Limitations 437
14.3.5 Clamp-on Meters 437
14.3.6 Signal Processing and Transmission Timing 439
14.3.7 Reported Accuracy 442
14.3.7.1 Reported Accuracy – Spool Piece Meters 442
14.3.7.2 A Manufacturer's Accuracy Claims 443
14.3.7.3 Clamp-on Accuracy 444
14.3.8 Installation Effects 447
14.3.8.1 Effects of Distorted Profile by Upstream Fittings 447
14.3.8.2 Pipe Roughness and Deposits 453
14.3.8.3 Unsteady and Pulsating Flows 454
14.3.8.4 Multiphase Flows 454
14.3.8.5 Flow Straighteners and Conditioners 455
14.3.9 Other Experience of Transit-Time Meters 456
14.3.10 Experience with Liquid Meters 456
14.3.11 Gas Meter Developments 457
14.3.12 Applications, Advantages and Disadvantages of the Transit-Time and Related Designs 463
14.4 Doppler Flowmeter 466
14.4.1 Simple Explanation of Operation 466
14.4.2 Operational Information for the Doppler Flowmeter 468
14.4.3 Applications, Advantages and Disadvantages for the Doppler Flowmeter 468
14.5 Correlation Flowmeter 469
14.5.1 Operation of the Correlation Flowmeter 469
14.5.2 Installation Effects for the Correlation Flowmeter 470
14.5.3 Other Published Work on the Correlation Flowmeter 471
14.5.4 Applications, Advantages and Disadvantages for the Correlation Flowmeter 472
14.6 Other Ultrasonic Applications 472
14.7 Conclusions on Ultrasonic Flowmeters 473
14.A Mathematical Methods and Further Research Relating to Ultrasonic Flowmeters 474
14.A.1 Simple Path Theory 474
14.A.2 Use of Multiple Paths to Integrate Flow Profile 477
14.A.3 Weight Vector Analysis 478
14.A.4 Development of Modelling of the Flowmeter 479
14.A.5 Doppler Theory and Developments 482

15 Acoustic and Sonar Flowmeters 484
15.1 Introduction 484
15.2 SONARtrac® Flowmeter 484

15.2.1 Basic Explanation of How the Passive Sonar Flowmeter Works 484
15.2.2 A Note on Turbulent Eddies and Transition to Laminar Flow in the Pipe 485
15.2.3 Flow Velocity Measurement 485
15.2.4 Speed of Sound and Gas Void Fraction (Entrained Air Bubbles) Measurement 486
15.2.5 Localised Velocity Measurements 487
15.2.6 The Convective Ridge 487
15.2.7 Calibration 489
15.2.8 Sound Speed Used to Obtain Fluid Parameters 490
15.2.9 Additional Sensors 491
15.2.10 Clamp-on System 491
15.2.11 Liquid, Gas and Multicomponent Operation 492
15.2.12 Size Range and Flow Range 493
15.2.13 Signal Handling 493
15.2.14 Accuracy Claims 494
15.2.15 Installation Effects 494
15.2.16 Published Information 496
15.2.17 Applications 496
15.3 ActiveSONAR™ Flowmeter 496
15.3.1 Single and Multiphase Flows 497
15.3.2 Brief Summary of Meter Range, Size etc. 497
15.4 Other Related Methods Using Noise Emissions 498
15.5 Chapter Conclusions 500

16 Mass Flow Measurement Using Multiple Sensors for Single-Phase Flows 501
16.1 Introduction 501
16.2 Multiple Differential Pressure Meters 502
16.2.1 Hydraulic Wheatstone Bridge Method 504
16.2.2 Theory of Operation 504
16.2.3 Industrial Experience 505
16.2.4 Applications 505
16.3 Multiple Sensor Methods 506
16.4 Chapter Conclusions 507

17 Multiphase Flowmeters 508
17.1 Introduction 508
17.2 Multiphase and Multi-Component Flows 509
17.3 Two-Phase/Component Flow Measurements 509
17.3.1 Liquid/Liquid Flows and Water-Cut Measurement 510
17.3.2 Entrained Solid in Fluid Flows 510
17.3.3 Metering Wet-Gas 511

17.4 Multiphase Flowmeters 514
17.4.1 Categorisation of Multiphase Flowmeters 514
17.4.2 Multiphase Flowmeters (MPFMs) for Oil Production 515
17.4.3 Developments and References Since the Late 1990s 519
17.5 Accuracy 527
17.6 Chapter Conclusions 528

18 Thermal Flowmeters 530
18.1 Introduction 530
18.2 Capillary Thermal Mass Flowmeter – Gases 530
18.2.1 Description of Operation 531
18.2.2 Operating Ranges and Materials for Industrial Designs 534
18.2.3 Accuracy 535
18.2.4 Response Time 535
18.2.5 Installation 535
18.2.6 Applications 536
18.3 Calibration of Very Low Flow Rates 536
18.4 Thermal Mass Flowmeter – Liquids 537
18.4.1 Operation 537
18.4.2 Typical Operating Ranges and Materials for Industrial Designs 538
18.4.3 Installation 538
18.4.4 Applications 538
18.5 Insertion and In-Line Thermal Mass Flowmeters 538
18.5.1 Insertion Thermal Mass Flowmeter 540
18.5.2 In-Line Thermal Mass Flowmeter 541
18.5.3 Range and Accuracy 542
18.5.4 Materials 542
18.5.5 Installation 542
18.5.6 Applications 543
18.6 Chapter Conclusions 544
18.A Mathematical Background to the Thermal Mass Flowmeters 545
18.A.1 Dimensional Analysis Applied to Heat Transfer 545
18.A.2 Basic Theory of ITMFs 546
18.A.3 General Vector Equation 548
18.A.4 Hastings Flowmeter Theory 549
18.A.5 Weight Vector Theory for Thermal Flowmeters 551
18.A.6 Other Recently Published Work 552

19 Angular Momentum Devices 553
19.1 Introduction 553
19.2 The Fuel Flow Transmitter 554
19.2.1 Qualitative Description of Operation 554
19.2.2 Simple Theory 557

19.2.3 Calibration Adjustment 558
19.2.4 Meter Performance and Range 558
19.2.5 Application 559
19.3 Chapter Conclusions 559

20 Coriolis Flowmeters 560
20.1 Introduction 560
20.1.1 Background 560
20.1.2 Qualitative Description of Operation 563
20.1.3 Experimental and Theoretical Investigations 564
20.1.4 Shell-Type Coriolis Flowmeter 566
20.2 Industrial Designs 566
20.2.1 Principal Design Components 569
20.2.2 Materials 572
20.2.3 Installation Constraints 574
20.2.4 Vibration Sensitivity 576
20.2.5 Size and Flow Ranges 577
20.2.6 Density Range and Accuracy 578
20.2.7 Pressure Loss 578
20.2.8 Response Time 579
20.2.9 Zero Drift 580
20.3 Accuracy Under Normal Operation 581
20.4 Published Information on Performance 582
20.4.1 Early Industrial Experience 583
20.4.2 Gas-Liquid 583
20.4.3 Sand in Water (Dominick et al. 1987) 584
20.4.4 Pulverised Coal in Nitrogen (Baucom 1979) 584
20.4.5 Water-in-Oil Measurement 584
20.4.6 Two- and Three-Component Flows 585
20.5 Calibration 585
20.6 Applications, Advantages, Disadvantages, Cost Considerations 587
20.6.1 Applications 587
20.6.2 Advantages 589
20.6.3 Disadvantages 590
20.6.4 Cost Considerations 591
20.7 Chapter Conclusions 591
20.A Notes on the Theory of Coriolis Meters 593
20.A.1 Simple Theory 593
20.A.2 Note on Hemp's Weight Vector Theory 595
20.A.3 Theoretical Developments 597
20.A.4 Coriolis Flowmeter Reviews 601

21 Probes for Local Velocity Measurement in Liquids and Gases 603
21.1 Introduction 603
21.2 Differential Pressure Probes – Pitot Probes 604

21.3 Differential Pressure Probes – Pitot-Venturi Probes 607
21.4 Insertion Target Meter 608
21.5 Insertion Turbine Meter 609
21.5.1 General Description of Industrial Design 609
21.5.2 Flow-Induced Oscillation and Pulsating Flow 611
21.5.3 Applications 612
21.6 Insertion Vortex Probes 612
21.7 Insertion Electromagnetic Probes 614
21.8 Insertion Ultrasonic Probes 615
21.9 Thermal Probes 616
21.10 Chapter Conclusions 616

22 Verification and In Situ Methods for Checking Calibration 617
22.1 Introduction 617
22.2 Verification 617
22.3 Non-Invasive, Non-Intrusive and Clamp-On Flowmeter Alternatives 620
22.3.1 Use of Existing Pipe Work 620
22.3.2 Other Effects: Neural Networks, Tracers, Cross-Correlation 622
22.3.3 Other Flowmeter Types in Current Use 622
22.4 Probes and Tracers 623
22.5 Microwaves 624
22.6 Chapter Conclusions 624

23 Remote Data Access Systems 625
23.1 Introduction 625
23.2 Types of Device – Simple and Intelligent 626
23.3 Simple Signal Types 627
23.4 Intelligent Signals 629
23.5 Selection of Signal Type 630
23.6 Communication Systems 630
23.7 Remote Access 630
23.8 Future Implications 631

24 Final Considerations 633
24.1 Is there an Opportunity to Develop New Designs in Collaboration with the Science Base? 633
24.2 Is Manufacture of High Enough Quality? 633
24.3 Does the Company's Business Fall within ISO 9000 and/or ISO 17025? 636
24.4 What are the New Flow Measurement Challenges? 637
24.5 What Developments Should We Expect in Micro-Engineering Devices? 638

24.6 Which Techniques for Existing and New Flow Metering Concepts Should Aid Developments? 639
24.7 Closing Remarks 641

References 643
Main Index 735
Flowmeter Index 739
Flowmeter Application Index 743

Preface

This is a book about flow measurement and flowmeters written for all in the industry who specify and apply, design and manufacture, research and develop, maintain and calibrate flowmeters. It provides a source of information both on the published research, design and performance of flowmeters, and also on the claims of flowmeter manufacturers. It will be of use to engineers, particularly mechanical and process engineers, and also to instrument companies' marketing, manufacturing and management personnel as they seek to identify future products.

I have concentrated on the mechanical and fluid engineering aspects and have given only as much of the electrical engineering details as is necessary for a proper understanding of how and why the meters work. I am not an electrical engineer and so have not attempted detailed explanations of modern electrical signal processing. I am also aware of the speed with which developments in signal processing would render out of date any descriptions that I might give.

I make the assumption that the flowmeter engineer will automatically turn to the appropriate standard and I have, therefore, tried to minimise reproducing information which should be obtained from those excellent documents. I recommend that those involved in new developments should keep a watching brief on the regular conferences which carry much of the latest developments in the business, and are illustrated by the papers in the reference list.

I hope, therefore, that this book will provide a signpost to the essential information required by all involved in the development and use of flowmeters, from the field engineer to the chief executive of the entrepreneurial company which is developing its product range in this technology.

In this book, following introductory chapters on accuracy, flow, selection and calibration, I have attempted a clear explanation of each type of flowmeter so that the reader can easily understand the workings of the various meters. I have, then, attempted to bring together a significant amount of the published information which enlightens us on the performance and applications of flowmeters. The two sources for this are the open literature and the manufacturers' brochures. I have also introduced, to a varying extent, the mathematics behind the meter operations, but to avoid disrupting the text, I have consigned some of this to the appendices at the end of the chapters.

However, by interrogating the appropriate databases for flowmeter papers it rapidly becomes apparent that inclusion of references to all published material is unrealistic. I have attempted a selection of those which appeared to be more relevant and available to the typical reader of this book. However, it is likely that, owing to the very large number of relevant papers, I have omitted some which should have been included.

Topics not covered in this book, but which might be seen as within the general field of flow measurement, are: metering pumps, flow switches, flow controllers, flow measurement of solids and granular materials, open channel flow measurement, hot wire local velocity probes or laser Doppler anemometers and subsidiary instrumentation.

In this second edition, I have left in much of the original material, as I am aware of the danger of losing sight of past developments and unnecessary reinvention. I have attempted to bring up to date items which are out of date, but am conscious that I may have missed some, and I have attempted to introduce the new areas and new developments of which I have become aware. In two areas where I know myself to be lacking in first-hand knowledge, I have changed the focus of the chapters and greatly reduced their length. Modern Control Methods has gone and been replaced by Remote Data Access Systems, and the chapter on manufacturing by a brief chapter entitled Final Considerations which touches on manufacturing variation and ISO quality standards and also takes in final comments.

I have included three new chapters covering magnetic resonance flowmeters, sonar and acoustic flowmeters and verification. They are brief chapters, but represent new developments since the first edition. I have also separated multiphase flowmeters into another new chapter, but have done so recognising that my knowledge of the subject is minimal and the coverage in the chapter is very superficial.

The techniques for precise measurement of flow are increasingly important today when the fluids being measured, and the energy involved in their movement, may have a very high monetary value. If we are to avoid being prodigal in the use of our natural resources, then the fluids among them should be carefully monitored. Where there might be pressure to cut corners with respect to standards and integrity, we need to ensure that in flow measurement these features are given their proper treatment and respect.

Acknowledgements

My knowledge of this subject has benefitted from many others with whom I have worked and talked over the years. These include colleagues from industry, national laboratories and academia, visitors and students, whether on short courses or longer-term degree courses and research. I hope that this book does justice to all that they have taught me.

In writing this book, I have drawn on information from many manufacturers, and some have been particularly helpful in agreeing to the use of information and diagrams. I have acknowledged these in the captions to the figures. Some went out of their way to provide artwork, and I am particularly grateful to Katrin Faber and Ruth O'Connell.

In preparing this second edition, I have been conscious of the many changes and advances in the subject, and so I have depended on many friends and colleagues, near and far, to read sections for me and to comment, criticise and correct them. In the middle of already busy lives they kindly made time to do this for me. In particular I would like to thank:

Anders Andersson, Matthias Appel, Andy Capper, Marcus Conein, Andrew Cowan, Paul de Waal, Steve Dixon, Mark England, Chris Gimson, John Hemp, Jankees Hogendoorn, Daniel Holland, Geoff Howe, Foz Hughes, Ian Hutchings, Masahiro Ishibashi, Edward Jukes, Peter Lau, Chris Lenn, Tony Lopez, Larry Lynnworth, George Mattingly, Gary Oddie, Christian O'Keefe, Michael Reader-Harris, Masaki Takamoto, Scott Pepper, Janis Priede, Michael Sapack, Steve Seddon, Takashi Shimada, Henk Versteeg, Takeshi Wakamatsu, Tao Wang, Ben Weager, Charles Wemyss and Xiao-Zhang Zhang.

I am extremely grateful to them for taking time to do this, and for the constructive comments which they gave. Of course, I bear full responsibility for the final script, although their help and encouragement is greatly valued.

I have had the privilege of being based back at my alma mater for the past 15 years, and they have been some of the most enjoyable of my working life. I am very grateful to Mike Gregory, who was key in making this possible; to Ian Hutchings, with whom I have collaborated; and to others of the Department of Engineering, particularly librarians and technical support staff, who have facilitated my experimental and theoretical research. I have also appreciated the friendship of the late Yousif Hussain, who provided a strong industrial link over this period.

I acknowledge with thanks the following organisations which have given permission to use their material:

American Society of Mechanical Engineers

Elsevier based on the STM agreement for the use of figures from their publications and for agreement to honour my right to use material from papers of my own for Chapters 10 and 20.

National Engineering Laboratory (NEL)

Professional Engineering Publishing for permission to draw on material from the Introductory Guide Series of which I was editor, and to the Council of the Institution of Mechanical Engineers for permission to reproduce material which is identified in the text from Proceedings Part C, *Journal of Mechanical Engineering Science*, Vol. 205, pp. 217–229, 1991.

Permission to reproduce extracts from British Standards is granted by BSI Standards Limited (BSI). No other use of this material is permitted. British Standards can be obtained in pdf or hard copy formats from the BSI online shop: www.bsigroup.com/shop.

I have also been grateful for the help and encouragement given to me by many in the preparation of this book. It would be difficult to name them, but I am grateful for each contribution.

I have found the support of my family invaluable and particularly that of Liz, my wife, whose patience with my long hours at the computer, her willingness to assist with her proofreading skills, her encouragement and help at every stage, have made the task possible and I cannot thank her enough.

Finally, I am grateful to Cambridge University Press for the opportunity of preparing this second edition.

Nomenclature

Chapter 1

c_i	sensitivity coefficient
$f(x)$	function for normal distribution
K	K factor in pulses per unit flow quantity
k	coverage factor
M	mean of a sample of n readings
m	index
$N(\mu,\sigma^2)$	normal curve
n	number of measurements, index
p	probability, index
$\overline{q}$	mean of n measurements q_j, index
q_j	test measurement
q_v	volumetric flow rate
r	index
s	index
$s(\overline{q})$	experimental standard deviation of mean of group q_j
$s(q_j)$	experimental standard deviation of q_j
t	Student's t
U	expanded uncertainty
$u(x_i)$	standard uncertainty for the ith quantity
$u_c(y)$	combined standard uncertainty
x	coordinate
x_i	result of a meter measurement, input quantities
$\overline{x}$	mean of n meter measurements
y	output quantity
z	normalised coordinate $(x-\mu)/\sigma$
μ	mean value of data for normal curve
ν	degrees of freedom
σ	standard deviation (σ^2 variance)

$\Phi(z)$	area under normal curve e.g. $\Phi(0.5)$ is the area from $z = -\infty$ to $z = 0.5$
$\phi(z)$	function for normalised normal distribution

Chapter 2

A	cross-section of pipe
c	local speed of sound
c_p	specific heat at constant pressure
c_v	specific heat at constant volume
D	diameter of pipe
d	diameter of flow conditioner plate holes
f_D	Darcy friction factor: $f_D = 4f_F$
f_F	Fanning friction factor
g	acceleration due to gravity
H	Hodgson number Equation (2.13)
K	pressure loss coefficient
L	length of pipe (sometimes given as a multiple of D e.g. 5D)
M	Mach number
n	index as in Equation (2.4)
p	pressure
p_0	stagnation pressure
Δp_{loss}	pressure loss across a pipe fitting
q_v	volumetric flow rate
q_m	mass flow rate
R	radius of pipe
Re	Reynolds number
r	radial coordinate (distance from pipe axis)
T	temperature
T_0	stagnation temperature
V	velocity in pipe, total volume of pipework used in Hodgson number
V_0	velocity on pipe axis, maximum axial velocity at a cross-section
V_{rms}	fluctuating component of velocity
$\bar{V}$	mean velocity in pipe
v	local fluid velocity
v_τ	friction velocity $v_\tau = \sqrt{\frac{\tau_w}{\rho}}$
x	distance from pipe axis in horizontal plane
y	distance from the pipe wall $= (R - r)$
y_1	viscous sublayer thickness
y_2	extent of buffer layer
z	elevation above datum

γ	ratio of specific heats
μ	dynamic viscosity
ν	kinematic viscosity
ρ	density
τ	shear stress
τ_w	wall shear stress: $\tau_w = f_F \dfrac{\rho \bar{V}^2}{2}$

Subscripts

1,2	pipe sections

Chapter 4

C_d	concentration of tracer in the main stream at the downstream sampling point
C_{dmean}	mean concentration of tracer measured downstream
C_i	concentration of tracer in the injected stream
C_u	concentration of tracer in the main stream upstream of injection point (if the tracer material happens to be present)
c_i	sensitivity coefficient
K_{fm}	mass flowmeter factor
k_S	factor for the weigh scale
M_n	net mass of liquid collected in calibration
M_D	weight of deadweight
M_s	conventional mass of material of density 8,000 kg/m^3
M_L	mass of water in weigh tank
M_G	mass of air displaced
ΔM_{LDV}	change in mass within the connection pipe between the flowmeter and the weir
m_{CAL}	reading of the weigh scale when loaded with deadweights
m_L	weigh scale reading
P	pulse count
p	pressure
q_v	volumetric flow rate in the line
q_{vi}	volumetric flow rate of injected tracer
R	gas constant for a particular gas
T	temperature
t	collection time during calibration
V	amount injected in the sudden injection (integration) method
v	specific volume
ρ	liquid density
ρ_D	actual density of deadweight

ρ_G	air density
ρ_{LW}	liquid density

Chapter 5

A	function of Re_D
a_1	expression in orifice plate-bending formula
a_ε	constant
b_ε	constant
C	discharge coefficient
C_{Re}	part of discharge coefficient affected by Re
C_{Taps}	part of discharge coefficient which allows for position of taps
C_∞	discharge coefficient for infinite Reynolds number
$C_{Small\ orifice}$	correction for small orifice sizes
c_1	expression in orifice plate-bending formula
c_ε	constant
D	pipe diameter (ID)
D'	orifice plate support diameter
d	orifice diameter
E	thickness of the plate, velocity of approach factor $(1-\beta^4)^{-½}$
E_T	total error in the indicated flow rate of a flowmeter in pulsating flow
E^*	elastic modulus of plate material
e	thickness of the orifice (Figure 5.3), Napierian constant
F	correction factor used to obtain the mass flow of a (nearly) dry steam flow
f	frequency of the pulsation
H	Hodgson number
h	thickness of orifice plate
K	loss coefficient, related to the criterion for Hodgson number
L_1	$= l_1/D$
L'_2	$= l'_2/D'$ signifies that the measurement is from the downstream face of the plate.
l_1	distance of the upstream tapping from the upstream face of the plate
l'_2	distance of the downstream tapping from the downstream face of the plate. ′ signifies that the measurement is from the downstream face of the plate.
M'_2	$= 2L'_2/(1-\beta)$
M_1	numerical value defined in text
n	index
p	static pressure
p_u	upstream static pressure

p_d	downstream static pressure
p_1	static pressure at upstream tapping
p_2	static pressure at downstream tapping
Δp	differential pressure, pressure drop between pulsation source and meter
q_m	mass flow rate
q_v	volumetric flow rate
Re	Reynolds number
Re_D	Reynolds number based on the pipe ID
r	radius of upstream edge of orifice plate
T_f	temperature of the fluid at flowing conditions
t	time
V	volume of pipework and other vessels between the source of the pulsation and the flowmeter position
$\bar{V}$	mean velocity in pipe with pulsating flow
V_{cl}	centre line velocity
V_{max}	maximum velocity
V_{rms}	rms value of unsteady velocity fluctuation in pipe with pulsating flow
x	dryness fraction, displacement of the centre of the orifice hole from the pipe axis (m)
α	$= CE$ the flow coefficient
β	diameter ratio d/D
γ	ratio of specific heats
δq_m	small changes or errors in q_m etc.
ε	expansibility (or expansion) factor
ε_1	expansibility (or expansion) factor at upstream tapping
κ	isentropic exponent
ρ	density
ρ_1	density at the upstream pressure tapping
ρ_g	density of gas
ρ_l	density of liquid
σ_y	yield stress for plate material
θ	angle defined in Figure 5.B.1 caused by deposition on the leading face of the orifice plate
Φ_{fo}^2	ratio of two-phase pressure drop to liquid flow pressure drop
ϕ	maximum allowable percentage error in pulsating flow

Chapter 6

C	coefficient of discharge
C_{tp}	coefficient for wet-gas flow equation
C_{dry}	discharge coefficient for fully dry gas

C_{fullywet}	discharge coefficient for fully wet gas $X \geq X_{\text{lim}}$ where $X_{\text{lim}} = 0.016$
D	pipe ID
d	throat diameter
E	velocity of approach factor $= 1\big/\sqrt{(1-\beta^4)}$
Fr_{g}	superficial gas Froude number
$\text{Fr}_{\text{g,th}}$	Froude number at the throat
g	gravitational acceleration
n	index
p_1	upstream pressure tapping
p_2	downstream pressure tapping
Δp	differential pressure
$q_{\text{m,g}}$	mass flow rate of gas
$q_{\text{m,l}}$	mass flow rate of liquid
q_{m}	mass flow rate
q_{tp}	apparent flow rate when liquid is present in the gas stream
q_{v}	volume flow rate
Ra	roughness criterion
Re	Reynolds number
V_{sg}	superficial gas velocity
β	diameter ratio d/D
ε	expansibility (or expansion) factor
κ	isentropic exponent
ρ_1	density at upstream pressure tapping
ρ_{l}	liquid density
$\rho_{\text{1,g}}$	gas density at upstream tapping point
τ	pressure ratio $\dfrac{p_2}{p_1}$
ϕ	defined in Equation (6.1)

Chapter 7

A_2	outlet cross-sectional area
A_*	throat cross-sectional area
a	constant
b	constant
C	discharge coefficient
C_{R}	$= C_*\sqrt{Z}$
C_*	critical flow function
$C_{*\text{i}}$	critical flow function for a perfect gas
c	speed of sound
c_{p}	specific heat at constant pressure
c_{v}	specific heat at constant volume
c_*	speed of sound in the throat

d	throat diameter
d_2	outlet diameter
l	dimension given in Figure 7.5
M	Mach number
M_1	Mach number at inlet when stagnation conditions cannot be assumed
$\mathbf{M}$	molecular weight
n	exponent in Equation (7.10)
p_o	stagnation pressure
p_1	pressure at inlet when stagnation conditions cannot be assumed
p_{2i}	ideal outlet pressure
p_{2max}	actual maximum outlet pressure
p_*	throat pressure in choked conditions
q_m	mass flow
$\mathbf{R}$	universal gas constant
Re_d	Reynolds number based on the throat diameter
r	toroid radius
T_o	stagnation temperature
T_*	throat temperature in choked conditions
Z	compressibility factor
Z_o	compressibility factor at stagnation conditions
β	$= d/D$
γ	ratio of specific heats
ε	error
κ	isentropic exponent
ν	kinematic viscosity
θ	angle given in Figure 7.5
μ_0	dynamic viscosity of gas at stagnation conditions
ρ_*	density of gas in the throat
ρ_o	density at stagnation conditions

Chapter 8

A	cross-sectional area of the pipe, constant
A'	constant
A_f	cross-sectional area of float
A_x	cross-sectional area of tapering tube at height x
A_2	annular area around float, annular area around target
a	area of target
B	constant
C	coefficient
C_0 to C_4	constants in curve fit for target meter discharge coefficient
C_c	contraction coefficient
C_d	coefficient of discharge

D	pipe diameter
d	diameter of ball float, throat diameter for pipe inlet
G	see Section 8.A.3
g	gravity
K	loss coefficient, bend meter coefficient
L	length of laminar flow tube
p_i	pressure on inside of bend
p_o	pressure on outside of bend
p_1	upstream pressure
p_2	downstream pressure
Δp	pressure difference between upstream and downstream V-cone pressure tappings
q_v	volumetric flow rate
q_{vG}	see Section 8.A.3
R	radius of bend or elbow
Re	Reynolds number
V	velocity
$\bar{V}$	mean velocity in tube
V_f	volume of float
W	immersed weight of the float
x	height of float in tube
β	for V-cone meters is the square root of the ratio of annulus area between V-cone and pipe wall/cross-sectional area of pipe
ε	expansibility factor
κ	isentropic exponent
μ	dynamic viscosity
μ_g	dynamic viscosity of calibration gas at flowing conditions
μ_{std}	dynamic viscosity of reference gas at standard conditions
ρ	density
ρ_f	density of float material

Chapter 9

E	Young's modulus of elasticity
F	friction force
g	acceleration due to gravity
L	length of clearance gap in direction of flow
l	axial length of clearance gap
l_{ax}	axial length of measuring chamber
M	mass of rotor
N	rotational speed of rotor
N_i	rotational speed of cam disc of calibrator (clutch system)
N_{IN}	rotational speed of shaft into calibrator (epicyclic system)
N_{max}	maximum rotational speed of rotor

N_o	rotational speed of outer ring of calibrator (clutch system)
N_{OUT}	rotational speed of shaft out of calibrator (epicyclic system)
N_s	slippage in N_{IN}
n	number of teeth on gear
p_d	downstream pressure
p_u	upstream pressure
p_{ref}	reference pressure
p_{RPFM}	value of pressure at the meter
q_{ideal}	as used in Equation (9.A6)
$q_{leakage}$	leakage flow rate
q_{BULK}	bulk volumetric flow rate
q_{slip}	volumetric flow rate through meter at no rotation
q_v	volumetric flow rate
$q_{v,RPFM}$	flow rate indicated by the meter
$q_{v,ref}$	true flow rate at reference meter
R	radius of friction wheel
r	radius of point of friction wheel contact
r_i	inner radius of measuring chamber
r_o	outer radius of measuring chamber
r_s	shaft radius
T	temperature, torque
T_{IN}	torque on input shaft to calibrator
T_{OUT}	torque out from calibrator
T_0	constant drag torque
T_1	speed-dependent drag torque
T_{ref}	reference temperature
T_{RPFM}	values of temperature at the meter
t	clearance between stationary and moving members
t_o	thickness of outer casing
U	velocity of moving component as in Figure 9.A.1
u	fluid velocity
y	position coordinate across clearance gap
α_m	coefficient of linear expansion of metal
α_1	coefficient of volume expansion of liquid
Δ	change in quantity, distance between centre of rotation of arm and centre of discs in calibrator
δ	reduction in area of the measuring chamber due to blade sections
ε	deviation being the fractional difference between indicated and true flow rates: due to leakage this is expected to be negative
θ_c	angular position of centre disc of calibrator
θ_i	angular position of cam disc of calibrator
θ_o	angular position of outer ring of calibrator

μ	dynamic viscosity
Φ	angular position of arm of calibrator

Other Subscripts

j	dummy suffix for summation
l	liquid
m	metal
1,2	different materials
1–8	epicyclic gears

Chapter 10

A	cross-sectional area of the effective annular flow passage at the rotor blades
A_1/A_2	area ratio
a_0, a_1, a_2, a_3	constant coefficients
B	axial length of rotor
b	turbine meter aerodynamic torque coefficient; bearing length; frequency response coefficient unless used as an exponent
C	proportional to 1/K
$C_{water}/C_{liquid\ nitrogen}$	calibration factor ratio to allow for thermal expansion
C_D	drag coefficient
C'_D	drag coefficient adjusted to allow for C_h, dimensionless drag coefficient
C_f	fluid drag coefficient
C_h	constant for a particular design. Behaves like an adjustment to the main aerodynamic drag term coefficient.
C_L	lift coefficient
C'_{B0}	coefficient defined following Equation (10.A.26)
C'_{B1}	coefficient defined following Equation (10.A.26)
C'_{B2}	coefficient defined following Equation (10.A.26)
C'_D	coefficient defined following Equation (10.A.26)
C_p	pressure coefficient
c	chord
D	drag force; pipe diameter; impeller diameter
D_1	rotor response parameter
D_2	fluid drag parameter
D_3	non-fluid drag parameter
f	frequency of blade passing
I	moment of inertia of rotor system about rotor axis
i	incidence angle; pulsation index
K	lattice effect coefficient; characteristic meter factor (pulses/unit volume)

K_h	constant used in equation for helical blade angle
K_i	ideal meter factor (rad/m^3)
L	lift force; helical pitch of blades
N	number of blades
m	ratio across the meter (inlet to cross-section at blade annulus)
n	rotational speed (frequency)
p_{loss}	pressure loss
p_{01}	stagnation pressure at inlet
p_{02}	stagnation pressure at outlet
p_1	static pressure at inlet
p_2	static pressure at outlet
Q_1	or Q_{min} minimum flow rate in standard documents
Q_2	or Q_{trans} flow rate between two zones where maximum error changes
Q_3	or Q_n normally recommended continuous flow rate
Q_4	or Q_{max} maximum flow rate for short periods in standard documents
Q_{max}	maximum flow rate for short periods
Q_{min}	lowest flow rate at which the meter operates
Q_n	normally recommended continuous flow rate
Q_{trans} or Q_t	flow rate between the two zones
q_1	initial flow rate
q_2	final flow rate
q_i	indicated average flow rate
q_t	flow rate at time t
q_v	volumetric flow rate
Re	Reynolds number
Ro	Roshko number
r	radial position; index in error equation
r_j	journal-bearing radius
r_h	hub radius
r_t	tip radius
S	slope of no-flow decay curve at standstill
St	Strouhal number
s	blade spacing
T	temperature; fundamental period of pulsating flow; torque
T_B	bearing drag torque
T_d	driving torque
T_{F0}	mechanical friction torque on rotor at zero speed
T_h	hub fluid drag torque
T_n	non-fluid drag torque
T_o	temperature at operation
T_r	drag torque
T_t	blade tip clearance drag torque

T_W	hub disc friction drag torque
t	time
t_B	blade thickness
t_R	relaxation time
V	non-dimensional fluid velocity
V_0	time average value of V_z over period T
V_1	inlet relative flow velocity
V_2	outlet relative flow velocity
V_{max}	maximum value of V_z
V_{min}	minimum value of V_z
V_z	axial velocity (instantaneous inlet fluid velocity of pulsating flow)
W	rotor blade velocity, non-dimensional instantaneous rotor velocity under pulsating flow
W_h	relative velocity at the hub
Y	tangential force
Z	axial force
α	wave shape coefficient; thermal coefficient of expansion; angle between inlet flow direction and far field flow
β	blade angle at radius r measured from axial direction of meter
β_1	relative inlet angle of flow
β_2	relative outlet angle of flow
$\sqrt{-1}$	mean of the inlet and outlet flow field
Γ	full-flow amplitude relative to average flow
δ	deflection of flow at blade outlet from blade angle
η	flow deviation factor
μ	dynamic viscosity
ν	kinematic viscosity
ρ	fluid density
τ	non-dimensional time (t/T); period of modulation
τ_0	rotor coast time to standstill
ω	instantaneous rotor angular velocity

Chapter 11

A	pipe full flow area, constant of value about 3.0 used in equation for avoiding cavitation in flows past vortex meters
A_{min}	flow area past bluff body
a	area when integrating vorticity
B	constant of value about 1.3 used in equation for avoiding cavitation in flows past vortex meters
D	pipe ID
f	shedding frequency
H	streamwise length of bluff body
h	parallel flats on sides of bluff body at leading edge

K	calculation factor for shedding frequency (Zanker and Cousins 1975), K factor = pulses/unit volume
L	length of bluff body across pipe between end fittings
Δp	pressure drop across vortex meter (about 1 bar at 10 m/s)
p_{atmos}	atmospheric pressure
p_{gmin}	minimum back pressure 5D downstream of vortex meter
p_v	saturated liquid vapour pressure
q_v	volumetric flow rate
Re	Reynolds number
St	Strouhal number
s	length along curve when integrating velocity around a vortex
V	velocity of flow
V_{max}	velocity of flow past bluff body
w	diameter or width of bluff body
δ	shear layer thickness
ω	vorticity

Chapter 12

A	used for area of electrode leads forming a loop causing quadrature signals in the electromagnetic flowmeter
a	pipe radius
B	magnetic flux density in tesla
$\mathbf{B}$	magnetic flux density vector
B_θ	θ component of magnetic flux density
D	diameter of pipe
$\mathbf{E}$	electric field vector
f	frequency of magnetic field excitation
i	$=\sqrt{-1}$
$\mathbf{j}$	current density vector
K	defined in Equation (12.A.25)
l	length of wire traversing magnetic field, typical length scale
M	a constant related to the polarizability of a single molecule of the medium
$\mathrm{Re_m}$	magnetic Reynolds number = $\mu\sigma Vl$
r	radial coordinate, radius of pipe bend
S	electromagnetic flowmeter sensitivity
U	electric potential
ΔU_{EE}	voltage between electrodes
ΔU_P	voltage across a wire P moving through a magnetic field (similarly for Q and R)
V	velocity
$\mathbf{V}$	velocity vector
V_m	mean velocity in the pipe in metres per second
v	volume element in integration

W	weighting function
$\mathbf{W}$	weight function vector
W'	rectilinear flow weight function
W''	axisymmetric weight function
W_z	axial component of W
Z	defined in Equation (12.A.27)
z	axial coordinate
α	void fraction
ε	permittivity of the medium
ε_0	permittivity of free space
θ	cylindrical coordinate
μ	permeability of the medium
μ_0	permeability of free space
ρ	fluid density
σ	conductivity
τ_n	sampling times for DC converter system
Φ	scalar potential
ω	excitation frequency

Chapter 13

$\boldsymbol{B}$	imposed magnetic flux density vector
$\lvert\boldsymbol{B}\rvert$	scalar magnitude of the magnetic flux density vector
E	energy level of photon
f	frequency of absorbed or emitted radiation
h	Planck's constant
M	magnetisation
M_0	equilibrium magnetisation state
S_g	signal amplitude for flow with 100% gas
S_l	signal amplitude for flow with 100% liquid
S_{meas}	measured signal amplitude for multiphase flow
T_1	longitudinal relaxation time or spin lattice relaxation time
T_2	transverse relaxation time
t	time
α	gas fraction (or gas hold-up) in the flow
γ	gyromagnetic ratio
τ	elapsed time

Chapter 14

A	weighting factors for Gaussian Quadrature
a	radius of pipe
c	velocity of ultrasound in the medium, coefficients of a polynomial
c_1	speed of sound in medium 1

c_2	speed of sound in medium 2
D	pipe diameter
F	factor to allow for difference between mean velocity in pipe cross-section and mean velocity along diametral path of ultrasound
f	frequency of stable quartz oscillator for ultrasonic measurement system
f_d	frequency for the downstream sing-around pulse train
f_t	transmitted Doppler frequency
f_r	reflected Doppler frequency
f_u	frequency for the upstream sing-around pulse train
Δf	difference between the sing-around frequencies, frequency shift in Doppler flowmeter
$f(t)$	function of t
H	defined in Section 14.3.8.1
h	displacement of ultrasonic beam from axis of pipe
I	driving current in ultrasonic flowmeter
i_1	angle of incident sound ray
i_2	angle of refracted sound ray
k_s	adiabatic compressibility
L	distance along path in transit-time flowmeter, distance between transducers in ultrasonic correlation flowmeter
N_d N_n	
N_t N_u	counters for ultrasonic measuring system
n	$1/n$ is the index for turbulent profile curve fit, index
P_R	ultrasound power reflected
P_T	ultrasound power transmitted
p	gas pressure
q_m	mass flow rate
q_v	volumetric flow rate
R	radius of pipe band
Re	Reynolds number
$R_{yx}(\tau)$	correlation coefficient
r	radial coordinate
T	correlation time for integration
t	pipe wall thickness
t_d	downstream wave transit time
t_m	mean transit time of up- and downstream waves
t_u	upstream wave transit time
Δt	difference between these two wave transit times
t_1,t_2	beam positions for ultrasonic flowmeter using Gaussian Quadrature
t_{umeas}	total measured time of upstream pulse in clamp-on flowmeter
t_{dmeas}	total measured time of downstream pulse in clamp-on flowmeter
t_{wedge}	delay time in wedge

t_{pipe}	delay time in pipe wall
$U_I^{(2)}$, $U_{II}^{(1)}$	received voltages for ultrasonic flowmeter
V	flow velocity in pipe
$V(x)$	flow velocity profile across the pipe
$\mathbf{V}_s$	undisturbed flow
$\mathbf{v}_o$	acoustic field for cases (1) and (2)
v	volume element in integration, relative flow velocity in the direction of the acoustic beam
v_{actual}	mean cross-sectional velocity in pipe section
$v_{measured}$	average of ultrasonic velocities measured on each beam
$\mathbf{W}$	vector weight function
X	axial length between transducers in ultrasonic flowmeter
Y	length of chord in the plane of the ultrasonic path
y	distance along chord used in ultrasonic flowmeter
Z	acoustic impedance
z	ultrasonic beam deflection distance on opposite wall
β	ultrasonic beam deflection angle
γ	ratio of specific heats for a particular gas
θ	ultrasound beam angle
λ	ultrasound wave length
λ_t	transmitted wavelength
ρ	density of material
ρ_m	mean density in ultrasonic meter
τ	measurement period for ultrasonic system, time period between ultrasonic wave peaks
τ_m	mean ultrasonic correlation time

Chapter 15

c	speed of sound in the mixture
c_l	speed of sound in the liquid
c_g	speed of sound in the gas
D	pipe diameter
f	frequency
k	wave number $= \frac{2\pi}{\lambda}$
L	spacing between sensors
Re	Reynolds number
$U_{convect}$	convection velocity or phase speed of the disturbance
V	flow velocity
α	gas fraction
λ	wavelength
μ	fluid dynamic viscosity

ρ	fluid density, density of the mixture
ρ_l	density of the liquid in the flow
ρ_g	density of the entrained gas
τ	transit time of eddies between sensors
ω	frequency (radians/second)

Chapter 16

A	area of duct
c	speed of sound
K	constant
k_s	adiabatic compressibility
p	pressure
Δp_A	pressure drop to throat of Venturi A
Δp_B	pressure drop to throat of Venturi B
Δp_{AB} etc.	differential pressure between Venturi throats, across limbs of the hydraulic Wheatstone bridge or across diagonals of the hydraulic Wheatstone bridge
q_m	total mass flow
q_{vp}	metering pump volumetric transfer flow
V	velocity in the meter
γ	ratio of specific heats
ρ	density of the fluid

Chapter 17

K	pressure loss coefficient
$q_{m,l}$	mass flow rate of liquid
$q_{m,g}$	mass flow rate of gas
X	Lockhart-Martinelli parameter
ρ_g	gas density
ρ_l	liquid density

Chapter 18

A	area of the duct, constant equal to $1/\pi$
B	constant equal to $(2/\pi)^{0.5}$
C,D	function of temperature
c_p	specific heat at constant pressure
c_v	specific heat at constant volume
D	pipe diameter
d	heating element diameter
g	gravitational constant

I	current through resistance R
K	constants
k	thermal conductivity of the fluid
k'	constant allowing for heat transfer and temperature difference at zero flow
L	finite difference dimension
Nu	Nusselt number
n	index
Pr	Prandtl number
p	pressure
Q_h	heat transfer
q_a	rate of heat addition per unit volume
q_h	heat flux
$\mathbf{q_h}$	heat flux vector
q_m	mass flow rate
q_v	volumetric flow rate
R	resistance of heating element
Re	Reynolds number
S	flow signal
S_0	flow signal at start
T	absolute temperature of the fluid
T_1, T_2, T_c	temperatures used in finite difference approximation
ΔT	measured temperature difference in °K
t	time from start of flow change
V	fluid velocity
$\mathbf{V}$	vector velocity
v	volume element in integration
$\mathbf{W}$	weight vector
ρ	fluid density
μ	dynamic viscosity of the fluid
ν	kinematic viscosity
τ	time constant of flowmeter

Chapter 19

q_m	mass flow
R	radius of the annulus of angular momentum meter
s	spring constant
X	angular momentum
θ	angular deflection
τ	time difference between markers on the angular momentum meter rotating assembly
ω	angular velocity of the rotor

Chapter 20

A	cross-sectional area of the pipe
A_f	tube internal cross-section
A_p	pipe wall cross-sectional area
c	speed of sound in fluid
D	internal diameter of pipe
d	width of U-tube
E	Young's modulus
F	force due to Coriolis acceleration
G	shear modulus
I_s	inertia in plane of twisting oscillation
I_u	inertia in plane of normal oscillation
I	pipe moment of inertia in direction of vibration
I_f	rotational inertia of fluid
I_p	rotational inertia of pipe
K	constant, allows for the fact that the twist of the tube will not form a straight integration
K_s	spring constant of the U-tube in twisting oscillation
K_u	spring constant of the U-tube in normal oscillation
k	shear correction factor
l	length of the U-tube, pipe length
δm	element of mass equal to $\rho A\delta r'$
q_m	mass flow
r	radius
δr	elementary length of tube
$\delta r'$	elementary length of fluid
T	torque, combined tension in pipe and fluid
t	time
u	transverse movement of pipe
V	flow velocity
$\mathbf{V}$	vector velocity
v	flow velocity, volume element in integration
v_0	velocity of fluid "string"
$\mathbf{v}^{(1)}$	oscillatory velocity fields set up in the stationary fluid by the driving oscillator
$\mathbf{v}^{(2)}$	oscillatory velocity field set up in the stationary fluid by the Coriolis forces
$\mathbf{W}$	vector weight function
x	axial coordinate
θ	twist angle of U-tube
θ_o	amplitude of twist angle of U-tube when in sinusoidal motion
ρ	density
ρ_f	fluid density

ρ_p	tube density
σ_0	initial stress in pipe
τ	difference in transit time of two halves of twisted U-tube
$\nabla\phi$	phase difference between the total velocities at the two sensing points
Ω	angular velocity of the pipe caused by the vibration
Ω_o	amplitude of the angular velocity of the pipe caused by the vibration in sinusoidal motion
ω	angular frequency, operating frequency, driving frequency
ω_s	natural frequency of U-tube in twisting oscillation
ω_u	natural frequency of U-tube in normal oscillation

Chapter 21

k	coefficient
M	Mach number
p	pressure
p_o	stagnation pressure
Δp	dynamic pressure
V	velocity
γ	ratio of specific heats
ρ	density

Chapter 22

K	loss coefficient
Δp	pressure drop
V	velocity
ρ	density

1

Introduction

1.1 Initial Considerations

Some years ago at Cranfield, where we had set up a flow rig for testing the effect of upstream pipe fittings on certain flowmeters, a group of senior Frenchmen was being shown around and visited this rig. The leader of the French party recalled a similar occasion in France when visiting such a rig. The story goes something like this.

A bucket at the end of a pipe seemed particularly out of keeping with the remaining high-tech rig. When someone questioned the bucket's function, it was explained that the bucket was used to measure the flow rate. Not to give the wrong impression in the future, the bucket was exchanged for a shiny, new, high-tech flowmeter. In due course, another party visited the rig and observed the flowmeter with approval. "And how do you calibrate the flowmeter?" one visitor asked. The engineer responsible for the rig then produced the old bucket!

This book sets out to guide those who need to make decisions about whether to use a shiny flowmeter, an old bucket, nothing at all or a combination of these! It also provides information for those whose business is the design, manufacture or marketing of flowmeters. I hope it will, therefore, be of value to a wide variety of people, both in industry and in the science base, who range across the whole spectrum from research and development through manufacturing and marketing. In my earlier book on flow measurement (Baker 1988a/1989, 2002b, 2003), I provided a brief statement on each flowmeter to help the uninitiated. This book attempts to give a much more thorough review of published literature and industrial practice.

This first chapter covers various general points that do not fit comfortably elsewhere. In particular, it reviews guidance on the accuracy of flowmeters (or calibration facilities).

The second chapter reviews briefly some essentials of fluid mechanics necessary for reading this book. The reader will find a fuller treatment in Baker (1996), which also has a list of books for further reading.

A discussion of how to select a flowmeter is attempted in Chapter 3, and some indication of the variety of calibration methods is given in Chapter 4, before going in detail in Chapters 5–20 into the various high- (and low-) tech meters available.

In this edition, I have introduced three additional chapters to cover new commercial meters and to allow a brief and superficial review of multiphase hydrocarbon flowmeters. Chapter 21 deals with probes. Chapter 22 covers general issues relating to verification and clamp-on meters. Chapter 23 provides a brief introduction to remote data handling and Chapter 24 provides final personal reflections relating to manufacture and future developments.

In this book, I have tried to give a balance between the laboratory ideal, manufacturers' claims, the realities of field experience and the theory behind the practice. I am very conscious that the development and calibration laboratories are sometimes misleading places, which omit the problems encountered in the field (Stobie 1993), and particularly so when that field happens to be the North Sea. This may be more serious for flowmeters than for some other instruments, and may require careful consideration of the increase in uncertainty which results. In the same North Sea Flow Measurement Workshop, there was an example of the unexpected problems encountered in precise flow measurement (Kleppe and Danielsen 1993), resulting, in this case, from a new well being brought into operation. It had significant amounts of barium and strontium ions, which reacted with sulphate ions from injection water and caused a deposit of sulphates from the barium sulphate and strontium sulphate that were formed.

With that salutary reminder of the real world, we ask an important – and perhaps unexpected – question.

1.2 Do We Need a Flowmeter?

Starting with this question is useful. It may seem obvious that anyone who looks to this book for advice on selection is in need of a flowmeter, but for the process engineer it is an essential question to ask. Many flowmeters and other instruments have been installed without careful consideration being given to this question and without the necessary actions being taken to ensure proper documentation, maintenance and calibration scheduling. They are now useless to the plant operator and may even be dangerous components in the plant. Thus, before a flowmeter is installed, it is important to ask whether the meter is needed, whether proper maintenance schedules are in place, whether the flowmeter will be regularly calibrated, and whether the company has allocated to such an installation the funds needed to achieve this ongoing care. Such care will need proper documentation.

The water industry in the United Kingdom has provided examples of the problems associated with unmaintained instruments. Most of us involved in the metering business will have sad stories of the incorrect installation or misuse of meters. Reliability-centred maintenance recognises that the inherent reliability depends on the design and manufacture of an item, and if necessary this will need improving (Dixey 1993). It also recognises that reliability is preferable in critical situations to extremely sophisticated designs, and it uses failure patterns to select preventive maintenance.

In some research into water consumption and loss in urban areas, Hopkins, Savage and Fox (1995) found that obstacles to accurate measurements were

- buried control valves,
- malfunctioning valves,
- valve gland leakage,
- hidden meters that could not be read and
- locked premises denying access to meters.

They commented that "water supply systems are dynamic functions having to be constantly expanded or amended. Consequently continuous monitoring, revisions and amendments of networks records is imperative. Furthermore, a proper programme of inspection, maintenance and subsequent recording must be operative in respect of inter alia:

- networks,
- meters,
- control valves,
- air valves,
- pressure reducing valves,
- non-return valves."

They also commented on the poor upstream pipework at the installation of many domestic meters.

So I make no apology for emphasising the need to assess whether a flowmeter is actually needed in any specific application.

If the answer is yes, then there is a need to consider the type of flowmeter and whether the meter should be measuring volume or mass. In most cases, the most logical measure is mass. However, by tradition, availability and industrial usage, volume measurement may be the norm in some places, and as a result, the regulations have been written for volume measurement. This results in a Catch-22 situation. The industry and the regulations may, reasonably, resist change to mass flow measurement until there is sufficient industrial experience, but industrial experience is not possible until the industry and the regulations allow. The way forward is for one or more forward-looking companies to try out the new technology and obtain field experience, confidence in the technology and approval.

In this book, I have made no attempt to alert the reader to the industry-specific regulations and legal requirements, although some are mentioned. The various authors touch on some regulations, and Miller (1996) is a source of information on many documents. An objective of the Organisation Internationale de Métrologie Légale (OIML) is to prevent any technical barriers to international trade resulting from conflicting regulations for measuring instruments. With regard to flow measurement, it appears to have been particularly concerned with the measurement of domestic supplies and industrial supplies of water and gas (Athane 1994). This is because two parties, the supplier and the consumer, are involved, and the consumer is unlikely to be able to ascertain the correct operation of the meter. In addition, the supplier does not continually monitor these measurements, the meters may fail without anyone knowing, the usage is irregular and widely varying in rate, the measurements are not repeatable, and the commodities have increased in value considerably in recent years.

In order to reduce discussions and interpretation problems between manufacturers and authorised certifying institutes, the European Commission was mandating the European standardisation bodies (CEN and CENELEC) to develop harmonised standards that would give the technical details and implementation of the requirements based on OIML recommendations. These would be such that a measuring instrument complied with essential requirements, assuming that the manufacturer had complied with them (Nederlof 1994).

The manufacturer will also be fully aware of the electromagnetic compatibility (EMC), which relates to electromagnetic interference. In particular, the EMC characteristics of a product are that

- the level of electromagnetic disturbance the instrument generates will not interfere with other apparatuses, and
- the operation of the instrument will not be adversely affected by electromagnetic interference from its environment.

In order to facilitate free movement within the European area, the CE mark was designed to identify products that conformed to the European essential requirements. For further details relating to the European Community (EC), the reader is referred to the Measuring Instrument Directive (MID 2004, DTI 1993, Chambers 1994).

First, we consider the knotty problem of how accurate the meter should be.

1.3 How Accurate?

Inconsistency remains about the use of terms that relate to accuracy and precision. This stems from a slight mismatch between the commonly used terms and those that the purists and the standards use. Thus we commonly refer to an accurate measurement, when strictly we should refer to one with a small value of uncertainty. We should reserve the use of the word *accurate* to refer to the instrument. A high-quality flowmeter, carefully produced with a design and construction to tight tolerances and with high-quality materials as well as low wear and fatigue characteristics, is a precise meter with a quantifiable value of repeatability. Also, it will, with calibration on an accredited facility, be an accurate meter with a small and quantifiable value of measurement uncertainty. In the context of flowmeters, the word *repeatability* is preferred to *reproducibility*. The meanings are elaborated on later, and I regret the limited meaning now given to *precision*, which I have used more generally in the past and shall slip back into in this book from time to time! In the following chapters, I have attempted to be consistent in the use of these words. However, many claims for accuracy may not have been backed by an accredited facility, but I have tended to use the phrase "measurement uncertainty" for the claims made.

Hayward (1977a) used the story of William Tell to illustrate precision. William Tell had to use his crossbow to fire an arrow into an apple on his little son's head. This was a punishment for failing to pay symbolic homage to an oppressive Austrian ruler. Tell succeeded because he was an archer of great skill and high accuracy.

An archer's ability to shoot arrows into a target provides a useful illustration of some of the words related to precision. So Figure 1.1(a) shows a target with all the shots in the bull's-eye. Let us take the bull's-eye to represent ±1%, within the first ring ±3%, and within the second ring ±5%. Ten shots out of ten are on target, but how many will the archer fire before one goes outside the bull's-eye? If the archer, on average, achieves 19 out of 20 shots within the bull's-eye [Figure 1.1(b)], we say that the archer has an uncertainty of ±1% (the bull's-eye) with a 95% confidence level (19 out of 20 on the bull's-eye: $19 \div 20 = 0.95 = 95 \div 100 = 95\%$).

Suppose that another archer clusters all the arrows, but not in the bull's-eye, Figure 1.1(c). This second archer is very consistent (all the shots are within the same size circle as the bull's-eye), but this archer needs to adjust his aim to correct the offset. We could say that the second archer has achieved high repeatability of ±1%, but with a bias of 4%. We might even find that 19 out of 20 shots fell within the top left circle so that we could say that this archer achieved a repeatability within that circle of ±1% with a 95% confidence. Suppose this archer had fired one shot a day, and they had all fallen onto a small area [Figure 1.1(c)], despite slight changes in wind, sunshine and archer's mood; we term this good day-to-day repeatability. But how well can we depend on the archer's bias? Is there an uncertainty related to it?

Finally, a third archer shoots 20 shots and achieves the distribution in Figure 1.1(d). One has missed entirely, but 19 out of 20 have hit the target somewhere. The archer has poor accuracy, and the uncertainty in this archer's shots is about five times greater than for the first, even though the confidence level at which this archer performs is still about 95%.

If the third archer has some skill, then the bunching of the arrows will be greater in the bull's-eye than in the next circle out, and the distribution by ring will be as shown in Figure 1.1(e).

We shall find that the distribution of readings of a flowmeter results in a curve approximating a normal distribution with a shape similar to that for the shots. Figure 1.1(f) shows such a distribution where 95% of the results lie within the shaded area and the width of that area can be calculated to give the uncertainty, ±1% say, of the readings with a 95% confidence level. In other words, 19 of every 20 readings fall within the shaded area.

With this simplistic explanation, we turn to the words that relate to precision.

Accuracy

It is generally accepted that *accuracy* refers to the truthfulness of the instrument. An instrument of high accuracy more nearly gives a true reading than an instrument of low accuracy. Accuracy, then, is the quality of the instrument. It is common to refer to a measurement as accurate or not, and we understand what is meant. However,

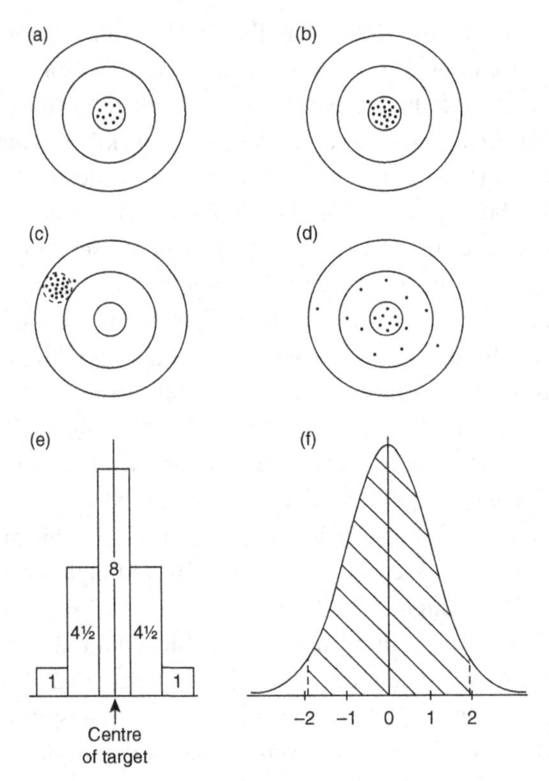

Figure 1.1. Precision related to the case of an archery target. **(a)** Good shooting – 10 out of 10 arrows have hit the bull's-eye. An accurate archer? **(b)** Good shooting? – 19 out of 20 arrows have hit the bull's-eye. An accurate archer and a low value of uncertainty (±1%) with a 95% confidence level. (**c**) Shots all fall in a small region but not the bull's-eye. Good repeatability (±1%) but a persistent bias of 4%. (**d**) Shots, all but one, fall on the target – 19 out of 20 have hit the target. A ±5% uncertainty with 95% confidence level. **(e)** Distribution of shots in (d) on a linear plot, assuming that we can collapse the shots in a ring semicircle onto the axis. **(f)** The normal distribution, which is a good approximation for the distribution of flowmeter readings.

the current position is that accuracy should be used as a qualitative term and that no numerical value should be attached to it. It is, therefore, incorrect to refer to a measurement's accuracy of, say, 1%, when, presumably, this is the instrument's measurement uncertainty, as is explained later.

Repeatability

In a process plant, or other control loop, we may not need to know the accuracy of a flowmeter as we would if we were buying and selling liquid or gas, but we may require repeatability within bounds defined by the process. *Repeatability* is the value below which the difference between any two test results, taken under constant conditions with the same observer and with a short elapsed time, are expected to lie with 95% confidence.

Precision

Precision is the qualitative expression for repeatability. It should not take a value and should not be used as a synonym for accuracy.

Uncertainty

Properly used, *uncertainty* refers to the quality of the measurement, and we can correctly refer to an instrument reading having an uncertainty of ±1%. By this we mean that the readings will lie within an envelope ±1% of the true value. Each reading will, of course, have an individual error that we cannot know in practice, but we are interested in the relationship of the readings to the true value. Because *uncertainty* is referred to the true value, by implication it must be obtained using a national standard document or facility. However, because it is a statistical quantity, we need also to define how frequently the reading does, in fact, lie within the envelope; hence the confidence level.

Confidence Level

The *confidence level*, which is a statement of probability, gives this frequency, and it is not satisfactory to state an uncertainty without it. Usually, for flow measurement, this is 95%. We shall assume this level in this book. A confidence level of 95% means that we should expect on average that 19 times out of 20 (19/20 = 95/100 = 95%) the reading of the meter will fall within the bracket specified (e.g. ±1% of actual calibrated value).

Linearity

Linearity may be used for instruments that give a reading approximately proportional to the true flow rate over their specified range. It is a special case of *conformity* to a curve. Note that both terms really imply the opposite. *Linearity* refers to the closeness within which the meter achieves a truly linear or proportional response. It is usually defined by stating the maximum deviation (or nonconformity e.g. ±1% of flow rate) within which the response lies over a stated range. With modern signal processing, linearity is probably less important than conformity to a general curve. *Linearity* is most commonly used with such meters as the turbine meter.

Range and Rangeability

An instrument should have a specified range over which its performance can be trusted. Therefore, there will be upper- and lower-range values. This reflects the fact that probably no instrument can be used to measure a variable when there are no limitations on the variable. Without such a statement, the values for uncertainty, linearity etc. are inadequate. The ratio of upper-range value and lower-range value may be called the *rangeability*, but it has also been known as the *turndown ratio*. The difference between upper- and lower- or negative-range values is known as *span*. It is important to note whether the values of uncertainty, linearity etc. are a percentage of the actual flow rate or of the full-scale flow [sometimes referred to as full-scale deflection (FSD), full-scale reading (FSR), maximum-scale value, or upper-range value (URV)].

1.4 A Brief Review of the Evaluation of Standard Uncertainty

Kinghorn (1982) points out the problem with terminology in matters concerning statistics and flow measurement. To the engineer and the statistician, words such as *error* and *tolerance* may have different meanings. The word *tolerance* was used for what is now known as uncertainty.

In providing an introduction to the terminology of uncertainty in measurement, I shall aim to follow the guidance in BIPM et al. (2008), which is usually known as the *Guide* or GUM, and also in a document consistent with the *Guide*, which provides the basis for uncertainty estimates in laboratories accredited in the United Kingdom (UKAS 2012). The reader should note that the *Guide* may also be available as ISO/IEC 98-3: 2008 and that a further valuable document is a guide on the vocabulary of metrology, ISO/IEC 99: 2007. The reader is strongly advised to consult this document, which is full of clear explanations and useful examples. Those wishing to pursue background arguments are referred to Van der Grinten's (1994, 1997) papers.

Random error, the random part of the experimental error, causes scatter, as the name suggests, and reflects the quality of the instrument design and construction. It is the part that cannot be calibrated out, and the smaller it is, the more precise the instrument is. It may be calculated by taking a series of repeat readings resulting in the value of the standard deviation of a limited sample n, and sometimes called the experimental standard deviation:

$$s(q_j)=\left\{\frac{1}{n-1}\sum_{j=1}^{n}\left(q_j-\bar{q}\right)^2\right\}^{1/2} \tag{1.1}$$

where $\bar{q}$ is the mean of n measurements q_j. The experimental standard deviation of the mean of this group of readings is given by

$$s(\bar{q})=\frac{s(q_j)}{\sqrt{n}} \tag{1.2}$$

Where too few readings have been taken to obtain a reliable value of $s(q_j)$, an earlier calculation of $s(q_j)$ from previous data may be substituted in Equation (1.2). In obtaining the overall uncertainty of a flowmeter or a calibration facility, there will be values of group mean experimental standard deviation for various quantities, and so UKAS (2012) defines a standard uncertainty for the *i*th quantity as

$$u(x_i) = s(\bar{q}) \tag{1.3}$$

where x_i is one of the input quantities. For those with access to UKAS (2012), this is, essentially, dealt with there as a Type A evaluation of standard uncertainty.

Systematic error, according to flowmeter usage, is that which is unchanging within the period of a short test with constant conditions. This is, essentially, dealt with in UKAS (2012) under the heading Type B evaluation of standard uncertainty. It is also called *bias*. However, in modern flowmeters and in calibration facilities, it is likely that this bias or systematic error will result in a meter adjustment, or a rig correction. The resulting uncertainty in that adjustment or correction will contribute to the overall uncertainty. The systematic uncertainty, therefore, may derive from various factors such as

a. uncertainty in the reference and any drift,
b. the equipment used to measure or calibrate,
c. the equipment being calibrated in terms of resolution and stability,
d. the operational procedure, and
e. environmental factors.

From these we deduce further values of $u(x_i)$.

There has been debate about the correct way to combine the random and systematic uncertainties. We can combine random and systematic uncertainties conservatively by arithmetic addition. This results in a conservative estimate. UKAS (2012) has followed the *Guide* in taking the square root of the sum of the squares of the standard uncertainties in consistent units. Thus the combined standard uncertainty is

$$u_c(y) = \sqrt{\sum [c_i\, u(x_i)]^2} \tag{1.4}$$

where y is the output quantity. To ensure consistent units, a sensitivity coefficient, c_i, will be required for each input x_i, although in practice this may be unity in most cases (as in Figure 4.3).

The final step (and we have glossed over many important details in UKAS 2012) is to deduce from u_c the bracket within which the reading of, say, the meter lies.

In the past, bearing in mind that u_c or its components have been derived from standard deviations, we have used Student's t value, which for a number of readings n is given by

n	t
10	2.26
20	2.09
>30	2.0

for a 95% confidence level. The *Guide* replaces this, in general, with a coverage factor, k, to obtain the expanded uncertainty

$$U = ku_c(y) \tag{1.5}$$

The recommended value is $k = 2$, which gives a confidence level of 95.45% taken as 95%, assuming a normal distribution. If this assumption is not adequate, then we need to revert to Student's t.

The net result is that the assumption of a factor of 2 has now been given a systematic basis. The reader who is interested in more details about the basis of normal and t distributions is referred to Appendix 1.A.

1.5 Note on Monte Carlo Methods

An alternative approach, which is an outcome of the speed and accessibility of personal computers, is based on the use of random number generators to model instrument errors, and on running many tests to obtain the overall uncertainty of the system, say, a flow calibration rig. This is known as the Monte Carlo method for assessing uncertainty.

Not being a statistician, my perception of these methods is that, essentially, a numerical model of the measurement system is set up on a computer, instrument and system errors are modelled using values obtained from a random number generator and the measurement procedure is, thereby, modelled. The program is then run very many times, to obtain the likely uncertainty by averaging all the results. The procedure may be less conservative in its assessment than the standard GUM (BIPM et al. 1993) approach.

Monte Carlo computer programs are available, some as freeware. One or more such programs may be specifically modelled on the latest GUM approach (e.g. GUM-Workbench may be available) (private communication from Peter Lau).

Some explanation of the procedure can be found in Coleman and Steele (1999).

1.6 Sensitivity Coefficients

Suppose that output quantity, a flow rate, has the relationship

$$y = x_1^p x_2^q x_3^r x_4^s \tag{1.6}$$

then if x_2, x_3 and x_4 are held constant, we can differentiate y with respect to x_1 and obtain the partial derivative. This is the slope of the curve of y against x_1 when the other variables are kept constant. It also allows us to find the effect of a small change in x_1 on y. This slope (or partial derivative) is the sensitivity coefficient c_1 for x_1 and may be found by calculation. It will have the value $c_1 = px_1^{(p-1)} x_2^q x_3^r x_4^s$, where the values of x_1, x_2, x_3 and x_4 will be at the calibration point and may be dimensional. In some cases, it may be a known coefficient (e.g. a temperature coefficient of expansion). For cases where it is difficult to calculate, it may be possible to find the coefficient by changing x_1 by a small amount and observing the change in y. In some cases, the sensitivity coefficient may provide a conversion between different sets of units (e.g. where output quantity or velocity may be obtained from a dimension, a pressure, a movement or a voltage).

1.7 What Is a Flowmeter?

We take as a working definition of an ideal flowmeter:

> A group of linked components that will deliver a signal uniquely related to the flow rate or quantity of fluid flowing in a conduit, despite the influence of installation and operating environment.

The object of installing a flowmeter is to obtain a measure of the flow rate, usually in the form of an electrical signal, which is unambiguous and with a specified expanded uncertainty. This signal should be negligibly affected by the inlet and outlet pipework and the operating environment. Thus the uncertainty of measurement of a flowmeter should be reported as $y \pm U$, where U, the uncertainty band, might have a value of, say, 0.5%, and it should be made clear whether this is related to rate, full-scale deflection (FSD) or other value that might be a combination of these [e.g. in the form $\pm a$ (rate) $\pm b$ (FSD)]. The range should be given (e.g. 1 m^3/h to 20 m^3/h).

The statement of performance should include the coverage factor $k = 2$ and the level of confidence of approximately 95%, and, if appropriate, the authority that accredited the calibration facility (national or international).

In addition, the ranges of properties for which it can be used should be specified, such as fluid, flow range (beyond calibration), maximum working pressure, temperature range of fluid and ambient temperature range.

It is useful to introduce two factors that define the response of flowmeters, although they are most commonly used for linear flowmeters with pulse output. The K factor is the number of pulses per unit quantity. In this book, we shall take it as number of pulses per unit volume when dealing with turbine and vortex meters:

$$K = \frac{\text{Pulses}}{\text{True volume}}$$

whereas the meter factor is usually defined as

$$\text{Meter factor} = \frac{\text{True volume}}{\text{Indicated volume}}$$

The reader should keep a wary eye for other definitions of meter factor such as the reciprocal of the K factor.

Let us take a specific example of a fictitious, but reasonably realistic flowmeter. In Figure 1.2(a), a typical flowmeter envelope is shown. It defines an approximately linear flowmeter with a 10:1 turndown and an uncertainty of ±1% of rate with a confidence level of 95% against a traceable standard calibration. This is a reasonable performance for a flowmeter and probably satisfies most industry requirements. This, let us assume, is the performance specification the manufacturer carries in its sales literature. Actually the characteristic of the flowmeter may be the curve shown in Figure 1.2(a). If the company works to a high standard of manufacture, then the company may know that this characteristic lies within close tolerances in all cases. It may, therefore, only be necessary for the manufacturer to calibrate each flowmeter at, say, 90% of FSD, or 50% and 90% of FSD, in order to make the claim that the characteristic falls within the envelope specified in the sales literature.

If the meters are actually of this standard, it may well be feasible to calibrate them in much greater detail so that a 5-, 10- or even 20-point calibration may provide a characteristic that ensures that the reading is known to, say, ±0.2%. The values obtained from the calibration will then be programmed into a flow computer, which will interpret each reading of the flowmeter against this look-up table. Since we are comparing the flowmeter's signal to a linear one, if it were without error it would also be linear. Consequently, companies sometimes record linearity within their literature. In this case, it would also be ±1% with a 10:1 turndown.

The envelope just discussed gives the uncertainty at each flow rate in terms of the actual flow rate. Because of the physical basis of some flowmeters, this method is not appropriate, and the uncertainty may then be given in terms of the full scale. Figure 1.2(b) shows such an envelope where the performance of the flowmeter would be defined as ±1% FSD. It is apparent that the uncertainty in the flowmeter's reading at, say, half scale is ±2% of rate and at 20% of reading will be as much as ±5% of rate. The problem often arises wherein the user has a particular flow range that does not match that of the actual instrument. The user's full flow may be at only 60% of the instrument's range, and so for the user the instrument has an uncertainty, at best, of 1.7% of rate.

A third type of envelope is shown in Figure 1.2(c). This is particularly common in the specifications for water and gas meters. In the example shown, the meter has an uncertainty of ±2% of rate from full flow down to 20%. Below this value of flow rate, the uncertainty is ±5% of flow rate down to 2% of range. In practice, a meter might have more steps in its envelope.

In many cases, the manufacturer's specification of uncertainty may be a combination of these. As indicated earlier, it is common to have an uncertainty that combines a value based on rate and another on full-scale deflection. In addition, there may be allowances to be added for zero drift, temperature change and, possibly, even pressure change. In some flowmeters, viscosity is important but is probably accommodated by charts showing the variation in performance with viscosity.

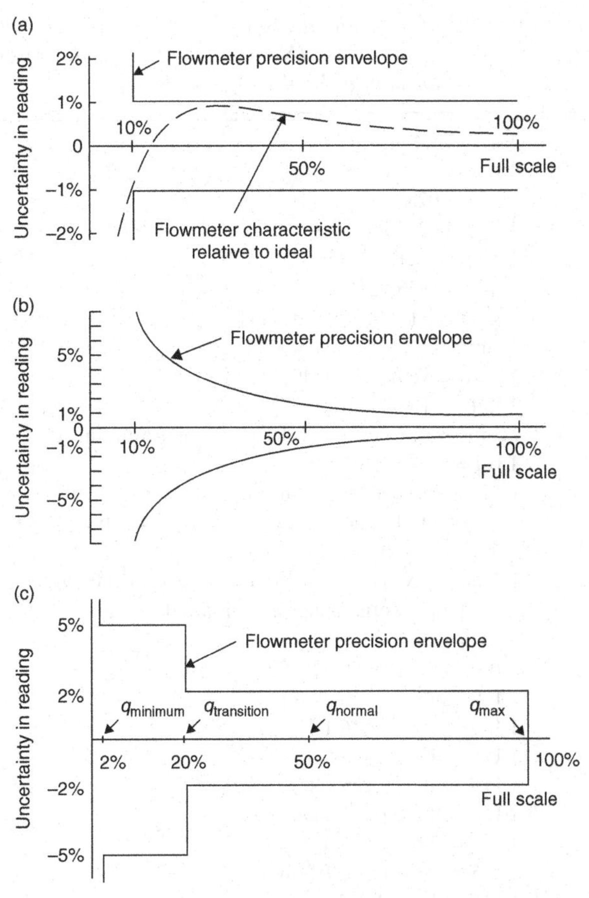

Figure 1.2. Required envelope for a flowmeter. **(a)** Envelope as a percentage of rate; **(b)** envelope as a percentage of FSD; **(c)** stepped envelope with increased uncertainty at low-range values. The Measuring Instruments Directive of the European Commission should be consulted for the latest definitions of Qmax etc. for the EC.

One note of caution! Clever electronics can take any signal, however nonlinear, and straighten the characteristic before the signal is output. Suppose such a procedure were used for the characteristic in Figure 1.2(a), below 10% of range. The characteristic is probably very sensitive to minor variations in this region, and any attempt to use the characteristic could lead to disguised, but serious, errors.

Variation of temperature and pressure can affect the performance of a flowmeter, as can humidity, vibration and other environmental parameters.

Often the units used in a manufacturer's catalogue are not those that you have calculated. For this reason, conversion factors that provide conversions to four significant figures of flow rate, velocity, temperature, pressure, length etc. have been included in Table 1.1. If not otherwise specified, the International System of units (SI) based on meter, kilogram, second is assumed.

Table 1.1. *Conversion for some essential measurements from Imperial, U.S. and other units to metric, to four significant figures (Note that the EC has time limits by which certain standard units must be used.)*

Length	1 in[a] = 25.4 mm
	1 ft = 0.3048 m
Volume	1 ft^3 = 0.0283 m^3
	1 ft^3 = 28.32 l (litre)
	1 bbl (barrel) = 0.1590 m^3
Mass	1 lb = 0.4536 kg
	1 long ton (2,240 lb) = 1,016 kg
	1 short ton (2,000 lb) = 907.2 kg
	1 metric tonne (2,205 lb) = 1,000 kg
Density	1 lb/ft^3 = 16.02 kg/m^3
Temperature	(Temperature in °F – 32)/1.8 = Temperature in °C
Pressure	1 psi = 6,895 N/m^2
Viscosity	Dynamic viscosity: SI (metric) unit is the Pascal second (Pas) to which the more common unit, the centipoise (cP), is related by 1 cP = 10^{-3} Pas.
	Kinematic viscosity: SI (metric) unit is m^2/s to which the more common unit, the centistoke (cSt), is related by 1 cSt = 10^{-6} m^2/s = 1 mm^2/s.
Velocity	1 ft/s = 0.3048 m/s
Volumetric flow rate	1 ft^3/s (1 cusec) = 0.02832 m^3/s
	1 Imp gal/s = 0.004546 m^3/s
	1 Imp gal/s = 4.546 1/s
	1 U.S. gal/s = 0.003785 m^3/s
	1 U.S. gal/s = 3.785 1/s
	1 Imp gal/h = 0.004546 m^3/h
	1 Imp gal/h = 4.546 1/h
	1 U.S. gal/h = 0.003785 m^3/h
	1 U.S. gal/h = 3.785 1/h
Mass flow rate	1 lb/h = 0.4536 kg/h
	1 lb/s = 0.4536 kg/s

[a] An approximate conversion is 4 in. to 100 mm.

1.8 Chapter Conclusions (for those who Plan to Skip the Mathematics!)

I have tried to bring together, within the compass of this book, essential information for all who may have dealings with flowmeters and flow measurement, although I have tried to avoid duplicating information available in other excellent books on the subject. For this reason, the chapters not only address the technical aspects but also the selection, maintenance, calibration and typical applications of the various meters. I hope that this book will provoke the prospective entrepreneur, the small and medium-sized enterprises (SME) or the major instrument company to assess the market needs and the relevant development and production needs of their companies for new devices.

The management of flowmeters, at all stages from selection through application in complex systems, to identifying malfunction, is clearly an area where modern information technology methods would be attractive. How does one select? How do we allow for the costs of ownership? How can we check performance and identify emerging problems? De Boom (1996) considered life-cycle analysis to help users select the most appropriate technology for their use. Menendez, Biscarri and Gomez (1998) used a model of a water supply net to deduce the errors in flow measurement from: analysis of the system, assignment to the meters of flow measurement uncertainty, estimate of flow distribution in the net and comparison of the estimated values with the measured values. Nilsson and Delsing (1998) and Nilsson (1998) considered malfunctions and inaccuracies in gas flowmeters. Scheers and Wolff (2002) saw the production measurement process from field data collection to final reporting as the entire chain.

As the reader moves into the following chapters, two sets of information may be useful. Table 1.1 lists the conversion factors for Imperial, U.S. and metric units, and Table 1.2 relates volumetric and mass flow rate to linear velocity in various sizes of tube (Baker 1988a/1989, 2002b, 2003). It is common in flow measurement to require the velocity of flow, and Table 1.2 provides an order of magnitude.

Finally, the whole matter of accuracy and the limits of accuracy, when related to all the parameters that influence a flowmeter's operation, remains an area with unanswered questions.

1.9 Mathematical Postscript

I have left this note to the end so that those who are not concerned with advanced mathematical concepts can ignore it.

I have included essential mathematics in the main text of this book. In certain flowmeters, the mathematical theory is more complex (e.g. the turbine meter), and the theory has, accordingly, been consigned to an appendix after the relevant chapter.

One important and interesting mathematical approach, which starts to develop a unified theory of flow measurement, was first suggested by Shercliff (1962) and significantly extended by Bevir (1970). Both applied it to electromagnetic flowmeters, where it has been highly successful. Hemp (1975) has also applied this theory to electromagnetic flowmeters, but he has developed the theory for other types of flowmeters: ultrasonic (1982), thermal mass (1994a) and Coriolis (1994b, and Hemp and Hendry 1995). In Chapter 12 on electromagnetic flowmeters, an appendix describes the essential mathematics. The weight function developed in this theory provides a measure of the importance of flow in each part of the meter with respect to the overall meter signal. The flow at each point of a cross-section is weighted with this function. Ideally, the weighting should result in a true summation of the flow in the meter to obtain a volume flow rate. It has been possible to approach this ideal for the electromagnetic flowmeter. For the other types of meters, the reader will be given only a brief explanation and will be referred to relevant papers.

A second mathematical physics theory, first (to my knowledge) applied by Hemp (1988) to flow measurement, is reciprocity. This, essentially, states that if you apply a voltage to one end of an electrical network and measure the current at the other end, you find that by reversing the ends and hence the direction you obtain the same relationship. Hemp has proposed this as a means of eliminating some errors in flowmeters to which the theory is applicable.

Table 1.2. *Velocity in pipes for various flow rates to two significant figures*

	m^3/h[a]	l/min	gal/min	gal/min	ft^3/min	Mean Velocity (m/s) in a Circular Pipe of Diameter							
						10 mm	25 mm	50 mm	100 mm	200 mm	500 mm	1000 mm	2000 mm
Very low	10^{-3}	0.017	3.7×10^{-3}	4.4×10^{-3}	5.9×10^{-4}	3.5×10^{-3}	5.7×10^{-4}	1.4×10^{-4}	3.5×10^{-5}				
	10^{-2}	0.17	3.7×10^{-2}	4.4×10^{-2}	5.9×10^{-3}	3.5×10^{-2}	5.7×10^{-3}	1.4×10^{-3}	3.5×10^{-4}	8.8×10^{-5}	1.4×10^{-5}		
	0.1	1.7	0.37	0.44	5.9×10^{-2}	0.35	5.7×10^{-2}	1.4×10^{-2}	3.5×10^{-3}	8.8×10^{-4}	1.4×10^{-4}	3.5×10^{-5}	
	1	17	3.7	4.4	0.59	3.5	0.57	0.14	3.5×10^{-2}	8.8×10^{-3}	1.4×10^{-3}	3.5×10^{-4}	8.8×10^{-5}
	10	170	37	44	5.9	35	5.7	1.4	0.35	8.8×10^{-2}	1.4×10^{-2}	3.5×10^{-3}	8.8×10^{-4}
	100	1,700	370	440	59	350	57	14	3.5	0.88	0.14	3.5×10^{-2}	8.8×10^{-3}
	1,000	1.7×10^4	3700	4,400	590	3,500	570	140	35	8.8	1.4	0.35	8.8×10^{-2}
	10^4	1.7×10^5	3.7×10^4	4.4×10^4	5,900	3.5×10^4			350	88	14	3.5	0.88
	10^5	1.7×10^6	3.7×10^5	4.4×10^5	5.9×10^4	3.5×10^5				880	140	35	8.8
Very high	10^6	1.7×10^7	3.7×10^6	4.4×10^6	5.9×10^5	3.5×10^6						350	88

Reproduced from Baker (1988a/1989) with permission of Professional Engineering Publishing.

[a] Since water has a density of 1,000 kg/m^3 (approximately), the mass flow rate in kilograms per hour of water may be obtained by multiplying this column by 1,000.

In preparing this second edition, I have added new material to the appendices. This is partly because much of the new material goes further into the mathematics or into experimental research, but also to accommodate many new references without disturbing the flow of the main text.

Appendix 1.A Statistics of Flow Measurement

1.A.1 Introduction

The engineer's main needs are to

- understand and be able to give a value to the uncertainty of a particular measurement;
- know how to design a test to provide data of a known uncertainty;
- be able to combine measurements, each with its own uncertainty, into an overall value; and
- determine the uncertainty of an instrument at the end of a traceable ladder of measurement.

The international and national documents set the recommended approach for flow metering. Most standard statistics books will provide the essentials (Rice 1988; cf. Campion, Burns and Williams 1973, which is often quoted but may not be easy to obtain), but good school texts may be more accessible (Crawshaw and Chambers 1984, Eccles, Green and Porkess 1993a, 1993b). Hayward (1977c) is an extremely well-written and elegant little book, which deserves to be updated and reprinted; Kinghorn (1982) provides a well-written and useful brief review of the main points; and Mattingly (1982) addresses some of the problems concerned with transfer standards. I would also draw the reader's attention to an excellent book on experimentation and uncertainty analysis by Coleman and Steele (1999).

1.A.2 The Normal Distribution

The normal distribution, Figure 1.A.1(a), is also known as the Gaussian distribution after Carl Friedrich Gauss, who proposed it as a model for measurement errors (Rice 1988). The notation used for the normal curve is $N(\mu, \sigma^2)$, which is the distribution under the curve

$$f(x)=\frac{1}{\sigma\sqrt{2\pi}}e^{-\frac{1}{2}[(x-\mu)/\sigma]^2} \tag{1.A.1}$$

where μ is the mean value of the data, and σ^2 is the variance. Alternatively, σ is the standard deviation for the whole population. We can simplify the curve (normalise it) by putting $z = (x - \mu)/\sigma$ and obtaining [Figure 1.A.1(b)]

$$\phi(z)=\frac{1}{\sqrt{2\pi}}e^{-\frac{1}{2}z^2} \tag{1.A.2}$$

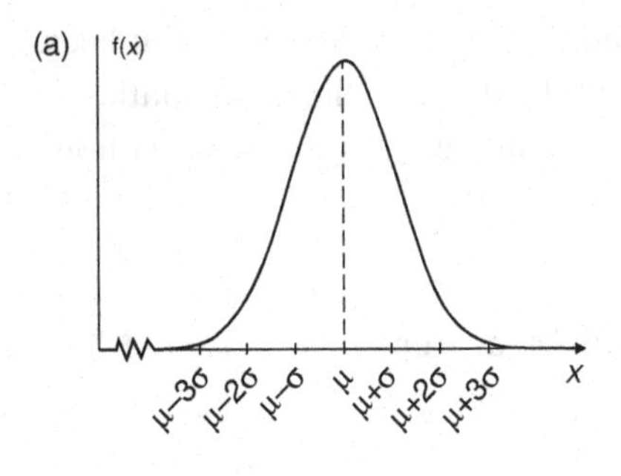

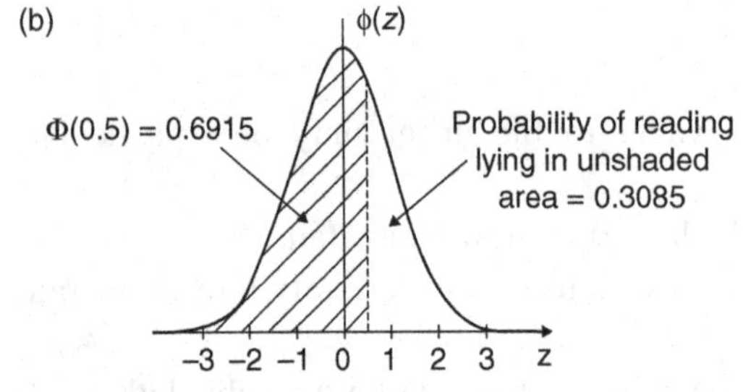

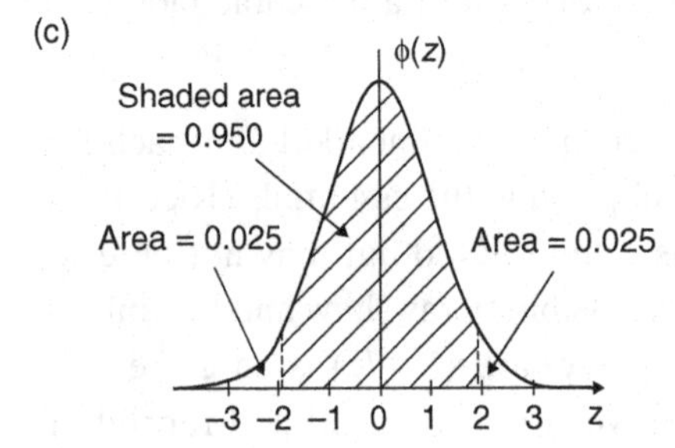

Figure 1.A.1. The normal distribution.

With the form of Equation (1.A.2), the curve does not vary with the size of the parameters μ and σ.

What the curve tells us (in relation to instrument measurements) is that the statistical chance of an instrument reading giving a value near to the mean μ is high, but the farther away the reading is from the mean, the less the chance is of its occurring (indicated by the curve decreasing in height the further one moves from the mean), and as values of the reading get farther still from μ, so the chance gets less and less.

The area under the curve of Equation (1.A.2) [Figure 1.A.1(b)], which reaches to infinity each way, is unity, and this is the probability of the reading lying within this curve (obviously). The area under the curve between $z = -\infty$ and some other value of z is given by

$$\Phi(z) = \frac{1}{\sqrt{2\pi}} \int_{-\infty}^{z} e^{-\frac{1}{2}t^2}\, dt \tag{1.A.3}$$

and is the probability that a reading will lie within that range and is obtained numerically and given in Table 1.A.1 in normalised form. For instance, if $z = 0.5$,

Table 1.A.l. *A selection of values from the normal distribution function $\Phi(z)$*

z	$\Phi(z)$	Symmetrical central area under curve
0	0.5000	
0.5	0.6915	
1.28	0.8997	
1.282	0.9000	0.80
1.29	0.9015	
1.64	0.9495	
1.645	0.9500	0.90
1.65	0.9505	
1.96	0.9750	0.95
2.57	0.99492	
2.576	0.99500	0.99
2.58	0.99506	
3.29	0.99950	0.999
3.30	0.99952	

After D. V. Lindley and W. F. Scott, *New Cambridge Statistical Tables*, 2nd ed., Cambridge: Cambridge University Press. Table 4, pp. 34, 35.

$\Phi(z) = 0.6915$, where $\Phi(z)$ is the area under the curve from $z = -\infty$ to $z = 0.5$ in this case. So the chance of a reading lying beyond this point is 0.3085, or about 30%.

We shall be interested in the chance that a flowmeter reading will fall between certain limits each side of $z = 0$, the mean value. A chance of 95% is often used and is called a 95% confidence level. This means that 19 times out of 20 the reading will fall between the limits. This requires that the central area of the curve [Figure 1.A.1(c)] has a value of 0.95, or 0.475 each side of the mean. To obtain z from this value, we need to add $0.475 + 0.5 = 0.975$, and this gives a value (Table 1.A.l) of $z = 1.96$. If we put this in terms of x, we obtain

$$x-\mu=1.96\,\sigma \tag{1.A.4}$$

or the band around the mean value of the reading within which 95% of the readings statistically should fall, is approximately $\pm 2\sigma$, or two standard deviations from the mean.

If we are interested, not in the spread of individual readings, but in the spread of the mean of small sets of readings, a statistical theorem called the Central Limit Theorem provides the answer. If a sample of n readings has a mean value of M, then the distribution of means like M is given by $N(\mu, \sigma^2/n)$. This is intuitively reasonable because one would expect that the scatter of means of groups of n readings would have a smaller variance, σ^2/n, than the readings themselves, as well as a smaller standard deviation, $\sigma/\sqrt{n}$. In this discussion, we have skated over the need to know the value of the standard deviation of the whole normal population. If we do not know σ, then we can approximate it with the value of the standard deviation s of the small set of n. So if $n \geq 30$, it is usually sufficiently precise to take $\sigma = s$. If $n < 30$, the standard deviation should be taken as $\sigma = s\sqrt{n}/\sqrt{n-1}$.

1.A.3 The Student *t* Distribution

We now need to look at one more subtlety of these estimates. The normal distribution assumed that we had obtained many readings and could with confidence know that they formed a normal distribution. We can agree that if the error is random, then it is a fair assumption that many readings would form a normal distribution. However, often we have only a few readings, and these may not be uniformly distributed within the curve of Figure 1.A.1. Too many may lie outside the 1.96σ limit. For this reason, we use the Student *t* distribution, which allows for small samples on the assumption that the distribution, as a whole, is normal. Figure 1.A.2 shows the effect of the small number of readings. Because, with a small number of readings, one has to be subtracted from all the others to obtain a mean, the number of independent values is one less than the number of readings, and so the statisticians say that there is one less degree of freedom than the number of readings. In Figure 1.A.2, ν is the symbol for the degree of freedom, and $\nu = n - 1$, where n is the number of readings. For $\nu \to \infty$, the *t* distribution tends to a normal curve with mean zero and variance unity.

Figure 1.A.2 shows clearly the larger area spreading beyond the normal curve in which the readings may lie and the reason for the greater uncertainty. The curves are used in a similar way to the normal curve, but, as an alternative, Table 1.A.2 provides the information we need. If we have 10 readings, say, and so select the value of $\nu = 9$ for the degree of freedom, and if we wish to find the limits for a confidence level of 95%, we shall need to use the 5% column. We obtain $t = 2.262$, which we can apply to obtain the limits for a 95% confidence of $\pm 2.262\,\sigma/\sqrt{n}$ on the mean values of groups of readings, where n is the number of readings in the group. We should note, however, that the 95% confidence level from Table 1.A.2 gives a *t* value that varies little from 2.0 if $\nu \geq 20$. The limits for 95% confidence will then be $\pm 2.0\,\sigma/\sqrt{n}$ on the mean values of groups of readings.

I have always been puzzled by the name Student, but Eccles et al. (1993b) explained that the originator of this technique was William S. Gosset, born in 1876, who used the pseudonym *Student.*

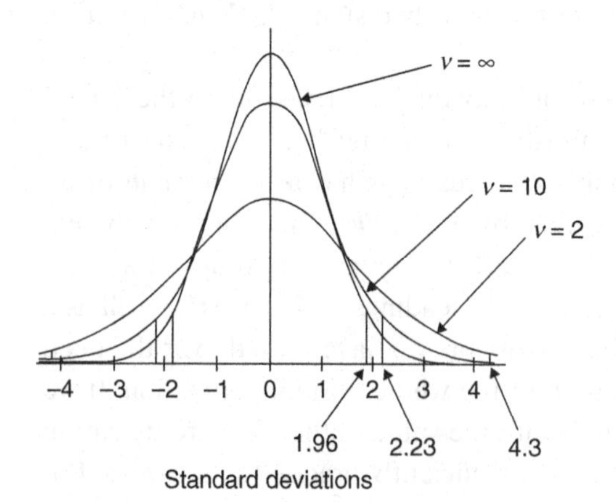

Figure 1.A.2. Student's *t* distribution curves compared with the normal curve. Note $p = 5\%$ as related to Table 1.A.2 for both tails.

Table 1.A.2. *A selection of values from the Student* t *function*

n	*v*	p(%) (Total)				
		20	10	5	1	0.1
		p/2 (%) (per tail)				
		10	5	2.5	0.5	0.05
2	1	3.078	6.314	12.71	63.66	636.6
3	2	1.886	2.920	4.303	9.925	31.60
5	4	1.533	2.132	2.776	4.604	8.610
10	9	1.383	1.833	2.262	3.250	4.781
20	19	1.328	1.729	2.093	2.861	3.883
30	29	1.311	1.699	2.045	2.756	3.659
61	60	1.296	1.671	2.000	2.660	3.460
121	120	1.289	1.658	1.980	2.617	3.373
∞	∞	1.282	1.645	1.960	2.576	3.291

After D. V. Lindley and W. F. Scott, *New Cambridge Statistical Tables*, 2nd ed., Cambridge: Cambridge University Press, Table 10, p. 45.

1.A.4 Practical Application of Confidence Level

The method described in Section 1.4 leads to the following steps (cf. Guide ISO/IEC 98-3: 2008, Hayward 1977b, 1977c):

i. Write down systematic uncertainties and derive the standard uncertainty for each component.
ii. Write down random uncertainties and derive the standard uncertainty for each component.
iii. Calculate the combined standard uncertainty for uncorrelated input quantities (and refer to UKAS 2012 if correlated).
iv. Obtain the expanded uncertainty using $k = 2$ for 95% confidence.

Taking a simple example, where we need to revert to *t*, suppose that we obtain a series of volumetric flow readings from a 50 mm ID flowmeter with the flow set at $10\text{m}^3/\text{h}$:

$$10.06, 10.01, 9.95, 9.99, 9.85, 10.02, 10.03, 10.12, 9.90, 9.98.$$

The results are plotted in Figure 1.A.3. The mean of these readings is 9.991, and the standard deviation is 0.07752. We could thus conclude that the true reading of this meter fell in the bracket $9.991 \pm 2.262 \times 0.07752/\sqrt{9} = 9.991 \pm 0.05845$, or between 9.93 and 10.05 with a 95% confidence. This fairly brackets the value of $10\ \text{m}^3/\text{h}$.

The actual readings should have all fallen within 9.991 ± 0.07752, or 9.91 and 10.07. In fact, three fell outside this bracket – rather higher than the 1 in 19 implied by the 95% confidence level. We might wish to look more closely at the procedure

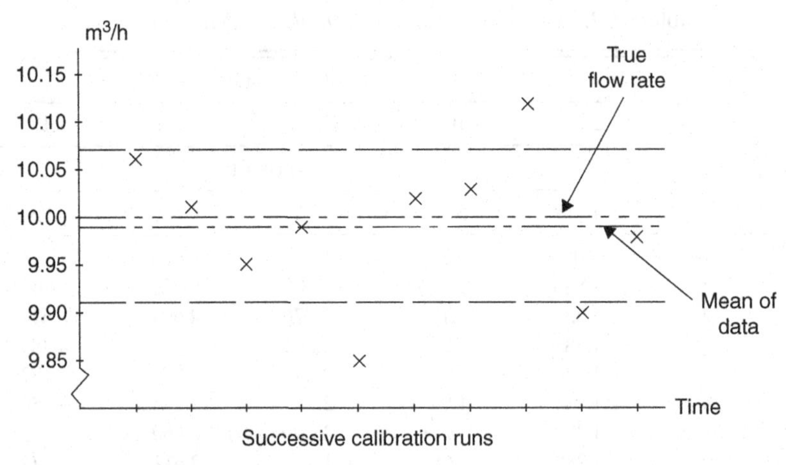

Figure 1.A.3. A set of flowmeter test readings (after Kinghorn 1982) for a fixed flow rate.

for obtaining these results since this suggests a possible problem with the means for obtaining the data.

1.A.5 Types of Error

There are essentially four types of error (Kinghorn 1982).

- *Spurious errors* result from obvious failures, obvious in the sense that they can be identified and documented. Readings with these should be eliminated.
- *Random errors* cause a variation in the output reading even when the input parameter has not changed.
- *Constant systematic error*, which is also called bias, may vary over the range but is constant in time, and could, in principle, be corrected out of the reading.
- *Variable systematic error* (bias) slowly varies with time, usually in a consistent direction, and may be caused by wear in bearings of a rotating meter, fatigue in components of a vibrating meter, erosion of geometry, etc.

Figure 1.A.4 illustrates these errors. Clearly one of the readings is so far out that there must be some explanation other than randomness. It is comforting to know that some of the most eminent experimentalists of the past have had cause to discard readings in critical experiments!

The scatter around the mean line will provide the basis of the calculation which we did in Section 1.A.4. The constant systematic error (bias) can be seen and could be built into a flow computer. The change in the mean value with time shows the changing systematic error, which is, in part, the reason for regular recalibration of meters.

The repeatability is related to the closeness of readings. If we expect a reading to lie within a band given by $\pm 2s$, the worst case difference between successive readings that fall within this band would be $4s$ $[(2 \times 2)s]$, but a less extreme working value is

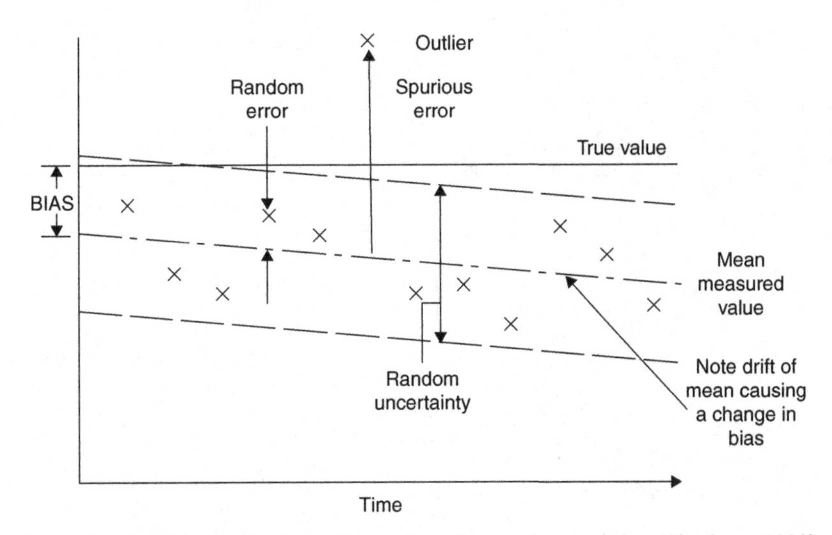

Figure 1.A.4. Diagram to show the various types of error (after Kinghorn 1982).

obtained from the root of the sum of the squares (or the quadrature; cf. Pythagoras and the length of the hypotenuse):

$$\sqrt{(2s)^2+(2s)^2} \text{ or } 2\sqrt{2}s$$

1.A.6 Combination of Uncertainties

If we combine uncertainties due to the nature of a flowmeter's operating equation, then we take the following approach. Suppose that the flowmeter has the equation

$$q_v = x_1 x_2^n x_3 x_4^m \tag{1.A.5}$$

To obtain the uncertainty in q_v, we need the partial derivative of q_v with respect to x_1, x_2 etc. The required result can be achieved, either by differentiating the equation as it stands or by first taking logarithms of both sides. We shall skip this and go straight to the result:

$$\frac{u_c(q_v)}{q_v} = \pm\frac{u(x_1)}{x_1} \pm \frac{nu(x_2)}{x_2} \pm \frac{u(x_3)}{x_3} \pm \frac{mu(x_4)}{x_4} \tag{1.A.6}$$

The problem with this equation is that the arithmetic sum of the uncertainties is usually overly pessimistic. It is, therefore, recommended that they be combined in quadrature, or by the root-sum-square (rss) method. This leads to the following equation:

$$\frac{u_c(q_v)}{q_v} = \sqrt{\left(\frac{u(x_1)}{x_1}\right)^2 + \left(\frac{nu(x_2)}{x_2}\right)^2 + \left(\frac{u(x_3)}{x_3}\right)^2 + \left(\frac{mu(x_4)}{x_4}\right)^2} \tag{1.A.7}$$

(a)

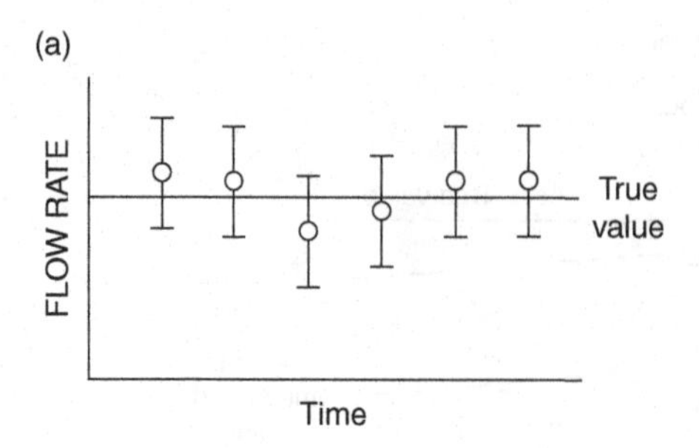

(b)

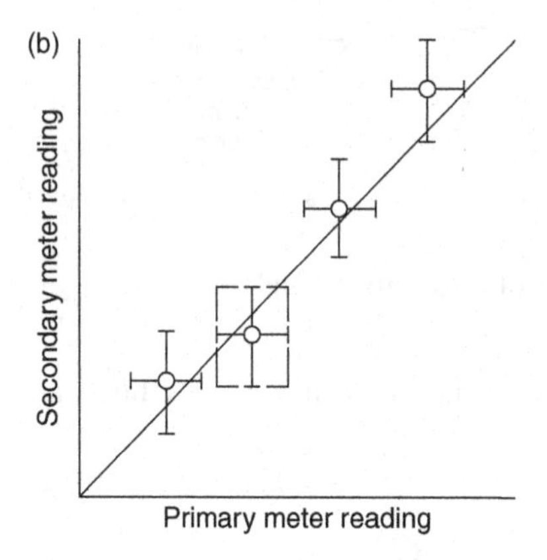

Figure 1.A.5. **(a)** Uncertainty limit bars on readings. **(b)** Uncertainty limit bars on both flow rate and readings.

There are complications beyond this equation. x_2 may appear in the equation as $x_5 + x_6$. In this case, $x_5 + x_6$ will need to be dealt with first and will require careful consideration as to whether the actual errors in these quantities are combining, cancelling or random.

1.A.7 Uncertainty Range Bars, Transfer Standards and Youden Analysis

It is sometimes useful to indicate the range of uncertainty estimated from the experimental method in each reading. This can be done by using bars that give uncertainty limits on each experimental point. This will then indicate, for a particular flow rate, the likely uncertainty in the reading. This is shown in Figure 1.A.5(a). In some cases where flow rate varies, there may be an uncertainty in both primary flow rate measurement and reading of second meters. In this case, uncertainty bars are needed in both directions, and the rectangle [Figure 1.A.5(b)] will define the limits of the possible uncertainty. In other words, the maximum uncertainty will have been obtained by quadrature.

Mattingly (1982) describes the procedure for checking the validity of different flow measurement laboratories using a Measurement Assurance Program (MAP) [or Proficiency Testing, as referred to by Mattingly in a draft report on the approach of the National Institute of Standards and Technology (NIST)] where a good-quality flowmeter acting as a transfer standard is exchanged between laboratories. Figure 1.A. 6(a) shows the results of such a cycle of checks; and the bars on

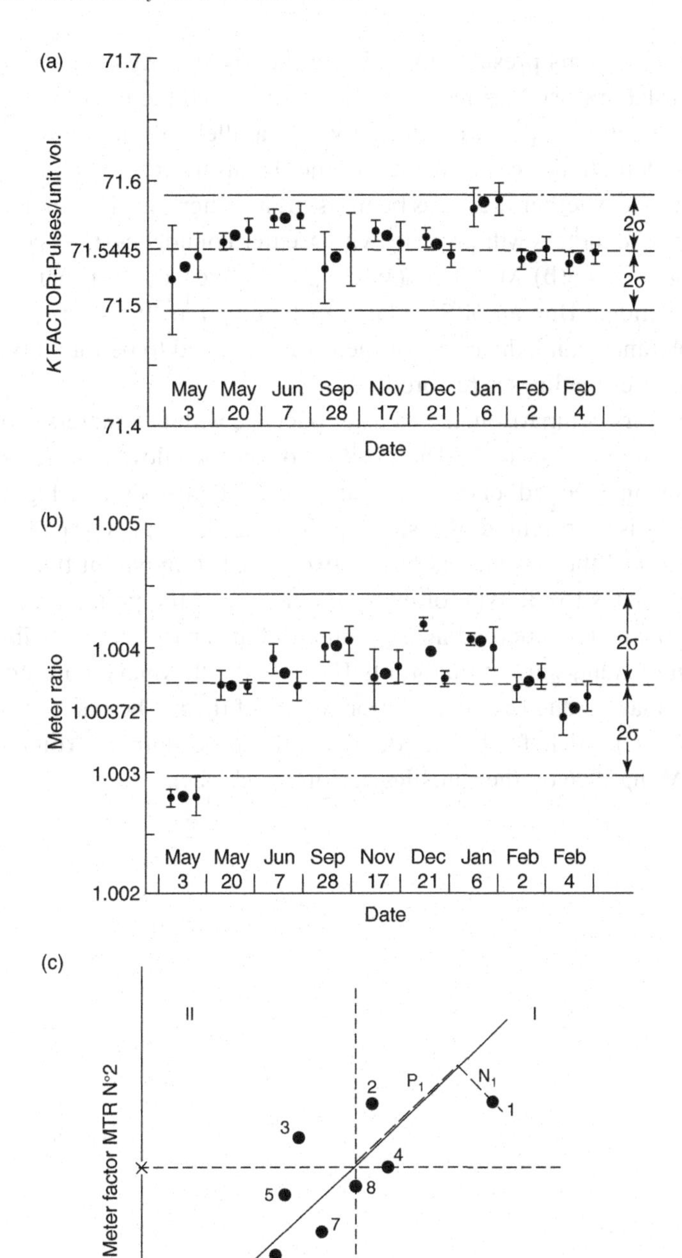

Figure 1.A.6. Turbine meter as a transfer standard (from Mattingly 1982; reproduced with the author's permission). **(a)** Typical turbine meter control chart for meter factor; **(b)** typical turbine meter control chart for ratio; **(c)** graphical representation of the Youden plot.

the experimental points presumably indicate the uncertainty of the turbine meter at a particular laboratory. Various people have suggested the use of two flowmeters usually in series, but as a possible alternative in parallel, to enhance the accuracy of a transfer standard. In this case, the ratio of the signals from the two meters will give an indication as to whether there has been a shift in either, and, if there has not been, the reading of the meters will provide, with greater confidence, the accuracy of the facility. In Figure 1.A.6(b), Mattingly (1982) gave a typical control chart for the ratio of a pair of National Bureau of Standards turbine meters. If the ratio falls outside an agreed tolerance band, the cause of the error will need to be found before confidence in the meter readings is restored.

For laboratory comparison, the transfer package with two meters should be run at one or two agreed flow rates. The position of the two flowmeters may be interchanged to obtain a second set of data. Using the Youden analysis in Figure 1.A.6(c), each laboratory is represented by a single point (1, 2, 3, etc.) resulting from plotting the meter factor of the two meters on the two axes. It is apparent that the position of the points relates to the type of error. Essentially, if the points lie in quadrants I or III, then the meters are reading the same, and the error can be attributed to the flow rig. If the readings are in quadrants II or IV, the flowmeters are not agreeing, and the error may be due to a malfunction in one of them. The reader is referred to Mattingly's (1982) article for further details of this procedure (cf. Youden 1959 and see Wu and Meng 1996 on the statistics of Youden circles).

2

Fluid Mechanics Essentials

2.1 Introduction

In an earlier book (Baker 1996), I provided an introduction to fluid mechanics and thermodynamics, particularly aimed at instrumentation. I do not, therefore, propose to repeat what is written there but rather to confine myself to essentials. In this book, I shall use the term *fluid* to mean liquid or gas and will refer to either liquid or gas only when the more general term does not apply.

2.2 Essential Property Values

Flowmeters generally operate in a range of fluid temperature from –200°C (–330°F) to 500°C (930°F), with line pressures up to flange rating for certain designs. Typical values of density and viscosity are given in Table 2.1.

It should also be noted that liquid viscosity decreases with temperature, whereas gas viscosity increases with temperature at moderate pressures. In common fluids, such as air and water, the value of viscosity is not dependent on the shear taking place in the flow. These fluids are referred to as Newtonian in their behaviour as compared with others where the viscosity is a function of the shear taking place. The behaviour of such fluids, known as non-Newtonian, is very different from normal fluids like water and air. Newtonian fluid behaviour is a good representation for the behaviour of the bulk of fluids.

2.3 Flow in a Circular Cross-Section Pipe

An essential dimensionless parameter that defines the flow pattern at a particular value of the parameter is the Reynolds number

$$\mathrm{Re} = \frac{\rho V D}{\mu} \tag{2.1}$$

where ρ is the density of the fluid, μ is the dynamic viscosity and, when applied to flow measurement, V is the velocity in the pipe and D is the pipe diameter. Typical

Table 2.1. *Typical values of viscosity (approximate values at 1 bar)*

	Temperature (°C)	Density, ρ (kg/m^3)	Viscosity	
			Dynamic, μ (cP)	Kinematic, ν (cSt)
Water	20	1,000	1.002	1.002
Benzene	20	700	0.647	0.92
Oxygen	0	1.43	0.019	13.3
Nitrogen	0	1.25	0.017	13.6
CO_2	0	1.98	0.014	7.1
Air	0	1.29	0.017	13.2

Adapted from Kaye and Laby (1966).

values of Reynolds number are: for water with $\mu/\rho = 10^{-6}$ m^2/s, $V = 1$ m/s (3.3 ft/s), $D = 0.1$ m (4 in.) Re $= 10^5$; and for air at ambient conditions with $\mu/\rho = 1.43 \times 10^{-5}$ m^2/s, $V = 10$ m/s (33 ft/s), $D = 0.1$ m (4 in.) Re $= 0.7 \times 10^5$.

If Re is less than about 2,000, and the fluid has had a sufficient length of pipe to reach a steady state, it all moves parallel to the axis of the pipe. At the pipe wall, the fluid "sticks" to the pipe in what is known as a nonslip condition. The velocity of the fluid increases, therefore, from zero at the pipe wall to a maximum at the centre, and the shape of the profile is parabolic. In this case, the flow is called laminar.

At about Re = 2,000, a major change takes place, and the smooth, parallel nature of the flow gives way to eddies in the flow. These eddies mix the high velocity at the pipe axis with the lower velocity near the pipe wall. The resulting profile is flatter (although it still goes to zero at the pipe wall). When fully developed, this is known as a turbulent profile. It is of a well-defined shape and with a known range of eddy sizes. It may, therefore, be misleading to use the term *turbulent profile* for any other profiles created by upstream disturbances due to bends etc. Between the turbulent flow and the laminar flow is a transition when the flow alternates randomly, in space and time, between laminar flow and turbulent flow.

Any flowmeter application to non-Newtonian fluids will need advice from the flowmeter manufacturer. It is probable that clear bore flowmeters, if applicable, will be most suitable. A flowmeter with relatively low sensitivity to profile, such as electromagnetic or multi-beam ultrasonic designs, will be worth consideration provided the fluid is conducting for the first type or transmits an adequate level of sound for the second (cf. Appendix 2.A.2).

It is apparent from the calculations of Reynolds number for water and air that, in the majority of industrial applications, the flows will be turbulent, and, therefore, the behaviour of the turbulent profile, and of flowmeters subject to it, is of primary importance to us.

For laminar flow, we can use the equation

$$V = V_0\left[1-\left(\frac{r}{R}\right)^2\right] \tag{2.2}$$

Table 2.2. *Approximate turbulent velocity profiles from Equation (2.4) (after Schlichting 1979)*

Re	4×10^3	2.3×10^4	1.1×10^5	1.1×10^6	2.0×10^6 to 3.2×10^6
n	6.0	6.6	7.0	8.8	10
$V/\bar{V}$ (at $r = 0.75R$)	1.003	1.004	1.004	1.005	1.005
$V/\bar{V}$ (at $r = 0.758R$)	0.998	0.999	1.000	1.002	1.002

where V_0 is the velocity on the axis of the pipe, r is the radial point at which we are measuring velocity, and R is the pipe radius and where V is zero at the wall of the pipe because of the nonslip condition for a fluid at a solid boundary.

The mean velocity in the pipe is then given by

$$\bar{V} = \frac{V_0}{2} \tag{2.3}$$

For turbulent flow an approximate curve-fit is

$$V = V_0 (1 - r/R)^{1/n} \tag{2.4}$$

where V_0 is, again, the centre line velocity, and we can relate n to Re from experimental data (Table 2.2). Note that pipe roughness, if the pipe is not hydraulically smooth, will also affect the profile.

This expression, therefore, provides not only a convenient, although approximate, representation of the profile shape but also an index adjustment to allow for the change in Reynolds number. The resulting profile shapes for the laminar and the turbulent regimes, based on Equation (2.4), are shown approximately in Figure 2.1. The turbulent profile should become flatter with increasing Reynolds number.

One interesting feature of the turbulent profiles based on Equation (2.4) is that the ratio of the velocity at about the 3/4 radius point to the mean velocity is approximately unity. The mean velocity is given by

$$\frac{\bar{V}}{V_0} = \frac{2n^2}{(n+1)(2n+1)} \tag{2.5}$$

Hence

$$V/\bar{V} = \frac{(n+1)(2n+1)}{2n^2}(1 - r/R)^{1/n} \tag{2.6}$$

Table 2.2 provides the values of this ratio for various values of Re and two radial positions. Thus if a single measurement of velocity in a pipe is to be used to obtain the mean of the velocity in a turbulent flow, a point at $0.758R$ may be found best. This figure will clearly be of importance when we consider where to position the probes described in Chapter 21.

A final important point about the profiles in Figure 2.1 is that there is much difference between the flows, even though they are drawn as smooth curves. In the

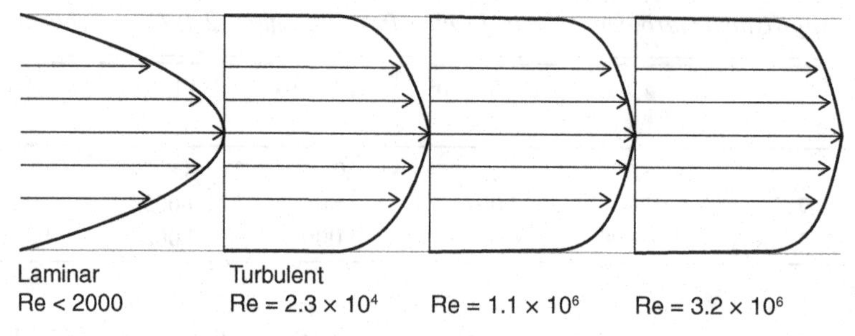

Figure 2.1. Laminar and turbulent pipe profiles.

laminar profile, the fluid shears smoothly over itself, but it all moves essentially parallel to the axis of the pipe. In the turbulent profiles, the curves represent a mean profile and ignore the turbulent eddies, which can be as much as 10% of the velocity of the mean flow and cause fluctuations in all directions. In the context of flow measurement, this is particularly significant in that the meter must measure accurately against this background fluctuation.

Equation 2.6 is an approximation to the actual turbulent profile in a pipe, although a widely used approximation. It may be in some cases that a better analytical expression is required, for instance in an ultrasonic flowmeter's software for signal correction. Ward-Smith (1980) sets out the laws for velocity distribution in terms of the law of the wall or the inner velocity law (comprising the viscous sublayer, the buffer layer and the logarithmic region), and the outer or core region, in a clear presentation, which those needing more precision are recommended to read as an introduction to more complicated mathematical approaches. A brief presentation of this will be found in Appendix 2.A.1.

So far we have assumed that the pipe flow profile is fully developed. To achieve such a profile requires an upstream straight length of pipe greater than may be available in many industrial flowmeter installations. 60D is sometimes quoted as a minimum requirement.

We, therefore, have a problem when installing a flowmeter in that the calibration conditions for a fully developed profile are unlikely to be achieved in the installation and the meter accuracy will be degraded. We have four options.

i. Find an installation point where fully developed flow occurs.
ii. Calibrate the flowmeter with the upstream pipework (downstream has little effect beyond a maximum of about 8D).
iii. Take extensive measurements of the effects of bends, valves, T-pieces, multiple bends, expansions and contractions on a particular flowmeter design and allow for the change in precision.
iv. Attempt to reorder the flow to recreate a turbulent profile of the sort used for calibration.

Outside a few laboratory situations, (i) is probably unlikely to be an option, but (ii) may not be realistic or financially viable unless in situ calibration is possible. Option (iii) will be a recurring theme through the discussion of most types

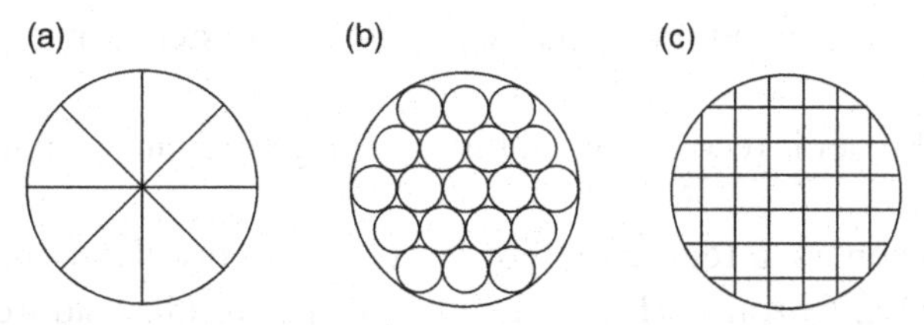

Figure 2.2. Approximate diagrams of flow straighteners: (**a**) Etoile; (**b**) tube bundle; (**c**) box (a honeycomb layout may also be used).

of flowmeters. An alternative approach is to compute the profiles resulting from pipe fittings. One example of such an approach is that of Langsholt and Thomassen (1991), who used a commercially available computer program to obtain the flow downstream of various piping fittings. Further discussion of computational applications is given in Section 2.A.5. Some indication that computer prediction and experiment gave similar swirl decay downstream of a double bend in perpendicular planes was given, and 41D reduced the angle to 3.3° for measured values and 2.4° for simulated values. Mottram and Rawat (1986) suggested that the swirl from two 90° bends in perpendicular planes is reduced by 50% in smooth pipes after L/D = 45 and in rough pipes after an L/D value in the range of 10 to 20 (cf. however Mattingly et al. 1987, who noted swirl effects on a turbine meter after 90D).

Langsholt and Thomassen (1991) also provided some flow contours at points in a complex pipe configuration. In the context of this chapter, it is important to give some attention to (iv).

The reader is also directed to an excellent book edited by Merzkirch (2005) which develops related areas.

2.4 Flow Straighteners and Conditioners

Flow straighteners and flow conditioners have been used, for many years, to attempt to reorder a profile that has been disturbed by upstream fittings. The flow straightener was designed to remove swirl from the flow. Swirl, the bodily rotation of the fluid in a pipe, takes a long length to decay and can introduce severe metering errors. The three types of straightener commonly used are shown in Figure 2.2.

It is essential that they are set straight or they will become the cause of swirling flow rather than the correction for it. Examples of the effectiveness of etoile straighteners of various lengths were given by Kinghorn et al. (1991). Their conclusions follow:

- An eight-vaned (on four diameters) straightener of 0.5D length virtually removed even the maximum swirl.
- The flat profiles downstream of the straighteners appear to have caused most of the residual negative errors in the orifice coefficient, as opposed to the positive errors due to swirling flow.
- To ensure an error of less than ±0.5%, a straightener 1D long should be placed upstream of the orifice meter at least 6D for plates with a beta ratio of 0.5 and 14D for plates with a beta ratio of 0.8.

- Even with 16D separation, virtually all tests suggested some effect from the straightener.
- A 1D straightener is recommended for removing swirl and minimising head loss.

Flow conditioners attempt to reorder the flow profile to recreate a fully developed turbulent profile. However, this reordering must address both mean profile shape and turbulent eddy pattern. Examples of conditioners are given in Figure 2.3.

Recent published work reviewed in Section 2.A.3 has thrown much helpful light on the effectiveness of these devices. Bates (1991) reported on field use by Total Oil Marine of a K-Lab flow conditioner. Strong swirl was found to be present in the gas metering system on a process platform in the Alwyn North field, which had been installed to ISO 5167 in three 14-in. lines with beta ratio of 0.6. The meter was separated from a pipe reduction of 0.86 by 30D, but upstream of this there were other fittings. Greasy deposits on the orifice plate gave an early indication of swirl being present. This was rectified by installing K-lab flow conditioners. This conditioner is machined from solid, with holes arranged approximately as in Figure 2.3(b). (See Erdal, Lindholm and Thomassen 1994 on the development of the K-Lab conditioner and Spearman, Sattary and Reader-Harris 1991 on LDV measurements downstream of a Mitsubishi flow conditioner and of an orifice plate in a combined package.)

Karnik, Jungowski and Botros (1991, 1994) addressed the important question of how closely the flow downstream of a flow conditioner represented fully developed flow as far as a flowmeter was concerned. For undisturbed fully developed flow, they found that, for Re of about 0.9×10^5, the profile was approximated by $n = 7.4$ in Equation (2.4), compared with $n = 7$ in Table 2.2. Measurements of velocity and turbulence profiles and of orifice plate performance were made to determine the effect of an elbow upstream of the tube bundle with various spacings between the elbow, the flow straightener and the metering position. They confirmed the point that, even though the profile may be approximately correct, the turbulence characteristics are unlikely to be correct, and the orifice plate appears to be influenced by both.

Laws and colleagues have done some useful work on straighteners and conditioners (Laws 1991; Laws and Ouazzane 1992, 1995b, 1995c). Her work initially, and rather unexpectedly, suggested that tube bundles of length recommended by ISO 5167 (hexagonal pack of 19 tubes of $L \geq 20d$), AGA(3) and ANSI/API 2530:1985 (circumferential pack with $L > 10d$) are little more, and possibly less, effective than one of only $L = 1.25d$. She found that this was as effective in swirl elimination and attenuation of profile non-uniformities. She also suggested that the AGA design was more effective and had a lower Δp. This led her to suggest that perforated plates may be a better and more easily constructed option.

This observation was reinforced by work on the Zanker flow conditioner with thicker perforated plate, which performed, in some cases, better without downstream honeycomb (Laws and Ouazzane 1992). A criterion of ±6% of the fully developed profile 100D downstream was used to assess whether the resultant Zanker profiles at

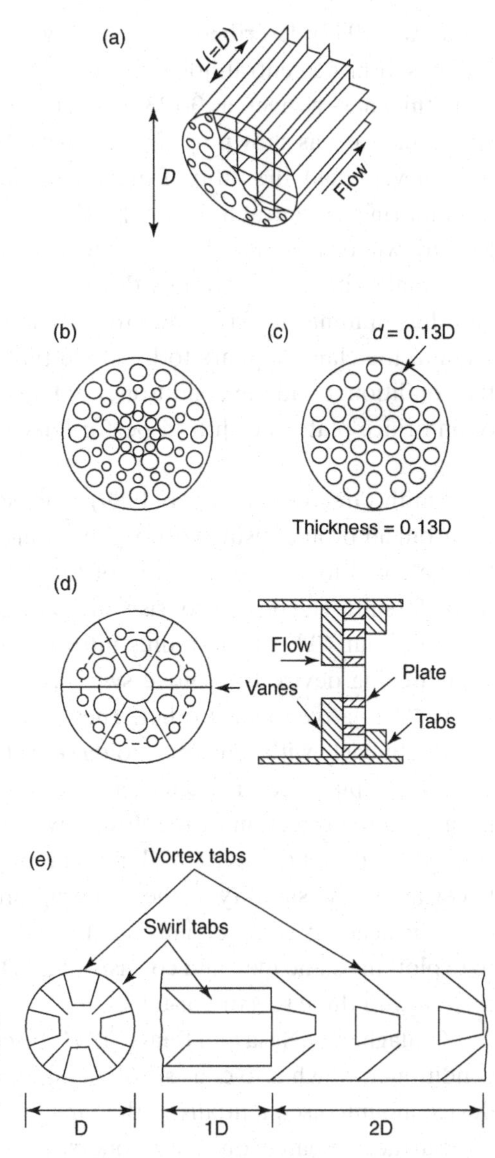

Figure 2.3. Approximate diagrams of flow conditioners: (**a**) Zanker; (**b**) K-lab Mark 2; (**c**) Mitsubishi; (**d**) Laws (after Laws 1990; Laws and Ouazzane 1995a); (**e**) Vortab [reproduced with permission of Laaser (UK) Ltd.].

8.5D were satisfactory. The reported experiments suggested that, for a 12-mm plate in a 100-mm line, the Zanker flow conditioner without downstream honeycomb performed better than with the honeycomb. Laws and Ouazzane (1995b) also suggested that a thick Zanker plate (10% of the pipe diameter) with upstream honeycomb flow straightener is preferable to the Zanker straightener with thinner plate and downstream straightener. Not surprisingly, they suggested that the Laws plate was superior. In further tests, Laws and Ouazzane (1995c) found that the etoile flow straightener was not a very good flow conditioner, although it was effective as a straightener. They suggested that if the etoile is made with the vanes stopping short of the pipe axis, the wake from the hub of the etoile is removed and the performance is improved.

Laws (1990) described her new flow conditioner, which benefitted from these findings. It has a central hole and two rings of holes on 0.4616D and 0.8436D. The plate thickness is given as 0.123D and, for best performance, has an arrangement of vanes and tabs as shown in Figure 2.3(d). The number of holes is 1:7:13 (or 1:6:12 in later versions), the ratio of open area outer to inner ring is 1.385, and the ratio of inner ring to central hole is 5.42. The total open area is 51.55%. Laws claims that the unit will be an enhancement to most flow installations, enabling flow conditions to be maintained close to fully developed irrespective of the operating conditions and thus minimising errors due to installation effects. The loss coefficient is about 1.4, and the claim appears to be made that with a flow distortion device 3D from the conditioner and the conditioner 3D upstream of the upstream pressure tapping of an orifice plate, the shift in performance is of order 0.2% (Laws and Ouazzane 1995a).

Smith, Greco and Hopper (1989) proposed a different approach to the recreation of turbulent profiles using vortex-generating devices. Their results looked promising for the conditioner 5D upstream of the flowmeter when they appear to claim less than 1% shift. The device consists of four axial swirl tabs of about 1D length, which project about 40% of the radius into the flow from the wall and are essentially a straightening device to remove swirl. Two sets are recommended; they should be set at 45° relative to each other. These are followed by, preferably, two sections of 1D length each, with four trapezoidal tabs (vor-tabs) spaced around the circumference of the pipe and projecting inward with an angle of about 30° to the pipe wall in the general direction of the flow. Laws and Harris (1993) tested this device and considered that it did quite well in recreating a turbulent profile, but they appeared uncertain of the stability of the resulting profile. The loss is claimed to be only 0.7 dynamic heads. A commercially available device, the Vortab [Figure 2.3(e)], appears to exploit the same idea and consists of a 1D straightening section followed by a 2D section with three sets of vor-tabs.

Gallagher, LaNasa and Beaty (1994) discussed the development of the Gallagher conditioner, which also consists of an anti-swirl device, settling chamber and profile device mounted sequentially in the pipe.

Further research on conditioners, including the Spearman (NEL) promising design, is given in Section 2.A.3.

2.5 Essential Equations

The *continuity equation* results from the classical physics concept that mass can neither be created nor destroyed and so the same mass must pass through each point in a pipe

$$\text{Mass flow} = q_{\text{m}} = \rho q_{\text{v}} = \rho VA \tag{2.7}$$

and we can relate the velocity at different pipe sections 1 and 2 (Figure 2.4)

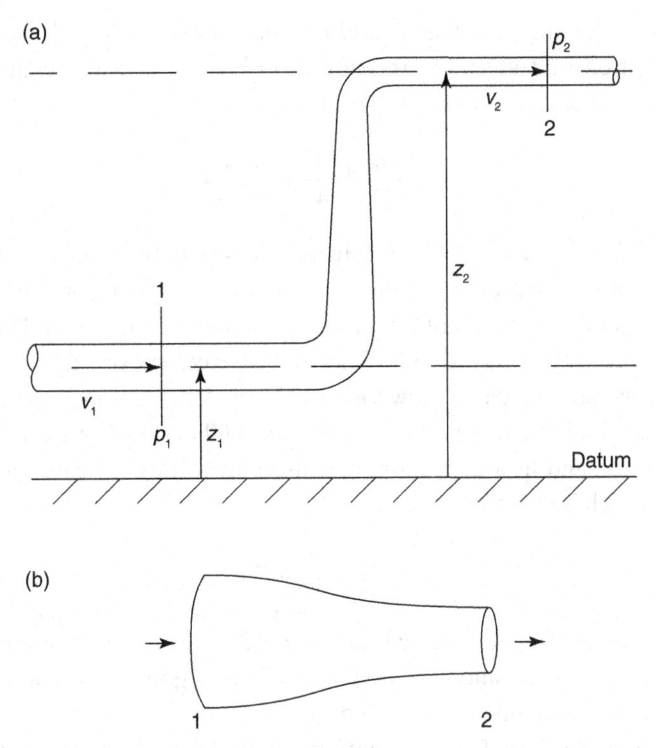

Figure 2.4. Duct sections. (**a**) Varying height and duct section; (**b**) varying area duct to show positions referred to in equations.

$$\rho_1 V_1 A_1 = \rho_2 V_2 A_2 \tag{2.8}$$

Bernoulli's equation can be derived from energy considerations (Baker 1996) and relates the pressure change to the velocity change. Using Figure 2.4(a), we can relate pressure, velocity and height above datum in a compressible fluid without flow losses as

$$\int_1^2 \frac{dp}{\rho} + \frac{(V_2^2 - V_1^2)}{2} + g(z_2 - z_1) = 0 \tag{2.9}$$

This equation is known as the compressible fluid form of Bernoulli's equation. If ρ is constant for an incompressible fluid with constant density, then we can rewrite the Bernoulli equation as

$$\frac{p_2 - p_1}{\rho} + \frac{(V_2^2 - V_1^2)}{2} + g(z_2 - z_1) = 0 \tag{2.10}$$

Note from this, assuming that we can neglect changes in z, that, if the flow is brought to rest, the resulting pressure, p_0, is the total or stagnation pressure, the pressure while fluid is flowing is the static pressure, and the difference, $\Delta p = \frac{1}{2}\rho V^2$, is the dynamic pressure.

If we neglect changes in height and combine Equations (2.8) and (2.10), we obtain an equation that gives the relationship between pressure change and flow rate through a duct such as in Figure 2.4(b)

$$p_2 - p_1 + \frac{q_m^2}{2\rho}\left(\frac{1}{A_2^2} - \frac{1}{A_1^2}\right) = 0 \tag{2.11}$$

Loss coefficients result, in real flows, from viscosity in the fluid, and it is essential to know these losses in a piping system to ensure that the pump, or whatever creates the flow, will be sufficient to maintain the required flow. To understand how to calculate losses, the reader is referred to Miller's (1990) excellent book.

Thus all flowmeters create a flow loss, although clear-bore meters do not significantly add to the loss due to an equal length of straight pipe. However, if the profile is highly disturbed and flow conditioning is used, these devices cause an additional pressure loss, which is given by

$$\Delta p_{\text{loss}} = K\left(\tfrac{1}{2}\rho V^2\right) \tag{2.12}$$

K may be up to order 5 for the devices in Figures 2.2 and 2.3, but for some flow conditioners it may be larger since the flow is effectively going through a contraction with very poor downstream pressure recovery.

For a fuller derivation of these equations, the reader is referred to Baker (1996) or other fluid mechanics books.

2.6 Unsteady Flow and Pulsation

In this section, we consider the effect of unsteadiness on flowmeters (see Svete et al. 2012 – flow pulsator). The use of fluid instability in flowmeters is covered in Chapter 11, where it is an essential part of the meter designs.

Most flowmeters are affected by unsteady or pulsatile flows for the following reasons.

- Pulsatile flow affects the velocity profile in the pipe, and the distorted profile may, in turn, affect the flowmeter response. Hakansson and Delsing (1994) found that flattening of the flow profile occurred due to the pulsating flows. This, in the case of the ultrasonic flowmeter, causes incorrect averaging of the diametral beam.
- The characteristic of the meter may be nonlinear, and the resulting output average will not correspond to that for the flow. The differential pressure flowmeters with their square law are examples.
- The inertia of parts of the flowmeter or of the fluid may not allow the meter to track the pulsating flow correctly. The gas turbine meter is highly affected by this. The importance of this effect in the orifice plate has been the cause of some discussion in the literature, for which the reader is referred to Mottram's papers.
- The natural operating frequency of the flowmeter may be near the pulsation frequency, and this may cause aliasing or other errors. Vortex meters are sensitive to such pulsations.

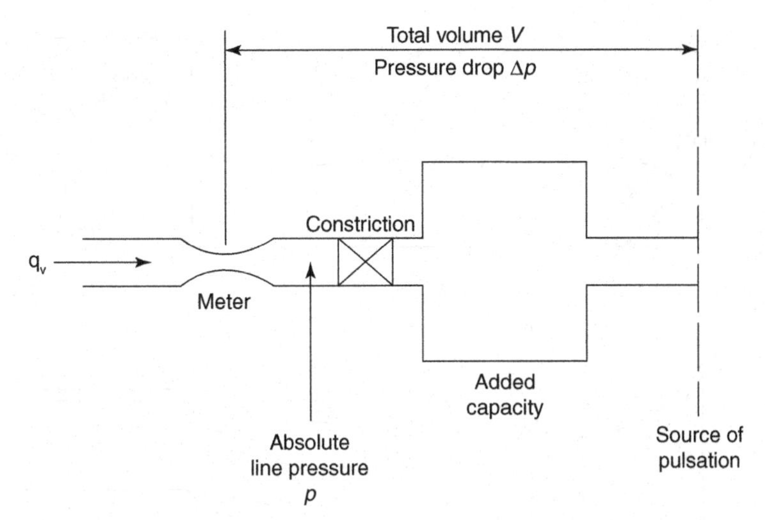

Figure 2.5. Diagram to indicate symbols used in the Hodgson number calculation.

- The secondary instrumentation may not be able to follow the pulsation, and, as a consequence, errors may be introduced. The manometer is the most obvious secondary device to be affected, but pressure transducers, connecting leads and other devices may also be affected.

One approach associated with Hodgson was developed to show how much damping was needed to reduce pulsation to acceptable levels.

Mottram (1989) reviews some of the work on Hodgson's number (cf. Mottram 1981), but makes the wise comment that "if you can't measure (the size of the pulsation) damp it" (Mottram 1992). The Hodgson number is given by (Figure 2.5)

$$H = \frac{Vf}{q_v}\frac{\Delta p}{p} \tag{2.13}$$

where V is the total volume of pipework and other vessels between the source of the pulsation and the flowmeter position, Δp is the pressure drop over the same distance, f is the frequency of the pulsation, q_v is the volumetric flow rate, and p is the absolute line pressure. With this number, it is possible to plot curves to show the likely error levels for various values of H (cf. Section 5.5 for orifice plates).

Mottram gives a simplified value for the damping based on the Hodgson number, allowing for the worst case where resonance occurs, of

$$\frac{H}{\gamma} = \frac{1}{4\pi}\frac{\left(V_{rms}/\overline{V}\right)_{ud}}{\left(V_{rms}/\overline{V}\right)_{d}} \tag{2.14}$$

where $\left(V_{rms}/\overline{V}\right)_{ud}$ is the undamped velocity ratio, $\left(V_{rms}/\overline{V}\right)_{d}$ is the damped velocity ratio, and γ is the isentropic exponent.

He cautions that the criterion resulting from using this with the Hodgson number is untested. He also suggests that the Hodgson number should be doubled where the pulsation amplitude is estimated and not measured.

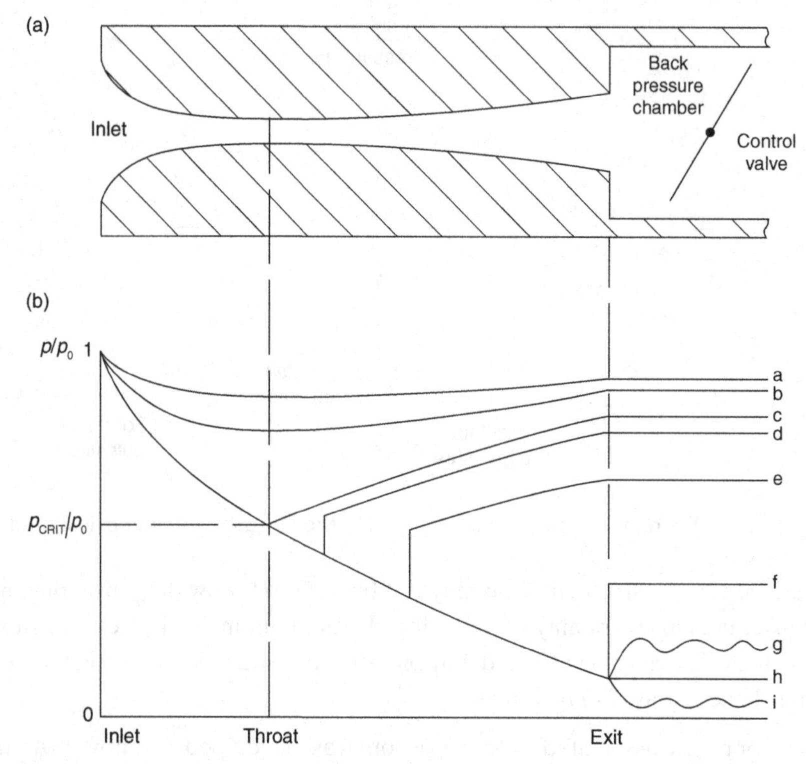

Figure 2.6. Convergent-divergent nozzle: (**a**) geometry; (**b**) p/p_0 against distance through convergent-divergent nozzle.

Some meters [e.g. differential pressure (cf. Section 5.5), turbine or vortex] are seriously affected by pulsating flow, whereas others (e.g. electromagnetic or ultrasonic time-of-flight) are probably little affected.

2.7 Compressible Flow

When a gas flows at velocities comparable to the speed of sound, its behaviour is markedly different from the behaviour of incompressible fluids. Such high-velocity flows occur in sonic or critical nozzles.

To appreciate the special features of compressible flow of a gas, we shall start by imagining that we have the nozzle in Figure 2.6(a) set up in an experimental rig with a compressor sucking air downstream of the control valve. We consider a convergent-divergent nozzle that reduces in area from the inlet to a minimum at the throat, and then increases in area to the exit, where it opens into the back pressure chamber. As the flow increases, there would be a steadily increasing noise emitted by the nozzle. A point is reached where the mass flow becomes constant, and the noise disappears irrespective of the downstream pressure.

What is happening? The normal information transfer mechanisms between the changes downstream and the upstream flow appear to have ceased. The loss of sound is clearly linked to this loss of information transfer, and it is intuitively reasonable to link the information transfer to the very small pressure waves that constitute sound and to suggest that, when the velocity is as fast or faster than sound speed, it is impossible for the waves to move upstream, and so impossible also to communicate upstream any flow change that happens downstream. In this case, sound speed has been reached at the throat of the nozzle.

It is beyond the scope of this book to derive the full equations for compressible flow, but we state three equations that will be useful in later discussions. These are for the special case of flow when conditions are adiabatic, isentropic and reversible. Although an idealisation, the flow for the inlet to the nozzle is a good approximation to such a flow.

$$\frac{T_0}{T} = 1 + \frac{\gamma - 1}{2} M^2 \qquad (2.15)$$

$$\frac{p_0}{p} = \left(1 + \frac{\gamma - 1}{2} M^2\right)^{\gamma/(\gamma-1)} \qquad (2.16)$$

$$\frac{q_m \sqrt{c_p T_0}}{A p_0} = \frac{\gamma}{\sqrt{\gamma - 1}} M \left(1 + \frac{\gamma - 1}{2} M^2\right)^{-\frac{1}{2}(\gamma+1)/(\gamma-1)} \qquad (2.17)$$

They give the value of T, p and q_m/A for a given γ and M, where

	Units
A = area	m^2
c = local sound speed	m/s
c_v = specific heat at constant volume	J/kgK
c_p = specific heat at constant pressure	J/kgK
M = Mach number V/c	dimensionless
p = pressure	Pa
p_0 = stagnation pressure	Pa
q_m = mass flow rate	kg/s
T = temperature	K
T_0 = stagnation temperature	K
V = gas velocity	m/s
γ = ratio of specific heats c_p/c_v (isentropic exponent)	dimensionless

Note that, in an isentropic flow without heat transfer, the stagnation temperature and pressure do not change and represent the values if the gas were to be brought to rest. However, stagnation pressure will change where irreversibility occurs (e.g. through a shock wave or in a flow with friction), and stagnation temperature will change where heat transfer takes place.

Figure 2.6(b) gives the plot for pressure variation through the convergent-divergent nozzle. As gas flows through the convergent-divergent nozzle in Figure 2.6(a), the velocity of the gas increases towards the narrowest point in the duct, the throat. The flow is created by a low pressure downstream of the nozzle. As the back pressure is reduced, so the pressures through the nozzle fall, and the velocity increases [a and b in Figure 2.6(a)]. In the sonic or critical condition, the pressure downstream is reduced until the velocity of the gas at the throat has increased to the speed of sound when the Mach number M is unity. Figure 2.6(b) shows that this condition can be achieved with the flow downstream of the throat subsonic (c), partly supersonic and partly subsonic with shock waves (d and e) or with the flow all supersonic (f, g, h and i with various types of flow in the plenum). Had the nozzle stopped at the throat, it would still be possible to obtain sonic conditions there, but there would be no possibility of the pressure recovery that occurs in the diffuser in c, d and e.

Equations (2.16) and (2.17) give the variation of pressure and area of the duct as we move towards the throat and M increases. At the throat where $M = 1$, we obtain two important relationships: the critical pressure ratio

$$\frac{p_*}{p_0} = \left(\frac{\gamma+1}{2}\right)^{-\gamma/(\gamma-1)} \tag{2.18}$$

and the mass flow rate at choked or critical conditions

$$q_m = \frac{A_* p_0}{\sqrt{c_p T_0}} \frac{\gamma}{\sqrt{\gamma-1}} \left(\frac{\gamma+1}{2}\right)^{-\frac{1}{2}(\gamma+1)/(\gamma-1)} \tag{2.19}$$

where * as a subscript indicates throat conditions. For air, with $\gamma = 1.4$, the critical pressure ratio for sound speed at the throat is $p_* / p_0 = 0.528$. It can be shown that the pressure recovery in the diffuser allows the exit/inlet pressure ratio to be much higher than this while still achieving critical conditions (sonic at the throat).

The importance of choked or critical conditions is twofold.

i. The flow rate is unaffected by downstream variations because the sonic throat condition acts as a block to downstream changes (see Baker 1996 for a fuller description).
ii. The mass flow of gas may be obtained from a knowledge of γ for the gas, the throat area A_* and the upstream stagnation values p_0 and T_0.

We shall consider the practical version of Equation (2.19) in Chapter 7.

2.8 Multiphase Flow

The term *multiphase flow* is somewhat misleading because it covers both multicomponent and multiphase. Because of the dearth of data, the fluid engineer attempts to learn from data from different sources.

We may consider first the flow from an oil well, making the assumption that this is vertical. This is not strictly true, but it will give us some basic concepts. This was described by Baker and Hayes (1985). The crude oil will reach the wellhead, having flowed up a pipe of about 100-mm bore for distances of several kilometres or about 30,000 pipe diameters. The flow will therefore, presumably, be fully developed. Initially, the flow will be single-phase, essentially oil only. As oil is removed from the well, the pressure in the reservoir will decrease, and the gas fraction in the

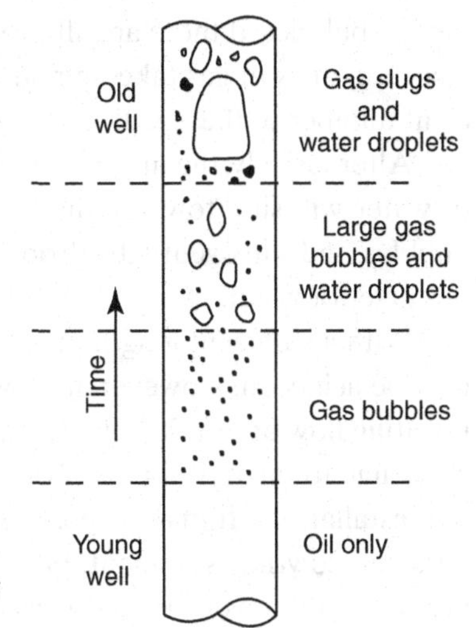

Figure 2.7. An example of three-phase vertical flow from an oil well.

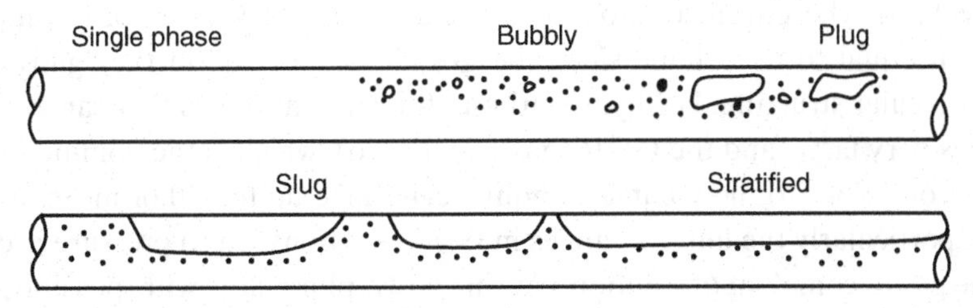

Figure 2.8. Horizontal two-phase flow.

well flow line will increase, and appear as gas bubbles. This is known as bubbly flow (Figure 2.7). With further ageing, the gas bubbles will become larger. An equilibrium size distribution will result from breakup due to turbulence and from coalescence due to the breakdown of the liquid film on close approach of two bubbles. In addition, the water content is likely to increase. Water will be present as droplets and will form a third phase. Yet further ageing will result in slugs of gas that travel up the centre of the pipe, leaving a slower layer of moving liquid on the wall (which, at times, may even reverse in direction during the passage of the slugs). These slugs tend to overtake each other, forming larger slugs many metres long. These slugs may be up to 20 metres long and may form an equilibrium size distribution, giving a balance between coalescence due to bubbles overtaking each other and breakup due to instabilities when they become too large. As a further complication, the oil flow may contain waxy deposits and sand. Even though separation of components is a standard process, the development of subsea systems requires multicomponent handling.

We next consider horizontal two-phase flows (Figure 2.8). The most obvious effect is the loss of axisymmetry. Gravity now causes the less-dense phase to migrate to the top of the pipe. Thus, in a gas-liquid flow, the gas will move to the top of the pipe as bubbles. If these are allowed to become large, plugs of gas result, and as these coalesce, slugs of gas take up regions against the top of the pipe. Eventually, a sufficient number of these will lead to stratified flow.

Alternatively, the mixtures may be of two liquids (e.g. water in oil). The droplets of water will sink towards the bottom of the pipe, mirroring the behaviour of gas bubbles, and will eventually drop out onto the bottom of the pipe, causing a continuous layer of water.

With a sufficient length of straight pipe (100D or more), a fully developed flow may be achieved. However, in most applications we shall not be able to predict the resulting flow or how it will affect equipment in the line. This presents a considerable challenge to the development of instrumentation for handling oil well flows in subsea installations. It may be possible to mix the fluid to create a more homogeneous fluid for flow measurement, but this will cause severe turbulence and a changing profile, conditions generally considered unsuitable for such a measurement, not to mention the pressure loss which may result.

Liquids may contain gases in solution. For water, the maximum amount is about 2% by volume. The gas in solution does not increase the volume of the liquid by an amount equal to its volume since the gas molecules "fit" in the "gaps" in the liquid molecular structure. For hydrocarbons, the amount of gas that can be held in solution is very large, and the GOR (gas-to-oil ratio), which is the volume of gas at standard conditions to the volume of liquid, can range up to 100 or more. In either case, but particularly the latter, changes in flow conditions (e.g. a pressure drop) can cause the gas to come out of solution, causing a two-phase flow with, possibly, severe effects on instrument precision.

In low head flows (e.g. in a sewage works), the flowing stream may entrain air, and a dispersion of air in water may result.

2.9 Cavitation, Humidity, Droplets and Particles

Cavitation may occur in certain liquid flows at pressures around ambient. Cavitation is the creation of vapour cavities within the liquid caused by localised boiling at low pressure. It can cause damage (e.g. in pumps, propellers, valves and other flow components) because the cavities can collapse very quickly with large, although localised, impacts and can erode solid surfaces. It can also cause errors in flowmeter readings because it results in a larger volume than for the liquid alone.

High humidity may also create problems if it results in a consequent large amount of water vapour changing to liquid droplets in the gas (e.g. in critical-flow Venturi nozzles).

Particulate matter can, in addition, cause wear and may need to be removed with a fine filter.

Much can be learned by computing the trajectories of droplets and particles (cf. Ahmad, Baker and Goulas 1986; Hayes 1988), and such an approach may be usefully applied to flowmeters. Droplets and particles will affect their performance in various ways.

a. They will cause flowmeters to read incorrectly, and most often this will be due to the flowmeter responding to the volume flow of the gas phase and not fully accounting for the greater mass carried in solid and liquid.
b. They may cause erosion of the flowmeter body and sensing element.
c. They may give false pulse readings in vortex flowmeters by hitting the pressure sensor.
d. They may become trapped in vortex structures around both stationary and moving parts of the meter.

The basic flow around most meter internals is now predictable from standard computer programs, and a bubble/droplet/particle trajectory model should be combined with such a prediction. From such studies, it may be possible to design internals less susceptible to bubble/droplet/particle effects or to understand better the effects on, and to optimise the design of, meters such as the turbines described by Mark, Sproston and Johnson (1990a) and Mark et al. (1990b).

For electromagnetic and ultrasonic meters, further work on the interaction of the field and bubble/droplet/particle may lead to more sophisticated sensors. Further study should also address the effect of large bubbles on the performance of such meters and on ways of flow-pattern mapping in general.

2.10 Gas Entrapment

Gas entrapment appears to occur in some flow geometries with important effects. Thomas et al. (1983) observed that "Transient large eddies (vortices) in turbulent free shear flows entrap and transport large quantities of bubbles, and may also force the coalescence of bubbles." At the conference where this paper was presented, the appearance of this phenomenon was discussed in relation to two papers in particular, which suggested the presence of the phenomenon. In one of these, Baker and Deacon (1983) had tested a turbine meter that appeared to exhibit hysteresis in its response to increasing and decreasing fractions of air in water. This may have been due to the particular meter or the flow circuit, but it raised speculation as to whether vortex structures in the vicinity of the ball-bearing fluid access path could be entrapping air and holding it in the bearing after the external air content in the flow had dropped. Presumably such entrapment could occur in the vortex downstream of an orifice plate [Figure 2.9(a)].

Hulin and Foussat (1983) observed the entrapment of a second phase in the shed vortices behind a meter bluff body. Figure 2.9(b) shows a diagram of this mechanism.

I have also observed, in some field data from an ultrasonic flowmeter, a behaviour that could result from entrapment of air. A meter in a low-head flow, where air was entrained with the water, periodically failed. The possibility that, in such a flow, the transducer cavities could cause small local vortices which could entrap the air and block the ultrasonic beam offered a possible explanation [Figure 2.9(c)].

There is considerable scope for investigating in other applications the appearance of such entrapment and the likely effect. In flow measurement, entrapment may

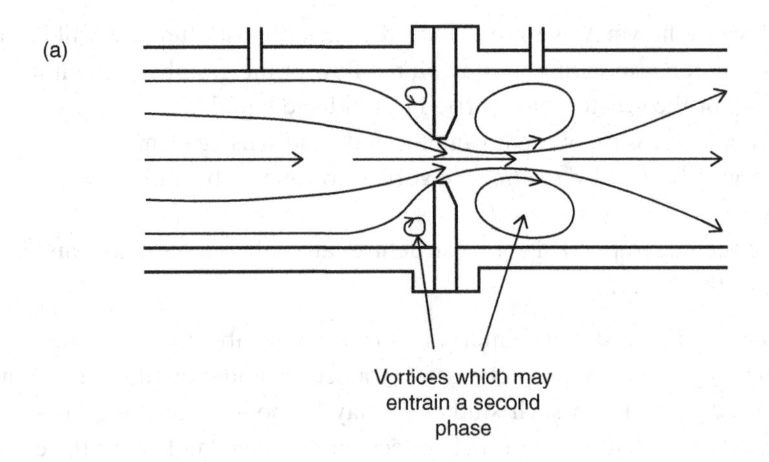

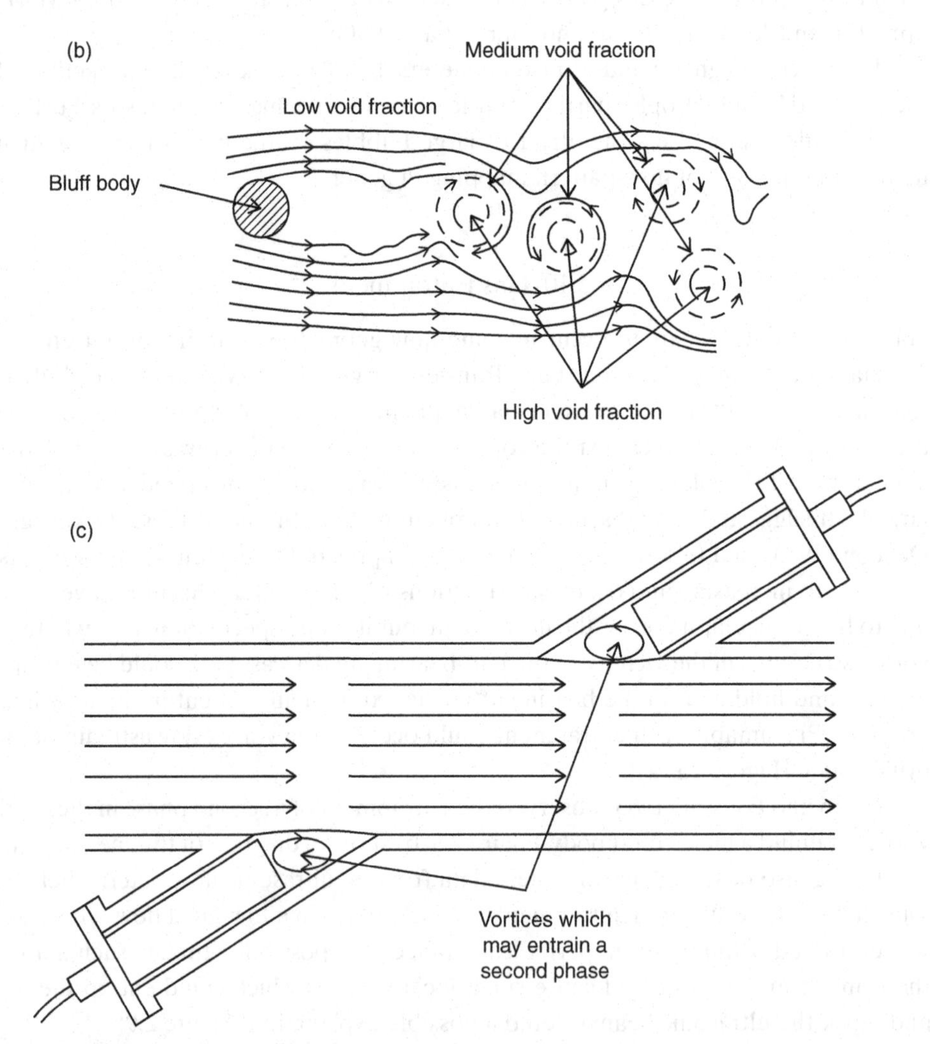

Figure 2.9. Possible mechanisms for gas entrapment: (**a**) Upstream and downstream of an orifice plate; (**b**) in vortices; (**c**) in an ultrasonic transducer cavity.

also occur around the throat of a Dall tube and behind an averaging pitot tube or other probes inserted in the flow. In all these cases, it is likely to affect the reading of the flowmeter. Entrapment may also occur in pumps, where it may reduce performance, and in pipework, where abrupt area changes exist.

2.11 Steam

One truly two-phase fluid is steam. Superheated steam may be treated as a gas, and its properties are well tabulated. However, it is increasingly important to measure the flow of wet steam made up, say, of about 95% (by mass) vapour and about 5% liquid. The droplets of the liquid are carried by the vapour but will not follow the vapour stream precisely. As with water droplets in oil, the liquid will drop through the vapour to land on the pipe wall, and we may obtain an annular flow regime until sufficient turbulence is created to re-entrain this liquid. The measurement of such a flow causes major problems because the pressure and temperature remain constant while the dryness fraction changes. It is therefore not possible to deduce the dryness fraction or density from the pressure and temperature, and, in addition, the droplets may cause an error when we attempt to measure the flow.

Some pointers from the literature will be referred to in relation to some of the types of flowmeter. However, the best and safest advice is to use a water separator upstream of the flowmeter and to measure the resulting flow of dry saturated steam.

2.12 Chapter Conclusions

I have been highly selective in choosing topics for inclusion in this chapter. Once again the reader is referred to fuller texts on fluid mechanics. Baker (1996) was written, particularly, with the industrial flowmeter user in mind and enlarges on some of the points made in this chapter.

The development of computational fluid dynamics provides a tool for analysing flowmeter behaviour, and references will be mentioned in the other chapters. I consider that perturbation methods should be explored. These methods would allow a greater precision in defining performance change between the ideal flow through a meter and the disturbed flow due to turbulence change, profile distortion or swirl.

The effects of small amounts of a second phase offer possibilities for investigating the behaviour of particle/bubble/droplet in flows through meters by developing, for instance, methods such as those of Ahmad et al. (1986) or Hayes (1988).

Other recent work on flow conditioning and flowmeter installation will be found in Appendix 2.A.

The critical-flow Venturi nozzle, as we shall see, appears to be capable of high precision. There may be value in fluid behaviour investigations in this and other areas to define the ultimate limits of accuracy.

Appendix 2.A Further Aspects of Flow Behaviour, Flow Conditioning and Flow Modelling

2.A.1 Further Flow Profile Equations

This section follows Ward-Smith's (1980) very useful exposition of the relevant flow profile equations. Equation (2.6) is an approximation to the actual turbulent profile in a pipe, and we will compare it with another equation. Ward-Smith (1980) sets out the laws for velocity distribution in terms of the law of the wall or the inner velocity law (comprising the viscous sublayer, the buffer layer and the logarithmic region), and the outer or core region.

The viscous sublayer next to the wall has a linear relationship when small-order terms are neglected

$$\frac{v}{v_\tau} = \frac{y v_\tau}{\nu} \tag{2.A.1}$$

where friction velocity $v_\tau = \sqrt{\frac{\tau_w}{\rho}}$, y is the distance from the pipe wall $= (R - r)$, ν is the kinematic viscosity and

$$\tau_w = f_F \frac{\rho \bar{V}^2}{2} \tag{2.A.2}$$

where f_F is the friction factor, and the viscous sublayer thickness, y_1, is given by

$$\frac{y_1 v_\tau}{\nu} \approx 5 \tag{2.A.3}$$

The Fanning friction factor, f_F, which we use here, following Ward-Smith (1980), can be obtained from experimental data charts based on Reynolds number and pipe roughness, e.g. Moody chart. If we take the mean velocity in a smooth pipe as $\bar{V} = 2$ m/s and the pipe diameter as $D = 0.05$ m for water flow, then Reynolds number is $\mathrm{Re} = 10^5$, and we can deduce $f_F = 0.00447$. Note that the Fanning friction factor, f_F, should be distinguished from the Darcy friction factor, f_D (e.g. used by Miller 1990), since $f_D = 4f_F$. Taking $f_F = 0.00447$, we obtain $V_\tau = 0.0944$ and hence $y_1 \approx 5 \times 10^{-5}$ m. Beyond this layer is a buffer layer which extends to a value of $y_2 \approx 30 \times 10^{-5}$ m.

The equation for the logarithmic region is (Ward-Smith 1980)

$$\frac{v}{v_\tau} = 2.5 log_e \left(\frac{y v_\tau}{\nu} \right) + 5.5 \tag{2.A.4}$$

Figure 2.A.1 compares these curves with Equation (2.6), with n = 7.

By substituting $v = v_0$ when $y = R$, he rewrites this equation in terms of the "velocity defect law" as

$$\frac{v_0 - v}{v_\tau} = 2.5 log_e \left(\frac{R}{y} \right) \tag{2.A.5}$$

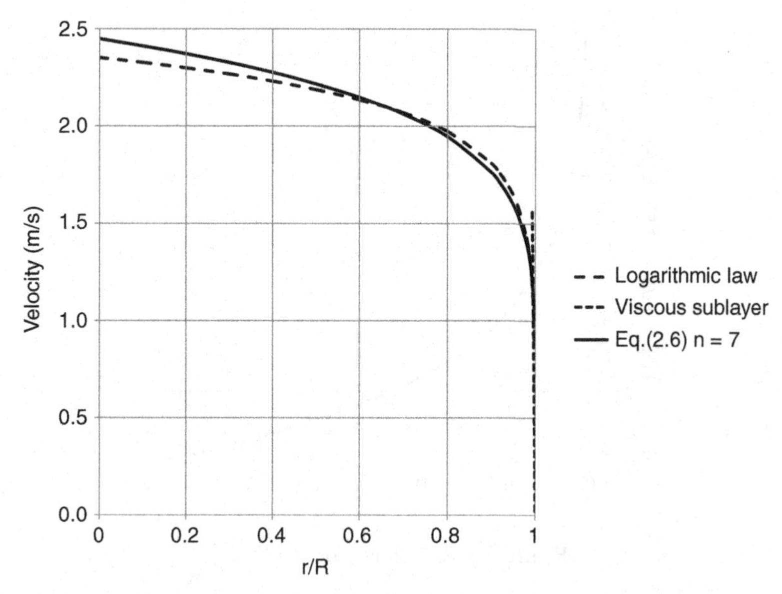

Figure 2.A.1. Comparison of Equation (2.6) with Equations (2.A.1) and (2.A.4).

It appears that the velocity defect law also describes the velocity in the outer region. Figure 2.A.2 plots this curve and Ward-Smith (1980) shows good agreement with experimental data for both smooth and rough pipes.

For those who wish to delve further into these matters, the book edited by Merzkirch (2005) is strongly recommended.

2.A.2 Non-Newtonian Flows

There appears to be a shortage of data on the effect of non-Newtonian flows on flowmeters. In some cases, of course, the effect should be predictable. However, Owen, Fyrippi and Escudier (2003) and Fyrippi, Owen and Escudier (2004) have presented data on the performance of Coriolis, electromagnetic and ultrasonic clamp-on meters when used for such flows. They found that the Coriolis meter behaved best, but that the other meters, which are recognised to be very sensitive to flow profile, were subject to errors, particularly the ultrasonic meter.

2.A.3 Flow Conditioning

Flow conditioning continues to have a significant development line based on the perforated plates (Gallagher and Laws). While important for differential pressure devices, they are also required in some installations for other meter technologies. Much of the work to date has been experimental, but some recent papers have used computational methods. A Laws plate which took some account of Reynolds number (Karnik 1995) appears to have given good results for single and two elbows out-of-plane for an orifice. Karnik noted the sensitivity of the performance of the

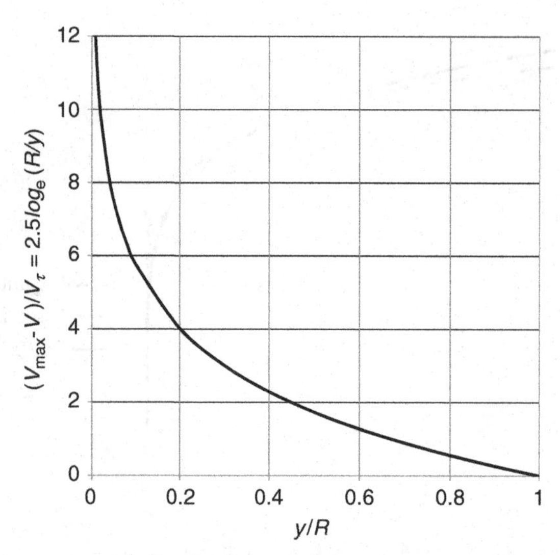

Figure 2.A.2. Velocity defect law.

plate to the size of the holes (see also Sanderson and Sweetland 1991; Spearman, Sattary and Reader-Harris 1996). He noted the need for the jets to have coalesced and the desired profile to have been achieved. Many of the modern plates continue to favour the perforated plate approach.

Schluter and Merzkirch (1996) undertook work on the performance of flow conditioners based on shape of profiles, and Spearman et al. (1996) compared velocity and turbulence profiles downstream of four perforated plate flow conditioners installed 4D downstream of the disturbing bends. This is an important paper with some interesting localised velocity and turbulence measurements. The factors evident from this research are the development of the velocity profile, the turbulence level and the angle caused by swirl and the head loss. The data seems to suggest that the Spearman (NEL) plate may be the best at dealing with both a single bend and two bends in succession, although the measured head loss appeared to be higher than for other plates (K was quoted as 2.9 for this plate compared with 0.8 for the unchamfered Laws plate). The Spearman (NEL) plate, 0.12D thick, appears to have three concentric rings of circular holes, Table 2.A.1.

As an example of the data obtained, Figure 2.A.3 is an approximate diagram to show how the data for the plates is set out in Spearman et al. (1996). The flow profiles downstream of the conditioner are set out on a graph with the ±5% tolerance limits for the baseline (fully developed turbulent) profile. The data is then added for various downstream spacings where I have shown the crosses in this diagram. The data should all fall between the ±5% tolerance lines to be acceptable.

The graphs for the r.m.s. fluctuating turbulence velocity profiles are similar except one baseline is included and no tolerance limits. Similar tolerance limits may be required for the turbulence curves. From the paper it appears that, for the Spearman (NEL) conditioner, the experimental measurements of the flow downstream of the conditioner all appear to fall within the ± 5% tolerance limits except

Table 2.A.1. *Spearman (NEL) conditioning plate layout (my interpretation of the values in Spearman et al. 1996)*

Ring of holes	Inner	Middle	Outer
Diameter of circular holes	0.10D	0.16D	0.12D
Pitch circle diameter for circle centres	0.18D	0.48D	0.86D
Number of holes	4	8	16
Angle between holes	90°	45°	22.5°

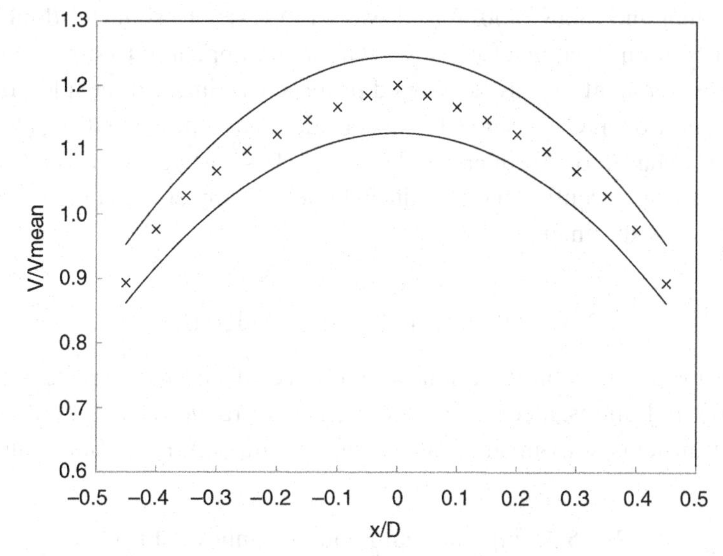

Figure 2.A.3. Approximate diagram to show how Spearman et al. (1996) set out profile data downstream of a conditioner installed 4D downstream from a bend. The curves are the ±5% bounds and the crosses are indicative of experimental results.

for the 3D downstream of the plate, and some appear to be marginally outside particularly in the pipe wall region. Data on other plates and turbulence data can be obtained from the paper.

The creation of a fully developed profile has always seemed to me a very elusive task, and the ultimate test as far as flow measurement is concerned will be with the measurement for a particular meter with a linked conditioner. The exception to this will be, of course, if an ISO or similar standard authorises its use within the uncertainty claims of the meter. Further data is provided by Morrison et al. (1997) on standard tube bundle and three different porous plate flow conditioners. Erdal (1997) appeared more confident of obtaining approximately fully developed velocity profiles a few diameters downstream by grading the porosity, but with reservations about turbulence profiles. Morrow (1996, 1997) compared tube bundles with the Gallagher flow conditioner and with no conditioner. The Gallagher conditioner appeared to reduce the error to less than ±0.2% for β ratios 0.2 to 0.75 and for 4D to 12D between conditioner and orifice plate.

It seems likely that standard documents will be updated, and may indicate acceptance tests for conditioners as further data is available, and the reader should check the relevant standards (Zanker and Goodson 2000). On the other hand, the value, in some applications, may be questioned (Merzkirch 1999) with the added disadvantage of pressure loss. Other relevant references are Benhadj and Ouazzane (2002) and Barton et al. (2002), who suggested that perforated plate conditioners developed and optimised in liquid flow would perform equally well in high pressure gas flow. Gallagher et al. (2002) noted that developments in flow conditioners had also affected the design of metering stations for natural gas [cf. Merzkirch (2005)].

Yang, Chen and Shaw (2002) used two CFD codes to simulate double-elbow pipe flows and found that an etoile flow straightener appeared to suppress swirl but delay the development of fully developed profiles. This implied the additional need for a flow conditioner. CFD is now being considered as a possible design tool, and a recent paper (Manshoor, Nicolleau and Beck 2011) suggested a "fractal" pattern of circular holes from a central one to reducing ones on concentric circles which made use of CFD and experiments.

2.A.4 Other Installation Considerations

Some other installation considerations for meter certification are (van Bokhorst and Peters 2002): pulsations, mechanical vibration (with ranges of amplitudes and frequencies), transients, start-up and shut-down, cavitation and gas bubbles in liquid.

2.A.5 Computational Fluid Dynamics (CFD)

The development of computational fluid dynamics (CFD) is increasingly providing a tool for analysing flowmeter behaviour, with increasing accuracy. The application of some form of perturbation method, where the error being sought is isolated by keeping all other aspects constant, may be yielding useful results. These would allow a greater confidence in defining performance change between the ideal flow through a meter, and the disturbed flow due to turbulence change, profile distortion or swirl. Hilgenstock and Ernst (1996) considered that CFD solutions might be capable of replacing testing in the near future. Recent developments appear to bring this possibility closer. Apart from the references below, there will also be references found under various flowmeters, where CFD has been used with some success.

Wilcox, Barton and Laing (2001) used CFD to check whether meter performance would meet ISO 5167 at a metering station due to upstream pipework. Barton and Peebles (2005) discussed the redesign of the installation of the ETAP gas export system. They undertook a CFD study which suggested that the Zanker plate would be effective and avoid an additional 0.5% uncertainty

von Lavante and Yao (2012) made numerical predictions of swirl decay, and they identified errors caused in a Venturi tube due to the swirl. Their results appear to indicate the significant effects of the radial distribution of the swirl. See also von Lavante and Yao (2010) on computation of flow upstream of meters.

CFD has been used in many other detailed designs and evaluations, of which some are: a flow diverter (Ho et al. 2005), upstream fittings (Barton, Gibson and Reader-Harris 2004, cf. Reader-Harris 1986), orifice plate back to front (Brown et al. 2000), eccentric orifice plates within carriers with damaged O-ring seals (Barton, Hodgkinson and Reader-Harris 2005), on the effect of contamination on orifice plates (Reader-Harris, Barton and Hodges 2010, 2012), low-pressure wet-gas metering (Kumar and Ming Bing 2011), effect of vertex angle and upstream disturbed flow on cone flowmeters (Singh, Singh and Seshadri 2009, 2010; Sapra et al. 2011), Venturi meter (Reader-Harris et al. 2005) and with low Reynolds number (Stobie et al. 2007), elbow meter (Meng, Li and Li 2010a), sonic nozzles (Bignell and Takamoto 2000), vortex meter (Chen et al. 2010; von Lavante et al. 2010), fluidic meter (Fang, Xie and Liang 2008), ultrasonic flowmeter effect of upstream profile (Frøysa, Lunde and Vestrheim 2001; Frøysa, Hallanger and Paulsen 2008, Hallanger 2002; Temperley, Behnia and Collings 2004; Gibson 2009; Zheng et al. 2010), thermal stratification effects in ultrasonic flowmeters (Morrow 2005), ultrasonic cross-correlation flowmeter (Lysak et al. 2008a, 2008b) and averaging Pitot tubes (Sun, Qi and Zhang 2010) etc.

3

Specification, Selection and Audit

3.1 Introduction

The object of this chapter is to assist the reader in specifying a proposed flowmeter's operational requirements as fully as possible and hence to enable the reader to select the right flowmeter. The reason for wishing to select a flowmeter may be obvious: a flow rate to be measured. However, the reason in some cases may be less obvious. If you are a flowmeter manufacturer, or prospective manufacturer, you may wish to identify a type and design which would meet a particular market niche. If you are a member of an R&D team, you may be exploring measurement areas which are inadequately covered at present.

This chapter is little changed from Baker (2000), which developed earlier ideas (Baker 1989; Baker and Smith 1990), and I have benefitted from the ideas of others such as Endress+Hauser (1989; cf. 2006). In Baker and Smith (1990), we took a different line from Baker (1989), providing the means to specify the user's needs in as much detail as possible. We then provided a form for communication with the manufacturer.

Many people have attempted to provide a means of selecting a flowmeter. An expert system to assist in selection has seemed to me the way forward, but the on-line or CD selection methods, available from some manufacturers, may essentially be providing a satisfactory approach. Such data is likely to reflect the manufacturer's latest developments.

3.2 Specifying the Application

The major manufacturers appear to have developed their choice of flowmeter types to cover most of the commonest technologies. I suspect that this has the useful effect of ensuring that the most appropriate type will be utilised for a particular application.

On flow measurement workshops we have run over the years at Cranfield University, at Cambridge and at Endress+Hauser, we divided up into working groups for selecting flowmeters for particular applications. Almost invariably there was more than one meter suitable to a lesser or greater extent for each application.

This will mean that more than one meter may be suitable for the job in hand, but the manufacturer may give an experienced view as to the best one.

I have, therefore, developed the theme of Baker and Smith (1990) in this chapter. We start with a form and discuss the implications of each item for user and manufacturer. Before the reader turns to the form and concludes that it is unrealistic for a user or manufacturer to complete or respond to such a form, I would hasten to defend its inclusion because:

i. it provides a checklist of features that might otherwise be forgotten in specifying the meter; and
ii. it is there to be modified, simplified and reduced in length and complexity, until it suits individual needs.

The form is set out in Appendix 3.A.1 under eight headings labelled alphabetically. It can be photocopied and/or adapted. It is aimed at the user and manufacturer, but it should provide a starting point for the prospective manufacturer, who may wish to identify a type and design for a particular market niche, or for a member of an R&D team which is exploring new development areas. In summary the eight headings are:

A. General information about the meter manufacturer
B. General description of the meter user's business and specific application
C. Details of fluid to be measured
D. Site details
E. Electrical and control details
F. Other user information such as preferred accuracy, materials, maintenance and cleaning
G. Manufacturer's additional information and requirements such as warranty, filtration requirements and price
H. Other considerations and final conclusion

The audit of a flowmeter should link closely with the specification and selection. To ensure consistency, the final section on audit reverts to the specification table and reviews the correctness of the original assessment (if there was one) or the appropriateness of the selection.

3.3 Notes on the Specification Form

A Information about the Manufacturer and/or Distributor

In approaching a manufacturer, it is important to know whether it will want to sell you its one and only flowmeter type – however appropriate or inappropriate for your application – or whether it has a range of meters, which will ensure that one is appropriate to your needs. It is also useful to know how many the manufacturer sells and the satisfaction of other customers. A manufacturer may be wise to include illustrations of a high-quality production line in its literature as well as indications of its international sales and servicing structure.

In making this point, I have no intention of excluding the smaller high-quality manufacturer whose products will be sold on their special features and adaptability. Such a manufacturer will have its own strengths. Indeed, the manufacturer should be clear as to what they are and base its selling strategy on them. In fact, many, if not most, manufacturers' outlets in a particular country will fall in the category of small and medium-sized enterprises and are likely to give a personalised service.

B User's Business and Specific Application

The object of this section is to provide the manufacturer with a clear overview of the application, site and essential background to maintenance and safety on site.

The relevant safety documents and those dealing with other relevant standards and legislation on electromagnetic compatibility (EMC) and Safety Integrity Level (SIL) should be referred to (note the CE marking requirement in Europe). Endress+Hauser (2006) provides an extensive discussion of the safety requirements, standards and approvals.

C Fluid Details

The user should endeavour to provide the manufacturer with the fullest information possible to avoid any incompatibility between fluid and meter. Of those properties requested, conductivity only applies to liquids where an electromagnetic flowmeter is to be considered. Opacity would be relevant only for meters that use optical sensing through the fluid. Compressibility may relate to the behaviour of differential pressure and ultrasonic flowmeters, and it may be useful for the manufacturer in advising on such applications.

Lubricity may be used in the flowmeter business, where the measured fluid is often expected to act as a lubricant for the flowmeter rotor or piston. *Lubricity* may be taken to mean boundary-lubricating properties (private communication from J. D. Summers-Smith). It is a term which may not be mentioned in books on lubrication or tribology. It generally is used to indicate the ability of the metered fluid to lubricate the bearings and sliding surfaces of a rotating or oscillating meter insert.

Summers-Smith (1994) states that: "Boundary lubrication can be demonstrated by the deposition of a single monolayer of soap on a steel surface. (A soap is the metal salt of a fatty acid, e.g., stearic acid reacts with the oxide film on steel to form iron stearate, i.e., a soap)." Long-chain hydrocarbon molecules may also provide this surface layer. Polar molecules which are chemically active at one end, react with the surfaces and are close packed rather like a carpet pile, are particularly effective as a boundary lubricant (Figure 3.1). Even if the surface layer is worn away the molecules will re-attach to replace the layer (Summers-Smith 1994). See also Hutchings (1992; Section 4.6) for a further description of boundary lubrication. It would, therefore,

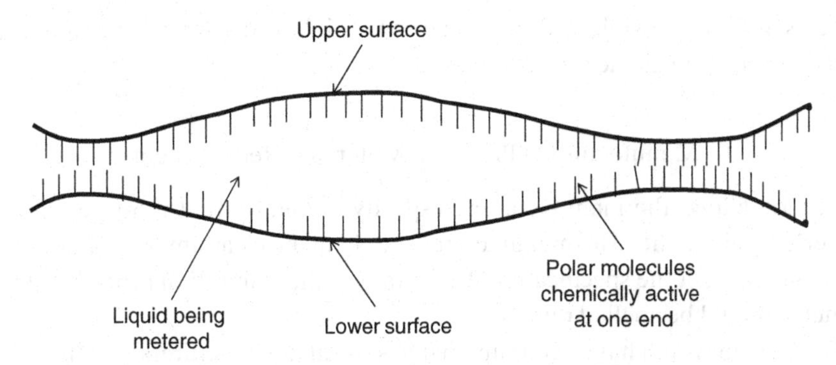

Figure 3.1. Simple diagram to explain boundary lubrication (after Hutchings 1992).

appear that if such a layer is not present or the liquid prevents such a layer forming, the "lubricity" will be poor.

Some gases (e.g. oxygen) require special lubrication. Because it oxidises steel pipes with the danger of chunks of rust detaching from the pipe wall, a maximum velocity is sometimes specified for oxygen flows.

Abrasiveness caused by solids in the flow will clearly be a problem for bearings and will lead to wear unless the particulate matter is filtered out.

D Site Details

Much of the data requested would be clarified and reinforced by a site plan. Details of access and environmental hazards are clearly essential. The access may be such that the meter has to be man-handled into position, placing severe limitations on weight.

E Electrical and Control Details

Whether the proposed instrument becomes part of a sophisticated modern telemetry system or is a stand-alone with local power – not to mention all the intermediate variants – will make a lot of difference to meter specification. In addition, all safety details for the site must be set out either here or under environmental hazards in D. Appropriate documents on safety should be consulted.

F Other user Information

In specifying accuracy, it is essential that the user does not overstate requirements or that the manufacturer does not oversell the product. The accuracy should be realistic when the total installation constraints are taken into account. Response time should be appropriate to the requirements of the plant. For a plant with, say, a dosing period of 10 s, the response will need to be rapid enough to achieve the required dosing precision. The instrument identified by the manufacturer should fit in with existing plant schedules as far as possible and should be compatible with extreme conditions (e.g. steam cleaning of hygienic process plant).

G Manufacturer's Proposed Meter and Requirements

This section allows the manufacturer to specify the meter that most nearly meets the user's requirements. The manufacturer should make clear any ways in which the meter will not meet the specification. The details of the calibration procedure for the flowmeter should be spelled out.

The form may well have been supplied to several manufacturers, and this section will provide the user with a comparison of the possibilities.

H Any other Considerations and Conclusions

This is an opportunity for the manufacturer to add further information and for the user to summarise the decision on selection.

3.4 Flowmeter Selection Summary Table

Table 3.1 lists the main types of flowmeter as set out in the chapters of this book and seeks to encapsulate ranges and other information. It is an attempt to provide the reader with an indication of a meter's capability. I am very conscious that in some cases the table may not adequately reflect the performance of some meters, and should be treated with caution, as ranges may be wider or narrower than indicated. The reader should refer to the manufacturers for actual ranges and other details.

For convenience, they are subdivided into momentum-, volume- and mass-sensing meters. This subdivision also emphasises that, for instance, a meter responding to momentum will be sensitive to density change. Where a less common meter of the same family has particular features that make it unique or particularly useful in certain applications, these may be indicated in notes in the table and in the discussion in subsequent chapters, where other points arising from this chapter will be elaborated. This table, therefore, provides a first cut at the selection for the task which has been identified by the specification questionnaire. The values given are my best estimates at the time of writing, but they probably fail to reflect the inherent variation in designs and extending size range and applications.

Nature of the Fluid to be Metered

The first choice, liquid or gas, will cause a few types of meters to be eliminated. The selection of suitable meters for slurries and two (or more) phase flows is severely restricted at present. In Table 3.1, X indicates unsuitable, and * indicates suitable except where some are more suitable (**) than others. A ? indicates that it is not common but might be appropriate in this application. The electromagnetic flowmeter has found a valuable role in slurry flow measurement. However, the Coriolis mass flowmeter is sometimes claimed to be suitable for slurry. The electromagnetic flowmeter is also useful in two-phase flows when the continuous phase is conducting. Venturis may be usable. The ultrasonic correlation flowmeter and the Doppler flow monitor may also be worth considering in such flows.

Flowmeters with rotating parts will be less suitable for dirty, abrasive fluids or for fluids that do not lubricate adequately. Abrasion may also detract from the performance of meters, such as the vortex meter. Cavitation should be avoided. The liquid must be conducting for current commercial electromagnetic flowmeters.

Flowmeter Range and Specification

Accuracy is perhaps the most difficult to determine because both user and manufacturer are prone to overstate their requirements and capabilities. It may be more useful to think of this as repeatability after calibration. In Table 3.1,

*** is for random error < 0.1% rate,
** is for random error < 0.5% rate, and
* is for random error < 2% rate.

The estimates reflect my view of the accuracy of flowmeters. These values should be achievable with a particular type of flowmeter. For a standard orifice the uncertainty before calibration due to the discharge coefficient is 0.5–0.75%, and overall uncertainty will be greater than this when errors in the measurement of dimensions, pressure and density are included. After calibration, the total uncertainty may be less than this, and it is reasonable to assume that the random error may be less than 0.5%. Data appear to confirm this. Some manufacturers may feel able to claim better values for the various meters than those in the table, but the reader should ask for justification for such claims. Commercial flowmeter claims of better than 0.5% indicate a level of accuracy that is probably only achieved in a few designs. Some manufacturers may not always achieve the values indicated. The repeatability of a variable area meter is often quoted as a percentage of full scale, and so on balance may be just outside the * rating for flow rates over, say, the top half of the range.

Precision relates to random error. It does not allow for changes in bias error caused by changes in method of measurement, location or long elapsed times.

Diameter, temperature and flow are my best estimates based on manufacturers' literature, but the reader should check latest claims. In all cases, the diligent reader can probably find examples of greater ranges than those given, and manufacturers are always seeking to extend the capability of their instruments. Note that, in the column for flow range, the values in parentheses indicate kilograms per hour rather than cubic meters per hour.

The maximum pressure has not been included because, apart from exceptionally high-pressure applications, when one might turn to an orifice plate, a turbine meter or a specialist meter such as the oscillating circular piston positive displacement meter designed for very high pressures, versions of most designs appear to be offered

Table 3.1. *Flowmeter selection summary tables*

1	2	3	4	5	6	7	8	9	10	11	12	13
Flowmeter type	Liquid or Gas	Slurry	Other Two-Phase	Accuracy	[d] Typical turndown	[e] Dia. Range (mm)	Temp.[f] Range (°C)	Flow Range[f] (m^3/h (kg/h))	Pressure Loss	Sensitivity to Installation[g]	Initial Cost	Notes
Momentum [a]												
Orifice	L	X	?	**	5:1	50–1,000		$1–3\times10^5$	H	H/M	L/M	Concentric ISO orifice with differential pressure cell assumed
	G			**	5:1	50–1,000		$1–4\times10^6$				
Venturi	L	*	*	**	5:1	50–1,200		30–7,000	M	M	M/H	
	G			**	5:1	50–1,200		$400–10^5$				
Nozzles	L	?	*	**	5:1	50–630		$2–1.7\times10^4$		H/M	M	Does not include critical nozzle
	G			**	5:1	50–630		$20–2.5\times10^5$				
Critical-flow Venturi nozzle	G	X	X	***	Reynolds Number range		2.1×10^4	to 3.2×10^7	M	L	M	**This is a mass flowmeter** Consult ISO 9300:2003 for details
Variable-area	L	X	X	?	10:1	15–150	–200 to 350	$10^{-4}–100$	M	L/M	L	Glass/plastic assumed, higher ratings for steel. Ranges dependent on designs
	G			?	10:1	15–150		$10^{-2}–2,000$				
Other Venturi-like	L	?	?	?	5:1	13–1,200			M	H/M	M/H	Consult manufacturers.
	G				5:1	13–1,200						
Averaging pitots	L	X	X	?	10:1	25–12,000	up to 450	$10–3\times10^4$	L	H	L	
	G				10:1	25–12,000	–100 to 450	$200–6\times10^5$				
Laminar	G	X	X	?	20:1	?		up to 120	H	L	M	Particularly appropriate for pulsating flows

Volume												
Positive displacement	L	X [b]	X [b]	***	10:1	4–200	–50 to 290	0.01–2,000	H/M	L	H/M	Turndown given may be achieved for rotary positive displacement gas meters
	G			***	80:1		–40 to 65	0.01–1,200				
Turbine	L	X	X	***	10:1	5–600	–265 to 310	0.03–7,000	M	H	L/M	High precision instruments assumed rather than water meters etc.
	G			***	30:1	25–600	–10 to 50	0.01–25,000				
Vortex	L	?	?	**	20:1	15–300	–200 to 400	3–2,000	H/M	H/M	L/M	Fluidic flowmeter suitable for wide range of utility flows. May be suitable for gas/solid flows.
	G			**	20:1	15–300		$50–10^4$				
Electromagnetic	L	**	**	**	100:1	2–3,000	–50 to 190	$10^{-2}–10^5$	L	M	M	Only available commercially for conducting liquids.
Magnetic resonance	L/G	?	**	**		100		L 1–100 G 2–400 actual	L	?	H/M	Suitable for multiphase flows.
Transit-time ultrasonic	L	? [b]	? [b]	**	20:1	10–2,000	–200 to 260	$3–10^5$	L	H/M	M/H	Single beam are more sensitive to installation.
	G			**	30:1	20–1,000	–50 to 260	$0.04–10^5$		H/M	M/H	Correlation and Doppler for two–phase flows
Sonar	L/G	**	**	**	10:1	50–1,500	–40 to 100	10–50,000	L	L?		Ranges may be exceeded
Mass												
Multisensor	L	** [c]	** [c]	?	?	?	?		H		H	
Wheatstone bridge	L	X	X	**	50:1	6–60	–18 to 150	$(0.05–2.3\times10^4)$	?	L	H	Suitable for engine testing
Multiphase Flowmeter	L+G		***	**	10:1	19–150	–40 to 120	0.1–10 m/s "L>G" <20 m/s "G>L"	?	?	H	"L>G" liquid dominated "G>L" gas dominated Figures depend on design and technology[h]

(continued)

Table 3.1 (*continued*)

1	2	3	4	5	6	7	8	9	10	11	12	13
Flowmeter type	Liquid or Gas	Slurry	Other Two-Phase	Accuracy	[d] Typical turndown	[e] Dia. Range (mm)	Temp.[f] Range (°C)	Flow Range[f] (m^3/h (kg/h))	Pressure Loss	Sensitivity to Installation[g]	Initial Cost	Notes
Thermal	L	X	X	*	15:1	2–6	0 to 65	(0.002–100)	M	L(CTMF)	M	
	G			*	50:1	6–200	–50 to 300	(2×10^{-4}–8,000)	L/M	M/H(ITMF)		for high gas flows CTMF+bypass or ITMF
Angular mom.	L	X	X	*	7:1	20–50	–40 to 150	(100–4.5×10^3)	M	M	M	suitable for aircraft fuel flow.
Coriolis	L	*	X	***	100:1	1–200+	–240 to 400+	(0.1–2×10^6)	M/L	L	H	Straight single tube has essentially no pressure drop unless smaller ID than pipe.
	G			**								Gas ranges are probably more limited.

Key: ***, Very high; **, More suitable/High; *, Suitable/Medium; ?/blank, Unsuitable/lower; X, Unsuitable. L = Low; M = Medium; H = High.
a Some proprietary devices offer special features: higher differential pressure, linear characteristics.
b Some designs have been produced for slurries.
c Multisensor systems have been developed specially for multiphase.
d Typical estimates that may be exceeded by using smart transducers and may be greater or less than the value attained by some designs.
e Larger or smaller sizes may be available. DP meters will require calibration outside ISO limits.
f Other ranges may be available.
g Flow conditioning may be used in some applications to reduce this effect.
h (Personal communication from Dr G Oddie 2015.)

for normal industrial ranges. In some cases, minimum pressures may be important because cavitation may be a danger, and the manufacturer should be consulted.

The pressure loss is given as H (high), M (medium) or L (low). Only meters with a clear bore have been categorised as L because there is virtually no more pressure loss than in a similar length of pipe. The orifice plate meter is H, and positive displacement meter and vortex are H/M. Losses will depend on such things as the instrumentation being driven by the positive displacement meter, but for modern meters with electronic sensors the drag should be negligible.

Installation effects caused by upstream pipework will be a recurring theme in most chapters. The positive displacement meter may be assumed to be insensitive to upstream installation effects and is, therefore, rated L. A few others, also, have low sensitivity to installation. However, most flowmeters are affected, and these have been categorised for a fitting, such as a bend, at 5D upstream.

L, negligible effect;
M, <2% increase in uncertainty;
H, >2% increase in uncertainty.

These categorisations are my best estimates, and manufacturers may be prepared to uprate them.

Most flowmeters are sensitive to pulsation over part of their range, with the possible exception of some positive displacement and some ultrasonic flowmeters and the laminar flowmeter, which was specially designed to cope with pulsating gas flows.

The flowmeters with essentially a clear straight bore (nonintrusive) are the electromagnetic, ultrasonic, sonar, magnetic resonance and some Coriolis designs. The only flowmeter that may fail and block the line is the positive displacement flowmeter. All the others have partial line blockage, although the Venturi has such a smooth change in section that its effect is probably of small concern.

Only the ultrasonic and sonar families offer clamp-on options (non-invasive) at present.

Response time will depend on fluid inertia, component inertia or electrical damping. The first two are inherent to the meter type and are not reducible. The electrical response may be speeded up in some meters provided stability and noise permit.

The installation of the flowmeter will also be affected by environmental considerations, such as ambient temperature, humidity, exposure to weather, electromagnetic radiation and vibration. These will need to be assessed in consultation with particular manufacturers. The area safety and explosion classification should also be checked.

Price and Cost

This is possibly the most important consideration and is difficult to tabulate because the range of price for any design is wide and the data are always changing! Therefore, I have attempted to rate the initial cost of the flowmeters as H (high), M (medium) or L (low). However, the medium bracket tends to become a "catch-all" and covers, for instance, electromagnetic and turbine flowmeters, which may actually have an initial cost range of 4:1 for a 100-mm diameter design. The move of companies to produce low-cost versions of more expensive meters has also affected this category. For this reason, L/M or M/H have been used where low- or high-cost versions are available.

Initial cost will, of course, include both purchase price and cost of installation. There will also be ongoing costs associated with maintenance, energy loss due to the presence of the flowmeter and savings based on the information provided by the flowmeter. If the flowmeter is clearly required, then the savings should outweigh the costs. This does not reduce the responsibility of the user to install the best and most economical instrument for the job!

3.5 Draft Questionnaire for Flowmeter Audit

In this chapter, we have taken a detailed look at the questions that a user and a manufacturer need to answer before selecting a flowmeter for a particular application. We have then looked at a selection table, which does little more than indicate the range of possible alternatives for most applications.

This final section sets out the essential information needed if the flowmeter installed is to be audited in the future. The objectives of such an audit should probably establish one or more of the following:

- The suitability of the flowmeter for the task,
- The current state of the meter,
- The optimum specification for a new flowmeter, and
- The suitability of the flowmeter's position for precise measurement and maintenance.

If in situ calibration is to be used, then the options need to be considered. The list in Appendix 3.A.2, based on Baker and Smith (1990), offers a starting point for the flowmeter owner. The list should be used either with the specification questionnaire if such exists or to create such a questionnaire for future meter actions.

3.6 Final Comments

To the company that uses flowmeters, note that the results of auditing flowmeters may vary widely. The replacement of meters because of age, wear or incorrect sizing may result in more precise measurements provided installation constraints allow it. However, this may not mean that the user reduces costs because correct measurement may mean higher costs. One is tempted to suggest that, for statistical reasons, the overall gain will be hardly worth the cost of the audit and replacement exercise. Of course, the need for greater accuracy will exist if an absolute flow rate is needed for control or fiscal reasons.

To the person responsible for purchase, make sure that you have specified the requirements carefully, and do not be content until you are sure that you have found a commercial device that matches your requirements. This will require a balanced view in assessing any concessions needed.

To the manufacturer, make sure that you have a clear understanding of the following:

a. what you are offering and why it is superior to your competitors' designs;
b. whether you expect to sell your products because you can offer a complete range for 95% of applications, or because you have the only device for particular applications;
c. the weaknesses in your products that need to be addressed and the gaps in your range that need to be filled.

To the research centre, identify the applications that have not been met and the technology most likely to meet them.

To the inventor, start from the application problem, look for features that may be inherently flow-dependent, and see if they can be used as the basis for a flowmeter. Remember that the power of modern signal processing should allow the mechanical design to be very simple.

The development of an expert system to satisfy all parties would clearly be valuable, but its production and updating may cost too much to make it worthwhile. I should like to prove myself wrong about this one day.

Appendix 3.A Specification and Audit Questionnaires

3.A.1 Specification Questionnaire

		Response
A	Information about the manufacturer and/or distributor	
A1	Address	
A2	Contact person and position in company	
A3	Company size and international standing	
A4	Flow product range and turnover	
A5	Other products	
A6	Accreditation (ISO 9000, ISO17025 etc.)	
B	User's business and specific application	
B1	Company's site address	
B2	Contact person and position in company	
B3	General description of application and telemetry links	
B4	Site safety requirements (Safety Integrity Level, SIL)	
B5	Current general policy on instrument maintenance	
C	Fluid details	
C1	Chemical composition, state (gas or liquid) and materials that are compatible or incompatible with the chemical	
C2	Purity of chemical and if contaminated or multicomponent, the nature of other components, likelihood of deposition	
C3	Commonly known properties	
	C3.1 Density	
	C3.2 Viscosity (if non-Newtonian, state type)	
	C3.3 Conductivity	
C4	Other features if known	
	C4.1 Lubricity and other features if known	
	C4.2 Abrasiveness	

C4.3 Opacity
C4.4 Compressibility
C4.5 Flammability and other safety considerations

D Site details (a site plan should be attached if possible)

D1 Flow range within an uncertainty value (e.g. ± 3%): maximum
minimum
normal flow
likelihood and size of over-range flows

D2 Pipework (a diagram should be attached if possible extending to at least 50D upstream, and should indicate the position of fittings such as bends, reducers and valves, particularly in cases where the fittings create swirl)
D2.1 Outside diameter
D2.2 Inside diameter
D2.3 Material of pipe and pipe lining if present
D2.4 Flange size and details or other connection method
D2.5 Nearest upstream pipe fitting
D2.6 Distance to upstream fitting
D2.7 Distance to nearest downstream fitting
D2.8 Orientation of pipe run

D3 Environment etc.
D3.1 Hazards including required safety and explosion-proof classification
D3.2 Access, size and weight limitations including vehicle access, machinery transport and shelter
D3.3 Flow source (gravity, pumped etc.)
D3.4 Flow stability
D3.5 Mounting requirements (including maximum weight)
D3.6 Ambient working pressure (where relevant external pressure subsea): maximum
minimum
normal
D3.7 Fluid working pressure: maximum
minimum
normal
D3.8 Ambient working temperature: maximum
minimum
normal
D3.9 Fluid working temperature: maximum
minimum
normal
D3.10 Maximum acceptable pressure drop
D3.11 Electromagnetic and radio frequency interference at site

E Electrical and control details

El Overall control system in use with communication protocol details

E2 Power supply availability and variability

E3 Preferred signal transmission if not defined previously:
0–20 mA
4–20 mA
0–1 kHz
0–10 kHz

	Pulse output:	frequency range
		pulse size
	Digital	
	Bus architecture	
	Wireless	
	2 wire/4 wire	
	Active/Passive	
	Other	
E4	Failure system:	
	fail soft	one operating and one stand-by channel
	fail passive	two operating channels
	fail operational	three operating channels and one decision element
	fail safe	output goes to safe condition
E5	Interelement protection (IP) AINSI/IEC 60529 describes the classification of the degree of protection provided by the enclosure of electrical equipment: 1) protection of people against access to hazardous parts and protection of equipment against access of solid foreign objects; 2) protection against ingress of water.	
F	Other user information	
F1	Accuracy (within a specified range)	uncertainty repeatability
F2	Response time	
F3	Preferred construction materials	
F4	Existing instrument maintenance schedules	
F5	Line-cleaning arrangements	
G	Manufacturer's proposed meter and requirements	
G1	Manufacturer's proposed meter specification	
G2	Special requirements resulting from this proposal (e.g. filtration, power supply, environment)	
G3	Calibration details, rig conditions and results	
G4	Warranty period and conditions	
G5	Delivery	
G6	Price	
H	Any other considerations and conclusions	
H1	Considerations and constraints by the manufacturer not covered elsewhere	
H2	Agreement of the user to the proposed meter	
H3	Conclusion and decision	

3.A.2 Supplementary Audit Questionnaire

This questionnaire should be read in conjunction with the specification questionnaire in Appendix 3.A.1.

		Answer
A	Information about the manufacturer	
B	User's business and specific application If the auditor is not part of the user's company, then this section may usefully be completed.	

C Fluid details

Is the meter of suitable design, construction and materials for this fluid?

D Site details (Is the site plan correct?)

On the plan, mark power and signal cables and the position of the main meter electronics.

What is the meter power consumption?

What is the flow range?

Are these within the meter specification?

Pipework

Is the site of the meter satisfactory? If not, should the meter be moved?

Environment etc.

Is the meter suitable for this environment?

Is the meter design suitable for these hazards?

Is the meter capable of operating within these parameters?

Is access sufficient for proper maintenance?

E Electrical and control details

Do power, control and signal systems match requirements for operating and protection?

Is there power and system access for audit and maintenance work?

What is the signal level?

Is the meter permanently energised and if not what is the warm-up procedure?

What is the control system for the meter?

Are there special: alarms?
low-flow cut-offs?

F Other user information

Has the meter accuracy been checked? If so, what is the uncertainty? repeatability?

What are the materials of the meter? Do they match preferred construction materials?

Have instrument maintenance schedules been adhered to?

If so, are there any apparent operational changes or peculiarities that may have caused problems etc.?

Has the meter reading changed unexpectedly?

What are the line-cleaning arrangements?

G Manufacturer's information and requirements

Meter specification:

Manufacturer?

Type?

Model?

Serial number?

Specification?

Where specification is unavailable, obtain key detail: as far as possible. Expected accuracy?

Construction date?

Installation date?

Subsequent checking/calibration/maintenance?

Is documentation available? If so, list.

Are all special requirements met (e.g. filtration, power supply, environment)?

Has the meter met the expected accuracy? If not, was it unrealistic/ unnecessarily stringent?

H Any other considerations and conclusions

Actions resulting from this audit?

4

Calibration

4.1 Introduction

The object of calibration is to benchmark a flowmeter to an absolute datum. Just as a benchmark tells us how a particular geographical point compares in height with sea level datum, so a calibration of a flowmeter tells us how the signal from the flowmeter compares with the absolute standard of a national laboratory. The analogy is not perfect at this stage. The national laboratory standard must also be compared with other more fundamental measures of time and mass, and it will be essential to check and compare different national standards in different countries from time to time.

There is a desire to reference back to fundamental measurements such as mass, length and time. Thus if we can measure mass flow on a calibration facility by using fundamental measurements of mass and time, this will bring us nearer to the absolute values than, say, obtaining the volume of a calibration vessel by using weighed volumes of water and deducing the volume from the density. The first is more correctly termed primary calibration, whereas the second fails strictly to achieve this. It is likely that the final accuracy will reflect this. Liquid calibration facilities can achieve a rather higher accuracy than is possible for most of the gas calibration facilities. In part, this difference will result from the lower density and the increased difficulties of handling a gas.

The result of a calibration will be both a comparison with the national standard and also a range within which the reading is likely to lie.

4.1.1 Calibration Considerations

Figure 4.1 shows three different approaches to obtaining the data. Although the selection of calibration points and their optimum distribution over the range is not obvious, most of the reference books seem to omit any comment on this. Figure 4.1(a) records data at one flow rate. This might be appropriate if the flowmeter were expected to operate at one rate. It shows the fact that there is variation

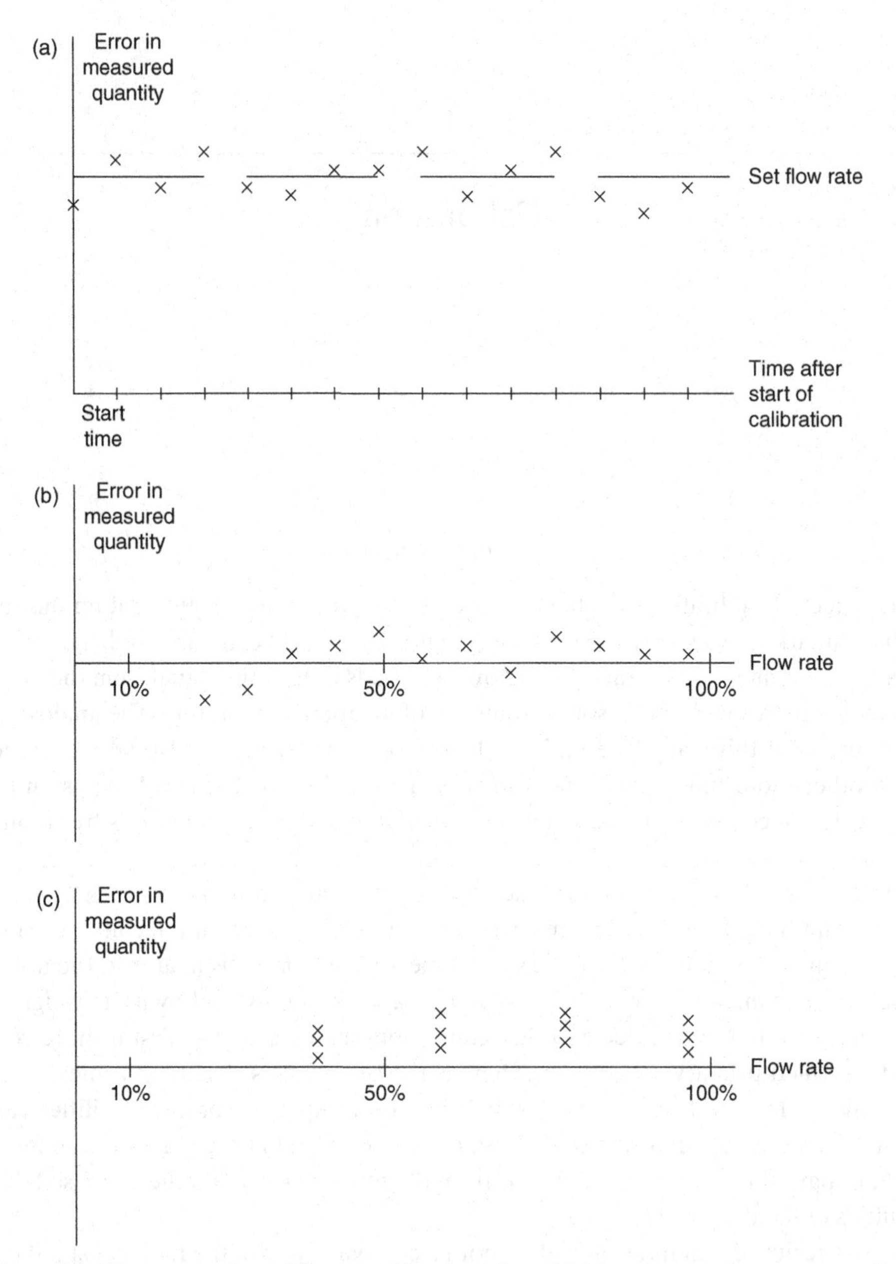

Figure 4.1. Diagram to show different calibration plots: **(a)** Measurements at a constant flow rate taken over a short period of time; **(b)** measurements of the full characteristic of the flowmeter; **(c)** measurements repeated at certain points in the flowmeter characteristic.

with time to which the repeatability refers. Such a plot might be used to compare the calibration facility itself with other facilities or to test the consistency of diverter operation. Figure 4.1(b) is probably the most usual plot. Here a number of points have been taken over the full range of the flowmeter. Figure 4.1(c) shows a characteristic that has been investigated in four positions, and three readings have been obtained in each.

The number of points will depend on the user's confidence in the calibration facility, the quality and linearity of the flowmeter and the statistical requirements for the required confidence level. Thus a manufacturer may have sufficient confidence in the quality of manufacture of its products to use, say, just two calibration points, and three readings might be taken for each to guard against one spurious reading. It might then be justified in claiming a ±1% envelope for its product. This is unlikely to be a nationally accredited calibration. On the other hand, 20 points or more along a complete characteristic of adequate linearity may be needed to ensure, for a high-quality meter, an acceptable value of uncertainty. At any point on the characteristic, there will need to be sufficient neighbouring points to achieve the required confidence level. (See Section 1.A.4 where there is such a calculation.)

Ideally the flowmeter would be calibrated under conditions identical with service ones. In practice, in situ calibration is not very accurate, and a calibration facility will provide the reference conditions for an instrument given that installation and other influences will increase the uncertainty. We shall therefore wish to consider the following factors.

Fluid: If the service fluid is water, then we shall have few problems, although it will still be necessary at least to check for gas in solution, for purity and for temperature. However, if the fluid is anything else, there will be a range of decisions. If we cannot match the fluid perfectly, we shall add to the final uncertainty of the calibration because the effect of, say, calibrating a meter for oxygen service on air will require a calibration adjustment. This adjustment will, itself, have an uncertainty in it.

Flow Range: It is obviously preferable if this is identical to that for the service. However, if the fluid conditions are not identical, then there may be a compromise between velocity through meter, mass/volumetric flow through meter, and Reynolds number. If the fluid for calibration differs from the fluid for service, the differing viscosity may cause the Reynolds numbers not to match. The importance of this is that the Reynolds number defines the shape of the flow profile and the turbulence spectrum (if the pipe is smooth). The profile shape (and the turbulence), as we noted in Chapter 2 and as we shall see for many of the flowmeters in this book, can affect performance. The viscosity may also change the nature of the flow over the internals of flowmeters, which in turn may alter their performance. Even if the fluid is the same, the temperature may cause changes in viscosity, which may be significant. There are also, of course, flow rate limitations due to the calibration facility.

Pipe Size and Configuration: The service pipe may not be of the same size as that on the calibration facility. Many flowmeters are sensitive to step changes in pipe diameter upstream of the inlet. It may therefore be necessary to calibrate the flowmeter with the upstream service pipework in place. It may be wise to spigot the flowmeter to the pipe, so that if they are separated, they can be reunited in precisely the same way as when they were calibrated as a unit. In cases where there are major changes in the diameter or fittings causing disturbance, and it is impractical to install enough upstream pipe in the actual facility, it may be necessary to use a flow conditioner package upstream, and the calibration, if possible, should be done with the complete conditioner package and flowmeter bolted together in the calibration facility. These requirements will be discussed in relation to the various flowmeter designs.

In addition, the upstream straight length of pipe should be sufficient to create a fully developed flow profile. This length is usually taken as of order 60D, although some suggestions have been made that, even after 70D, the flow has not settled, and it is known that swirl is very persistent.

Pipe Material and Finish: One aspect of this, which may be important (although possibly of secondary effect in many applications), is the roughness of the inside surface – the pipe should be smooth to ensure that the flow profile is predictable. Roughness may be defined as the arithmetic mean deviation of surface contour from a mean line based on the minimum sum of the square of the deviation. The maximum permitted roughness will probably be in the range 10^{-3}–10^{-4}D. In some cases, there may be questions of compatibility with the service conditions. For instance, the electromagnetic flowmeter can be sensitive to surrounding pipework if it has special electrical or magnetic properties. Also if the flowmeter is transferred with a length of pipe, this may result in incompatibility with the flow calibration facility.

Steady Flow: Most flowmeters are affected by unsteady conditions. There are several reasons for this; they are set out in Section 2.6. However, the calibration facility should be designed to ensure that flow is steady.

Environmental Conditions: Even though ideally the flowmeter should be unaffected by its environment, in practice this is unlikely to be the case. It is important that any effects are controlled and recorded. Where the service is very likely to be subject to external effects, temperature and pressure variation, humidity, immersion in water, vibration or hazardous environment, the selection of the meter prior to calibration will need to allow for these. If the meter is in a vulnerable position, there is always the possibility of accidental damage. And, of course, Joe Bloggs, who always taps his spanner on any convenient piece of pipe as he walks through the plant, may find a tender part of the flowmeter!

It should be remembered that, having taken great trouble to calibrate the flowmeter in the correct flow conditions, the actual service installation will probably have many of the shortcomings that we have been at pains to eliminate from the calibration facility and will, therefore, raise questions about the value of the calibration. In subsequent chapters, we shall review much of the available advice in these circumstances.

4.1.2 Typical Calibration Laboratory Facilities

Information on international flow calibration facilities is now available on the Web from BIPM [http://kcdb.bipm.org/AppendixC/default.asp].

As examples of such facilities I have listed in Table 4.1 some obtained from this website and from elsewhere. The selection given were of some for which the calibration rig was defined, information which did not appear to be available for all.

Table 4.1. *Examples of calibration facilities obtained from websites in November 2014*

Fluid	Calibration method	Approximate flow range	Relative expanded uncertainty ($k = 2$) of 95% confidence level
water[a]	constant water flow and weigh tank	0.67 to 65 kg/s	0.05% or less
water[b]	Static, gravimetric, (flying start / finish)	1.39 to 83.3 kg/s	0.042 %
water[b]	Static, gravimetric, (flying start / finish)	13.9 to 833 kg/s	0.060 %
liquid hydrocarbon[b]	Static, gravimetric, (flying start / finish)	0.022 to 67 kg/s	0.020%
Dry air or nitrogen[b]	Static, gravimetric, (flying start / finish)	0.1 to 180 g/min	($0.001Qm + 0.05$), Qm mass flow rate in g/min %
Dry air[b]	Critical nozzles	1.7 to 1670 g/s	0.28%
Dry air[b]	Static, PVTt, (flying start / finish)	1.7 to 333 g/s	0.17%

[a] NIST (The National Institute of Standards and Technology, USA): http://www.nist.gov/calibrations/mechanical_index.cfm

[b] NMIJ (National Metrology Institute of Japan) based on BIPM data in December 2014 (http://kcdb.bipm.org/appendixC/search.asp?branch=M/FF)

The reader should consult the websites for further details of temperature ranges etc. and for other flow ranges and national laboratories.

It appears from this sample that uncertainties may be better than ±0.1% for liquid flow facilities and better than ±0.3% for dry air flow facilities.

We discuss in the remainder of this chapter some of the typical designs of calibration facilities. Section 4.3.1 gives a typical design of a water flow facility as may be found in many calibration laboratories. The resulting calibration could be such as that in Figure 4.1(c) with four or five sets of readings of five average values at each flow rate. The whole procedure may then be rerun on a subsequent occasion to check repeatability.

Other rigs, of various designs, may be available for liquids other than water.

In many calibration facilities, critical nozzles and provers will provide secondary standards. The area of multiphase flow calibration is, also, increasingly likely to be present.

4.1.3 Calibration from the Manufacturer's Viewpoint

Weager (1993/4), who approached the subject from the standpoint of a manufacturer, described aspects of development of a flow facility. He commented that "the most used and abused term in any brochure for a measuring instrument is 'accuracy', and even the phrase 'typical accuracy' has been employed." Standards committees continue to address the specification of flowmeters. European co-operation for Accreditation publication EA – 4/02 also deals with uncertainty. Table 4.2 shows the typical quality of an approved laboratory, with uncertainty in the range ±0.1–0.2%.

Table 4.2. *Summary of accreditation of a manufacturer's flow calibration facility derived from inter-laboratory comparison over whole flow range with the national laboratory and not based on calculation of uncertainty budgets (reproduced with permission of Danfoss Flowmetering Ltd.)*

Test rig	Accredited flow range l/s	Measurement uncertainty for flow rate	Measurement uncertainty for quantity passed
100 kg	0.5–2.5	± 0.1%	± 0.1%
1,200 kg	5–25	± 0.2%	± 0.1%
8 tonnes	6–200	± 0.1%	± 0.1%
46 tonnes	200–1,200	± 0.1%	± 0.1%

The quantity passed was derived as in Figure 4.2. The national authority should ensure that round-robin tests of a transfer standard package are regularly organised between commercial laboratories and national and international standard laboratories to ensure that no undiagnosed changes have occurred. It is essential that the national laboratories are as aware of their own unexpected changes as are other laboratories and are open about the way they address these problems.

Table 4.2 is an example of the derivation of uncertainty in a commercial laboratory. An example of traceability for a laboratory is shown in Figure 4.3 for a water facility. In addition, flow rate uncertainty is assessed by inter-comparison between calibration laboratories, for instance with the national laboratory, using a transfer standard, and hence providing a further traceability. Table 4.2 indicates the importance of this. See "Inter-comparisons" in Section 4.A.2 and Karnik (2000).

The reality is that many flowmeters may be sold with a production calibration and without a fully traceable calibration certificate. The precision claimed for some flowmeters may be in conflict with the reality of the precision of the calibration facility. A flow rig with ±0.2% uncertainty cannot be used to credit a flowmeter with a lower uncertainty of, say, ±0.15%. It would be useful if manufacturers justified, on the basis of a calibration facility accuracy, their uncertainty values.

4.2 Approaches to Calibration

There are several approaches to calibration, and in the remainder of this chapter they are reviewed in turn.

Dedicated Flow Calibration Facility: This is at present the surest way to achieve a high accuracy. The facility will usually be traceable back to fundamental measurements of time and mass.

Master Meters: For a simple facility to compare two meters, the master meter offers a suitable transfer standard.

In Situ Calibration: Although there are various ways of calibrating on site, the ultimate accuracy is usually far less than on a calibration facility, and the skill and

CALCULATION OF UNCERTAINTY BUDGETS

Test Rig:- 8 Tonnes [Low Flowrate]
Date:- 21.05.97

Flowrate:-	**6**	**l/s**
Target Weight:-	**4000**	**kg**
Typical Density:-	**998.0000**	**kg/m³**
Diversion Time:-	**667**	**secs**

Measurement quantity:-

Mass

Symbol	Source of uncertainty	Value ± g	Probability Distribution	Divisor	c_i	$u_i(W_x)$ ± g
W_S	Calibration of standard weights	50	normal	2.0	1.0	25
S_D	Discrimination of weighscale	1000	rectangular	1.73	1.0	578
S_L	Linearity of weighscale	100	normal	2.0	1.0	50
S_H	Hysteresis of weighscale	0				
S_T	Temperature stability of weighscale	200	rectangular	1.73	1.0	116
S_R	Repeatability of weighscale	1000	normal	2.0	1.0	500
$u_c(Wx)$	Combined uncertainty		normal			775
$U(Wx)$	Expanded uncertainty		normal (k=2)			1550

Density (densitometer method)

Symbol	Source of uncertainty	Value ± kg/m³	Probability Distribution	Divisor	c_i	$u_i(D_x)$ ± kg/m³
D_{TX}	Calibration of Density Transducer	0.200	normal	2.0	1.0	0.100
D_T	Temperature compension	0.025	rectangular	1.73	1.0	0.014
D_P	Pressure compensation	0.005	rectangular	1.73	1.0	0.003
D_S	Long term stability	0.200	rectangular	1.73	1.0	0.116
$u_c(Dx)$	Combined uncertainty		normal			0.154
$U(Dx)$	Expanded uncertainty		normal (k=2)			0.307

Time

Symbol	Source of uncertainty	Value ±	Probability Distribution	Divisor	c_i	$u_i(T_x)$ secs
T_C	Calibration of Timer	0.001 %	normal	2.0	1.0	0.003
T_D	Discrimination of Timer display	0.1 ms	rectangular	1.73	1.0	0.000
T_T	Temperature stability of Timer	0.001 %	rectangular	1.73	1.0	0.004
T_{DIV}	Diverter operation	126 ms	rectangular	1.73	1.0	0.073
$u_c(Tx)$	Combined uncertainty		normal			0.073
$U(Tx)$	Expanded uncertainty		normal (k=2)			0.146

Summary - Flowrate measurement

Symbol	Source of uncertainty		Probability Distribution	u_i(FLOW) ± l/s	u_i(FLOW) ± %
$u_c(FLOW)$	Combined uncertainty (FLOW)		normal	0.0016	0.027
$U(FLOW)$	Expanded uncertainty (FLOW)		normal (k=2)	0.0032	0.054

Summary - Quantity passed measurement

Symbol	Source of uncertainty		Probability Distribution	u_i(FLOW) ± l/s	u_i(FLOW) ± %
$u_c(FLOW)$	Combined uncertainty (QUANTITY)		normal	0.0015	0.025
$U(FLOW)$	Expanded uncertainty (QUANTITY)		normal (k=2)	0.0030	0.049

Figure 4.2. Example of calculation of uncertainty budgets in flow rate measurement. Note that the symbols, not included in the nomenclature of this book, are defined in the table. Summation of uncertainties is by means of root-sum-squares (reproduced with permission of Danfoss Flowmetering Ltd.).

experience needed is greater. It may, therefore, be preferable to use the term in situ *verification*.

Dry Calibration: Dry calibration is not strictly calibration. An orifice meter, if correctly constructed according to the standards, can be measured, and from the measurements it should be possible to deduce the flow for a particular pressure drop. The

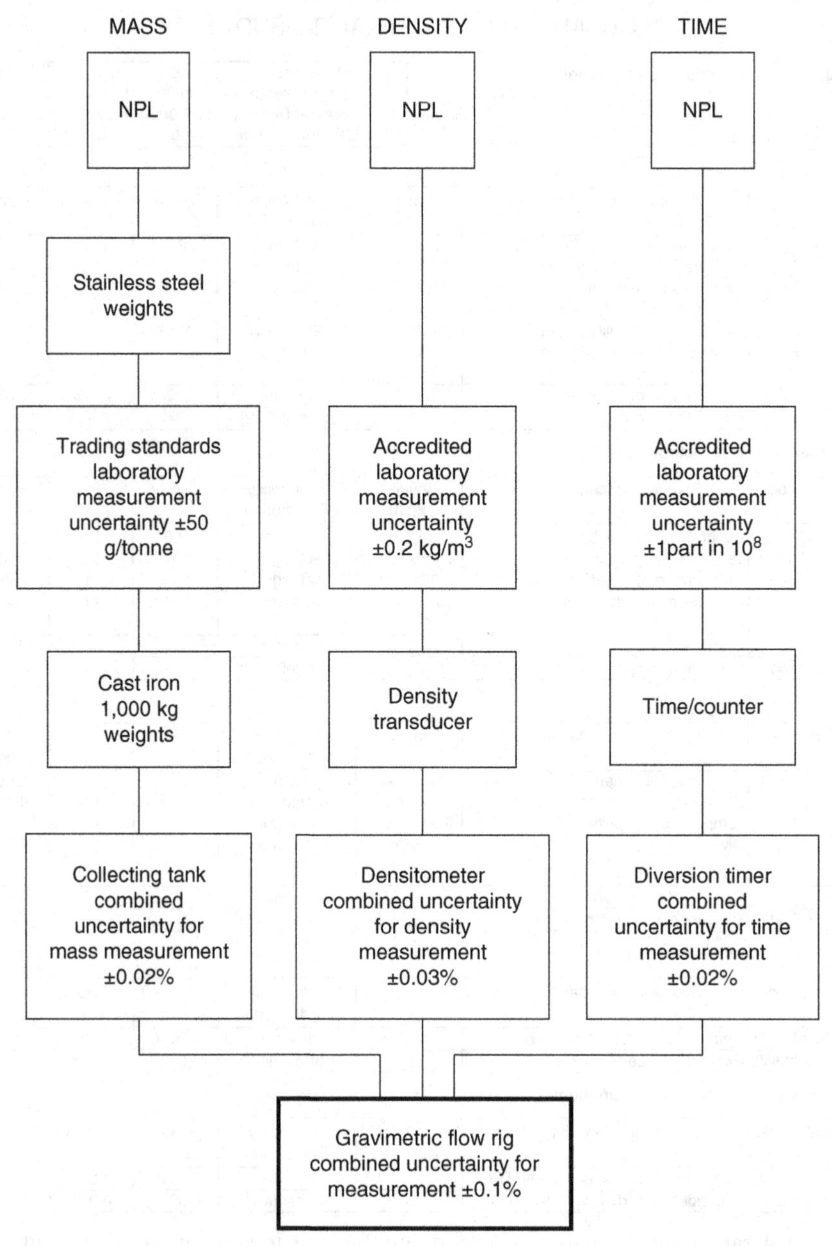

Figure 4.3. Example of traceability of measurements made on a UK commercial 8-tonne test facility with achievable uncertainties (after Weager 1993/4).

same is true of other standard meters such as the Venturi, nozzle and critical flow Venturi nozzle. Other meters are less predictable. No flow test is used, and so *dry calibration* is something of a misnomer, since we are really talking about the ability of standards or theory to predict performance. The reader is referred to the appropriate chapter for the likely predictability of particular flowmeters such as the electromagnetic and ultrasonic.

4.3 Liquid Calibration Facilities

4.3.1 Flying Start and Stop

Figure 4.4(a) shows the outline of such a facility, which is most commonly used for water. This is known as the static weighing method because the water is diverted into a tank for a fixed period, and the tank can then be measured with the water static in the tank. In its ideal form, there would be a header tank of large enough size so that change in level was negligible. A header tank, with a weir to retain constant head, is shown in Figure 4.4(b). The flow from this, by gravity, would pass through the flowmeter under test and thence to a control valve before discharging into a diverter valve. This valve would allow the flow to drop either into a weighing tank or into the sump tank. Water in the weighing tank would be drained into the sump tank once the weight had been taken. The beauty of this design is that the flow would be extremely steady due to the header tank and errors would be restricted to the measurement of the time during which the flow was diverted into the weighing tank and to the weight itself. Figure 4.5 shows a diagram of the diversion process and the source of timing error. Mattingly (1982) elaborated on the error, and the reader is referred to the original article for full details.

The essential problem is that it is physically impossible to divert the flow to and from the weighing tank instantaneously. The time at which the diversion takes place is, therefore, not at a precise moment, and the changing flow is illustrated in Figure 4.5. The timer is set to start when volume a equals volume b. This is when the lost water shown by volume b is just balanced by the gained water in volume a, which passed before timing began. The same compromise is needed at the end of the timing period with c equal to d. The setting is often, but not necessarily, the same at the beginning and end of the diversion (cf. Buttle and Kimpton 1989 for the design of a diverter). See "Flow diverters" in Section 4.A.2.

In order to check the validity of this compromise method, one test is to divert many times after short periods, comparing the result with one diversion over a period equal to the total of the short runs. This will tend to emphasise the errors in the diversion, and an adjustment can be made to minimise them. However, there may be variation in the diversion flow depending on flow rate, and this will require a more sophisticated compensation method.

It is reasonable to assume that the time measurement can now be several orders of accuracy higher than the other measured quantities. The errors will therefore come from the weight measurements and the setting of the diverter. Scott (1975a) suggested that the true weight of water is about 1/1,000 greater than the apparent weight (0.12% may sometimes be used as an approximation). For accurate weighing, therefore, the upthrust due to immersion of steel test weights and the water in the weigh tanks in air will need to be allowed for (see 4.A.6). The typical overall uncertainty of such a facility is currently of order 0.1%.

One variant on the weighing method is to use a dynamic weighing system where the flow enters the one tank, and a fast drain is available in this tank. The flow

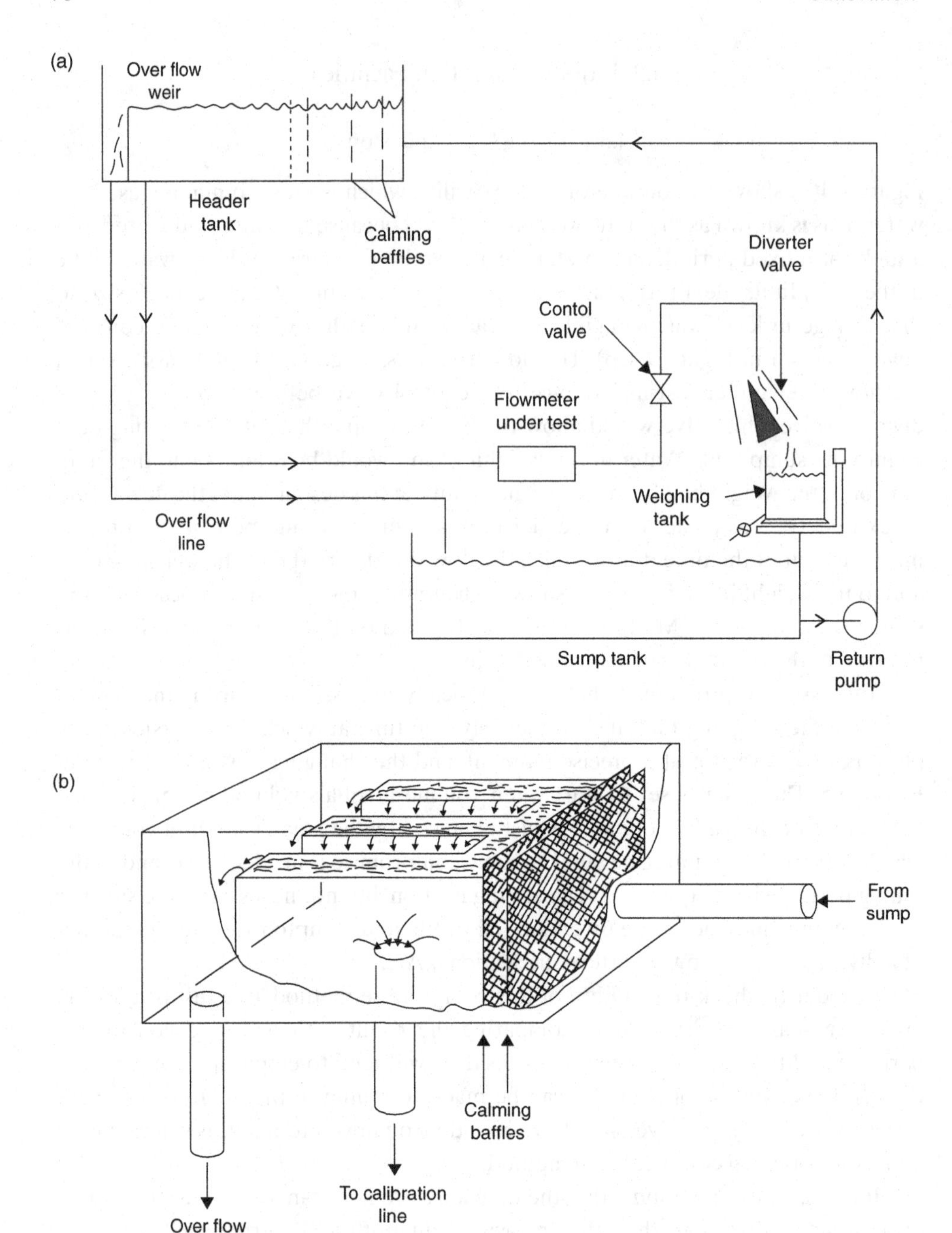

Figure 4.4. Flying start and stop liquid flow calibration facility with static weighing: **(a)** Outline diagram; **(b)** typical constant head tank design (after Harrison 1978a).

entering the empty tank rises in the tank, and the weight is measured (usually electronically) at the start of flowmeter output measurement. At the point at which sufficient water has been collected, the second weight is measured and the drain valve is opened. Pursley (1986) advised that the tank should have vertical sides and that the

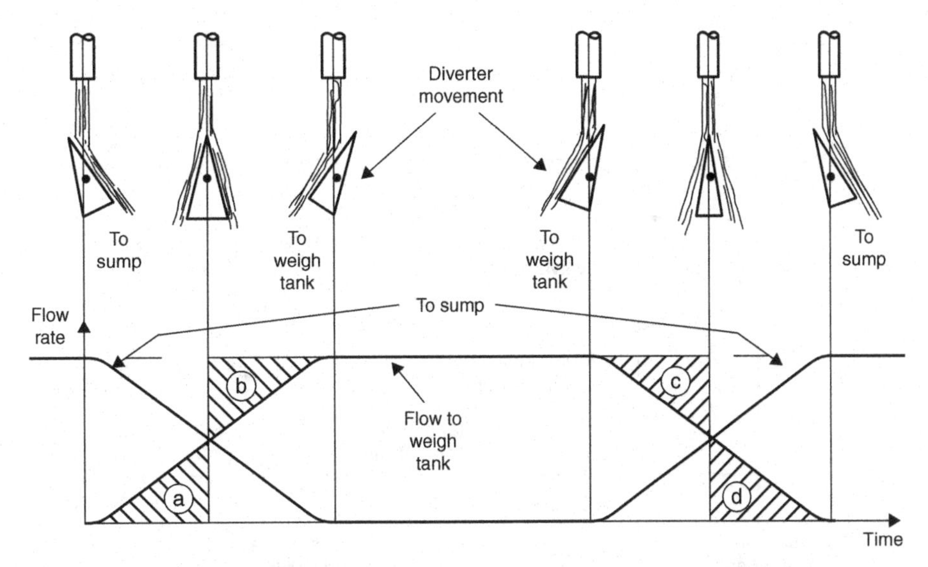

Figure 4.5. Diagram of the flow division for a diverter valve in a flying start and stop facility (after Harrison 1978a).

liquid jet should enter vertically. There are problems with the dynamic nature of this at high flow rates: Does the inertia of the weighing tank affect the measurement? Does the jet momentum have any effect?

Another variant is known as the substitution weighing method. At the start, the weighing machine has the nearly empty tank and a calibrated weight of the same value as the liquid that will enter the weigh tank. The weight is removed, and the water is allowed to enter until the balance is restored. If standing start and stop is used, then the final balance can be perfected with additional small weights (Pursley 1986).

In practice, it is often inconvenient to use a header tank and necessary, therefore, to use some means (a pump) for recycling the liquid. This in turn is likely to make it more difficult to ensure that the flow is absolutely steady and great care will be required in the design and specification of components such as pipework, pump, flow conditioners etc.

Scott (1975a) encouraged protection of the calibration weights and their integrity by keeping them in soft material.

It should be noted that, for very high-precision facilities, even air circulation may affect the readings.

4.3.2 Standing Start and Stop

The standing start and stop amounts to allowing fluid to pass through the meter for a fixed time and then comparing the total registered by a reference meter with the amount collected. The amount collected could be measured either as a volume or a weight. The problem with this method is that the meter must follow the changing flow rate accurately, and many meters are unable to do so. Figure 4.6 is a diagram

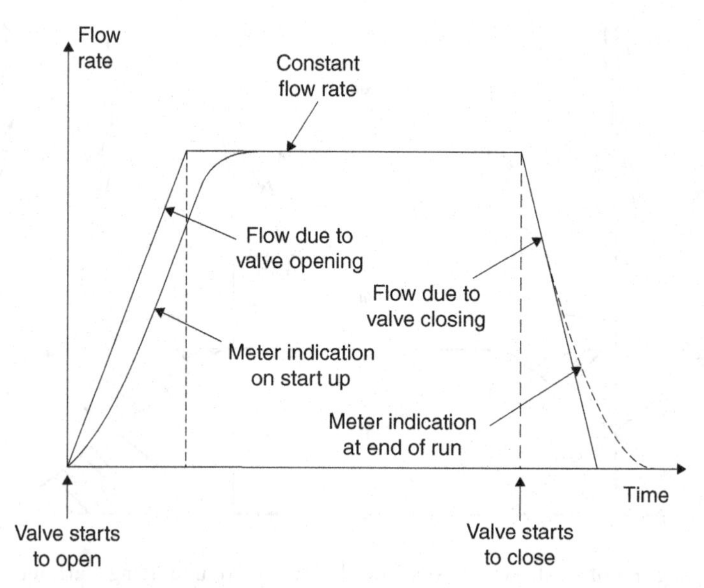

Figure 4.6. Discrepancy between changing flow rate and the meter's response.

to indicate the variation between the meter's response and the actual changing flow rate. Each meter will differ according to how it measures the varying flow. There will, therefore, be an error in the measurement in most cases, particularly if the meter is of poor accuracy at low flows.

For standing start and stop, it is clearly essential to be sure that the same amount of fluid has passed through the meter as arrives in the mass or volume measuring tank. This usually requires particular attention to the design of weirs and sight glasses, which confirm that the level in the tank is the same both before and after the run and that the flow stops quickly. Examples of weirs are shown in Figures 4.7(a, b) and Figure 4.A.4. It also requires care to ensure that valves do not leak. The block and bleed is one system to achieve this [Figure 4.7(c)].

A typical measuring vessel for volumetric measurement is shown in Figure 4.8. The vessel is calibrated using weighed quantities of distilled water at a known temperature. The top and bottom of the vessel are of small diameter so that small volume changes cause measurable level changes compared with the large volume contained in the centre portion of the vessel. The initial fluid level is usually fixed by a weir and sight glass. The fluid is then allowed to enter the vessel and is stopped when it rises into the narrow top section. Against this, there is either a calibrated sight glass or a means to measure the surface level to obtain the final volume. It has been found that the vessel will need to be filled and emptied three or four times before it has wetted sufficiently on the inner surface to ensure that a consistent and correct reading is obtained. When a consistent reading is obtained, the value can be used to calibrate the flowmeter.

The viscosity of the liquid will also affect the emptying of the vessel, and as a result Pursley (1986) recommended a maximum viscosity of 5 cSt for this calibration method.

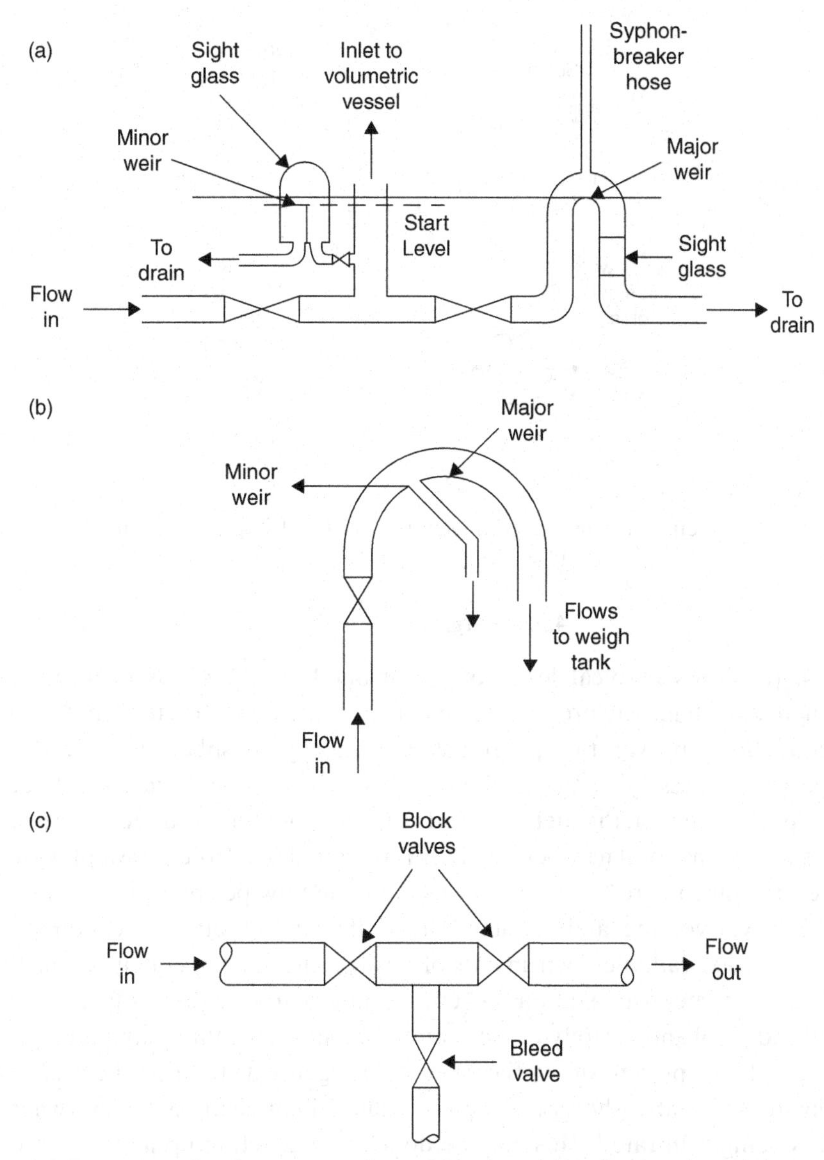

Figure 4.7. Examples of weirs, swan necks and block-and-bleed valves in calibration (after NEL). **(a)** Arrangement for precise starting level for volumetric tank; **(b)** arrangement for standing start and stop gravimetric system; **(c)** block-and-bleed valving system.

Proving vessels of this type are used to calibrate meters to uncertainties of ±0.1% for custody transfer of hydrocarbon liquids. A typical vessel could have a capacity of 500 l. Such vessels, of various sizes, have been used for a variety of liquids, including foods.

Another type of volumetric facility uses the falling head method. A tall tank is filled with liquid. The liquid is allowed to flow through an outlet pipe to a flowmeter. The volume between two level switches is known, and the meter is calibrated against this volume. The disadvantage of this method is that the flow rate may change slightly as the head falls.

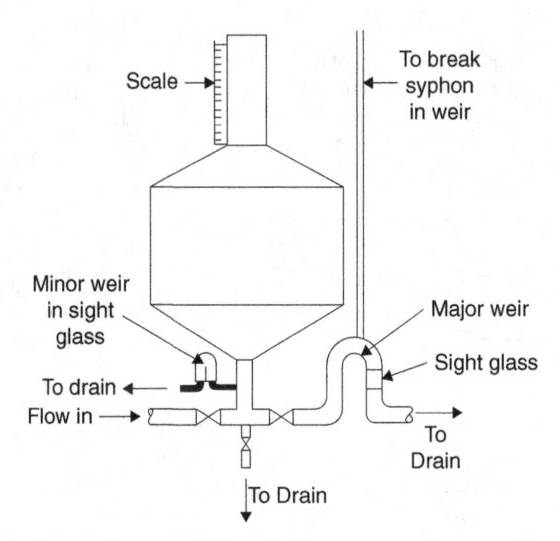

Figure 4.8. Volumetric proving vessel for standing start and stop calibrations (after NEL).

4.3.3 Large Pipe Provers

Figure 4.9(a) shows a typical design of prover, and Figure 4.9(b) is a photograph of a commercial system. The prover consists of a carefully constructed length of pipe. In a bidirectional prover, the sphere passes both ways. A sphere is projected into the pipe, and its passage is recorded past two set points where there are detectors. When it passes the first, the meter-pulsed output is counted until the sphere passes the second, and the total number of pulses is obtained. The volume swept is known, and the time taken is recorded. With this data, the flow per pulse can be checked with the prover volume, and the actual flow rate can be obtained. Unidirectional provers are also available. Instruments of this type have been commonly included in hydrocarbon metering stations, where oil and hydrocarbon products are transhipped and fiscal and custody transfer requirements come into play. Figure 4.9(b), a photograph of a prover, shows the valving arrangements to bring the prover into play. Figure 4.9(a) also shows the separate flanged connections for use when the prover is being calibrated. This may be done by connecting up a mobile proving vessel or transfer standard flowmeter. The calculation of the internal volume of the prover by measurement, even if the accuracy is achievable, would seem unlikely to be an economical method (cf. Paulsen 1991 on prover ball material problems and materials).

Claims are made that these devices can achieve uncertainties of the order 0.1% (or less) with calibration.

4.3.4 Compact Provers

The large bulk of the pipe prover, which leads to weight (e.g. in oil platform installations) and space considerations, and also the amount of liquid needed to complete the calibration have led to the design of very compact piston provers [Figure 4.10(a)].

(a)

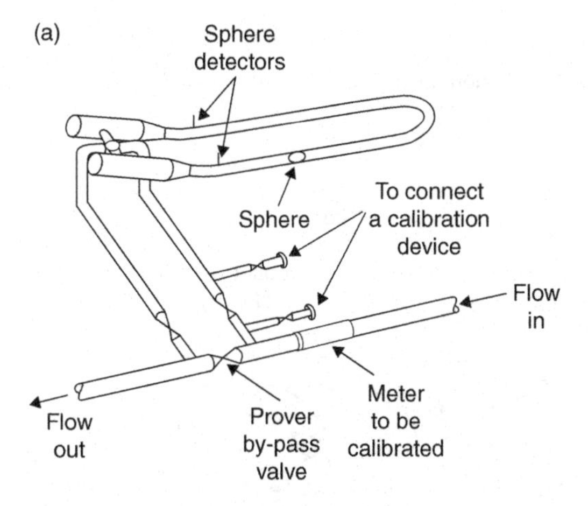

(b)

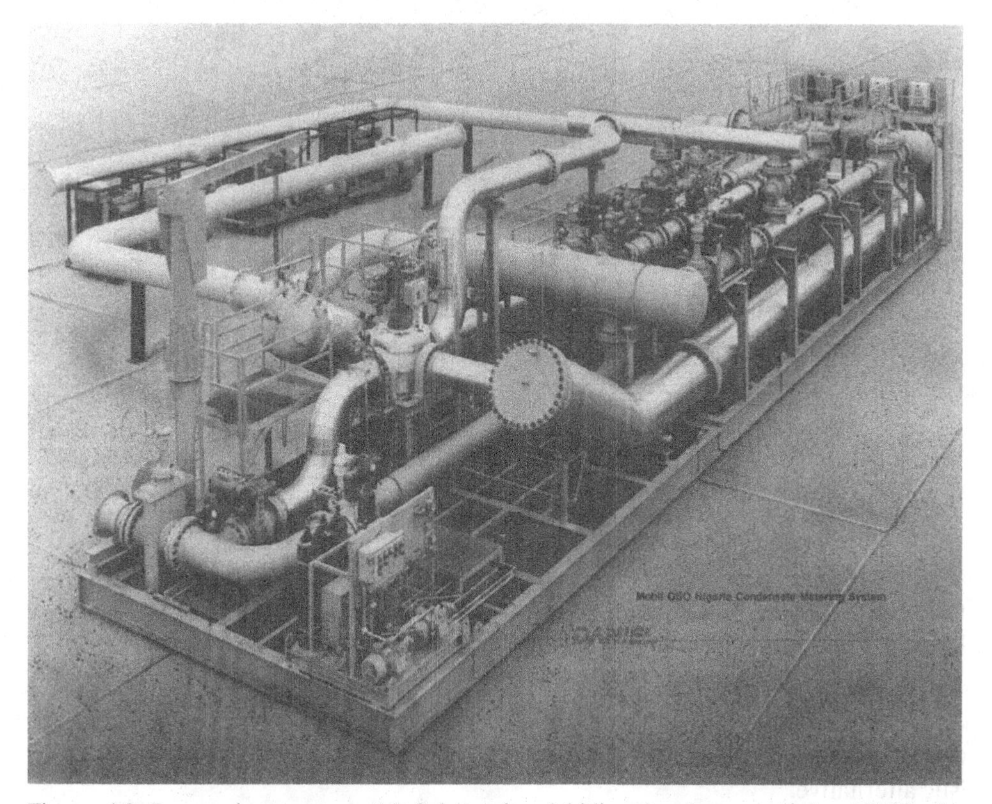

Figure 4.9. Large pipe provers: **(a)** Schematic of bidirectional prover (after NEL). **(b)** Photograph of a bidirectional prover (reproduced with permission of Emerson Process Management Ltd.).

These devices still depend on the swept volume of a tube but use a piston moving over a much shorter distance. Figures 4.10(b–d) show the stages in the measurement cycle. Thus, with great manufacturing precision, it appears that the performance of these compact provers can rival that of the large pipe prover. Compact provers will,

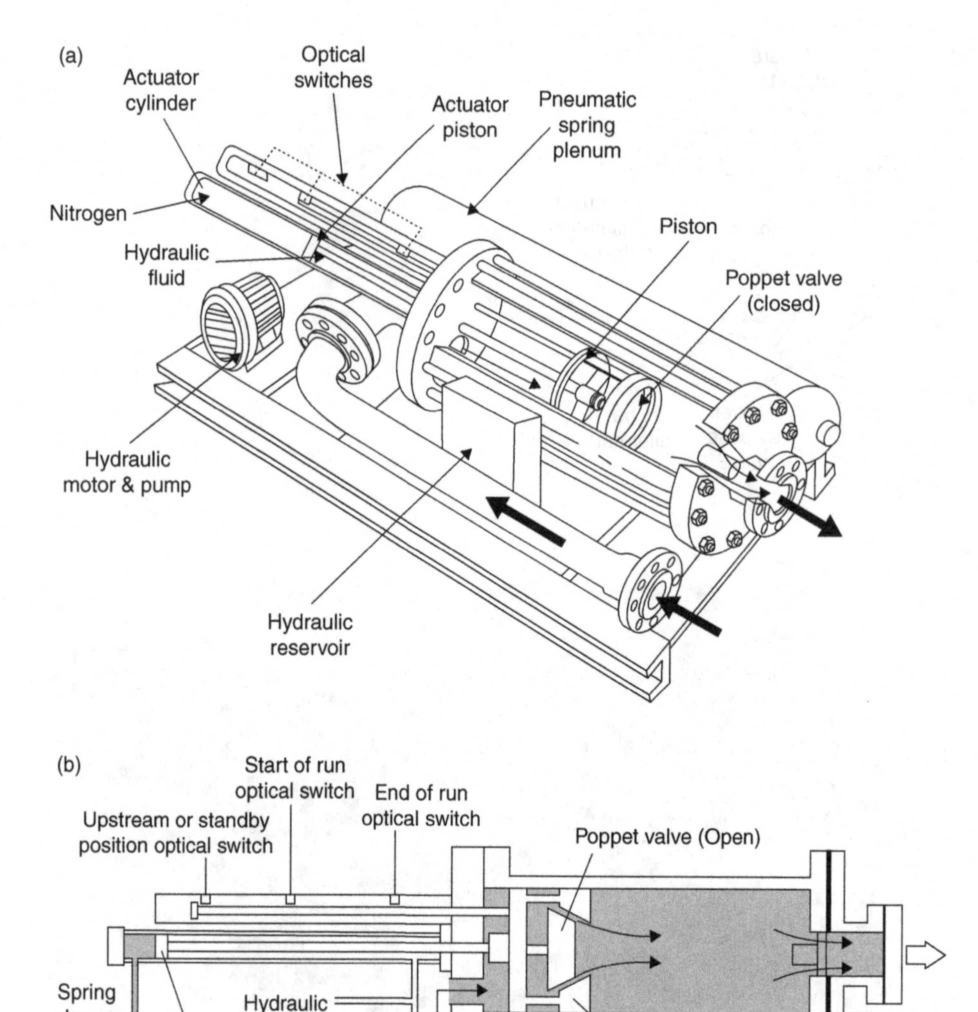

Figure 4.10. Compact prover (reproduced with permission of Emerson Process Management Ltd.): **(a)** Principal components; **(b)** standby mode; **(c)** start of run; **(d)** end of run.

certainly, target the same level of accuracy as pipe provers in order to provide a realistic alternative.

One problem that arises in the use of these compact prover results is the small amount of fluid that passes. Thus, if a flowmeter produces a pulse for a certain volume of fluid passed, the number of pulses resulting from the fluid passed by a compact prover may be very low, causing a discrimination error. This has led to the special techniques to overcome the problem, such as pulse interpolation (Furness and Jelffs 1991) in order to achieve the required 0.01% resolution (based on 10,000 pulses for the pipe prover). Reid and Pursley (1986) used a double chronometry

(c)

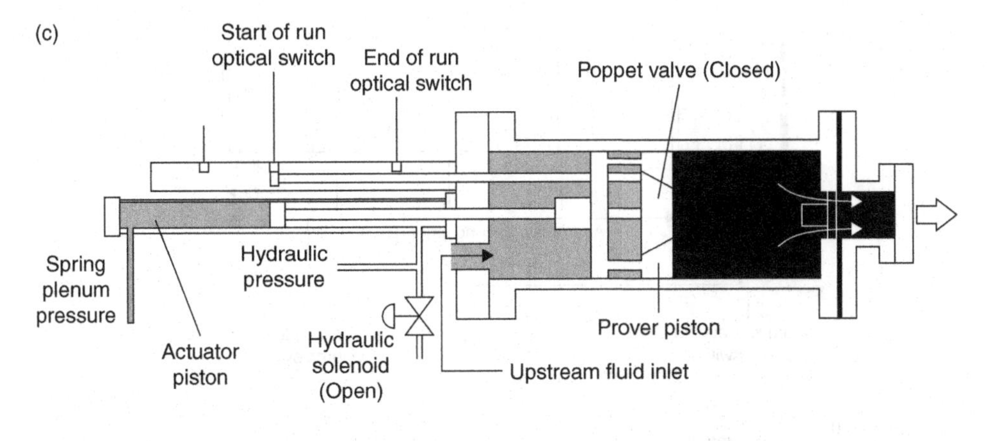

(d)

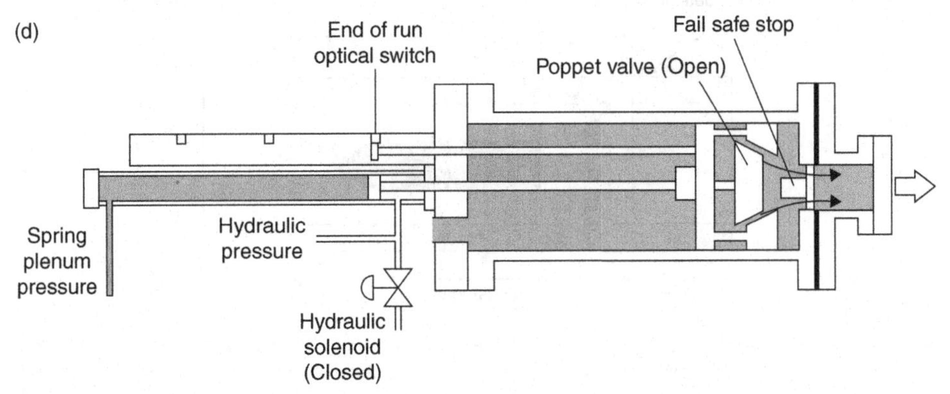

Figure 4.10 (*continued*)

method for pulse interpolation [Figure 4.11(a)]. As the first prover volume switch is activated, two timers are started. The first timer (A) runs for the full period of piston movement, from activation of the first prover volume switch until the second prover volume switch is activated (volume D). The second timer runs until the first turbine pulse and starts again at the second prover volume switch, running until the subsequent turbine pulse. Alternatively, the second timer (B) runs from the first turbine pulse after the first volume switch until the first pulse after the second volume switch (C pulses in total). Thus the full time, the number of pulses and the fractional period between pulses are all obtained. The turbine pulses per unit volume K can be found by multiplying (Time A/Time B) by the total pulses C and dividing by the displacer volume D [Figure 4.11(b)].

An advantage of the compact prover (Furness and Jelffs 1991), especially for meters delivering products to ships, is that it can be easily cleaned between grades and so eliminate contamination. It can also be skid mounted. Wherever possible, comparisons between ship and meter quantities should be monitored continuously to provide warning of meter factor drift (cf. Hannisdal 1991 on alternative metering concepts to reduce space and weight).

Shimada et al. (2015) investigated the performance of a small volume prover (SVP) and concluded that they should consider the performance of each type of

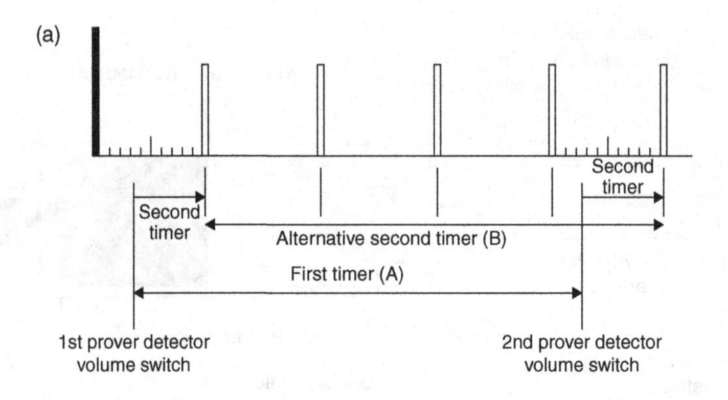

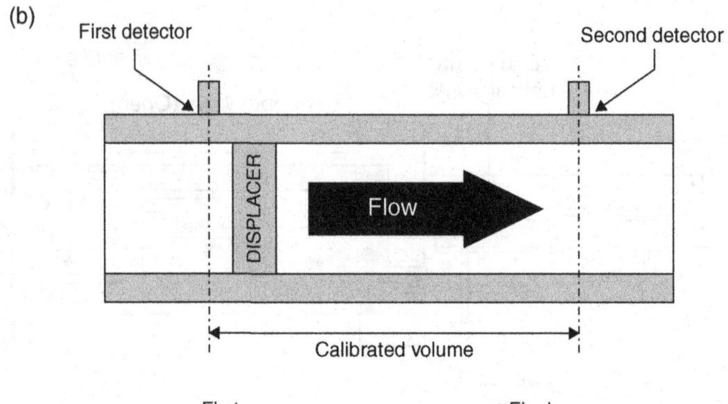

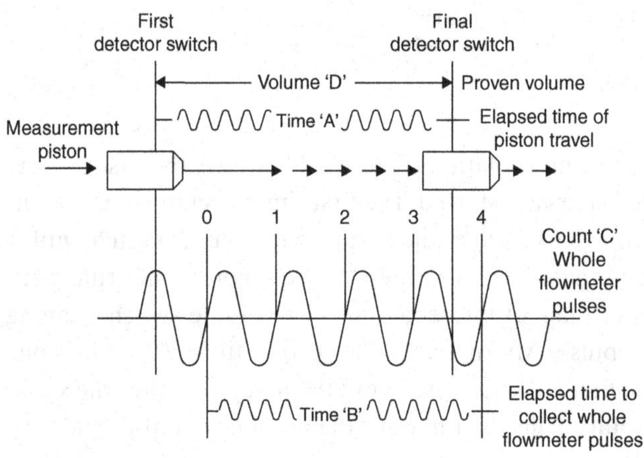

$$K = \frac{\text{Time A}}{D} \times \frac{C}{\text{Time B}}$$

Figure 4.11. Interpolation of flowmeter pulses for compact prover. **(a)** General concept and double chronometry method for pulse interpolation (after Reid and Pursley 1986). **(b)** Double chronometry method as used by Emerson Process Management Ltd. (reproduced with permission).

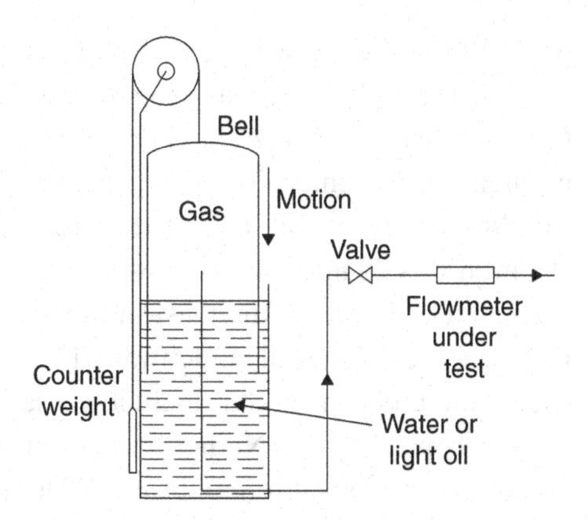

Figure 4.12. Bell prover.

SVP in future since some computer-based flowmeters such as Coriolis and ultrasonic may be affected by the proving action.

4.4 Gas Calibration Facilities

4.4.1 Volumetric Measurement

The bell prover uses a standing start and stop. Figure 4.12 shows the main features. An inverted vessel – the bell – is held by a cable and counterweight so that it dips into a sealing fluid. This is usually water or a light oil. The flowmeter to be calibrated is connected into a pipe that allows the gas in the gas cavity under the bell to flow out. The bell is allowed to sink into the liquid so that the gas is expelled from the cavity through the outlet pipe and so through the flowmeter under test. It is therefore possible to take the height of the vessel at start and end and to deduce the volume of gas that has passed. For high accuracy, various other factors need to be taken into account. The volume of gas in the vessel will be affected by the pressure and temperature within the vessel. To a lesser extent, it will be affected by change in the volume of the vessel. The humidity of the gas will be important if water is used instead of oil. The movement of the vessel, if continuous monitoring of the flow is taking place, may need to be allowed for. This method is particularly suitable for small domestic gas meters and is capable of achieving an uncertainty of as little as ±0.2% if particular care is taken. A number of elderly bell provers are still in use, but some new ones are also providing high accuracy in national laboratories. See also Choi et al. (2010).

A similar approach, using a large bag of known volume, accommodates larger volumes.

Bellinga and Delhez (1993, cf. Bellinga et al. 1981) of Nederlandse Gasunie, Groningen, described the conversion of a portable piston prover for liquids to use with gases; the prover was capable of flows from 20 to 2000 m^3/h. The cylinder length was 12 m with a measuring section of 5.22 m, diameter of 0.584 m and volume of 1.3988 m^3. They also described the traceability chain before the prover came into operation. During the initial calibration of the piston prover, the volume was determined by weighing the quantity of water displaced by the piston when moving through the measuring section. A dynamic calibration was achieved by using a master meter. As a check, the volume displaced was obtained from measurement of the dimensions of the prover. They claimed that these different methods should agree to within 0.03%. The pressure difference needed to drive the piston at 150 m^3/h was 20 mbar, but at lower flow rates the motion became unsteady and was corrected by reducing the mass of the piston and adding a dynamic damping system. Bellinga and Delhez also mentioned other small changes to valves, valve operating speed and electronics that include pulse interpolation. The aim was for the prover to be accepted by the Netherlands Metrology Institute as the primary standard for high-pressure gas meter calibration. The meters calibrated from this prover should have an uncertainty in meter factor of as good as 0.1–0.15%. Reid and Pursley (1986) considered the use of nitrogen as a pressure-balancing gas to reduce the differential pressure across the piston and hence to reduce the leakage and to provide stability of motion at low flow rates.

Reid and Pursley claimed 500:1 operation in air for pressures greater than 30 bar with agreement to 1% and repeatability of order 0.25% up to 8 bar.

Jongerius, Van der Beek and Van der Grinten (1993) described a low-flow test facility consisting of four mercury seal piston provers at the Nederlands Meetinstituut with an operating range of 2×10^{-5}–3.5 m^3/h at approximately ambient conditions and operable, in principle, on any gas. The low-flow facility is traceable to the primary length standard, and uncertainty is about 0.2% for the smallest prover and 0.15% for the others.

See 4.A.4 for further discussion of recent references on piston provers.

4.4.2 Mass Measurement

Figure 4.13 shows the scheme for a gravimetric calibration facility. The system shown depends on the use of critical nozzles (explained in Chapter 7) as transfer standards. The facility is used to calibrate the sonic nozzles, then to calibrate the flowmeter under test, and, if appropriate, to recalibrate the nozzles. The problem with gravimetric measurement in gases is the low density of the gas and the consequent problems of mass measurement. In addition, the flow control in the calibration facility must be such that the flow rate through the meter under test is the same as that into the weighing vessel. The facility at the National Engineering Laboratory, Scotland, has been used for flows of up to 5 kg/s at pressures of up to 50 bar to within an estimated uncertainty of ±0.3% (Brain 1978).

4.4.3 Gas/Liquid Displacement

The problems of calibration of gas flow have led to the use of a device that exchanges liquid flow by volume for gas flow by volume, say, by means of a piston. The liquid is metered to a high accuracy, and the gas at controlled temperature and pressure is used to calibrate the flowmeter.

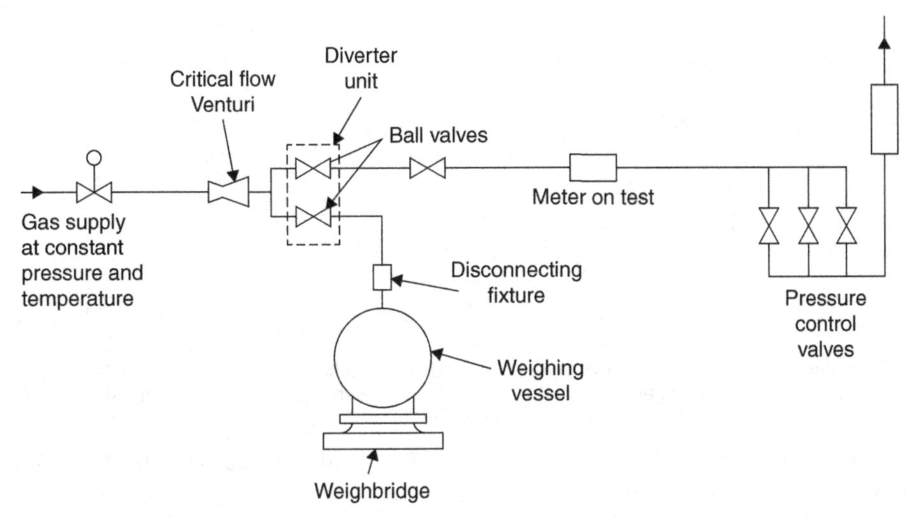

Figure 4.13. Scheme for a gravimetric gas calibration facility (Pursley 1986 reproduced with permission of NEL).

Lapszewicz (1991) described a flowmeter for gas mixtures in which liquid is displaced, and the time and volume between two points on a tube is measured. A valving system allows the gas to enter and be exhausted.

4.4.4 *pvT* Method

Provided that the gas under calibration obeys a well-defined relationship between the pressure p, the specific volume v and the temperature T (e.g. for an ideal gas),

$$pv = RT$$

where R is the gas constant, the change in quantity in a pressure vessel can be deduced from the change in pressure and temperature, and the volume or mass passing through the meter under test can be deduced. A simple facility is shown in Figure 4.14. The flow and temperature of the gas are controlled by heat exchanger and pressure regulation, and the flow rate may be stabilised by a critical nozzle followed by a further heat exchanger.

4.4.5 Critical Nozzles

The design and operation of these devices will be described in Chapter 7. Here we note only that they are rather inflexible in that the flow through them is, essentially, defined by the inlet pressure and the area of the throat of the nozzle. Thus to accommodate a range of flows, they need to be set up as a bank of flowmeters. A typical bank of such flowmeters is shown in Figure 7.2. The approach will be to set these in series with high-quality master flowmeters. In order that the flow range is as complete as possible, the mass flows through the bank need to be in a progression such

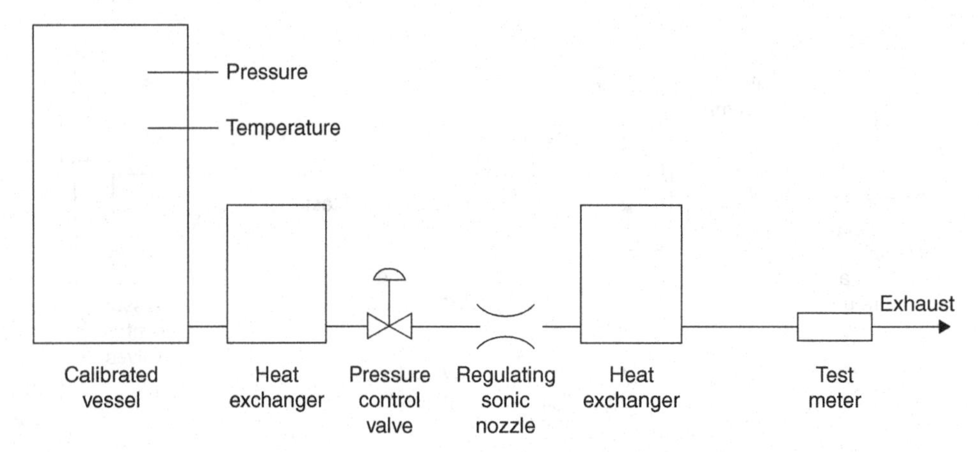

Figure 4.14. Diagram of a typical facility for the *pvT* calibration method (Pursley 1986 reproduced with permission of NEL).

as 1, 2, 4, 8, 16. This will allow flow rates of 1, 2, 3, 4, 5, …, 31, by suitable combination of the nozzles (cf. Aschenbrenner 1989, who described the calibration of a bank of 16 nozzles). In such a system, it will, clearly, be essential to have a means of checking the integrity of the valving system so that accidental leaks do not affect the accuracy. With this range, it is then possible to calibrate the master meters over the same range and to use these either on the facility or as transfer standards (cf. Bignell 1996a on comparison techniques for small sonic nozzles).

4.4.6 Soap Film Burette Method

Figure 4.15 shows a simple arrangement of the soap film burette calibration system (Pursley 1986). This is suitable for very low flow rates. The gas enters the burette after leaving the flowmeter, and a soap film, created within the gas flow, is driven up the burette by the flow of gas. The movement can be observed and timed by eye, or sensing methods, probably optical, can be used to measure its transit time. Clearly the burette will need to be calibrated for volume. Flow range is 10^{-7}–10^{-4} m^3/s at conditions near to ambient (Brain 1978).

4.5 Transfer Standards and Master Meters

Calibrated master meters may also be used to measure the flow in a specially constructed flow facility in which the masters are in series with the meter to be calibrated. This may be the most economical approach for many manufacturers. The cost will, of course, include the regular recalibration of the master meters. Such a facility will need careful design to achieve:

- steady flow,
- fully developed flow profile upstream of both the reference (master) meter and the meter under calibration,

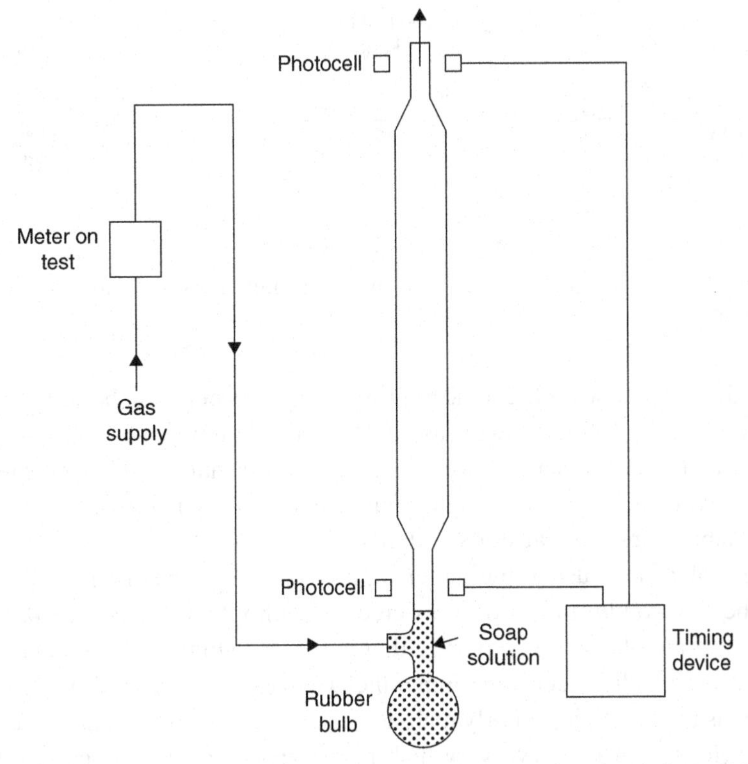

Figure 4.15. Soap film burette (Pursley 1986 reproduced with permission of NEL).

- adequate range for all meters to be tested,
- flexible means to accommodate flowmeters of different diameters and axial lengths and to mount them in the line, and
- high-quality sensing and recording systems to provide the comparison between the meter under calibration and the master meter.

A good transfer standard should have the following characteristics (Harrison 1978b):

- highly repeatable,
- stable with time,
- wide flow range,
- insensitive to installation,
- simple and robust,
- compact,
- low head loss,
- suitable for a variety of fluids and
- available in a wide range of sizes.

Meters suitable for use as masters for transfer standard work are, therefore, the high-accuracy meters, which will be described in later chapters. In particular, these have in the past included positive displacement meters and turbine meters. For conducting liquids, electromagnetic meters have reached a sufficient standard to be

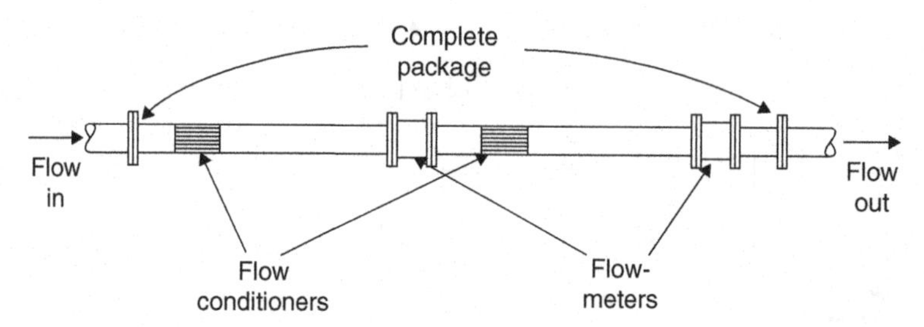

Figure 4.16. Transfer package allowing two meters, conditioners and pipework to be calibrated together.

considered in some cases. Ultrasonic multi-beam meters may also be suitable, as may Coriolis meters. In gas flows, the ultrasonic multi-beam may be suitable, and the critical nozzle meters, as indicated earlier, are very appropriate despite their inflexible flow range. Wet gas meters, although a fairly elderly design, are still used by some standards laboratories as transfer standards.

The problem with using one meter on its own is that the stability of its signal may not be observable, and errors may creep in without any check. For this reason, master meters are often used in pairs, either in series, so that the consistency of their readings is continually checked, or in parallel when one is used most of the time and the second is kept as a particularly high-accuracy meter for occasional checks.

A development to achieve very high performance and to reduce sensitivity to installation is the calibration of the meter with the upstream pipework and a flow straightener permanently in position, and the use of two of these packages in series (Figure 4.16; this arrangement was suggested by Mattingly et al. 1978, with or without the initial flow straightener). In this way, any deviation caused by installation or by drift of one meter will appear as a relative shift in calibration between the two meters. Such meters or meter pairs are referred to as transfer standards in that they allow a calibration standard to be transferred (with only a small increase in uncertainty) from a nationally certified facility to a manufacturer's facility or a research laboratory facility. The components of the transfer standard package should be spigotted together to ensure correct alignment and repeatable contact joints between components. It may also be useful to use two flowmeters of different technologies to provide an additional check.

The meters selected as transfer standards must be capable of retaining a high performance with removal, transport, calibration, subsequent removal, transport, reinstallation, removal, storage, etc.! Several of the main flowmeter designs will meet this requirement, and the actual selection will depend on other factors. The line liquid should be without gas bubbles or cavitation. The meters should be selected for optimum range; in some cases, overlapping ranges may be an advantage.

Apart from flow profile, the pipework for transfer standard installation would have the following features:

- uniquely linking meters with convenient valves to control flow path and for ease of access;
- suitable environment, meeting safety requirements;

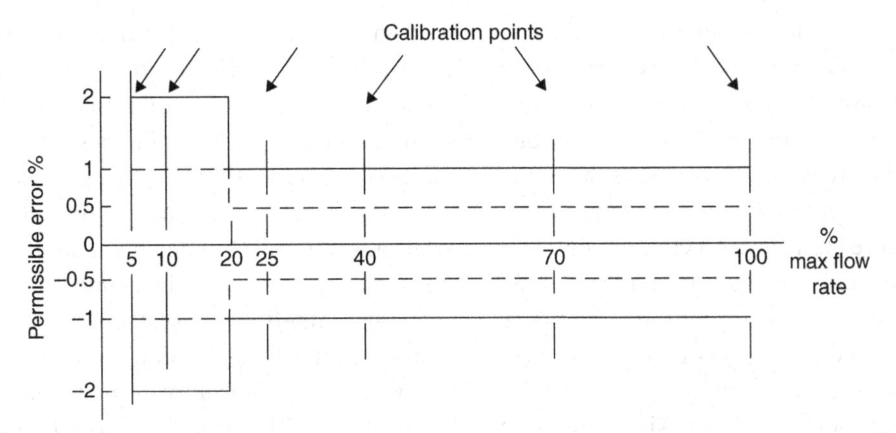

Figure 4.17. Maximum permissible errors at time of paper: solid line, ISO 9951; broken line, Gasunie specification (reproduced with the agreement of Elsevier from Van der Kam and Dam 1993).

- freedom from vibration and cathodic protection; and
- surrounding pipework clean, in good condition and of the same diameter as transfer package.

Van der Kam and Dam (1993) described the replacement of orifice meters by turbines, in export stations of Nederlandse Gasunie. Even though ISO 9951 specified a Maximum Permissible Error (MPE) as ±1% in the upper range at time of paper, van der Kam and Dam reported that, in their experience, turbines can operate within an envelope of ±0.5% down to about 25% maximum flow and of ±1% in the lower range (Figure 4.17). They suggested that the four-path ultrasonic meter would probably be best of all.

Pereira and Nunes (1993) described a facility with an array of ISA 1932 nozzles as the calibration source. The sizes were such as to allow a 50:1 turndown in air, and where nozzles were not needed, rubber balls were inserted in the inlets to seal them off and prevent flow. Uncertainty was claimed to be in the range 0.5–0.8%.

4.6 In Situ Calibration

A new flowmeter should have some performance specification. The need to keep documentary evidence on flowmeters should be reinforced. It may start well with a manufacturer's minimal or full calibration. The flowmeter should, therefore, have a known performance under reference conditions. Let us assume that this is ±1% uncertainty with 95% confidence, a reasonable and not uncommon value. What happens when we install this flowmeter on site?

In many cases, there may not be an adequate upstream straight length to retain the calibration, and this may be further degraded by misalignment etc. Let us assume that this will cause a total uncertainty of ±2%. The meter then starts its working life, and we cease to have any satisfactory means of predicting the likely drift that results. Experience from the water industry suggests that this may range from 5% to as much as 50%. Because of this uncertainty, it is necessary to determine a documentation, maintenance and recalibration schedule for each meter. One way to decide the period between recalibration is by experience. If, after the first recalibration, the shift is unacceptably large, then the next period to recalibration should be half the length. If the calibration is essentially unchanged, then the period may be increased. Descriptions such as "initial period", "essentially unchanged" and "unacceptably large" are rather subjective, but they may be given a value based on lost revenue against costs of maintenance and calibration. The outcome of all this is the need to recalibrate, and one choice is in situ calibration with a likely uncertainty of, say, ±3%, considerably poorer than originally obtained on a test stand.

In deciding between the merits of test stand and in situ calibration for a specific case, the reader should develop a table with pros and cons in terms of ultimate uncertainty and total costs (including removal, transport, calibration, lost revenue and reinstallation). Other factors such as damage to the meter should also be allowed for in the decision.

Methods of in situ calibration include

- provers (sphere, piston and tank);
- other meters (transfer standards);
- pipework features (inlets and bends);
- drop test;
- in situ meter measurement and inspection (dry calibration);
- clamp-on flowmeters (see Chapters 14, 15 and 22); and
- probes (see Chapter 21) and tracers.

Traceability for the provers and transfer standards will be through a calibration facility. For other methods, traceability will be more tenuous, and in situ verification will be a more appropriate term.

4.6.1 Provers

Provers have already been covered but are mentioned here for completeness. The sphere prover (Figure 4.9) is widely used in permanent systems with banks of turbine meters in the oil industry. The piston prover can be trailer mounted. Claims of order 0.1% or better are sometimes made for these devices (cf. Eide 1991, who claimed ±0.03%). The volumetric tank prover (Figure 4.8) discussed earlier is reckoned capable of the same sort of accuracy level.

Other Meters

An example of a transfer standard used in a survey of domestic water consumption in Scotland (Harrison and Williamson 1985) is shown in Figure 4.18 and was capable of calibrating meters from 0.004 to 35.7 m^3/h. This was achieved by using three meters within the transfer standard:

- an Avery-Hardoll PD meter in series with a Kent 50-mm Helix 3000 flowmeter – range 35.7–3.7 m^3/h,

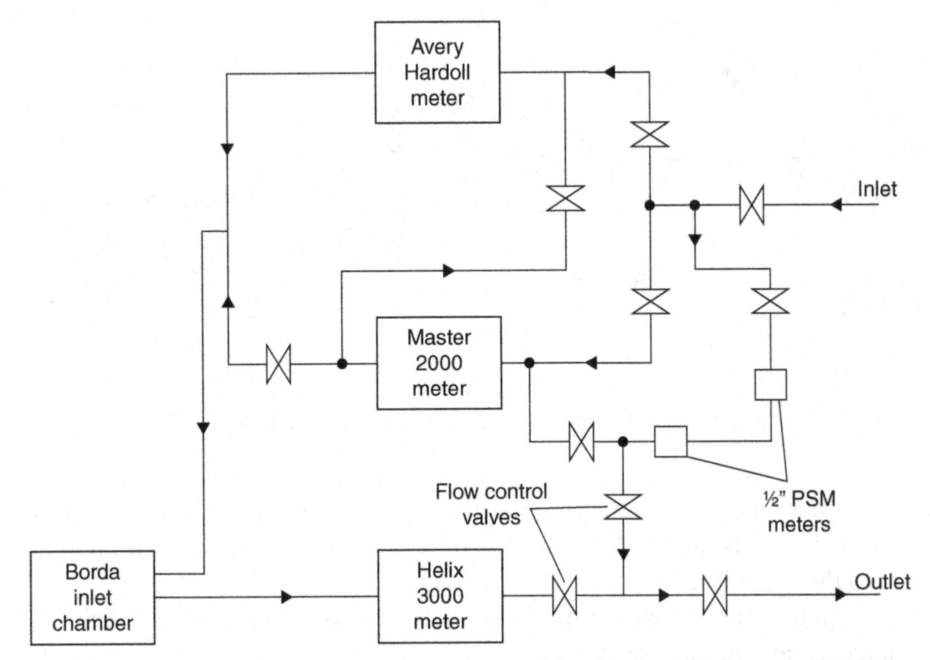

Figure 4.18. Wide-flow range reference meter transfer standard (Harrison and Williamson 1985 reproduced with permission of NEL).

- a Kent 50-mm Master 2000 flowmeter in series with a Kent 50-mm Helix 3000 flowmeter – range 10.8–0.7 m^3/h, and
- two Kent 12-mm PSM meters in series – range 2.7–0.004 m^3/h.

The Borda inlet chamber allowed the flow to settle prior to the Helix meter, which is sensitive to installation.

It is obviously necessary for provers, such as this one, to be coupled into the site pipework in series with the meter to be tested.

Pipework Features – Inlets, Bends/Elbows

The inlet from a large tank to a pipe may be used to obtain a measure of the flow rate. The pressure difference between the inside and the outside as flow goes round a bend or elbow may be used to obtain flow rate. The accuracy of these methods will depend on how well-formed the pipework features are. The equations and methods are described in Chapter 8.

Drop Test

In the drop test method, the volumetric flow through the meter is compared with the volume change due to a level change in a reservoir (Figure 4.19). To achieve this, we require the following:

a. A suitable system with the pipework leading from the reservoir so as to allow closure of all flows except that which leads from the reservoir to the meter under

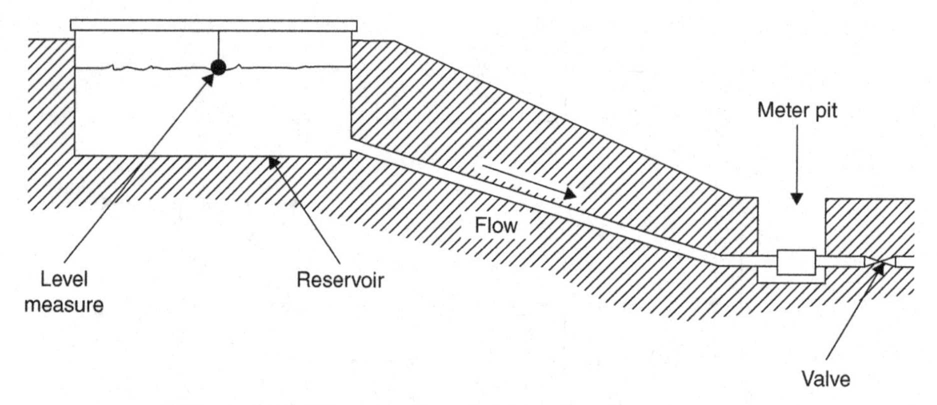

Figure 4.19. Diagram of a suitable system for a drop test.

test. There must be confidence that the water leaving the reservoir all passes through the test meter. There must also be a valve suitable for controlling the flow through the meter.

b. Strapping tables, which give the relationship between water level and the volume of water in the reservoir.
c. A means to obtain the level of water in the reservoir.
d. Satisfactory means of communication between the person who is taking the level measurements and the person who is taking the volumetric readings on the meter. Of course, this process may be automated.

Typical uncertainty for such a test may be

Strapping tables (say 1:10,000 dimensional precision)	0.02%
Level measurement (say 2 mm in 10 cm)	2.0%
Timing errors for rate (say 0.2 s in 100 s)	0.2%

Within this group of errors, although the volume of the reservoir may be affected by tank supports etc., the level is likely to be predominant. The measurement of level, often in places with poor accessibility, small changes and surface waves, may result in such an error, and the preceding example may be overly optimistic for some sites. Taking the root of the sum of the squares of these errors gives 2.01%, which may be optimistic.

In Situ Measurement of Meter Dimensions (Dry Calibration)

By measuring the dimensions of a differential pressure flowmeter, it may be possible to deduce the performance. If the measurements confirm that the design meets the ISO requirements, then the flow rate should be predictable with an uncertainty of about 1.0–2.5%.

The term *dry calibration* was particularly coined for the electromagnetic flowmeter where it is possible to measure the magnetic field strength and the meter bore and so deduce the meter performance (Al-Rabeh and Baker 1986). Although the methods have increased in sophistication, it has been found to be not a very satisfactory approach except under very carefully controlled conditions. However, an uncertainty of order 5% may be achievable. This will, of course, require checking all secondary instrumentation as well as measurements of the meter geometry.

The use of retrofitted ultrasonic transducers also depends on the measurement of pipe dimensions and the positions of the transducers, The author is not aware of experience on whether this may offer genuine calibration in an in situ mode [cf. Drenthen and de Boer (2001)].

Clamp-on Flowmeters

The possibility of using clamp-on transit-time ultrasonic meters is being explored. Apart from the inherent uncertainties of using such meters on pipework of unknown condition, it is also essential to avoid installations that are seriously affected by upstream pipe fittings. These meters are discussed in more detail in Chapter 14.

Sonar clamp-on flowmeters (Chapter 15) might be considered as an alternative in some applications.

Probes

Probes will be covered in Chapter 21. However, the use of probes for in situ calibration is common, and a few points need to be made here.

- If a single-point measurement is required, the reader is referred to Table 2.2 where it is shown that, for a well-developed turbulent profile, if the probe is placed near the three-quarter radius position ($0.76 \times$ radius from the pipe axis), the velocity at that point is approximately equal to the mean velocity in the pipe. However, this is a region of strong shear and turbulence, and incorrect positioning of the probe may result in measurement errors. If the probe is positioned on the axis of the pipe, the actual positioning is less critical, but the velocity at that point is not proportional to the mean velocity in the pipe as the Reynolds number changes. Allowing for a probe calibration of $\pm 1\%$ uncertainty, the best that is likely to be achieved for the total flow is of order 3% and may be much poorer than this.
- If the velocity profile across the pipe is required, then multiple measurements will be necessary. These can be equally spaced, but to obtain mean velocity in the pipe it may be quicker and preferable to use an integration method that selects fewer positions, which may or may not be equally weighted. The log

linear positioning of measurements (with each measurement equally weighted) follows:

Number of points radially	Position, r/R		
3	0.359	0.730	0.936
4	0.310	0.632	0.766
	0.958		
5	0.278	0.566	0.695
	0.847	0.962	
6	0.253	0.517	0.635
	0.773	0.850	0.973

Salami (1971) recommended that at least six traverse points on each of six radii were needed to keep errors down to 0.5% for non-axisymmetric profiles. If we assume that the profile is axisymmetric, then we may take one diameter and select the number of points according to the time available. The specification of the positions, the recording of the data from the probe and the calculation of the mean flow in the pipe are often done today by a portable flow computer designed for the purpose. We shall also need a value of the pipe internal diameter (ID). For further information, the reader is referred to the standards. Realistically we are unlikely to achieve an uncertainty of better than 3%.

- The act of inserting a probe will alter the flow in the pipe, and the probe's response will vary with insertion position. This is mentioned in Chapter 21, but the manufacturer of the probe will need to supply detailed information. In addition, the possibility that the probe will vibrate should also be taken into account.

See Thomas et al. (2004a) and Thomas, Kobryn and Franklin (2004b) who used an electromagnetic probe (Chapter 21).

Tracers

Tracer methods for obtaining flow in an unknown system are an elegant and attractive approach, but they have a disadvantage. Those undertaking them really need extensive experience; otherwise, the resulting accuracy may not be very high. The method can conveniently be categorised as constant rate injection, sudden injection or time of transit.

In the constant rate method, a solution of the tracer material in the same fluid as that in the main stream is injected from a branch pipe, which has a valve, at a known rate q_{vi} to the line (Figure 4.20). The concentration in the injected stream C_i is known; the concentration in the main stream, if the tracer material happens to be present, is measured as C_u; and, far enough downstream, the concentration resulting from the injection and thorough mixing C_d is also measured. The equation that gives conservation of tracer material and hence the volumetric flow rate in the line q_v is

$$C_u q_v + C_i q_{vi} = C_d \left(q_v + q_{vi} \right) \tag{4.1}$$

and rewriting this to obtain the unknown flow rate in the line we have

$$q_v = q_{vi} \frac{C_i - C_d}{C_d - C_u} \tag{4.2}$$

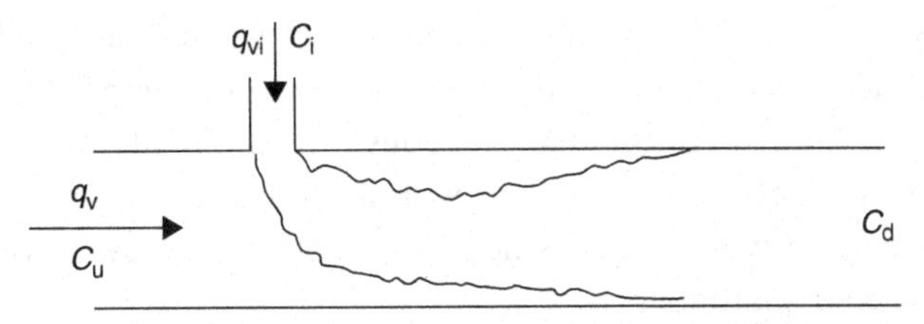

Figure 4.20. Constant rate injection method (dilution method).

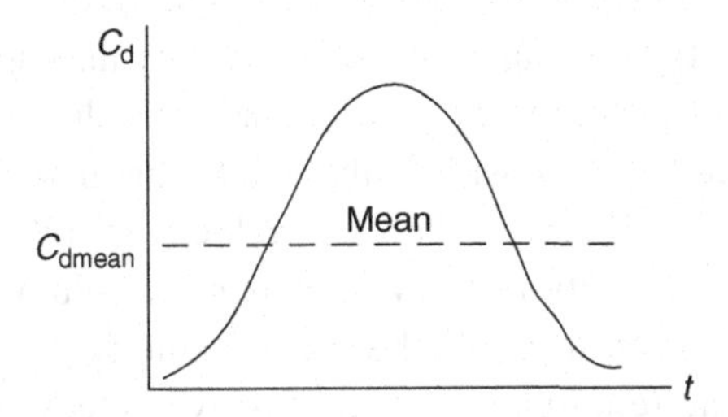

Figure 4.21. Concentration variation downstream with sudden injection method (integration).

For this method to be successful, we do not need to know the size of pipe, but we do need to ensure that all the tracer material goes downstream, and that there is a genuine steady state, without accumulation of tracer in trap areas in the pipework. It is probably necessary for the upstream concentration C_u to be less than 15% of the downstream concentration after injection C_d. Clearly the tracer must be such as can be injected into the line without any health or environmental problems, and it must be cheap enough to use in continuous injection. If we make the assumption, to obtain an idea of the accuracy, that $C_i \gg C_d \gg C_u$, then with a high-quality flowmeter for the injection and a precise concentration in the injected flow, the downstream value C_d is likely to be the least reliable of the dominant measurements and will affect the accuracy. Hermant (1962) in the discussion of his paper appears to claim ±1% as the uncertainty compared with use of current meters. Clayton et al. (1962a), and Clayton, Clark and Ball (1962b), using a radioactive tracer, were able to claim better than 1% for the method including using a portable system. They suggested that 110–220D should be allowed for mixing.

In the sudden injection method (integration), one known amount of V is injected with a known concentration C_i, and the variation in the downstream concentration is measured with time as in Figure 4.21. If this curve is integrated, and we assume that the flow rate is constant during the measurement period, we obtain the total of tracer passing, or the mean over the measuring period. Thus

$$q_v = \frac{VC_i}{\int_0^t (C_d - C_u)\,dt} \tag{4.3}$$

$$= \frac{VC_i}{t(C_{dmean} - C_u)} \tag{4.4}$$

where C_{dmean} is the mean concentration measured downstream. In addition to the constraints for the constant rate method, the accuracy of sampling the concentration is likely to be lower because instantaneous measurements will be needed and the accuracy of the concentration sensor will introduce uncertainty. However, the amount of tracer used will be less, which may, therefore, allow a wider choice of tracer (cf. Spragg and Seatonberry 1975).

In addition to these two methods, it may be possible to sense the passing of tracer at two points in the line, or, indeed, the variation in some naturally occurring component that might allow the use of a correlation method. It should be remembered that, in this case, the measure will be of velocity and not necessarily of volumetric flow (cf. Aston and Evans 1975, who estimated likely uncertainties within ±1% for steady flows and ±2% for normally fluctuating flows).

Scott (1982) also described the Allen salt-velocity method, which measures the time of transit between two stations between which the volume of the pipe is known. Again, he mentions the need to inject the tracer explosively, in this case a minimal amount of brine, and the requirement for a sensitive detection system. The marker point as the brine passed each station was taken as the midpoint of the half height trace [Figure 4.22 (cf. Hooper 1962 and Clayton et al. 1962b, who suggested uncertainties of order 1%)].

Various tracers have been used. For the constant rate method, sample analysis is likely to be the most accurate method of obtaining C_u and C_d, the values of concentration. In the case of sudden injection, a continuous detector will be needed. Examples follow:

Main fluid	Tracer	Detector
Water	Brine	Electrical conductivity
Other liquids	Dyes or chemicals	Colour or light sensing
Liquids or gases	Temperature	Temperature sensor
	Radioactivity	Scintillation or Geiger counter
Gases	CO_2, N_2O, He, methane	Infrared spectrometer

More recently Lovelock (2001) reviewed the use of alcohol tracers for two-phase flow measurement from geothermal fields. Ali et al. (2003) used tracers for multiphase flow and Kuoppamäki (2003) discussed a calibration method based on a radiotracer transit time measurement.

4.7 Calibration Uncertainty

Mattingly (1990/1991) gave a very useful run down of calibration uncertainty. Before the facility is built, the uncertainty can be estimated. For a static gravimetric liquid facility,

$$q_v = \frac{M_n}{\rho t} \tag{4.5}$$

where q_v is the volumetric flow rate, M_n is the net mass of liquid collected, ρ is the liquid density, and t is the collection time. The uncertainties (cf. Equation 1.A.7 for alternative symbols) can be combined as

$$\frac{\Delta q_v}{q_v} \leq \left\{\left(\frac{\Delta M_n}{M_n}\right)^2 + \left(\frac{\Delta\rho}{\rho}\right)^2 + \left(\frac{\Delta t}{t}\right)^2\right\}^{1/2} \tag{4.6}$$

or

$$\frac{\Delta q_v}{q_v} \leq \left|\frac{\Delta M_n}{M_n}\right| + \left|\frac{\Delta\rho}{\rho}\right| + \left|\frac{\Delta t}{t}\right| \tag{4.7}$$

By inserting percentages of reading, an initial estimate can be obtained. This presupposes that Equation (4.5) is correct. However, other factors are likely to be involved. At the National Institute of Standards and Technology (NIST), Gaithersburg, Maryland, for liquid flow, the three values to three standard deviations were

$$\frac{\Delta M_n}{M_n} = 0.02\% \quad \frac{\Delta\rho}{\rho} = 0.02\% \quad \frac{\Delta t}{t} = 0.01\%$$

and combine to ±0.03% for Equation (4.6) and ±0.05% for Equation (4.7). For gas, the volume was 0.04%, density pressure effects were 0.13%, temperature effects were 0.05%, and collection time for the device was 0.01% and for switching was 0.02%, giving totals of ±0.15% and ±0.25%, respectively. The next stage is to obtain data from the facility, and the third stage is to obtain systematic errors from round-robin tests. These round-robin tests are known as Measurement Assurance Programs (MAPs). The systematic error is either root-sum-squared with the random error or, preferably, obtained by straight addition.

A satisfactory transfer standard is formed by using two meters, possibly turbines or more recent designs such as ultrasonic or Coriolis, with a flow conditioner placed between them in a tandem configuration (Figure 4.16), as already mentioned for laboratory flow facilities. Adequacy of the data is established by specifying the number of repeat calibrations done for each flow rate and meter

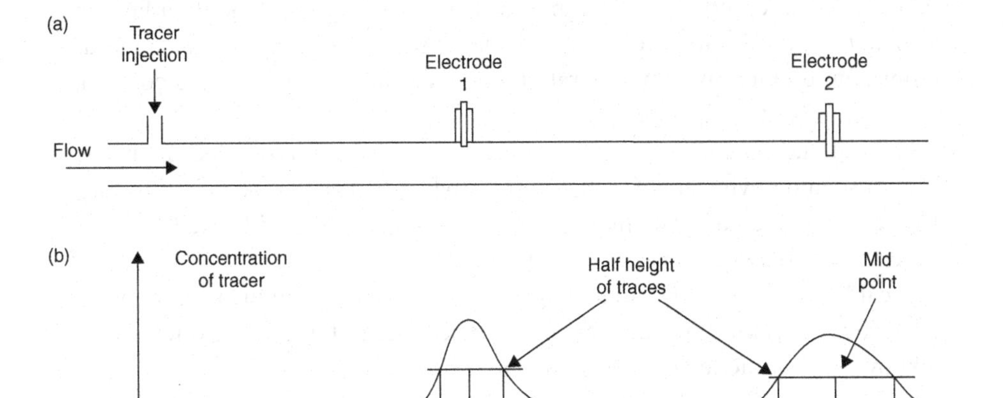

Figure 4.22. The Allen salt-velocity method for time of transit between two stations: **(a)** Pipe for which volume is known; **(b)** marker points as the brine passes each station (after Scott 1982).

configuration. These will ensure statistical significance. The Youden (1959) procedure is recommended (Chapter 1).

4.8 Traceability and Accuracy of Calibration Facilities

For a calibration to be acceptable, the ultimate source of the measurement must be known, and the calibration must be traceable to that standard, as in Figure 4.3. Laboratories holding national standards exist in several countries (Benard 1988), and the address of the nearest should be available from national government information or from laboratories such as:

Country	Laboratory
United Kingdom	National Engineering Laboratory, East Kilbride, Scotland
United States of America	National Institute of Standards and Technology, Gaithersburg, MD

(See Section 4.1.2 for weblink to BIPM.)

National standards should themselves be traceable back to more fundamental measures of mass, time and length. As a result of this, a traceability chain is formed. Each link is formed from a facility, or a flowmeter calibrated on a facility or against a flowmeter of lower uncertainty. The chain needs to be rechecked with sufficient frequency to ensure continuing confidence.

It should also be clear that, if a transfer standard with an uncertainty of, say, ±0.25% is used to calibrate a flowmeter, the uncertainty in the flow rate measurement made by that flowmeter will be greater than ±0.25% due to additional random errors in the flowmeter.

Harrison (1978b) clarified the meaning of traceability with the following three points:

a. Each standard used for calibration purposes has itself been calibrated against a standard of higher quality up to the level at which the higher-quality instrument is the accepted national standard. This is usually a unique item held in a national standards laboratory, but it could, in some cases, be a local standard of equivalent quality built and operated to a national specification and confirmed as operating to that specification.
b .The frequency of such calibration, which is dependent on the type, quality, stability, use and environment of the lower-quality standard, is sufficient to establish reasonable confidence that its value will not move outside the limits of its specification between successive calibrations.
c. The calibration of any instrument against a standard is valid in exact terms only at the time of calibration. and its performance thereafter must be inferred from a knowledge of the factors mentioned in b.

The standards for mass and time are national, derived from equivalent international standards. The international one for mass is in Paris and for time is based on a fundamental frequency, which may in turn be broadcast nationally.

Thus calibration uncertainties achievable at present appear to be: for liquids about 0.1% and for gases 0.2–0.3% or possibly better. When total mass or volume, rather than flow rate, is required, these values may be improved. On the other hand, in situ calibration is likely to be at best 2% and often 5% or more and should be seen as a last resort in most cases. The exception is where a meter installation is equipped with off-takes so that a high-quality transfer standard prover or meter can be coupled in series and will result in an accuracy approaching that of a dedicated facility (cf. Johnson et al. 1989 for further useful discussion of component uncertainties for a gas facility).

4.9 Chapter Conclusions

For most readers, the essential value of this chapter will be knowledge of calibration methods and accuracy levels, as well as signposts pointing to where to obtain further detailed information.

For those in the business of designing, installing, and commissioning flow calibration facilities, there are some fundamental questions to be addressed, not all of which may yet be answerable. This chapter has attempted to flag some of these questions and to point to relevant literature. Some of these questions concern:

- limits of accuracy for a meter in a turbulent flow;
- limits of accuracy for a test stand;
- the nature of flow diversion and its repeatability;
- the design of provers and the interpolation of pulse trains;
- ultimate in situ accuracy including
  - more user-friendly tracer methods,
  - more experience with retrofitted ultrasonic meters,
  - improved confidence in clamp-on ultrasonic meters,
  - improved confidence in sonar clamp-on meters,
  - improved probe measurements;
- theoretical optimum accuracy as traced from national standards and round-robin meter exchange.

Because, therefore, existing methods may not meet all needs (Paton 1988), new developments continue to appear. We may question whether other factors resulting from environment, impurities or other components in the fluid may affect the calibration of certain flowmeter types or the behaviour of the test stand.

It is also appropriate to emphasise the importance of achieving a fully developed flow profile at the test section of the calibration rig. Chapter 2 has indicated the requirement to achieve this.

In addition, the sensitivity of the various designs of flowmeter to upstream profile has been set out in the relevant chapters.

Appendix 4.A Calibration and Flow Measurement Facilities

4.A.1 Introduction

In this appendix, some publications are briefly reviewed to indicate the scope of new developments in flow calibration facilities and related matters. The appendix then concludes with a case study of a flow rig suitable for a manufacturing plant, and in contrast, a very large flow calibration facility in China. The reader may find it useful to peruse the recent conference proceedings, such as ISFFM and FLOMEKO, for reports on recent developments.

4.A.2 Flow Metrology Developments

Flow Facilities

There is a continuing concern with calibration and the requirements for traceability. Pöschel and Engel (1998) described their proposals for a new primary standard for liquid flow measurement at PTB Braunschweig with a claimed total standard uncertainty of 0.01%. Wright and Johnson (2000) considered one source of uncertainty in gas flow standards, the origin of which is rapid valve movement causing transients. Other papers covered particular facilities, specific devices, very stable transfer standards and dynamic traceability.

Delsing (2006) brought together various papers on flow measurement facilities for *The Journal of Flow Measurement and Instrumentation* (Volume 17 Issue 3). The papers covered liquid calibration rigs, air flow facilities and a three-phase hydrocarbon system. Doihara et al. (2004, 2006) described an important development in the diversion of liquid flows in "flying start and stop" facilities, achieving impressively low uncertainties. Ishibashi and Morioka (2006) described the updated closed-loop calibration facility with a constant volume tank, in Japan, for airflow of 5 to 1,000 m^3/h which achieve uncertainties of 0.1% or better. Robøle, Kvandal and Schüller (2006) described the Norsk Hydro three-phase flow loop. Nakao (2006) detailed a new pvT system.

In a useful paper on calibration, Griffin (2009) commented that: "*when calibrating a measuring device, the uncertainty of the calibration standard should be a factor of 10 less than that of the device itself*," which brings realism into uncertainty determination, from which it followed that parameters contributing less than one-tenth of overall uncertainty might be ignored. The author also observed that for multiphase meters, test separators may not be an order of magnitude better than the meters.

Shimada et al. [2004, 2007] discussed the development of a new primary standard for hydrocarbon flow measurements and the uncertainty analysis of the facility. The expanded uncertainty was estimated to be 0.03% for volumetric flow rate and 0.02% for mass flow rate.

Sato et al. (2005) described the design and Furuichi et al. (2009) described the resulting calibration facility for water flow at NMIJ for flowmeters in nuclear power plants. It allowed high Reynolds numbers (maximum flow rate 12,000 m^3/h at 70°C and $Re = 1.8 \times 10^7$ is achievable). Its expanded uncertainty was claimed as 0.077%.

As suggested earlier, the reader is encouraged to access recent proceedings from conferences such as ISFFM and FLOMEKO where reports on new or updated calibration facilities are often presented. Shinder and Moldover (2009, 2010) reported

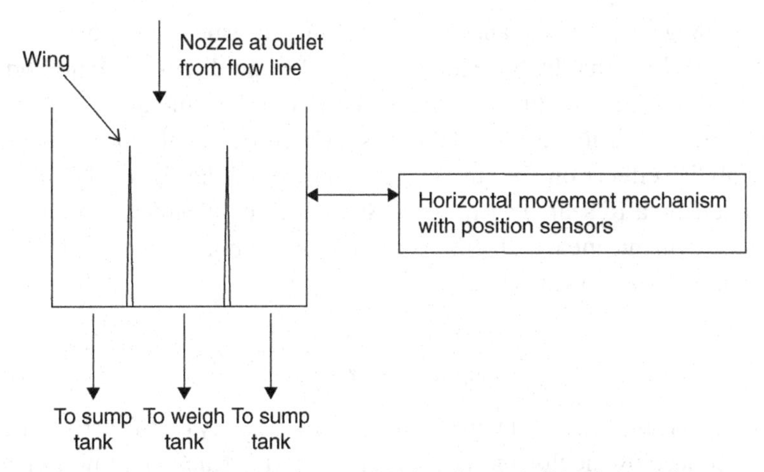

Figure 4.A.1. Diagram to indicate concept of Shimada et al. (2003) new diverter system for liquid flows.

progress in testing a dynamic gravimetric standard using water flows at NIST, developments at PTB are mentioned later, and further reports from China, France, Japan, Taiwan and elsewhere may be included at such conferences.

Flow Diverters

Shimada et al. (2003) described the development of a new diverter system for liquid flows which they referred to as double diverting wings. The unit moves horizontally with the two diverter "wings" (Figure 4.A.1) rather than rotate as in the conventional diverter design (Figure 4.5). The flow appears to return to the storage tank before and after the pair of "wings" and to go to the weigh tank between the "wings". To improve on this design and overcome one of the disadvantages, Doihara et al. (2004, 2006) developed the concept into a half cylinder which rotates around its axis and in which it appears that the wings are radial closures of the half cylindrical box (cf. Shimada et al. 2004, 2007).

Ho et al. (2005) have also described a novel flow diverter for water flow measurement. They used CFD to facilitate development and optimise flow patterns. Marfenko, Yeh and Wright (2006) described a new diverter design such that the nozzle moved in the same direction at start and end of a flying start/stop calibration. They claimed an uncertainty less than 0.01% for water flow calibrations.

The improvement which these offer appears to be due to the fact that the jet from the nozzle is always moving in the same direction relative to a divider or wing, with a more consistent division of liquid between weigh tank and sump.

Expansion and other Techniques

Shimada et al. (2010a) described the development of expansion techniques from the primary hydrocarbon standard to achieve wider flow ranges and to the use of other liquids, by adding an advanced analysis to the characteristic of the flowmeter. Other contributions include that by Spazzini et al. (2010), who discussed a method for the comparison of calibration curves, for instance in the determination of a reference curve from the calibration curves of various laboratories. Lim (2005) discussed the detailed stability effects on the uncertainty analysis of the light oil flow standard system at the Korea Research Institute of Standards and Science. Catherine (2002) described the dynamic measurement of volumes in a major network of multiproduct hydrocarbon pipelines in Europe.

Signal Processing

Elliott (2004) discussed a testing programme focused solely on the errors and uncertainties introduced by the flowmeter electronics used to calculate flow and generate flow proportional pulses. This paper was limited to Coriolis and ultrasonic flowmeter technologies. See also Section 4.A.6 for a case study (Baker et al. 2013).

Provers

Martin (2009) discussed pipe prover uncertainty and suggested that it may be possible to achieve ±0.02%. Johnson et al. (2010) described analysis of the National Institute of Standards and Technology's (NIST) bi-directional 20-l piston prover, its primary standard for measuring hydrocarbon liquid flows ranging from 1.86×10^{-5} m^3/s (0.3 gpm) to 2.6×10^{-3} m^3/s (40 gpm), and indicated an uncertainty of 0.074 % ($k = 2$).

Manufacturing

Meter calibration is a time-consuming process in the manufacture of flowmeters, and an addition to the fabrication costs (Engel 2002). An alternative approach, with an electronic balance or force-measuring device as a dynamic weighing system in a gravimetric flow calibrator, was discussed in the paper.

Inter-Comparisons

Paik, Lee and Mattingly (2005) gave an explanation of the procedure to check uncertainties for an inter-comparison of water flow calibration facilities. A transfer standard was used consisting of three flowmeters: ultrasonic, Coriolis and turbine (www.bipm.org). The pilot comparison, performed with flowmeters in series, established about ±0.1% agreement in four different flow conditions which appeared to be high and low flow rates, Coriolis or turbine flowmeter upstream, and in various test laboratories.

Gersl and Lojek (2010) reviewed the results of supplementary comparisons of European national water flow laboratories. Eleven laboratories took part. Two electromagnetic flowmeters were used as transfer standards, for water flow rates between 1 and 10 m^3/h.

In Situ Calibration

At a different level of precision is the need to do in situ calibrations. For these the use of tracers is one possibility where other methods are unavailable. Le Brusquet and Oksman (1999) combined CFD predictions with tracer methods to obtain an improved prediction of the flow rate.

Drysdale, Frederiksen and Rasmussen (2005) discussed the use of laser Doppler velocimetry for on-site calibration of large (100 to 1,000 mm) flowmeters using specially installed optical access windows near to the flowmeter. An uncertainty of ±0.9% may be achievable, and in addition, flow conditions may be mapped.

Density

Engel (2010) addressed the problem of knowing, by measurement or computation, precise liquid density in a gravimetric calibration rig used to obtain volumetric flow. The fluid density is one of the quantities which affect the uncertainties. The author also considered the effect of temperature variation (see also Engel and Baade 2012).

4.A.3 Multiphase Calibration Facilities

There has been development of multiphase flow facilities at some national laboratories to meet the demand for calibration of multiphase flowmeters. Britton, Seidl and Kinney (2002) described a test loop for both hydrocarbon gases and hydrocarbon liquids. Britton, Kinney and Savidge (2004) presented an overview of preliminary data from the facility at Colorado Engineering Experiment Station, Inc. (CEESI). A small 50 mm (2 inch) pilot facility was constructed to understand the characteristics and problems associated with handling multiphase fluids.

4.A.4 Gas Calibration Facilities

Factors affecting uncertainty of a volumetric primary standard for gases were considered by Kegel (2002a). Kegel (2002b) discussed the CEESI natural gas calibration facility for calibration up to 760 mm, 830 bar.

Jousten, Menzer and Niepraschk (2002) described an automated gas flowmeter at PTB for flows from 10^{-13} mol/s to 10^{-6} mol/s. The meter operates by means of a volume displacer which brings the pressure back to the datum value so that the volume flowing at constant pressure and temperature can be deduced. They claimed relative standard uncertainties from 1.45% to 0.14%.

Caron (2001) described a flow measurement system for air mass flow rates based on critical Venturi nozzles. It was checked at various laboratories and was aimed at use in development and manufacturing, for example in the measurement of air flow in IC engines. Caron et al. (2002) also described a primary flow standard for compressible flow using both gravimetric and volumetric methods.

Nakao et al. (2002) described the development of the primary flow standard at NMIJ for gas flow rates less than 1 mg/min using the dynamic gravimetric method, which aimed for an expanded uncertainty of better than 0.5%. Cignolo et al. (2002) discussed a primary standard for measurement of gas flows (1 ml/min to 1 l/min) using a piston prover. They claimed standard uncertainty of 0.03% or better.

Berg and Tison (2002) described two primary standards for low flows of gases: one using a precision piston which transferred oil from a bellows which in turn allowed a calibrated volume of gas to be measured; the other weighed a small cylinder to determine the change in gas contained within the cylinder. A laminar flowmeter was used to provide a transfer standard to compare the two methods. The error of both standards was considered to be less than 0.06%.

Altfeld (2002) discussed the PTB *pigsar*™ test rig, the German national primary standard for high pressure natural gas, which appeared to achieve (Mickan et al. 2002) an uncertainty for meter calibration, with $k = 2$, of 0.15%. Bremser et al. (2002) discussed the traceability and uncertainty of this German national standard. Müller et al. (2004) discussed the optical laser Doppler velocimeter (LDA) method for natural gas flow measurement under high pressure in Germany. The velocity was measured by LDA across the outlet flow from a carefully designed set of nozzles. These were followed by a diffuser and then a bank of critical flow nozzles to stabilise the flow and provide a transfer standard. Mickan et al. (2009) discussed the operation of *pigsar*™ as one of the leading high-pressure gas calibration facilities worldwide and the reduced uncertainty of *pigsar*™ from initially 0.16% to 0.13% and their aim to reduce this down to 0.1%. It includes a high-pressure piston prover, and gas turbine meters as secondary or transfer standards. Mickan and Kramer (2009) discussed PTB's two new primary volumetric standards: a conventional bell prover with operating range 1 to 65 m^3/h, and a new actively driven piston prover operating mainly between 0.04 and 4 m^3/h, for which they claimed an uncertainty of 0.04%.

Espina (2005) presented results of the North American natural gas flow calibration laboratory comparison. The transfer package consisted of turbine and multipath ultrasonic meters. It had day-to-day reproducibility of about 0.37%. The laboratories showed a consistency of about 0.3% to 0.4%, near the limits achievable allowing for the transfer package uncertainty and the testing protocols used.

Jian (2004) described the setting up of a bell prover as a mid-range standard in Singapore. Dongwei (2004) described the updating of the Chinese National Institute of Metrology's pvTt facility (cf. Meng and Wang 2004). Wang et al. (2004) provided some data for bell prover design.

Barbe et al. (2010) discussed the achievement of calibration using a "traced gas method" developed at the Laboratoire National de Métrologie et d'Essais (LNE), Paris, France. Its application appears to be, for example, for very small flows of a gas such as helium, in a chemical analysis laboratory.

Chunhui and Johnson (2010) reported that data from laboratory comparisons using critical flow Venturis, between China (NIM) and the United States (NIST), were in agreement with theoretical models to within 0.07% and with the ISO 9300 empirical equation to within 0.3% expanded uncertainty.

Chahine (2005) compared a bell prover and a five-tube mercury sealed piston prover, with glass tubes and optical sensors, for low-pressure gas flow standards. For the latter, cross-sectional area, length travelled and possibly timing contributed the highest components to the uncertainty. Sonic nozzles were used as transfer standards between these two standards.

At NMIA (National Measurement Institute Australia) sonic nozzles, in additive combination, were calibrated against a bell prover, with an uncertainty of 0.13% ($k = 2$) to cover the range 1 to 25 m^3/h. This allowed estimation of the nonlinearity of the bell prover (Chahine and Ballico 2010) and tabulation of uncertainties. An assessment of repeatability and reproducibility of sonic nozzle calibrations by the NMIA bell prover gave an apparently low value of 0.015% ($k = 1$). This reproducibility has allowed the estimation of nonlinearity of the bell prover at high flow rates (maximum value 0.07%). Benkova et al. (2010) also described calibration and measuring of the geometry of a bell prover.

A new 1m^3 bell prover was developed by the flow laboratory in China's National Institute of Metrology (Cui 2010). Laser tracking and laser interferometry were applied to measure the volume of the bell prover. The result suggested that this kind of geometric method is feasible and the uncertainty of the result appears to have been less than 0.1%.

It should be noted that various parameters such as temperature and flow rate may affect the measured values.

Calibration – Gases at Very Low Flows

Lashkari and Kruczek (2008) used commercially available soap flowmeters to measure gas flow rates in the μl/min range. They developed a fully automated soap flowmeter for micro-flow measurements (cf. Calcatelli et al. 2003).

Calibration – Other Considerations

Wiklund and Peluso (2002a, 2002b, 2002c) described a method for determining and specifying the dynamic response of flowmeter characteristics. To quantify dynamic response, they suggested that the frequency response test method was superior to the step response test. They suggested that it also provided easier quantification of the flowmeter dead time. They then discussed effects in various flow applications. They suggested that for custody transfer applications the transient response of flowmeters does not affect the totalising of the flow. For flow monitoring applications they suggested that the magnitude of transients would be underestimated, becoming worse for slower-response flowmeters and for higher-frequency transients. Underestimation is made worse with increased damping of the flowmeter. For flow control there may be an advantage in using faster response flowmeters. Berrebi et al. (2004a) and Berrebi, Van Deventer and Delsing (2004b) discussed how to reduce errors resulting from pulsating flows.

Van den Heuvel and Kemmoun (2005) reported on some unexpected flow measurement errors caused by stratified flow conditions due to temperature variation.

Calibration – Piston Provers

Padden (2002, 2004) discussed a piston prover that used clearance seals and for three sizes, designed for 10 ml/min to 50 l/min, uncertainty range was 0.064% to 0.073%. This design eliminates mercury seals for low piston speeds for gases at low flow rates. Maginnis (2002) explained how devices like piston provers, where the motion of the piston is not at constant velocity, may be subject to an error due to acceleration of the piston.

In order to minimise the uncertainty arising from measurements of volume provers, Ilha, Doria and Aibe (2010) considered the thin liquid films that form on the inner surfaces of provers. Jeronymo and Aibe (2010) presented an electronic circuit for quadruple-timing pulse interpolation applied to compact piston provers.

Masri, Lin and Su (2010) reported on a primary standard with the capacity of 0.005 l/min to 24 l/min at the National Institute of Metrology (Thailand), NIMT, consisting of a mercury-sealed piston prover with three precision-machined glass cylinders. Gas flow rate was determined from displaced volume, time, pressure and temperature of the gas with an uncertainty of less than 0.13%. See also Choi et al. (2009).

4.A.5 Gas Properties

Johansen (2010) explained that constant values for isentropic exponent and absolute viscosity were commonly used for the calculation of gas flow rates, and may introduce significant errors. REFPROP 8.0 was recommended as a source of accurate thermodynamic and transport property values. Wright (2010) described calibrations conducted by NIST's Fluid Metrology Group using the NIST-supported database REFPROP version 8.1.

4.A.6 Case Study of a Water Flow Calibration Facility Which Might Be Used in a Manufacturing Plant or a Research Laboratory from the Author's Experience

The author moved to the Department of Engineering at the University of Cambridge, and joined the Institute for Manufacturing. Research, therefore, had a manufacturing focus, and it was decided to build a flow calibration rig both for measuring manufacturing variation in flowmeters, and also as a source of guidance for manufacturers and others who need to install a calibration rig. The background and some initial work is set out in a paper by the author and colleagues (Baker et al. 2006b).

The flow circuit was designed and a calculation undertaken based on Miller's (e.g. 1990 edition) book on internal flow systems to obtain the pressure losses in the system and so to be able to size the pump. A pump and inverter set was selected with sufficient spare pressure range and flow to allow subsequent development of the rig. This system gave very good control and avoided some of the valve control which would otherwise have been necessary to adjust the flow. The maximum flow rate of the system was about 300 kg/m. The energy thus applied to the water resulted in heating, and a tank of sufficient size was selected to ensure that the temperature change was small. Of course a cooler could be added to provide this temperature control. We used a simple scheme of adding melt from ice in sufficient quantities to balance the heat addition.

In the original design, we provided a concrete base to support the flowmeters and to eliminate any vibration to or from the Coriolis meters. In a later redesign of the rig, we were able to simplify this and provide a framework to support the pipe-work and meters which were, by then, much less sensitive to vibration etc. Pipework was a combination of ABS and stainless steel. Each had advantages, and the type will need to be selected with other considerations in mind.

Since the main rig testing was likely to be linear flowmeters, it was decided to install a standing start and stop system with a weigh tank and scales. This reduced the uncertainties due to flow diverters and timing by ensuring that the total in the weigh tank should be the total passed through the system. Continuous flow tests were not excluded, since transfer standard flowmeters could be calibrated and then used for continuous flow calibrations. The details of the weighing system are described by Baker et al. (2006b). Two electro-pneumatic valves controlled the flow direction, either around the recirculation flow circuit or for batch mode into the weigh tank. The graphical user interface (GUI) which was developed showed the circuit diagram and for each instrument gave the output data. The software allowed the data to be modified for units or operational changes or for setting up the current parameters. Figure 4.A.2 shows the GUI with recirculation mode selected and with additional meter test sections.

The upthrust due to the buoyancy of the air on the weigh tank was calculated from the scale reading, m_L, the mass of water in the weigh tank, M_L, and the mass of air displaced, M_G, where

$$m_L = M_L - M_G \tag{4.A.1}$$

Assuming that ρ_{LW} is the density of the liquid (water) and ρ_G is the density of the air,

$$m_L = M_L\left(1 - \frac{\rho_G}{\rho_{LW}}\right) \tag{4.A.2}$$

resulting in a correction factor for the value given by the weigh tank read-out of

$$1 + \frac{\rho_G}{\rho_{LW} - \rho_G} \tag{4.A.3}$$

The consequent adjustment to the value is about 0.12%.

The paper discusses the effect of the start and stop periods and the error these could cause, and also considers the effect of continuing dripping from the outlet nozzle, and other matters such as air in the water. Our preliminary estimates of uncertainty, which, we recognised, did not cover all factors, gave a value for the combined uncertainty of the weigh tank system of about 0.03%.

This was followed by a paper giving a very thorough assessment of the uncertainty and causes thereof (Shimada, Mahadeva and Baker 2010b). This gave

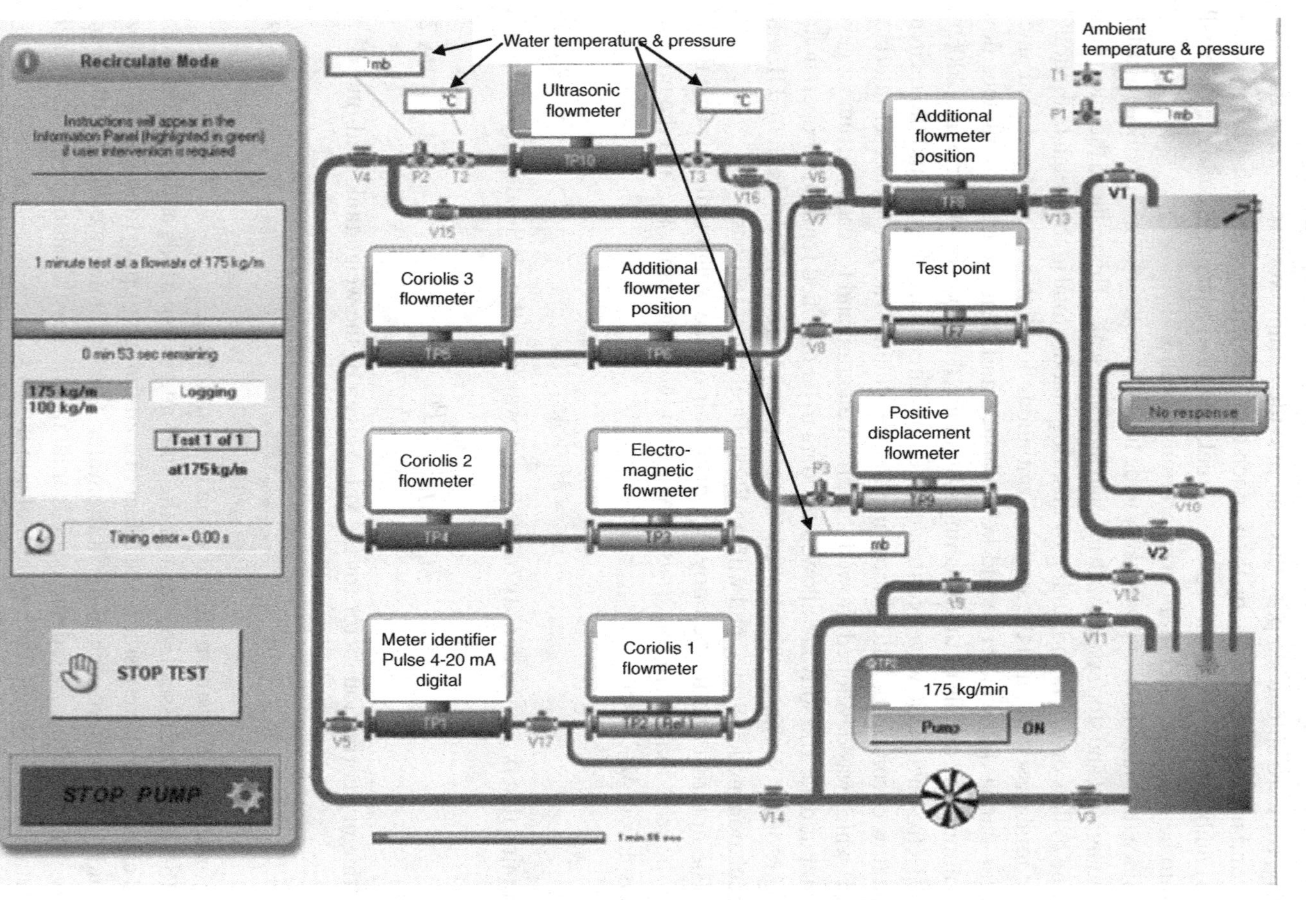

Figure 4.A.2. Screen shot of graphical user interface (GUI) edited.

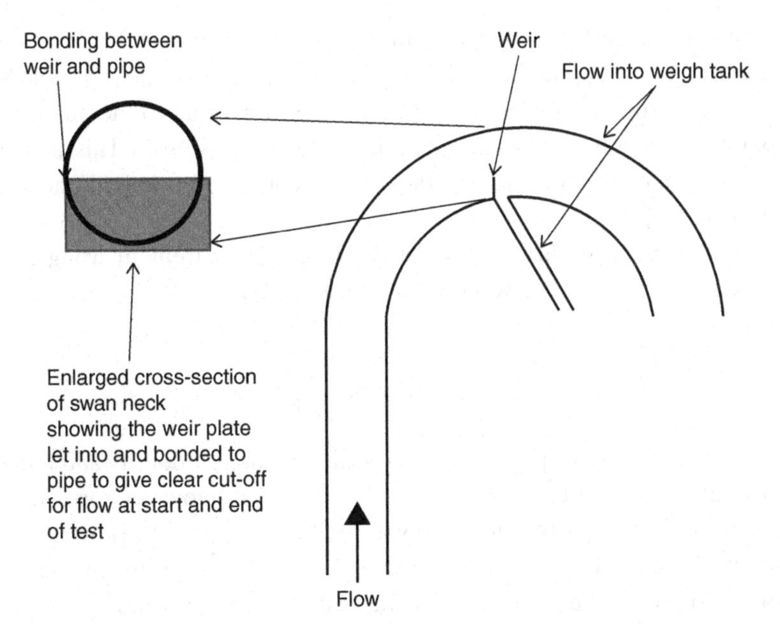

Figure 4.A.3. Diagram of weir in inlet pipe to weigh tank. (After Shimada et al. 2010b)

further details of the rig, including the weir in the inlet pipe to the weigh tank which ensures that the level was the same for each test, Figure 4.A.3. There were three main stages in the assessment of uncertainty attainable by the rig: the uncertainty of the weigh scale, the uncertainty of the fluid flow and the uncertainty of the meter under test. We first considered the uncertainty and traceability of the weigh scale.

The deadweight calculation is instructive in reflecting the detailed considerations required for precise assessment. The upthrust due to the displacement of air by the deadweight will depend on the local gravity and the density of the material of the deadweights. OIML (2004) (cf. BIPM 2008) the reader should check for the latest version of all standards referenced) defines a conventional value of mass (conventional mass = M_s) of material of density ρ = 8,000 kg/m^3 and expresses the weight of the deadweight as

$$M_D = M_s \frac{\left(1 - {}^{1.2}\!/_{8000}\right)}{\left(1 - {}^{1.2}\!/_{\rho_D}\right)} \tag{4.A.4}$$

where ρ_D is the actual density of the deadweight. This adjustment allows for the small change in upthrust. The k_S factor for the weigh scale is given by:

$$k_S \approx \frac{M_S}{m_{CAL}} \left(1 - \frac{1.2}{8000}\right) \tag{4.A.5}$$

where m_{CAL} is the reading of the weigh scale when loaded with the deadweights, M_D, and the formula for the weigh scale uncertainty is set out.

The standard uncertainty of the calibration factor is obtained, allowing for the effects of temperature, eccentric loading, repeatability, linearity and reproducibility. In addition, the uncertainties of the weights, the uncertainty due to the air density and the temperature and temporal variation of the scales were assessed. This allowed, with reading resolution etc., an estimate of the relative combined standard uncertainty of the weigh scale of 7.5×10^{-5}.

The paper covers the uncertainty of the liquid measurement using the weigh scale. The mass of liquid in the weigh tank was given by

$$M_L = \frac{k_S\, m_L}{\left(1 - \rho_G / \rho_{LW}\right)} \tag{4.A.6}$$

where m_L, ρ_G and ρ_{LW} are respectively weigh scale reading, air density and liquid density. The resolution is as for the weigh scale calibration, but buoyancy correction uncertainty, and any difference between the flow through the pipe and the flow into the tank requires consideration. These additional factors raised the uncertainty to 0.013%.

In order to obtain the factor K_{fm} for the mass flowmeter against pulse count, P, we used the following equation

$$\left\{\frac{u(K_{fm})}{K_{fm}}\right\}^2 = \left\{\frac{u(P)}{P}\right\}^2 + \left\{\frac{u(M_L)}{M_L}\right\}^2 + \left\{\frac{u(\Delta M_{LDV})}{M_L}\right\}^2 + \left\{\frac{u(\Delta K_{fm})}{K_{fm}}\right\}^2 + \left\{\frac{u(\delta K_{fm})}{K_{fm}}\right\}^2 \tag{4.A.7}$$

where the first term on the r.h.s. is the uncertainty of the pulse count, the second term is the uncertainty of the mass flow measurement from the rig, the third term relates to the change in mass, ΔM_{LDV}, within the connection pipe between the flowmeter and the weir, the fourth term relates to the effect of diversion on the flowmeter characteristic and the final term relates to the random effect of the flowmeter.

The paper discussed the possible effects of pipework expansion, trapped air, initialising the rig and valve behaviour. The effect of valve closure produced some very interesting effects. The pulsation set up caused oscillation in the pipework, which introduced an increased uncertainty for the Coriolis meters which could be reduced by adjusting the time delay settings in the meter. For a positive displacement meter, while these oscillations were generally not important, if the magnets in the rotors were close to the pick-up when rotation stopped and the oscillation took place, the meter appeared to sense additional pulses.

Allowing for all these considerations, the authors estimated that the expanded uncertainty for the mass flowmeters was 0.023% and, when density uncertainty was introduced for volumetric meters, the value was 0.12%.

If, however, we heed the comment of Griffin (2009): "*when calibrating a measuring device, the uncertainty of the calibration standard should be a factor of 10 less than that of the device itself*," we may be more cautious in our estimates.

The final paper on the rig concerned the software and hardware of the signal processing (Baker et al. 2013). There are essentially three aspects to this paper: the communication between the PC and the rig, the software aims and outcome including

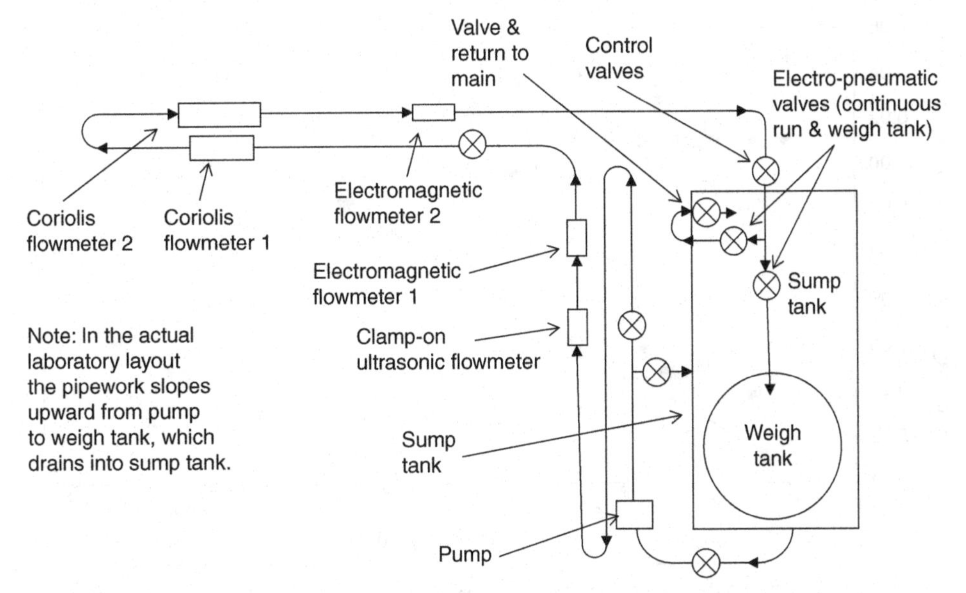

Figure 4.A.4. Rebuilt flow rig (reproduced with the agreement of Elsevier from Baker et al. 2013).

the GUI and the problems with obtaining signal integrity. In addition the data from the flowmeters, when tested on the rebuilt rig, Figure 4.A.4, were shown to be within a figure of order ±0.05% compared with their calibrations obtained from national standards, Figure 4.A.5.

4.A.7 Example of a Recent Large Water Calibration Facility

A very different scale of facility is described by Ben Weager, who has provided the following brief description of a major flow rig development with which he has been involved (Figure 4.A.6).

"The large flow calibration rig of Siemens Sensors and Communications Ltd. in Dalian, China, has a capacity of 4000 l/s and can calibrate flowmeters from 600 up to 3,000 mm in diameter. It is based on the closed loop concept in which water is continuously circulated in a loop which contains the measuring section and reference line[s].

"The reference lines are designed with lengthy upstream straight pipework which includes flow straighteners and conditioners. Each line contains three electromagnetic reference meters which are calibrated in situ against two static weighing systems of 8 tonnes and 30 tonnes capacity, selected according to flow rate. During the reference system calibration the lines are individually selected and data from all three reference meters is collected.

"When a test line is built with the device under test in position, the complete system comprising reference lines, manifolds and test section are filled from the reservoir using the line filling pumps and all air is vented.

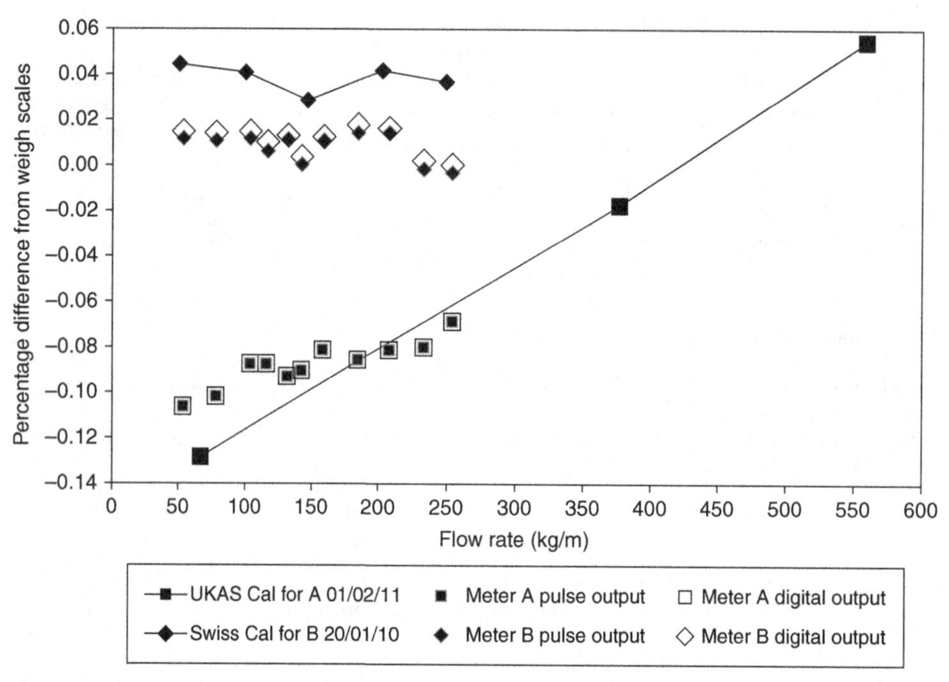

Figure 4.A.5. Calibration of two meters on laboratory flow rig compared with their calibrations on national facilities (reproduced with the agreement of Elsevier from Baker et al. 2013).

Figure 4.A.6. Large-flow calibration rig in Dalian, China (reproduced with permission of Siemens Sensors and Communications Ltd).

"Each reference line has its own closed loop circulation pump and the lines are selected to match the test flow rate required. The pumps have good matched characteristics to give stable flow with no interaction. In order to reduce the build in length for very large test flowmeters, attention has been given to ensure a good flow profile at the test point by means of a specially designed upstream flow conditioner.

"The calibration process can then be undertaken with duration to suit the device being tested. This is very advantageous with large electromagnetic flowmeters with low field excitation frequencies.

"The estimated uncertainty is stated to be less than 0.1%" (Ben Weager private communication 2015).

5

Orifice Plate Meters

5.1 Introduction

The orifice plate flowmeter, the most common of the differential pressure (DP) flowmeter family, is also one of the most common industrial flowmeters. It is apparently simple to construct, being made of a metal plate with an orifice that is inserted between flanges with pressure tappings formed in the wall of the pipe. It has a great weight of experience to confirm its operation. However, it is far more difficult to construct than appears at first sight, and the flow through the instrument is complex.

Some key features of the geometry of the orifice and of the flow through it are shown in Figure 5.1. The behaviour of the orifice plate may be predicted, but the predictions derive from experimental observation and data. The inlet flow will usually be turbulent and will approach the orifice plate where an upstream pressure tapping (one diameter before the orifice plate) will measure the pressure at the wall, that is in this case the static pressure. (Flange and corner tappings will be discussed later.) The flow close to the orifice plate will converge towards the orifice hole, possibly causing a recirculation vortex around the outside corner of the wall and orifice plate. The inward momentum of the flow at the orifice hole will continue downstream of the hole, so that the submerged jet coming out of the hole will reduce to a smaller cross-section than the hole, which is known as the *vena contracta*, the narrowest point of the submerged jet. Outside this submerged jet is another larger recirculation zone, and the downstream pressure tapping (the one after the orifice) set in the wall senses the pressure in the *vena contracta* across this recirculation zone. Downstream of this point, diffusion takes place with considerable total pressure loss.

The data on which the orifice predictions are based may be presented in three ways (cf. Miller 1996):

1. The most accurate method is to use a discharge coefficient-Reynolds number curve for the required geometry, which includes all dimensional effects and other influences.
2. To reduce the number of curves, a datum curve is used in conjunction with correction factor curves. This was essentially the procedure adopted for the British Standard 1042: Part 1:1964.

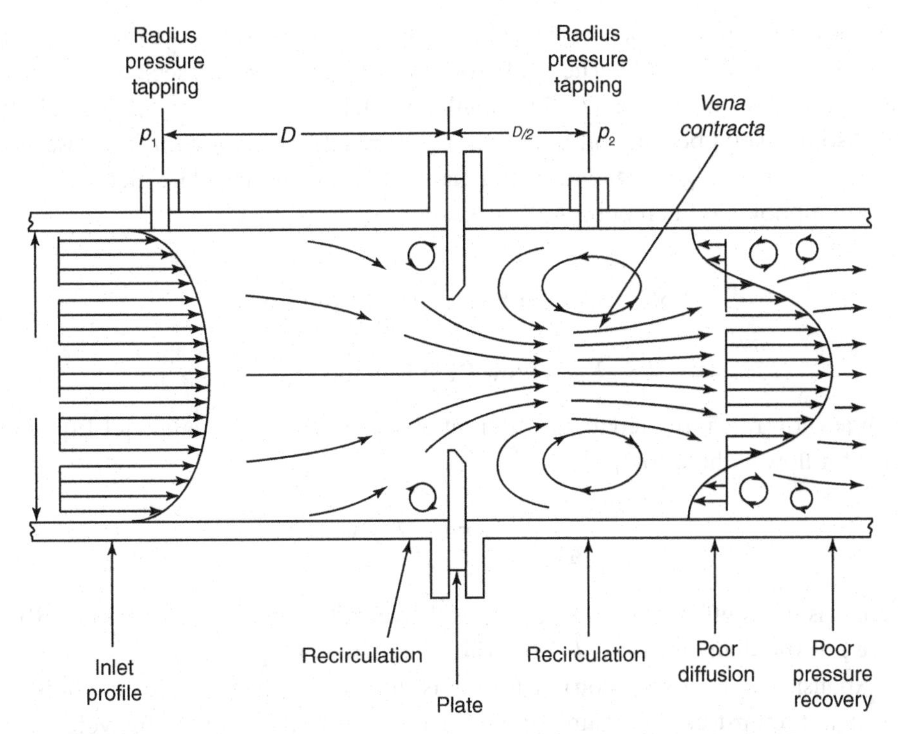

Figure 5.1. Diagram to show geometry and flow patterns in the orifice plate flowmeter.

3. For convenience with the advent of modern flow computers, the data are reduced to a best-fit equation. This is essentially the procedure with the most recent versions of the international standard, and its form will be dealt with in the following pages.

The most common orifice plate is a metal disc spanning the pipe with a precisely machined hole in the centre of the plate; it is usually mounted between flanges on the abutting pipes, with pressure tappings fitted in precisely defined positions and to precise finishes. The differential pressure is measured by manometer, Bourdon tube or a pressure transducer, and the flow is deduced from the equations and probably computed using a flow computer.

The importance of the orifice is its simplicity and predictability, but to achieve high accuracy it is essential that the detailed design of the meter is the same as that from which the original data were obtained, and that the flow profile entering the meter is also the same. To ensure that the details are correct, the national and international standards lay down the precise requirements for constructing, installing and operating the orifice meter. It must be stressed that departing from the standard requirements removes the predictability and prevents the standard from being used to obtain the flow prediction.

In this chapter, I have sought to avoid duplicating the standards or the very thorough presentation of Miller (1996) and instead have attempted to present recent published information on the performance of the meter. It is, therefore, essential that the reader have a copy of the relevant standards from which to work and, in particular, ISO 5167-2: 2003. The reader is also strongly advised to seek out a copy of Miller (1996). Access to Bean (1971), Spink (1978), Danen (1985) and Spitzer (1991) may also be useful because all have extremely valuable information for those concerned with the design, installation and maintenance of orifice plate meters. A very important book has been published by Reader-Harris (2015).

5.2 Essential Background Equations

Mass Flow Rate Equation

The mass flow rate is related to the differential pressure by the equation [cf. Equation (2.11) for flow without loss]

$$q_{\mathrm{m}} = \frac{CE\varepsilon\pi d^2 \sqrt{2\rho_1 \Delta p}}{4} \tag{5.1}$$

where C is the coefficient of discharge, E is the velocity of approach factor $(1-\beta^4)^{-1/2}$ where β is the diameter ratio d/D of orifice diameter to pipe internal diameter, ε is the expansibility (or expansion) factor, Δp is the differential pressure, and ρ_1 is the density at the upstream pressure tapping cross-section. The value of the volumetric flow can be obtained from this using the relationship:

$$q_{\mathrm{v}} = q_{\mathrm{m}}/\rho \tag{5.2}$$

where ρ is the density of the fluid at the appropriate conditions of pressure and temperature.

Coefficient of Discharge

A simple expression for the discharge coefficient is

$$C = C_\infty + \frac{C_{\mathrm{Re}}}{\mathrm{Re}^n} \tag{5.3}$$

where C_∞ is the coefficient for infinite Reynolds number, C_{Re} is a constant for a particular installation, Re is the Reynolds number based on the pipe ID, and n is the index to which this is raised. The relative simplicity of Equation (5.3) is not matched by the most commonly used expressions for C, C_∞ or C_{Re} in either their complexity or variety.

Three versions of the coefficient are set out in Appendix 5.A: Stolz and the two versions of the Reader-Harris/Gallagher equation from which the current U.S. and ISO standards were developed. These cover much of the usage currently in various countries. American practice appears to be to use API Chapter 14 Section 3 (14.3). The ISO equation supersedes the Stolz equation. However, it is essential that the reader refers to the appropriate standard document to obtain all the conditions necessary to its valid use, and any updates. These include ranges for parameters, detailed design, pipe smoothness etc.

The discharge coefficient, C, used in Equation (5.1) is given in slightly differing forms in various standard documents. In general it takes the form

$$C = C_\infty + \frac{C_{Re}}{Re^n} + C_{Taps} + C_{Small\ orifice} \tag{5.4}$$

where all the coefficients are functions of β. C_∞ is the basic part of the coefficient, C_{Re} reflects the slope of the characteristic due to change of Re, the Reynolds number, C_{Taps} varies with the position of the tapping points be they D-D/2, flange or corner tappings. It includes values of $L_1 = \ell_1/D$ and ℓ_1 is the distance of the upstream tapping from the upstream face of the plate, $L_2' = \ell_2'/D$ and ℓ_2' is the distance of the downstream tapping from the downstream face of the plate and the ′ signifies that the measurement is from the downstream and not the upstream face of the plate. $C_{Small\ orifice}$ gives a correction for small orifice sizes.

The form of this equation as used in ISO 5167-1981 was known as the Stolz equation, and this is given in Appendix 5.A.1. It will be noted that this equation did not have a special small orifice term.

As a result of extensive international tests, two newer equations have been adopted, and both are attributed to Reader-Harris and Gallagher. The first of these, and the one which is now generally adopted in the United States, is that given in API Manual of Petroleum Chapter 14–3, and the equation from which it was derived is reproduced in Appendix 5.A.2 in a rearranged form to match the terms in Equation (5.4).

The second of these newer equations is that which is set out in ISO 5167-2:2003, and also in Reader-Harris and Sattary (1996). It is given in Appendix 5.A.3. It will be noted that there are differences in the coefficients between these two equations. The arrangement of the small orifice term in this version differs slightly from that in Appendix 5.A.2, but the actual value of diameter for which the term becomes non-zero is $D = 71.12$ mm in each equation.

The data for the ISO equation appears to have been obtained for orifice plates with diameter ratios 0.1–0.75, throat Reynolds numbers from 1,700–5×10^7 and pipe diameters from 50–600 mm. Data points for orifice diameters less than 12.5 mm were very scattered and were not included.

Miller (1996) gave the proposed Stolz II and NEL/TC 28 equations and tabulated the values for these compared with ISO and ANSI/API.

A sample of values to show the extent to which the value of C differs between these equations is given in Table 5.1, from which it can be seen that in most cases the difference in the value of C between the equations is less than ±0.05%. However, for the high Reynolds number it is within ±0.09% and for the small pipe diameter it is within ±0.14%.

Whichever equation is used, it is important to ensure that the specified constraints are met on d, D, β and Re. Outside the specified limits the standards will not necessarily define the uncertainty in the value of C. Within the constraints the value of C will have the uncertainty indicated in the relevant standard. For instance, for the ISO 5167-2:2003, the value in Appendix 5.A.3 is 0.5% at best. The reader should refer to the appropriate standard document for detailed information.

Table 5.1. *Typical values of the coefficient of discharge for orifice plates with flange tappings with β = 0.5 derived from three equations*

D	Re	Stolz	API 14.3	ISO 5167-2:2003
101.6	100 000	0.60579	0.60625	0.60620
101.6	1 000 000	0.60342	0.60327	0.60313
101.6	10 000 000	0.60300	0.60228	0.60204
50.8	1 000 000	0.60347	0.60399	0.60503
203.2	1 000 000	0.60319	0.60347	0.60315

Cristancho et al. (2010) suggested an alternative formulation of the standard orifice equation for natural gas which avoids the need for iteration resulting from the presence of Reynolds number in the ISO equation.

Expansibility Factor

The expansibility (expansion) factor provides an adjustment factor to allow differential pressure devices to be calibrated on water, an essentially incompressible fluid for these purposes, for use on compressible gas. It, essentially, provides an adjustment factor to the coefficient of discharge that allows for the compressibility of the gas. Reader-Harris (1998) provides a review of the past development of this factor and the current recommended equation for it. It is found to be virtually independent of Reynolds number. For many years, it was given by an equation of the form:

$$\varepsilon_1 = 1 - (a_\varepsilon + b_\varepsilon \beta^4)\frac{\Delta p}{\kappa p_1} \tag{5.5}$$

where a_ε and b_ε are constant coefficients, κ is the isentropic exponent, which for an ideal gas is equal to γ, the ratio of specific heats, and if κ is not known γ should be used. p_1 is the pressure at the upstream tapping. However, according to Kinghorn (1986), the coefficients in common use in Equation (5.5) were probably in error.

In ISO 5167–2: 2003, the equation for ε_1 is given in the form

$$\varepsilon_1 = 1 - (a_\varepsilon + b_\varepsilon \beta^4 + c_\varepsilon \beta^8)\left\{1 - \left(\frac{p_2}{p_1}\right)^{1/\kappa}\right\} \tag{5.6}$$

provided that $p_2/p_1 \geq 0.75$. c_ε is another constant coefficient. The final bracket may be approximated by $\Delta p/\kappa p_1$ if p_2/p_1 is very close to unity, but the work of Reader-Harris (1998) suggests that the expression in the final bracket is preferable. He gave the following equation after careful analysis of the data, which is in ISO 5167-2: 2003:

$$\varepsilon_1 = 1 - (0.351 + 0.256\beta^4 + 0.93\beta^8)\left\{1 - \left(\frac{p_2}{p_1}\right)^{1/\kappa}\right\} \tag{5.7}$$

He gave a relative uncertainty for ε_1 of 3.5 $\Delta p/\kappa p_1$ %, which differs slightly in detail from that in ISO. He also gave a very useful theoretical derivation and obtained an equation that differs from Equation (5.7) by about the uncertainty in Equation (5.7).

Pressure Loss

The expression in the ISO standard for the pressure loss across the orifice plate now appears to be

$$\text{Pressure loss} \approx \frac{\sqrt{1-\beta^4(1-C^2)}-C\beta^2}{\sqrt{1-\beta^4(1-C^2)}+C\beta^2}\Delta p \tag{5.8}$$

Taking a typical value of C of about 0.6, this results in a pressure loss for a plate with $\beta = 0.5$ (and $E = 1.033$) of about $0.73\Delta p$ [cf. Urner (1997), who suggested that the previous ISO equation (for nozzles and orifice plates) gave anomalous results if used outside the limits of applicability of the ISO standard].

An alternative for orifice plates allowed by a recent version of the standard is

$$\text{Pressure loss} \approx (1-\beta^{1.9})\Delta p \tag{5.9}$$

which, for the preceding case, also gives $0.73\Delta p$.

The most recent version appears to have provided a loss coefficient K given by

$$K = \frac{\text{Pressure loss}}{\frac{1}{2}\rho_1 V^2} \tag{5.10}$$

where

$$K = \left(\frac{\sqrt{1-\beta^4(1-C^2)}}{C\beta^2}-1\right)^2 \tag{5.11}$$

5.3 Design Details

Design details are set out fully in the ISO 5167 standard. It is important that the detailed requirements are followed so that performance can be predicted, but also so that new data can be added to existing data, to increase our overall knowledge. In this book, some of the requirements will be touched on, but the standard should be referred to for full details.

The design is, of necessity, an iterative procedure starting from the known requirements of, typically, design flow rate, pipe size and differential pressure. Danen (1985) gives some helpful calculation flowcharts for each type of differential meter.

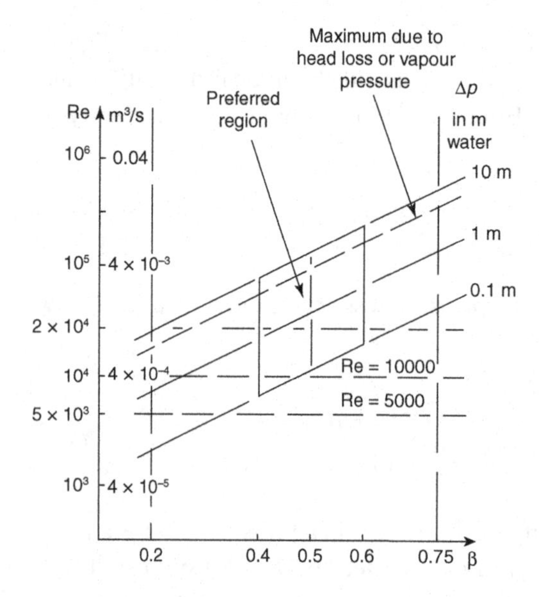

Figure 5.2. Compromise decision for sizing an orifice plate for water with D and $D/2$ tappings with a line size of $D = 50$ mm.

The first educated guess will relate to the value of β. With this and the selected values, a new value of β is found via the flow rate equation. Iteration continues until a satisfactory value of β is achieved. However, it is likely that most users will have access to a computer program for this procedure (either from a manufacturer or in their own software library) that will enable the optimum size of plate to be obtained. These will ensure a balance between too great a pressure drop through the plate and too small a differential pressure for measurement. It is often suggested (Miller 1996) that the pressure drop across the orifice plate should be of the order of 100 in. of water (2.5 m or 25 kPa) and that the beta ratio should be close to 0.5. In some cases, this may not be achievable. The pressure loss can never exceed the difference between supply and demand pressures, and the minimum pressure should be kept above the liquid's vapour pressure to avoid cavitation. Figure 5.2, which has been suggested in different forms by various people, illustrates the compromise decision that needs to be made between these factors.

For wide-range metering, Miller suggests the use of two transmitters ranged for 180 in. of water (4.5 m and 45 kPa) and 20 in. of water (0.5 m and 5 kPa).

a. The plate. Some details of the plate geometry are shown in Figure 5.3 for a pipe of diameter D and with an orifice diameter of d. The plate must be flat with parallel faces, and concentric. Flatness is defined in the standards. The upstream face is the most critical because this has a strong effect on the flow entering the orifice. In addition to the flatness, the plate must have a roughness of less than $10^{-4}d$. It is useful to indicate the correct flow direction on the plate in a position that can be seen when the plate is installed. The plate may have to be rather thick to withstand the forces due to the flow, and in this case the downstream edge of the orifice is bevelled at an angle of 45° ± 15°, so that the final

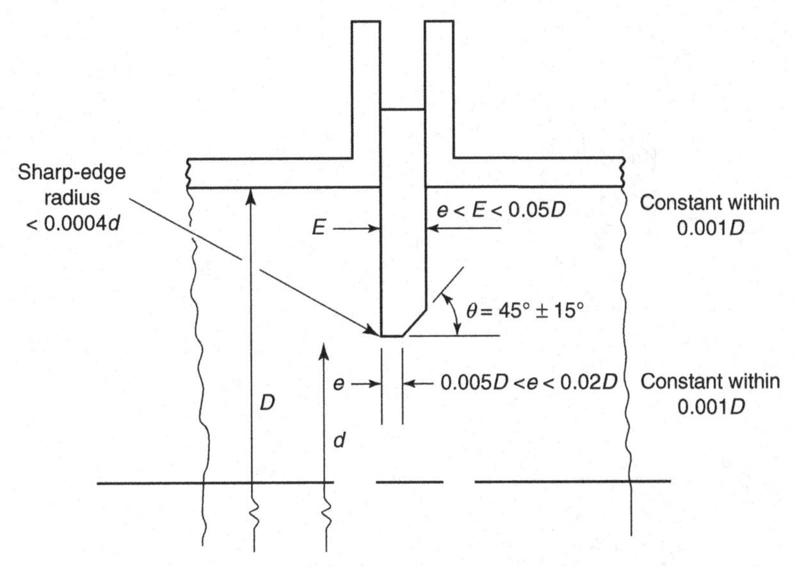

Figure 5.3. Plate geometry: E is thickness of the orifice plate, and e is thickness of the orifice.

thickness of the actual orifice opening is between $0.005D$ and $0.02D$, and constant to within $0.001D$. All edges of the orifice must be clear of burrs etc., and the upstream edge must be sharp, defined as of radius less than $0.0004d$. If d is less than 25 mm (cf. 125 mm in the previous version of the standard), it is necessary to measure the radius to ensure that this condition is met. Sun et al. (1996) gave an equation for change in discharge coefficient with edge radius, r

$$\frac{\Delta C}{C}\% = 0.85 \ln\left(10^3\, r/d\right) + 1.74 \qquad (5.12)$$

for $0.0002 < r/d < 0.0035$. This appears to give a rather larger change than most of the data in BS 1042: Section 1.5:1997, although some data for $D = 150$ mm lines indicate changes greater than Equation (5.12). If these are valid, then even for $r/d < 0.0004$, $\Delta C/C$ could be nearly 1% high. (Sun et al. also suggested an orifice chamfered at entry and referred to the Russian National Code RD 50-411-83.)

Two methods have been used to measure the edge radius (Hobbs and Humphreys 1990): a stylus that followed the contour and allowed magnifications of up to 500, and a cast around the edge that could be sliced and polished after removal from the edge. Results to 0.005 mm are claimed (cf. Jepson and Chamberlain 1977, who described British Gas's orifice radius inspection system).

b. Pressure tappings. There are three standard methods of sensing the pressure drop across the orifice plate (Figures 5.1 and 5.4): D and $D/2$ tappings (Figure 5.1), flange tappings (Figure 5.4) and corner tappings (Figure 5.4).

Again, the standards set out the precision of positioning of the tappings and the details of their manufacture (cf. Zedan and Teyssandier 1990, who concluded that

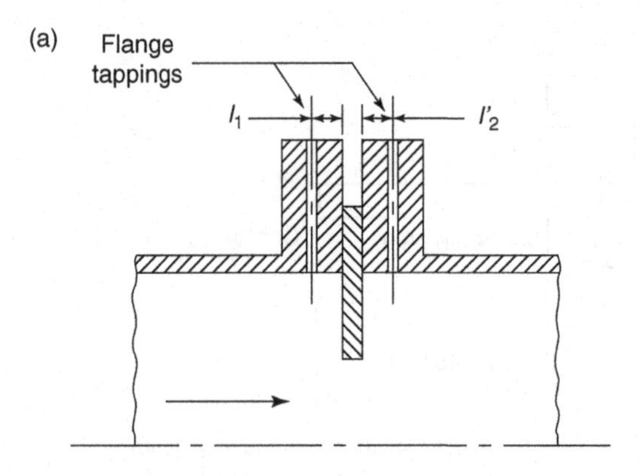

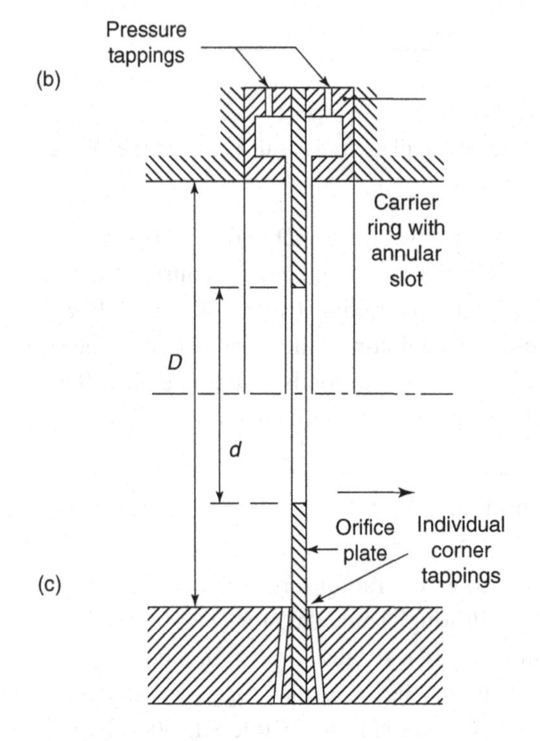

Figure 5.4. Layout of pressure tappings (after ISO 5167): **(a)** Flange; **(b)** corner – ring with annular slot; **(c)** corner – individual tappings.

the tolerances that ISO and ANSI/API 2530 allowed for pressure tap locations were conservative).

5.4 Installation Constraints

More recent research on this topic may be found in Appendix 5.B.

The upstream and downstream pipe lengths are set out in some detail in ISO 5167-2: 2003. A sample of the information is given in Table 5.2. The values given are claimed to retain the predicted performance of the orifice plate, whereas the values in parentheses increase the measurement uncertainty by 0.5%. For instance, a meter with $\beta = 0.5$ mounted within 22D of a 90° bend but with more than 9D clear

Table 5.2. *Examples of installation requirements for zero additional uncertainty from the standard. (Bracketed values are ±0.5% additional uncertainty.)*

β	Upstream straight lengths				Downstream straight lengths
	Single 90° bend	Two 90° bends in perpendicular planes with less than 5D spacing between *	Concentric reducer 2D to D over a length of 1.5D to 3D	Full-bore ball valve or gate valve fully open	
≤0.2	6 (3)	34 (17)	5 (**)	12 (6)	4 (2)
0.4	16 (3)	50 (25)	5 (**)	12 (6)	6 (3)
0.5	22 (9)	75 (34)	8 (5)	12 (6)	6 (3)
0.6	42 (12)	65 (25)	9 (5)	14 (7)	7 (3.5)
0.67	44 (20)	60 (18)	12 (6)	18 (**)	7 (3.5)
0.75	44 (20)	75 (18)	13 (8)	24 (12)	8 (4)

* Not a satisfactory installation, as it is likely to create swirl and would require a flow straightener. The unexpected reduction of spacing in this column for large β (from the ISO) may be accounted for by the complex nature of the flow.
** Data not available.
Note that some of these values, from ISO5167-2:2003, are greater than in earlier versions (and the first edition of this book), and may be a response to concerns mentioned in the references.

pipe upstream and 3D downstream will have an uncertainty in the coefficient of 0.5 + 0.5 = 1.0.

Reader-Harris and Keegans (1986) suggested that 2.5D of smooth pipe upstream of corner tappings reduced the effect of pipe roughness further upstream. Sindt et al. (1989) found up to 1% change in coefficient for $\beta = 0.74$ and $Re = 7 \times 10^6$ between smooth (2.8 μm) and rough (8.9 μm), but for $\beta = 0.5$, the change was not measurable.

Constraints are also placed on thermometer pockets near to the flowmeter.

The validity of these values is based on many tests to obtain data. Some recent measurements are compared later.

Bends

Some tests raised questions as to whether the required upstream straight length of pipe is adequate (Branch 1995). Decrease in bend angles does not necessarily reduce the disturbing influence on the meter (Himpe, Gotte and Schatz 1994). There does not appear to be an optimum angle between the tapping position and the plane of the bend (Branch 1995), although a considerable influence has been noted by Himpe et al. (1994). Mattingly and Yeh (1991) gave the effect of distorted profile caused by elbows upstream of orifice meters (Figure 5.5). They also gave change in coefficient for upstream elbows with a tube bundle straightener between 3.8D and 5.7D from the bend outlet but concluded that this flow conditioner may introduce errors. Conditioner/flowmeter combinations should preferably be calibrated as one unit.

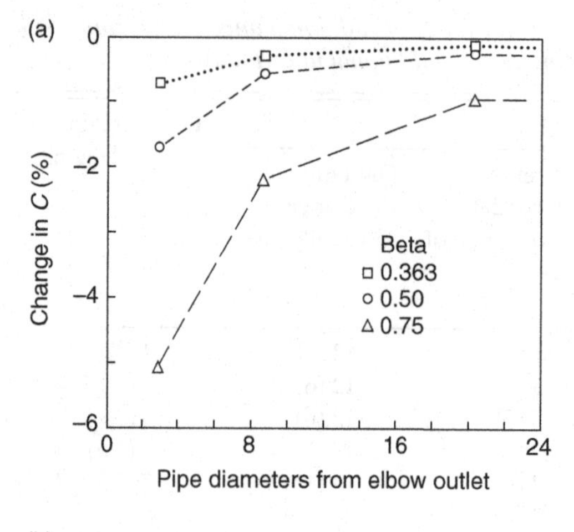

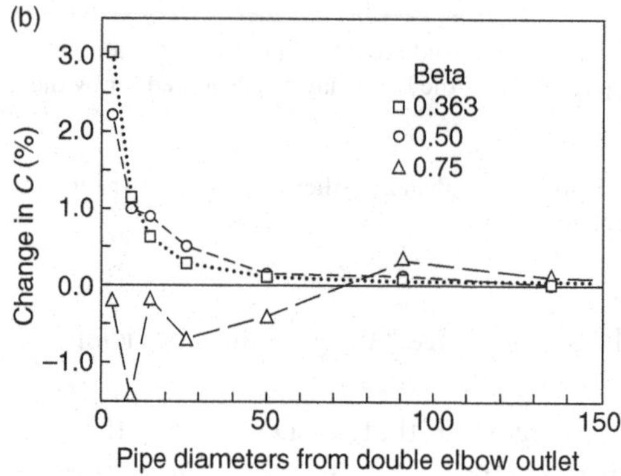

Figure 5.5. Change in orifice coefficient when downstream of: **(a)** single elbow; **(b)** closely coupled double elbows out of plane (reproduced with the agreement of Elsevier from Mattingly and Yeh 1991).

Fittings that Induce Swirl

Mattingly and Yeh (1991) also showed that, for Reynolds numbers between 10^4 and 10^5 in a 50-mm water flow facility, the angle of swirl caused by the double elbow out of plane configuration decreased from about 18° to about 6° after 40D and to about 3° after 80D. They questioned the 2° swirl criterion for safe operation of orifice installations (cf. Brennan et al. 1989 and Morrison, Hauglie and DeOtte 1995, whose work may require a knowledge of the level of swirl in a particular pipe installation).

Morrison et al. (1990a) gave measurements that confirmed that larger beta ratio orifice plates are more sensitive to swirl. They noted that expansion of the orifice jet (*vena contracta*) leaving the plate, resulting from conservation of angular momentum, would be expected to have a greater effect on small beta ratios, contrary to experience. On the other hand, the results of Brennan et al. (1991) appear to be contrary to expectation with increased beta ratios in most cases improving the performance downstream of tees and two elbows in perpendicular planes. In addition, the results show a much better performance by the straighteners than found by Sattary (1991), which is referred to later.

Other Profile Distortions

Yeh and Mattingly (1994) measured the profile downstream of a reducer (approximately 3:2), and the effect of the reducer on the coefficient of an orifice plate flowmeter. Their results appeared to confirm the ISO values for $\beta = 0.5$ and less, but not for $\beta = 0.75$ [cf. Morrison et al. 1992, who distorted the inlet profile to an orifice plate by varying the ratio of flow in the outer (wall) area of the pipe to that flowing through the centre of the pipe].

Flow Conditioners

McFaddin, Sindt and Brennan (1989) found that, for a 100-mm (4-in.) line, a 2.5D-long tube bundle flow conditioner (19 × 12 mm tubes) 7D upstream of the orifice with $\beta = 0.75$, the minimum distance specified in the ANSI/API 2530 standard, the error can be as much as 1%.

Kinghorn et al. (1991) found that the flat profiles downstream of etoile straighteners appeared to cause most of the residual negative errors in the orifice coefficient, as opposed to the positive errors due to swirling flow. To ensure an error of less than ±0.5%, a straightener 1D long should be placed upstream at least

- 6D for plates with a beta ratio of 0.5 and
- 14D for plates with a beta ratio of 0.8.

Even with 16D separation, virtually all tests suggested some effect from the straightener.

Karnik, Jungowski and Botros (1994) tested the effect of a 19-tube bundle in a 101.6-mm line for good flow conditions and downstream of an elbow and looked at the effect on an orifice plate. Their results are very interesting in showing that even though the profile from the tube bundle at about 20D was very close to the fully developed turbulent profile, approximating to Equation (2.4) with $n = 7.4$, the turbulence intensity pattern is markedly different, suggesting that this is a factor in the response of orifice plates. They showed that the orifice coefficient downstream of the elbow and the tube bundle, compared with the correct value, was high by about 0.2% between 10D and 15D and fell to about 0.1% high by about 20D. There was a point closer to the orifice plate where the coefficients were approximately the same, but it would be unwise to rely on this for precise measurements (cf. Morrow et al. 1991, who used an arrangement that allowed them to slide a tube bundle conditioner along the pipe between a 90° long-radius bend and an orifice plate to obtain the variation of discharge coefficient with position for a gas flow. Overall the results appeared to suggest that, if possible, the conditioner should be next to the bend).

Sattary (1991) concluded from the EEC orifice plate programme on installation effects that the ISO 5167 and AGA Report No 3 minimum straight length specifications needed to be revised. The results for conditioners 19D upstream of an orifice

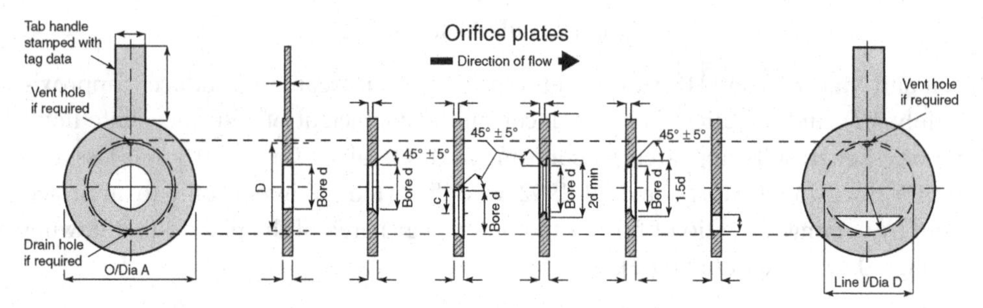

Figure 5.6. Other orifices (reproduced with permission from ABB).

in fully developed flows showed shifts of up to 0.33–0.5% for $\beta = 0.57$, rising to about 0.57–0.75% for $\beta = 0.75$. For a $\beta = 0.2$, there was negligible effect from a tube-bundle conditioner at 5D. The reader concerned with this topic is encouraged to refer to the original paper, which has extensive experimental data and indications that there is variation between the performances of the various straighteners. There may be some discrepancy between the various standards. For two bends in perpendicular planes, by far the most serious disturbance, the ISO requirement for no loss in accuracy is that the orifice should be at least 75D downstream for a $\beta = 0.5$ and also for a $\beta = 0.75$. In the past the requirement for a flow straightener may have resulted, in certain cases, in total lengths of upstream pipework (allowing for that up- and downstream of the straightener and the length of the straightener) greater than or equal to the requirements without a straightener.

5.5 Other Orifice Plates

Figure 5.6 gives a range of orifices that are useful for special purposes, data for some of which may be obtained from BS 1042 Section 1.2:1989.

The quarter circle (quadrant) orifice plate is for low Reynolds number flows, which result from high-viscosity liquids. The entry is more difficult to make precisely than that of the square-edged orifice, and manufacturing tolerances are therefore critical. An expression for the discharge coefficient in the standard has an uncertainty in the region of 2–2.5%.

The conical entrance design is for very low Reynolds number flows, which result from very high-viscosity liquids. The entry is again more difficult to make precisely than that of the square-edged orifice, and manufacturing tolerances are therefore critical. Provided that the standard is followed, $C = 0.734$ with uncertainty of 2%.

Eccentric and chordal (segmental) orifices are more suitable for flows that have a second component in them. Of these, the eccentric orifice plate is preferred, presumably on grounds of more accurate manufacture and more flexible sizing and of being defined in the standard. Thus for a liquid with solid matter, one would choose an eccentric orifice with the hole at the bottom of the pipe, whereas for a liquid with gas entrained, an orifice at the top of the pipe would be chosen. For β in the range 0.46–0.84, the standard gives the coefficient in the range 0.597–0.629 with an uncertainty in the region of 1–2%. In cases where the position of the second component is not clear, it may be necessary to go to a chordal orifice with the edge vertical to allow the second component to pass wherever it is in the pipe.

Dall orifices (see Chapter 8).

Slotted orifices (see Chapter 8).

Further information on "Guidelines for the specification of orifice plates, nozzles and Venturi tubes beyond the scope of ISO 5167" can be found in ISO/TR 15377:2007.

5.6 Deflection of Orifice Plate at High Pressure

Jepson and Chipchase (1975) provided a formula to calculate the deflection in the downstream direction and consequent error for plates operating at high pressure.

The original research was carried out at the British Gas Research Station in Killingworth, England. It used plates rigidly mounted between flanges and tested at differential pressures between 0 and 120 mbar.

Fulton et al. (1987) undertook further tests to determine the effect of seal ring mountings as shown in Figure 5.7 and differential pressures up to 1,000 mbar. Their results appear to suggest that the deflection with the Teflon seal ring was generally about 15% greater than the values from Jepson and Chipchase's (1975) formula, whereas the results for the No. 1 metal seal were close or about 20% less and for the No. 2 metal seal were close or about 30–40% less.

Simpson (1984) provided a useful digest of the papers by Norman, Rawat and Jepson (1983, 1984) in which this problem has been developed and experimental work undertaken to validate the results. Simpson suggested that Norman et al.'s equation (essentially that in BS 1042: Sect. 1.5 1997)

$$\frac{\Delta q_{\mathrm{m}}}{q_{\mathrm{m}}} = -\left(\frac{\Delta p}{E^*}\right)\left(\frac{D'}{h}\right)^2\left(\frac{a_1 D'}{h} - c_1\right) \tag{5.13}$$

where

q_{m} = mass flow rate

Δp = differential pressure across the plate

E^* = elastic modulus of plate material

D' = orifice plate support diameter

h = thickness of orifice plate (=E in Figure 5.3, following ISO 5167)

could have simpler expressions for a_1 and c_1 than given by Norman et al. because the plate thickness will usually be fixed by the next most appropriate thickness of material. Simpson, therefore, gave (as in the standard)

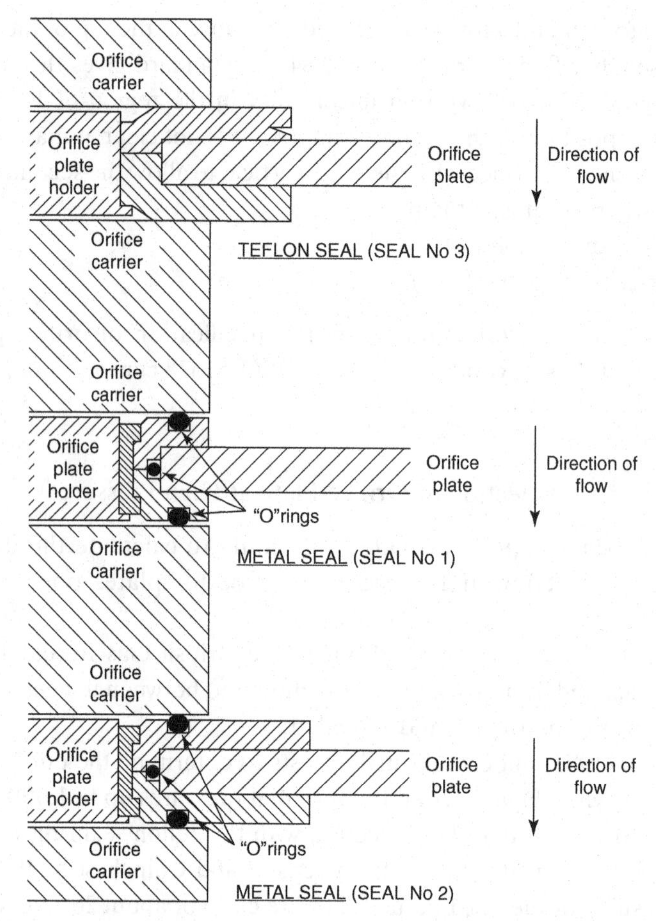

Figure 5.7. Differential seal ring arrangements (from Fulton, Hammer and Haugs 1987 reproduced with permission of the authors, the Norwegian Society of Chartered Engineers and Christian Michelsen Research).

$$a_1 = \beta(0.135 - 0.155\beta)$$
$$c_1 = 1.17 - 1.06\beta^{1.3}$$

To avoid plastic deformation, the differential pressure must be below:

$$\Delta p_y = \sigma_y \left(\frac{h}{D'}\right)^2 \frac{4}{6(0.454 - 0.434\beta)} \tag{5.14}$$

where

σ_y = yield stress for plate material
β = orifice plate diameter ratio *d/D*

With Equations (5.13) and (5.14), it is possible to design for an error of, say, not more than 0.1% for a certain working pressure and for safe operation without plastic deformation with fault conditions of flow and differential pressure drop.

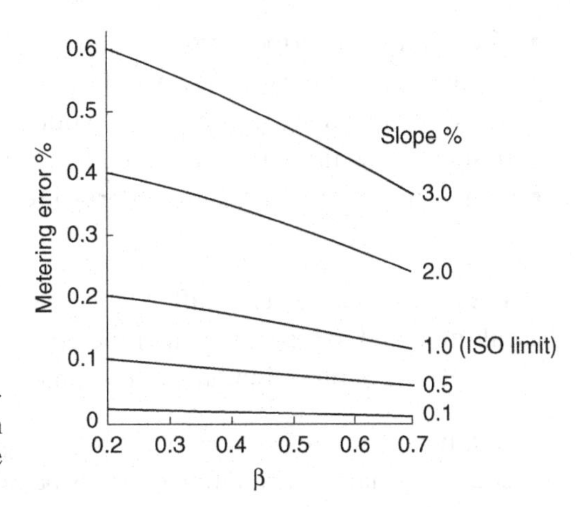

Figure 5.8. Metering error due to initial lack of plate flatness (from Simpson 1984; reproduced with permission of the Institute of Measurement and Control).

For a working differential pressure of up to 500 mbar (and up to 1,000 mbar under fault conditions without yielding) for $\beta = 0.4$ with 304SS plate material having $E^* = 195.4 \times 10^9$ Pa and $\sigma_y = 215 \times 10^6$ Pa, Simpson obtained values of

$h/D' \geq 0.016$ for required maximum error
$h/D' \geq 0.014$ to avoid plastic deformation

h must, therefore, be equal to or greater than $0.016D'$. It should be noted that, in extreme cases of high differential pressure, the allowed thickness according to the standards may be exceeded in meeting these plate-bending requirements. The standard also implies that a safety factor on σ_y (of three) is a wise precaution.

Figure 5.8 gives curves for errors due to flatness slope against β ratio.

5.7 Effect of Pulsation

In considering the installation of a flowmeter, the assumption is usually made that the flow is steady. In many cases, this is probably a fair assumption. However, there are situations when it is not the case (e.g. when a reciprocating compressor, an internal combustion engine or some form of rotary valve is in the line). It is often very difficult to decide whether the flow is indeed steady or pulsating in some way.

We might idealise the pulsating flow as consisting of a sinusoidal ripple superimposed on a steady flow. This idealisation is probably seldom valid, and little will actually be known about the amplitude, frequency or flow profile. Mottram (1992, cf. BS 1042 Section 1.6, which uses some of his work), after the wise comment, "If you can't measure it, damp it!" made some additional useful points. See also ISO/TR 3313.

- If the frequency is above about 2 Hz, differential pressure transducers will, generally, be too heavily damped to pick it up. (Transducers with a response up to 2 kHz or more are likely to be specified and to be more expensive.)

- The standard describes some detection techniques including the use of a thermal probe to sense the presence of pulsation.
- Some flowmeter signals show indication of pulsation.
- In some cases, the error may be deduced from the raw flowmeter signal.
- Connecting leads in differential pressure meters can cause resonance and confusing effects.

The orifice plate meter is affected by pulsation, and so it is necessary to reduce the pulsation and to have an idea of the error likely from any residual pulsation.

The effects of pulsation on an orifice meter are:

a. Nonlinear (square root) error. If a signal is averaged to obtain a mean flow, it is necessary that the instantaneous flow be proportional to the signal. If this is not so, an error will be introduced.
b. Inertia error. If the fluid does not follow the changes in the flow rate instantaneously, an error due to the inertia of the fluid will result.
c. Velocity profile effect. This effect is a result of the change in profile from fully developed to the unsteady flow.
d. Resonance. Resonance occurs because some component of the system is resonating at the pulsation frequency.
e. Limitations in the pressure measurement device. If the pressure measurement device does not respond correctly to pulsating flow, the pressure measurement may be incorrect.

Of these, the most important are (a), (d) and (e), the last two of which are related. The latter two are particularly possible in the connecting tubes between the flow tube and the transducer, and because of the rate of response of the transducer. A manometer will possibly be of little value in an application with pulsation.

Thus for (a), Gajan et al. (1992) reviewed work that showed that, for quite high values of the ratio of rms fluctuating pressure difference to steady pressure difference, up to 0.5 at least, the main error is due to the square root effect and can be eliminated with a high-speed response transducer. Even with a slow-response transducer, the error for the ratio, if within 0.1, may be within about 1%.

Botros, Jungowski and Petela (1992) looked at (d) and (e) and found that the combination of lines and transmitter form a resonator that can amplify or attenuate. They recommended the use of very short lines of essentially constant diameter and a transmitter with a small volume and a high frequency response.

Commercially available differential pressure transmitters are generally unsuitable for dynamic measurement (Clark 1992). The dynamic response of differential pressure transducer systems is significantly affected by length of transmission line and whether the measurement medium is liquid or gas. Gas and liquid lines appear to be susceptible to oscillations, and in liquid systems there is a need for proper gas venting arrangements. Considerable care is required in order to make reliable dynamic measurements particularly when using a liquid medium (Clark 1992).

The square root problem is shown by

$$\frac{1}{t}\int_0^t \sqrt{\Delta p}\, dt \leq \sqrt{\frac{1}{t}\int_0^t \Delta p\, dt} \tag{5.15}$$

A transducer with pressure connections that is capable of responding within a time that is short compared with the period of the fluctuation, and the signal of which is square-rooted immediately and then averaged, will give the correct average value. This is essentially the left-hand side of the inequality (5.15). On the other hand, if the pressure is averaged and then the square root is taken, this is essentially the right-hand side of the inequality and the apparent flow rate will be high compared to the actual flow rate. Mottram's (1981) results for pulsation error are shown in Figure 5.9(a), and they suggest that the experimental results confirm the dominance of the square-root error when compared with errors (b) and (c). However, Mainardi, Barriol and Panday (1977) appeared to suggest that (b) and (c) cause a change in the coefficient so that the effect of pulsation for a fluctuating ripple of 6% of the mean flow will cause an error if pulsation is measured with a high-frequency response transducer of 1.5%, whereas the error is 4% for a 15% ripple. (Frequencies appeared to be up to about 50 Hz.)

Williams (1970) identified the possible sources of error when manometric devices are used to indicate mean differential pressure in a pulsating flow as

- wave action and resonance effects in the connecting leads,
- volumes of ducts of varying section within the manometer and leads, and
- restrictions in the leads, which cause a nonlinear relationship between flow and pressure drop and result in nonlinear damping.

The effect of pulsation on a manometer is shown in Figure 5.9(b).

Before leaving the subject of pulsation effects, it is important to include a further note (cf. Section 2.6) on the Hodgson number:

$$H = \frac{Vf}{q_v}\frac{\Delta p}{p} \tag{5.16}$$

where V is the volume of pipework and other vessels between the source of the pulsation and the flowmeter position (Figure 2.5), Δp is the pressure drop over the same distance, f is the frequency of the pulsation, q_v is the volumetric flow rate, and p is the absolute line pressure. With this number, it is possible to plot curves to show the likely error levels for various values of H. Oppenheim and Chilton (1955) have given plots of error levels for various types of flow. In Figure 5.10(a), one such set of curves is shown.

Mottram (1989) suggested that an expression for the total error in the indicated flow rate of a differential pressure flowmeter is

$$E_T = \left(1 + \left(V_{rms}/\overline{V}\right)^2\right)^{1/2} - 1 \tag{5.17}$$

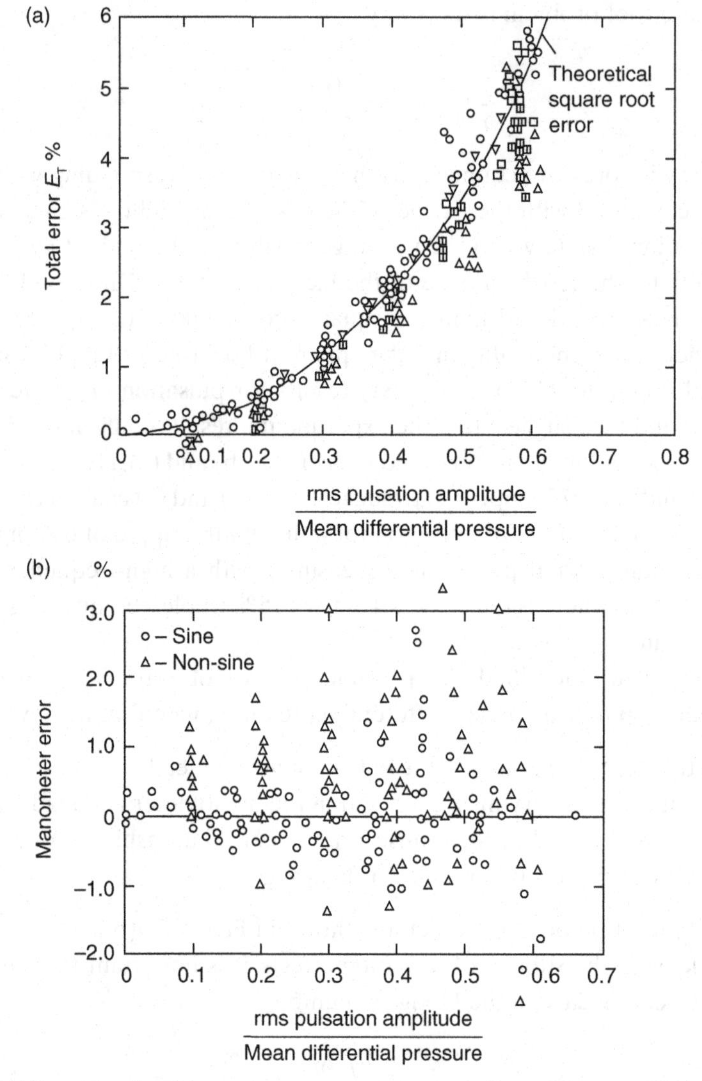

Figure 5.9. Effect of pulsation on pressure measurement (from Mottram 1981; reproduced with permission of the author and of the Instrument Society of America): **(a)** Total error at low pulsation amplitude; **(b)** U-tube manometer error.

which for small amplitude ratios can be written as

$$V_{rms}/\overline{V} = (2E_T)^{1/2} \tag{5.18}$$

where $\overline{V}$ is the mean velocity and V_{rms} is the rms value of the unsteady velocity fluctuation in the pipe. He then suggested that if a maximum allowable percentage error is ϕ where $\phi = 100E_T$, and using the criterion for damping in Equation (2.14), then

$$\frac{H}{\gamma} \geq \frac{1}{4\pi} \frac{\left(V_{rms}/\overline{V}\right)_{ud}}{\sqrt{2\phi/10}} \tag{5.19}$$

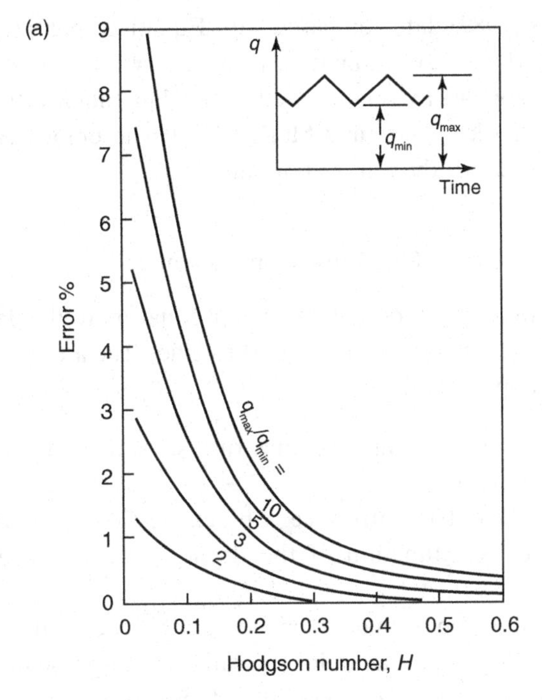

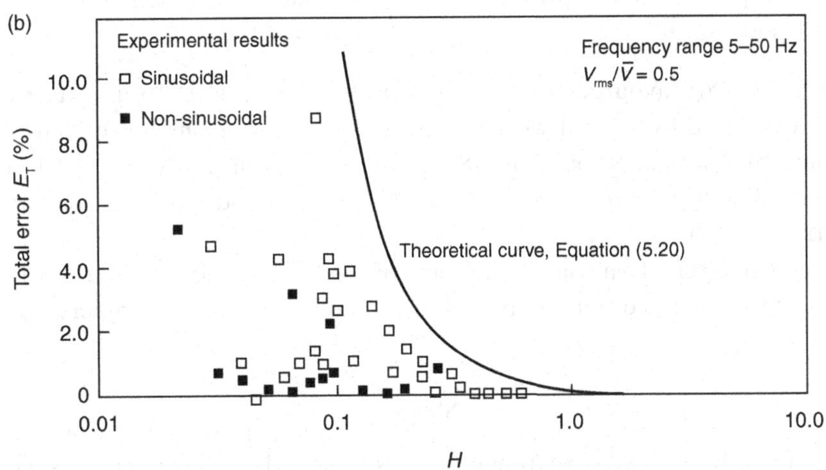

Figure 5.10. Hodgson number, **(a)** pulsation error curves for a triangular pulsation of varying amplitude (from Oppenheim and Chilton 1955; reproduced with permission of ASME); **(b)** comparison between Hodgson number and experimental results (reproduced with the agreement of Elsevier from Mottram 1989).

where subscript ud implies undamped and γ is the isentropic index for the gas or

$$\frac{H}{\gamma} \geq K \frac{\left(V_{rms}/\bar{V}\right)_{ud}}{\sqrt{\phi}} \tag{5.20}$$

where $K = 10/\left(4\sqrt{2}\pi\right) = 0.563$. He claimed that experimental evidence is available to validate the safety of this criterion provided the flow pulsation amplitude can be measured. Some of Mottram's (1989) data compared with Equation (5.20) with $K = 0.563$ is shown in Figure 5.10(b). Where estimation is necessary, he suggests a safety margin by using $K = 1$. This appears to be confirmed by the data that Mottram (1989) included in his paper (cf. Sparks, Durke and McKee 1989, who emphasised the problems and the only safe solution – the removal of pulsation).

5.8 Effects of More than One Flow Component

Attempts have been made to relate the differential flowmeter equation to the data for flow of a fluid mixture, most commonly steam, through the orifice. The attempted adjustments have followed three approaches:

1. Adjusting the value of density to reflect the presence of a second component (cf. James 1965/1966);
2. Adjusting the discharge coefficient to introduce a blockage factor for the other components, expressed as a function of the dryness (cf. Smith and Leang 1975);
3. Relating the two-phase pressure drop to that which would have occurred if all the flow were passing either as a gas or as a liquid. The ratio of two-phase pressure drop to liquid flow pressure drop Φ^2_{fo} is equated to a function of x, the dryness fraction.

Miller (1996) favoured the third approach as the easiest to use. It has also been investigated by several workers, and various expressions for Φ^2_{fo} have been tabulated by Grattan, Rooney and Simpson (1981) from the work of Chisholm (1977), Collins and Gacesa (1970), James (1965/1966) and Watson, Vaughan and McFarlane (1967).

The main reservation concerning some of these correlations relates to whether they can be applied to different pipe geometries, fluids, Reynolds numbers etc.

Steam

Grattan et al. (1981) gave experimental pressure loss data for orifice tests with an empirical curve:

$$\Phi^2_{fo} = 1.051 + 291x - 3{,}796x^2 + 74{,}993x^3 - 432{,}834x^4 \quad \text{for } 0.00005 \leq x \leq 0.1 \qquad (5.21)$$

and this is shown in Figure 5.11. The scatter is an indication of the limited value of such an empirical curve and of attempts to predict flowmeter performance in two-phase flows. Rooney (1973) also obtained values for the parameter Φ^2_{fo} (cf. Chisholm 1967; Chisholm and Leishman 1969; Chisholm and Watson 1966).

Owen and Hussein (1991) pointed out that if the water content was very small, the droplets would probably have a negligible effect on the flowmeter response, and so the mass flow should be corrected to allow for the mixture density rather than the vapour density. For wet steam of dryness x greater than about 0.9, this correction is

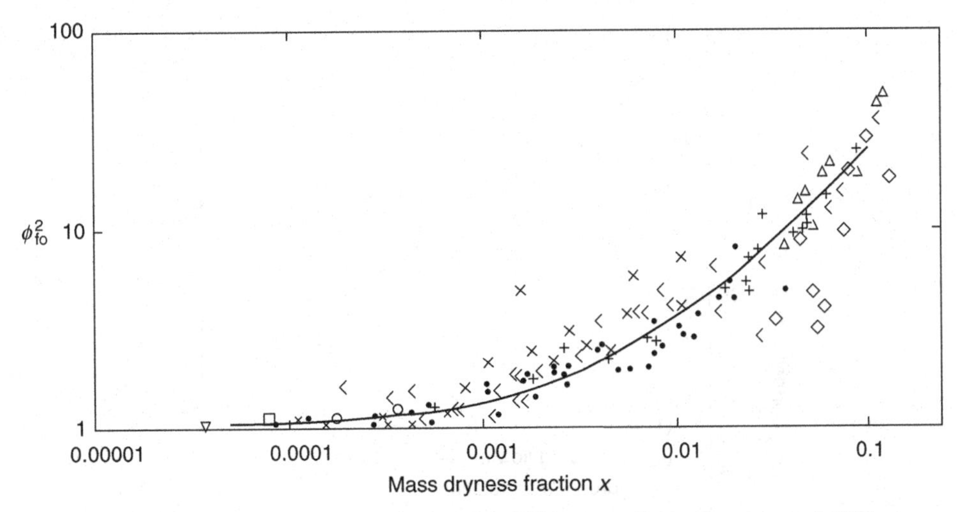

Figure 5.11. Experimental pressure loss data for orifice tests (from Grattan et al. 1981; reproduced by permission of NEL).

approximately $F = 1/x$. Owen and Hussein compare this value with other corrections. In particular, BS 3812 suggested

$$F = \sqrt{2 - x} \tag{5.22}$$

and the *Shell Flowmeter Engineering Handbook* proposed

$$F = 1.74 - 0.74x \quad \text{for } x > 0.95 \tag{5.23}$$

Hussein and Owen (1991) referred to two correlations: James's giving a correction factor of

$$F = \left\{ \frac{\rho_l}{x^{1.5}\rho_l + (1 - x^{1.5})\rho_g} \right\}^{1/2} \tag{5.24}$$

and Murdock's

$$F = \frac{1}{x + 1.26(1 - x)(\rho_g/\rho_l)^{1/2}} \tag{5.25}$$

Figure 5.12 for seven alternative correlations suggests that the data scatter around those of James and Murdock. Hussein and Owen (1991) showed that pressure had little effect on correction factors, and that the *Shell Flowmeter Engineering Handbook's* correlation is the best compromise for $x > 95\%$ for the orifice. They wisely advocated an upstream separator and indicated that several types of upstream separators give a value of $x > 95\%$ with reasonably low pressure loss. Owen, Hussein and Amini (1991) provided further convincing experimental evidence for the

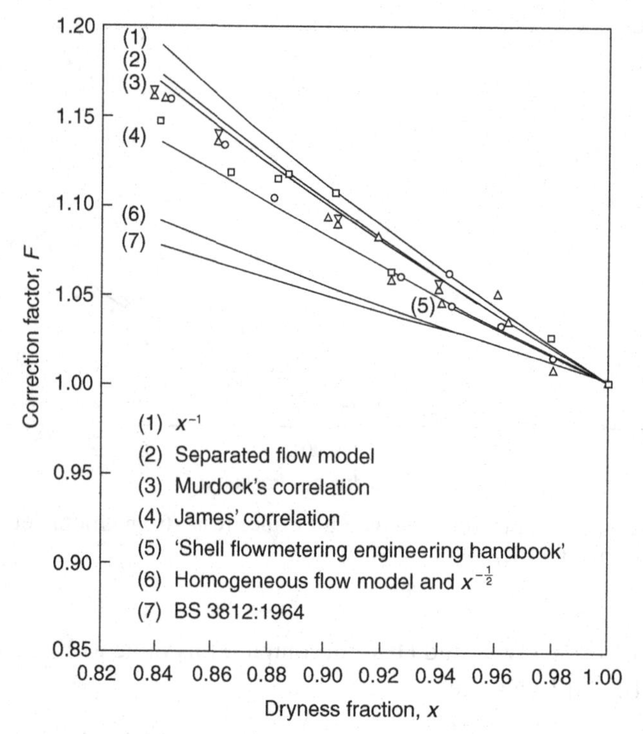

Figure 5.12. Wet steam flowmeter correction factor. Comparison between data for wet steam correction factor and proposed factors from the literature (reproduced with the agreement of Elsevier from Hussein and Owen 1991).

benefits of fitting a separator in a steam line before an orifice plate. The separator, of course, should make more accurate flow measurement possible. They showed that slugs of water in a gas flow, trapped by valves and then released, can travel at 50 m/s or more and result in impact pressures at 5 bar tank pressure of, in some cases, well over 50 bar. One of the orifices they used was made of 3-mm-thick mild steel and had deformed very substantially after 20 impacts.

Wenran and Yunxian (1995) suggested a simple model for deducing flow rate and phase fraction for steam flows in orifice plates, and they claim that errors were, respectively, 9% and 6.5% for their model. The reader is referred to the original article for more details. Pressure noise has also been suggested as a means of obtaining more information in steam-water flows (Shuoping, Zhijie and Baofen 1996). (cf. Fischer's 1995 calculation for gas-liquid annular mist, see also Chien and Schrodt 1995 on steam quality).

Oil-in-Water Flows

Pal and Rhodes (1985) reported results of tests of orifice meters in horizontal oil-water mixtures. Figure 5.13 shows the effect of increasing oil content on each meter.

Pal (1993) tested orifices with a range of oil-in-water emulsions ranging from 30% to 84% (inversion took place at 78% oil) and found that stable emulsions introduced

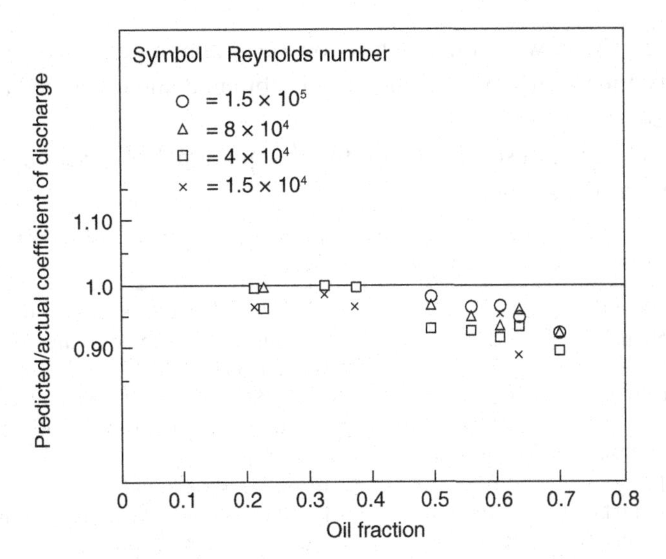

Figure 5.13. Performance of orifice meter in an oil-in-water emulsion (from Pal and Rhodes 1985; reproduced with permission of BHR Group).

coefficient changes within about 3% using a generalised Reynolds number, which is defined in the paper, but for unstabilised emulsions the changes were greater.

Other Applications

Murdock (1961) reported correlations to within ±1.5% for two-phase flow. Others are reported by Lin (1982) and Mattar et al. (1979). Washington (1989) claimed good predictions using wet-gas. Majeed and Aswad (1989) appear to have achieved agreement between measured and predicted flows based on a correlation by Ashford-Pierce when using oil field data. They included a computer listing.

5.9 Accuracy Under Normal Operation

To achieve the accuracy set out in the standards, it is necessary to construct and install the meter to the detailed specification of the standards.

From the equation

$$q_m = \frac{CE\varepsilon \pi d^2 \sqrt{2\rho_1 \Delta p}}{4} \tag{5.1}$$

the standard provides a formula for the total uncertainty (cf. Equations (1.A.7) and (7.14) for alternative symbols) in the mass flow rate:

$$\frac{\delta q_m}{q_m} = \left\{ \left(\frac{\delta C}{C}\right)^2 + \left(\frac{\delta \varepsilon}{\varepsilon}\right)^2 + \left(\frac{2\beta^4}{1-\beta^4}\right)^2 \left(\frac{\delta D}{D}\right)^2 + \left(\frac{2}{1-\beta^4}\right)^2 \left(\frac{\delta d}{d}\right)^2 + \frac{1}{4}\left(\frac{\delta \Delta p}{\Delta p}\right)^2 + \frac{1}{4}\left(\frac{\delta \rho_1}{\rho_1}\right)^2 \right\}^{1/2} \tag{5.26}$$

where α, (=CE) the flow coefficient, has now been eliminated. To obtain the overall uncertainty, the uncertainty in each term is obtained and combined according to Equation (5.26).

It is instructive to put some values into this equation. Let us set out some values and the likely errors in them:

Parameter	Value	Uncertainty %	Comment
D	100 mm	0.2	Will depend on pipe specification.
d	50 mm	0.1	Machine to 0.002 in. or 50 μm.
β	0.5	–	Equal to d/D.
C	0.6051	0.5	From ISO formula or table (value of Re is assumed, but iteration is normally required).
E	1.0328	negligible	
ε	1	–	Incompressible liquid.
ρ	1,000 kg/m^3	0.1	Possible variation due to temperature or purity.
Δp	31.54 kPa	1	

From these values, we obtain

$$q_m = 9.75\,\text{kg/s} \qquad (\text{Re}=1.24\times10^5)$$

$$\frac{\delta q_m}{q_m} = \left\{(0.5)^2+(0)^2+(0.133)^2(0.2)^2+(2.133)^2(0.1)^2+\frac{1}{4}(1)^2+\frac{1}{4}(0.1)^2\right\}^{1/2}$$

$$= 0.89\%$$

The estimate of uncertainty for gas metering is likely to be considerably larger than this figure. In a paper describing the specification, installation, commissioning and maintenance of typical orifice metering stations for the British Gas national high-pressure transmission system, with flows up to 1.4×10^6 std m^3/h at pressures up to 69 bar, Jepson and Chamberlain (1977) concluded that, to achieve uncertainty of less than ±2%, a very rigorous checking procedure was needed to ensure that the complete measuring system remained in specification all the time.

It should be noted that 1% is normally required as the uncertainty for fiscal gas measurement and is achieved if measurements are made correctly (Dr M Reader-Harris, private communication 2014).

Miller (1996) provided a number of special corrections: for steam quality with gas-liquid flows, for saturated liquids with up to 10% saturated vapour, for drain and vent holes, for water vapour and for indicated differential when the pressure-measuring device is at a different elevation to the differential pressure device and the pressure lines are filled with fluid of a different density.

Ting and Shen (1989) commented that the majority of natural gas flow measurements in the United States were determined by orifice meters. They led a systematic study of measurement by 152.4-mm (6-in.), 101.6-mm (4-in.) and 50.8-mm (2-in.) meters in the Reynolds number range 1 million to 9 million.

Ting and Shen used Honeywell smart static and differential pressure transducers, which had microprocessor-based built-in pressure and temperature compensation, higher span-turndown ratio, improved accuracy and easy rangeability. Bias was quoted as better than ±0.15% for differential pressure and precision as ±0.015%. A precision aneroid barometer of 0.1% full-scale uncertainty was used for atmospheric pressure. Careful calibration of these instruments took place, and the paper implied careful experimentation, the only weakness being that the definition of the inlet pipework was not entirely clear. Their tests showed that the 101.6- and 50.8-mm orifice discharge coefficients agreed with the ANSI and ISO standards within the estimated uncertainty levels of 0.35% with 95% confidence level. However, the 152.4-mm meter agreed for low and medium beta ratio, but for a beta ratio of 0.74, ANSI and ISO gave a discharge coefficient about 2% low. Their plots also, helpfully, show the divergence between the two previous standards, which is not more than about 0.2% over the range.

5.10 Industrially Constructed Designs

Orifice Plates

Orifice plates may be available from various manufacturers machined to the requirements of ISO 5167 (BS 1042) in stainless steel 316, 321, 304, Hastelloy, tantalum, Inconel and plastics such as polytetrafluoroethylene (PTFE), polyvinylchloride (PVC) and polyvinylidene fluoride (PVDF). Also possibly available are orifices that are removable from the holding ring, with integral gaskets that may be spiral wound in stainless steel or of PTFE or graphite filled. Size availability extends beyond the common standard's range going down to a 25-mm bore or less. The orifices should be stamped permanently with detailed information (e.g. tag number, pipe size, rating, orifice diameter, material), usually on the upstream side (Figure 5.14). The information will also state which is the upstream side of the plate.

Orifice Plate Carrier Assemblies

Some examples of orifice carrier assemblies are shown in Figure 5.15(a). The materials of construction are generally carbon steels, but stainless steels and duplex steels can be provided. Carriers may be split-ring or one-piece integral design.

Orifice Metering Run Assemblies

Orifice metering run assemblies may be constructed to the appropriate standards and may incorporate flange tappings [Figure 5.15(b)]. Although plates are relatively easy to install and remove, there are occasions when a plate needs to be inspected without disrupting the flow or depressurising the line. For this reason, at least one manufacturer (Figure 5.16) developed a system for removing and replacing plates without shutting the line down. This consists of a means for sliding the orifice plate in and out, possibly with a rack and pinion mechanism, and a dual chamber pressure sealing system while the plate is removed.

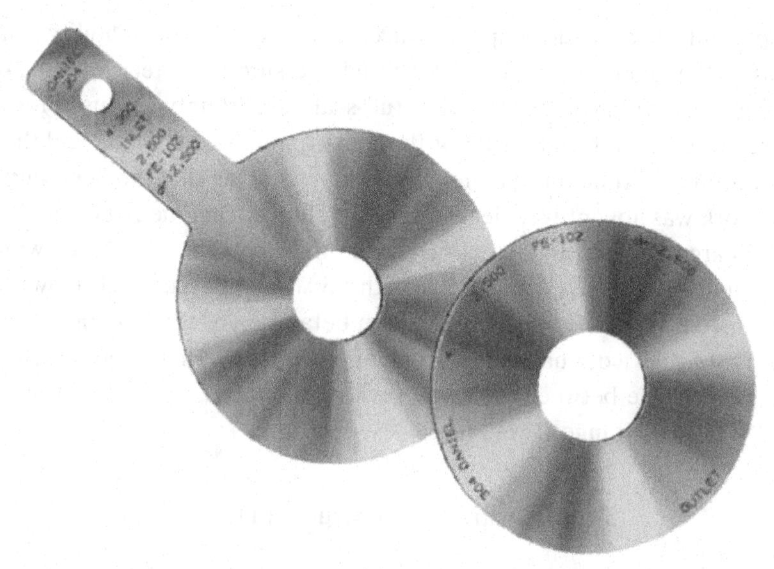

Figure 5.14. To indicate the main features of industrial orifice plates (reproduced with permission of Emerson Process Management Ltd.).

Sizing

The plate manufacturer may offer a sizing service based on a computer program that undertakes the iteration and optimisation, an otherwise laborious procedure.

5.11 Pressure Connections

Tapping Orientation

The tapping positions should be such that any unwanted component of the line fluid or any second phase in the line fluid will not enter the pressure tapping of the impulse line or become trapped in the impulse line connections. For vertical lines, any position is satisfactory (Miller 1996). Where the line is not vertical, the following recommendations should be noted to reduce the chances of accidental blockage (BS 1042: Part 1:1964 and Miller 1996):

- For liquids, within an angle of 45° above or below the horizontal, but preferably on the horizontal;
- For dry gases, between the horizontal and vertical upward with a suggestion that for clean noncondensable gases the tapping should be in a vertical position;
- For moist gases, between an angle of 30° above the horizontal and vertically upward; and
- For steam and other vapours, horizontal only.

These requirements are illustrated in Figure 5.17. Where a drain hole is provided through the orifice plate, the single tapping should be orientated so that it is between 90° and 180° to the position of the drain hole.

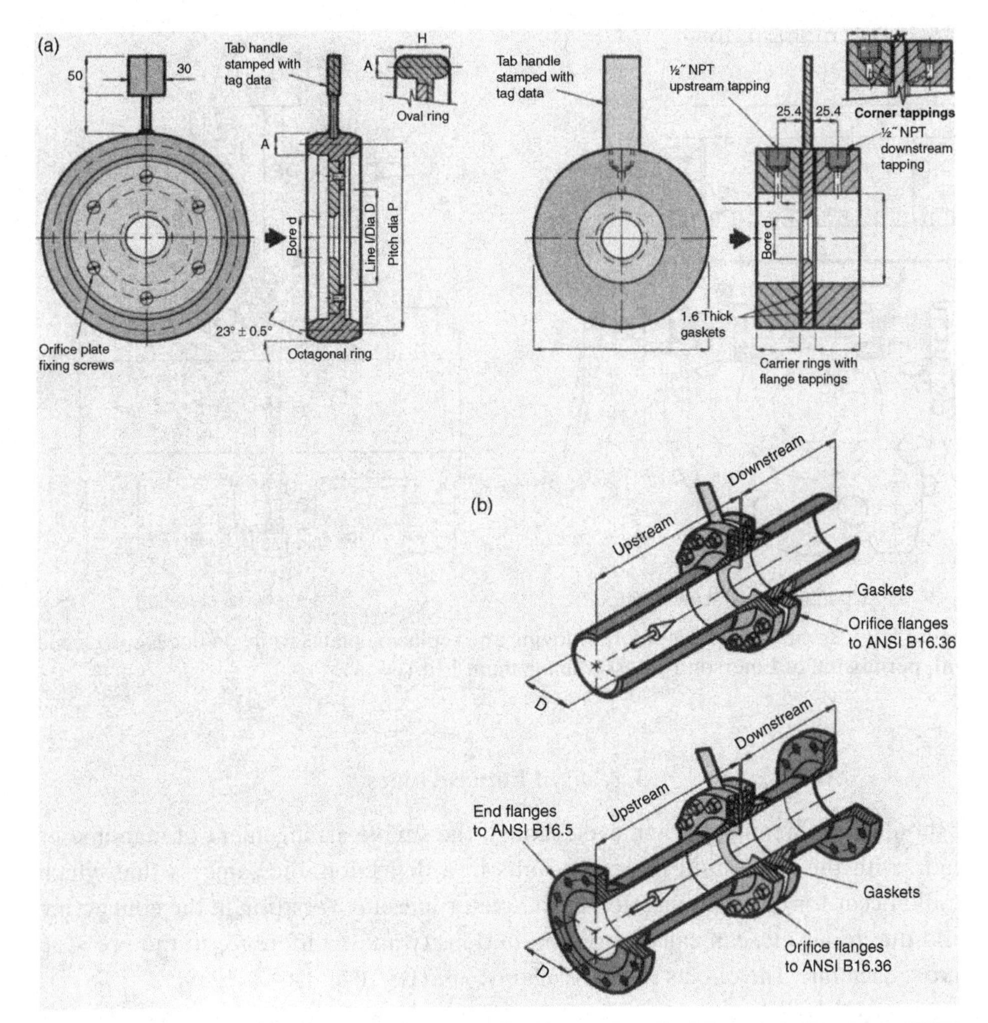

Figure 5.15. Commercial orifice plate assemblies (reproduced with permission from ABB): **(a)** Carriers; **(b)** metering runs.

The standard also recommends that, where there is a danger of blockage, provision should be made for rodding out the tapping from outside the line.

Piezometer Ring Arrangements

In pipelines of 150 mm (6 in.) or greater, or in lines where one hole may become blocked, it may be useful to use a piezometer ring, as shown in Figure 5.18. In the presence of steam or other vapour flows where condensation may occur, a piezometer ring may form an unwanted trap for condensate. The manifold cross-section should always be equal to or greater than the total area of the individual tappings which it serves. The triple T arrangement may be preferred, and, in any case, traps for condensate and adequate bleed points should be provided. In laboratory applications, and at low pressures, the use of transparent tubing is recommended.

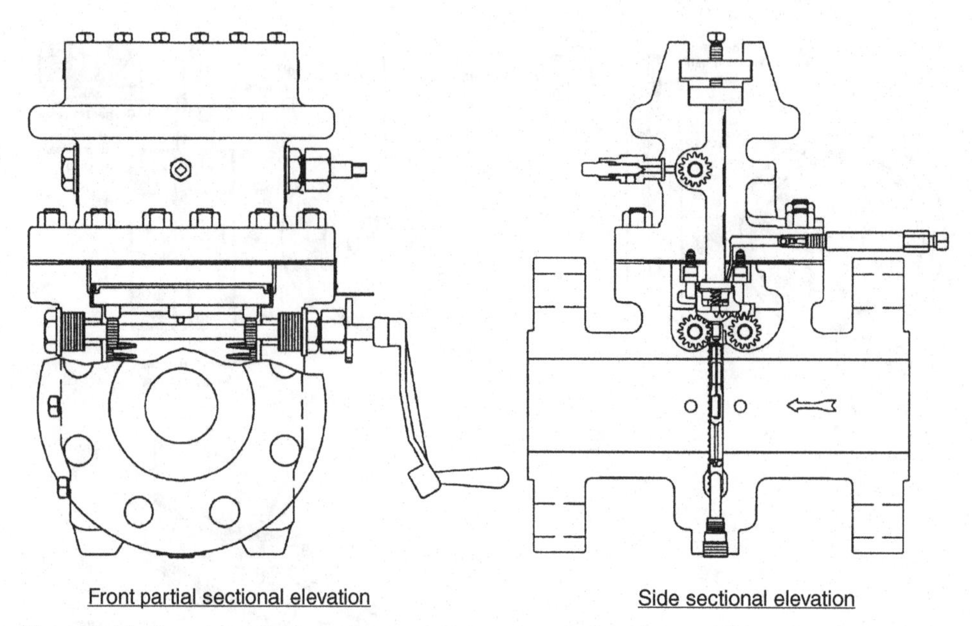

Figure 5.16. Commercial system for removing and replacing plates while on line (reproduced with permission of Emerson Process Management Ltd.).

Layout of Impulse Lines

It should be remembered that, even though the simple arrangement of manometer leads, with the same fluid in each, results in a deflection, the same as that which would occur for direct connection to the meter line, any variation in the connecting fluid due to air etc. can cause variation in density and a difference in the pressure across each line. The effects to guard against are (BS 1042: Part 1: 1964):

- Temperature difference between the lines leading to a density difference;
- Discontinuity in the impulse line fluid due to a second component, be that solid, liquid or gas; and
- Blockage of any sort including solidification of the impulse fluid.

The impulse lines should be as short as possible, at the same temperature (which may require lagging together), and at a sufficient gradient to allow drainage etc. as required.

Miller (1996) should be consulted for detailed guidance on impulse lines, sealant liquids and transmitter positions. A typical arrangement for steam is shown in Figure 5.19. (ISO 5167-1: 2003 provides information on this and other related topics.)

5.12 Pressure Measurement

The main methods used for pressure measurement are described in Noltingk (1988) and are manometers, mechanical devices that deflect under pressure and electromechanical pressure transducers.

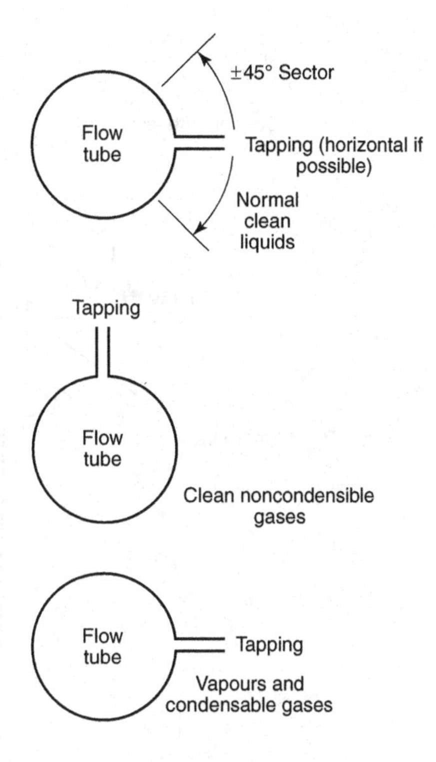

Figure 5.17. Diagram to show positioning of the pressure tapping for nonvertical lines.

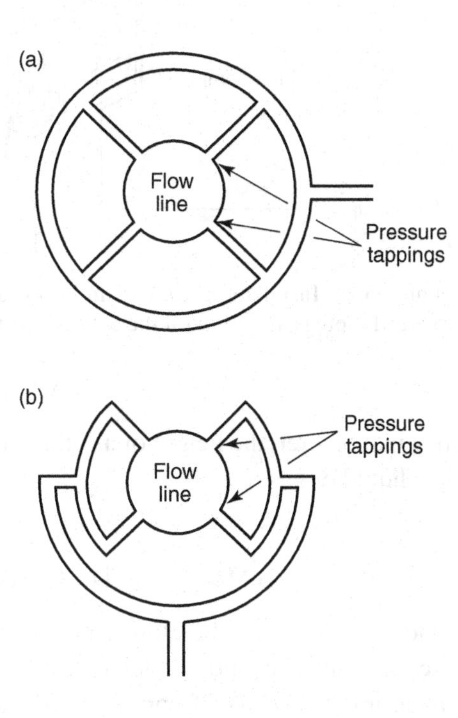

Figure 5.18. Diagram to show piezometer ring connections: **(a)** Conventional; **(b)** triple T.

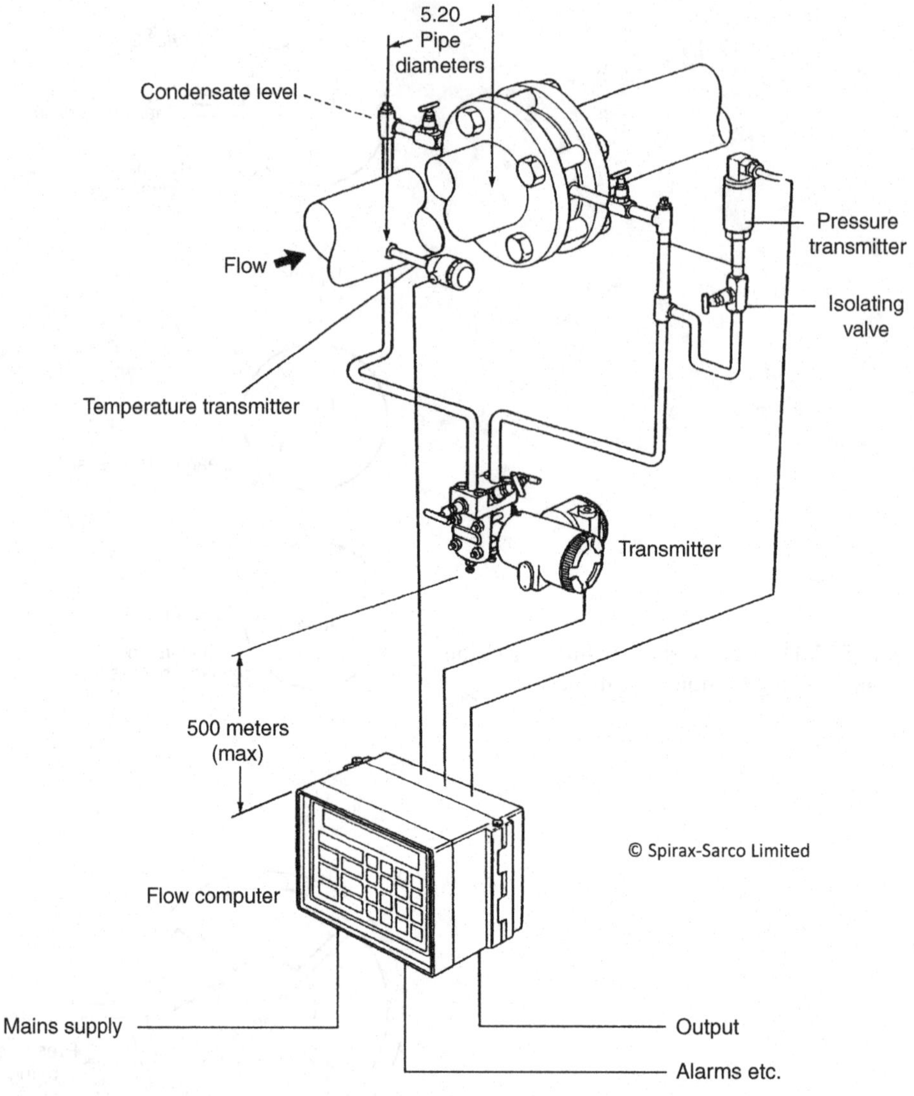

Figure 5.19. Installation for steam (reproduced with permission of Spirax-Sarco Ltd.). [Note: Tappings are often on the same side of the pipe.]

A Manometer

In a manometer, the height of a column of liquid provides a measure of the pressure in a liquid or gas.

B Mechanical Devices that Deflect Under Pressure

One such device is the Bourdon tube, which has a flattened tube wound into an arc. The tube unwinds under pressure. Another design uses diaphragm elements made up from pairs of corrugated discs with spacing rings welded at the central

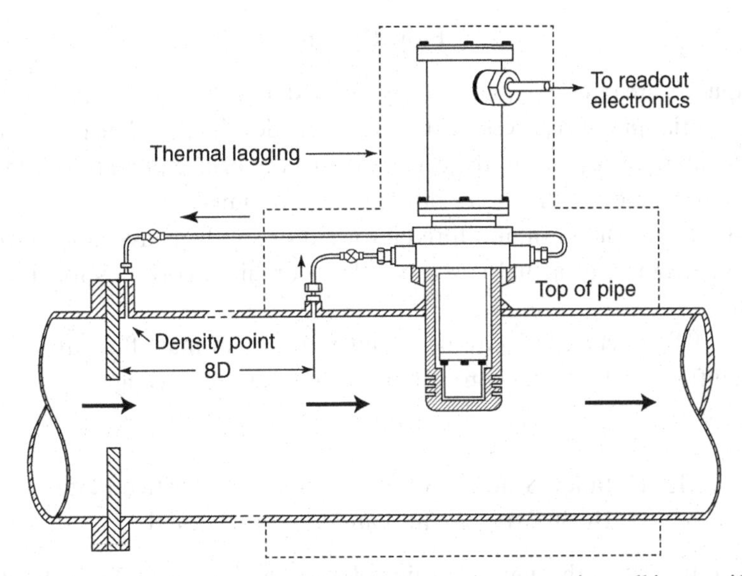

Figure 5.20. Typical example of the installation of a density measuring cell in an orifice metering line (courtesy of the Roxboro Group PLC).

hole. Pressure will cause the assembly to elongate, and this will in turn be used as a method of registering the pressure change. Other devices based on expanding bellows or diaphragm elements are also used.

C Electromechanical Pressure Transducers

This is the most important and accurate device for the future and is in a state of continual development. Various methods have been used to obtain an electrical signal from a mechanical deflection. Some of these are capacitance transducers, piezoresistive strain gauges and piezoelectric and resonant devices. In one design, pressure applied to a diaphragm causes it to deflect and to exert a force on a cantilever beam that has a strain gauge bonded to it and that provides an electrical signal. Smart devices can provide condition monitoring, self-diagnosis of temperature, memory and reference voltages and will provide ranging and temperature compensation. Measurement uncertainty can be as low as, or lower than, 0.1% of span. Regular recalibration at 1–6 month intervals is recommended.

5.13 Temperature and Density Measurement

Methods of temperature and density measurement are described by Noltingk (1988). A typical installation arrangement for a density cell in a metering line is shown in Figure 5.20. The cell is arranged to project into the line so that it is at the gas temperature, and the gas is extracted from the line, passed through the cell and returned to the line.

5.14 Flow Computers

Flow computers are microcomputers programmed to obtain flow from sensors measuring differential pressure, pressure, temperature, density or other parameters, and are of particular importance in the calculation of the orifice plate flows. When used with smart transmitters, they may offer on-line optimisation of differential pressure measurement. They contain stored information on the conversion of frequency and analogue signals to actual units, and they contain algorithms for interpreting density etc.

Taha (1994) described electronic circuitry to obtain mass flow from an orifice plate with differential pressure, pressure and temperature sensing.

5.15 Detailed Studies of Flow through the Orifice Plate, Both Experimental and Computational

An early study of flow through an orifice plate using laser doppler techniques was reported by Bates (1981) for a BS 1042 orifice plate with $\beta = 0.5$ in the range of Reynolds number 67,600–183,000. The shape of the profiles upstream at 0.535D and 0.32D show no recirculation zone in the plate corner, but this may be because the region of any such recirculation is closer to the plate than was investigated. Downstream at 0.5D the recirculation region has a zero velocity point at about 25% radius in from the pipe wall, and this moves to about the 20% radius position at 0.9D downstream.

Morrison et al. (1990b) reviewed past work on flow patterns in orifice meters, both theoretical and experimental. They used a laser doppler anemometer (LDA) to obtain detailed flows in an orifice. They also obtained (Morrison et al. 1990a) wall pressure distributions, and some of these are shown in Figure 5.21(a) for a $\beta = 0.5$ at Re = 122,800 both for conditioned flow and for three angles of swirl. Figure 5.21(a) shows the pipe wall pressure distributions with a minimum pressure at about $D/2$ downstream of the orifice plate. Note that data were not obtained between 0.125D upstream and 0.25D downstream of the plate. Figure 5.21(b) shows the variation of the velocity on the centreline of the orifice meter, the high peak value at the *vena contracta* and possibly a suggestion of poor diffusion.

Erdal and Andersson (1997) provided a brief review of some computational fluid dynamics (CFD) work. They emphasised the sensitivity to the use of a particular code (cf. Spencer, Heitor and Castro 1995). Davis and Mattingly (1997) claimed agreement between computed and experimental discharge coefficients to within about 4% for β in the range 0.4–0.7 and Re between 10^4 and 10^6. Sheikholeslami, Patel and Kothari (1988) and Barry, Sheikoleslami and Patel (1992) may have claimed an agreement of about 2%. Patel and Sheikholeslami (1986) computed the discharge coefficient for Re = 2,500,000 with $\beta = 0.4$ to within 1.5%.

Reader-Harris and Keegans (1986) and Reader-Harris (1989) claimed that discharge coefficients could be calculated to within 0.64%. Figure 5.22 shows some of Reader-Harris's computations using the $k - \varepsilon$ turbulence model with a computer

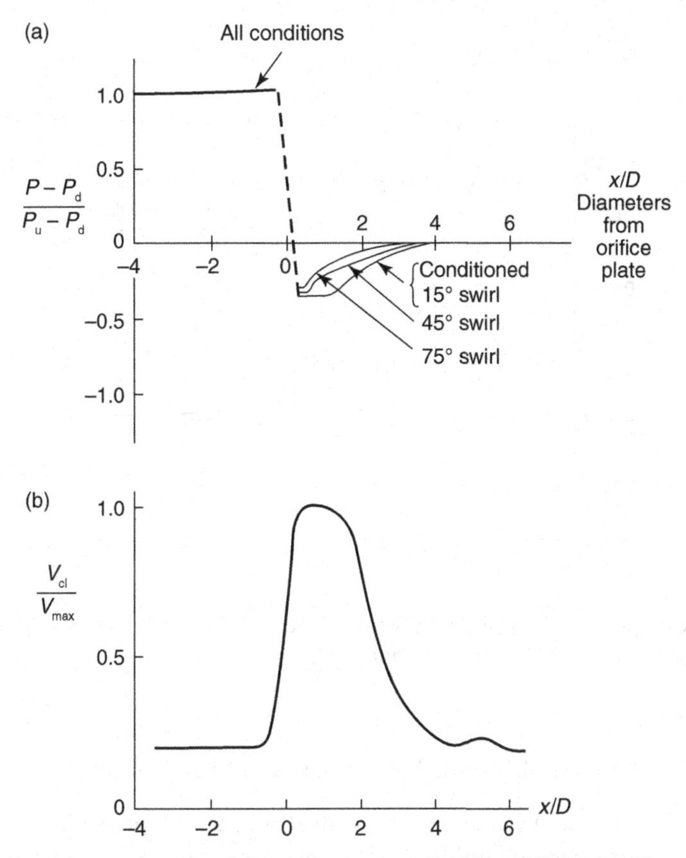

Figure 5.21. Experimental curves (after Morrison et al. 1990a, 1990b). **(a)** Pressure distributions for conditioned flow and for three angles of swirl from pipe wall tappings for an orifice with $\beta = 0.5$ and Re = 122,800; **(b)** centreline velocity distribution using a laser doppler anemometer for an orifice with $\beta = 0.5$ and Re = 18,400.

package (QUICK). In Figure 5.22, computed results for a smooth pipe and a $\beta = 0.7$ are compared with the equation in the standard at that date (Stolz) and with an improved equation (cf. Reader-Harris and Sattary 1990) from which the current ISO equation has developed. Where two symbols occur at one value of Reynolds number, the higher is for the more refined grid. Reader-Harris only considered that he had obtained independence from grid size for Re $\geq 3 \times 10^6$. His results give encouraging agreement with the modified equation.

Spearman, Sattary and Reader-Harris (1991) described laser Doppler velocimeter (LDV) measurements downstream of a Mitsubishi flow conditioner and of an orifice plate in a combined package.

Erdal and Andersson (1997) indicated the problems which still exist in obtaining adequate modelling of the flow through an orifice plate. They commented that the predictions revealed the need for considerable expertise and care before results, especially for pressure drop, could be predicted with confidence. (A further note on CFD related to orifice plates will be found in Appendix 5.B.)

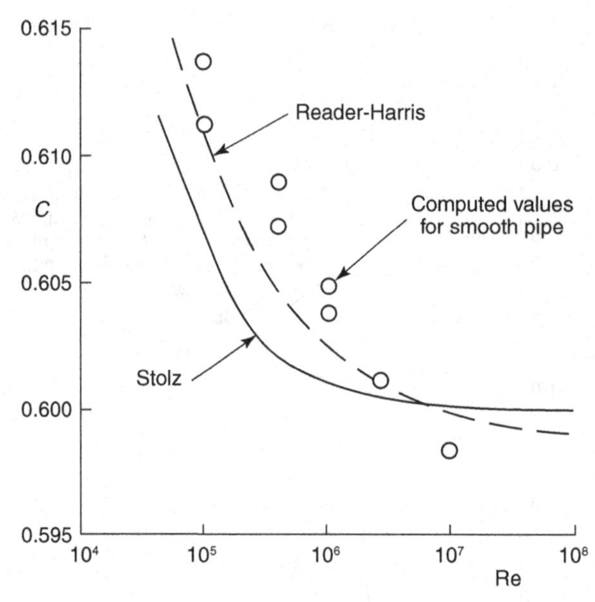

Figure 5.22. Computed values of discharge coefficient C, for an orifice with $\beta = 0.7$, corner tappings and smooth pipe (after Reader-Harris 1989).

5.16 Application, Advantages and Disadvantages

The standard orifice can be used with almost any single-phase Newtonian flow, whereas for high-viscosity fluids, a quadrant or conical orifice may be used. Abrasive fluids are likely to change the shape of the leading edge of the plate and therefore the calibration of the orifice meter and are not to be advised. The application of the orifice plate to two-phase flows should be attempted with great caution. However, the orifice plate in wet-gas with appropriate wet-gas correlation performs well, provided that the plate is undamaged (Dr M Reader-Harris, private communication 2014).

An advantage of the orifice plate flowmeter is that the experience over many years has been well documented and is distilled in the standards, allowing a device to be designed and constructed with confidence. The uncertainty is, also, calculable from the standards. Thus without calibration of the instrument, it is possible to set a value against the uncertainty of the measurements. Few other meters offer this possibility.

Its ease of installation in the line has to be balanced by the need for very careful construction in order to obtain the calculable uncertainty. Some features of the design (e.g. the edge radius of the plate inlet for small plate diameters) are very difficult to achieve.

Another disadvantage is the nonlinear characteristic of the meter due to Equation (5.1). This, in turn, leads to a restricted range since the differential pressure measurement device will need to respond to, say, 100:1 turndown for a 10:1 flow turndown. With smart transducers, this is more possible than in the past.

Another problem with the nonlinear characteristic is the effect of pulsation on the reading.

The high pressure loss due to the poor pressure recovery after the plate may be a disadvantage in many applications and will certainly lead to energy losses.

The cost of installation of the orifice will not be markedly less than other full-bore flowmeters, when allowance has been made for pressure transducers, flow computer etc. Maintenance will be aided if one of the commercial designs which allow the orifice plate to be withdrawn from the line, even under operating conditions, is used. Pressure transducers should be regularly checked. This does not ensure that the total system has retained operational integrity or that the calibration is unchanged by debris in the impact pipes or by fouling or rusting of the flow pipe causing changes in roughness and hence of the profile entering the meter. However, it allows a fuller record to be kept without removal for calibration and, with other operating data, may give greater confidence in the integrity of the instrument.

5.17 Chapter Conclusions

The orifice continues to have an important and valued place in flow measurement because of its longevity and its documentation in the standards and because custom and practice require its use in some industries. The relatively recent update of the standard is evidence for this.

It will, no doubt, continue to have an important niche, and manufacturers of orifice meter components will continue to provide a useful service particularly for fiscal metering of natural gas and oil products (Reader-Harris 1989).

The measurement of C, of C with upstream fittings causing distortion to the inlet profile, of C with upstream conditioners etc. has been, and will continue to be, the subject of experimental programmes. However, there appears to be reasonable confidence in the statements in ISO 5167-2: 2003.

Users should be conscious of the degrading of precision that results from corrosion, wear and deposits on the plate (Appendix 5.B), in the pipe and in the pressure tappings. The effect of wear on the inlet edge of the plate may be substantial.

Steven and Hodges (2012) described the background to the development of an orifice diagnostic method which is now available as a product (called Prognosis) which appears to be providing a valuable source of verification for orifices. The essence appears to be the use of a third downstream pressure tapping and the relationship between the differential pressures between the three tapping points.

The computation of the flow, particularly if coupled with upstream distortion, offers a considerable challenge, and the results can be sensitive to the detailed application of computer programs. It would seem to be a prime candidate for a perturbation approach, where the computer program could be used to calculate changes from the norm rather than attempt to obtain the absolute discharge coefficient. Such an approach might shed further light on the shape of the empirical equations and the trends in discharge coefficient for variation in D, β and Re (cf. Reader-Harris 1989).

Appendix 5.A Orifice Discharge Coefficient Equation

I have included three important examples of discharge coefficients for information and comparison. I have attempted to ensure that the equations are accurate, but the reader should ensure that he/she works from the relevant standard.

Appendix 5.A.1 Stolz Orifice Discharge Coefficient Equation as Given in ISO 5167: 1981

In the version of ISO 5167 dated 1981, the equation due to Stolz (1978, cf. 1988) was given, and has been used widely since then. This equation was retained in the standard until 1998. The form of the equation is:

$$
\begin{aligned}
C = {} & 0.5959 + 0.0312\,\beta^{2.1} - 0.1840\,\beta^{8} && C_\infty \text{ term} \\
& + 0.0029\,\beta^{2.5}\left(\frac{10^6}{\mathrm{Re}}\right)^{0.75} && \text{Slope term} \\
& + 0.0900\,L_1\,\beta^4(1-\beta^4)^{-1} && \text{Upstream tap term} \\
& - 0.0337\,L_2'\,\beta^3 && \text{Downstream tap term}
\end{aligned}
\qquad (5.A.1.1)
$$

where C is the discharge coefficient, β is the diameter ratio and is equal to d/D, Re is the Reynolds number based on the pipe diameter, and:

$$L_1 = \ell_1/D$$

$$L'_2 = \ell'_2/D$$

where ℓ_1 is the distance of the upstream tapping from the upstream face of the plate and ℓ'_2 is the distance of the downstream tapping from the downstream face of the plate. For spacings which follow the standard:

tapping position	L_1	L'_2
corner	0	0
flange	25.4/D	25.4/D (D in millimetres)
D and D/2	1	0.47

When $L_1 \geq \dfrac{0.0390}{0.0900}$ (= 0.433 3), use 0.039 0 for coefficient of $\beta^4(1-\beta^4)^{-1}$.

In addition to the requirements for the value of C, the reader should refer to the appropriate standard for all other constraints on diameter, β, Reynolds number, roughness etc.

Uncertainties

Provided the ranges for Re etc. are observed, the uncertainties in C are:

$\beta = d/D$	Corner taps	Flange taps	D and D/2 taps
$\beta \leq 0.6$	0.6%	0.6%	0.6%
$0.6 \leq \beta < 0.8$	β%	–	–
$0.6 \leq \beta \leq 0.75$	–	β%	β%

Appendix 5.A.2 Orifice Discharge Coefficient Equation as set out by Gallagher (1990)

The following version of the equation for the discharge coefficient for orifice meters was given by Gallagher (1990a&b, reproduced here with permission of the author). There appear to be two versions of the equation credited to Reader-Harris/Gallagher of which this is the first.

It is applicable to:

- nominal pipe sizes of 50 mm (2 in.) and larger;
- diameter ratios of 0.1–0.75 provided the orifice plate bore diameter is greater than 11.4 mm (0.45 in.);
- pipe Reynolds numbers greater than or equal to 4,000

The Reader-Harris/Gallagher equation which is set out in the form of Equation (5.4), is:

$$\begin{aligned}
C = {} & 0.5961 + 0.0291\beta^2 - 0.2290\beta^8 && C_\infty \text{ term} \\
& + 0.000511\left(\frac{10^6\beta}{\mathrm{Re}}\right)^{0.7} && \text{Slope term 1} \\
& + (0.0210 + 0.0049A)\beta^4\left(\frac{10^6\beta}{\mathrm{Re}}\right)^{0.35} && \text{Slope term 2} \\
& + (0.0433 + 0.0712e^{-8.5L_1} - 0.1145e^{-6.0L_1})(1 - 0.23A)\frac{\beta^4}{1-\beta^4} && \text{Upstream tap term} \\
& - 0.0116(M_2' - 0.52M_2'^{1.3})\beta^{1.1}(1 - 0.14A) && \text{Downstream tap term} \\
& + 0.003(1-\beta)M_1 && \text{Small orifice term}
\end{aligned} \tag{5.A.2.1}$$

Note that Gallagher gives an alternative for slope term 2 for Re < 3,500

Also

$$M_1 = \max\left(2.8 - \frac{D}{25.4}, 0.0\right)(D : \mathrm{mm})$$

$$M'_2 = \frac{2L'_2}{1-\beta}$$

$$A = \left(\frac{19{,}000\beta}{\mathrm{Re}}\right)^{0.8}$$

Gallagher comments that the terms in A are significant only at small throat Reynolds numbers. (Note that I have used M'_2 and L'_2 to be consistent with nomenclature in this book.)

where:

β = diameter ratio
= d/D
d = orifice plate bore diameter calculated at T_f
D = meter tube internal diameter calculated at T_f
e = Napierian constant
= 2.71828
L_1 = 0 for corner taps
= $25.4/D$ when D is in millimetres (for flange taps)
= 1 for radius taps
L'_2 = 0 for corner taps
= $25.4/D$ for flange taps
= 0.47 for radius taps
Re = the pipe Reynolds number based on the pipe diameter D
T_f = temperature of the fluid at flowing conditions

For further background to the development of this equation the reader is referred to Bean (1971), Spitzer (1991), Reader-Harris and Sattary (1990), Reader-Harris (1989), Gallagher (1990a, b), Reader-Harris, Sattary and Spearman (1995), Reader-Harris (2015).

This equation was subsequently used by ANSI 2530/API 14.3/AGA-3/ GPA 8185 in their document "Orifice metering of natural gas and other related hydrocarbon fluids Part 1 General equations and uncertainty guidelines", dated 1990.

Appendix 5.A.3 Orifice Discharge Coefficient Equation as Given in ISO 5167–2: 2003

For details of some of the thorough work which has provided the data for this standard the reader is referred to Reader-Harris et al. (1995) and Spencer (1993) (cf. Fling and Whetstone 1985).

The form of the Reader-Harris/Gallagher equation for C set out in ISO 5167-2:2003 is as follows.

Ranges of applicability:

$$d \geq 12.5 \text{ mm}$$

$$50\,\text{mm} \leq D \leq 1{,}000\,\text{mm}$$

$$0.1 \leq \beta \leq 0.75$$

for D $D/2$ and corner tappings	for flange tappings
$\text{Re}_D \geq 5{,}000$ for $0.1 \leq \beta \leq 0.56$	$\text{Re}_D \geq 5{,}000$ and $\text{Re}_D \geq 170\beta^2 D$ where D is in mm
$\text{Re}_D \geq 16{,}000\beta^2$ for $\beta > 0.56$	

For $D \geq 71.12$ mm (2.8 in.)

$$
\begin{aligned}
C = {} & 0.5961 + 0.0261\beta^2 - 0.216\beta^8 && C_\infty \text{ term} \\
& \left.\begin{aligned} & +0.000521\left(\frac{10^6\beta}{\mathrm{Re}}\right)^{0.7} \\ & +(0.0188+0.0063A)\beta^{3.5}\left(\frac{10^6}{\mathrm{Re}}\right)^{0.3} \end{aligned}\right\} && \text{Slope term} \qquad (5.A.3.1) \\
& \left.\begin{aligned} & +(0.043+0.080e^{-10L_1}-0.123e^{-7L_1}) \\ & \qquad \times(1-0.11A)\beta^4(1-\beta^4)^{-1} \end{aligned}\right\} && \text{Upstream tap term} \\
& -0.031(M'_2 - 0.8M'^{1.1}_2)\beta^{1.3} && \text{Downstream tap term}
\end{aligned}
$$

Where $D < 71.12$ mm (2.8 in.) the following term should be added:

$$+\,0.011\,(0.75-\beta)(2.8-D/25.4)\ (D\text{: mm}) \qquad \text{Small orifice term}$$

where Re is based on the pipe diameter D,

$$M'_2 = \frac{2L'_2}{1-\beta}$$

$$A = \left(\frac{19000\beta}{\mathrm{Re}}\right)^{0.8}$$

where $L_1 = \ell_1/D$ and ℓ_1 is the distance of the upstream tapping from the upstream face of the plate, and $L'_2 = \ell'_2/D$ and ℓ'_2 is the distance of the downstream tapping from the downstream face of the plate. The ′ signifies that the measurement is from the downstream and not the upstream face of the plate.

The terms in A are only significant for small throat Reynolds number. M'_2 is in fact the distance between the downstream tapping and the downstream face of the plate divided by the dam height.

For the purposes of Equation (5.4) the values of the upstream and downstream lengths are given here:

	L_1	L'_2	
D and $D/2$	1	0.47	
Flange	25.4/D	25.4/D	(D in mm)
Corner	0	0	

The uncertainty associated with Equation (5.4) is:

$(0.7-\beta)$%	for	$0.1 \leq \beta < 0.2$
0.5%	for	$0.2 \leq \beta \leq 0.6$
$(1.667\beta - 0.5)$%	for	$0.6 < \beta \leq 0.75$

and for $D < 71.12$ mm (2.8 in) the following should be added arithmetically:

$$+\,0.9(0.75-\beta)(2.8-D/25.4) \qquad (D\text{: mm})$$

If $\beta > 0.5$ and Re < 10,000 an uncertainty of 0.5% should be added arithmetically to the uncertainty in C.

It appears that the API document was published before the 24″ (610 mm) and some additional 2″ (50.8 mm) data had been analysed, and while some data were still being collected. The Reader-Harris/Gallagher equation was amended to take account of the 24″ (610 mm) and additional 2″ (50.8 mm) data and ISO considered it necessary to use the latest information.

Appendix 5.B Review of Recent Published Research on Orifice Plates

Appendix 5.B.1 Installation Effects on Orifice Plates

Work on installation effects continues as shown by Studzinski and Karnik (1997). They comment that "the results of the tests indicate that the interaction of the orifice meter with various flow disturbances is a complex process and deviations of the orifice reading from baseline measurements are sometimes not as expected." The measurement of:

- C (Coefficient of Discharge),
- C with upstream fittings causing distortion to the inlet profile,
- C with upstream conditioners etc.

has been, and will continue to be, the subject of experimental programmes. The assessment of error which results from corrosion, wear (particularly on the inlet edge of the plate) and deposits on the plate, in the pipe and in the pressure tappings is also important.

Weiss, Studzinski and Attia (2002) evaluated the performance of orifice meter standards for installation downstream of T-junction and elbow in a low-pressure air facility, in the light of modified requirements in API/ANSI-2530 and ISO-5167. They found them to be on the conservative side, with a small exception, with data outside the acceptance range, for a rounded edge T-junction with, it appears, a 10D blind extension on the branch of the T in line with the meter run for one β-ratio.

Reader-Harris and Brunton (2002) examined the effect of diameter steps in upstream pipework. With pipes of Schedules 80, 120 and 10 installed at various distances upstream of an orifice meter which was of Schedule 40, the measured shifts in discharge coefficient showed that the requirements in ISO 5167-1: 1991 were excessively conservative, and the data was used to recommend maximum steps for the revision of ISO 5167.

Reader-Harris et al. (2003) looked at the effect of damage to a Zanker flow conditioning plate and the effect of roughness in the upstream pipe where this is restricted to the bottom of the pipe. For the conditioner, blocked perforations were more serious than rounding the upstream edge of the conditioner, and the shift in discharge coefficient was approximately proportional to the fraction of the pipe roughened.

Barton, Hodgkinson and Reader-Harris (2005) identified from inspection that installed orifice plates for gas flow rate measurement (in 16″ runs) may be eccentric within the carriers with damaged O-ring seals. The authors investigated and estimated the likely error from published test data and CFD. The eccentricity was taken as:

$$\text{Eccentricity} = x/D$$

where x is the displacement of the centre of the orifice hole from the pipe axis (m), D is the pipe diameter (m). Plots suggested that 0.025 eccentricity may be negligible but may allow leakage. Eccentricity might cause errors between −1% and 4% and in addition leakage may add to this.

Other research on installation has been reported by Laribi, Wauters and Aichouni (2003) and Kim et al. (2002).

Some axisymmetric internal configurations were numerically simulated to investigate the development of turbulent swirling and non-swirling flows (von Lavante and Yao 2010). The authors referred to Ferron (1962), Akashi, Watanabe and Koga (1978), Parchen and Steenbergen (1998), Steenbergen and Voskamp (1998), Reader-Harris (1994), Reader-Harris et al. (1997) and Sofialidis and Prinos (1996).

Appendix 5.B.2 Pulsation

The effect of pulsation also continues to generate test data (See also Section 5.7). McBrien (1997) reported tests at high gas pressures. For pressure in the range 46 to 51 bar and Re between 10^6 and 3×10^6, the error is expected to be less than ±0.5% for rms to average pressure ratio ≤ 0.25. The error is expected to be less than ±0.2% for rms to average pressure ratio ≤ 0.1 (see Norman et al. 1995 for higher frequencies).

Appendix 5.B.3 Contamination

Pritchard, Marshall and Wilson (2004) noted that evidence from various studies had identified that contamination of the orifice plate tended to result in under registration of gas flow. The experimental work behind ISO 12767 was, apparently, undertaken primarily by British Gas in the 1970s at atmospheric pressure with air as the gas and revealed metering errors ranging from 0 to 24% for various degrees of contamination, caused by pipeline sludge, oil, grease, liquids etc. These test conditions were reviewed under current conditions to assess the validity of the estimated errors when used under actual operating conditions. Hence, further experimental study looking at different contamination patterns have been undertaken. Reader-Harris et al. (2010) reported on research on the effect of contamination on orifice plates caused by oil, grease etc. (cf. ISO/TR 12767). Their work combines experiment and

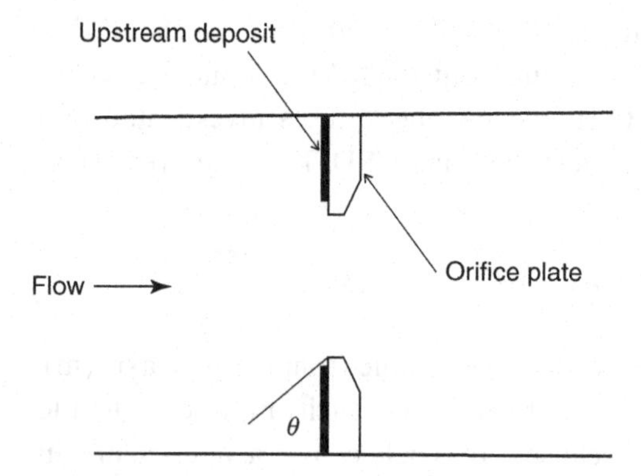

Figure 5.B.1. Contamination on leading face of orifice plate showing the approach angle, θ, caused by deposit (after Reader-Harris et al. 2010).

CFD for which they reported good agreement. They noted that grease spots could cause increases in the C. Their findings suggested consequent changes in discharge coefficient up to about 1.3% in some cases (see also Reader-Harris et al. 2012). If the contamination does not extend right up to the sharp edge (i.e. there is a portion of the orifice plate cleaned by the flow), they identified the change in flow angle on approach to the plate caused by the upstream contamination layer which will result in an angle, θ (Figure 5.B.1). This angle will be of increasing size as the contamination approaches the sharp edge and becomes thicker. This, in turn, changes the shape of the *vena contracta*, leading to an increase in C. Table 5.B.1 gives a few values which have been obtained.

Appendix 5.B.4 Drain Holes

Reader-Harris et al. (2008) and Reader-Harris and Addison (2013) have discussed the percentage change in C, the discharge coefficient, due to drain holes. Their findings suggest that the previous estimate for the effect of the holes (ISO/TR 15377) was not accurate, and in their latter paper they have proposed an approach which they consider would be preferable as an amendment to the standard. See also Reader-Harris and Addison (2014).

Appendix 5.B.5 Flow Conditioning for Orifice Meters

Karnik (1995) described the development of a Laws plate which took account of Reynolds number. The perforated plate was sandwiched in a 13D meter run. For a single elbow and two elbows out-of-plane the meter error appeared to be within ±0.1%. Sensitivity of the performance of the perforated plate conditioners to the size of the holes (cf. Spearman, Sattary and Reader-Harris 1996) was mentioned. He made the following observations:

Table 5.B.1. *Some values of the increase of discharge coefficient with θ, due to contamination, for a pipe of 300 mm (from Reader-Harris et al. 2010)*

β	0.6	0.6	0.6	0.2	0.4, 0.6, 0.75	0.2	0.6
θ (degrees)	5	10	20	20	30	32	40
Approximate maximum increase shift in discharge coefficient C (%)	1	2.4	2.7	5	7.5	13	10

i. The orifice plate should be placed where the jets have coalesced and the desired profile has been achieved.
ii. If the flow is allowed to develop further, then the pipe roughness may become important and may begin to affect the profile shape.
iii. The package should be used at receipt and delivery.

Schluter and Merzkirch (1996) undertook work on the performance of flow conditioners based on shape of profiles, and Spearman et al. (1996) compared velocity and turbulence profiles downstream of perforated plate flow conditioners, including a Spearman (NEL) one (Appendix 2.A.3). Morrison et al. (1997) undertook more tests of the development of flow downstream of a standard tube bundle and of three different porous plate flow conditioners. Erdal (1997) commented that it was possible to obtain approximately fully developed velocity profiles a few diameters downstream of the plates with different overall porosities if grading of porosity is appropriate. But it also revealed that the turbulence profile downstream of the plates could be quite different. This would affect the distance before the flow is fully developed. Morrow (1997) reported a set of tests (full data in Morrow 1996). He compared results for tube bundle with cylindrical and hexagonal tubes, the Gallagher flow conditioner and no conditioner. Although the tube bundles limited error to ±2% for β ratios of 0.2 and 0.4, this was not so for β greater than 0.5 for 17D from disturbance to orifice plate. However, for the Gallagher conditioner the error was less than ±0.2% for all β ratios from 0.2 to 0.75 for 4D to 12D between outlet of conditioner and orifice plate.

Zanker and Goodson (2000) commented that the new AGA 3/API 14.3 standard on concentric square edged orifice meters allows other flow conditioners to be qualified for use by meeting certain type approval requirements. They listed tests proposed: baseline, good flow conditions, two 90° elbows in perpendicular planes, gate valve 50% closed, high swirl (24°). A modified Zanker flow conditioner of perforated plate type was tested and qualified.

Barton and Peebles (2005) undertook a CFD study of the flows in a gas export system header to identify whether a conditioner would be necessary to meet ISO requirements, and they suggested that the Zanker plate would be effective and avoid an additional 0.5% uncertainty.

Appendix 5.B.6 Plate Thickness for Small-Diameter Orifice Plates

Kim et al. (1997) tested effects of cavitation and plate thickness on small diameter ratio orifice meters. The work may be of interest for specialist uses of orifice plates.

Appendix 5.B.7 Variants on the Orifice Plate

A "smart-orifice" meter (Krassow, Campabadal and Lora-Tamayo 1999) has been suggested which makes use of silicon technology to incorporate the DP sensor within the meter. The resulting orifice being outside the ISO parameters was modelled with CFD.

Appendix 5.B.8 Impulse Lines

Strom and Livelli (2001) made suggestions on the benefit of eliminating impulse lines and improving accuracy by mounting DP cells on the meter. Livelli (2002) considered best practice for process plant safety, efficiency and reduction of variability. This paper also suggested the advantages of eliminating impulse lines by making use of new DP instrumentation which allowed direct mounting to the primary element. Transmitters and control systems can be programmed to reduce bias errors.

Appendix 5.B.9 Lagging Pipes

Kimpton and Niazi (2008) examined the value of lagging orifice plates, upstream and downstream straight lengths and temperature fittings as well as impulse lines. They appeared to find the effects were small, and suggested that CFD showed local insulation around a thermowell or a surface-mounted sensor to be sufficient, without the need to insulate the entire meter runs. This finding confirmed reports by Ingram (British Gas internal reports and memorandum) that local insulation is sufficient for accurate measurement of gas temperature.

Appendix 5.B.10 Gas Conditions

The reader requiring detailed information on the gas properties required for differential pressure meter calculations is referred to various references, a selection of which follow. Starling and Luongo (1997) discussed the electronic implementation of AGA Report No 8 for the compressibility factor of natural gas. Johansen (2010) discussed the effect of using real gas absolute viscosity and isentropic exponent on orifice flow measurement and considered adoption of REFPROP 8.0 as a standard for the natural gas industry. Marić (2005) appears to have provided some valuable work on the Joule-Thomson effect which is suitable for application in flow measurements using differential pressure devices. With colleagues, he has (Marić, Galović and Šmuc 2005) provided a calculation of the natural gas isentropic exponent and

(Marić 2007) a numerical procedure for calculation of the natural gas molar heat capacity, the Joule-Thomson coefficient and the isentropic exponent, by applying the fundamental thermodynamic relations to AGA-8 (1992) extended virial-type equations of state. Marić and Ivek (2010) also discussed compensation for the Joule-Thomson effect in flow rate measurements by the "Group Method of Data Handling" (GMDH) polynomial.

Appendix 5.B.11 Emissions Testing Uncertainty

Frøysa et al. (2007) described an uncertainty analysis of the sum of the flows through orifice plates used for emission monitoring from the Statoil Mongstad oil refinery. As well as temperature and pressure, gas density was measured. The authors describe the uncertainty model in detail, and the issue of correlation between various measured quantities. With improved instrumentation an expanded uncertainty within 1.5% was achieved.

Appendix 5.B.12 CFD Related to Orifice Plates

The development of solutions for real internal flows of the sort encountered in pipe flows by commercial flowmeters continues, see Morrison (1997) and Holm, Stang and Delsing (1995). Wilcox, Barton and Laing (2001) used CFD to check the flow in a metering station where there were, upstream of the header, bends etc. which could have caused flows and meter performance outside that expected from ISO5167.

Mokhtarzadeh-Dehghan and Stephens (1998) presented computational analysis of fixed and variable area orifice meters. For the standard orifice they obtained discharge coefficients for corner taps about 1% low and for D and $D/2$ taps about 0.8% high. It would be interesting to have explored this difference further, and to find the effect of, say, changes in the turbulence model, edge treatment, upstream flow profile and turbulence distribution. The work on the variable area meter is interesting, but not so easily compared with a known performance.

Barton, Gibson and Reader-Harris (2004) reviewed the value of CFD modelling of orifice plate flows and upstream fittings. They considered that it provided a useful complement to experimental data, particularly when applying information derived from tests to field conditions and in circumstances where a high degree of control is required which may be difficult to achieve in conventional tests. The authors compared simulation results from various projects undertaken at NEL with test and field data to provide a feel for the accuracy of the CFD method. To predict the flow measurement error associated with some non-ideal operational conditions, CFD simulations were usually run of the orifice plate in both ideal and non-ideal conditions, effectively to eliminate most of the inaccuracies associated with the simulation method. As a result a large number of simulations have been run of orifice plates in ideal flow conditions. The authors expected most modern simulations of orifice plates in good flow conditions to agree with ISO 5167 to within 1%. CFD methods could also give reasonable predictions of orifice-plate installation errors provided that an appropriate turbulence model was used. CFD had proved useful in the study of rough pipes upstream of orifice plates and contamination of plate surfaces with dirt or grease. The authors considered that results of CFD simulations were most useful when the source of errors was well understood and when predictions could be validated against experimental data. They suggested that as computing power increased and as more validation work was performed, CFD was likely to be increasingly used to assess and to understand orifice-plate metering applications (see also Reader-Harris 1986; Brown et al. 2000).

6

Venturi Meter and Standard Nozzles

6.1 Introduction

In this chapter, we are concerned with devices defined in international standard documents. Among these, the Venturi meter is one of the oldest industrial methods of measuring flow, although other methods may be more common. Chapter 8 deals with other devices such as variable area meters, target meters, averaging pitot tubes, Dall meters, V-cone meters and other Venturi, nozzle and orifice-like devices.

As in the case of the orifice plate, the reader should have access to a copy of the latest version of the ISO standard (5167). I have refrained from repeating information that should be obtained from the standard, and include only the minimum information to provide an understanding of the discharge coefficients, the likely uncertainty of the various devices and the type of material from which they may be constructed. There is a growing interest in the use of some of these devices for difficult metering tasks and a re-evaluation of their behaviour. Recent publications are briefly reviewed in Appendix 6.A.

Spink (1978) suggested that a standard orifice should be used unless:

- The velocity is such as to require the value of β to exceed 0.75: if higher differential pressure or larger pipe is not possible, he suggested considering a flow nozzle.
- The fluid contains a second phase of some sort – particulate, bubble or droplet – in suspension: either a Venturi or a flow nozzle with vertically downward flow, an eccentric or segmental orifice, or a different type of device such as a target meter or electromagnetic flowmeter should be considered.
- The fluid is very viscous or results in low Reynolds number flows: the meters already mentioned may be suitable, or a quadrant or semi-circular orifice may be worth considering.
- The flow is very low and the fluid very clean: an integral orifice or a small calibrated meter run may be appropriate.
- Pumping costs are a major consideration: in which case go to a lower-loss device.

If a liquid or a gas flows down a pipe of decreasing cross-section, the velocity (provided, in the case of a gas, that it is below the speed of sound) increases

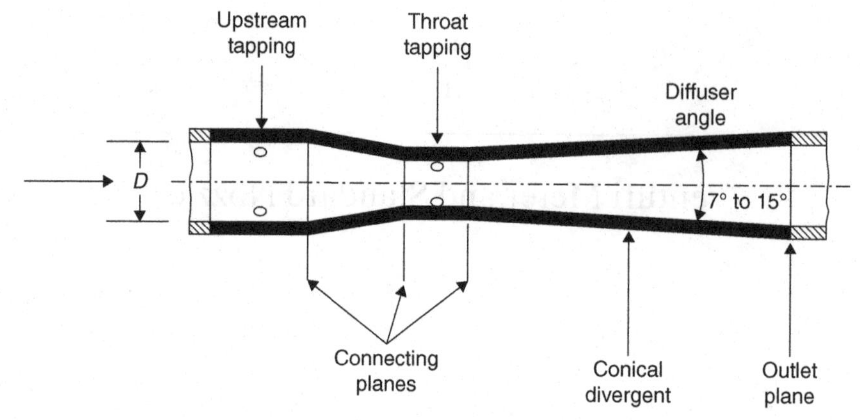

Figure 6.1. Classical Venturi meter (after ISO 5167-1 1997).

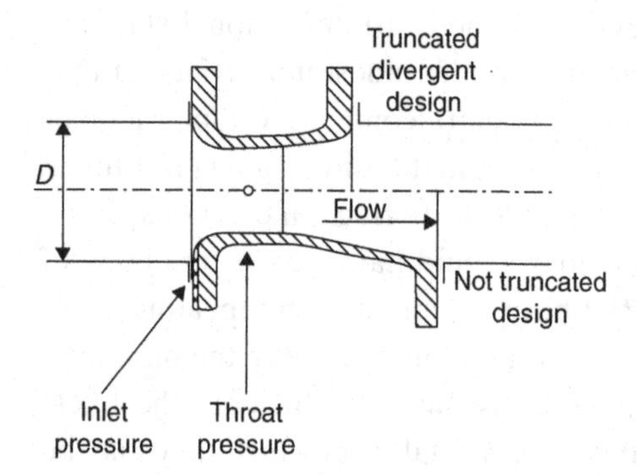

Figure 6.2. Venturi nozzle (after ISO 5167-1 1997).

as the area decreases, and the pressure also decreases with the increase in velocity. This behaviour allows the measurement of the change in pressure as a means of obtaining the flow rate in the duct. If the duct has a smoothly varying cross-section and slowly changing area, the pressure change for a particular flow rate can be predicted from Equation (2.11). This is particularly so for the case of the Venturi and the inlet to the standard nozzles, and in contrast to the standard orifice plate, the flow rate can be predicted to within about 1–1.5% as shown from experimental data.

The inlet flow will usually be turbulent and will approach the Venturi (Figure 6.1) or nozzle (Figures 6.2–6.4) where an upstream pressure tapping (one at entry to the flowmeter) will measure the static pressure. The convergent flow at the inlet of the Venturi or nozzles passes smoothly to the throat, except that it is possible that a small recirculation zone may exist upstream at the corner of nozzle entry. The downstream pressure tapping for the Venturi is in the parallel throat section and is set in the wall. In the case of the nozzles, the downstream tapping is either in the throat or in the pipe wall. In the latter case, it senses the pressure of the jet as it leaves the outlet of the nozzle. Downstream of the throat in the Venturi, there is a controlled diffusion, the small angle ensuring that good pressure recovery is achieved with a

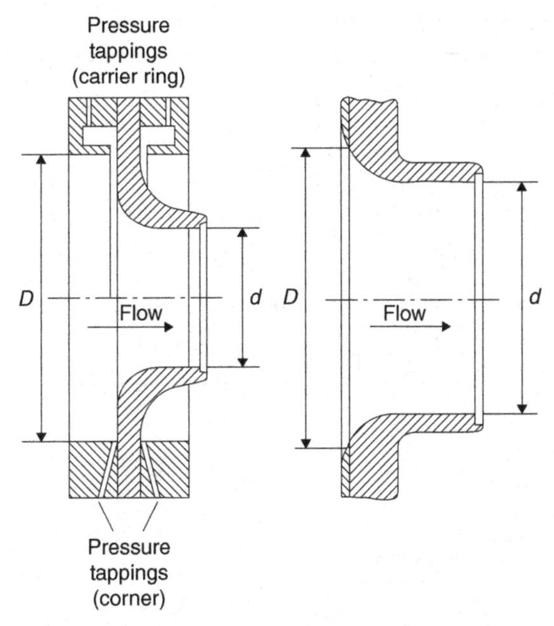

Figure 6.3. ISA 1932 nozzle (after ISO 5167-1 1997).

low total pressure loss. The losses at the outlet of the nozzles vary according to the particular design.

The comparison between the flow in the Venturi (Figure 6.1), a duct of (relatively) smoothly changing area, and the flow through an orifice plate (Figure 5.1) is marked.

The essential equations governing the behaviour of the Venturi and the nozzles have been given in Chapter 2 (cf. Baker 1996), and many of the ideas are touched on in Chapter 5. In this chapter, we shall, as far as possible, avoid repetition.

6.2 Essential Background Equations

The equation relating mass flow to the differential pressure for Venturi and nozzles is reproduced from Chapter 5:

$$q_m = \frac{CE\varepsilon\pi d^2\sqrt{2\rho_1\Delta p}}{4} \tag{5.1}$$

where C is the coefficient of discharge, E is the velocity of approach factor $(1-\beta^4)^{-1/2}$ where β is the diameter ratio d/D of orifice diameter to pipe internal diameter, ε is the expansibility (or expansion) factor, Δp is the differential pressure, ρ_1 is the density at the upstream pressure tapping cross-section, and q_v is related to q_m by Equation (5.2).

The expansibility (or expansion) factor ε is given in the standard for each device, Venturi or nozzle, which should be consulted for details. The expressions differ from those in Chapter 5 (see Equation 6.A.3).

Table 6.1. *Coefficient of discharge for the classical Venturi*

Type of Convergent	Constraints	C	Uncertainty in C (%)
Rough-cast (or as-cast)	$100\text{ mm} \le D \le 800\text{ mm}$ $0.3 \le \beta \le 0.75$ $2 \times 10^5 \le \text{Re} \le 2 \times 10^6$	0.984	0.7
Machined	$50\text{ mm} \le D \le 250\text{ mm}$ $0.4 \le \beta \le 0.75$ $2 \times 10^5 \le \text{Re} \le 1 \times 10^6$	0.995	1.0
Rough-welded sheet iron	$200\text{ mm} \le D \le 1200\text{ mm}$ $0.4 \le \beta \le 0.7$ $2 \times 10^5 \le \text{Re} \le 2 \times 10^6$	0.985	1.5

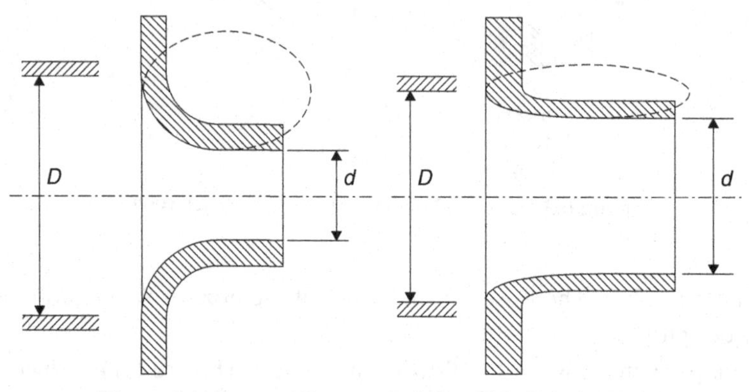

Figure 6.4. Long radius nozzle (after ISO 5167-1 1997).

For the classical Venturi meter, the coefficients are given in Table 6.1. It is important to note that, although the flow in the meter is a flow that approaches the ideal and results in a coefficient within about 1% of unity, and therefore very close to the result from Bernoulli's equation, the uncertainty is greater than for the orifice. This reflects not on the stability of the flow but on the consistency with which the devices can be machined and the pressure tappings formed. The experience that has been amassed of the behaviour of these instruments will also relate to this.

Shufang, Yongtao and Lingan (1996) reported tests outside the ISO limits with a 1,400-mm design using a maximum flow rate of 18,000 m^3/h with a system uncertainty of 0.2% and for $1.85 \times 10^6 < \text{Re} < 5.78 \times 10^6$ obtained discharge coefficients:

for machined convergent 0.983 ± 0.003
for rough-welded 0.987 ± 0.006

For the nozzles, the standard sets out the discharge coefficient formulas. The reader is referred to ISO 5167-3:2003 for the complete formulas, the conditional requirements and the limits of applicability.

Note that in all the equations, Re refers to the Reynolds number based on the upstream pipe internal diameter. However, Re_d based on the throat diameter is sometimes used in the standards, and the reader should be aware of this.

6.3 Design Details

As for the orifice plate, these are set out fully in ISO 5167, which provides the detailed requirements. The reader must refer to the standard to confirm that these are the latest values and to obtain all other details for each of these devices.

For the Classical Venturi Meter

The main features of the Venturi design are shown in Figure 6.1. In addition the standard deals in matters such as the smoothness of the finish of the throat and the adjacent curvature. These are required to have $Ra/d < 10^{-4}$ where Ra is the roughness criterion and d is the throat diameter. For the machined Venturi, the entrance cylinder of length not less than D and the convergent section are required to have a finish to the same quality. The radius of curvature between the sections must be less than the following values:

$0.25D$ between entrance cylinder and convergent section,
$0.25d$ between the convergent section and the throat,
$0.25d$ between the throat and the divergent section.

In each case, the standard would prefer that the radius of curvature be zero. The requirements for the other two methods of fabrication of the Venturi are also given in the standard.

Pressure tappings are shown in Figure 6.1, and the standard requires that four (at least) shall be provided at both the throat and the upstream positions and specifies their size, finish and precise positions.

The value of the pressure loss across the Venturi meter compared with the differential pressure between inlet and throat can be taken to be in the range 5–20%.

For the Venturi Nozzles

The nozzle is shown in Figure 6.2 and must have a roughness less than $10^{-4}d$. It can be made of any material provided it meets the requirements set out in the standard. Obviously the material must be stable enough to retain its shape and to resist erosion and corrosion by the fluid.

For the ISA 1932 Nozzles

The nozzle is shown in Figure 6.3 and has a roughness as for the Venturi nozzle. Again, any unchanging material is acceptable.

For the Long Radius Nozzle

The nozzle is shown in Figure 6.4 and has the same roughness as the Venturi nozzle. Again, any unchanging material is acceptable.

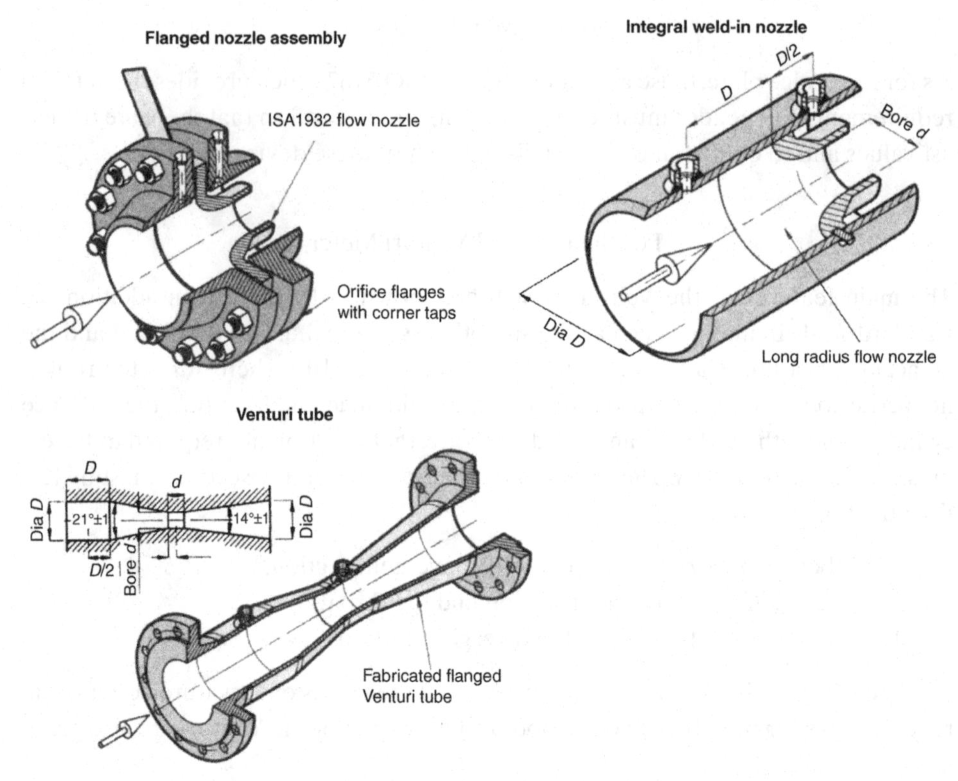

Figure 6.5. Commercial devices (reproduced with permission from ABB).

6.4 Commercially Available Devices

The standards allow prospective users to design and make their own differential pressure device. However, various manufacturers also produce standard equipment. Figure 6.5 shows examples of such devices.

6.5 Installation Effects

Upstream Fittings

A sample of upstream installation spacings from ISO 5167:2003 is given in Table 6.2.

Himpe, Gotte and Schatz (1994) described results for the influence of upstream bends on the discharge coefficient of classical Venturi tubes which suggest that the influence of disturbances does not necessarily decrease with decreasing bend angles, and the influence of the angle between the plane of the pipe bend and the position of the tapping is considerable.

Al-Khazraji et al. (1978) used an eccentric orifice upstream of a Venturi to assess the error that results from the flow disturbance. The orientation relative to the single

Table 6.2. *Examples of installation requirements for the classical Venturi meter for zero additional uncertainty from the standard. (Bracketed values are ±0.5% additional uncertainty.)*

β	Upstream straight lengths			
	Single 90° Bend	Two or more 90° bends in the same or different planes	Concentric reducer 3D to D over a length of 3.5D	Full-bore ball valve or gate valve fully open
0.4	8 (3)	8 (3)	2.5 (*)	2.5 (*)
0.5	9 (3)	10 (3)	5.5 (2.5)	3.5 (2.5)
0.6	10 (3)	10 (3)	8.5 (2.5)	4.5 (2.5)
0.75	16 (8)	22 (8)	11.5 (3.5)	5.5 (3.5)

At least four throat diameters should separate the throat tapping from downstream fittings.
* Data not available
Note that some of these values, from ISO5167-4:2003, differ from earlier versions (and the first edition of this book).

pressure tapping had a negligible effect, but a spacing of 5D reduced the error to about 1%. At 10D spacing, there was negligible effect, showing the same trend as for a single bend with $\beta \leq 0.6$ in Table 6.2.

Two-Phase Flows

Baker (1991a) provided several references relating to the behaviour of Venturi meters in multiphase flows. de Leeuw (1997, cf. 1994) proposed a correlation for over-reading in wet-gas horizontal flow which appeared to give results within 2% of the data. He claimed that the improvement was, in part, because it allowed for the gas velocity. Taking q_{tp} as the mass flow rate deduced from the differential pressure when liquid is present in the gas stream, $q_{m,l}$ as the liquid mass flow rate and $q_{m,g}$ as the gas mass flow rate, the over-reading ratio is given by

$$\phi = \frac{q_{tp}}{q_{m,g}} = \sqrt{1 + C_{tp}X + X^2} \tag{6.1}$$

where X is the Lockhart-Martinelli parameter and

$$C_{tp} = \left(\frac{\rho_l}{\rho_{1,g}}\right)^n + \left(\frac{\rho_{1,g}}{\rho_l}\right)^n \tag{6.2}$$

the liquid and gas densities are given by ρ_l and $\rho_{1,g}$ and

$$n = 0.606\left(1 - e^{-0.746\mathrm{Fr_g}}\right) \quad \text{for } \mathrm{Fr_g} \geq 1.5 \tag{6.3}$$

and

$$n = 0.41 \quad \text{for } 0.5 \leq \mathrm{Fr_g} < 1.5 \tag{6.4}$$

where the superficial gas Froude number is given by

$$\mathrm{Fr_g} = \frac{V_{sg}}{\sqrt{gD}}\sqrt{\frac{\rho_{1,g}}{\rho_l - \rho_{1,g}}} \tag{6.5}$$

where V_{sg} is the superficial gas velocity, and the Lockhart-Martinelli parameter is given by

$$X = \frac{q_{m,l}}{q_{m,g}}\left(\frac{\rho_{1,g}}{\rho_l}\right)^{\frac{1}{2}} \tag{6.6}$$

The relationship is valid for gas densities above 17 kg/m^3 and up to the liquid density, for gas Froude numbers above 0.5, and for X up to 0.3 (cf. Chisholm 1977).

A recent wet-gas correlation is given is given in Appendix 6.A.3.

6.6 Applications, Advantages and Disadvantages

Early work on the classical Venturi by Herschel was aimed at use in the water industry, and it has continued to be used in measurement of water flows. Where energy conservation is an important consideration in large pumped flows, the low head loss of the Venturi becomes attractive.

Rivetti, Martini and Birello (1994) have described a quite different application of a small-size Venturi for liquid helium service, which was essentially a scaled-down ISO 5167 design. Differential pressure must not be so high as to cause cavitation, and pressure ducts should have a diameter between 1/10 and 1/5 of the diameter of the respective Venturi section and be four or more (an even number). Flow rates ranged from 0.5–4 g/s to 12.5–100 g/s for upstream/throat diameters of 6/2.6 and 16/12 mm, respectively, and differential pressures of 34–2,190 and 33–2,140 Pa, respectively.

The classical Venturi has also been used for slurry and two-phase flows for which some data are available. Baker (1991a) has reviewed work on this application of the Venturi. In multiphase flows the Venturi meter has become very widely used, although the behaviour of the flow and the appropriate models of the multiphase flow may not be straightforward. One such example is the performance of the Venturi meter in an oil-in-water emulsion in Figure 6.6 (Pal and Rhodes 1985).

The initial cost of the Venturi is higher than the orifice because it is a much larger device to make, and precision in manufacture may not be easy. Various manufacturers offer ready-made designs (Figure 6.5).

The nozzles tend to be more stable than the orifice for high temperatures and high velocities, experiencing less wear (due to the smooth inlet contour as opposed to the sharp orifice upstream edge) and being less likely to distort. They are particularly applicable to steam flows, where they have been widely used.

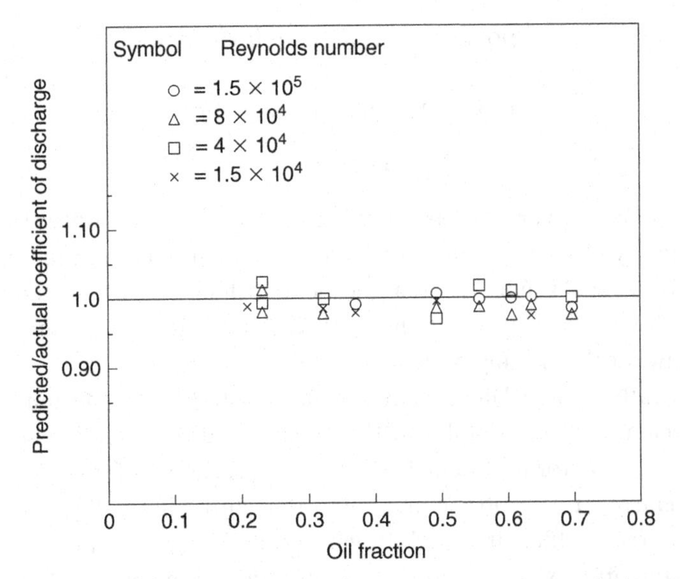

Figure 6.6. Performance of a Venturi meter in an oil-in-water emulsion (Pal and Rhodes 1985; reproduced with permission from BHR Group).

6.7 Chapter Conclusions

While this chapter has been brief, the Appendix indicates the increasing importance of this technology. There is substantial interest in the oil and gas industry for using Venturi meters for single and multiphase (including wet-gas) metering, and further work is taking place which is reported in Appendix 6.A and Chapter 17. Three areas of interest are the following.

a. Jamieson et al. (1996) observed discharge coefficients for a 150-mm (6-in.) diameter Venturi in high pressure air (70 bar) at Re up to 8×10^6 and throat velocities up to 125 m/s, which were several percent higher than those obtained in water calibration. The reason appears to lie in the tapping chambers and the possible acoustic effects there.
b. Computational fluid dynamics (CFD) work has been done (cf. Sattary and Reader-Harris 1997) for Venturi and nozzles. For Venturi with $\beta = 0.4$, 0.6 and 0.75, CFD tended to underestimate relative to experimental and ISO by up to about 1.2%. Sattary and Reader-Harris (1997) suggested that the difference was due to the actual method of measuring static pressure compared with the CFD model. They also looked at compressibility effects. There is likely to be growing interest in developing the CFD work to explore the ultimate precision of these instruments.
c. An inside-out nozzle or short Venturi in the form of a bullet-shaped target meter could yield a highly predictable meter with direct electrical output via a force sensor. This could be combined with wall differential pressure measurements to provide a condition-monitoring function.

Appendix 6.A Research Update

6.A.1 Design and Installation

Venturis

Reader-Harris, Brunton and Sattary (1997) gave installation distances for zero additional uncertainty for Venturi meters which were at variance with the values in the ISO. They also found values for multiple bends which seemed to require shorter lengths than for single bends, but they still recommended the longer installation length to allow for the spacing of the two bends.

More recently Reader-Harris et al. (2000a cf. 2001) have reported tests for discharge coefficients of Venturi tubes with non-standard convergent angles, and they obtained an equation for the coefficient with an uncertainty of 0.71%.

Reader-Harris et al. (2002) commented that until about 1995 it was assumed that the discharge coefficient at high Reynolds number in high-pressure gas would be constant and approximately equal to that obtained in water at Reynolds numbers greater than 2×10^5. However, further data indicated that the performance of Venturi tubes at high Reynolds number in gas is very different from that in water. Some discharge coefficients in gas were greater than would have been expected by 3% or even more. They suggested that one factor was the internal shape of the Venturi. They reported on four Venturi tubes calibrated in water and high-pressure gas: 100 mm diameter, diameter ratio 0.6, but with different convergent profiles. The best convergent profile was found to be one with 10.5° included angle and sharp corners since it gave the lowest standard deviation of data (cf. Reader-Harris et al. 2000c). Two further Venturi tubes with the same profile gave similar results in water and gas. Further work was reported by Reader-Harris, Rushworth and Gibson (2004) to understand data where discharge coefficients, at Reynolds numbers greater than 2×10^5, appeared to be 3% or more in gas flows than expected based on water tests. Reader-Harris et al. (2005) undertook work to confirm the performance of Venturi tubes with a 10.5° convergent angle (half that in the traditional standard Venturi). They developed a discharge coefficient equation and they imply that this made use of CFD to compute the effect of roughness. They implied that it was important to take pipe roughness into account. (See also Jamieson et al. 1996; Van Weers, Van der Beek; and Landheer 1998; Reader-Harris et al. 1999, 2000a, 2000b, 2001, 2002.) The effect on discharge coefficient of a single bend, and of two bends in perpendicular planes, was investigated. CFD was used to compute the effect of a single bend on Venturi tubes with convergent angles of 10.5° and 21°. For small β the required upstream straight lengths for Venturi tubes of convergent angle 10.5° were smaller than those for a standard Venturi tube; for large β they were larger than those for a standard Venturi tube. The straight lengths in Table 1 of ISO 5167-4:2003 appear sufficient for Venturi tubes of convergent angle 10.5° for β up to about 0.67.

Steven (2010) discussed additional pressure sensing in Venturi meters to provide additional diagnostics for monitoring the meter under assumed single phase flow. The method used an additional downstream tap after the Venturi and to use the three differential pressures, inlet to throat, throat to outlet and inlet to outlet, to obtain three flow measurements and from these to identify metering problems. He referenced the following papers: Steven (2008b), Steven (2009a, 2009b) and Geach and Jamieson (2005).

Numerical Analysis

CFD has been used to analyse the effect of flow on meter reading. The interested reader is referred to von Lavante and Yao (2010) for a detailed description of the flow analysis and the effect of swirl on discharge coefficients, and to other references cited by them: Reader-Harris (1994), Reader-Harris et al. (1997).

6.A.2 Meters in Nuclear Core Flows

Gribok et al. (2001) described a method of inferential sensing applied to nuclear reactor flows, which allowed a model to be developed to predict feedwater flows, and compared the predicted values with those obtained from a Venturi meter. By a technique of regularisation, for which the reader is referred to the original paper, the actual feedwater flow rate was predicted from other variables. This resulted in a value of feedwater flow rate lower than that obtained from the Venturi meter, which suffered from fouling. This, in turn, resulted in a loss of power due to derating. Problems relating to the drift of Venturi meters, particularly resulting from fouling in nuclear plant feed water systems, have also been highlighted by others (Roverso and Ruan 2004). Estrada (2002), in a paper on the effect of in-service velocity profiles on flow measurement in nuclear power plant feedwater systems, commented on the effect of swirl and the consequent effect on flow nozzles. Neural networks and cross-correlation techniques were considered as means to deal with this problem. Na, Shin and Jung (2005a) and Na, Lee and Hwang (2005b) also discussed a software approach and used actual data to verify its potential.

6.A.3 Special Conditions

Low Reynolds Number

Stobie et al. (2007) obtained some interesting experimental data for a Venturi meter with Reynolds numbers covering the laminar flow range down to 80 for discharge coefficient measurements and compared the data with other data and made use of CFD predictions. There was a slight "hump" on the discharge coefficient curve at transition. They also examined the effect of a viscous fluid stream, containing sand, on the erosion of the meter and presented data on the change in the discharge coefficient. A water/sand slurry resulted in significant erosion.

Extreme Conditions

Jitschin, Ronzheimer and Khodabakhshi (1999) and Jitschin (2004) commented that in vacuum technology flow measurement, flow constrictions are convenient, and they appear to have tested thin orifices and Venturis at critical flow conditions. At Re of about 10^5 they commented that the Venturi coefficient differed from the expected value by about 2%.

Wet-gas Correlation

In Section 6.5, the correlation for wet-gas due to de Leeuw (1997) was mentioned. Reader-Harris and Graham (2009) gave a useful review of wet-gas correlations and derived an improved model for Venturi over-reading in wet-gas (equations reproduced with permission). This approach has influenced PD ISO/TR 11583:2012. (See Reader-Harris (2012) on review of the standard cf. de Leeuw et al. 2011). The equation for gas mass flow in wet-gas is given as

$$q_{m,g} = \frac{C}{\sqrt{1-\beta^4}} \varepsilon \frac{\pi}{4} d^2 \frac{\sqrt{2\Delta p \rho_{1,g}}}{\phi} \tag{6.A.1}$$

The gas density, $\rho_{1,g}$, is that at the upstream tapping point.

We shall require the Lockhart-Martinelli parameter

$$X = \frac{q_{m,l}}{q_{m,g}} \sqrt{\frac{\rho_{1,g}}{\rho_l}}$$

where $q_{m,l}/q_{m,g}$ is the ratio of the mass flow rates.

The authors observed that the wet-gas discharge coefficient was not the same as that for the dry gas, and this change could result from a very small amount of liquid. An interesting change which they noted was that a Venturi tube can often emit a singing tone with dry gas, but never with wet-gas. They also noted that the over-reading appeared strongly linked to the surface tension of the liquid component. The discharge coefficient from a data fit was given as

$$C = C_{fullywet} \qquad X \geq X_{lim} \tag{6.A.2}$$

$$C = C_{dry} - (C_{dry} - C_{fullywet})\sqrt{\frac{X}{X_{lim}}} \qquad X < X_{lim}$$

where $X_{\text{lim}} = 0.016$ and

$$C_{\text{fullywet}} = 1 - 0.0463e^{-0.05\text{Fr}_{\text{g,th}}} \tag{6.A.3}$$

giving the summary equation (since $C_{\text{dry}} = 1$ gives the best fit for an uncalibrated Venturi tube)

$$C = 1 - 0.0463e^{-0.05\text{Fr}_{\text{g,th}}} \min\left(1, \sqrt{\frac{X}{0.016}}\right) \tag{6.A.4}$$

The Froude number is given by

$$\text{Fr}_{\text{g}} = \frac{4q_{\text{m,g}}}{\rho_{1,\text{g}} \pi D^2 \sqrt{gD}} \sqrt{\frac{\rho_{1,\text{g}}}{\rho_{\text{l}} - \rho_{1,\text{g}}}} \tag{6.A.5}$$

and, for the throat, is

$$\text{Fr}_{\text{g,th}} = \frac{\text{Fr}_{\text{g}}}{\beta^{2.5}} \tag{6.A.6}$$

The expansibility factor, according to ISO 5167-4:2003 (which appears to be for both the Venturi and for the nozzles), is

$$\varepsilon = \sqrt{\left(\frac{\kappa \tau^{2/\kappa}}{\kappa - 1}\right)\left(\frac{1-\beta^4}{1-\beta^4 \tau^{2/\kappa}}\right)\left(\frac{1-\tau^{(\kappa-1)/\kappa}}{1-\tau}\right)} \tag{6.A.7}$$

(obtained from data for air, steam and natural gas but with the suggestion that it could be used more widely if the isentropic exponent, κ, were known), where τ is the pressure ratio

$$\tau = \frac{p_2}{p_1} \tag{6.A.8}$$

and the actual values of p_1 and p_2 should be used.

The Chisholm approach was used for over-reading:

$$\phi = \sqrt{1 + C_{\text{tp}} X + X^2} \tag{6.A.9}$$

where the Chisholm expression for density ratio is:

$$C_{\text{tp}} = \left(\frac{\rho_{\text{l}}}{\rho_{1,\text{g}}}\right)^n + \left(\frac{\rho_{1,\text{g}}}{\rho_{\text{l}}}\right)^n \tag{6.A.10}$$

They proposed a new value for n:

$$n = \max(0.583 - 0.18\beta^2 - 0.578e^{-0.8\text{Fr}_{\text{g}}/H}, 0.392 - 0.18\beta^2) \tag{6.A.11}$$

where $\beta = d/D$ and H depends on the liquid type, and has a value of about unity for hydrocarbon liquids and 1.35 for water.

This correlation is claimed to be suitable for the following Venturi tube parameters

$0.4 \leq \beta \leq 0.75$
$0 < X \leq 0.3$
$3 < Fr_{g,th}$
$0.02 < \rho_{1,g} / \rho_l$
$D \geq 50$ mm

with an uncertainty of:

3% for $X \leq 0.15$
2.5% for $0.15 < X \leq 0.3$

If X is not known, the reader should refer to Reader-Harris and Graham (2009) and/or the ISO standard where a method of obtaining X is set out. The standard should, in any event, be perused for further details of the method and good practice such as avoiding placing the Venturi at a low point in the flow circuit, orientation of the tube etc.

Numerical modelling by He and Bai (2012) of wet-gas flows appeared to support the suggestion that a liquid film formed on the inlet wall of the Venturi accounting for the over-reading observed (see other references in Chapter 17).

Multiphase Flows

The application of the Venturi to two-phase and multiphase flows has been growing over the past 10 years. The reader is referred to papers by Hall (2001) at the UK National Engineering Laboratory, Elperin, Fominykh and Klochko (2002), who modelled the release of gas in the liquid phase and obtained the pressure drop, and Paladino and Maliska (2002, 2011) who introduced slip velocity into a homogeneous widely used model for estimating pressure drop for multiphase flows. They obtained the effect of void fraction and slip velocity, and computed models for Venturi and nozzles with bubbly flows.

The reader is referred to Chapter 17 where further work is reported.

7

Critical Flow Venturi Nozzle

7.1 Introduction

This meter makes use of a fascinating effect which occurs when gas flows at very high velocity through a nozzle. As the gas is sucked through the nozzle the velocity increases as the cross-section of the nozzle passage decreases towards the throat. At the throat the maximum velocity which can be achieved is sonic – the speed of sound. Downstream of the throat, the velocity will either fall again returning to subsonic, or will rise and become supersonic. In normal operation of the nozzle, the supersonic region is likely to be small, and to be followed by a shock wave which stands across the divergent portion of the nozzle, and causes the gas velocity to drop, very suddenly, from supersonic to subsonic. The existence of the shock wave does not mean that one will hear a "sonic boom"! Such booms are usually caused by moving shock waves carried forward by high-speed aircraft.

The fascinating effect of sonic conditions at the throat is that changes in the flow downstream of the throat have no effect on conditions upstream. The sonic or critical or choked condition, as it is called, appears to block any information which is trying to penetrate upstream. A simple picture of this is that the messengers carrying such information travel at the speed of sound, and so they are unable to make any headway over the fast-flowing gas stream.

Two important effects of this are that:

i) the mass flow rate is a function of the gas properties, the upstream stagnation temperature and pressure and the area of the throat, provided that the nozzle is actually running at critical conditions (which can be ascertained by checking the downstream pressure);
ii) the nozzle acts as a flow controller, creating steady conditions upstream even though conditions downstream are unsteady.

Figure 2.6(a) shows a simple illustration of a convergent-divergent nozzle with a throttle valve downstream. Figure 2.6(b) shows the variation of pressure through the nozzle, and the critical flow conditions are those bounded by curves c and i for which the flow becomes sonic at the throat.

For a fuller description of the fluid flow behaviour the reader is referred to Baker (1996). There are two essential equations which apply when the velocity at the throat is equal to the speed of sound (Mach number equals unity) and the conditions are critical. The first equation gives the pressure ratio for choked flow [Equation (2.18)]:

$$\frac{p_*}{p_0} = \left(\frac{\gamma+1}{2}\right)^{-\gamma/(\gamma-1)} \tag{7.1}$$

where p_* is the pressure at the throat and p_0 is the stagnation pressure and γ is the ratio of specific heats c_p/c_v.

The second equation gives the ideal mass flow, q_m, for choked flow from Equation (2.19), making use of the relationships:

$$\boldsymbol{R}/\boldsymbol{M} = c_p - c_v$$

and

$$\sqrt{\frac{\gamma}{c_p(\gamma-1)}} = \frac{1}{\sqrt{\boldsymbol{R}/\boldsymbol{M}}}$$

to obtain

$$q_m = \frac{A_* p_0}{\sqrt{T_0(\boldsymbol{R}/\boldsymbol{M})}}\sqrt{\gamma}\left(\frac{2}{\gamma+1}\right)^{\frac{1}{2}(\gamma+1)/(\gamma-1)} \tag{7.2}$$

where $\boldsymbol{R}$ is the universal gas constant (8.314 5 kJ/kmol K, see NIST[1]), $\boldsymbol{M}$ is the molecular weight, c_p and c_v are the specific heats at constant pressure and constant volume, A_* is the throat cross-sectional area and T_0 is stagnation temperature.

Thus if we can measure upstream stagnation pressure and temperature, p_0 and T_0, and if we know the throat area, A_*, and the gas properties we have a mass flowmeter. The flow must be choked. Also this equation is limited to a perfect gas and assumes an idealised and one-dimensional flow.

We are interested in the use of this device for real gases. For precise work we have to recognise that gases are NOT perfect. We also find that boundary layer growth and curvature of the flow through the nozzle result in a flow which is not truly one-dimensional.

Thus, as with many flowmeters, there is a coefficient of discharge, to be discussed later in this chapter. First we consider a practical flowmeter installation.

7.2 Design Details of a Practical Flowmeter Installation

The critical flow Venturi nozzle is fully specified by the most recent version of the standard document ISO 9300: 2005. The reader is strongly advised to obtain a copy

[1] http://physics.nist.gov/cgi-bin/cuu/Value?r

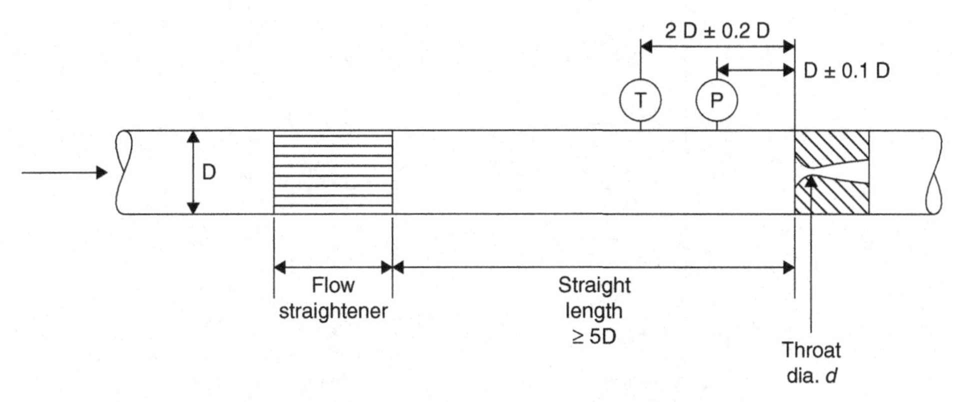

Figure 7.1. Details of Venturi nozzle installation.

of the standard for reference if planning to build a facility using this device, and to view this chapter as complementary to the standard.

Figure 7.1 shows a diagram of a critical flow Venturi nozzle installation. Note that the temperature tapping can be upstream of flow conditioners since the temperature change is likely to be negligible. However, the pressure tapping must be sensing a true inlet stagnation value. There may also be a downstream measure of pressure to ensure that critical conditions are maintained.

If the upstream pressure allows only limited variation, then the range of q_m will be limited, and only by using a bank of nozzles of different areas can we achieve a satisfactory calibration range, Figure 7.2. Caron (1995) mentioned a test stand which used a bank of sonic nozzles in binary sequence. Nozzle 2 has twice the flow rate of nozzle 1 etc. Thus by selecting the required nozzles and isolating the remainder, a range of discrete flows can be set.

Nozzles may have a convergent portion only, in which case there will be a substantial pressure loss at exit and the pressure ratio of outlet to inlet stagnation must be less than 0.528 for air. On the other hand, the convergent-divergent Venturi nozzle in Figure 7.3 may achieve a pressure ratio (exit/inlet) for critical conditions at the throat of $p_2 / p_0 \approx 0.9$. The one shown is the toroidal throat Venturi nozzle. An alternative, not favoured by some, is the cylindrical throat type where a constant area throat of one diameter in length separates inlet contraction and outlet diffuser.

The diagram in the standard provides the details of design finishes and tolerances which differ slightly from, but cover, those in Figure 7.3. It is also important to note other features of the nozzle in Figure 7.3. D must be greater than 4d, and referring to Figure 7.1 with a pipe diameter of 50 mm upstream, this would put a maximum on the throat diameter of 12.5 mm. Also the preferred distance to the upstream pressure tapping is between 0.9D and 1.1D. Other spacings are allowed provided that it can be shown that the tapping gives a correct measure of the nozzle inlet stagnation pressure. The standard allows for a larger space upstream with its own requirements.

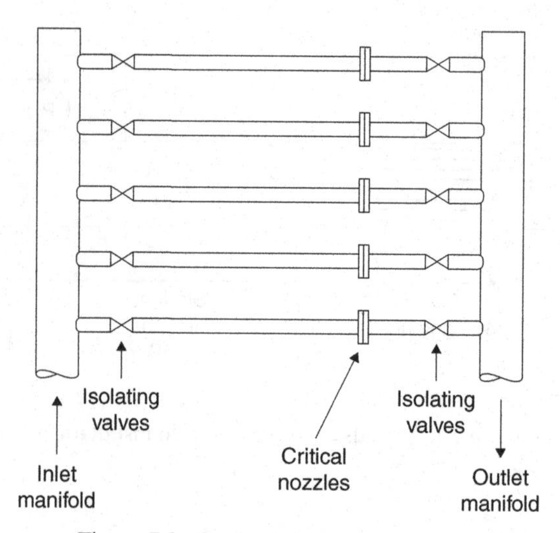

Figure 7.2. Bank of critical flowmeters.

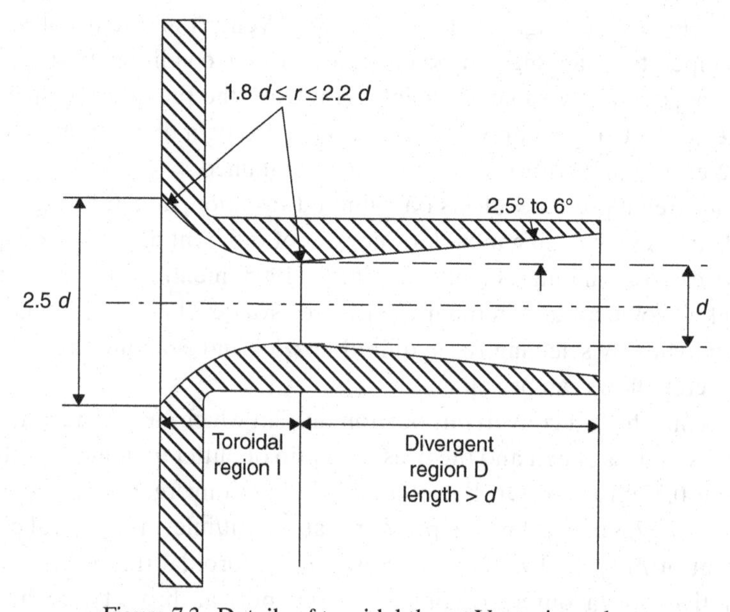

Figure 7.3. Details of toroidal throat Venturi nozzle.

The standard also sets out requirements on temperature measurement which cover the position in Figure 7.1, but allow a greater spacing if it gives, reliably, the stagnation temperature at nozzle inlet. The standard also warns that care needs to be taken if the surroundings of the installation differ in temperature by more than 5°C from the gas.

Where there is the possibility of condensate, drain holes are allowed provided there is no flow through them when flow measurement is taking place, and they are spaced upstream of the pressure tapping by at least D and not in the same axial plane as the pressure tapping, provided the hole diameter is smaller than 0.06D.

The standard requires that the nozzle be made from material capable of finish to the required smoothness, resistant to corrosion in the intended application and dimensionally stable so that temperature corrections can be used for throat area.

7.3 Practical Equations

The practical use of critical flow Venturi nozzles is set out in ISO 9300: 2005. Apart from some nomenclature differences, this section gives the method set out in the standard. Using the expression of Equation (7.2) we introduce a discharge coefficient, C, to allow for the deviation from the ideal equation, and we also introduce a coefficient for the gas, C_*, the critical flow function, obtainable from tables and formulae, and for a perfect gas written as:

$$C_{*\mathrm{i}} = \sqrt{\gamma}\left(\frac{2}{\gamma+1}\right)^{\frac{1}{2}(\gamma+1)/(\gamma-1)} \tag{7.3}$$

With these coefficients we can rewrite the equation as:

$$q_\mathrm{m} = \frac{A_* C C_* p_0}{\sqrt{T_0\left(\boldsymbol{R}/\boldsymbol{M}\right)}} \tag{7.4}$$

Alternatively, the equation may be written in terms of p_0 and ρ_0

$$q_\mathrm{m} = A_* C C_\mathrm{R}\sqrt{p_0\rho_0} \tag{7.5}$$

where

$$C_\mathrm{R} = C_*\sqrt{Z_0} \tag{7.6}$$

and

$$\frac{p_0}{\rho_0} = \frac{Z_0 \boldsymbol{R} T_0}{\boldsymbol{M}} \tag{7.7}$$

In this case we require values of C_* and Z_0, the compressibility factor at stagnation conditions.

The design should, preferably, ensure that the values of p and T are obtained where the Mach number is small enough to assume that they are equal to p_0 and T_0. Otherwise, the standard gives equations to obtain the corrected values.

In order to ensure that the nozzle operates critically, the standard requires that the actual maximum outlet pressure, $p_{2\mathrm{max}}$, when compared with the ideal value, $p_{2\mathrm{i}}$, for which critical conditions are achievable at throat Reynolds numbers greater than 2×10^5 and having an exit cone longer than d, is determined from the relationship

$$p_{2\mathrm{max}} - p_* = 0.8\left(p_{2\mathrm{i}} - p_*\right) \tag{7.8}$$

Using Equations (2.16) and (2.17), it is possible to obtain values of $p_{2\mathrm{i}}$ for various values of outlet area, A_2, assuming that the gas behaves sufficiently closely to

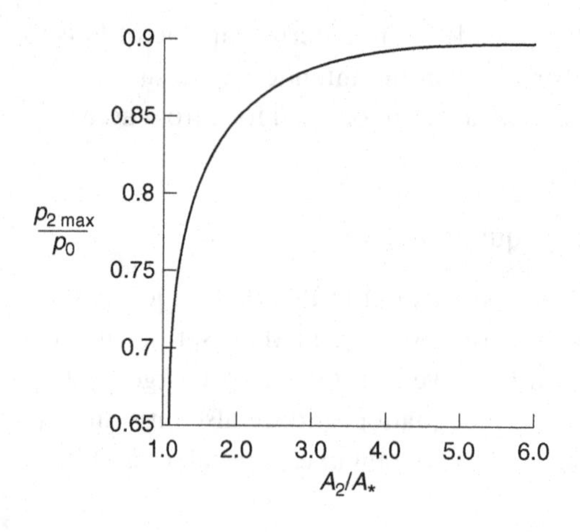

Figure 7.4. Maximum value of back-pressure for $\gamma = 1.4$ to ensure critical operation based on the criterion in ISO 9300:2005.

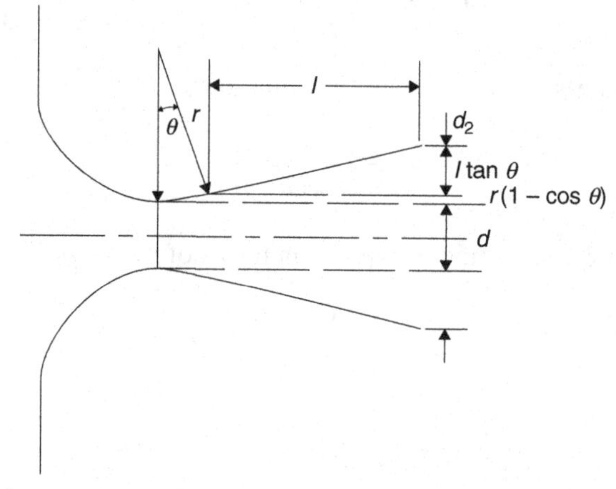

Figure 7.5. Geometrical calculation to obtain the outlet diameter from the diffuser geometry.

an ideal gas to use the ratio of the specific heats γ. Figure 7.4 shows the values of p_{2max}/p_0 for $\gamma = 1.4$. The standard provides a plot for other values of the isentropic exponent.

It should be noted that γ is the ratio of specific heats for an ideal gas, and κ is referred to in the standard as the isentropic exponent and applies to real gases. Miller (1996) gives plots of κ. For recent information on Z and κ the reader is referred to ISO 12213 and ISO 20765, respectively.

It is relatively simple to relate the area ratio between the throat and the outlet based on design angles of the diffuser. If the half-angle of the diffuser is θ, the radius of the toroid is r and the length of the diffuser is l beyond the toroid, the increase in diameter is (Figure 7.5) $2[r(1 - \cos\theta) + l\tan\theta]$ so that the outlet diameter is

$$d_2 = d + 2\left[r\left(1 - \cos\theta\right) + l\tan\theta\right]$$

or

$$\frac{d_2}{d} = 1 + \frac{2r(1-\cos\theta)}{d} + \frac{2l\tan\theta}{d} \tag{7.9}$$

from which the expression in the standard for A_2/A_* can be simply obtained.

7.4 Discharge Coefficient C

(I am indebted to Takamoto and Ishibashi (2014)* for the update of this section (7.4), based on ISO 9300:2005.)

The basic design of the standard geometry for toroidal throat critical flow Venturi nozzles (Figure 7.3) was originally proposed by Smith and Matz (1962) and Stratford (1964) in order to have better accuracy in theoretical calculations. The first version of ISO 9300 published in 1990 presented a single correlating equation of the discharge coefficient covering the Reynolds number from 10^5 to 10^7 with the uncertainty of ±0.5% based on calibration data including works by Arnberg et al. (1973, 1974), Brain and MacDonald (1975) and Brain and Reid (1978, 1980). Its revised version published in 2005 (the current version) presented two correlating equations for the same geometry based on a review by Arnberg and Ishibashi (2001). Figure 7.6 shows the universal curve from their paper, to which the reader is referred for details of the data. All but about 4% of the data referred to in the paper fell within the ±0.3% band and the remaining data points fell below the universal curve in the range from Re = 10^6 to 10^7 and within about −0.5% of the curve.

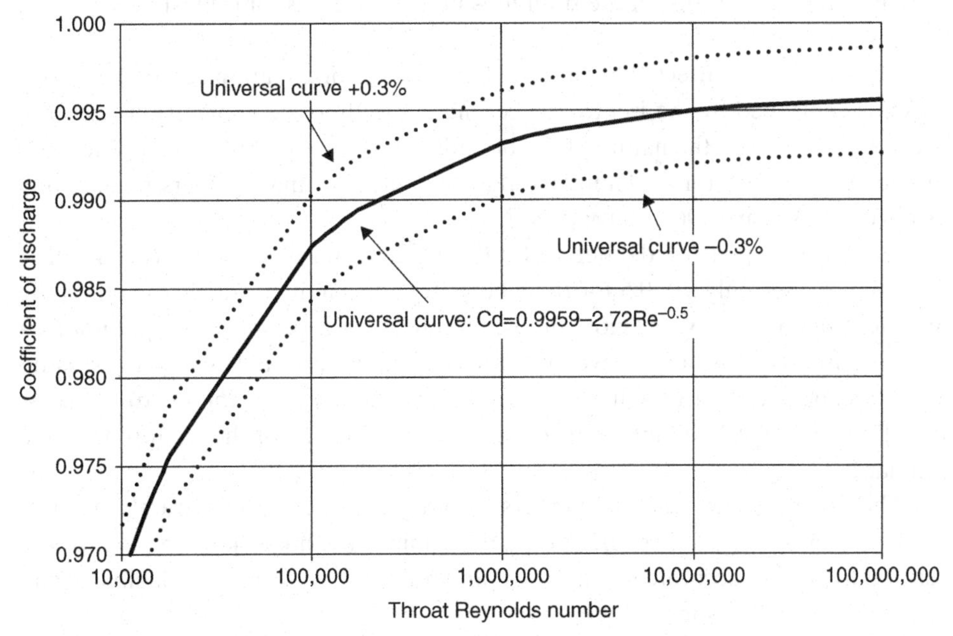

Figure 7.6. Universal curve with ±0.3% error bounds. For the full data points see Arnberg and Ishibashi (2001).

* Personal communication from Dr M. Takamoto and Dr M. Ishibashi, 2014.

Table 7.1. *Values of constants in Equation (7.10) for toroidal throat Venturi nozzles**

Re_d range	a	b	n	Reference
2.1×10^4–3.2×10^7	0.995 9	2.720	0.5	EN ISO 9300:2005
2.1×10^4–1.4×10^6	0.998 5	3.412	0.5	Accurately machined EN ISO 9300:2005

* Other values are available for cylindrical throat Venturi nozzles in ISO 9300:2005.

One of the equations defined in the new version covers the whole Reynolds number range from 2.1×10^4 to 2.3×10^7 and the other covers only the laminar boundary layer region from 2.1×10^4 to 1.4×10^6. Both the equations have the common form of

$$C = a - b\mathrm{Re_d}^{-n} \tag{7.10}$$

where $n = 0.5$. Values of the coefficients are found in Table 7.1. It should be noted that the standard always uses the Reynolds number defined by

$$\mathrm{Re_d} = \frac{c_* d\rho_*}{\mu_0}, \tag{7.11}$$

where the throat gas velocity is equal to the speed of sound

$$c = \sqrt{\gamma(\boldsymbol{R}/\boldsymbol{M})T_*} \tag{7.12}$$

and ρ_* is the density of the gas in the throat. But what is somewhat strange is that μ_0 is the value of the viscosity at stagnation although the other values refer to the critical condition. It is one of the traditions in the standards for critical flow Venturi nozzles.

The value of a is affected by the core flow velocity distribution across the throat, which is controlled by the isentropic exponent κ and nozzle geometry. Those of b and n are affected by the nature of the boundary layer in the throat. Theoretically, n should be 0.5 and 0.2 for the laminar and the turbulent boundary layers, respectively (Stratford 1964; Tang 1969; Geropp 1971).

One of the equations defined in ISO 9300: 2005, termed the Universal Curve by Arnberg, uses only $n = 0.5$ for the whole Reynolds number range including the turbulent boundary layer region and is, therefore, an experimental equation as a compromise, but it is very convenient for the standard users because there is no need to choose one of two different equations depending on the upstream pressure, flow rate, nozzle quality and so on, which is also one of the traditions in the standard.

The other equation defined in the standard is, on the contrary, purely theoretical by using $n = 0.5$ and covering only the laminar boundary layer region. It was verified to be very accurate by analytical and numerical calculations (Ishibashi and Takamoto 1997; Johnson et al. 1998).

The uncertainties of the Universal Curve and the higher accuracy curve in the standard are ±0.3% and ±0.2%, respectively. For the latter one assumes the use of accurately machined nozzles, which have been machined by a super-accurate lathe to have mirror surface finish without being polished, and that results in very small geometric errors.

Boundary layer transition is clearly observed between $Re_d = 1 \times 10^6$ and 2×10^6 in accurately machined nozzles (Ishibashi and Takamoto 2001). Data are also being accumulated that indicate that the normally machined ones, if carefully made, also obey the same curve (Morrow 2004; Mickan, Kramer and Dopheide 2006c, Johnson and Wright 2008). It will therefore be possible to define a new single equation to represent the discharge coefficient more accurately in the whole Reynolds number range by unifying the accurate curve and the Universal Curve (Morrow 2004; Mickan et al. 2007; Ishibashi and Funaki 2013).

For smaller critical flow Venturi nozzles such as d<2 mm, the effect of geometry error becomes significantly large; consequently, it is difficult to investigate the nozzle behaviours based on the discharge coefficient since it is only an individual parameter inherent to each nozzle in this range. However, being calibrated one by one, small critical flow Venturi nozzles down to d<0.1 mm have been widespread in laboratories and industries because they still act as precise flowmeters and precise flow controllers (Nakao, Yokoi and Takamoto 1997; Wendt and Von Lavante 2000; Mickan, Kramer and Dopheide 2006b; Wright et al. 2008).

Too small inlet curvature will induce separation, but Ishibashi and Funaki (2013) confirmed experimentally that the toroidal throat critical flow Venturi nozzles with the inlet curvature down to $0.5d$ can perform as stable flowmeters as the standard nozzles. The boundary layer becomes thinner and more robust in nozzles of smaller inlet curvatures, resulting in delay of the boundary layer transition. There may be a possibility of improving their performance by defining a new standard geometry apart from the traditional one, which was defined half a century ago based on theoretical considerations.

7.5 Critical Flow Function C_*

The calculation of C_* has been changed in ISO 9300: 2005, and the reader is recommended to consult the standard if C_* is to be calculated.

7.6 Design Considerations

Referring to Figures 7.3 and 7.7, some details are given from the standard (cf. ISO 9300 for full details). Wall smoothness, curvature and axial distances from the inlet plane are defined. The inlet plane is that plane perpendicular to the axis which cuts the inlet toroid at a diameter of $2.5d$.

Swirl and other upstream disturbances (cf. Yoo et al. 1993) may need to be eliminated with a straightener. Drain holes to remove condensation may also be necessary to avoid an accumulation of water reducing the upstream plenum.

These points are summarised below:

Region I

- Average roughness ≤ $15\times10^{-6}d$ and ≤ 0.04 μm for normally machined and accurately machined nozzles, respectively
- Free from deposits of any sort
- Radius of toroid (in plane through the axis) $1.8d \le r \le 2.2d$
- Contour toroidal to $\le 0.001d$

Region D

- Diffuser half angle between 2.5° and 6°
- Length of diffuser from toroid at least d
- No discontinuities exceeding 1% of local diameter
- Average roughness $\le 10^{-4}d$

Inlet plane

- Defined by $2.5d \pm 0.1d$ intercept

Swirl free

- Inlet flow to be swirl free, if necessary, by use of a straightener at $>5D$

Distance to the wall

- For a circular inlet pipe, $\beta = d/D \le 0.25$
- For a large space (chamber), no wall closer than $5d$ to the axis of the nozzle or to its inlet plane. Multiple nozzles can be used in a large space.

Drain holes

- $<0.06D$ at greater than one diameter upstream of p_0 tapping, and away from plane of tapping.

The design of the pressure tappings is shown in Figure 7.7, where d_t is the diameter of the tapping. The inlet tapping should be situated $0.9D$ to $1.1D$ upstream of the inlet plane, or in a large inlet chamber it should be within $10d \pm 1d$ from the inlet plane.

The downstream pressure tapping (if required) should be less than 0.5 times the conduit diameter from the nozzle exit plane. In some cases, downstream pressure will be known, for instance, when the nozzle exhausts to the atmosphere.

The inlet temperature tapping should be positioned to give a reliable measure of stagnation temperature.

The accuracy of the method will depend heavily on the accuracy of measurement of p_0. Takamoto et al. (1993a) showed that cross flows can cause errors due to the lack of a measure of stagnation. If we cannot assume that the flow is negligible in the upstream pipe, then p_0 cannot be directly measured at the wall tapping, and we shall have to deduce it from p_1 using

$$\frac{p_0}{p_1} = \left(1 + \frac{\gamma - 1}{2} M_1^2\right)^{\gamma/(\gamma-1)} \approx 1 \tag{7.13}$$

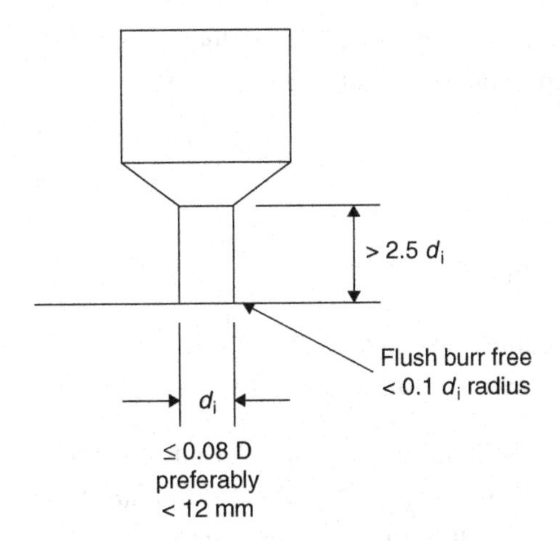

Figure 7.7. Detail of pressure tap.

where M_1 will be small, but the additional step is likely to reduce the precision of the final measurement.

A similar correction may be established also for T_0, but considering that any temperature sensor fixed on the pipe wall will not measure the static temperature of the flow but some value close to the stagnation temperature, the correction will be overly generous. As far as the standard's recommendation of $\beta<0.25$ is kept, the static temperature drop is smaller than 0.02%, so there is practically no need to apply the correction on T_0. On the contrary, as for p_0, the maximum drop reaches 0.06%, so the correction may not be necessary depending on the required accuracy. It is, therefore, preferable to ensure that stagnation conditions do exist in the inlet plenum.

Pereira, de Pimenta and Taira (1993) suggested making the nozzle throat and diffuser in two parts with an O-ring between and discussed the effect of a step between throat and diffuser which they found was not critical within certain limits.

7.7 Measurement Uncertainty

The equation for the error summation is:

$$\varepsilon(q_m) = \pm[\varepsilon(A_*)^2 + \varepsilon(C)^2 + \varepsilon(C_*)^2 + \varepsilon(p_0)^2 + \tfrac{1}{4}\varepsilon(M)^2 + \tfrac{1}{4}\varepsilon(T_0)^2]^{1/2}) \quad (7.14)$$

If we apply some of the possible uncertainties mentioned earlier, we have:

$\varepsilon(C) = \pm 0.3\%$ for a normally machined nozzle ($k = 2$)
$\varepsilon(C) = \pm 0.2\%$ for an accurately machined nozzle ($k = 2$)
$\varepsilon(C_*) = \pm 0.2\%$ based on the value declared in the standard with a certain safety margin ($k = 2$)
$\varepsilon(A_*) = \pm 0.4\%$ for a normally machined nozzle of 5 mm diameter throat by measuring at uncertainty of 0.010 mm ($k = 2$)

$\varepsilon(A_*) = \pm 0.1\%$ for an accurately machined nozzle of 5 mm diameter throat by measuring at uncertainty of 0.002 mm ($k = 2$)
$\varepsilon(M)$ guess ± 0.5%
$\varepsilon(p_0)$ guess ± 0.1%
$\varepsilon(T_0)$ guess ± 0.2%

With these values we obtain an overall uncertainty of

$\varepsilon(q_m) = \pm 0.6\%$ for a normally machined nozzle ($k = 2$)
$\varepsilon(q_m) = \pm 0.4\%$ for an accurately machined nozzle ($k = 2$)

Caron (1995) gave some details of the critical nozzle test stand that the Ford Company designed and used to calibrate air flow sensors. The ASME Standard toroidal design was selected, and it was reckoned that after calibration an uncertainty of ±0.25% in the coefficient could be achieved. The paper gives the following values for uncertainties (the area is calibrated out):

Source	Uncertainty	Sensitivity coefficient	Product
Pressure	±0.05%	1	±0.05%
Temperature	±0.37%	0.5	±0.185%
Discharge coefficient	±0.25%	1	±0.25%
Critical flow function	±0.0125%	1	±0.0125%
Combined uncertainty			**±0.315%**

Since this omits, due to calibration, the measurement of the throat area, and uses higher precision values, it is not out of line with the value obtained earlier.

7.8 Notes on the Calculation Procedure

Making use of the data given in this chapter and that in ISO 9300: 2005, we identify the steps to calculate mass flow of gas, q_m, given the throat diameter, d, inlet stagnation pressure and temperature, p_0 and T_0, and the calculation of the outlet conditions necessary to operate sonically.

To obtain an initial value of C we require Re_d from Equation (7.11) using Table 7.1 and the speed of sound from Equation (7.12) using Equation (2.15) for T_0, noting the use of μ_0. A value of γ is required for the gas e.g. $\gamma = 1.4$ for air.

The value of q_m from Equation (7.4) requires C_* from ISO 9300: 2005. The precise velocity through the nozzle throat resulting from iteration on C and Re_d, making use of q_m and the throat area, will provide the correct values.

Using Equation (2.16) with p_2 will provide the value of M_2, the outlet Mach number, and Equation (2.17) will allow the ideal areas to be calculated, and Equation (7.8) will then allow p_{2max} to be obtained. This should ensure that the outlet diameter and the outlet pressure allow the throat to be running at sonic velocity.

7.9 Industrial and Other Experience

The use of the nozzle in saturated steam with a dryness fraction down to 84% requires a wet steam correction factor (Amini and Owen 1995, cf. ASME/ANSI MFC-7M 1987). However, Amini and Owen suggest that a more practical approach might be to precede the nozzle with a steam/water separator, assume the steam to be then dry saturated, and accept an uncertainty in the mass flow of 3%. An uncertainty of 1% arises from the lack of perfect separation.

The Gas & Fuel Corporation of Victoria, Australia, had a gas meter test facility with 11 critical nozzles (Wright 1993) and a range of 6 to 5600 m^3h^{-1} for meters from 80mm to 450 mm and another with six nozzles for 2.4 to 30 m^3h^{-1}. The coefficient had ±0.5% uncertainty. The flow rate sequence was such that the flow rate of each nozzle was close to the sum of flows through the next three smaller nozzles. This was achieved by using a flow rate ratio for each nozzle compared to the one below of 1.839:1. The paper claims that the use of two nozzles to match the higher flow rate one requires more nozzles to cover the range. The paper also mentions a Wheatstone Bridge technique to calibrate nozzles.

Nakao, Yokoi and Takamoto (1996) described a calibration system for sonic nozzles which used the weight of a collecting vessel to obtain the flow rate. Four lines were available for the Venturi nozzles, and the gas was drawn through with a vacuum pump until the required flow conditions were established. The flow line was then changed to the vessel which had been evacuated and weighed, and the flow continued until a sufficient mass had been collected, when the diversion time ended. Care was needed to allow for dead volumes, and the changeover mechanisms. The authors claimed an overall uncertainty for the rig of about ±0.1%.

Bosio et al. (1990) observed a lack of repeatability during gravimetric primary calibration of nozzles in natural gas containing approximately 3 mg/m^3 of hydrogen sulphide. They identified the problem as deposition of sulphur in the throat. Chesnoy (1993) also discusses the problem of sulphur deposition in the nozzle throat, causing change in nozzle performance. To reduce the possibility of sulphur formation either the content of H_2S needs to be reduced, or the oxidant eliminated. If this is not possible, work supported by Chevron suggested that the position of the deposition moved downstream due to temperature increase (Mottram and Ting 1992). Chesnoy claimed that K-Lab's primary calibration was within ±0.3% between 20 and 100 bar. A zinc oxide bed had been used to treat the inlet gas successfully up to 100 bar(absolute), but other actions were under consideration above this pressure.

Bignell (1996a) described comparison techniques for the performance of a number of small sonic nozzles and Kim and O'Neal (1995) used critical flow models to estimate two-phase flow of HCFC22 and HFC134a through short tube orifices.

Nozzles with throat diameters of 10–1,000 μm have been reported (Takamoto 1996). Recent calculations of discharge coefficient using finite elements (Wu and Yan 1996) gave very impressive agreement with ISO. See Park (1995) on inlet shape.

7.10 Advantages, Disadvantages and Applications

The advantages of this device are that it allows mass flow measurement of gas, and it ensures that, provided the flow is choked and, therefore, operating in its correct regime, flows upstream are not affected by changes in downstream conditions.

It is, however, very inflexible, in that flow rate can only be increased by changing the upstream stagnation values or by increasing the throat area. This leads to the requirement for a bank of carefully sized nozzles to span the working range.

Its primary application is in gas flow measurement of high precision, particularly in flow calibration.

7.11 Chapter Conclusions

This meter appears to have the potential for very high precision. It has been suggested that the ISO uncertainty figure is too great (cf. Gregor et al. 1993).

It is a meter where upstream conditions should be capable of control and precise replication so that the profile and turbulence distribution approaching the meter is always the same.

There are clearly sources of reduced accuracy such as deposits and possible humidity or small quantities of contaminants in the flow.

It also should be capable of CFD modelling with increasing precision. The precise positioning, design and consequent effect of pressure and temperature tappings may be amenable to such modelling. Such an approach may suggest, where possible, a most repeatable and most stable overall layout for the upstream pipe, conditioners etc.

Other directions in which experimental and theoretical studies could go, suggested by Takamoto and Ishibashi,* are in the miniaturisation of nozzles and in overcoming the instability by the jump of the discharge coefficient during the boundary layer transition. The so-called premature unchoking phenomenon is another issue to be solved in which a critical flow Venturi nozzle sometimes does not choke especially at low Reynolds number although it is operating under a high enough pressure difference to choke in theory.

Appendix 7.A Critical Flow Venturi Nozzle – Recent Published Work

Reviews

The critical flow Venturi nozzle has the potential for very high precision. Paik, Park and Park (1998) gave data on discharge coefficients which suggest that the

* Personal communication from Dr M. Takamoto and Dr M. Ishibashi, 2014.

ISO uncertainty is too cautious. It should also be capable of CFD modelling with increasing precision. The precise positioning, design and consequent effect of pressure and temperature tappings may be amenable to such modelling. Such an approach may suggest, where possible, a most repeatable and most stable overall layout for the upstream pipe, conditioners etc. It may also indicate the effect of deposits.

A special issue of *The Journal of Flow Measurement and Instrumentation* (Volume 11, Number 4, December 2000) was devoted to sonic nozzles. Papers deal with numerical and experimental investigations, correlations for gases, uses as standards and provers and various effects which influence the discharge coefficient. The editors of the issue, Bignell and Takamoto (2000), brought together contributions from around the world, demonstrating the continued high level of interest in the critical (sonic) nozzle. They suggested that the reasons for this are:

a) the increased use of natural gas, and the need to monitor its flow, but also to ensure that there are adequate metrological standards;
b) advances in CFD allowing better modelling of the flows;
c) improved absolute gas flow measurement standards;
d) increasing importance of uncertainties in nozzle dimensions;
e) behaviour of very small nozzles including oscillations at low Reynolds numbers;
f) discharge coefficients greater than unity for some gases.

The reader is referred to the issue, which covers many other aspects including a prover system, a transfer standard, use as a secondary standard and inter-laboratory comparisons.

Design and Manufacture

Delajoud, Girard and Blair (2005) described design details and calibration for their toroidal throat Venturi nozzles to maximise precision in gas flow transfer standards.

Hu and Lin (2009) investigated KOH-etched silicon sonic nozzles with throats of about 90 μm and in the range $6 \times 10^2 < Re < 8 \times 10^3$. They appear to have achieved good performance and stability.

Mickan et al. (2006b) appeared to claim that micro-nozzles at both sonic and subsonic conditions could be handled using the same non-dimensional relationships of flow rate to differential pressure. They viewed these devices as small and of low cost. See Li and Mickan (2012a) on the flow characteristic of small MEMS nozzle.

Operational Aspects

Chien and Schrodt (1995) reported the use of empirically derived constants to obtain the steam quality and flow rate using a critical flowmeter. To ensure nozzles remain choked Britton and Caron (1997) suggested:

- diffuser angle $\leq 4°$;
- area ratio (diffuser exit)/(throat) > 4;
- choking pressure ratio should be determined experimentally at operating condition to check for and ensure choking exists.

Choi, Park and Park (1997) tested a sonic nozzle bank with seven toroidal critical Venturi nozzles to evaluate the influence of adjacent nozzles and the tube wall (see also Choi et al. 1999).

For low Reynolds numbers Park et al. (2001) developed a relationship between critical pressure ratio for small nozzles and Reynolds number (based on throat diameter, Re_d) given by:

$$\text{Critical pressure ratio} = 0.9801 - 39.046 \times Re_d^{-0.5}$$

They recommended this for Reynolds numbers below 10^5, but advised the use of a safety factor also.

Calibration and Standard Predictions

Stewart, Watson and Vaidya (2000) provided a new correlation for natural gas mixtures based on the equation of state AGA8-92.

Ishibashi and Takamoto (2000a) demonstrated several methods to calibrate critical nozzles against reference nozzles. By using several reference nozzles of various throat diameters, the Reynolds number dependence was obtained for low Reynolds numbers, and they reported (Ishibashi and Takamoto 2000b) on theoretical and experimental values which agreed to better than 0.1%. Ishibashi (2002) discussed the possibility of a standard using a super-accurate lathe with small machining error to machine the nozzle (see discussion in Section 7.4).

von Lavante et al. (2000, 2001) obtained reasonable agreement between theoretical and experimental results using the nozzle design based on the ISO standard design. The unsteady phenomena they identified appeared to result from pressure waves propagating upstream and causing unsteadiness in the throat. Since some effects appeared to be non-axisymmetric, a 3-D numerical solution might be required.

Lim et al. (2010) discussed a step-down procedure of sonic nozzle calibration at low Reynolds numbers. This included using a piece of test equipment which allowed two or more lower-range nozzles to be installed downstream of a higher-range calibrated nozzle to "step down" the calibration [cf. Lim et al. (2009) on the effect of humidity in a critical flow Venturi nozzle and a correction equation (2011) for humidity effects. See also Li and Mickan (2012b) on humidity effect].

CFD and Numerical Modelling

von Lavante and Mickan (2005) used a Navier-Stokes solver to examine the laminar/turbulent transition in critical Venturi nozzles. For a Re = 8.75×10^5 the transition was found to be well downstream of the throat, but for a Re = 4.15×10^6 the transition was predicted to be in the intake part of the nozzle influencing the flow rate (see also von Lavante, Mickan and Kramer 2004a). Mickan et al. (2004) proposed a useful approach to obtaining the discharge coefficient without knowledge of the throat diameter, by developing a relationship between parameter *b* and parameter *a* and a new parameter. Mickan, Kramer and Dopheide (2006a) appear to have been successful in using displacement thickness theory to allow for the thickness of the boundary layer on the nozzle. Johnson and Wright (2006) also allowed for boundary layers in their theoretical critical Venturi calculations. The best agreement with measured values was found to be for the laminar regime.

Cruz-Maya, Sánchez-Silva and Quinto-Diez (2006) developed a theoretical discharge coefficient for a critical Venturi nozzle with turbulent boundary layer, which was about 0.2% higher than the ISO9300 value. Li, Peng and Wang (2010) appear to have predicted that the influence on sonic nozzle discharge coefficient, of diffuser angle variation in the range 2.5° to 6°, was negligible for throat diameters greater than 1 mm, but may have an effect when the diameter is less than 1 mm.

Kim, Kim and Park (2006) undertook computational modelling and experiments on a nozzle, the throat cross-section of which could be varied by inserting a rod axially up the centre of the nozzle along the axis of the flow. Their results appear to have been promising.

Hu et al. (2010a, 2012) fabricated small sonic nozzles with throat diameters of about 100 μm and with Reynolds numbers ranging from 580 to 4,500. Simulation results revealed that, besides flow separation, the first set of oblique shocks which appeared in the nozzle jet could lead to great pressure loss; the weaker the oblique shocks, the higher the critical back pressure ratio. Further work by this group is reported by Hu, Lin and Su (2011). Kramer, Mickan and Schmidt (2010) described an arrangement of two micro nozzles in series.

Experimental Research

Ishibashi and Morioka (2010) measured flow fields in the nozzles with observed interaction of oblique shocks and a strong shock moving along the nozzle axis possibly affected by premature unchoking, and they suggested a design improvement in the supersonic region (Ishibashi and Morioka 2012). Mickan, Kramer and Li (2012) developed a correlation for critical back pressure ratio allowing for various parameters including diffuser design and gas characteristics. Kramer and Mickan (2012) discussed requirements for small sonic nozzle used to test meters in natural gas, in particular where meters with thermal mass flow elements (apparently an on-chip flow sensor based on thermal mass flow measurement principles) were used, so that calibration gas would be as similar as possible to the application.

Applications

Vulovic, Vallet and Windenberger (2002) discussed advantages of the meter for the high-pressure gas based on experience of two French laboratories, noting their stability and repeatability.

Nakao (2005) discussed its application for a high-pressure hydrogen gas dispenser at a hydrogen gas station. Dahlström (2003) appeared to explain how the outlet geometry of a flare stack could be used as a sonic nozzle by adjusting an internal restrictor to provide a convergent nozzle. Temperature and pressure readings were estimated based on losses etc., and, provided the nozzle ran at critical conditions, it should have been possible to obtain the mass flow rate within the limitations of precise knowledge of area etc.

8

Other Momentum-Sensing Meters

8.1 Introduction

The previous chapters have been mainly concerned with devices, the designs of which have been defined by various standard specifications. In this chapter, we consider other mainly proprietary meters, which depend on the changing flow pattern and which essentially sense momentum of the flow. The output is in some cases given by a differential pressure measurement and in others by a position or force measurement. This means that the user must be aware of the possible effect of density on the results. Viscosity will also affect some readings.

The somewhat arbitrary order of consideration and the uneven amount of detail in this chapter reflect the available literature and my experience. Thus the variable area meter takes up a large proportion of this chapter.

The meters considered in this chapter are:

- variable area (VA) meters, which depend on gravity to oppose the movement of the float and consist of two main types:
  - those with a tapered tube and a float with a fixed metering edge and
  - those with an orifice with a fixed metering edge and a moving tapered plug;
- spring-loaded profiled plug in an orifice;
- target (drag plate) meter sometimes spring loaded;
- Venturi-type meters claiming a low loss, such as Dall, Epiflo, Gentile and Low-Loss;
- wedge meter;
- V-cone meters;
- meters using a bypass with an oscillating vane, a Pelton or a VA meter in it;
- slotted orifice meter;
- pipework features used as meters such as inlets and bends;
- averaging pitots under various names (Annubar, Torbar etc.); and
- laminar (viscous) flowmeters.

8.2 Variable Area Meter

The variable area (VA) meter is sometimes known as a Rotameter, but this is a trade name of a particular company and I shall, therefore, refer to it as the variable area meter. In the following sections on the VA meter, I have avoided history, theory or other items, which have been consigned to Appendix 8.A. One way to view the VA meter is as a variable orifice meter with a fixed pressure drop resulting from the weight of the float.

8.2.1 Operating Principle and Background

The VA meter consists of a float that rises up a conical tube due to the upward flow (Figure 8.1). *Float* is a misleading term because it rises not from any buoyant effect but as a result of the upward drag of the fluid, which, in turn, results in a pressure difference across the float. As the flow increases, the float rises higher in the tube to a point where the annulus formed between the float and the conical tube wall is sufficient to allow the drag on the float to balance the float's immersed weight. The flow is then deduced from the height to which the float has risen. (In Germany, the term *swimmer* is used rather than *float*, which may be more appropriate, but I shall stick with *float* in this book.)

8.2.2 Design Variations

Coleman (1956) provided useful information on the development of the float and tube. However, we will concentrate mainly on the current practice. Figure 8.2 shows some of the shapes of floats and of a tube.

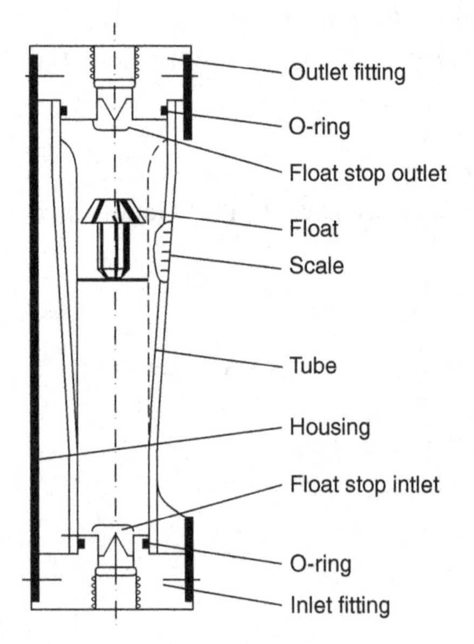

Figure 8.1. Variable area flowmeter (reproduced with permission from ABB).

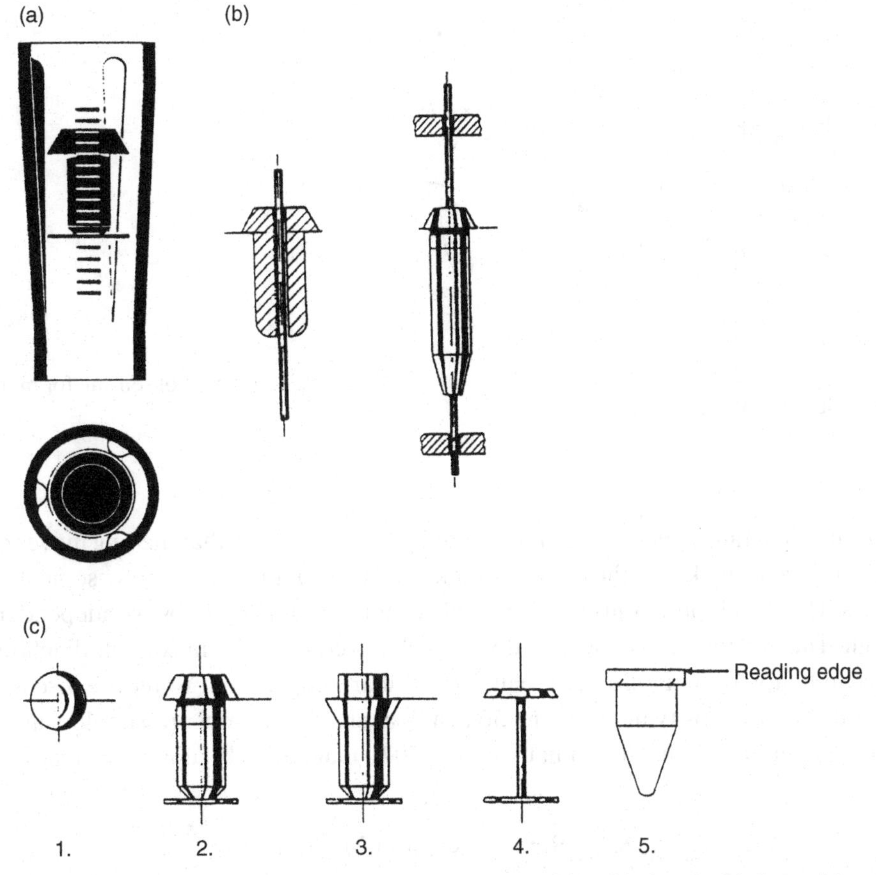

Figure 8.2. Tubes and floats: **(a)** Cross-section of a tube with guide beads; **(b)** float with guide rod; **(c)** typical float shapes (1, ball float; 2, viscosity immune float; 3, viscosity nonimmune float; 4, float for low pressure losses; 5, rotating float) (reproduced with permission from ABB).

Tubes

Tubes are often made of borosilicate glass. In addition to the plain taper tubes, there are tubes with three longitudinal beads moulded into the glass (beaded or fluted glass) to guide the tail of the float (rib-guided). For very low flow rates a conical tube with a plumb-bob float or a tri-flat tube with a spherical float is used. For plain, tapered tubes, a central guide is often used (rod-guided). If the fluid or the operating conditions are unsuitable for glass, a metal tube is used with a fixed orifice and tapered plug or a tapered tube and disc-head float. Figure 8.2(a) shows the cross-section of a tube with guide beads. In manufacture, the induction-heated glass tube is vacuum drawn onto a metal mandrel so designed that several tubes are made at once in-line.

Floats

There are differing views as to whether centrally guided floats with possible friction effects offer more repeatable results than unguided floats [Figure 8.2(b)]. An

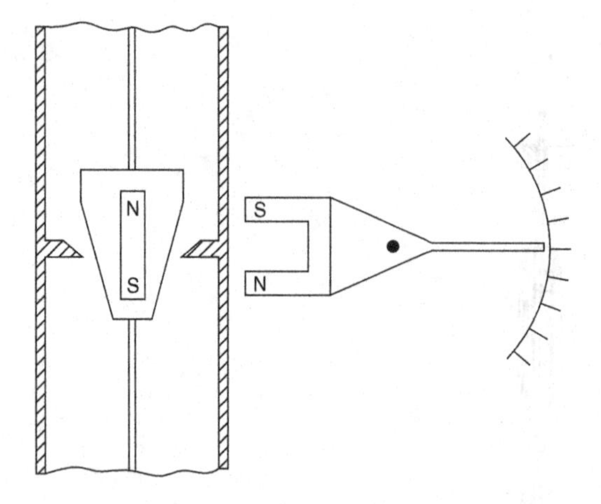

Figure 8.3. Diagram of sensing for metal tubes.

alternative is a tail-guided float, where a ring is held by webs so that the ring moves on the tube beads and keeps the float central in the tube. Figure 8.2(c) shows some float shapes. The development history included attempts to find a viscosity-independent design. This is most nearly achieved with a sharp edge at the maximum diameter. The five floats in the figure are a ball float, a sharp-edged float to reduce viscosity dependence, a viscosity nonimmune float, a float shaped for low pressure loss and a rotating unguided float. Cf. Neuhaus et al. (2007): magnetically supported float.

8.2.3 Remote Readout Methods

The most common remote readout method used is magnetic sensing with a metal tube. A diagram of a typical mechanism is shown in Figure 8.3. However, designs using newer field-sensing methods are being introduced.

ISA RP16.4 (1960)[1] dealt with extension devices that translate the float motion into a useful secondary function for indicating, alarming, transmitting etc. An extension tube may be used, and magnetic or electrical sensing coils appear to have been most commonly used.

Liu et al. (1995) described a sensing method for variable area flowmeters that used the capacitance between two conducting strips on the inner surface of the tapered tube and the float of good electrical conducting material. The capacitance between the strips will vary as the float rises in the tube. They claimed that analysis and experiment have shown that this sensor system can be used for non-conducting fluids. However, tests reported were for a prototype in a laboratory rig and wider testing would appear necessary.

[1] The standard ISA RP16 described the basics but has been withdrawn after several decades possibly because its content would be accepted as good engineering practice today.

Table 8.1. *Sample of dimensions for glass tube–type VA meters (after ISA)*

Pipe Size	A	B
$\frac{1}{2}''$ (12.7 mm)	$16\frac{1}{2}''$ (420 mm)	$3\frac{1}{2}'' - 4''$ (89–102 mm)
1″ (25.4 mm)	$17\frac{1}{2}''$ (445 mm)	$4'' - 4\frac{1}{2}''$ (102–114 mm)
2″ (51 mm)	21″ (533 mm)	$5'' - 5\frac{1}{2}''$ (127–140 mm)
4″ (102 mm)	28″ (711 mm)	6″–7″ (152–178 mm)

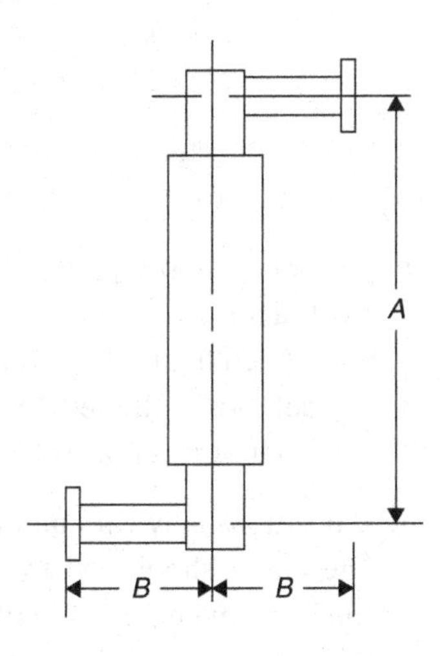

Figure 8.4. Connection dimensions for glass tube–type variable area meters (after ISA RP16.1.2.3 1959).

Parker (1990) described an optical transducer. In this, an array of nine infrared LEDs was mounted on one side of the flow tube, and a corresponding array of phototransistors was mounted on the other but displaced by half the LED pitch. The reading on each detector, or pair of detectors, is taken for each LED, and the responses are normalised. When a float is present, the transmission will vary, and a position-finding algorithm is claimed to give a resolution of 1 part in 2,048.

8.2.4 Design Features

ISA RP16.1.2.3 (1959) gave the recommended main dimensions of the connections for the meters, and some of these are shown in Figure 8.4 for the glass tube type, and a sample of the dimensions is given in Table 8.1.

ISA RP16.1.2.3 (1959) gave a uniform terminology, a small part of which has been reproduced in Figure 8.5. It also gives some guidance on precision and on safe working pressures of borosilicate glass tubes.

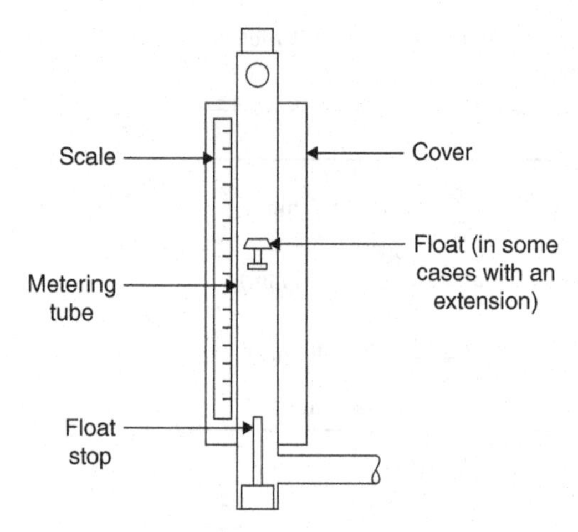

Figure 8.5. Diagram with a selection of the terminology used (after ISA RP16.1.2.3 1959).

The manufacturer's instructions on where to read the float level should be obtained (ISA RP16.5 1961a). If no other information is available, the following may be the case;

- For plumb-bob-type floats, their position is usually taken as the largest uppermost diameter.
- For guided floats, the level may be given by a shoulder on the float.
- For ball floats, the level is taken as the mid-position (equator) of the ball (but note that others may specify the top of the ball).

These positions may vary among manufacturers.

The meter should, where possible, be installed with a bypass arrangement so that the line can be flushed without affecting the meter and also so that the glass can be replaced without closing down the line. Sudden pressure caused, for instance, by water hammer, liquid close to boiling point or entrained gas can cause damage and should be prevented.

8.2.5 Calibration and Sources of Error

If the instrument is to be used as a repeatability measure (ISA RP16.6 1961b), it can usually be calibrated on a liquid such as water, and corrections can be made for density. Corrections for viscosity may also be possible. However, in some cases, the use of a different liquid for calibration may not be acceptable for viscosity reasons. If the absolute accuracy of the flowmeter is required, calibration should be with the actual fluid. If in doubt, refer the question to the manufacturer.

VDI/VDE 3513 (1978) suggests various reasons for error:

- positioning of the scale, or removing, replacing or changing of the scale for different ranges;
- backlash of the float indicator;
- nonlinearity of the scale, which may accentuate error if incorrectly positioned;
- friction, which affects the free movement of the float; or
- unsteadiness of the float.

This does not allow for changes caused by:

- density and viscosity beyond the scale allowance;
- temperature effects on the conical tube and the float;
- non-vertical installation, deposits on the equipment, corrosion to the tube or float, vibration or flow pulsation.

8.2.6 Installation

The meter must be set vertically, with the inlet at the bottom and the outlet at the top. Head (1946–1947) claimed that there was no need for long pipes at inlet or for flow straighteners. Coleman (1956) suggested that, if the tube inlet is less than the annulus area, there was increased sensitivity to inlet flow pattern. He suggested, therefore, that the maximum tube diameter should not be more than 1.41 times the float diameter ($1.414^2 = 2$ so that this figure ensures that the annulus area is less than the float cross-sectional area). Because of the entry arrangements and the low accuracy at the bottom of the scale, one might expect that installation would have a negligible effect. This appeared to be Head's (1946–1947) view. However, some manufacturers recommend at least 5D upstream installation length and may suggest a downstream straight length, which may, in some cases, be greater than the upstream length. A safe rule may be, in the absence of guidance, to take at least 5D of straight pipe both up- and downstream. Installation is more critical with short-stroke VA meters where turbulence leads to float instability and reading problems, but modern detector systems can partially overcome this by averaging. Installation into a large-diameter gas pipeline can result, at low flows, in vertical oscillation of the float (bounce), often with increasing amplitude and limited only by the float stops.

8.2.7 Unsteady and Pulsating Flows

Dijstelbergen (1964) reported extensive mathematics and experimental work on the behaviour of the VA meters in a pulsating flow and claimed that "the response of the instrument and the value of the error in the mean float position with pulsating flow can be predicted." Chatter and bounce can be reduced with guided floats or float rod extensions with a pneumatic damper. Nevertheless, Dijstelbergen's experience of applying these meters in pulsating flow would suggest the need for caution.

8.2.8 Industrial Types, Ranges and Performance

Upward flow is, of course, specified, and the meter should be mounted to minimise strain on the glass tube.

Typical uncertainty claimed for these instruments is 1–3% of full-scale deflection as supplied, and probably 0.5–1% repeatability.

Ranges for liquid may be from 0.001 l/min to 100 m³/h, and ranges for gas, from 0.1 l/min to 1,800 m³/h. A meter, typically, has a 10:1 turndown.

Meter tubes are made of glass, acrylic and special transparent materials where visual reading is used. There is a move to use metal to avoid breakage etc. for safety reasons.

Transparent tubes are now used mainly for inert liquid and gas applications. Where a metal tube is used, sensing is usually through a magnetic link. For magnetic sensing, the float may be in a tapered tube, or the float may be tapered and, with a fixed orifice, form a variable annulus. The float will probably contain a magnet, and the sensing head will form as closely coupled a magnetic circuit as possible (Figure 8.3). These designs are giving way to newer field-sensing methods. The remainder of the meter may be of aluminium, nylon, stainless steel, PVC etc.

Temperature can range from –180 to 400°C for metal tubes and rather less for others.

The manufacturer may give advice on flushing the line before the meter is installed and may recommend filters (magnetic if particles are ferromagnetic) for use with devices using metal tubes and magnetic position sensing.

8.2.9 Manufacturing Variation

Baker and Sorbie (2001) reviewed the manufacturing process at one manufacturer, where a PC-based computer system provided a control for each order passing through the factory. The meter was given a serial number. The calibration and subsequent scale production carried the serial number for the particular meter. As the meter proceeded around the assembly line, the code with the PC-based information system provided the details needed for constructing the meter. The paper then discussed various aspects of manufacturing variation and precision for: floats, tubes, scales, calibration. Detailed float shape may vary, particularly affecting the relationship between drag and viscosity. Thus change in liquid or in temperature may have effects which are difficult to predict and may lead to a larger uncertainty than normal in the performance of the meter. Possibly the most obvious variation may be in the production of the scale and its positioning on the tube. However, the authors suggested that the process, including firing, should be much less than ±1 mm, or less than 1% on a 100 mm length of tube.

Meters of this type which result from typical manufacturing methods have an uncertainty, likely to be in the range 1.6% to 5% of full scale deflection.

The paper was particularly concerned with higher-quality meters and suggested for these that for an uncalibrated meter the uncertainty range, depending on size, was from 0.8% to 3.1% combined with the calibration uncertainty of the prototypes on which the meter was based. For a calibrated meter, the uncertainty will, essentially, be that which the calibration facility can deliver. For a meter calibrated prior to positioning the scale, the calibration uncertainty would be combined with the scale positioning variation of 0.2% to 3.0%. In collaboration with a manufacturer of variable area (VA) meters flow tests of predicted meters taken from the company's production line were run (Baker 2004a) to measure the effect of manufacturing variation. The tests made use of a series of glass tubes and a series of floats. In addition small weights were used in one of the floats (inserted in a cut-out in the top of the float) to test the effect of variation in the weight of the float, while, at the same time, keeping other parameters fixed. Equation (8.A.6) was used to obtain an expression for the effect of parameter variation [symbol as for Equation (8.A.6)]:

$$\frac{\delta q_v}{q_v} = \frac{\delta A_x}{A_x} + \frac{1}{2}\frac{\delta V_f}{V_f} - \frac{1}{2}\frac{\delta A_f}{A_f} \tag{8.1}$$

Thus for the float at a fixed height the second term on the right would increase because of the added mass, and the left-hand side term would therefore increase, giving an apparent under reading as predicted from the theory. The tests confirmed that, for this source of meters, the variation was small. Results suggested that flowmeter variation was well within the predicted uncertainty of 2.5% FSD. The importance of correct positioning of the scale was confirmed by these tests.

8.2.10 Computational Analysis of the Variable Area Flowmeter

Using a commercial meter, Buckle et al. (1992, cf. Leder 1996) have studied the flow around the flowmeter by LDA methods and CFD calculations (Figure 8.6). The work gave quite good agreement between the computation and the experimental results apart from the region of the recirculation zone above the float, which, nevertheless, was shown well by the experimental results. One reason given for the lack of agreement was the small rotation of the float. However, in their subsequent paper (1995), while noting the need to allow for rotation, Buckle et al. concluded that the discrepancies were probably due to asymmetry in the experimental flow. Unfortunately, this work was done at a maximum Reynolds number of 400 so that all the flows were in the laminar regime. See also Xu et al. (2004).

8.2.11 Applications

These meters have been used on a wide variety of fluids. Tubes have been made with heating elements, insulating tube arrangements and armoured inspection glass with a pressure-balanced inner tube. Where metal tubes are used, the sensing is via a magnetic coupling, and this can link to pneumatic or electrical signal transmission. The meter has also been used as a bypass for an orifice plate.

Applications include chemicals, pharmaceuticals, refining, food and beverages, power plants, water and wastewater, pulp and paper, mechanical engineering and

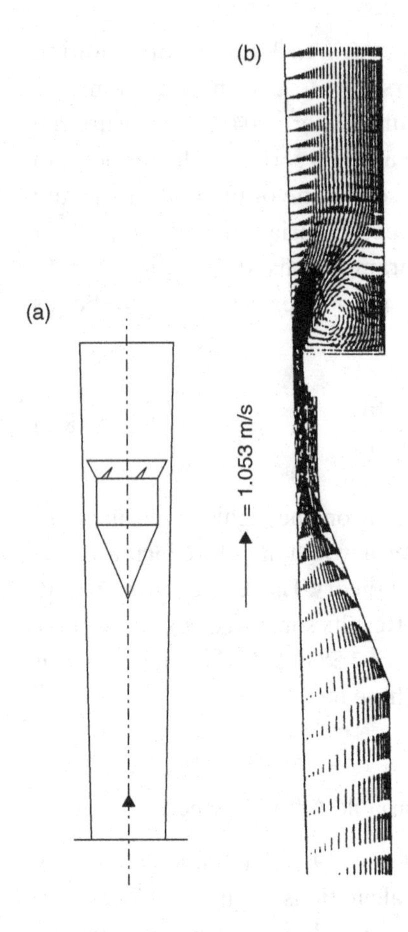

Figure 8.6. Computational results for the flow in a variable area meter: **(a)** Flowmeter geometry; **(b)** computed velocity vectors (reproduced with the agreement of Elsevier from Buckle et al. 1992).

plant construction. Fluids include air, acetylene, ammonia, argon, butane, chlorine, natural gas, helium, steam, oxygen and various liquids.

8.3 Spring-Loaded Diaphragm (Variable Area) Meters

One design is shown in Figure 8.7. Wide turndown and linear response are often claimed as the particular strengths of this device. Figure 8.7(a) clearly shows the mechanism of an early version, the Gilflo-B meter, which was taken over by Spirax-Sarco. The Gilflo-B had a spring-loaded profiled plug in an orifice which was forced back by the flow, causing a variable area orifice. The differential pressure (DP) across the meter was sensed between inlet and outlet and this turned out to be close to linear in relation to flow, with a claim that the operating range was up to 100:1 for gas, steam, liquid natural gas, cryogenic and other liquids. It required a differential pressure cell, temperature and pressure sensor and a flow computer.

This meter has been superseded by the Gilflo ILVA (In Line Variable Area) shown in Figure 8.7(b) which is a development of the Gilflo B, using the same cone/orifice and measuring differential pressure, but designed to fit between existing

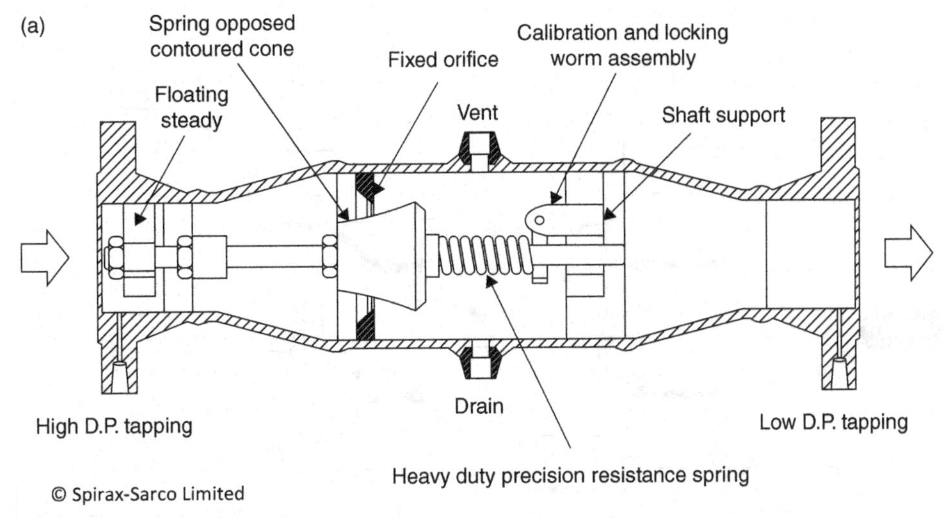

Figure 8.7. **(a)** Longitudinal section of the earlier version of the Gilflo-B spring-loaded diaphragm meter (reproduced with permission of Spirax-Sarco Ltd).

installation pipe flanges, in order to make the device more compact and cost-effective. It still requires a DP cell, temperature/pressure sensors and flow computer. Its turndown is also claimed as 100:1.

Spirax-Sarco also markets two target meter designs, one with similarities to the ILVA, but an important development in this meter is that, instead of measuring the differential pressure across the meter, the force on the cone is measured by strain gauges in the supporting structure, thus giving an electrical signal in place of the differential pressure. The on-board software also obtains a value for mass flow using measurements of temperature and pressure for superheated steam and, for saturated steam, the temperature measurement with a dryness fraction which is entered into the software. Turndown is claimed as 50:1. The other design is a fixed target meter for smaller pipe sizes with a 25:1 turndown.

Other variable area spring-loaded meters have been used for measuring the flow rate of water, paraffin, gasoline, oil, tar, distillates etc. They can handle both high and low flows. The manufacturers' claims for factory calibration may be of the order of 1–2% and may be based on either rate or maximum flow, although occasionally claims are made for very large turndown such as 100:1 with ranges up to 30,000 l/min, uncertainty of order ±1.0% of full-scale deflection, and repeatability of order ±0.3%, and with versions for liquid, steam and gas. Such claims should be treated with caution. A meter capable of performing to this specification based on rate would be outstanding. On the other hand, if not based on rate, the envelope of measurement uncertainty should be provided for the whole range (cf. Turner 1971).

Whitaker and Owen (1990) also obtained data for a spring-loaded variable area orifice meter in a horizontal water-air flow with void fraction up to 40%. For tests in the range 2–20 l/s, the meter retained its linearity, but the meter factor

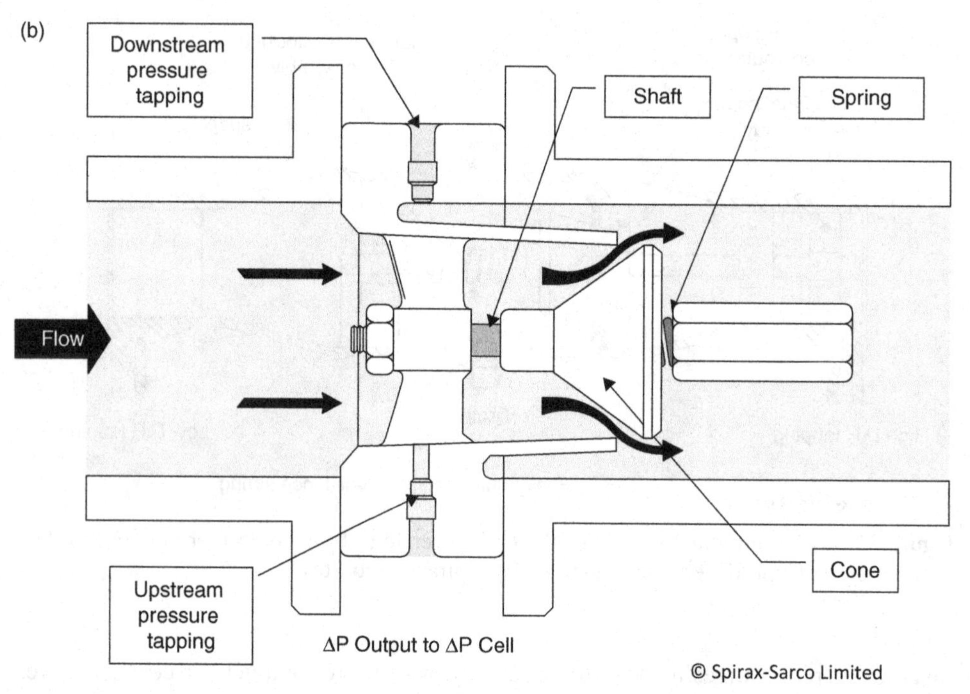

Figure 8.7. **(b)** Current design of Gilflo spring-loaded in-line variable area (ILVA) flowmeter (reproduced with permission from Spirax-Sarco Ltd).

changed with increasing void fraction by about 25%. Suggestions appear to imply that the meter plug/orifice homogenises the flow, removing the need for upstream conditioning.

An alternative approach is for the diaphragm or flap to rotate on a hinge and to close due to a spring. Increasing flow forces the flap open, and its angle gives a register of flow rate. Figure 8.8 shows such a design, which was applied to steam flow but has been discontinued.

The outlet signal was electrical from a rotary transformer, which converted the rotation of the flap to an electrical signal.

In Figure 8.9, a typical line arrangement of fittings is shown for steam service. Note that

- the separator with trap is set to allow condensate through but not steam;
- the strainer is set horizontally to avoid condensate collecting or particulate dropping into the line;
- the upstream and downstream spacings, in a reduced line size, are 6D and 3D, respectively; and
- a nonreturn valve at the end of the line may be necessary.

Hussein, Owen and Amin (1992) used temperature and pressure measurements, a condensate separator with a condensate flowmeter and finally a steam flowmeter

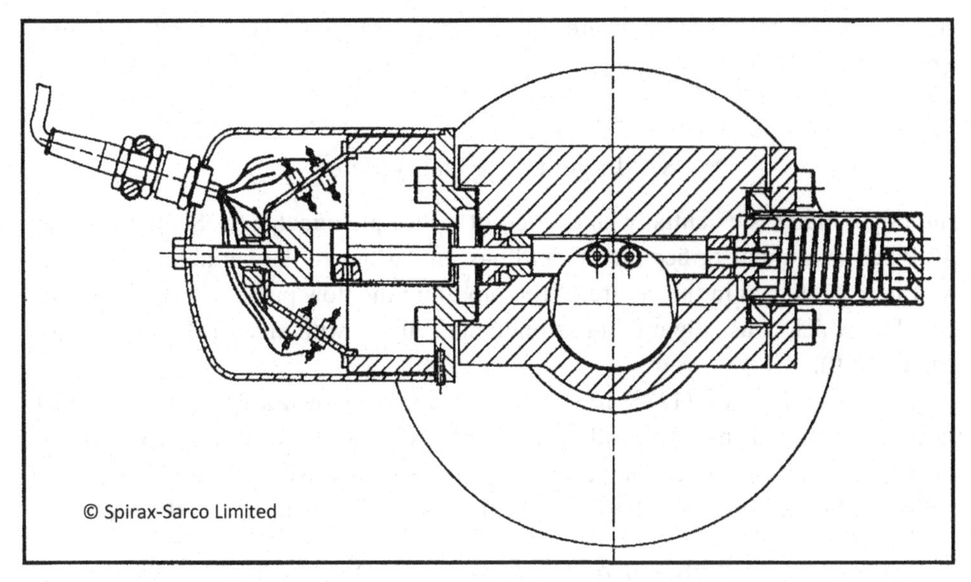

Figure 8.8. Cross-section of spring-loaded rotating diaphragm or flap-type meter (reproduced with permission of Spirax-Sarco Ltd).

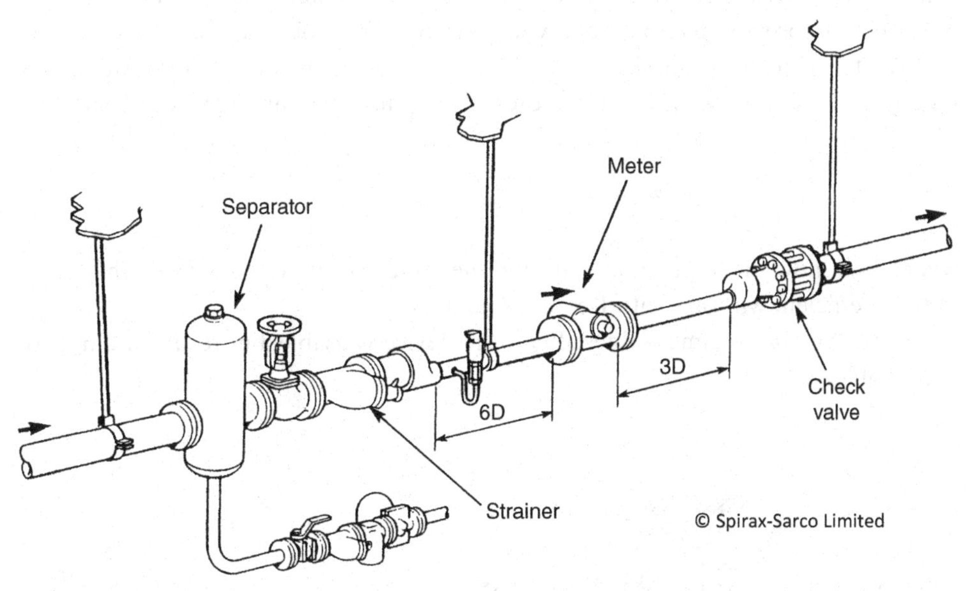

Figure 8.9. Diagram of the line arrangements for measurement of steam flow (reproduced with permission of Spirax-Sarco Ltd).

to obtain the mass flow rate and steam quality. The flow rate was obtained to about 1%, and the dryness fraction to better than 0.3%. The flowmeter was a spring-loaded variable area Spirax-Sarco meter. The separator efficiency was about 92%.

Hussein and Owen's (1991) work was mentioned in Chapter 5. They found that the *Shell Flowmeter Engineering Handbook's* correlation was the best compromise for $x > 95\%$ and for the variable area meter. They wisely suggested that an upstream separator was advisable.

8.4 Target (Drag Plate) Meter

The target meter is essentially an inside-out orifice plate meter (cf. Scott 1982, who describes an annular orifice meter) and has similarities to the variable area meter. A drag plate is held in the centre of the pipe, and the flow past it exerts a force on the plate which is usually measured pneumatically or with electrical strain gauges (Figure 8.10).

Hunter and Green (1975) reported early work on such a flowmeter in which they demonstrated that with a drag coefficient C_d based on the mean velocity in the gap around the target the variation in C_d over a Reynolds number range of 2,000–250,000 is from about 0.97 to 1.83. They proposed a curve fit of the form

$$C_d = C_0 + C_1\left(\frac{a}{A}\right) + C_2\left(\frac{a}{A}\right)^2 + C_3\left(\frac{a}{A}\right)^3 + C_4\left(\frac{a}{A}\right)^4 \tag{8.2}$$

where a is the area of the target, and A is the cross-section of the pipe, but they did not appear to propose a term to allow for change in Reynolds number.

The target meter (Ginesi 1991) for Re > 4,000 has a drag force proportional to the square of the velocity. The force on the plate may be obtained from Equation (8.A.3)

$$F = A(p_1 - p_2) = K\tfrac{1}{2}\rho q_v^2\, A/A_2^2 \tag{8.3}$$

where A_2 is, now, the annular area around the target. We may assume that the value of the coefficient will vary with Reynolds number.

In the laminar regime (Ginesi 1991) the device is usable, but results are not so predictable.

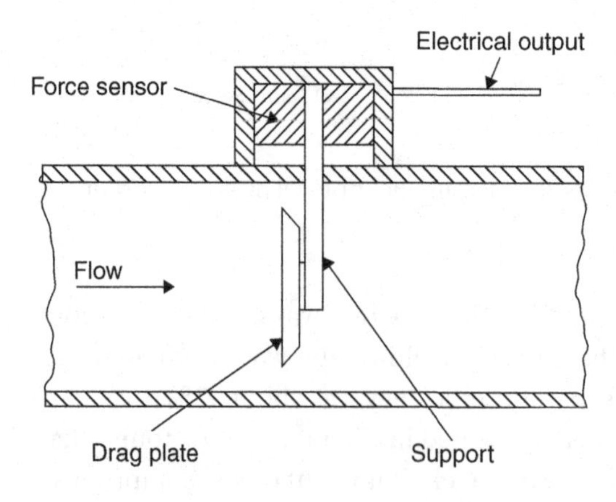

Figure 8.10. Target (drag plate) meter.

Because it allows gas or solids, entrained in the fluid, to pass, the meter has been used for two-phase flows. It needs to be used with care and understanding in this application.

Uncertainty is between $\frac{1}{2}$ and 2% of full scale (Ginesi 1991). The target can be sized for the flow so that for a 2-in. (50-mm) meter full span could be 20 or 200 gal/min (5.5 or 55 m^3/h), whereas turndown is limited to 4:1 or 5:1. The expansion of the target due to temperature change will change the response [e.g. with stainless steel 100°F (56°C) can cause about 0.1% error]. The target can also be subject to coating and build-up and will be affected by edge sharpness. Installation requirements may generally be similar to that for an orifice plate. Although the accuracy is not as high as other meters, reliability, repeatability, turndown and speed of response are important characteristics which may be achieved in some designs.

Figures for a commercial device using force balance arm with either electrical or pneumatic transmission claimed that flows from 0.4 to 1350 m^3/h for water and from 12 to 40,500 m^3/h for air could be measured by meters of this type with diameter range 25–300 mm and uncertainty of ±3% FSD. Such ranges should be treated with caution.

Yokota, Son and Kim (1996) appeared to suggest that in unsteady flow measurement up to 10 Hz the uncertainty was within 2%.

Peters and Kuralt (1995) described a flowmeter that was essentially a target flowmeter with such a small gap that the flow was laminar in the gap and the force was proportional to flow. It is likely to be of limited use (cf. Wojtkowiak, Kim and Hyun 1997).

Sun et al. (2000) described an optical interference system for a target meter. To obtain the drag-force on a bluff body made of rubber, Philip-Chandy, Scully and Morgan (2000) used optical fibres with grooves attached to the bluff body so that when it deflected there would be an intensity modulation.

See also a note on a target meter in Section 8.3.

8.5 Integral Orifice Meters

For small-diameter pipes and small flows of very clean liquids and gases, it may be possible to use an integral orifice. These devices, essentially, in some arrangement form a part of, or are closely coupled to, the differential pressure cell and can achieve repeatability (Miller 1996) of $\pm 1\frac{1}{2} - 6\%$ (cf. Cousins 1971).

8.6 Dall Tubes and Devices that Approximate to Venturis and Nozzles

The devices in this section all depend on Equation (5.1), with appropriate coefficients defined by the manufacturers.

The Dall tube looks somewhat like a Venturi in its largest version [Figure 8.11(a)]. An early description of this device is given by Dall (1962), who also refers to a

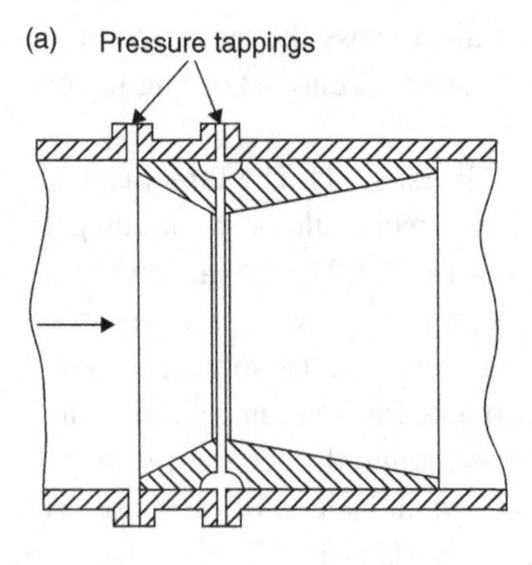

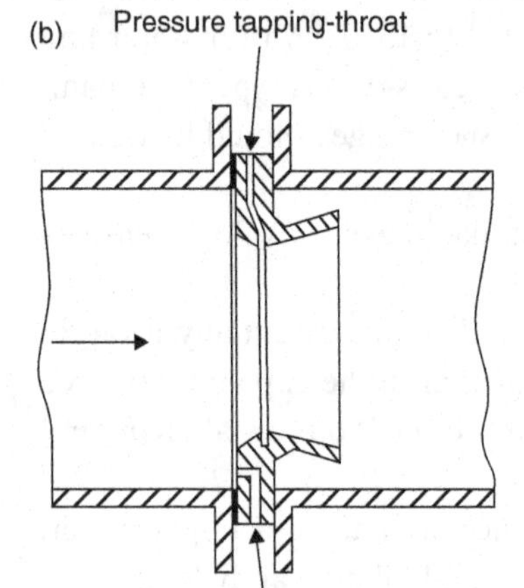

Figure 8.11. Dall tubes (approximate diagrams): **(a)** Dall Venturi; **(b)** Dall orifice.

segmental hump Venturi. However, it has certain design features that result in a higher differential pressure and better diffusion than a Venturi of equivalent size. These features follow.

i. The inlet pressure is raised as the flow is almost brought to rest on the inlet shoulder.
ii. The throat groove probably results in a toroidal vortex, which sits in the groove, causes the flow to contract and curve to pass around it, and so creates a slightly higher velocity and lower pressure than would otherwise occur.
iii. The vortex will presumably help the flow to negotiate the changing area at the throat and to reattach in the diffuser.

However, one drawback of this enhanced performance is that any gas in liquid flowing through the Dall tube may tend to get caught in the vortex and to cause substantial performance changes. Cousins (1975) discusses these and other effects caused by slot width, surface roughness, cone angle and some geometrical details. In a

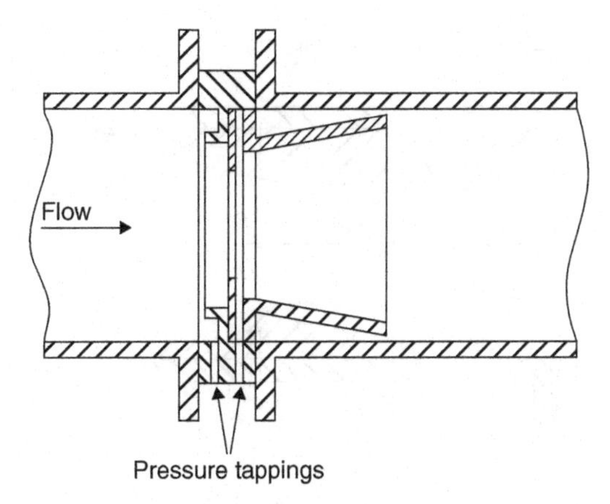

Figure 8.12. Epiflo tube (approximate diagram).

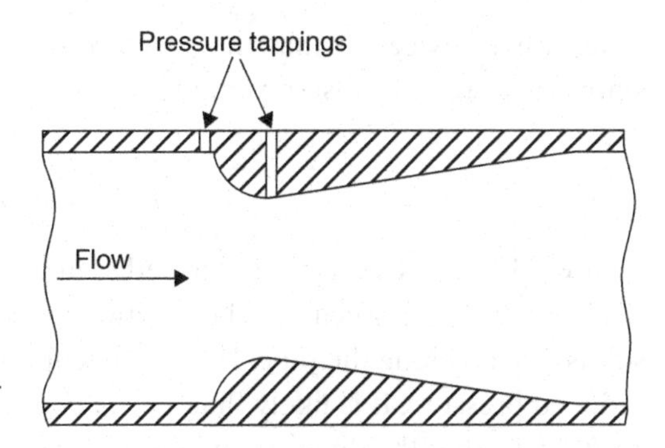

Figure 8.13. Low-Loss flow tube (approximate diagram).

subsequent discussion of this paper, it was suggested that the coefficient could shift by as much as 8% with changing Re from 0.4×10^6 to 1.0×10^6. The reason for this may have been cavitation or some other mechanism that prevented reattachment of the flow in the diffuser.

A Dall nozzle and orifice have also been made, and Figure 8.11(b) is an approximate diagram of the orifice. In each case, their performance data will need to be obtained from the manufacturer.

Lewis (1975) reported on another device (Epiflo Figure 8.12) in which, at inlet, flow is brought to rest in an upstream-facing annulus. Downstream of this is an orifice followed by a slot and then a short diffuser. The downstream tapping is in the slot. Thus it appears to have some of the features of the Dall tube.

Figure 8.13 shows a device of a Venturi type but which has a rounded (bell-mouth) inlet and may have a wider diffuser angle than ISO allows. It is sometimes referred to as a Low-Loss tube. The manufacturer may have claimed ±1% uncalibrated and ±0.25% calibrated with a repeatability of ±0.1% and a turndown of 10:1. Sizes ranged from 13 to 1,200 mm. Installation was claimed to require 12D upstream and 2D downstream for all but partially shut valves and elbows in different planes.

Vincent Gentile Jr. described a tube which, in one form, is illustrated in Figure 8.14 and is known as a Gentile tube. It is claimed that it differs from other

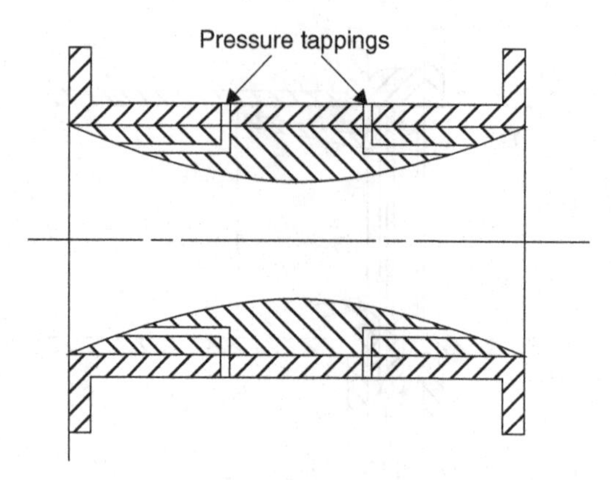

Figure 8.14. Gentile tube (approximate diagram).

differential devices in that the pressure difference is due to dynamic rather than static pressures. It is also reversible (Scott 1982).

8.7 Wedge Meter

Figure 8.15 shows a diagram of a wedge meter. They are designed with an asymmetrical constriction, which may be positioned at the top of the pipe, thus allowing any solids to pass along the pipe. The constriction will result in an increased velocity and reduced pressure. It is likely that a *vena contracta* will form downstream of the constriction so that the downstream tapping will sense the static pressure near the *vena contracta* (cf. Dall 1962, who refers to a segmental hump Venturi).

It is, therefore, apparent that the orifice equation with a suitable coefficient will be appropriate, but the size of the coefficient and its variation with Reynolds number will depend on the manufacturer's calibration data. Uncertainty after calibration of ±0.5% is suggested.

Oguri (1988) suggested that a wedge flowmeter was suited to solid-liquid and gas-liquid flows. This is probably the reason for its preference over a segmental orifice for dirty or solids-bearing fluids – it is less likely to be damaged by them. It is claimed to have a long life without maintenance and without fouling. It may have a smaller pressure loss than the segmental orifice because of the slightly greater fairing of the wedge.

It was available in sizes from about 13 mm to about 600 mm and for a Reynolds number range up from 500.

The V-shaped restriction to contract the flow is characterised by H/D (Ginesi 1991). H is the height of the opening below the restriction (Figure 8.15). There are no critical dimensions or edge sharpness to be retained, and the element can withstand wear and tends to keep clear of build-up due to the slanted upstream face. It is claimed to retain a square law down to about Re = 500, regardless of the state of the flow (laminar, transition or turbulent). With few published recommendations and no standardised geometry, the meter has to be calibrated, and suggestions of 0.5% uncertainty have been made. For Re < 500, the calibration should be on a fluid

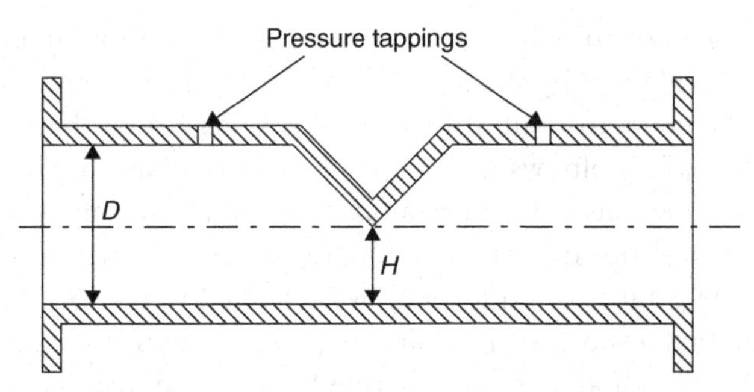

Figure 8.15. Wedge meter (approximate diagram).

of the correct viscosity. The non-clogging feature with the low-flow capability and ruggedness have made this a realistic choice for fluids such as dewatered sludge, black liquor, coal, flyash and taconite slurries. Ginesi claims that it is less sensitive to installation than other differential pressure meters. The worst case reported by him, 1.74%, was with three short radius bends in different planes at 5D upstream. To eliminate deviation, 20D is claimed as sufficient.

With calibration, ±0.5 to ±0.75% of rate may have been claimed with repeatability of order ±0.2%. The pressure tappings are presumably symmetrical because the device is claimed as bidirectional. H/D was in the range 0.2–0.5 (Figure 8.15).

Steven et al. (2009) discussed the potential for using the wedge meter with wet natural gas flows.

8.8 V-cone Meter (Cone Meter)

The V-cone meter (Venturi cone, which may also be referred to as a McCrometer tube) was introduced in the late 1980s and uses a conical flow restriction centrally mounted in the metering tube (Ginesi 1991; cf. Ginesi 1990) (Figure 8.16). Sizes may range from 1 to 16 inches (25 mm to 400 mm). It has resistance to wear and may be suitable for dirty liquids and gases.

Data and applications of the V-cone meter have been many over the past 10 years or so, and it is an interesting example of the development of flow metering even making use of long-established technology where a new design can establish a niche in a crowded market!

The V-cone meter is claimed to have a typical turndown of 10:1 or more. The differential pressure is measured between an upstream tapping and the trailing face of the cone. The β ratio is defined as the square root of the ratio of the annular area to pipe area.

Ifft (1996) claimed that for a β = 0.5, the performance was ±0.5%, and this was supported by a current pdf on the Web (http://www.mccrometer.com/library/pdf/24517-16.pdf), which claimed uncertainty of ±0.5% of rate for a spool piece design (±1% for wafer design), with repeatability of ±0.1% or better.

It has been claimed for installation with nearby pipe fittings, and depending on the specific installation, 3D or less of straight pipe upstream of the spool piece and 1D or less of straight pipe downstream results in minimal installation effects. This was explained by the suggestion that the profile entering the meter is flattened by the cone. Ifft and Mikklesen (1993) suggested that a single elbow and double out-of-plane elbows with R/D = 1.5 had little or no effect. The reader should check these claims with the manufacturer for confirmation. Prahu et al. (1996) undertook a series of tests to compare an orifice plate with a cone flowmeter. In both the β ratio, or equivalent, was claimed to be 0.75. However, they appeared to find that a short straight pipe of 10D with an open upstream end, the details of which do not appear to be described, caused a percentage change in C_d, given by

$$\Delta C_d = \frac{C_d \text{ in disturbed flow} - C_d \text{ in undisturbed flow}}{C_d \text{ in undisturbed flow}} \times 100 \tag{8.4}$$

for the orifice plate of –1%. It is not clear to this reader what effect the open-ended upstream fittings would have had when compared with an inlet flow profile more expected in turbulent pipe flow.

Singh et al. (2006) tested the V-cone meter downstream of a gate valve and claimed that the discharge coefficient was not affected by the valve with gate open from 25% to 75% and fully open, if the meter was 10D or more downstream. Singh, Singh and Seshadri (2009) used CFD to study the effect of vertex angle and upstream swirl on the performance characteristics of the cone meter. They appeared to predict an error of about 3% for 30° of swirl. Singh, Singh and Seshadri (2010) also used CFD predictions of the effects of the upstream elbow fittings on the performance of these flowmeters.

Peters, Reader-Harris and Stewart (2001) obtained an equation for the expansibility for a V-cone meter based on data from tests at NEL:

$$\varepsilon = 1 - (0.649 + 0.696\beta^4)\frac{\Delta p}{\kappa p_1} \tag{8.5}$$

They confirmed that the expansibility factor was not significantly affected by pipe diameter or by Reynolds number. The value for the wafer-cone meter was found to be:

$$\varepsilon = 1 - (0.755 + 6.787\beta^8)\frac{\Delta p}{\kappa p_1} \tag{8.6}$$

Peters et al. (2004b, 2006) tested a wafer V-cone meter with water and air and with upstream disturbances. Their findings suggested that the tests of the V-cone meter, according to API (API Chapter 5.7), demonstrated that it met the manufacturer's claims, that the expansibility equation was appropriate, and that the meter appeared to be able to cope with upstream flow disturbances.

Toral et al. (2004a, 2004b) presented results from the development of the neural nets with the V-cone meter in wet-gas, and results from the NEL facility and from K-Lab. The interested reader should refer to the original paper where apparently good predictions were possible with closed data sets, but rather less good ones with data sets not used for "training" the software. Steven, Britton and Kinney (2010) discussed the need for a wet-gas correlation factor to correct for liquid in cone meters. Lloyd, Guthrie and Peters (2002) reported on a calibration facility for steam meters which made use of a V-cone meter as the standard. They reported that as far as that project was concerned the best overall meter was "the Cone with the intelligent DP transmitter".

Steven (2008b, 2009) proposed the use of additional pressure measurements for DP meters to show up problems such as an orifice plate installed the wrong way round. He also considered other meters such as V-cone in the paper.

From an earlier patent (Lisi 1974) Storer and Steven (2010) developed the idea of combining vortex and cone meters to make a mass and volume flowmeter and a densitometer. An initial 100 mm prototype meter was claimed to have 0.5% uncertainty while the gas density could be predicted to 1% uncertainty. Predicting the gas density also gave redundancy to an external density calculation system. Furthermore,

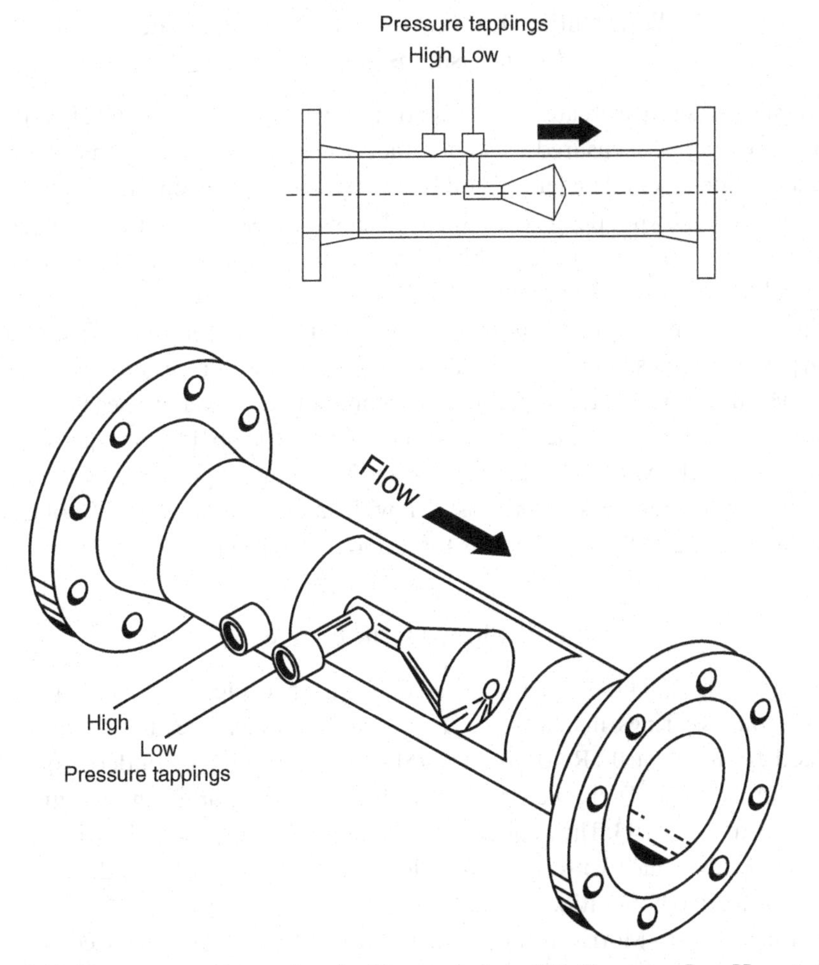

Figure 8.16. V-cone meter (reproduced with permission of McCrometer, Inc., Hemet, CA).

with the addition of a downstream pressure tap the DP meter could have three flow rate equations, thereby adding extra redundancy to this metering system.

Hodges et al. (2010) reviewed a large amount of data relating to cone meters. Test data included single phase and wet-gas from several sources. Meters ranged from about 50 mm to 350 mm with β ratios from 0.45 to 0.85 and Reynolds numbers from 50,000 to 5,000,000. The authors appeared to find differences between nominally identical meters. They warned against extrapolation of data from low Reynolds numbers to higher values. They noted that cone assemblies might suffer distortion with consequent calibration change due, for instance, to slug impingement. Small misalignments of 1° might be significant, and damage during sizing should also be avoided. See also Borkar et al. (2013).

This author wonders whether the cone meter, or a similar design, could yield a highly predictable meter with direct electrical output via a force sensor to obtain the flow rate, and this could be combined with wall differential pressure measurements to provide a condition monitoring function.

8.9 Differential Devices with a Flow Measurement Mechanism in the Bypass

The pressure difference created across an orifice can be used to create a bypass flow, which in turn can be measured by some other flowmeter. One which has been used is a small Pelton wheel. The jet created by the bypass impacts on the wheel, causing it to rotate and allowing the ferrites in the blades to cause pulses in the sensing coil. Claims are for ±1% of full scale on linearity or ±0.5% with 40:1 turndown using linearizing electronics and repeatability ±0.25% for 90% of range.

The oscillating vane meter uses an orifice with the vane in the orifice structure as a bypass (cf. Ginesi 1991). It may be available as a wafer design to go between flanges. The flow through the orifice creates a pressure drop that causes flow through the oscillating vane. The orifice appears to be concentric apart from one segmental region in which the vane is situated. An electric pickup or proximity switch senses the oscillation. Turndown is claimed as 10:1, with an uncertainty of ±0.5% of rate and a repeatability of ±0.2%. The meter is for liquid applications.

8.10 Slotted Orifice Plate

Morrison et al. (1994a, 1994b) suggested using an orifice that had radial slots arranged on two rings instead of a single central hole. The inner ring of slots extended from about 0.1R to about 0.45R, and the outer extended from about 0.55R to about 0.9R. There were 8 slots in the inner ring and 24 in the outer, making an effective $\beta = 0.43$. They concluded from tests in a 50.8-mm line that the slotted orifice is less sensitive to upstream flow conditions than a standard orifice of the same effective beta ratio.

In further tests Morrison and Hall (2000) compared it with a conventional orifice plate of β ratio 0.5 and they reported that their tests suggested a superior performance to the conventional plate for a range of profiles and swirl. Further tests (Morrison et al. 2001) on a slotted orifice plate with an equivalent β ratio of 0.5 in an air and water flow suggested that it responded to two-phase flow in a well-behaved manner. The authors measured the product of flow coefficient and expansion factor for various quality values. A proposal was made to use two meters of different designs in series to obtain other parameters in two phase flow such as density (Morrison et al. 2002a, 2002b), an approach other workers have explored. Geng, Zheng and Shi (2006) studied the metering characteristics of a slotted nozzle for wet-gas flow, as did Li, Wang and Geng (2009). Kumar and Ming Bing (2011) undertook a CFD study of low pressure wet-gas metering using slotted orifice meters.

8.11 Pipework Features – Inlets and Pipe Lengths

If a well-formed inlet is provided to a pipe from a stilling tank or large gas vessel, then the pressure drop from the tank or vessel into the pipe can be used to obtain the flow rate. Equation (6.1) is simplified since $\beta = 0$, the flow coming from a large, essentially still, container, and so $E = 1$. The coefficient is also very close to unity.

The *Shell Flowmeter Engineering Handbook* (Danen 1985) gives values for the coefficient. The volumetric flow rate will be given by

$$q_v = C(\pi/4)d^2\sqrt{2\Delta p/\rho} \tag{8.7}$$

where d is the throat diameter (at the downstream tapping) and may equal D, the pipe diameter downstream; Δp is the differential pressure between inlet and throat; and ρ is the density. The coefficients of discharge for various designs may be obtained from the *Shell Flowmeter Engineering Handbook* (Danen 1985). For example, for a well-formed bell-mouth intake (cf. Figure 8.17), the values of C may be of order

Re number	5×10^3	10^4	10^5	3×10^5
C	0.914	0.940	0.987	0.991

The use of such an inlet will clearly depend on the condition of the installation and the upstream flow. The latter should be undisturbed and have had sufficient stilling time. Even so, the method should be used with caution.

Ito, Watanabe and Shoji (1985) obtained values of about 0.95 at Re = 20,000 to nearly 0.99 at Re = 600,000, and also gave a useful list of references.

An interesting idea put forward by Oddie et al. (2005), described a new differential pressure concept making use of a helical energy dissipator attached to the inside of a straight pipe wall, which causes a pressure drop, but also mixes the flow in a way which is particularly suitable for multiphase flows.

Crainic, Cornel and Ilie (2000) used the flow along a piece of straight pipe to deflect a manometer filled with ferrofluid and for the displacement to be read from the magnetic coils outside the limbs of the U-tube.

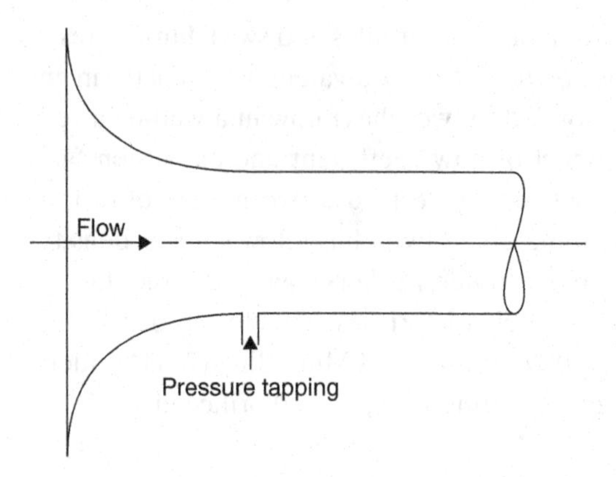

Figure 8.17. Bell-mouth intake.

8.12 Pipework Features – Bend or Elbow used as a Meter

In Chapter 4, the elbow meter is described as an additional means of in situ calibration. However, Xu-bin (1993) reported that the device was widely used in China for the measurement of primary coolant flows in pressurised water reactors, and that it has also been used for measurement of steam, gases, clean liquids, slurries, sludges and corrosive liquids. Xu-bin also pointed out that it is easy to install and maintain, can cope with large pipe sizes, and is less prone to damage.

Bean (1971) gave a full derivation of the relationship between pressure and flow. Here we give a simple derivation, which results in the same answer. Figure 8.18 shows the geometry of the bend meter. If we make the grossly simplifying assumptions that

- V is uniform across the bend (clearly this is not the case and there will be flow profile and secondary velocity effects as the flow negotiates the bend), and
- V^2/R is the centrifugal acceleration for all the fluid in the bend (it won't be since the radius of each fluid element differs, and their velocities differ),

we can write that (Figure 8.18):

Force across bend on unit area = Mass of fluid of unit cross-section × acceleration

or

$$p_o - p_i = D\rho \times V^2/R \tag{8.8}$$

Thus

$$V = \sqrt{\frac{R}{D}\frac{\Delta p}{\rho}} \tag{8.9}$$

We are interested in the volumetric flow, and we introduce a flow coefficient to allow for the approximations to real flow. We obtain

$$q_v = KA\sqrt{\frac{R}{D}}\sqrt{\frac{\Delta p}{\rho}} \tag{8.10}$$

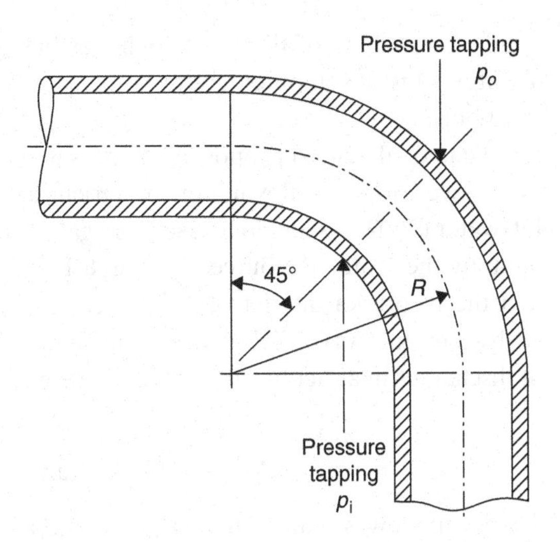

Figure 8.18. Bend used as a flowmeter.

Bean gave the coefficient as

$$K = 1 - 6.5/\sqrt{\text{Re}} \tag{8.11}$$

for $10^4 \leq \text{Re} \leq 10^6$ and $R/D \geq 1.25$ and suggested that an uncertainty of ±4% should be allowed for. (Note that there appears to be a disagreement with Xu-bin's equations.)

The use of pipe bends is also explained in the *Shell Flowmeter Engineering Handbook* (Danen 1985), where the coefficient includes the $\sqrt{R/D}.\sqrt{R/D}$ for a normal 100-mm short radius bend schedule 40 is approximately unity (Danen 1985).

This method should be used with extreme caution because the condition of the bend will be unknown and the uncertainty is unlikely to be as good as 4%.

Further work on the use of an elbow as a flowmeter is given by Silva, Velazquez and Ruiz (1997). The equation and coefficient appear to differ from some of the previous forms recorded earlier.

Li and Wang (2004) gave an important addition to the information on elbow flowmeters with experimental results and a statement that about 5,000 elbow meters had been applied, possibly mainly in China, by 2003.

Malinowski and Rup (2008) reported on numerical and experimental investigation into flow rate measurement with an elbow with oval cross-section.

Meng, Li and Li (2010a) studied the measurement characteristics of the V-type elbow meter (pipe bends out of line with the inlet pipe, arches back and symmetrically rejoins the outlet pipe). They used CFD to obtain the velocity and pressure distributions in the elbow meter. Experimental data were from a DN100 pipe, and demonstrated that the flow coefficient of the meter was stable, and agreed with CFD simulation to within ±1.0%. They referred to a reference from Hebei University of Technology by two of the authors.

Morris et al. (2001) discussed a measurement system specially designed to measure, for instance, flows from an axial fan. A torque resulted from flow through a 90°

bend, downstream of the fan, which was balanced by a force transducer. This was an interesting development, but not likely to be of general interest in the context of this book.

Yuan et al. (2003) proposed a bypass flowmeter using, for example, a pipe bend to achieve a low-cost flowmeter for agricultural applications. Einhellig, Schmitt and Fitzwater (2002) discussed its use in irrigation schemes. An evaluation of elbows was done by the Water Resources Research Laboratory of the Bureau of Reclamation and the Nebraska district of the Natural Resources Conservation Service. Trung, Nishiyama and Anyoji (2007) suggested a possible low-cost option for measurement of discharges in agricultural production using a bypass with a 45° bend flowmeter.

8.13 Averaging Pitot

The Annubar was claimed to be the first averaging pitot (Britton and Mesnard 1982), but many others have followed. The averaging pitot is, strictly, neither averaging nor pitot. In Chapter 21 we shall discuss pitot tubes, which are tubes coaxial with the flow. Although the averaging pitot has a series of upstream-facing holes, they are unlikely to provide a true average of the axial velocities across the pipe. The averaging pitot consists of a bar that spans the pipe and has holes facing upstream and downstream [Figure 8.19(a)]. It is claimed to cope with disturbed profiles due to upstream fittings. The flow will be brought to rest locally on the leading edge of the bar and at the upstream holes unless interconnecting flows occur. The resulting pressures in each of the holes, which will approach the local stagnation pressures, will be linked by the manifold that runs inside the bar. The rear-facing hole or holes will experience a pressure that approximates to the static pressure in the flow. These two pressures, the approximate average stagnation and the approximate static, are then carried by the internal tubes out of the bar and into a suitable pressure transducer where the pressure difference $p_u - p_d$ will be calibrated against the velocity. Essentially, we assume that this pressure difference will approximate to the dynamic pressure obtained from Equation (2.10). The calculation of the actual upstream-sensed pressure will depend on internal flows in the connecting manifold from the centre holes with higher stagnation pressure to, and possibly out of, the holes nearer the pipe wall where the stagnation pressure will be lower. In addition, the squared relationship

$$\Delta p = \tfrac{1}{2}\rho V^2 \qquad (8.12)$$

may result in an underweighting of the outer pressures where most of the flow passes. Calculation may suggest that the bar is a near-centreline measurement device with correction for profile shape from the outer tappings.

Figure 8.19(b) shows some of the cross-sections of the commercially available bars. Because flow around a circular cylinder has a changing flow pattern depending on the boundary layer, other shapes have been used to create a less varying pattern. Britton and Mesnard (1982) gave data comparing round and diamond shaped sensors, which clearly demonstrated the benefit of the latter over the former. While

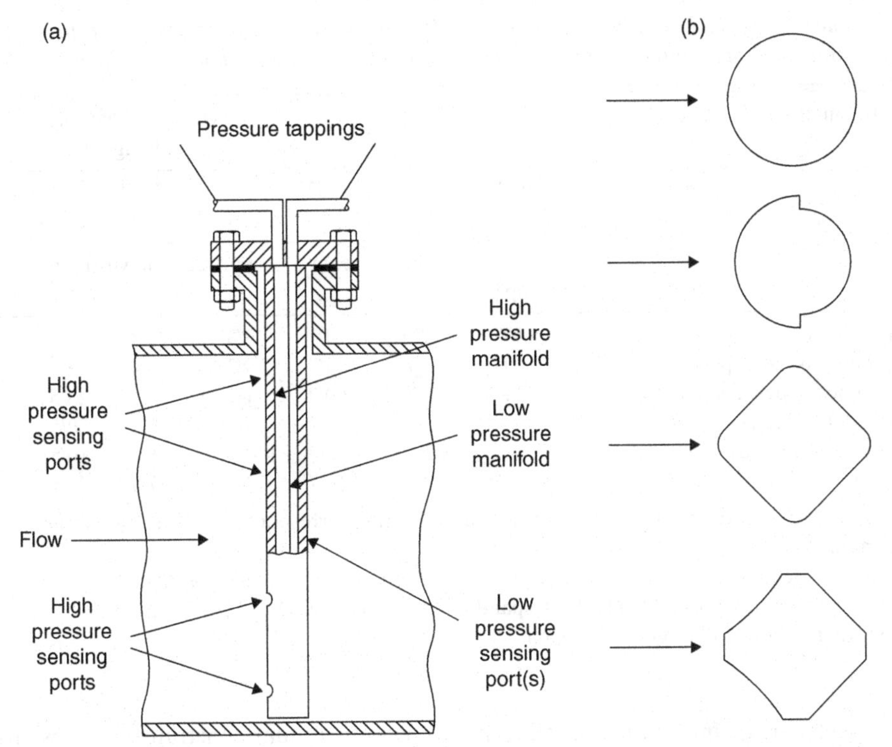

Figure 8.19. Averaging pitot tube/bar: **(a)** Diagram of a typical meter; **(b)** cross-sections of various designs.

the coefficient for the round cylinders varied by about 10% for a Reynolds number range of 2×10^5 to 10^6, the diamond cylinder varied by about 2%. The final version of the cross-section for one manufacture was a square section with flattened corners set so that it presented a corner to the flow direction. In addition to the coefficient, a blockage factor that allows for the reduced pipe cross-section at the probe is required. They claimed a measurement uncertainty of ±1% and a repeatability of ±0.1%. It has been suggested that a hexagonal bar with faces perpendicular to the flow overcomes the problem of a circular bar, in forcing the flow separation at the edge vertices, as well as the fabrication problem of diamonds and ellipses, which require a weld at the pipe entry point. By using a hexagon, if necessary just at the actual taps, the integrity of a round bar is retained. No doubt different manufacturers will have their own reasons for cross-sectional superiority! The cross-sectional dimension of the bar may be in the range 10–50 mm.

Typical performance claims for these devices are ±1% of reading with ±0.1% repeatability for turndown of 10:1 or more and they are suitable for pipe sizes of 25 mm (1 in.) to 12 m (about 40 ft.). Manufacturers should be consulted on flow ranges. Flows of 10–30,000 m^3/h for liquids and 200–600,000 m^3/h for gases may be possible. Materials may be stainless steel, Monel, Hastelloy, titanium, Inconel, PVDF etc. Temperature range may be from –100 to 450°C. Static pressure may be up to 70 bar for some designs and at least one manufacturer claims 250 bar. Some

Table 8.2. *Typical installation spacing (in diameters) for averaging pitots as has been indicated in manufacturers' literature, compared with orifice plate requirements*

Upstream fitting	Orifice (β)			Typical estimates for averaging pitot		
	0.2	0.5	0.67	A	B	C (worst case)
	upstream/downstream spacing			upstream/downstream spacing		
Single bend	6/4	22/6	44/7	7/3	24/4	10/5
Two bends in same plane	10/4	22/6	44/7	9/3	11/4	15/5
Two bends in perpendicular planes	34/4	75/6	60/7	17/4	-	28/5
Reducer (tapered)	5/4	8/6	12/7	7/3	9/4	10/5
Expander (tapered)	6/4	20/6	28/7	7/3	9/4	10/5
Valve	12/4	12/6	18/7	24/3	27/4	28/5

Orifice values are from ISO 5167-2:2003 and are for lengths corresponding to "zero additional uncertainty" for:

- Two 90° bends are separated by: ≤ 10D (same plane); < 5D (perpendicular planes),
- Reducer: 2D to D over 1.5D to 3D, Expander: 0.5D to D over D to 2D
- Valve: full-bore ball or gate fully open

manufacturers claim that installation should be with similar constraints to those for the orifice plate, and the connections should also be similar to avoid condensation problems etc. Some typical values for spacing are given in Table 8.2.

This suggests that, in the absence of clear guidance from the manufacturer, taking the spacing for an orifice with β between 0.5 and 0.67 may be the wisest precaution, with the exception of the upstream valve. It should be remembered that the accuracy of any insertion device is subject to the uncertainty with which the pipe cross-sectional area can be measured. In the case of averaging-pitots, there is a blockage caused by the bar, the effects of which must be obtained from calibration, although some careful computation may be useful for this meter, which should also take in the blockage effect of the device (cf. Walus 2000). Dobrowolski, Kabaciński and Pospolita (2005) undertook computational studies of the behaviour of an averaging pitot tube.

Emerson has introduced a development of the Annubar using a T-shaped sensor (http://www2.emersonprocess.com/siteadmincenter/PM%20Rosemount%20Documents/00803-0100-6113.pdf). It is claimed to have an uncertainty of 0.75% of rate, presumably for the probe itself in a specified flow profile and not for the installed value in a pipe of uncertain cross-sectional area, and also a repeatability of 0.1%. Numerical and experimental studies of a range of cross-sections have been undertaken (Kabaciński and Pospolita 2008, 2011), using twin sections which apparently create a higher differential pressure. Further experimental and numerical studies have been reported on eight cross-sections (Węcel, Chmielniak and Kotowicz 2008) which appear to indicate installation effects, for example that for a single bend, spacings of or greater than $L/D = 11$ should cause negligible addition to uncertainty. Sun, Qi and Zhang (2010) modelled averaging pitot tubes with CFD and also carried out tests. Cross-sections were circular, diamond and with "flow conditioning wings", projections laterally which may have given very specific separation positions. The wings appear to have shown an improvement over the diamond in some cases, presumably due to the well-defined shedding point on the pitot bar.

It is not clear how manufacturers overcome the uncertainty of the pipe ID and cross-section into which these devices are inserted. If installation of the pipe and bar are at the same time, no doubt this measurement will be possible. The manufacturer will then only need to know the allowance for blockage caused by the bar. However, if the pipe is in position, the measurement is considerably more problematic.

In some cases, it is possible that these devices may also create vortex shedding, a phenomenon that we shall encounter and use in the vortex-shedding flowmeter. Unfortunately, such shedding causes lateral forces on the bar and, if these are close to the natural frequency of the bar, may cause vibrations of an unacceptable level. The manufacturer should be asked about this when considering such a device.

We shall discuss pitot tubes in Chapter 21 and shall find a very different shaped device. Cutler (1982), in a letter commenting on averaging-pitots, questions whether multi-hole sensors do, in fact, generate an average differential pressure. Cutler made the important point that, for fully developed flow profiles, a single-point measurement may suffice. We show in Chapter 2 that one measurement at about three-quarters radius will suffice for turbulent profiles. He gave an example of a device consisting of upstream pointing and downstream pointing tubes.

A meter of this type has been marketed as a flow/no-flow indicator. Impact (upstream-facing) and suction (downstream-facing) ports supply a differential pressure that is displayed on a gauge attached to the pipe or may be used to obtain an indication of flow rate via differential pressure measurement.

8.14 Laminar or Viscous Flowmeters

The viscous flowmeter for gases uses very small flow passages so that the pressure drop is due to the viscous as opposed to the inertial losses and thus is proportional to the flow rate. The reader is also referred to Chapter 18 where capillary thermal mass flowmeters are described. These are frequently used in conjunction with a laminar flow bypass for higher flow rates. The design with differential pressure measurement may have been developed to measure gas flows for research into internal combustion engines where there is a high level of pulsation and unsteadiness. As observed in Chapter 5, pulsation causes an overestimate of the flow if an orifice meter is used. The research engineers, presumably, realised that if viscous effects could dominate the flow behaviour, the mean differential pressure across such an element would be proportional to the mean flow without errors due to a squared term.

The laminar flow in a circular tube can be shown to obey the equation (Baker 1996)

$$\bar{V} = \frac{1}{8}\frac{\Delta p D^2}{4\mu L} \tag{8.13}$$

where V is the mean velocity in the tube, Δp is the pressure drop along the tube, D is the diameter of the tube, μ is the dynamic viscosity, and L is the length of the tube. From this, we obtain the volumetric flow rate

$$q_v = \frac{\pi}{8}\frac{D^4}{16}\frac{1}{\mu L}\Delta p \tag{8.14}$$

In practice, we shall probably need to include, for a particular device, a calibration coefficient C. For very small flow passages viscous effects ensure that, for most practical purposes, the equation will be valid for time-varying flows.

Figure 8.20 is a diagram of such a device with a laminar flow element through which the flow passes, and with pressure measured in the upstream inlet chamber and in the downstream chamber.

Flow ranges for such devices are claimed to be from 0–2 ml/min to 0–2,000 l/min with a pressure difference up to about 10 mm H_2O and a Reynolds number of about 500. The meter is designed to be unaffected by installation and pulsation. Turndown of 20:1 or, with a mass flow computer, of 100:1 are claimed.

Claims may suggest it to be suitable for any dry, clean gas. Calibration may be claimed to 2% of reading for 100:1 turndown with 0.2% repeatability and pressure in the range 0–7 bar. Construction may be of aluminium, brass and epoxy. The design allows capacity to be changed by altering the number of parallel passages.

Weigand (1994) referred to the Hagan-Poiseuille law (Poiseuille 1842) for flow in a capillary

$$\Delta p = A q_v \mu L / D^4 + B \rho q_v^2 \tag{8.15}$$

where A and B are constants, L is the length of the capillary, and D is its diameter. Other symbols have their usual meanings.

Stone and Wright (1994) gave consideration to this meter and obtained the first term in Equation (8.15) and accounted for the second term:

a. due to entry length effects, which introduced into Equation (8.14) a factor of $(1 + C_1\text{Re})$ where $C_1 \approx 0.03$;
b. due to entry and exit losses, which they estimated as being of order 0.6% of total pressure drop;
c. due to compressibility of the gas (Fanno line cf. Baker 1996) giving an error of about 1%.

If B is small compared with A and L/D^4 is large, Equation (8.15) becomes

$$\Delta p = A' \mu q_v \tag{8.16}$$

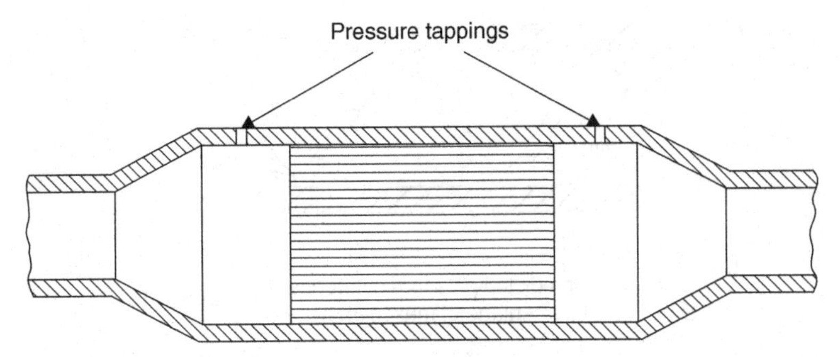

Figure 8.20. Diagram of a viscous flowmeter.

where $A' = AL/D^4$ (Bean 1971). This is essentially Equation (8.14) in an alternative form. If we include the factor in (a), we obtain

$$q_v = B(\Delta p/\mu)(1 + C_1 \,\mathrm{Re}) \tag{8.17}$$

which is similar in form to the equation Jones (1992) obtained, except that Jones also had a negative constant term.

Weigand claimed that the laminar flow elements that he described could operate over turndown ratios of up to 100:1. The main requirement was that viscosity was known. The laminar flow element (LFE) consisted of a housing and capillary element. The housing of stainless steel or aluminium provided inlet and outlet connections, structural support and connections (e.g. to measure differential pressure). The capillary element may have consisted of one capillary tube with internal diameter as small as 0.228 mm, a bundle of tubes or a matrix array consisting of a series of triangular passages made from 0.025 mm stainless steel stock and triangles with dimensions of 0.58 mm or less. The length of the capillaries were 76 mm (25 mm for the largest devices and maximum flow rates to reduce pressure loss) and were rigidly held to ensure geometric integrity and consequent repeatability. The 1.5 mm-diameter sensing ports were at the inlet and outlet and within 1.5 mm of the capillary element. The differential pressure was usually less than 254 mm water. Single tubes were used for 0–5 cm^3/min up to 0–230 cm^3/min. Bundles were used up to 1,300 cm^3/min, and arrays, up to 425 m^3/min. The normalised flow equation was given as

$$\mathrm{SCFM} = \mathrm{ACFM}\mu_g/\mu_{std} \tag{8.18}$$

where SCFM is the standardised flow rate, ACFM is the actual flow rate from calibration, μ_g is the viscosity of the calibration gas at flowing conditions, and μ_{std} is the viscosity of the reference gas at standard conditions.

The entrance and exit of the capillary caused a slight nonlinearity. Repeatability was claimed as of order ±0.1%. Mass could be obtained using the density. For moist air, there was a humidity correction. Provided the temperature was less than about 26°F and the relative humidity less than 80%, the correction was less than 0.4%. The meter should have been kept as warm as the flowing air. By combining temperature measurement, pressure measurement and differential pressure measurement and

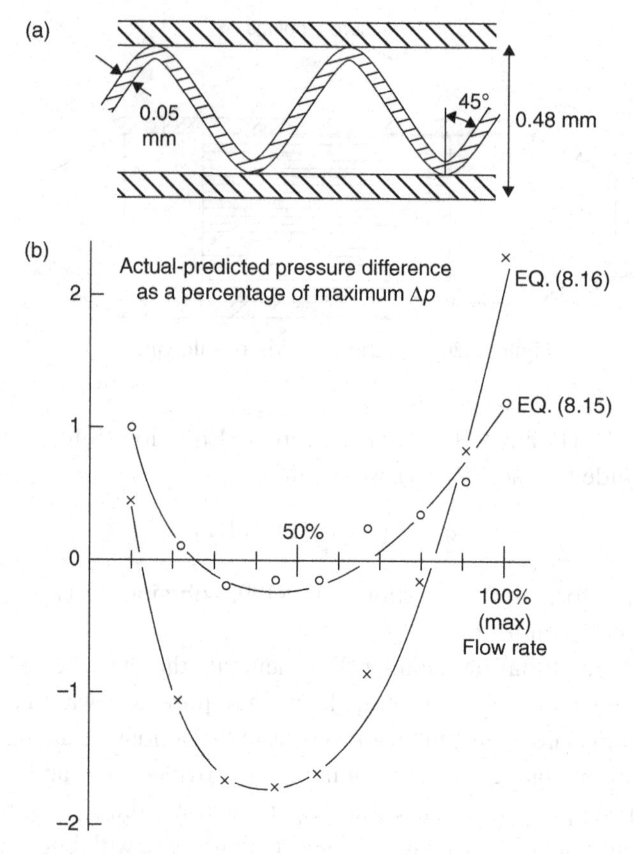

Figure 8.21. Test of viscous flowmeter (after Stone and Wright 1994): **(a)** Section through flow passages of meter; **(b)** comparison between experimental data for pressure drop and equations of the form of Equation (8.15) (∘) and Equation (8.16) (×). Maximum flow rate was about 32 l/s, and maximum Δ_p was about 250 mm water.

using a microprocessor, gas mass flow rate was obtained. An uncertainty of ±1% of reading for 10:1 turndown was claimed to be achievable.

Abe and Yoshinaga (1991) also described what appears to be essentially a laminar flowmeter with etoile-like resistors but with the centre blocked. Their slits were recommended to be 0.2 mm wide and 40–60 mm long. With two in parallel, they claimed flow rates up to about 0.5×10^{-3} m^3/s were possible, but nonlinearity appeared to set in below this figure. The device may be used as a flowmeter or a means to measure viscosity.

Stone and Wright (1994) tested a meter with passages as shown in Figure 8.21(a) with length 76.2 mm, about 18,000 passages and a flow area of 3507 mm^2. The hydraulic mean diameter was 0.347 mm, and the mean velocity at rated flow was 1.73 m/s. Figure 8.21(b) shows the flowmeter characteristic compared with a quadratic assumption as in Equation (8.15) and a linear assumption as in Equation (8.16). Both curves were obtained using regression analysis and not through evaluation of the constants.

Tison and Berndt (1997) reported calibrations of a laminar flowmeter using nitrogen and argon and obtaining typical long-term uncertainties of about 0.1%. A development of the laminar flowmeter which used a quartz capillary was reported by Berg (2004, 2005), and had been developed as a transfer standard by NIST. The device was modelled to allow for various corrections, including non-ideal gas behaviour, slip and tube curvature. It was designed for gas flow rates of 0.07 to 1,000 μmol/s in three sizes (see also Berg 2008). Nishimura, Kawashima and Kagawa (2008) used CFD on a laminar flowmeter and showed in tests that it is capable of measuring oscillatory flow up to 9 Hz. Cobu et al. (2010) calibrated three designs of laminar flowmeter with nitrogen at four pressures (100 kPa, 200 kPa, 300 kPa and 400 kPa) over a 10:1 flow range using NIST's primary flow standards and a physical model. Without additional calibration, each laminar flowmeter was used to measure the flow of three process gases (Ar, He and CO_2) over the same pressure and flow ranges with a maximum error of only 0.5%. The calibration and flow measurements used the gas-property data from NIST's database REFPROP 8.0 and a physical model for each meter that accounted for the viscous pressure drop, compressibility and non-ideal gas behaviour, slip flow effects, kinetic energy effects, gas expansion effects and thermal effects (see also Wright et al. 2012). The development of a laminar flow element consisting of a single or multiple straight glass capillaries in parallel as a transfer standard was described by Feng et al. (2010). Two gauges and one thermometer measured the inlet/outlet pressure and inlet temperature. The differential pressure was in the range 2 kPa to 100 kPa. The glass capillaries were manufactured by laser machining, resulting in consistent inner diameter and straight flow path. The reproducibility was within 0.03 %, and it was claimed that the meters could span flow rates of 0.8 to 986 μmol/s to within the ±0.15 %. Additional measurement with nitrogen demonstrated the feasibility of measurement with multiple gases.

8.15 Chapter Conclusions

This chapter has covered a number of types of meter, among which the variable area meter is possibly the most common. The great benefit of the transparent-tube designs of the variable area meter is the immediate visual reassurance that flow of a certain level is taking place. However, in many applications, a metal tube may be considered necessary for safety reasons. Work has started to analyse the flow and to identify reasons for the instability that has been observed since the first use of the meter. There may be value in pursuing these studies further to provide an answer to the reasons for limits of accuracy for the meter:

- Does upstream disturbance including swirl actually affect performance?
- Why is it necessary to quote an uncertainty as a percentage of full flow?
- Does turbulence level affect the meter reading?

Many of the devices depend on proprietary information on performance. The user is, therefore, heavily dependent on the information provided by the manufacturer.

The averaging-pitots clearly have a useful role where a full-bore meter is not possible, but whether the claims for installed accuracy are reproduced in practice is dependent mainly on manufacturers' data.

It seems likely that devices such as those in this chapter that use one or other of

- contoured tube walls,
- various shapes of insert, and
- existing pipe features

will continue to have a place. Several of the designs described have been introduced relatively recently. The V-cone is one such which appears to be finding a role.

There could be other differential meters proposed. One, for instance, might be an inside-out nozzle or short Venturi in the form of a bullet-shaped target meter, with some resemblance to a V-cone meter, but with direct electrical output via a force sensor and this could be combined with wall differential pressure measurements to provide a condition monitoring function. Ong et al. (2007) described an inverted Venturi design. Instead of a constricted section in the tubing, the inverted Venturi had an expanded section. Beaulieu et al. (2011) developed a meter for unsteady liquid flows for a biomedical application and based it on a symmetrical Venturi so that unsteady measurements might be possible. It appeared successful in its application.

If a manufacturer is considering the introduction of new designs, the user will be dependent on manufacturer's data for most of the designs that fall outside a standard's specification. The quality of the manufacturer's total operation will, therefore, be a key consideration. The effect of installation should be plausible compared with similar devices from elsewhere. The biggest uncertainty will be in the special conditions that are unforeseen: small traces of gas in a liquid becoming trapped in flow vortices, deposition of small quantities of a solid or liquid film and materials problems. Increasingly, however, the question will need to be asked, "Is there a linear meter that would be more appropriate?"

Appendix 8.A History, Equations and Maximum Permissible Error Limits for the VA Meter

8.A.1 Some History

In 1873, G. F. Deacon produced a design that resembled a variable area meter. This was very similar to an earlier device patented in 1868 and made by E. A. Chameroy of Paris (Coleman 1956). Sir J. Alfred Ewing (1924–1925) devised something similar about 1876 which was initially known as the Ewing flowmeter. Awbery and Griffiths

(1926–1927) carried out further experiments. They found that the float, if a ball, started chattering at high flows and caused an overreading. Three solutions were suggested:

i. Inclining the tube so that the ball rolled on the wall. This was Ewing's preferred solution.
ii. Using a float of a different shape. Awbery and Griffiths (1926–1927) mentioned that the meter used a cylindrical float with helical channels to cause slow rotation.
iii. Stretching a fine wire along the tube close to the wall, to prevent the chattering motion (Advisory Committee on Aeronautics 1916–1917). Awbery and Griffiths verified this in their observations.

One possible reason for this instability (Leder 1996) is a saddle point (free stagnation point) that forms in the wake downstream of the ball and off-axis and that rotates and induces a large-scale rotating helical vortex together with small-scale turbulence motions.

Ewing suggested that the range of the instrument could be extended by incorporating more than one float. The largest would be on top and would record low flows, and the smaller ones underneath would rise when the flow rate was sufficient. Ewing also used a dye trace to show the nature of the flow. An alternative could be to have two floats of similar size but with the top one of lower density.

Awbery and Griffiths showed that their experimental results collapsed onto a single curve for a given diameter ratio of D/d, where D is the diameter of the tube at a certain height, and d is the diameter of the ball. They plotted $\log[q_v D/(D^2 - d^2)]$ against $\log[d^3(\rho_f - \rho)/v^2\rho]$, where q_v is the volumetric flow rate, ρ is the density of the fluid, ρ_f is the density of the float, and v is the kinematic viscosity of the fluid.

British Patent 2428 granted to G. Joslin of Colchester (1879) covered a device for ascertaining rate of consumption of gas, which resembles a VA meter, and this was followed in 1908 by a patent granted to K. Kuppers of Aachen for a tube and float very similar to Joslin's. Felix Meyer of Aachen obtained the patent and started manufacturing the tubes. In 1921 Trost Brothers Ltd. of England took up Meyer's agency for the United Kingdom, and in 1931 Schutte and Koerting secured the manufacturing rights for the United States. Coleman (1956) gave a very useful historical review from which some of these notes are taken.

Head (1946) refers to a downdraught instrument in which the float is lighter than the fluid.

8.A.2 Equations

The basic theory of the flowmeter depends on the upthrust on the float resulting from the pressure loss across it

$$p_1 - p_2 = K\tfrac{1}{2}\rho V^2 \tag{8.A.1}$$

where p_1 and p_2 are the upstream and downstream pressures, K is a loss coefficient, ρ is the density, and V is the velocity in the annular passage past the float. Replacing velocity by volumetric flow rate q_v such that

$$V = q_v / A_2 \tag{8.A.2}$$

where A_2 is the annular area around the float, we obtain

$$p_1 - p_2 = K \tfrac{1}{2} \rho q_v^2 / A_2^2 \tag{8.A.3}$$

This equation assumes that the dynamic head at inlet is negligible, and that the pressure difference across the float is caused by the loss of the dynamic head downstream of the float. Dijstelbergen (1964) used a contraction coefficient C_c for the ratio A_2/A_x, where A_x is the area of the tapering tube at height x.

The immersed weight of the float is given by

$$W = V_f (\rho_f - \rho) g \tag{8.A.4}$$

where V_f is the volume of the float, ρ_f is the density of the material of the float, and g is the acceleration due to gravity. Thus, the balance is given by

$$K \tfrac{1}{2} \rho q_v^2 A_f / A_2^2 = V_f (\rho_f - \rho) g \tag{8.A.5}$$

where A_f is the maximum cross-sectional area of the float.

This is usually rewritten as

$$q_v = \frac{C_c}{\sqrt{K}} A_x \sqrt{\frac{2V_f}{A_f} \frac{(\rho_f - \rho) g}{\rho}} \tag{8.A.6}$$

with $1/\sqrt{K}$ replaced by C (e.g. Martin 1949; Coleman 1956; cf. Head 1946–1947).

Coleman (1956) gave values of C for three float shapes (Figure 8.A.1), which suggested that at full flow the value of C ranged from 0.6 for a float of low Re sensitivity to about unity. The plots also showed a decrease in C at low flow rates, and this would broadly reflect the fact that the loss coefficient for an orifice increases with reduced Re (Miller 1990). Coleman also plotted the effect of viscosity on the various float shapes and confirmed the implications of Figure 8.A.1, that the plumb-bob is the most viscosity dependent.

Schoenborn and Colburn (1939) gave a curve of C against Re/C, where Re is the Reynolds number based on D_{eq}. They called it the equivalent diameter but defined it as the difference between D and d. This curve is reproduced in Figure 8.A.2. It was obtained for a large, high-pressure design using a steel tube with extension rod observable through a graduated glass window. This curve could be used as a first approximation if no other is available. They used a balance of forces of the weight of the float in liquid to the upward force due to the pressure reduction across the float to provide the basis for their experimental correlation. They also noted that the curve for large meters appeared to be less fluid-dependent than for smaller instruments.

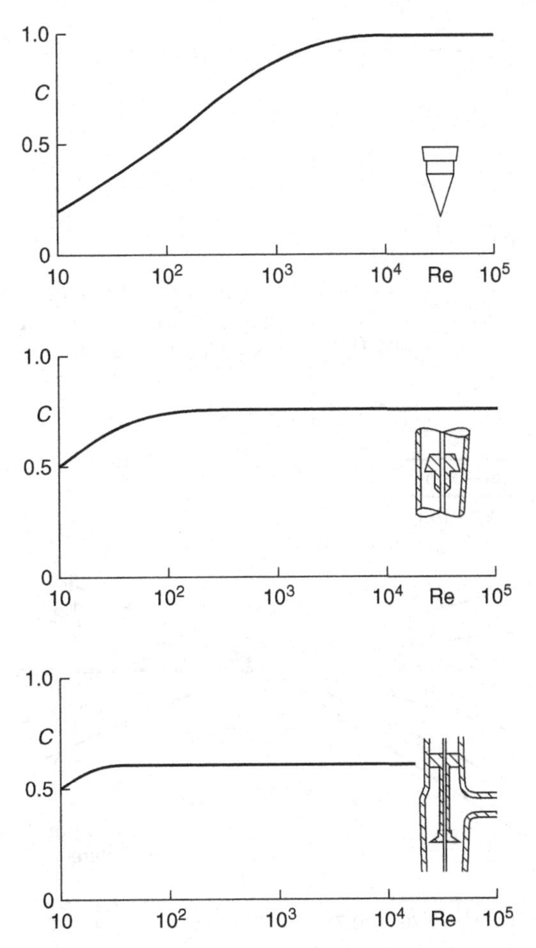

Figure 8.A.1. Effect of float shape on flow coefficient (after Coleman 1956).

Although it may be difficult to predict flow rates, it is possible to convert from one fluid to another at the same reading. Schoenborn and Colburn (1939) gave an equation similar to that given by ISA (1961b):

$$\frac{q_{v2}}{q_{v1}} = \frac{C_2\left[(\rho_f - \rho_2)/\rho_2\right]^{1/2}}{C_1\left[(\rho_f - \rho_1)/\rho_1\right]^{1/2}} \tag{8.A.7}$$

The difference is that ISA (1961b) omits C_2/C_1, thereby making the assumption that the ratio is unity. Coleman (1956) also used this equation with C_2/C_1 equal to unity, noting that variation of viscosity is ignored. He referred to the fact that if Equation (8.A.6) is rewritten for mass flow rate and differentiated for variation in fluid density, it can be shown that a float density of twice the fluid density gives a zero change and, therefore, is least sensitive to variation in density (cf. Head 1946). Figure 8.A.3 shows values of the density correction factor with change of fluid density and for various float densities, based on calibration for aviation gasoline of specific gravity 0.72. Head (1946) also raised the possibility that, with a downdraught

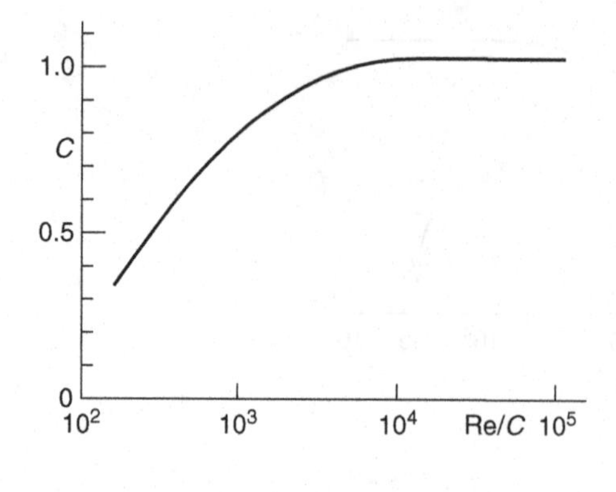

Figure 8.A.2. Plot of C against Re/C for a typical variable area meter (after Schoenborn and Colburn 1939).

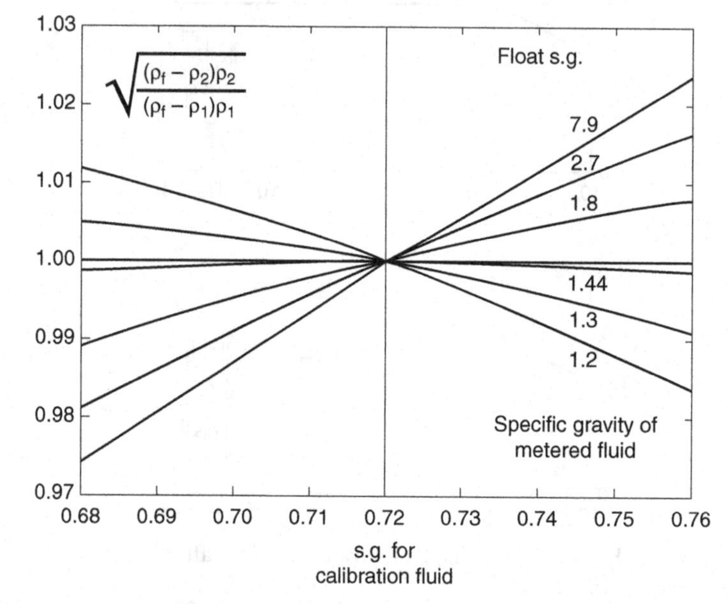

Figure 8.A.3. Density compensation float comparison curves (after Coleman 1956).

design, the requirement from differentiating the volume flow equation of a float of zero density could be achieved approximately. Head also suggested that float automatic compensation might be used to allow for temperature variation and other variables. Several authors have pointed out that the value of C is constant if $\mu/\sqrt{\rho(\rho_f - \rho)}$ is constant.

8.A.3 Maximum Permissible Error Limits

The error limits given by VDI/VDE 3513-2:2008 have been revised from earlier versions of this document. It now makes use of two parameters.

q_{vG} specifies the point in the flowmeter characteristic, as a percentage of full scale, above which the meter is approximately linear;

G is a permissible maximum percentage error for flow rates above q_{vG} for the linear region.

The maximum error limits, as a percentage of rate, may then be given by:

$$G\frac{q_{vG}}{q_v} \quad \text{for} \quad q_v < q_{vG}$$
$$G \quad \text{for} \quad q_v \geq q_{vG}$$

Note 1: All VA manufacturers, that apply the VDI/VDE directive, use consistently $q_{vG} = 50\%$.

Note 2: VDI/VDE 3513 is a directive about Variable Area Flowmeters and consists of 3 parts:

VDI/VDE 3513-1: Calculation methods
VDI/VDE 3513-2: Accuracy definition
VDI/VDE 3513-3: Installation recommendations

9

Positive Displacement Flowmeters

9.1 Introduction

The material in this chapter is based on the paper by Baker and Morris (1985) on positive displacement (PD) meters, which in turn has been updated with more recent industrial and published material and extended to gases. P. D. Baker's (1983) paper provided some additional useful information on these meters. The main types of liquid meter were given by Barnes (1982), Hendrix (1982), Henke (1955), and Gerrard (1979) (cf. Mankin 1955). The reader might refer to API (2005) and similar documents. Morton, Hutchings and Baker (2014a) provided additional references.

At least four of the meter designs to be discussed have been around for over 100 years. The nutating disc flowmeter for liquids was developed in 1850. The oscillating circular piston meter appeared in the late 19th century (Baker 1998).

The measurement of gas has depended, from an early date, on two types of positive displacement meter: the wet gas meter of high accuracy and credited to Samuel Clegg (1815), and the diaphragm meter of lower performance but greater range for which William Richards (1843) should take the credit and which has been widely adopted for the domestic gas metering market. The rotary PD gas meter appears to have shown good performance.

9.1.1 Background

The concept of carrying known volumes of fluid through a flowmeter is a short step from the use of a discrete measure such as a bucket or measuring flask. Thus in each of the designs described later, the flow enters a compartment that is as tightly sealed as is compatible with relative movement of adjacent components. A knowledge of how many of these compartments move through the flowmeter in one rotation of the shaft leads to a knowledge of the flow rate for a certain rotational speed. A simple theory is developed in Appendix 9.A.1. Clearly every leakage path will reduce or increase the amount carried and cause an uncertainty in the measurement. We can make an approximation of the leakage flows in the simple model. In addition, pressure and temperature will distort the volume and may cause small errors, and we may need to develop compensation for these. Compensation for pressure can be provided in the form of a double-walled meter in which the precision-measuring chamber is unpressurised since the pressure is the same inside and out, and the pressure drop to the outside world is taken across the second and outer wall, the shape of which will not affect the accuracy. The temperature compensation, if necessary, can be incorporated into the calibration adjustment by a temperature-sensitive element.

Very often these devices provide a direct sales ticket that is produced by the rotation of the meter and the starting and stopping of the flow. On the other hand, it is very difficult to make small adjustments as needed in calibration where the signal is from a mechanical rotation. We, therefore, need to introduce complicated mechanical calibrators where a frequency output is not acceptable. The combination of mechanical counter and ticket machine can cause a substantial drag on the meter and may cause slippage errors in certain types of calibrators. Recent electrical systems should overcome these problems.

Many of the points made in this chapter relate primarily to liquid measurement, have some relevance to gases, and should be noted by those mainly interested in gas measurement.

9.1.2 Qualitative Description of Operation

The essential of any positive displacement meter, as indicated earlier, is that a discrete and well-defined portion of the fluid is carried from inlet to outlet without loss or mixing with the remainder of the fluid. This applies to all the fluid that passes through the meter. The skill and ingenuity of the designer is shown, therefore, in achieving this as simply and precisely as possible. If, for instance, we know the volume of each compartment of a reciprocating pump, then, knowing how many cylinder-full portions are transmitted in a revolution, we can relate the amount passed to the speed or revolutions of the pump. This is essentially the principle of a metering pump. Taking this a step further, we could reverse the reciprocating pump. Instead of providing power externally to drive the pump, the fluid moves the pistons, and the valves ensure that each portion is transmitted, without loss, to the outlet. Of course there will be some loss due to imperfect sealing between the pistons and the cylinders, but provided this can be kept very small, we have the basis of a very accurate instrument. The reciprocating principle may not be satisfactory for high-throughput meters, and so other designs have been devised. It will be easiest to explain the working of each in turn with respect to diagrams.

The theory of the device is worked out in Appendix 9.A.1. Here we need only note that the simplest equation of flow through such a device is

$$q_v = \text{Volume per revolution} \times \text{Rotational speed} - \text{Leakage} \tag{9.1}$$

Leakage is unlikely to be a constant value for one revolution and is more likely to be speed-dependent. The output from some meters is a rotating shaft, and this needs to be related to the flow rate and total passed. If the shaft passes through the pressure containment of the meter, then a rotary-shaft-sealing arrangement will be needed. In one type of meter, an externally lubricated packing gland is used to

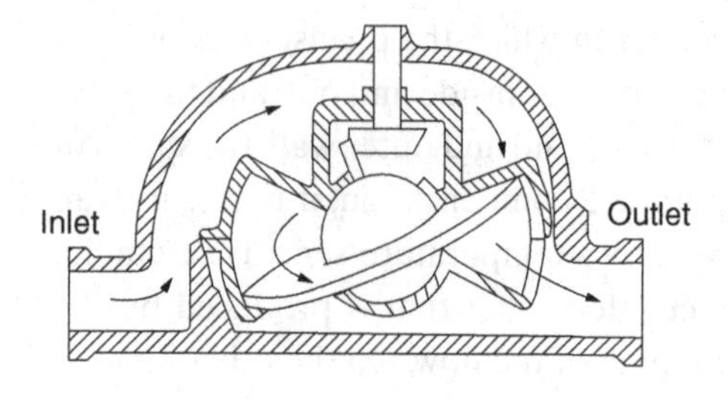

Figure 9.1. Nutating disc meter (reproduced from Baker and Morris 1985; with permission of the Institute of Measurement and Control).

isolate the dynamic shaft seal from the product (Baker 1983), thereby increasing the life of the packing gland. The external lubricant must, of course, be chemically compatible with the product being metered. In some cases, the problem is avoided by using a magnetic drive coupling.

In the past, calibration and transmission methods have tended to be by mechanical linkage. Such a linkage loads the rotor and may alter the characteristic of the meter. It is also difficult to design a mechanical system to adjust the calibration of the meter. Some ingenious methods have been adopted and will be described later in this chapter, largely for historical interest and for those who have dealings with such devices that remain in service. Today the possibility of using electrical or optical transmission while retaining intrinsic safety means that an output pulse train can be used, and signals can be manipulated electronically.

9.2 Principal Designs of Liquid Meters

This section reviews the main designs and their special operational features and, where possible, reviews the published information that is available about them.

9.2.1 Nutating Disc Meter

In the nutating disc meter (Figure 9.1), the disc is constrained by the central spherical bearing and by the transmission to nutate. Nutation is a movement rather like a spinning top that is slowing down. The disc is prevented from rotating by a partition separating the inlet and outlet streams. The incoming flow causes the disc to nutate and in doing so the disc forms a closed compartment trapping a fixed volume of liquid. There are effectively two such compartments, one above and one below the disc.

The disc element has the shape of the planet Saturn. The important difference is that the disc has a slit in its "rings", which makes a close fit with a radial partition. Flow enters, say, under the disc on one side of the partition and causes the disc to rise and fill the region under the disc, and so to nutate. When it has nutated half a turn, the entry port becomes blocked by the disc, and the other entry port above the disc has started to allow fluid above the disc. At the same time, the exit port for the liquid below the disc opens to discharge. Thus in terms of Equation (9.1), in one rotation, the cavity on one side of the disc is both filled and discharged. It is unlikely that the leakage can be kept very small, and the overall accuracy is not expected to be very

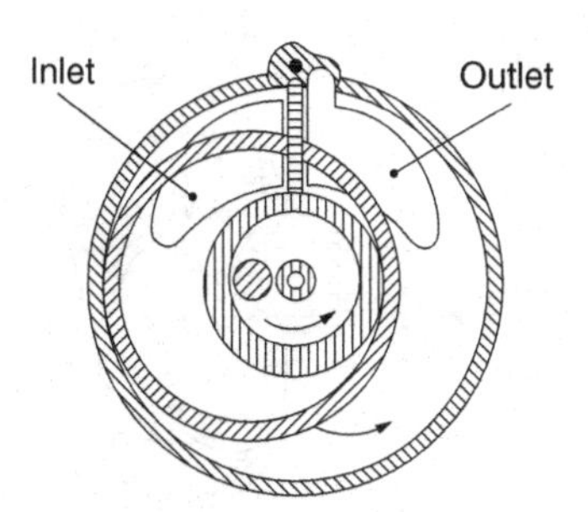

Figure 9.2. Oscillating circular piston meter (reproduced from Baker and Morris 1985; with permission of the Institute of Measurement and Control).

high. Although rather difficult to describe, it is a simple meter of low cost, rugged and capable of uncertainty within ±1.5% (Henke 1955).

Typical industrial designs, some of which may be for water flow measurement, may have a built-in filter. Claims for measurement uncertainty may be of order 1% for flow rates of 10–100 l/min with minimum pressure of 0.1 bar, viscosity up to 1,000 mPa and temperature up to 80°C. Materials may be stainless steel, polypropylene or ethylene-tetra-fluor-ethylene. Modern designs may include a microcomputer to convert flow rate into the correct units.

9.2.2 Oscillating Circular Piston (Also known as Rotary Piston) Meter

The oscillating circular piston meter (Figure 9.2) is similar to the nutating disc meter in that the rotation of the piston is constrained by a partition, so that the piston, in fact, oscillates. Its centre is constrained to move in a circle so that the radius of the cavity is essentially the sum of the radii of the oscillating piston and the circle on which its centre moves. When the centre of the oscillating piston is at the top of its travel, the piston forms a closed compartment with the cavity. One rotation will cause the oscillating piston to return to its starting place and so to discharge the volume of one compartment. For a hollow piston liquid will also be carried inside it. In some designs, the leakage may not be negligible. These meters have wide potential for use and, as well as use with water and chemicals, these meters may be made in designs suitable for liquid foods and other nutritional fluids (Anon. 1966a). Hamblett (1970) suggested that they could achieve a measurement uncertainty within ±0.25% for flow rates from 0.018 to 110 m^3/h (4–25,000 gal/h). This accuracy is rather higher than would be expected. He also listed the various types of registers and controls which were suitable for a range of applications. Recent research on the performance of these meters is described briefly in Appendix 9.A.2.

9.2.3 Multirotor Meters

There are several designs of meter with multiple rotors. Figure 9.3(a) has two rotors that seal against each other; each rotor carries liquid through the meter as it rotates. Figure 9.3(b) shows another two-rotor meter, but in this case the main rotor consists of four vanes that form metering compartments. The second rotor is a sealing

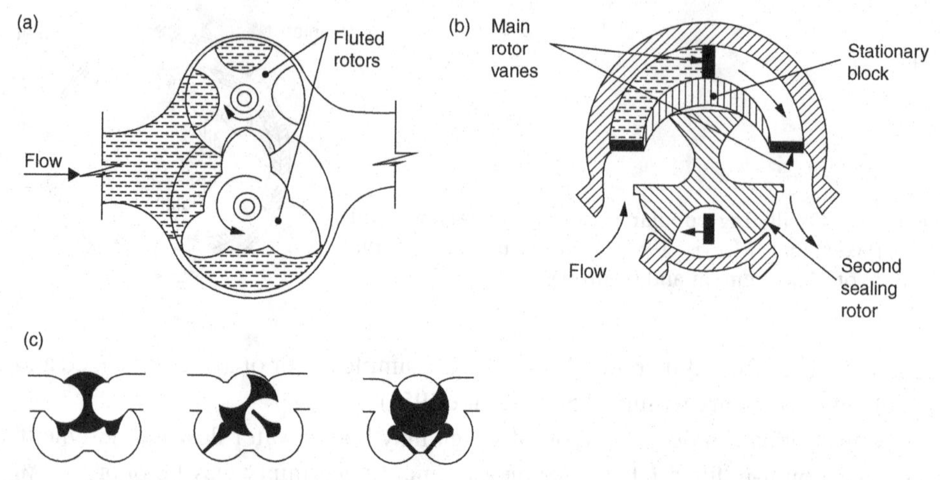

Figure 9.3. Multirotor meters: **(a)** Two rotor (reproduced from Baker and Morris 1985; with permission of the Institute of Measurement and Control); **(b)** two rotor, main rotor has four vanes (reproduced from Baker and Morris 1985; with permission of the Institute of Measurement and Control); **(c)** multirotor: large rotor with two small rotors (after Baker 1983, 1989).

rotor that returns the vanes to the inlet side of the meter. The second rotor may be two-or three-lobed, and its rotation will be precisely linked to that of the main rotor. It transmits a net fluid flow across the meter equal to the volume of the vane that it returns to the inlet. The mechanism, its design and engineering, is complex and, unless manufactured to a very high standard of precision, may lead to high pressure loss and interference between rotors and housing. Figure 9.3(c) shows a design with a large rotor and two small rotors.

A further design is shown in Figure 9.A.4.

9.2.4 Oval Gear Meter

The oval gear meter (Figure 9.4) is a special form of a multiple rotor meter in which each oval rotor is toothed, and sealing between the rotors is enhanced by the resulting labyrinth. Each rotor transmits fluid from inlet to outlet and forms a closed compartment when its major axis is aligned with the flow direction. The volume passed per revolution of each rotor is four times the volume between the rotor and the oval housing when the rotor is confining liquid [Figure 9.4(a)]. In place of the leakage paths between the rotors of multirotor meters, there will be, for this meter, extremely small leakage where the rotors mesh, and the tolerances for the other surfaces are likely to be to a high standard, giving a very small value for the overall leakage.

Small examples of this device are found in oil-flow lines in hydraulic systems, whereas larger high-precision stainless steel versions are also available. They are claimed to have a low pressure drop, to have low maintenance, to be of simple construction, and to achieve uncertainties within ±0.25% (Henke 1955). Meters are

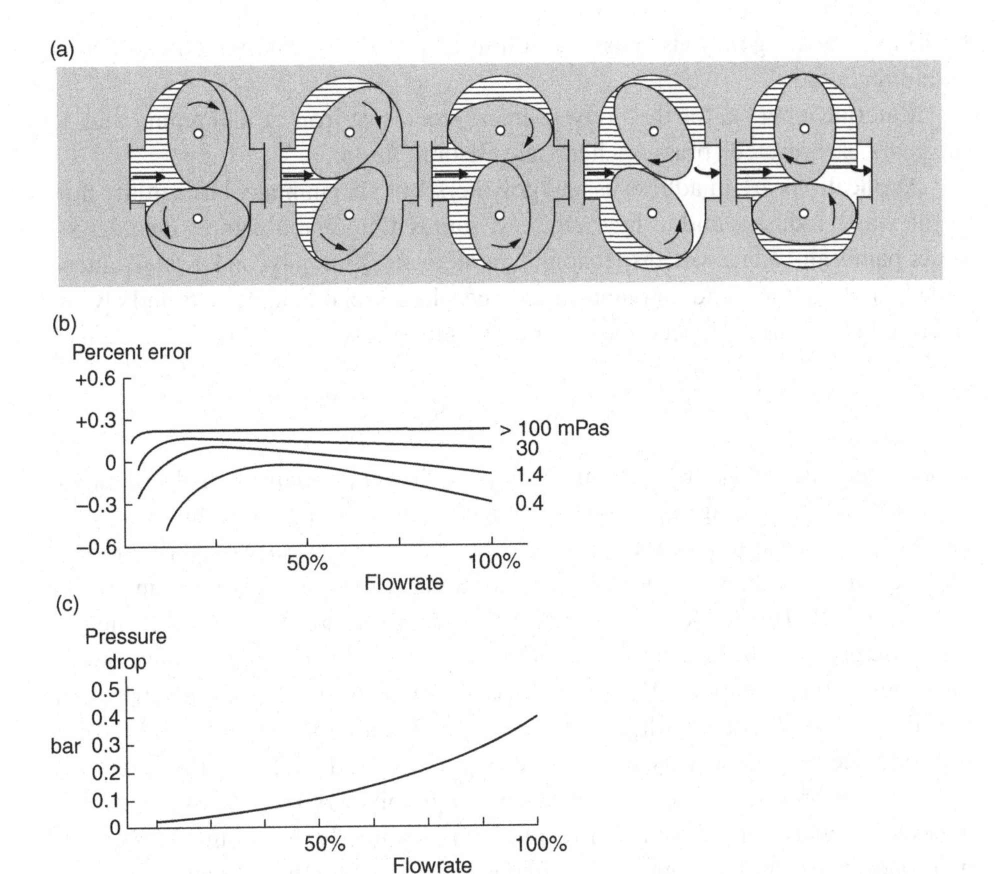

Figure 9.4. Oval gear meter: **(a)** Working principle (reproduced with permission of Bopp and Reuther); **(b)** error curves for a range of viscosity values (after Haar Messtechnik); **(c)** typical pressure drop for operating range (after Haar Messtechnik).

available with a stress-free measuring chamber sometimes achieved by using a double-walled case for high-pressure applications.

The meter can be subject to viscosity variation. Figure 9.4(b) shows the error, and Figure 9.4(c) shows the pressure drop for a typical commercial design. Measurement uncertainty is in the range ±0.1–2%. Sizes are from 6 to 400 mm, with flow ranges from 0.1 l/m or less up to 1,200 m^3/h, but many are for smaller flows. Turndown is of order 10:1 (Endress et al. 1989 see also 2006), and operation is possible with, and largely unaffected by, viscosity in the range 0.4–20 cP. However, the possibility of a calibration shift (of order 0.5%) should be checked with the manufacturer. One manufacturer suggests that operation is possible up to very high viscosities. Pressure loss is 0.5–1 bar, and pressure may be up to 40 bar or higher for certain designs, with temperature up to 290°C.

Transmission may be through a permanent magnet coupling or through magnets embedded in the oval wheels that generate a voltage in an external inductive sensor.

Calibration gearing may also pass the output to a totalising counter. Operation may be bidirectional.

Materials may be, for the body, stainless steel, cast iron or aluminium and, for the gears, carbon steel, brass or chemical-resistant plastic.

Applications are in adhesives and polymers, but also in many basic utility flows in the water industry and in hydraulic test stands. One manufacturer includes solvents, paints and adhesives; dispersions, polymerisates and polycondensates; glucose and alcohols; organic and inorganic liquids; gasoline, fuel oils, lubricants and raw and intermediate liquid products; and other liquid chemicals.

9.2.5 Sliding Vane Meters

Sliding vane meters (Figure 9.5) are, probably, the most accurate of this family of meters. Closed metering compartments are created by means of sliding vanes that move out from a slotted rotor to meter the fluid through a passage of constant shape. The movement of the vanes is caused, in the example in Figure 9.5(a), by an internal stationary cam. The inside edges of the vanes have rollers that follow the cam. The vanes are designed to have a small clearance on all fixed surfaces. The movement of the vanes, in the example of Figure 9.5(b), is achieved by an external cam, which the outside edge of the blade, where it meets the end faces, follows. There is therefore some surface contact, which can result in wear and a need for calibration checks.

The elaboration of Equation (9.1) in Appendix 9.A.1 is focussed on these devices. The measuring volume per revolution is essentially the volume between the two concentric cylinders, which make up the outer case and the rotating inner wheel with vanes. It is reduced by the volume of the vanes themselves. These designs can be highly accurate, and much of the following discussion is concerned with the limits of accuracy. For instance, they are expected to measure with great precision the transfer of valuable liquids, controlled by agencies such as H.M. Customs and Excise (1995). Such precision is likely to be demonstrated by short-term repeatability, such as three successive measurements of 500 l within 0.1%, and long-term repeatability over a six-month period within 0.15%.

Walles (1975) gave figures from his own tests indicating that these figures are achievable with this type [which he refers to as a rotary piston meter and, as shown by his illustration, is in our category of sliding vane meter, Figure 9.5(b)]. He divided the errors in a meter with a double-walled case into three classes – short, medium and long term. Experimental tests with kerosene at the UK National Gas Turbine Establishment, in which the flow was diverted into calibration vessels, suggested that the following figures were applicable: leakage, 0.005% of rated flow; short-term errors, 0.005–0.05% for different flow rates (affected by environmental and fuel temperatures and hence by weather and conditions); medium-term errors, 0.05%; and long-term errors, 0.2%. This gave a year-to-year repeatability of within 0.15%. Systematic (long-term) errors (e.g. in calibration equipment) were difficult to estimate but might be expected to add an additional 0.2% so that the absolute uncertainty was estimated at ±0.35%. Another paper (Walles and James 1985) suggested

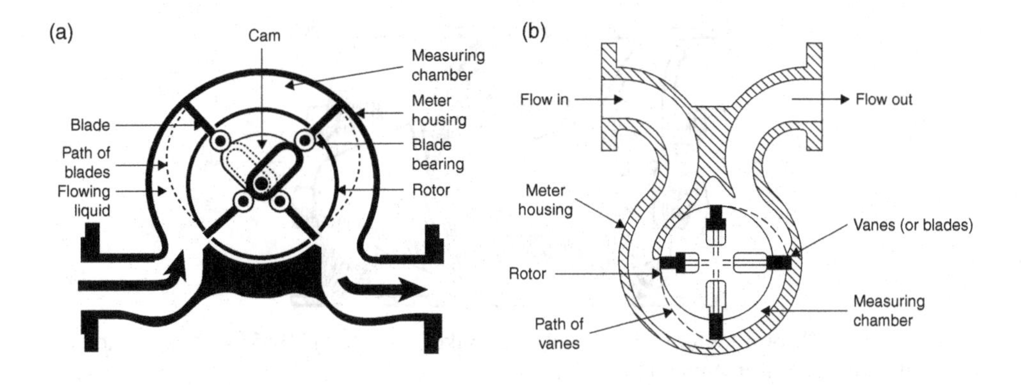

(c)

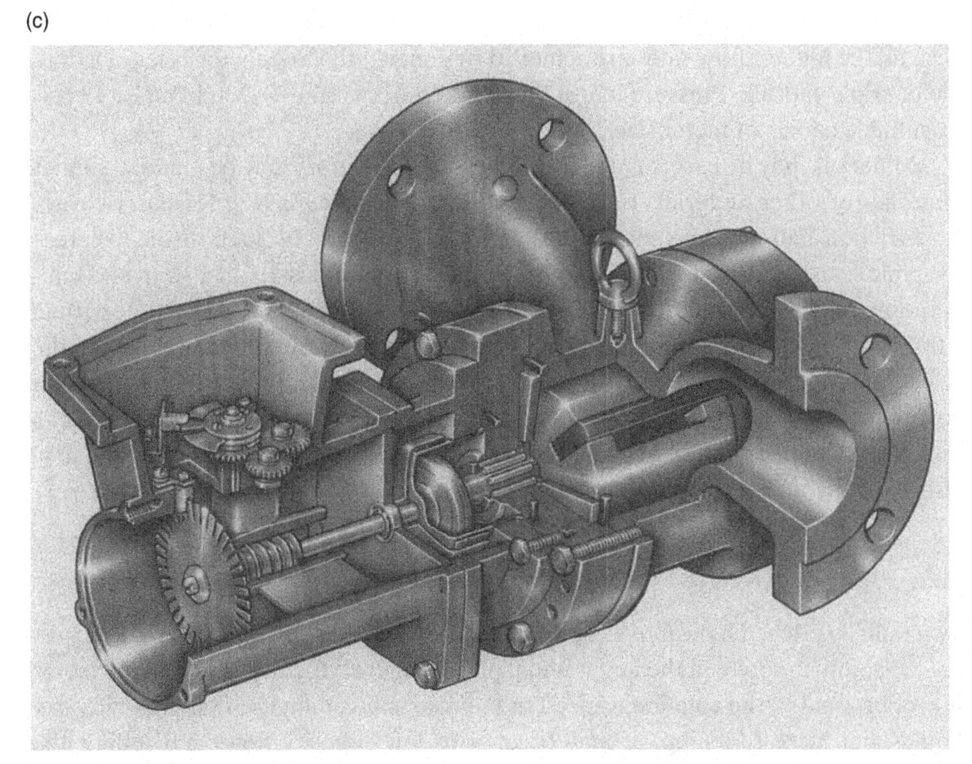

Figure 9.5. Sliding vane meter: **(a)** Smith type showing blade (vane) path (reproduced from Baker 1983; with permission of the author); **(b)** Avery Hardoll type (after the manufacturer's literature); **(c)** VAF J3000 (reproduced with permission of Flowquip Flowmeter Systems.).

that the performance of positive displacement flowmeters can be represented by an ideal meter of arbitrary but constant volume in parallel with a small leakage that is independent of flow rate.

Manufacturers claim very high performance in line with these published figures of 0.1–0.3%. A repeatability of 0.01–0.05% for flow ranges from about 3 m^3/h up to 2,000 m^3/h with turndown ratio of 20:1 and size range of 64–152 mm (2.5–6 in.) is typical. One penalty of this type of meter is its bulk and weight (60–136 kg). In some

Figure 9.6. Helical rotor meter (reproduced from Baker and Morris 1985; with permission of the Institute of Measurement and Control).

designs, the highest flow rates are achieved by ganging the meters together in a double or triple capsule. Pressure loss should be kept to within 70–100 kPa (10–15 psi) by suitable choice of meter size.

Materials may be cast iron for the bodies, and the rotors may be made of similar materials or other materials such as aluminium. The same is true for outer covers. Low-friction ball bearings may be used. Static seals may be high nitrile, whereas dynamic seals may also be fluorocarbon. Working pressure is about 10 bar, and temperature ranges from about –25 to 100°C. Temperature changes cause less than 0.0015%/°C, which is small compared with the theoretical values because of the selection of materials. The measuring chamber is fitted with pressure-balanced end covers to ensure that there is no pressure difference across the end covers of the measuring chamber and so to avoid distortion. Figure 9.5(c) shows a device that may also be available for 10- to 200-mm pipe bore sizes.

9.2.6 Helical Rotor Meter

Two radially pitched helical rotors trap liquid as it flows through the flowmeter, causing the rotors to rotate in the longitudinal plane (Figure 9.6). Flow through the meter is proportional to the rotational speed of the rotors. It can be used on high-viscosity liquids, but increased slippage may occur with low-viscosity flows and reduce the accuracy. The rotors form a seal with each other and with the body of the flowmeter so that these parts must be manufactured to a high degree of precision. One manufacturer (Gerrard 1979) claimed that its positive displacement meter with helical elements had an uncertainty of ±0.5% over a 150:1 flow range and a repeatability of ±0.1%. Performance of 0.1% rate uncertainty and 0.01% repeatability with low effect from viscosity were later claimed. Some designs may withstand pressures up to 230 bar, compensate for temperature changes, and allow the passage of small solids. The use of Equation (9.1) will require a knowledge of how many closed compartments the two helical rotors have created and how many of them pass in one rotation.

One manufacturer confirmed the precision level, suggesting that a higher pressure rating is possible, with maximum flow of about 13 m^3/h, an operating temperature up to about 290°C and viscosity up to 10^6 cP.

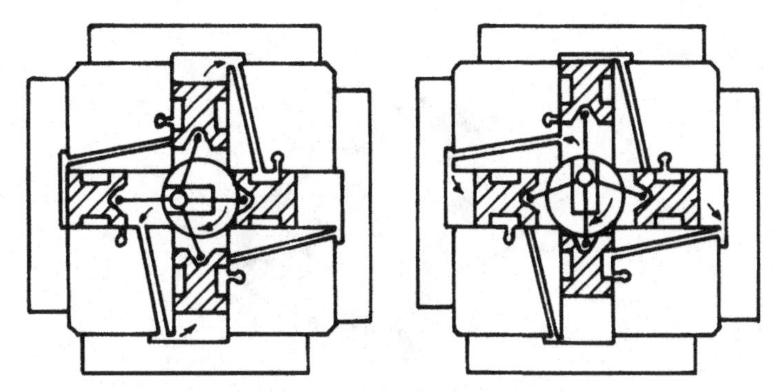

Figure 9.7. Reciprocating piston meter (reproduced from Baker and Morris 1985; with permission of the Institute of Measurement and Control).

Applications are to polymers and adhesives, fuel oils, lubricating oil, blending, hydraulic test stands, high-viscosity and thixotropic fluids. This type of meter has also been developed for multiphase flows in North Sea oil applications (Gold, Miller and Priddy 1991).

9.2.7 Reciprocating Piston Meters

One design in which four pistons trap liquid as it passes through the flowmeter is shown in Figure 9.7. The crankshaft rotates with a rotational speed proportional to the flow through the meter. The liquid that passes through each cylinder in one rotation of the shaft is equal to the swept volume of the cylinder.

Meters of this type have high accuracy claims and may also have a high pressure drop. The Aeroplane and Armament Experimental Establishment (Anon. 1966b) developed a three-cylinder meter for low flows. Uncertainties of ±0.5% were quoted for kerosene flows of 9–182 l/h (2–40 gal/h) with ±2% for gasoline at rates below 32 l/h (7 gal/h). Increased slippage made high accuracy difficult to achieve at low flows, particularly for less viscous liquids.

One commercial producer quoted up to 33 m^3/h (120 gal/min) with ±0.5% of reading and with an 1,800:1 turndown and another down to 1 l/h. Viscosity may be possible up to 30,000 cP or greater and temperatures up to about 280°C. Magnetic couplings may be used to reduce drag and avoid a rotating seal.

Endress et al. (1989 see also 2006) mentioned uncertainty of ±0.5 to 1% of rate, turndown of 10:1, maximum differential pressure of 5 bar, diameter range of 25–100 mm and maximum pressure of 40 bar. Others may offer pressure ratings up to 200 bar.

Materials of construction may be predominantly stainless steel with seals of Viton, Teflon, neoprene, Nordel, Buna-N etc. Filtration to 10 μm may be required, and magnetic coupling may be appropriate.

Applications include gasoline pumps, pilot plants, chemical additives, engine and hydraulic test stands, fuel consumption analysis, blending and ratio control and mercaptan odorant injection.

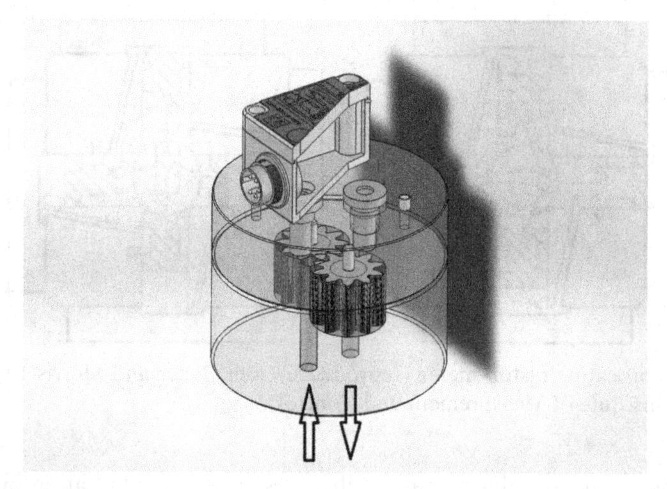

Figure 9.8a. Precision commercial gear meter of KEM Küppers (reproduced with permission of Litre Meter) (**a**) showing the general layout.

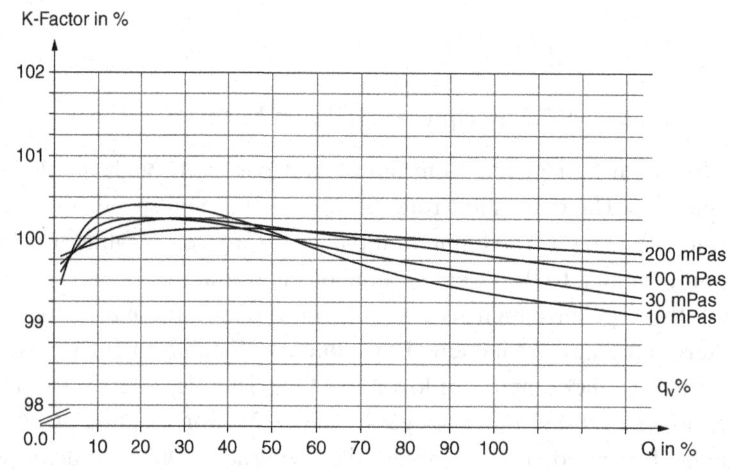

Figure 9.8b. Precision commercial gear meter of KEM Küppers (reproduced with permission of Litre Meter) (**b**) typical characteristics.

9.2.8 Precision Gear (Spur Gear) Flowmeters

Figure 9.8(a) shows an example of a commercial device. The spaces between the gears and the chamber wall form the fluid transfer compartments. Rotation may be sensed by electromagnetic sensors. Two sensors may be used to allow better resolution than one and to determine flow direction. The characteristic curves are given for a meter of this type in Figure 9.8(b). Meters for flow ranges of as low as 0.001 l/min and as high as 1,000 l/min may be available with temperature ranges of –30 to 150°C and pressure up to 300 bar or more. Figure 9.8(c) shows typical pressure losses for various viscosities. Plain bearings or ball bearings may be used depending on the liquid monitored.

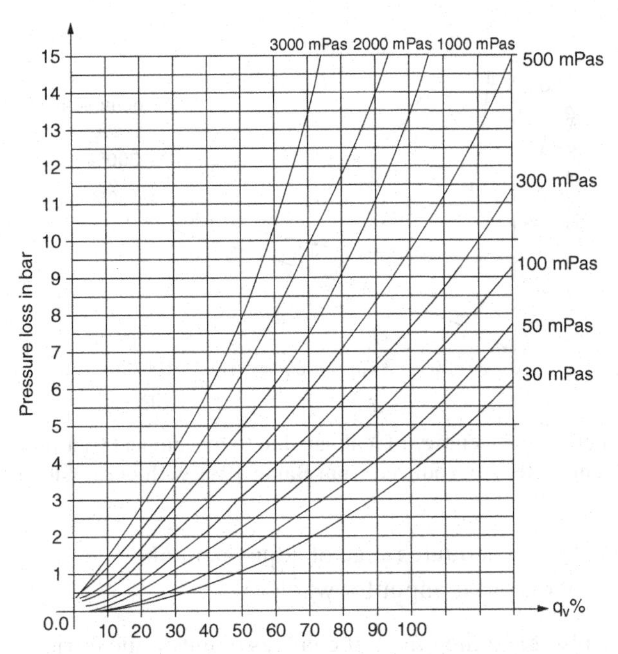

Figure 9.8c. Precision commercial gear meter of KEM Küppers (reproduced with permission of Litre Meter) (**c**) typical pressure loss curves for various viscosities.

Applications may include gasoline, paraffin, kerosene, diesel, mineral oil, hydraulic oil, ink, dyes, paint, greases, polyurethane, polyol-iso-cyanate, glue, paste, cream, resin and wax. For the auto industry, applications are hydraulic systems, fuel consumption, paint spraying, batch processes, adhesive coatings and hydraulics; and for the chemical industry, applications are process plant, batching, leakage and mixing ratios.

Designs are available using a servomotor to maintain zero pressure drop across the meter (cf. Conrad and Trostmann 1981; Katz 1971 regarding servo-control). A vehicle fuel flow measurement system (Anon. 1999) based on a servo-controlled meter used a density sensor to obtain high-accuracy mass flow calculations. Flow range was up to 240 kg/h.

9.3 Calibration, Environmental Compensation and Other Factors Relating to the Accuracy of Liquid Flowmeters

This section is primarily concerned with sliding vane meters. It should be remembered that the drive from these meters, when mechanical, may have to be used on a mechanical calibration device that loads the drive and causes a drag on the rotors. Manufacturers will clearly aim to reduce this as far as possible by the use of low-friction ball bearings etc. The mechanical methods of calibration adjustment are discussed first, and the theory behind them is elaborated in Appendix 9.A.5. Baker (1983) gave the characteristics of a good calibrator as

- ability to drive a high torque,
- long service life and low repair or replacement cost,

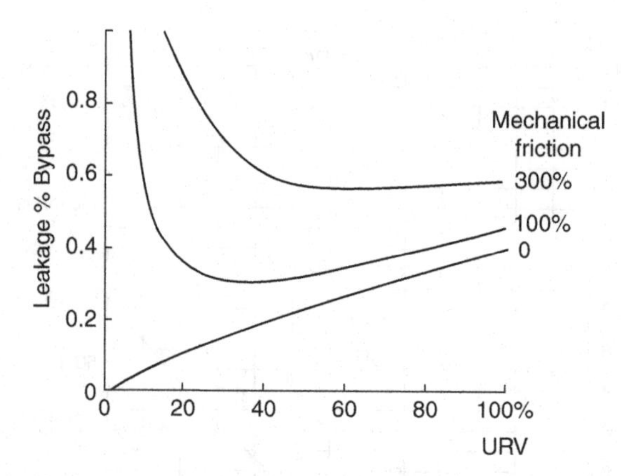

Figure 9.9. Typical effect on accuracy of leakage due to mechanical friction for a sliding vane positive displacement meter (reproduced from Baker 1983; with permission of the author).

- fine adjustment and adequate range of adjustment and
- constant ratio of output to input for whole cycle.

Baker (1983) also gave the effect of changes in mechanical friction in Figure 9.9 showing that greater friction leads to greater leakage past the vanes (cf. Keyser 1973 on the prediction of calibration shift).

9.3.1 Calibration Systems

While retaining this section on mechanical calibrators, it should be obvious that they will be superseded by electrical output systems.

Disc and Friction Wheel

In this system a friction wheel runs on the surface of a disc, the two shafts being perpendicular in the same plane (Figure 9.10). By adjusting the radius at which the friction wheel runs on the disc, a change in shaft rotation ratio can be obtained. This system is usually incorporated into an epicyclic gearbox to allow most of the torque to be transmitted via meshing gears. The disadvantage is clearly the problem of ensuring that slip does not occur while, at the same time, the disc is not worn. The advantages are that the output is proportional to the input, and the system may be used for temperature compensation.

Clutch System

In this method (Figure 9.11), a system of clutches transfers the rotation from one shaft to another by means of an offset shaft. The advantage is that frictional transmission by a friction wheel is replaced by the tight frictional lock of the clutch system,

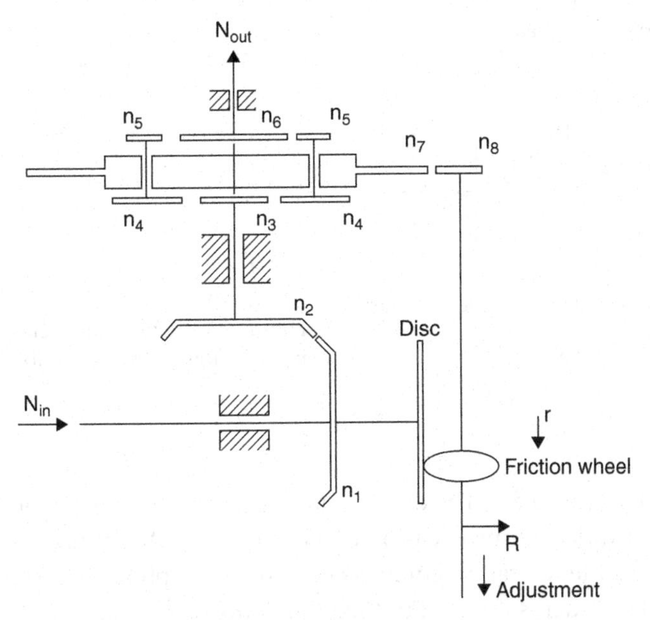

Figure 9.10. Epicyclic calibration adjustment system (reproduced from Baker and Morris 1985; with permission of the Institute of Measurement and Control).

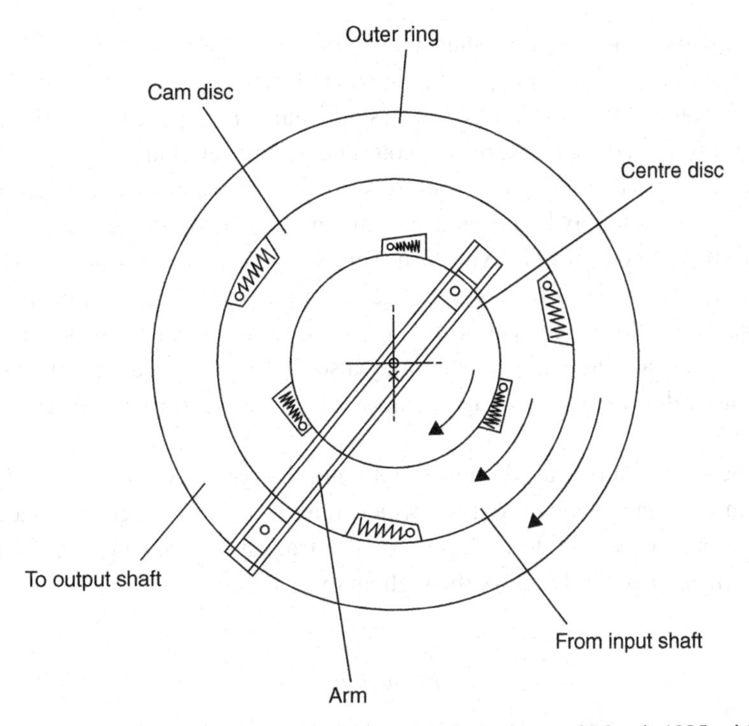

Figure 9.11. Clutch calibration system (reproduced from Baker and Morris 1985; with permission of the Institute of Measurement and Control).

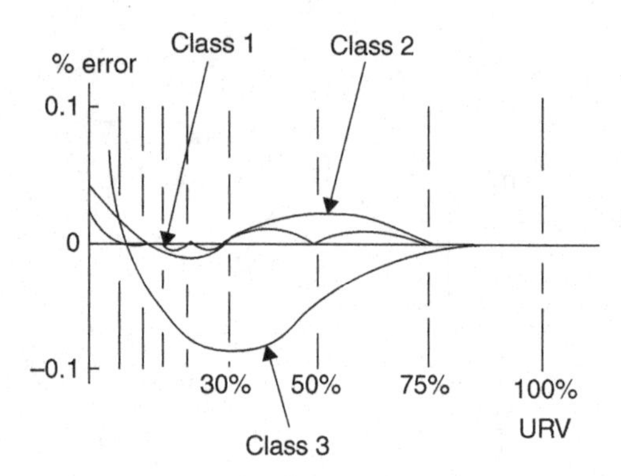

Figure 9.12. Calibration classes (based on information in Avery Hardoll brochures).

enabling a high torque drive. The disadvantage is that the output is not proportional to the input but varies during each rotation of the shaft (Appendix 9.A.5). It is also less well suited to temperature compensation than the previous system. For more information, the reader is referred to Baker and Morris (1985).

Gear box

The output drive from the rotor shaft passes through a gear box, which allows considerable freedom for interchanging the gears to obtain a required calibration ratio. The disadvantage of this method is that it is not simple to change calibration, but the advantages are that, once done, there should be no further change and also the output is directly proportional to the input. It is not suitable for temperature compensation. Some gear train may be necessary, in any meter, to adjust the rotational speed to a required volume flow per revolution. This will precede the calibrator, which can provide a fine adjustment. It may not be easy to provide a geared adjustment better than about 0.5%, and the calibrator will need to improve on this for a meter of this quality. The calibrator mechanism may also require that the output of the gear train has an under- or overreading to allow for the calibrator's physical behaviour (Baker 1983).

Whichever method is used, the output will pass typically through a calibration mechanism to a mechanical counter which may include a batch recorder, ticket printer and preset counter. In addition, a pulse transmitter may be included on the output shaft, and even a data link through an optical fibre.

Electronic

Electronic calibration adjustment, shown in Figure 9.12, is achievable using calibration data storage methods where class 3 has calibration at one flow rate, class 2 has calibration at four flow rates, and class 1 has calibration at eight flow rates.

9.3.2 Clearances

There are four general types of clearance in these meters, which will be sources of leakage:

a. clearance between moving blades and stationary members,
b. clearance between retracting blades and rotors,
c. clearance between two rotors moving relative to each other and
d. clearance between rotors and end housings.

A typical mesh size for a filter for one type of precision positive displacement meter was 0.250 mm, which is, presumably, of the same order of size as the clearances. If deposits form on the meter wall, they may affect this value, and the meter performance should be monitored carefully.

9.3.3 Leakage Through the Clearance Gap Between Vane and Wall

If the rotor is stationary, the leakage is obtained from the following equation (derived in Appendix 9.A), which is also the same equation as Baker (1983):

$$q_{\text{slip}} = \frac{\ell}{12\mu} \frac{(p_{\text{u}} - p_{\text{d}})}{L} t^3 \tag{9.2}$$

Ideally, the liquid in the gap would be carried at the same speed as the rotor blade, and so the leakage when the vanes are moving is

$$q_{\text{leakage}} = \left\{ \frac{Ut}{2} - \frac{1}{12\mu} \frac{(p_{\text{u}} - p_{\text{d}})}{L} t^3 \right\} \ell \tag{9.3}$$

We can introduce values into this equation for the meter dimensions in Figure 9.13. Assuming an axial length of 225 mm and clearances of 0.1 mm, this gives a value of the leakage as

$$q_{\text{leakage}} = 0.156tN - 3.4 \frac{(p_{\text{u}} - p_{\text{d}})}{\mu} t^3 \; m^3/s \tag{9.4}$$

Baker and Morris (1985), who used these values in their estimate of errors due to changes in clearance, showed that changes are of order 0.1% for a 0.1 mm clearance.

Baker (1983) appeared to suggest clearances of 0.003–0.005 in. (75–125 μm), less than the filter mesh size mentioned earlier. However, Baker (1983) appeared to provide experimental values that gave a higher value than those obtained by Baker and Morris [Figure 9.14(a)]. This is an area of necessary compromise in design. Consistent clearance requires high-quality production methods. However, a meter with extremely tight clearances will have reduced service time before inevitable bearing wear causes some clearances to close, subsequently resulting in increased mechanical friction and possibly severe damage to the meter.

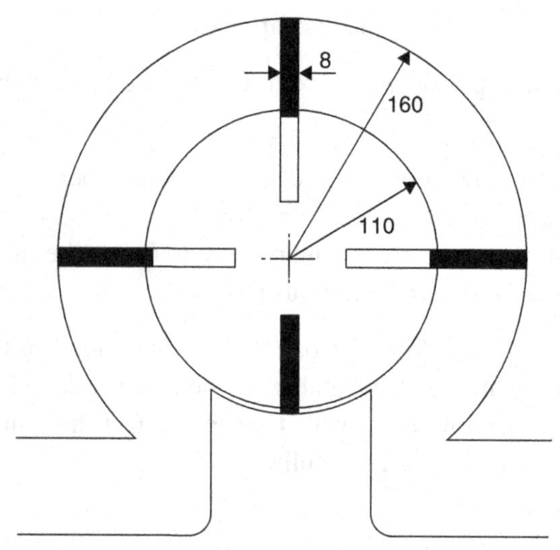

Figure 9.13. Dimensions of a typical sliding vane meter to obtain leakage (reproduced from Baker and Morris 1985; with permission of the Institute of Measurement and Control).

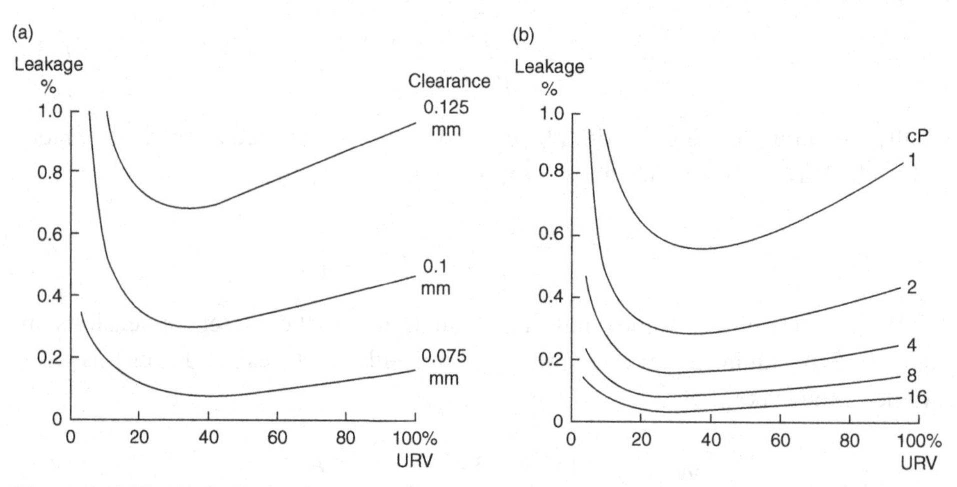

Figure 9.14. Typical effect on accuracy of leakage for a sliding vane positive displacement meter (reproduced from Baker 1983; with permission of the author): **(a)** With clearance width; **(b)** with viscosity.

The other point to note from Equation (9.4) is that the larger the viscosity, the smaller the second term, which is responsible for the part of the leakage past the vanes in the meter, which is not proportional to speed of rotation. This appears to be borne out by Figure 9.14(b) (the figure does not distinguish the direction of leakage which is shown in the equation by a negative sign). Baker (1983) made the point that the positive displacement meter is the ideal meter for liquids with viscosity above about 4 cP such as No. 2 Fuel Oil or 40° API gravity crude oil.

9.3.4 Slippage Tests

The existence of clearances will result in leakage of fluid past the vanes. This, in turn, will mean that the fluid metered may vary as a result of variation in leakage. This is known as slippage and will normally result, particularly at low rotational speeds, in more fluid passing than is registered, an effect that is anathema to the customs and excise authorities (cf. Plank 1951). In rare cases, back slip can occur, which means that the meter registers more fluid than has passed, and the customs and excise benefit. Thus slippage tests are often required, particularly where fuel is dispensed, because high slippage may result in low measure of liquid quantity and is sometimes exploited by using a slow flow rate to fill a vessel, thereby causing an under-registration of the liquid transferred. The last term in Equation (9.4) is not dependent on N, the rotational speed; hence, if N decreases, q_v will not decrease as fast, and there may be a small value of q_v when $N = 0$.

At $N = 1$ revolution per second the final term in Equation (9.4) is 0.01% of the speed-dependent first term. If on a slippage test the meter were to revolve as slowly as 0.01 revolutions per second, the relative size of the final term becomes about 1% and any changes in $(p_u - p_d)$, t, or μ will also be emphasised.

Back Slip

Although slip occurs when liquid passes through the meter in greater quantities than the rotation indicates, back slip is the apparent rotation of the meter while less liquid passes than expected. The former can be due to high bearing friction at low flows, increased clearances, increased pressure drop, etc. The reasons for the latter are less clear, although its occurrence during calibration tests appears to be observed occasionally. Possible causes of the observed behaviour may be categorised as back slip due to meter effects, system effects and calibration vessel.

Back Slip Due to Meter Effects

a. The back slip observations are for very small quantities of liquid passed (15–20 l). For these small quantities, the clutch-type calibrator's varying rotational speed could result in an apparent high rotation. Errors could be ±3%.
b. A meter in which two adjacent vanes had a slight volume difference could also exhibit rotational variation as in (a).
c. The temperature compensation system, where fitted, or other temperature effects due to the slow movement of liquid may cause errors.
d. The digital readout may not allow measurement to within a litre.
e. The low flow rate and rotation may markedly alter the liquid flows on sliding and stationary members. The bearing and clearance drag may be changed.

Back Slip Due to System Effects

f. The quantity of liquid in the pipework may not be precisely the same before and after the slip test. For instance, 20 l of liquid in a slip test fill 2.6 m of a 100 mm-diameter pipe.
g. In tests using manual and automatic valves, one may be closed at the start of the test and both closed at the end, slightly altering the pipe volume.
h. Upstream valve control may cause pressure changes leading to volume changes.
i. Gas may come out of solution during the test.

Back Slip Due to Calibration Vessel

j. The measurement uncertainty of the vessel is not better than about ±0.2% for so short a run.

It appears that various effects occasionally may come together to cause the observation of back slip.

Observations of back slip appear likely to result from uncertainties in operation of the meter at low rotation. Variation during rotation can become important if flows of about 15 l are involved because this may result, for a 100 mm meter, in fewer than two rotations. Thus the non-uniform calibrator transmission or variation in vane clearance may occasionally result in an increased recorded rotation for a given volume passed.

9.3.5 The Effects of Temperature and Pressure Changes

The effects of temperature and pressure variation are given in Equations (9.A.11) to (9.A.18), using values in Figure 9.13. These values are, therefore, for a sliding vane type meter but will give some indication of the effect on others. Making use of values of α_m of 0.23×10^{-4} for aluminium, 0.11×10^{-4} for mild steel and cast iron, and of α_ℓ of 9×10^{-4} for oil, Table 9.1 gives values for temperature changes of 10°C and pressure changes of 1 bar taking E for mild steel as 207 GN/m^2. One manufacturer gave the effect of changes in pressure and temperature on the meter excluding the effect of liquid changes, and this is reproduced in Figure 9.15. Where pressure is sufficient to cause internal distortion of components, significant errors (1 to 2%) might be caused by pressures of the order of 500 bar or more if the meter has no double wall or other pressure balance/protection.

9.3.6 The Effects of Gas in Solution

Another effect not considered so far will be the presence of dissolved gas in solution. The dissolved gas may cause no change in liquid volume, although it may have a large volume in gaseous form. The mass of gas involved will often be small enough to neglect.

Table 9.1. *The effects of temperature and pressure changes*

Description	Metal/ Liquid	$\Delta q_v/q_v$ (%)	$\Delta t/t$ (%)
For temperature change of 10°C			
Flowmeter expansion	Al	0.07	–
	M.S.	0.03	–
Differential expansion	Al	–	37
	M.S.	–	18
	Al/M.S.	–	19
Liquid-density change	Oil	0.9	–
For pressure change of 1 bar			
Flowmeter expansion	M.S.	0.003	–
Leakage gap radial	M.S.	–	1.2
Leakage gap axial	M.S.	–	0.4

However, under sudden pressure reduction, the gas may come out of solution. This might occur during a slippage test, if the test is controlled on a valve upstream of the meter, and might cause the rotor to turn because of the presence of the gas alone.

The problem can also arise when metering liquefied petroleum gas (LPG) if the back pressure on the meter is insufficient.

9.4 Accuracy and Calibration

Accuracy achievable from these meters will be highly dependent on the particular design. Values have been suggested earlier. At the top end, the sliding vane meters can achieve an accuracy that may be limited by the means of calibration. It is highly likely that precision devices such as the oval gear meter will approach the level of performance of the sliding vane meters.

The calculations of the effect of the pressure and temperature in Table 9.1 indicate that close tolerance control is necessary to achieve the required performance. However, it should be remembered that, although liquid density may change the reading of mass flow measurement, the requirement in many cases is for volume measurement.

The effect of changes in Δp and μ are indicated by Equation (9.4), and this clearly illustrates the errors due to change in clearances, pressure drop or viscosity. Wear in a meter will cause t and Δp changes, whereas temperature variation will cause changes in μ.

Apart from the published values already mentioned, Hayward (1977b) developed a technique for measuring the repeatability of flowmeters which involved comparing readings from two identical flowmeters in series. Tests at National Engineering Laboratory, Scotland, suggested that repeatabilities better than 10 ppm (0.001%) could be obtained with vane-type flowmeters provided that the drift in readings, which seemed to arise in the shaft encoder, could be eliminated.

Hayward (1979) commented that the rangeability was about 20:1, linearity was ±0.05%, and uncertainty when newly calibrated was ±0.2% of volume over the

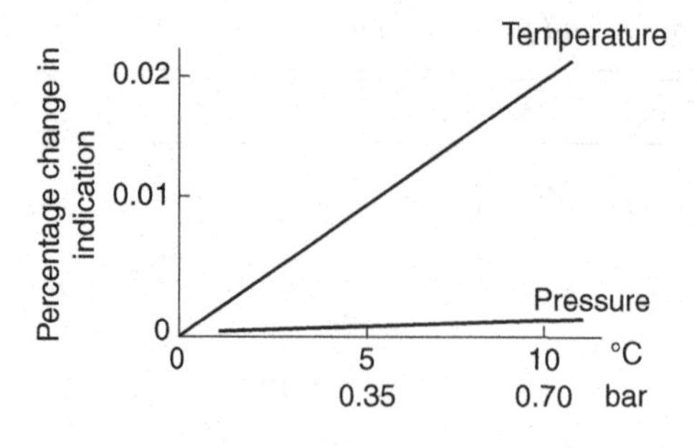

Figure 9.15. Sensitivity of a sliding vane meter to pressure and temperature change excluding the effect of liquid changes (based on Avery Hardoll data).

range. He suggested that these meters perform best in the maximum flow rate range of 0.002–0.05 m^3/s.

Barnes (1982) underlined the superior accuracy, repeatability (within 0.05%) and reliability of positive displacement flowmeters but drew attention to inaccuracies that arose if the liquid contained free or entrained gases. Data from fuel oil, gasoline and propane were presented to show that the accuracy was highest when the meters were used to measure flow at about 25% of the rated flow capacity. At both higher and lower flows, increased slippage occurs and causes the meter to under-register. The meters were also relatively insensitive to changes in viscosity. Barnes suggested that if a meter initially calibrated for a liquid of 1 cP was used with a liquid of 100 cP, the resulting shift in uncertainty was of order 1.2%.

Reitz (1979) quoted the following figures: repeatability within 0.02%, uncertainty ±0.25% over a 20:1 flow range and pressure loss 3 psi (0.2 bar) for liquids of viscosity 1 cP. He suggested that fluids with viscosities in the range 0.1 cP (gas) to 10^6 cP (liquid) could be metered in this way. He noted that errors arise from excessive wear, which increases slippage and causes under-registration, and entrained gas bubbles, which occupy part of the measuring chamber and cause the meter to give too high a reading.

Kent Meters (Scanes 1974) quoted an uncertainty of ±0.5% for a domestic oil meter designed for rates as low as 0.1–10 l/h. The temperature range was quoted as −10°C to 35°C, although temperatures as high as 90°C had apparently been recorded at meter boxes sited in direct sunlight.

One manufacturer gave meter calibration curves for the company's range of meters (Figure 9.16) and claimed linearity in the range 0.1–0.3% with repeatability for all but one of its product sizes in the 0.01–0.02% range.

9.5 Principal Designs of Gas Meters

As indicated in Section 9.1, the wet gas meter and the diaphragm meter have been in use for more than 100 years, and both still provide service for many applications. However, the predominant use of the wet gas meter has changed to that of

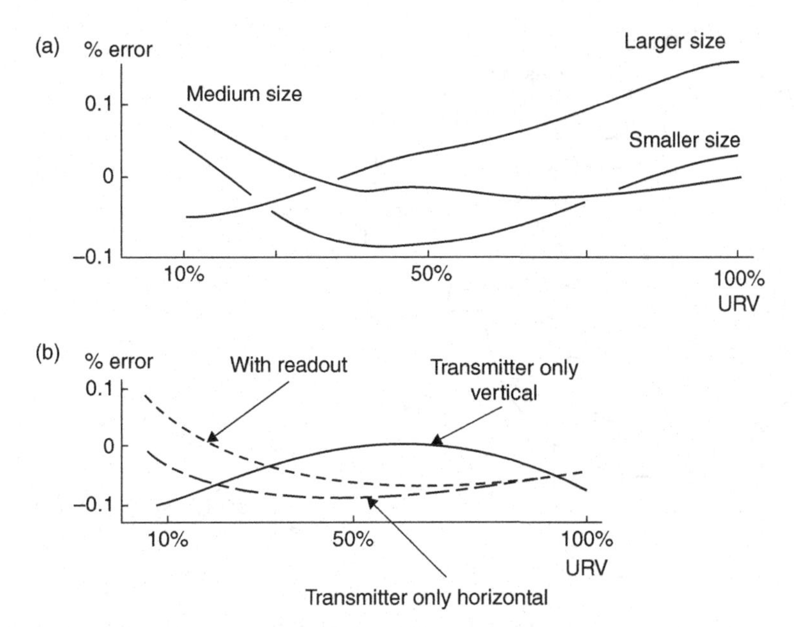

Figure 9.16. Typical calibration curves for a sliding vane PD meter (based on Avery Hardoll data): **(a)** For various models; **(b)** for various registration loads on the meter shaft output.

a secondary calibration standard, whereas mechanical meters offer an important option for commercial gas metering.

9.5.1 Wet Gas Meter

Figure 9.17 shows the wet gas meter, which uses a water (or other suitable liquid) bath as the gas seal to create closed compartments for the transfer of the gas. It is a high-precision meter capable of an operating range of about 10:1, but it is rather bulky. It was originally developed for the measurement of gas usage in industrial premises but recently has found a niche as a transfer standard for gas measurement. Modern designs may be capable of ranges up to 180:1 and uncertainty of the order of 0.2%.

It needs careful setting up to achieve the highest accuracy. Dijstelbergen (1982) made the following points.

- With careful use, the calibration curve is stable within 0.1%.
- The use of water has been replaced by Caltex Almag (Other fluids available may be: Ondina, Autin, Silox and Calrix.). This avoids the problem of gas take-up of the water. Evaporation into the gas stream (Park et al. 2002) may cause an error of about 2%.
- The meter can be used as a secondary standard at atmospheric conditions.

I should, perhaps, point out to the reader that this meter is not designed to measure wet-gas, despite its name!

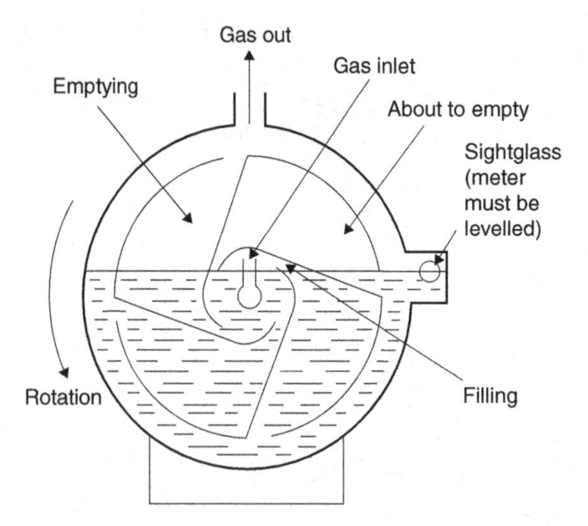

Figure 9.17. Wet gas meter.

9.5.2 Diaphragm Meter

Figures 9.18(a, b) show a diagram of the diaphragm meter. It is essentially a piston meter in which the compartments A, B, C and D form the cylinders and the bellows provide a common piston for A and B and similarly for C and D. The flow into and out of the compartments is controlled by the sliding valve. In Figure 9.18(a), A has emptied and B has filled, and the valve is moving to allow A to fill and B to empty. At the same time C is filling from the inlet chamber due to the valve's position, and D is emptying into the outlet manifold. In Figure 9.18(b), gas is entering the inlet chamber above the valves and is routed into compartment A, while gas leaves compartment B. At the same time compartments C and D have reached the change-over position so that C has filled and D has emptied, and the valve is changing over to allow C to empty and D to fill. Dijstelbergen (1982) made the following points:

- If the bellows are made either of a special type of leather or of fibre-reinforced rubber, they may dry out with low water content gases, causing over-registration.
- Most meters are manufactured in accordance with OIML recommendations (Organisation Internationale de Métrologie Légale) and should operate within 2% at high flows and 3% at low flows.
- Rangeability may be up to 150:1.

The typical measurement uncertainty is ±1% [Figure 9.18(c)]. The maximum allowable inlet pressure on some designs is about 0.1 bar above atmospheric, but it may be up to 1 bar on other special versions. The coupling from the mechanical meter to the output may consist of a magnetic coupling to the display. Operating temperature range is typically –20 to +60°C. The meters operate to DIN 3374 and BS 4161. The pressure loss through the meter is about 1.5 mbar [Figure 9.18(d)]. See also Bennett (1996).

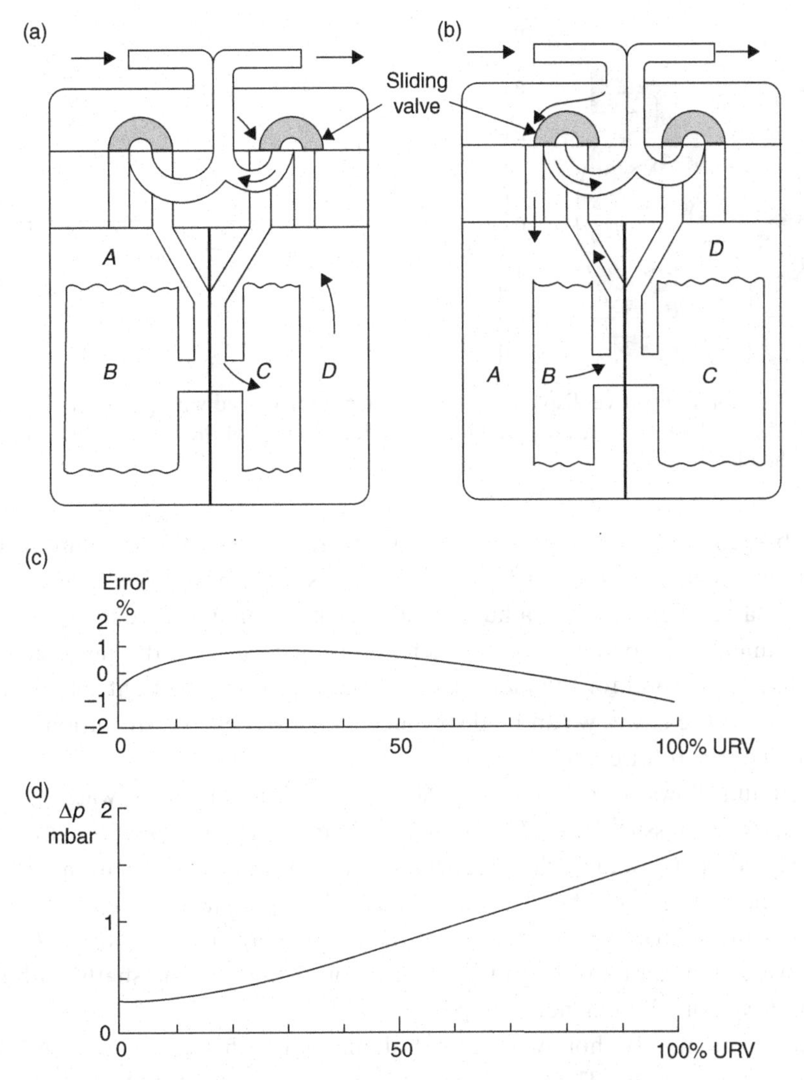

Figure 9.18. Diaphragm meter. **(a)** A, empty; B, full; C, filling; D, emptying; **(b)** A, filling; B, emptying; C, full; D, empty; **(c)** typical error curve (reproduced with permission of GMT and IMAC Systems Ltd.); **(d)** typical pressure loss curve (reproduced with permission of GMT and IMAC Systems Ltd.).

9.5.3 Rotary Positive Displacement Gas Meter

The rotary positive displacement gas meter is sometimes referred to as a Roots meter, but this name, in a particular style, is the registered trademark of a particular manufacturer's design (Dresser). Figure 9.19(a) shows a diagram of another design of this meter. Two rotors mesh and rotate within an oval body contour so that leakage is at a minimum between the rotors, but the gas is transferred at the outer point of the oval. In recent years they have performed to an increasingly high specification and offer one of the best options for high-precision gas flow measurement. There is a possible problem due to pulsating flow resulting from the use of these meters

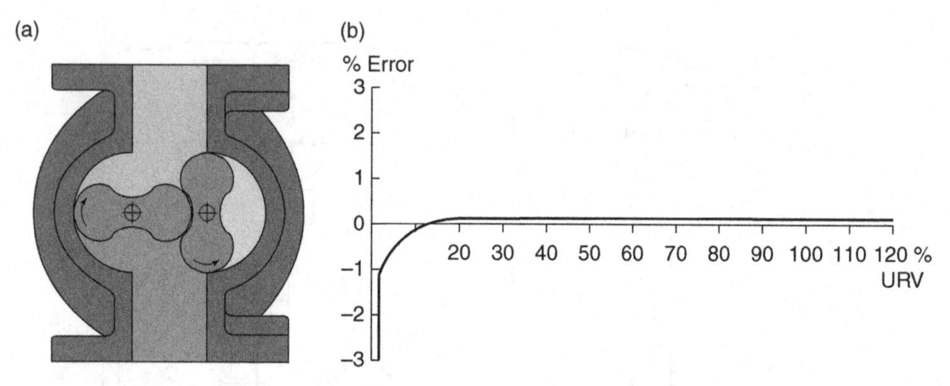

Figure 9.19. Rotary positive displacement gas meter (reproduced with permission of Romet Ltd. and IMAC Systems Ltd.): **(a)** Diagram of meter; **(b)** plot of the performance characteristic for the meter.

(Dijstelbergen 1982), and rapidly changing flow rates may damage them or downgrade their accuracy. The two rotors have a very small gap between (described by one manufacturer as less than a human hair), and the manufacture is to tight tolerances on impellers and meter body. It is, therefore, necessary for the gas to be free of particulate matter and for the meter to be mounted so that the tight tolerances are not distorted. The gas flow can be distributed to give an even distribution over the full length of the rotor entry.

Maximum flows have been from 28 m^3/h at 0.5 in. (13 mm) water gauge to 360 m^3/h, with a pressure loss of 0.5 to 1 in. (13–25 mm) of water gauge. Figure 9.19(b) shows a typical performance characteristic with a performance well within 0.5% over a 10:1 turndown. In fact, the claims are that 80:1 may be achievable. Temperature range is from –40 to 65°C. Greater ranges than these may now be obtainable.

Transmission can be through timing gears on the end of the shafts, with direct counter driven or with magnetic couplings.

The rotors may be hollow extruded aluminium with self-cleaning tips and a body of cast aluminium. The surfaces may be anodised. See also Dijstelbergen and van der Beek (1998), and the work of von Lavante et al. (2006) briefly reviewed in Appendix 9.A.2.

The CVM meter was of similar design to Figure 9.3(b) with a main rotor that consists of four vanes or moving walls that form metering compartments. The second rotor was a sealing rotor that returns the vanes to the inlet side of the meter. The second rotor had two or three lobes, and its rotation was precisely linked to that of the main rotor. According to Dijstelbergen (1982), it was suitable for high-pressure metering, did not cause pulsations in the flow, handled flows up to 1,200 m^3/h, and a repeatability of 0.05% has been achievable.

9.6 Positive Displacement Meters for Multiphase Flows

In a patent (1981), Arnold and Pitts proposed temperature and pressure sensing and a gamma-ray densitometer for sensing density inside a positive displacement

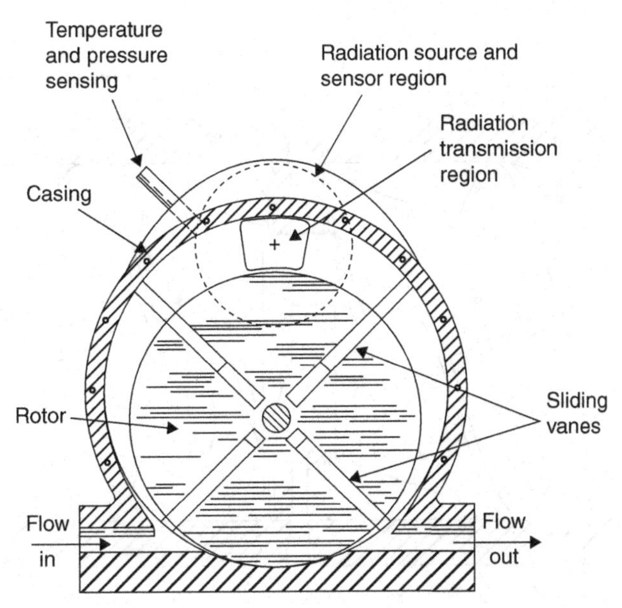

Figure 9.20. Diagram based on Arnold and Pitts' patent (1981) (reproduced from Baker 1991a; with permission of the Council of the Institution of Mechanical Engineers).

flowmeter (Figure 9.20), but they gave no experimental data. However, the patent presents an attractive device, which has the attribute of forcing the flow into a closed volumetric space regardless of the flow components. A major disadvantage is the possibility of solids causing it to seize up and block the line, requiring designs to have large clearances and consequent loss of precision.

An alternative approach has been taken by Gold et al. (1991), who used a helical rotor meter (Figure 9.21). In this design, two rotor sections are used in series, and between them is a section where the condition of the multiphase flow is measured.

Gold et al. (1991) described their solution, which controlled the flow being measured and so avoids the problem of slip between phases. The helical rotors are also claimed to be more resilient, with suitable bearings, to the rigours of multicomponent flow, reducing the possibility of damage due to transients. Between the two rotors, the density was measured. The preferred orientation was with flow vertically downward. The bearings were simple low-friction journals but with thrust bearings that gave long service.

The meter was seen as suitable for both surface and subsea service and aimed at a specification of ±5% rate for total flow and ±5% rate for oil flow rate with ±10% for both gas and water. The field trial meters were designed for up to 20,000 barrels per day (130 m^3/h) with at least 10:1 turndown and ±5% uncertainty. A nominal 4 in. (100 mm) bore meter was specified. The body had about 0.3 m outer diameter. The meter length was 1.2 m, and with nucleonic densitometer it weighed about 450 kg. The scatter on test was a maximum of about 20% in mid values of void fraction. Priddy (1994) reported tests at Prudhoe Bay, Alaska, which were extremely promising in terms of the specific aims.

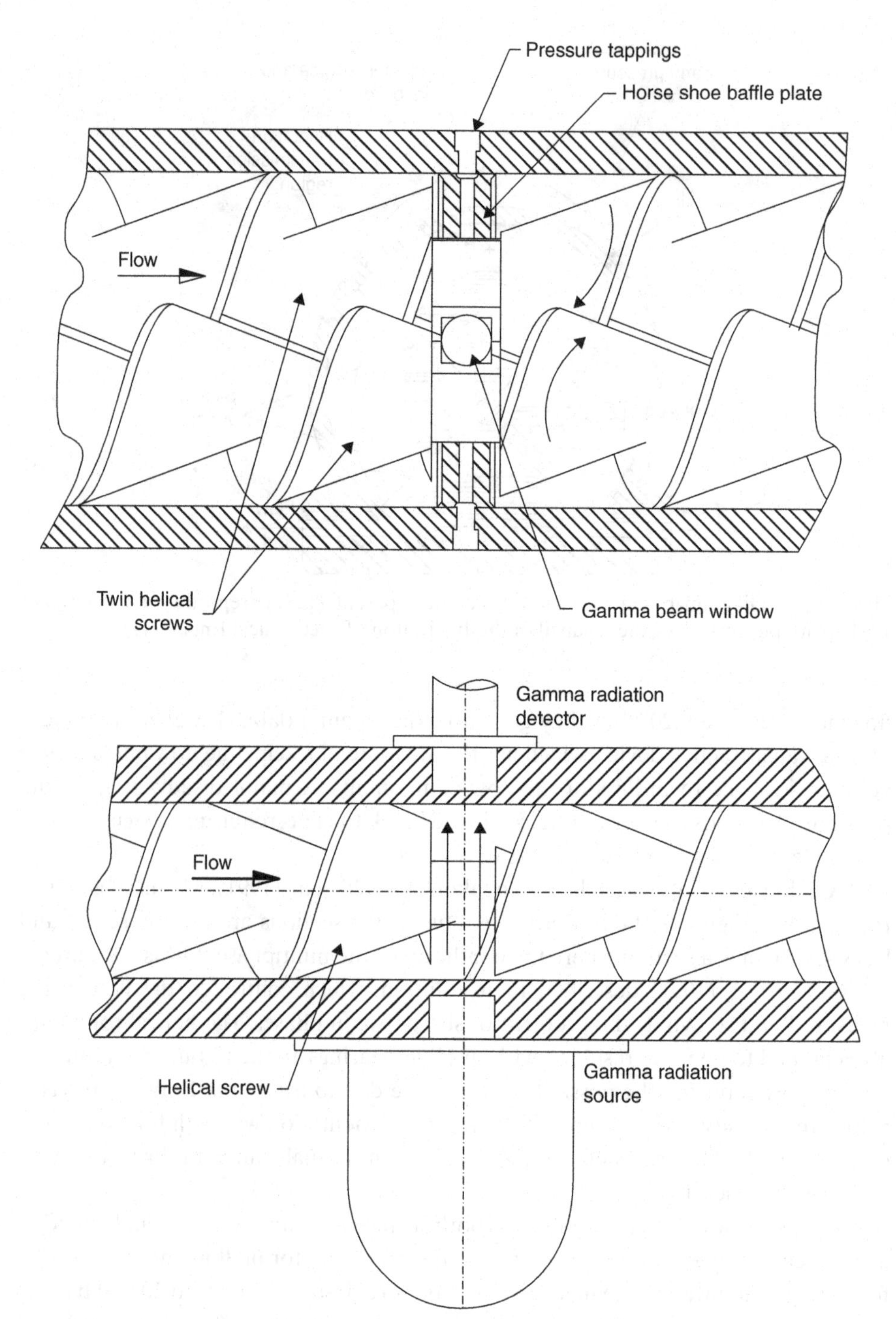

Figure 9.21. Helical multiphase meter (Gold, Miller and Priddy 1991, Priddy 1994; reproduced with permission of ISA Controls Ltd.; design is the property of BP International).

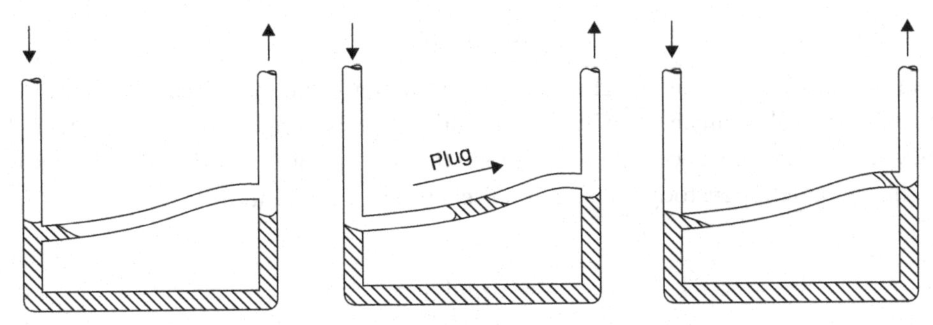

Figure 9.22. Meter using liquid plugs to measure low flows.

9.7 Meter Using Liquid Plugs to Measure Low Flows

Figure 9.22 shows a diagrammatic representation of a meter that uses liquid plugs of a sealing fluid such as distilled water to measure the flow of gas. The plugs are carried through the tube as shown, and the passage and velocity are sensed optically. Uncertainty claims are ±1% at the optimum rate, and the range is 8–3,000 ml/h. In this device, dry gases pick up moisture. It is not clear how the size of plug, which presumably varies, is allowed for. It may be that the optics provide a compensation. See also Guo and Heslop (2004).

Liu, Olsson and Mattiasson (2004) described the design and performance of a gas meter for biogas. A volumetric cell consists of a U-tube of glass. A three-way valve controlled flow into the U-tube. The flow displaced liquid and the movement of the liquid meniscus was sensed by optical sensors. Flow rates of 1 to 950 ml/h were claimed to an uncertainty better than ±3.3%.

9.8 Applications, Advantages and Disadvantages

Applications

The applications will vary to some extent for the various designs; in particular, the accuracy and cost of the meters will make them more suitable for particular applications. In general, they operate on clean or well-filtered fluids. The clearances are so fine that dirty or aggressive fluids may prevent operation and cause total blockage in the line. For the same reason, the instruments must be correctly mounted to avoid mechanical distortion, which could affect the clearances.

In particular, for the sliding vane meter, one manufacturer claimed: blending, batching, dispensing, inventory control, custody transfer of oils, solvents, chemicals, paints, fats and fertilisers.

For high-viscosity fluids, the flow passage may create too great a pressure drop. For some oils, the temperature of the liquid may need to be sensed and compensation used.

Presumably, the design developed for multiphase flows will eliminate the need for well-filtered fluids.

Advantages

The advantages are seen in high-precision instruments, and their accuracy after calibration is possibly only limited by the calibration facility used. Most are little affected by viscosity. The most accurate ones among them, therefore, provide a device that is highly suitable for custody transfer and fiscal transfer applications.

Disadvantages

The meters may create pulsations and may be subject to damage from rapidly varying flows. Damage to the meters causing them to stop rotating may also cause total line blockage. They may have a high initial price and may require careful maintenance with high associated costs. The meters may create a substantial pressure loss.

9.9 Chapter Conclusions

New Concepts

What is likely to be the future of such meter designs? No doubt other clever designs will be introduced such as that selected as a utility gas meter (Kim et al. 1993b; O'Rourke 1993, 1996), which used an undulating membrane (Figure 9.23). The sensor consisted of such a membrane, which trapped discrete volumes of gas as they moved through the flow chamber in pulses. Movement was sensed by piezoelectric elements bonded to the membrane. The design was patented. Tests suggested that it was capable of operating over a range of 0.007–7 m^3/h with uncertainty of ±1–2% for −40 to 140°F (−40 to 60°C), and with a life of 20 years (10-year battery life).

More generally, these essential workhorses of the industry will continue to benefit from improvement in materials, bearings and manufacture, so that they can be used with more difficult fluids.

Servo-Assisted PD Meter and Automation

Electronic control will also ensure that their standard applications are possible with greater automation and precision. Dilthey, Scheller and Brandenburg (1996)

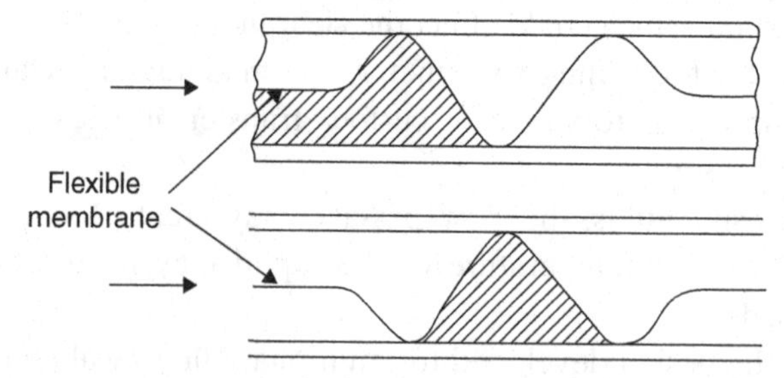

Figure 9.23. Undulating membrane meter (after Kim et al. 1993b and O'Rourke 1993,1996).

studied the performance of gear meter, helix-type meter and sliding vane meter for flow measuring systems in automated adhesive application. The helix-type meter appeared to be most suitable.

Shimada et al. (2002) developed a servo PD flowmeter to achieve wide rangeability, long-term stability and high accuracy in oil flow measurement. They claimed errors less than 0.2% over 1:10 of flow range. It was selected as a transfer standard for the hydrocarbon flow facility at the National Metrology Institute of Japan (NMIJ). The flowmeters had 50, 100 and 150 mm diameter with flow rates 3 to 30, 7.5 to 75 and 30 to 300 m³/h, respectively.

Adverse Conditions

The apparent possibly successful use of positive displacement meters in such a hostile environment as oil-well flows suggests that other difficult applications may be within range.

A Problem to be Overcome

Total blockage of the line in the event of failure is, perhaps, the greatest disadvantage, and ways of dealing with this through an automatically controlled bypass, in the event of blockage, may overcome this problem. One PD meter which may not cause total blockage is the oscillating circular piston PD meter.

Appendix 9.A Basic Analysis and Recent Research

9.A.1 Theory for a Sliding Vane Meter

The theory of the meter was set out by Hahn (1968). In his paper he considered the motion of the vanes and calculated the radial velocity and acceleration if the internal controlling cam gave a sinusoidal motion. He noted that the resultant forces between vane and wall contributed to the wear of the vanes. Avoidance of jerky motion and shock stresses is clearly important.

He also calculated the volumetric flow rate (cf. Baker and Morris 1985, whose work was in ignorance of that due to Hahn). Hahn pointed out that the radial motion of the vanes results in a sinusoidal fluctuation superimposed on the volume versus rotation characteristic and that the amplitude of this fluctuation can be of order 0.05–0.1% for various designs. This, in turn, would be shown in an irregularity of the rotation of the shaft of as much as 1.25%. However, in practice there will be a balance between varying flow, pressure variation and rotor velocity variation. Hahn also gave characteristics showing variation of within 0.2% over a range of 14:1 (slight drop of 1.5–2% with increasing flow) and changes of +0.15% for a change in viscosity from 0.75 to 4 cP.

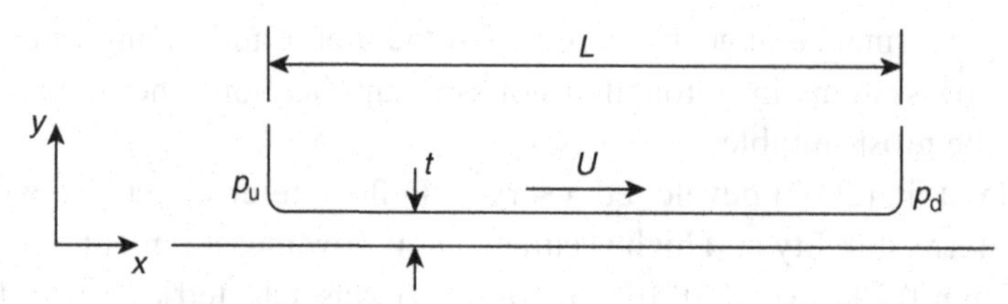

Figure 9.A.1. Geometry of leakage path between fixed and moving parts of the meter (after Baker and Morris 1985).

9.A.1.1 Flowmeter Equation

The ideal behaviour of the meter is given by the product of transferred volume, the number of volumes transferred per revolution and the rotational speed. For the actual behaviour allowance has to be made for the leakage that takes place past the moving rotors. The bulk volumetric flow for a meter of the sliding vane type is given by

$$q_v = \left[\pi\left(r_o^2 - r_i^2\right) - \delta\right] N \ell_{ax} - \text{Leakage flow} \tag{9.A.1}$$

where ℓ_{ax} is the axial length of the chamber. The flow in each gap (Figure 9.A.1) will be governed by

$$\frac{p_u - p_d}{L} = -\mu \frac{d^2u}{dy^2} \tag{9.A.2}$$

where μ is the fluid velocity at y in the gap, t is the clearance, p_u and p_d are the upstream and downstream pressures, L is the length of the clearance gap or, in the case of the vanes, the thickness of the vanes, and μ is the viscosity. This results in a velocity distribution of

$$u = \frac{1}{2\mu}\frac{(p_u - p_d)}{L}(t - y)y + \frac{Uy}{t} \tag{9.A.3}$$

and a volumetric flow through the gap of

$$q_v = \left\{\frac{1}{12\mu}\frac{(p_u - p_d)}{L}t^3 + \frac{Ut}{2}\right\}\ell \tag{9.A.4}$$

where ℓ is the axial length of the gap.

If the rotor is stationary, this quantity becomes the slippage;

$$q_{slip} = \frac{\ell}{12\mu}\frac{(p_u - p_d)}{L}t^3 \tag{9.A.5}$$

Ideally, the liquid in the gap would be carried at the same speed as the rotor blade:

$$q_{ideal} = Ut\ell \tag{9.A.6}$$

From Equations (9.A.4) and (9.A.6)

$$q_{\text{leakage}} = q_{\text{ideal}} - q_{\text{v}} \tag{9.A.7}$$

so

$$q_{\text{leakage}} = \left\{ \frac{Ut}{2} - \frac{1}{12\mu} \frac{(p_{\text{u}} - p_{\text{d}})}{L} t^3 \right\} \ell \tag{9.A.8}$$

From Equations (9.A.1) and (9.A.8) and assuming that the leakage can be taken as the sum of leakage through clearances t_j of length in the direction of leakage L_j and in the direction perpendicular to the leakage ℓ_j and with velocity of moving component U_j,

$$q_{\text{v}} = \left[\pi(r_{\text{o}}^2 - r_{\text{i}}^2) - \delta \right] N \ell_{\text{ax}} - \sum_{j} \left\{ \frac{U_j t_j}{2} - \frac{1}{12\mu} \frac{(p_{\text{u}} - p_{\text{d}})}{L_j} t_j^3 \right\} \ell_j \tag{9.A.9}$$

An expression for the pressure drop across the vanes is obtained from the torque balance on the rotor. The turning force due to $(p_u - p_d)$ will act on a vane and will precisely balance the frictional torques caused by bearings, vane clearance spaces and readout mechanism:

$$\frac{1}{2}(p_{\text{u}} - p_{\text{d}})(r_{\text{o}}^2 - r_{\text{i}}^2)\ell_{\text{ax}} = T_0 + \frac{T_1 N}{N_{\text{max}}} \tag{9.A.10}$$

where T_0 is a constant torque and $T_1\, N/N_{\text{max}}$ is a speed-dependent torque. Part of the torque $T_1\, N/N_{\text{max}}$ will be due to the clearance drag torque:

$$\mu \frac{U_j}{t_j} L_j \ell_j r_j$$

where r_j is the radius from the rotational axis of the *j*th clearance.

9.A.1.2 Expansion of the Flowmeter Due to Temperature

If α_{m} is the coefficient of linear expansion of the metal, then a change in temperature ΔT_{m} of the metal of the flowmeter will cause a change in length

$$\frac{\Delta \ell}{\ell} = \alpha_{\text{m}} \Delta T_{\text{m}} \tag{9.A.11}$$

so that

$$\frac{\Delta q_{\text{v}}}{q_{\text{v}}} = 3\alpha_{\text{m}} \Delta T_{\text{m}} \tag{9.A.12}$$

giving the fractional change in measured volume per revolution of the flowmeter rotor.

Differential Expansion. If the expansion of the rotor differs from that of the casing due either to different temperatures or to different materials, then the clearances will be designed for the greatest relative movement between these members. The change in the value of the clearance t will be

$$\frac{\Delta t}{t} = \alpha_{\mathrm{m}} \Delta T_{\mathrm{m}} \frac{r_{\mathrm{o}}}{t} \tag{9.A.13}$$

between uniform temperature conditions and conditions when the temperature of the internal rotor and the casing differ by ΔT_{m}. For the case where the temperatures are the same but the coefficients of expansion differ, this becomes

$$\frac{\Delta t}{t} = \Delta T_{\mathrm{m}} \frac{r_{\mathrm{o}}}{t} \left(\alpha_{\mathrm{m1}} - \alpha_{\mathrm{m2}} \right) \tag{9.A.14}$$

Liquid Density Change. The change in the bulk flow due to expansion of the liquid will be

$$\frac{\Delta q_{\mathrm{v}}}{q_{\mathrm{v}}} = \alpha_{\ell} \Delta T_{\ell} \tag{9.A.15}$$

where α_{ℓ} is the coefficient of volume expansion of the liquid.

Because many liquids are bought and sold by volume, this change is of a different significance from those given before. The volume passed by the meter may, therefore, be important. In some cases, the change in liquid density may be ignored.

9.A.1.3 Pressure Effects

Provided that the meter is supplied with an outer pressure vessel, there should be no net pressure causing distortion of the inner meter. However, this is not invariably so. The meter will then suffer from two distortions, hoop stress and axial stress. For convenience, we will consider these effects separately and ignore the effects of one on the other.

Bulk Flow Error. If Δp is the difference between internal and external pressure and E is Young's modulus of elasticity, then the change in the outer radius will be

$$\frac{\Delta r_{\mathrm{o}}}{r_{\mathrm{o}}} = \frac{r_{\mathrm{o}}}{E} \frac{\Delta p}{t_{\mathrm{o}}} \tag{9.A.16}$$

where t_{o} is the thickness of the outer casing. The axial stress will cause an elongation of

$$\frac{\Delta \ell_{\mathrm{ax}}}{\ell_{\mathrm{ax}}} = \frac{r_{\mathrm{o}}}{2E} \frac{\Delta p}{t_{\mathrm{o}}} \tag{9.A.17}$$

where ℓ_{ax} is the axial length of the meter's measuring chamber. Together these lead (neglecting Poisson's ratio effects) to an increase in measuring volume, which causes an error in the volumetric flow rate of

$$\frac{\Delta q_v}{q_v} = \frac{r_o}{2Et_o}\left\{\frac{5r_o^2 - r_i^2}{r_o^2 - r_i^2}\right\}\Delta p \tag{9.A.18}$$

Leakage Gap. In certain designs, the hoop stress will cause an increase in the clearance of

$$\frac{\Delta t}{t} = \frac{\Delta r_o}{t} = \frac{r_o^2 \Delta p}{tt_o E} \tag{9.A.19}$$

whereas the axial stress will cause an increase in end clearance at each end of

$$\frac{\Delta t}{t} = \frac{\Delta \ell_{ax}}{t} = \frac{r_o \ell_{ax}}{4tt_o}\frac{\Delta p}{E} \tag{9.A.20}$$

Clearly these calculations are simplifications of the actual expansion that will occur, but they allow some estimate of likely errors to be made.

9.A.1.4 Meter Orientation

Bearings at Each End of Shaft. Assuming that the shaft behaves like a beam simply supported at each end and that the rotor weight is evenly distributed along the shaft, then the maximum deflection at the point midway between the bearings will be given by

$$\text{Deflection} = \frac{5Mg\ell_{ax}^3}{96\pi E r_s^4} \tag{9.A.21}$$

where M is the mass of the rotor, g is acceleration due to gravity, and r_s is shaft radius.

Bearing(s) at One End of Shaft Only. Assuming that, in this case, the bearing(s) hold the shaft without deflecting, the shaft bending will be equivalent to a uniformly loaded encastré beam. The end deflection of this, which will be the maximum value, is given by

$$\text{Deflection} = \frac{Mg\ell_{ax}^3}{2\pi E r_s^4} \tag{9.A.22}$$

It is apparent that, other things being equal, the deflection of the end-bearing shaft will be nearly 10 times greater than the shaft with bearings at each end. This will require greater clearances and, consequently, allow greater leakages. To avoid this problem, it may be recommended that meters be mounted with shafts vertical.

9.A.1.5 Analysis of Calibrators

The rotation of these meters gives the volumetric flow. If every rotation causes a pulse, it is a simple matter with modern electronics to scale the pulse rate to give any required calibration factor or combine the electrical output from a temperature sensor to give a temperature compensation. However, in many applications, mechanical calibration and temperature adjustment is required as a result of existing installation constraints, and three main methods have been used or are possible.

Gearbox. The gearbox with interchangeable gears will not be discussed in this section. The drawback of the gearbox is clearly the limited adjustment possible. We are concerned with two calibrators, which may introduce special features into the meter performance.

Epicyclic System. A schematic diagram of this system is given in Figure 9.10. The drive from the flowmeter shaft enters with rotation N_{IN}. The drive then divides. The main drive is via gears n_1, n_2, n_3, and so to the epicyclic unit. The other is via the disc and adjustable friction wheel to n_8, n_7. The recombined drives leave the epicyclic unit via n_6 with rotation N_{OUT}. The integrity of the unit depends heavily on the wheel and disc contact. There must be no slip between these components, but they must also not wear and must be adjustable (r the radius of frictional contact being changed by the adjustment).

The ratio of N_{OUT}/N_{IN} is given by

$$\frac{N_{OUT}}{N_{IN}} = \frac{r}{R}\frac{n_8}{n_7}\left\{1 - \frac{n_3}{n_4}\frac{n_5}{n_6}\right\} - \frac{n_1 n_3 n_5}{n_2 n_4 n_6} \tag{9.A.23}$$

where R is the radius of the friction wheel.

Putting typical values of the numbers of teeth (n) into Equation (9.A.23)

$$n_2 = n_1$$
$$n_4 = \frac{20n_3}{15}$$
$$n_6 = \frac{20n_5}{15}$$
$$n_7 = 8n_8$$

then

$$\frac{N_{OUT}}{N_{IN}} = \frac{r}{R}\frac{1}{8}\left\{1 - \left(\frac{15}{20}\right)^2\right\} - \left(\frac{15}{20}\right)^2 = 0.055\frac{r}{R} - 0.563$$

If $R = 10$ mm and we let $r = 10$ mm,

$$\frac{N_{OUT}}{N_{IN}} = -0.508$$

If $r = 11$ mm,

$$\frac{N_{OUT}}{N_{IN}} = -0.503$$

or a change in calibration of 1%.

We may also write down the modified version of Equation (9.A.23), assuming that a small amount of slip occurs between disc and wheel equivalent to a rotation N_s in N_{IN}. Then

$$\frac{N_{OUT}}{N_{IN}} = \frac{r}{R}\frac{n_8}{n_7}\left\{1 - \frac{n_3}{n_4}\frac{n_5}{n_6}\right\}\left(1 - \frac{N_s}{N_{IN}}\right) - \frac{n_1}{n_2}\frac{n_3}{n_4}\frac{n_5}{n_6} \tag{9.A.24}$$

In this case, a relation between the torque in T_{IN} and the torque out T_{OUT}, the friction force between wheel and disc F, is given by

$$T_{IN} = \frac{N_{OUT}}{N_{IN}} T_{OUT} + \frac{N_s}{N_{IN}} rF \tag{9.A.25}$$

At the start of rotation if N_{OUT} were zero and N_{IN} were nonzero, then

$$T_{IN} = \frac{N_s}{N_{IN}} rF$$

and

$$\frac{n_s}{N_{IN}} = 1 - \frac{\dfrac{n_1}{n_2}\dfrac{n_3}{n_4}\dfrac{n_5}{n_6}}{\dfrac{r}{R}\dfrac{n_8}{n_7}\left(1 - \dfrac{n_3}{n_4}\dfrac{n_5}{n_6}\right)}$$

Using the values given earlier,

$$\frac{N_s}{N_{IN}} = -9.2$$

$$T_{IN} = -9.2\, rF$$

If slip does not occur, then

$$T_{IN} = \frac{N_{OUT}}{N_{IN}} T_{OUT} = -0.508\, T_{OUT}$$

With these relationships and the coefficient of friction, the required contact force between wheel and disc can be estimated from a knowledge of the torque required to operate the counter unit.

Clutch System. The clutch (or cam) system consists of three concentric discs and rings (Figure 9.11). The central ring is driven by the input shaft and carries a series of clutches that lock on the centre disc and the outer ring, causing them to rotate at least as fast as the central ring. An arm rotates on a separate shaft, which is not concentric with that of the disc and rings. In this arm are two sliders, one fixed to, and rotating with, the centre disc and the other fixed to, and rotating with, the outer ring. Because of the eccentricity of the arm shaft, the centre disc and the outer ring are forced to rotate at different speeds. The difference in speeds can be adjusted by manually altering the amount of eccentricity. Thus, the speed at which the output shaft revolves relative to the input shaft is adjusted to achieve compensation for use of the measuring element under a particular set of operating conditions. It can be shown that if Φ is the angular position of the arm and θ_c and θ_o are, respectively, the angular positions of the centre disc and outer ring [Figure 9.A.2(a)]

$$\Phi = \theta_c - \tan^{-1}\left(\frac{\Delta \sin\theta_c}{R_c + \Delta\cos\theta_c}\right) \tag{9.A.26}$$

and

$$\Phi = \theta_o + \tan^{-1}\left(\frac{\Delta \sin\theta_o}{R_o + \Delta\cos\theta_o}\right) - \pi \tag{9.A.27}$$

where Δ is the distance between the centres of rotation of the arm and the discs.

These are plotted in Figure 9.A.2(b), and it can be seen that at $\Phi = 0, \pi$, and 2π, the angular positions differ by π. The angular velocities will be equal when, for a given value of Φ, the slopes of θ_c and θ_o are equal.

In Figure 9.A.2(c), the relationship between output rotational speed N_o and input rotational position θ_i (which for constant flow rate is proportional to time) is plotted. It is important to note the non-uniform rotation of the output and the mean value in relation to actual rotational speed at various points in the cycle.

9.A.1.6 Application of Equations to a Typical Meter

Using the meter with dimensions shown in Figure 9.13 and Equation (9.A.9), we obtain for the main coefficients, assuming a total leakage length of 325 mm,

$$\begin{aligned} q_v = {} & 0.00918\,N \\ & -0.0000156N \ \text{(leakage)} \\ & +0.034\times10^{-10}\frac{\Delta p}{\mu} \ \text{(leakage due to pressure change)} \end{aligned}$$

Taking

$$\begin{aligned} \Delta p &= 20000\,\text{N/m}^2 \\ \mu &= 0.1\,\text{poise} \\ q_v &= 0.00916N + 0.677\times10^{-6}\,\text{m}^3/\text{s} \end{aligned}$$

The sensitivity to t and Δp is given by

$$q_v = 0.00918N - 0.156tN + 3.4\frac{\Delta p t^3}{\mu}\,\text{m}^3/\text{s}$$

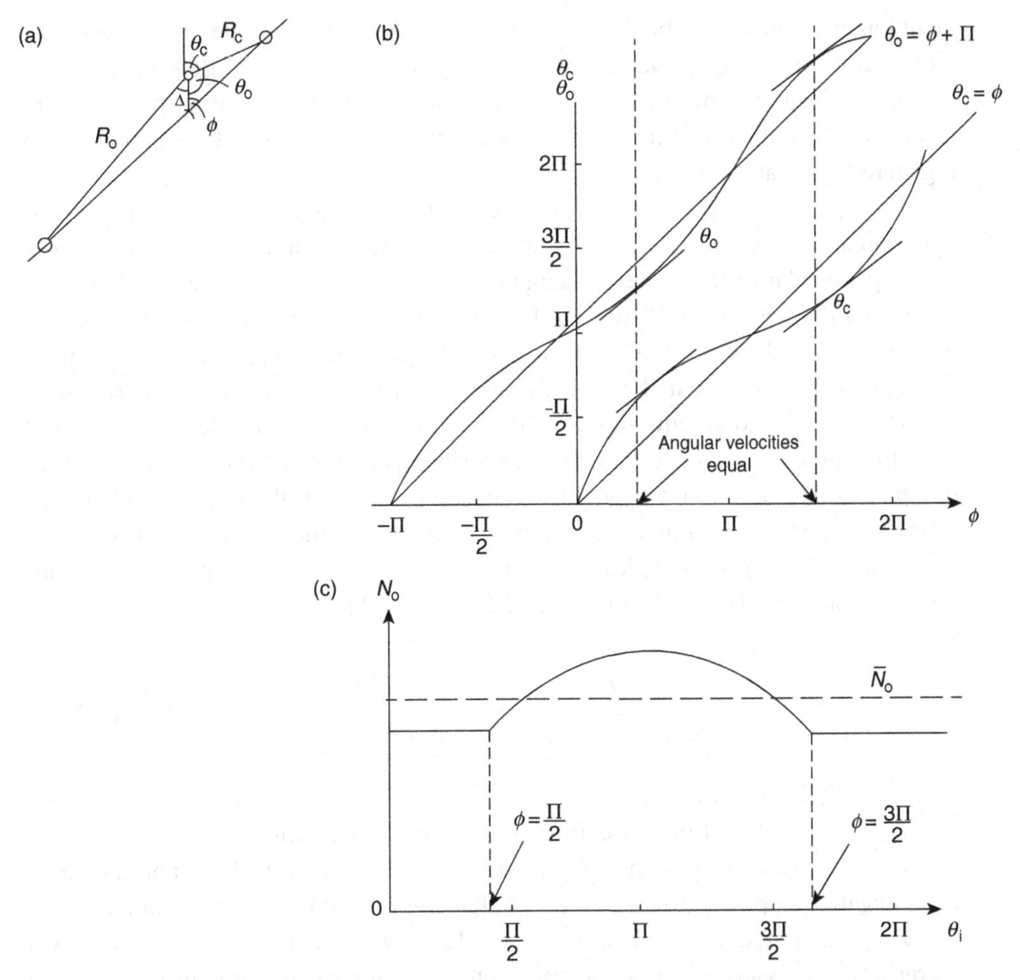

Figure 9.A.2. Clutch calibration adjustment system (reproduced from Baker and Morris 1985; with permission of the Institute of Measurement and Control): **(a)** Geometry; **(b)** rotational position of cebtre disc and outer ring against arm position; **(c)** rotational output speed as a function of input rotational position.

Taking the datum values of $t = 0.1$ mm, $\Delta p = 0.2 \times 10^5$ N/m² and $\mu. = 0.1$ poise, the effect of changes of $\pm 0.5\Delta p/\mu$ for $N = 1$ revolutions per second and $t = 0.2$ mm is about ±0.05% (Baker and Morris 1995).

9.A.2 Recent Theoretical and Experimental Research

Rotary Piston Positive Displacement Flowmeter for Gas

Von Lavante et al. (2006) described their research on the rotary piston positive displacement flowmeter (RPFM) which has become an important high-precision instrument for monitoring gas flows. They sought to explain some of the physical effects which have an influence on the accuracy of the meter. In their paper they gave a useful description of previous work on this type of meter by various researchers including previous theoretical and experimental work, work on the operation of the meter, understanding of leakage, accuracy and measurements of downstream pulsation. Some of the parameters they investigated were the operating pressure, the position of the inflow and the position of the pressure sensor needed to convert the registered flow rate to standard flow.

Their paper included a plot of the cyclical variation in flow rate which is shown approximately in Figure 9.A.3, where the flow rate is at a minimum when the two rotors are perpendicular to each other as in Figure 9.19. Leakage flow will occur between the rotors and the housing and between the rotors. They appear to have used a form of equation for the leakage similar to Equation (9.A.8) (see Figure 9.A.1). Their theoretical pressure loss curve had a similar shape to that shown for liquid in Figure 9.A.7

They developed an equation for the theoretical prediction of leakage flow and deviation, and they claimed that their experiments confirmed the effect of various parameters. However, they seemed to suggest that the results relate to what they term as a "simple" meter and may not be relevant to the most modern versions.

Figure 9.19(b) gives a typical performance characteristic. Their work on the simple model appeared to suggest that deviation, defined as

$$\varepsilon = \frac{q_{v,RPFM} \dfrac{p_{RPFM}}{p_{ref}} \dfrac{T_{ref}}{T_{RPFM}} - q_{v,ref}}{q_{v,ref}} \tag{9.A.28}$$

where $q_{v,RPFM}$ is the flow rate indicated by the meter, $q_{v,ref}$ is the true flow rate, p_{RPFM} and T_{RPFM} are values of pressure and temperature at the meter and p_{ref} and T_{ref} are the reference meter values, should always be negative due to leakage and increases in its negative slope due to an increase in density (caused by pressure change).

They suggested that decreasing the inlet flow area in a simple RPFM will increase the pressure drop across the meter and apparently, in consequence, the leakage. Changes in viscosity may be significant, but the indicated flow appeared to be relatively insensitive to the position of the pressure measurement point.

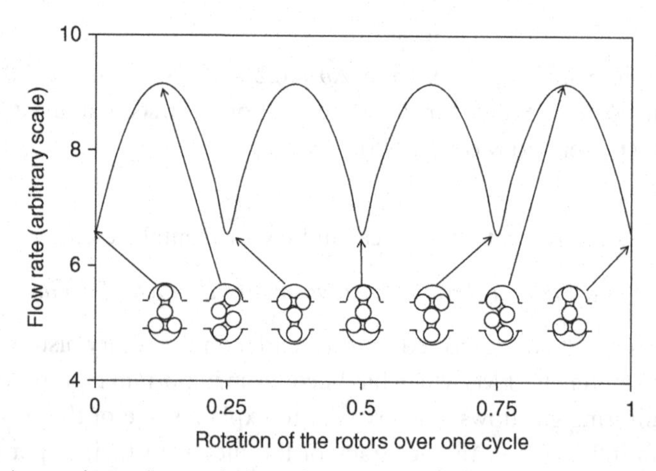

Figure 9.A.3. Approximate flow variation with rotation of rotary piston meter [after Nath and Löber (1999); von Lavante et al. (2006)].

Seals in a Fuel Dispensing Flowmeter

Yajun et al. (2012) looked at the wear rate of PTFE O-ring piston seals in a fuel-dispensing flowmeter. They also discussed the transfer film from the seal to the stainless steel liner of the cylinder and appear to suggest that this can break up, causing increased wear.

Leakage in an Oval Gear Flowmeter

Lim (2004) undertook tests on clearances to see the effect of leakage on the performance of oval gear meters.

Pressure Distribution in a Twin Helical Rotor Positive Displacement Flowmeter

Dr Shimada obtained some very interesting plots of pressure distribution in an OVAL gear meter and made measurements of the leakage caused by various clearances of the rotor ends, while working in our laboratory at Cambridge in 2007. Figure 9.A.4(a) shows the meter under test with the end cover removed, and Figure 9.A.4(b) shows the rotors which intermesh. He took pressure measurements at 15 positions around the upper chamber wall Figure 9.A.4(c). The average pressure distribution is shown in Figure 9.A.4(d) where the average pressure levels appear to follow the contours of the helical rotor.

The parameter ϕ relates to the position on the rotor axially by linking this to the angular position of one of the vanes which is facing the inlet flow at the central point of the meter.

Considering the plot, the line from about $\phi = -30°$ and $\theta = 45°$ to about $\phi = 60°$ and $\theta = -45°$ is, approximately, the edge of the compartment filled with liquid before it discharges on the outlet side of the flowmeter. So this should be the highest pressure and near stagnation pressure, and the averaging which occurs due to the movement of the vanes will reduce this from the stagnation pressure somewhat. On the other hand, the position $\phi = -60°$ and $\theta = 15°$ will approximately represent the edge of the next vane where the flow is still filling the channel, and will average over the movement to give a static average pressure of about -14×10^{-3} MPa.

The stagnation region in the bottom left corner is presumably due to streamlines which follow round the corner. The top right-hand corner, on the other hand, is where the liquid is flowing out at the outlet pressure.

Experimental and Theoretical Research on a Circular Oscillating Piston Positive Displacement Flowmeter

Morton (2009) wrote a very thorough and enlightening thesis on the behaviour of a circular oscillating piston positive displacement flowmeter while working with us at Cambridge. Her research was done in collaboration with Litre Meter, and she was able to make use of Litre Meter flowmeters. Her experimental work obtained data on: the movement of the piston, the pressure drop across the meter and during

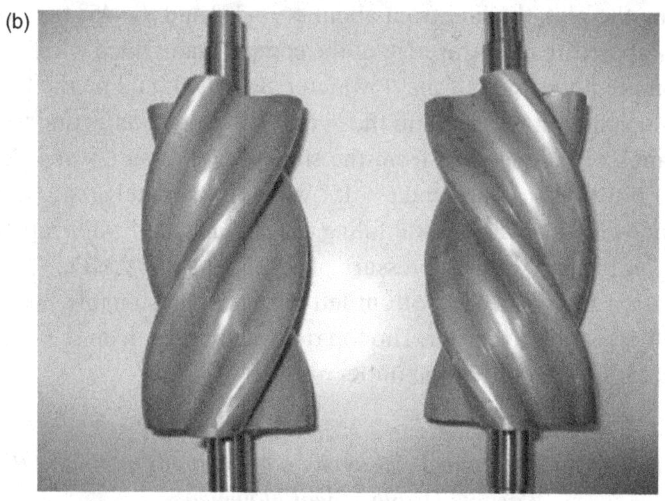

Figure 9.A.4. Pressure distribution in a positive displacement flowmeter (reproduced with permission of Dr Shimada).
(a) OVAL helical rotor positive displacement flowmeter **(b)** Helical rotors **(c)** Pressure tappings in the upper chamber wall **(d)** Average pressure distribution (pressure in MPa relative to upstream pressure)

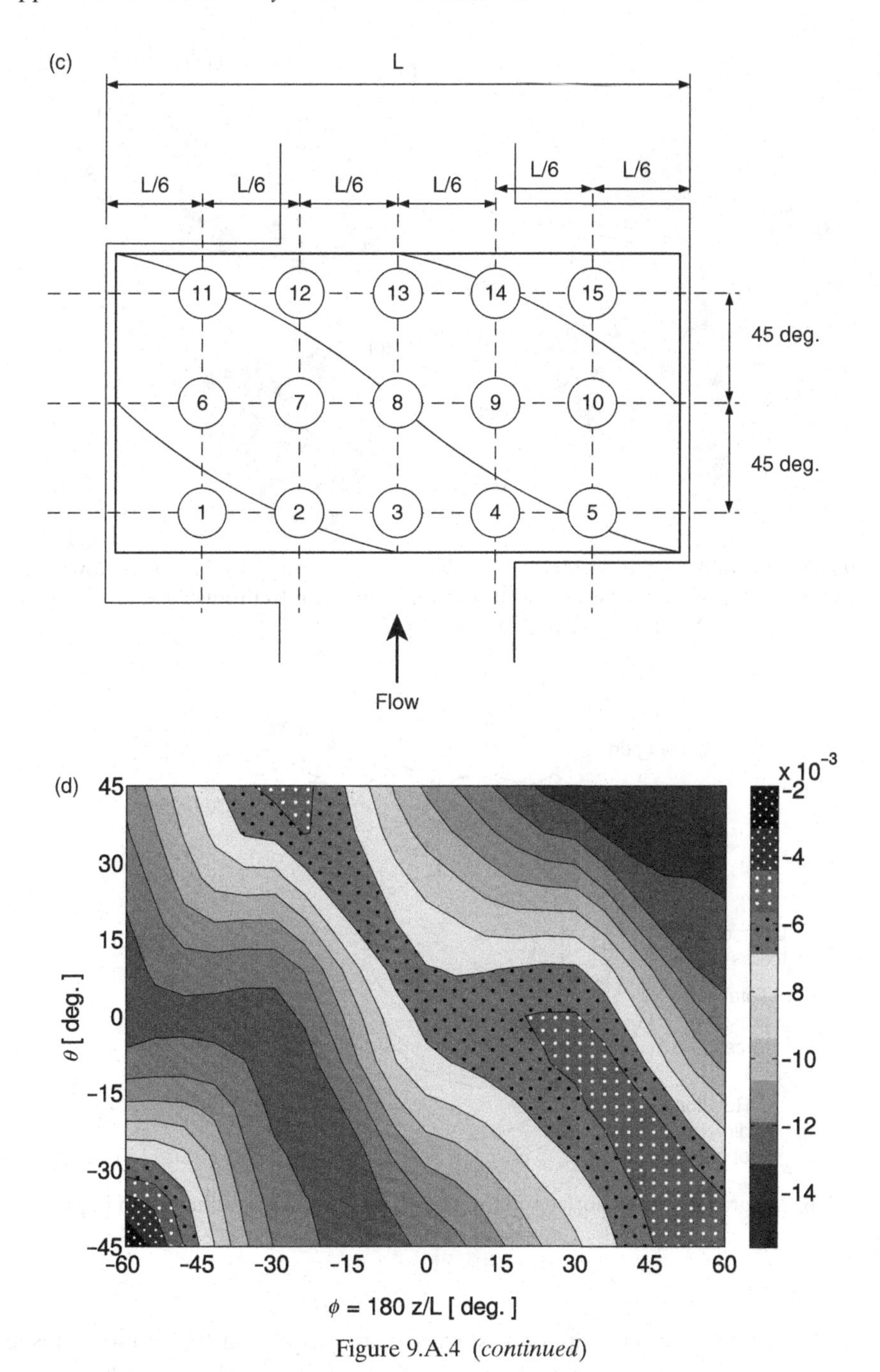

Figure 9.A.4 (*continued*)

oscillation and the leakage flows. These experiments were undertaken in parallel with the development of an analytical model of the piston movement and the resulting pressure losses and leakage. Figure 9.A.5 shows the main components of the meter. The chamber was of stainless steel and one of the two main pistons used was carbon and another stainless steel.

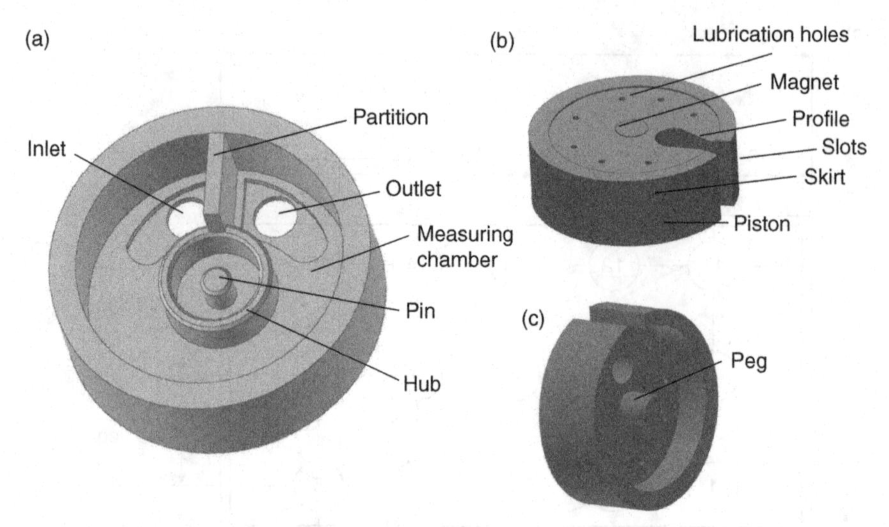

Figure 9.A.5. Diagram of the oscillating circular piston positive displacement flowmeter: **(a)** Chamber; **(b)** top view of piston; **(c)** bottom view of piston [reproduced with the agreement of Elsevier from Morton, Hutchings and Baker (2014a)].

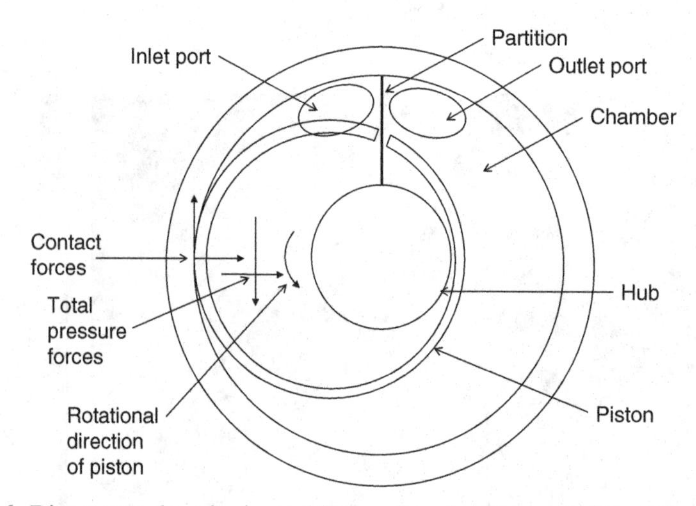

Figure 9.A.6. Diagram to show horizontal forces on and movement of the piston in plane of chamber.

The theoretical approach applied Newton's laws to obtain the motion caused by the forces acting on the piston; Figure 9.A.6 shows the predominant horizontal forces. The small vertical motion is caused by pressure and squeeze films under the piston which caused tilting and lifting. A numerical time-marching method was used to follow the changing forces and positions of the piston in its motion. The theory is set out by Morton (2009).

The measurement of the vertical movement using a fluorescent film was set out in Morton, Baker and Hutchings (2011). The experimental measurements of

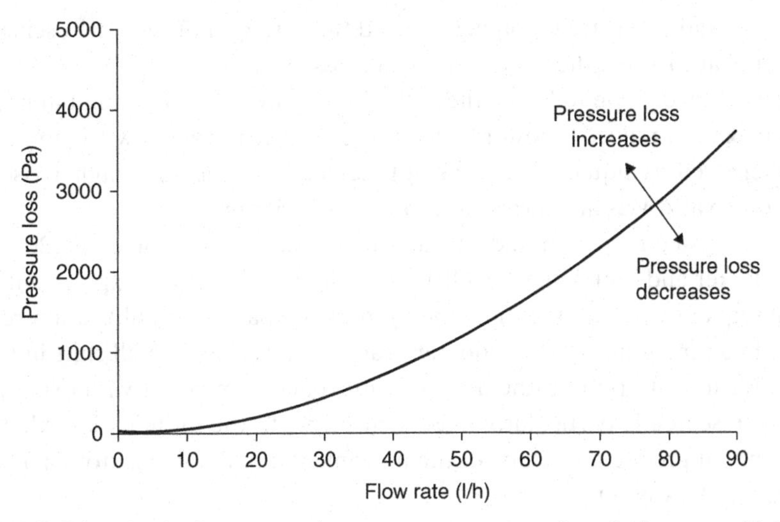

Figure 9.A.7. Diagram to show approximate average pressure loss during an oscillation and how it varies (after Morton et al. 2014a).

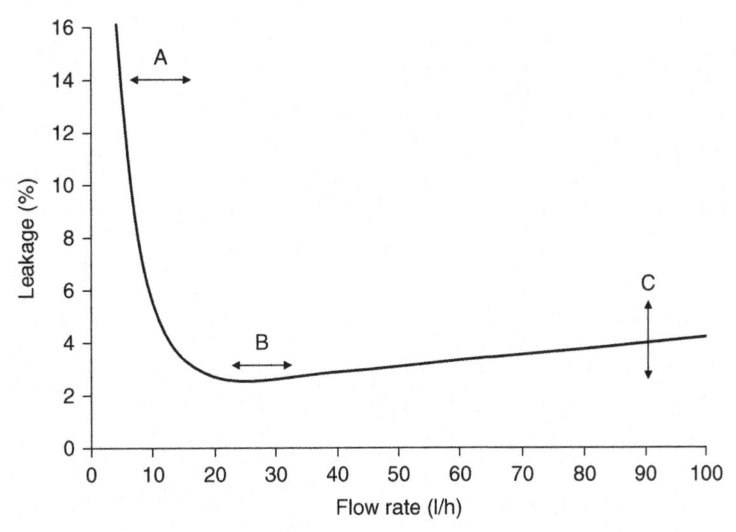

Figure 9.A.8. Diagram to show approximate average leakage variation (after Morton et al. 2014b).

movement and pressure loss have been given in Morton et al. (2014a). It appears that while the lighter pistons tilt, contact with the bottom of the chamber is not lost.

They showed that there was substantial variation in the velocity and in the pressure loss during one oscillation of the piston, and that the average pressure loss follows a curve of the form shown in Figure 9.A.7. The causes of increased and decreased pressure loss are set out by Morton et al. (2014a). The pressure losses increase for heavier pistons, higher-density liquid and higher viscosity. An interesting finding was the benefit of a low-friction coating on the stainless steel piston which reduced its pressure loss towards that of the carbon piston. They also found that while the length of upstream and downstream piping affected the variation during one oscillation, it did not appear to affect the average pressure loss.

Their leakage graph followed the pattern of Figure 9.A.8. The minimum point in the curve, B, was reduced to lower flow rates for lighter pistons, low-friction coatings and higher-viscosity liquids. The slope at C seemed to be less for higher viscosities, and the curve at A was at a larger flow rate for heavier pistons.

The interested reader should consult Morton (2009), Morton et al. (2014a) and Morton, Baker and Hutchings (2014b) for further details of the experimental findings. Subsequent work on the model in preparing a paper for publication revealed an error in an integral. Its correction appeared to result in a small shift in the predictions. We hope to publish the updated model and compare it with experimental data. This research is particularly important being, to the author's knowledge, the first attempt at precise modelling of the movement of a floating piston in a positive displacement flowmeter.

10

Turbine and Related Flowmeters

10.1 Introduction

10.1.1 Background

Spirals, screws and windmills have a long history of use for speed measurement. Robert Hook proposed a small windmill in 1681 for measuring air velocity, and later one for use as a ship's log (distance meter). A Captain Phipps, in 1773, employed the principle that a spiral, in turning, moves through the length of one turn of the spiral to create a ship's log. Many centuries earlier than this, it appears that a Roman architect, Vitruvius, suggested a more basic form of this device.

In 1870 Reinhard Woltmann developed a multi-bladed fan to measure river flows (Medlock 1986). The device was a forerunner of the long helical screw–type meter still called after him and used widely for pipe flows in the water industry. The first modern meters, of the type with which we are mainly concerned, were developed in the United States in 1938 (Watson and Furness 1977; cf. Furness 1982). These were attractive for fuel flow measurement in airborne applications. They consisted of a helically bladed rotor and simple bearings. Improved sleeve bearings were developed for longer life with hardened thrust balls or endstones to withstand the axial load. An alternative developed over several years and patented by Potter (1961) was to profile the hub of the rotor. The pressure difference caused by the hub shape, rather than the thrust on the bearings, may have held the rotor against the axial drag forces due to:

- the pressure balance across the rotor and/or:
- the spinning of the rotor on a film of the liquid, flowing upstream through the annular passage between the stationary axle and the moving rotor.

This allowed the rotor to run on a single journal bearing.

10.1.2 Qualitative Description of Operation

The turbine consists of a bladed rotor which turns due to the flow in the pipe. In most of the designs to be discussed, the rotor is designed to create the minimum disturbance as the oncoming flow passes round it. Ideally it cuts perfectly through the fluid in a helix so that every revolution of the helix represents one complete axial length of the screw and hence a calculable volume of the fluid. In practice drag forces slightly retard the rotation. These result from: frictional drag on the blades, the hub, the faces of the rotor and the tip of the blades; bearing drag, also a frictional effect; and drag due to the means by which the rotation is measured, which is usually a magnetic drag. These drag forces affect the otherwise ideal relationship of constant fluid volume for each rotor revolution and lead to nonlinearity.

The rotor has to be held in the stream and so supports are invariably necessary to position the bearings centrally in the pipe. Virtue is made of necessity and these supports are used to provide some flow straightening, since the turbine meter is of a type particularly susceptible to swirl. The oncoming fluid will therefore need to flow over the upstream support, and the naturally occurring profile, where the velocity of the fluid is lower at the pipe wall than at the centre of the pipe, will be redistributed into the annular passage past the blades and, perhaps most important, there will be a reduction in any swirl component due both to conservation of angular momentum during the redistribution and to the straightening effect of the supports. The velocity in this passage will vary a small amount with radial position, and a well-designed turbine rotor will generally have helical blades to match the axial and tangential velocity of the rotor at each radial position.

Dual rotor meters appear to have become more common recently.

The main variants from this description are the Pelton and vane or paddle wheel designs. In the former a tangential jet of fluid hits the buckets or blades of the wheel and causes it to rotate. This design has two alternative virtues: it can be made as a simple insertion probe to obtain an estimate of the flow rate; or it can be used in very small flows to register where other designs fail. Other designs using larger paddle wheels and angled propellers are also available.

10.1.3 Basic Theory

Many turbine meters use flat section blades. Ideally these will cut smoothly through the flow in a perfect helix. Using this basic concept the value of V_z can be obtained from the frequency of blade passing f (Figure 10.1). Thus:

$$V_z = W / \tan\beta \tag{10.1}$$

and:

$$f = N \tan\beta V_z / 2\pi r \tag{10.2}$$

where V_z is the axial velocity of the flow approaching the blades, W is the blade velocity, β is the blade angle, N is the number of blades and r is the radius of the blades at the measurement point.

Equation (10.2) can be used to obtain

$$f = Kq_v \tag{10.3}$$

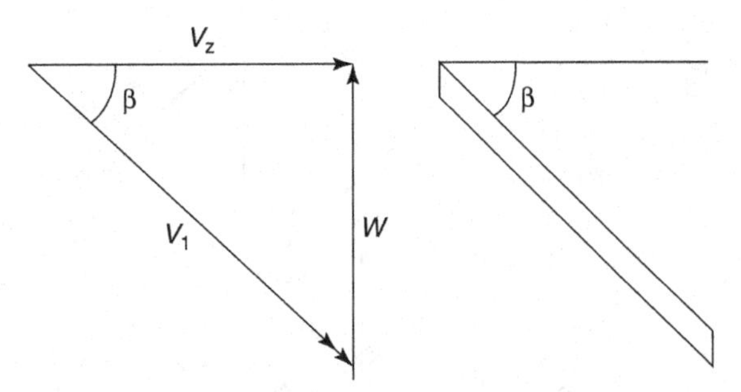

Figure 10.1. Relationship between velocities and angles for ideal flow over blades.

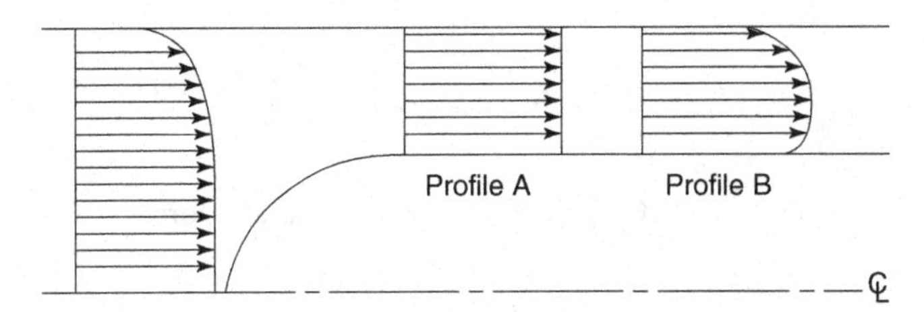

Figure 10.2. Velocity profiles at inlet to, and in the annulus of, a turbine meter.

which provides the basis for the constant which in various units and dimensions is known as meter registration, meter constant, meter coefficient, calibration factor etc., and is measured in pulses per unit volume or radians per unit volume. We shall call this the *K* factor (in pulses per m^3), as in Chapter 1.

Tan (1973; cf. Fakouhi 1977) gives an equation of the form

$$\frac{f}{q_v} = a_0 + \frac{a_1}{q_v} - \frac{a_2}{q_v^{\,2}} \tag{10.4}$$

where the second term is affected by the viscosity and leakage, and the third term is affected by the mechanical and aerodynamic drag. Adjustment of these two terms is found to affect the familiar hump in the turbine characteristic.

However the theory is not, unfortunately, as simple as this, as the blades do not cut the fluid perfectly, and the value of r and β must change to accommodate the profile across the pipe. Suppose we consider the variation of blade angle for a uniform profile A (Figure 10.2); then it can be seen that with no variation in axial velocity across the annulus the correct blade angles will be given by:

$$\tan\beta / r = constant \tag{10.5}$$

If the midspan blade angle is 41° as in Xu's (1992a, 1992b), papers then the values at hub and tip will be 30° and 49° as in Xu's meter, suggesting that this was the method applied in the design he used. If, however, profile B existed, then the flow will not meet the blades at the correct angle and the designer may adjust the blade twist to allow for this. The reader is also referred to Hochreiter's (1958) paper where he allows for blade obstruction.

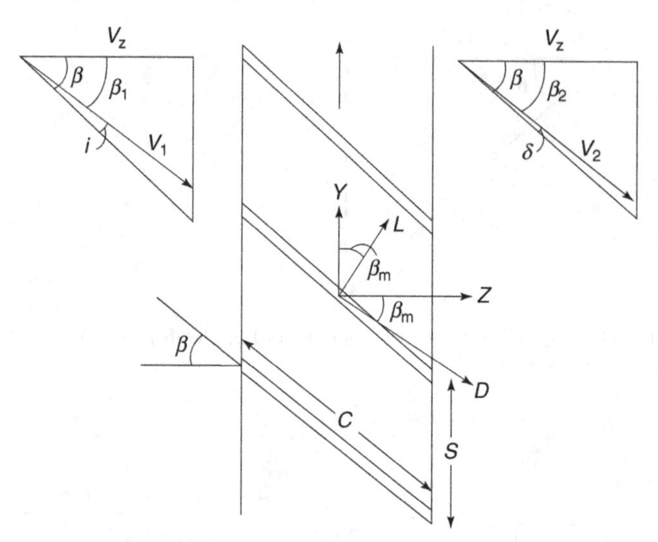

Figure 10.3. Cascade angles, dimensions and forces.

It thus becomes apparent why the flowmeter will be susceptible to incorrect installation, since this will cause a flow profile which results in the wrong incidence angle for some parts of the annulus. It also suggests why a turbine wheel which is optimised for flow profile range will give a better performance than one with, say, constant angle blades. To develop a fuller theory it is necessary to take account of the lift and drag on the blades, bearings etc. Figure 10.3 sets out the main angles and forces for a cascade of blades. The analysis is given in Appendix 10.A.1

Tsukamoto and Hutton (1985) predicted the complete characteristic of a turbine meter with reasonable accuracy, allowing for viscosity, inlet flow profile and meter geometry. Blows (1981) developed a model of a turbine flowmeter which allowed the effects of viscosity and tip clearance to be examined.

The equation for the driving torque Equation (10.A.1) given by Tsukamoto and Hutton (1985), Figure 10.4 and equivalent to the equation used by Blows (1981), is equated to the drag torque. Tsukamoto and Hutton (1985) gave four components of the drag torque:

T_B the bearing drag;
T_W the hub disc friction drag;
T_t the blade tip clearance drag;
T_h the hub fluid drag.

Blows (1981) discussed pick-up torque and thrust pad drag torque, but dismissed the former as difficult to calculate and likely to be significant only in small meters. He suggested that the thrust pad force might vary, not only in sign, but by a factor of 100 and hence, since effects of this significance are not observed, it could also be ignored. The consensus appears to be that the dependence of all drag forces on speed, apart from aerodynamic drag on the blades and the hub fluid drag, is a power less than two and therefore of decreasing significance with increasing speed.

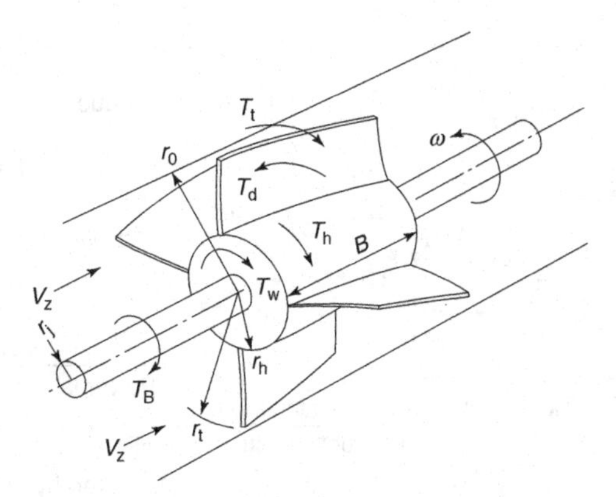

Figure 10.4. General view of a turbine rotor and torques on the rotor (based on Tsukamoto and Hutton 1985; with the authors' permission).

Tsukamoto and Hutton's (1985) results were impressive, giving a characteristic showing the hump and the effect of the blade boundary layer transition at different blade positions at different radii, and the linear range when the boundary layer is turbulent (Figure 10.5). This hump has been attributed to skin friction and to unsteady wake flow caused by straighteners. Tsukamoto and Hutton (1985) also obtained good qualitative agreement for the effect of change from uniform to fully developed turbulent inlet profile, and for tip clearance between Tan (1973), Jepson and Bean (1969) and their own theory.

Thompson and Grey (1970) derived expressions for rotor hub fluid drag and blade tip clearance drag, and introduced the bearing drag and the pick-up drag. Using calculated annulus flow profiles they demonstrated that even with helical blades the profile results in a variation of angle of attack for a 50 mm meter of +7° to −8° and since the actual incidence angle will be greater than the angle of attack, it is possible that blade stall may be occurring. The effect of this on the driving torque is well shown in a plot of torque against radius. It is also useful to note the point that change of fluid for the same velocity results in a change in Reynolds number and hence in profile, although the effect on angle of attack is comparatively small. The annulus profiles and the blade-driving torque are shown in Figures 10.6 and 10.7.

Jepson and Bean (1969) in their earlier work essentially reproduced Equation (10.A.1) with a slightly different constant. In their analysis they made three simplifying assumptions:

a) They assumed that the exit angle was the blade angle and for helical blades would be given by (cf. Equation 10.5):

$$\beta = \tan^{-1}\left(\frac{2\pi r}{K_h}\right) \tag{10.6}$$

where K_h is a constant;

b) They assumed that the retarding torque due to bearing friction was negligible compared with the aerodynamic forces and they justified this on the ground that

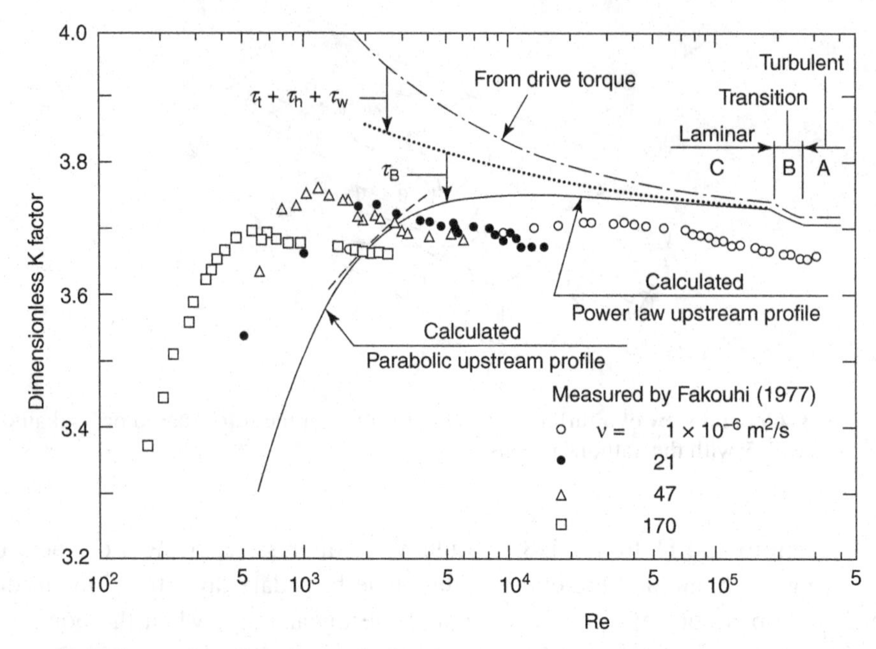

Figure 10.5. Comparison between calculated and measured meter performance. Relevant dimensions: r_0 = 18.288 mm. r_t = 18.008 mm. r_h = 9.703 mm. B = 14.224 mm, t_B = 1.245 mm, L = 95.280 mm, N = 6. r_j = 2.477 mm, o.$\nu = 1 \times 10^{-6}$ m^2/s. •, $\nu = 21 \times 10^{-6}$ m^2/s. Δ. $\nu = 47 \times 10^{-6}$ m^2/s, □. $\nu = \times 10^{-6}$ m^2/s (based on Tsukamoto and Hutton 1985: with the authors' permission).

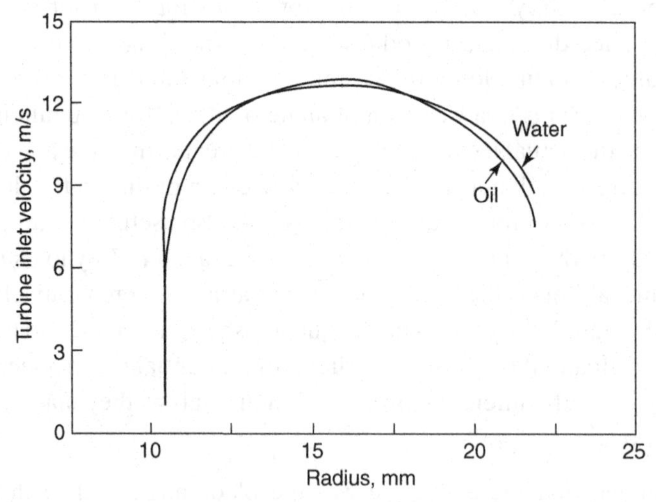

Figure 10.6. Turbine inlet velocity profiles for water and oil (Thompson and Grey 1970; reproduced with permission from ASME).

it is constant. They pointed out that the assumption was only valid for commercial meters with low-resisting torques;

c) They did not consider retarding torques apart from bearing friction and blade aerodynamic drag.

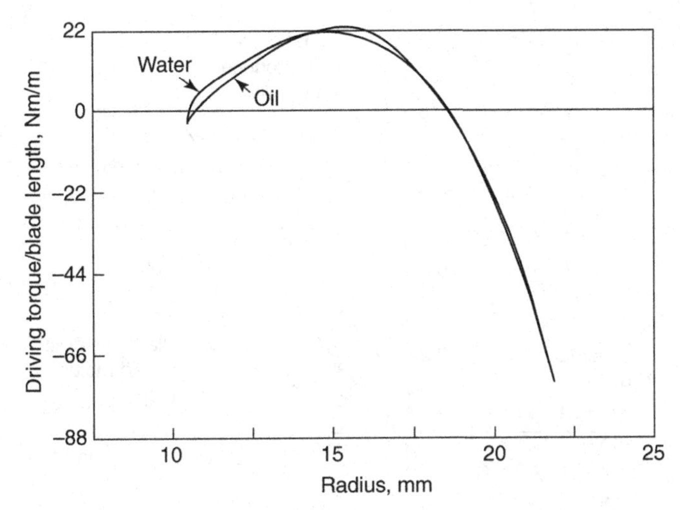

Figure 10.7. Driving torque per unit blade length for water and oil (Thompson and Grey 1970; reproduced with permission from ASME).

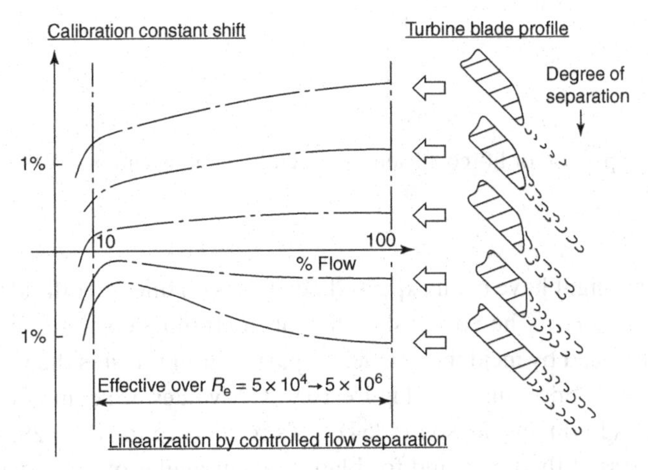

Figure 10.8. Effect of trailing edge on calibration (after Griffiths and Silverwood 1986; reproduced with permission of ONIX Measurement).

They used this simple approach to examine the effect of flow profile and, knowing the profile downstream of a cone exit, to calculate meter error with distance from the cone exit.

Hutton has made the point that a basic difficulty in applying blade theory to turbine meters is that their blades are not very like flat plates because the parallelogram cross-section due to the chamfer on leading and trailing edges results in an S-shaped camber line. The effect of the trailing edge is confirmed by Griffiths and Silverwood (1986) in Figure 10.8.

Xu (1992a) presented a very interesting study of the flow around a typical turbine meter blade and calculated the flow at leading and trailing edges. The flow at the

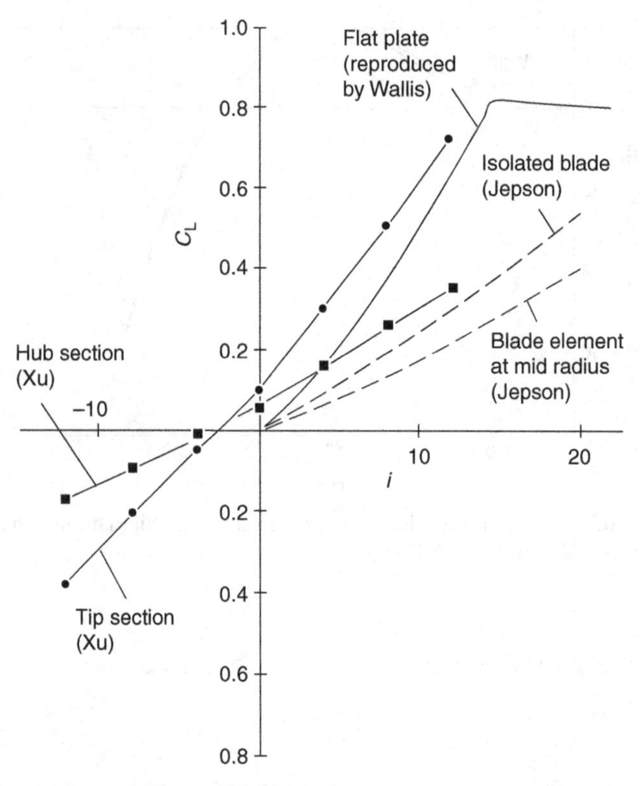

Figure 10.9. Comparison of lift coefficient as given by various authors (Wallis 1961; Jepson 1967; Xu 1992a).

leading edge, as might have been expected, separates on one or both sides depending on the incidence angle. The vortex shedding at the trailing edge does not appear to be greatly influenced by incidence angle. Of particular interest is the variation of lift and drag with incidence angle. In Figure 10.9 Xu's values are compared with those given by Wallis (1961) and Jepson (1964). Xu's results are for Re = 29,500 based on the half thickness of the blade and for blade stagger angles of: tip 49°; midspan 41°; hub 30°.

Xu (1992b) applied his results to the performance of a 100 mm turbine meter with and without swirl. He measured the inlet velocity profile which he then appeared to apply to a theoretical calculation of the meter characteristic starting essentially from Equation (10.A.8) (apart from a sign difference). The agreement he showed between prediction and measurement was impressive.

Tan and Hutton (1971) found that a full-diameter rotor had a rising characteristic while a turned-down rotor had a falling characteristic, and concluded that the tip clearance could account for this.

Hochreiter (1958) used non-dimensional groups to correlate experimental data but found that for a Reynolds number based on rotational speed below 6,000 to 10,000, scatter may have resulted from poor lubricating properties of gasoline.

Recent modelling is reviewed in Appendix 10.A.2.

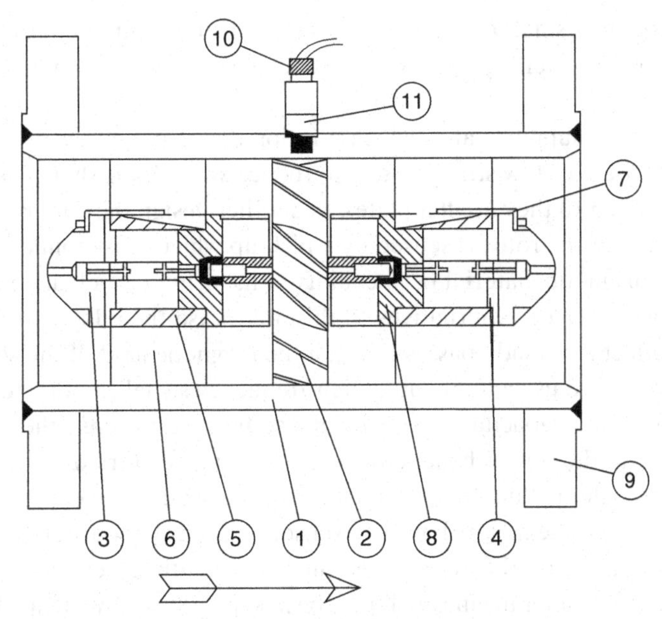

Figure 10.10. Precision commercial turbine flowmeter design for liquids for 100–500 mm diameter (1, body; 2, rotor; 3, support nut; 4, support plate; 5, support tube; 6, support vane; 7, lock tab; 8, end support assembly; 9, flange; 10, pickup assembly; 11, pickup stub) [reproduced with permission of ONIX Measurement].

10.2 Precision Liquid Meters

10.2.1 Principal Design Components

Figure 10.10 gives an example of a commercial design which introduces most of the essential features of these meters. The first essential component is the rotor itself, which is like a small turbine. This is formed from either straight or twisted (helical) blades. For good design it is suggested that: the hub radius should be approximately half the tip radius; the blades, when viewed axially, should just block the cross-section of the pipe; and the axial length of the blades should be approximately equal to the blade span, which is the length of the blade from hub to tip.

It is important to remember that unlike the power turbine, the meter aims to cut through the fluid without disturbing the flow. This clearly is an impossible ideal, but nevertheless the power extracted from the fluid is very small and the incidence of the flow on to the blades is consequently also very small. The reason for the popularity of helically twisted blades is that the relative angle of approach of the fluid on to the blades varies with radius and so flat plate blades will not allow a constant angle of attack and will lead to unnecessarily large incidence angles, both positive and negative, and thus will impart unnecessary flow disturbance and drag.

The rotor is usually machined from solid, and the flat plate blades often have leading and trailing edges which are milled in a constant end plane. The cross-section of the blade at a constant radius will therefore be essentially a parallelogram. Sharp edges should be avoided since they may cause cavitation. Griffiths and Silverwood

(1986) describe how small changes in the shape of the trailing edges can cause a change of the blades' aerodynamic drag and consequently of the meter calibration (Figure 10.8).

The hangers or support vanes which position the bearings centrally in the pipe are commonly used as flow straighteners to reduce swirl. Their shape varies considerably: in Figure 10.10 they are flat plates. In another design they are parallel tubes.

The rotation of the rotor is sensed by a pick-up which most commonly senses a change in the magnetic material permeability in its vicinity as a blade passes (magnetic reluctance change). Alternatives to this are to use the eddy current effect or the dynamo effect as a blade passes through the magnetic field of the sensor (magnetic inductive), or to use a modulated high-frequency signal (modulated carrier or radiofrequency). The capacitance effect can also be used to sense the proximity of a passing blade to the sensor head. Optical fibres have been tried, but if the optical path passes through the fluid there is a danger of the windows becoming fouled. For increased integrity some meters have two sensors spaced at about 120° and the pulse trains from both are sensed and the phase angle constantly checked.

In small sizes of meter the internals may be held in position by circlips, the removal of which allows the internals to be withdrawn. In other designs and larger sizes the bearing assemblies may be removed from the body by releasing a taper-locking nut.

10.2.2 Dual Rotor Meters

Dual rotor turbine meters are available and the internals of one are shown in Figure 10.11. A brief review of a paper by Pope et al. (2012), in which a dual rotor meter was used, is given in Appendix 10.A.3.

Tests of turbine flowmeters with single and dual rotors were reported by Jalbert (1999). The tests were particularly aimed at testing aircraft engines, which required wide dynamic range. The dual rotor meter appeared to perform well even in swirling inlet flow conditions. Olivier (2002) discussed a dual rotor turbine flowmeter that appeared to be insensitive to changes in fluid viscosity.

Zhang et al. (2004b) set out some equations and experimental results which they claimed demonstrated the performance of a turbine meter with two rotors for mass flow. I am not clear, from the theory they provide, how their approach will, indeed, provide precise mass flow rate. A design which does achieve this is described in Chapter 19.

Figure 10.11 shows an industrial dual rotor meter. It is suggested that they may be less sensitive to upstream flow profile than single-rotor designs, have a larger turndown than single-rotor meters and be less sensitive to viscosity change (see Section 10.A.3). In some designs the rotors turn in the same direction and in others in opposite directions. Turndown claims as much as 500:1 have been made. Dual frequency pickups are used, and the ratio may provide diagnosis of bearing condition.

10.2.3 Bearing Design Materials

Minkin, Hobart and Warshawsky (1966) attributed calibration unpredictability to bearing variation as a major factor.

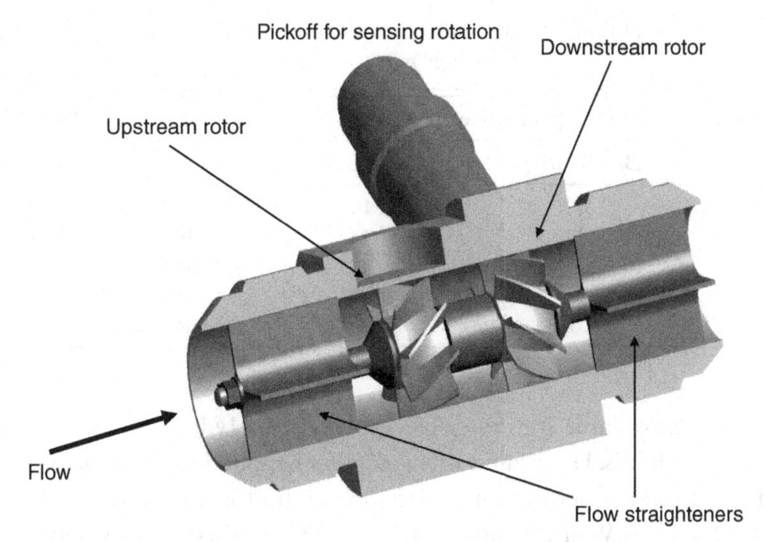

Figure 10.11. Dual rotor turbine meter (reproduced with permission of Badger Meter Europa GmbH).

Clean Liquids

Open ball bearings of stainless steel or other compatible material are suitable for liquids with lubricating properties in the temperature range –50°C to 250°C but a filter with a mesh size small enough to prevent solid particle ingress into the bearings should be used.

Liquids in General

Tungsten carbide or high-chrome/high-cobalt journal bearings possibly with carbide pinions and a stellite sleeve are suitable in the range –50°C to 400°C unlubricated. The bearings are lubricated by the metered fluid and made up of tungsten carbide with a hardness of Rockwell C-94 and a surface smoothness of 0.05 micrometres. Tungsten carbide with cobalt bonding is most common, but the cobalt is leached out by acidic solutions used for cleaning. Stellite may then be used as a less durable alternative for more corrosive fluids. A more recent alternative is nickel-bonded carbide bushes with titanium carbide shafts. An alternative design used rotor pins of hard metal in sapphire bearings with thrust plate and rings made with curved contact surfaces for line contact. PTFE is used in some applications. Ceramic bearings, such as Al_2O_3, may also be an option.

A ball and sleeve design has been used. The shaft was tipped with an ellipsoid and ran in a ceramic sleeve so that line contact was achieved. Typical bearing life was claimed as 12 to 24 months when used with 50 to 75% duty cycle in clean particulate-free gasoline. For ball race bearings correctly lubricated, a life of 4,000 to 6,000 hours was claimed, and for journal bearings 20,000 hours or more, depending on the properties of the measured fluid and rate of flow.

For some larger sizes (above about 80mm) only an upstream bearing is used, consisting of a tungsten carbide (or other suitable materials) bush and shaft. Special

Table 10.1. *Typical values for strainers from manufacturers' literature with maximum allowable particulate sizes*

Meter Diameter (mm)	Particulate Sizes (μm)		Mesh	Hole Size (mm)
	Ball Bearings	Journal Bearings		
13	60	100	150	0.100
20	90	142	100	0.150
25–80	100	185	80	0.175
100–150	142	251	60	0.200
200–300	251	401	40	0.300

bearings may be available for cryogenic applications. Hydrostatic bearings may be suitable for dirty fluids. House and Johnson (1986) applied hydrostatic bearings and obtained a remarkable performance (turndown ratio for water of 1000:1) and suitability for dirty fluids. Wemyss and Wemyss (1975) described the development of the Hoverflo bearingless meter. Some manufacturers provide application charts.

New materials have resulted in bearings which are tolerant to a wider range of liquids and wider temperature ranges. One manufacturer indicated the advantages of ceramic balls in terms of roundness, hardness, non-attraction of particles and temperature tolerance which may be from cryogenic to in excess of +400°C, and combined with new retainer and raceway materials.

10.2.4 Strainers

Typical values for strainers are given by manufacturers with maximum allowable particulate sizes (Table 10.1).

10.2.5 Materials

Because the signal is usually obtained from magnetic sensing it is common to make the tube of austenitic stainless steel to allow the magnetic field to penetrate. However, carbon steel is used for flanges and sometimes also for the bodies, and in this instance provision is made for the pickup assembly to function despite the magnetic body.

10.2.6 Size Ranges

Meter sizes are typically:

for flanged, 6 to 500 mm
for threaded, 6 to 50 mm

In some designs the bore reduces so that at the propeller the bore radius is a minimum. Flow ranges are typically 10:1 turndown (but may be up to 30:1) for flow rates of 0.03 m^3/h to 7,000 m^3/h. Overspeed up to 1.5 times the maximum flow rate may be permitted for short periods, but if it takes place for prolonged periods the bearing life is reduced.

Pressure losses at maximum flow rate, with water as the flowing medium, range up to 0.2 bar ($2 \times 10^4 N/m^2$) in 12mm sizes and up to 0.25 bar ($2.5 \times 10^4 N/m^2$) in 200mm sizes.

Maximum pressures range up to 25 bar for hygienic designs, up to 240 to 400 bar for threaded designs and according to the flange rating for flanged designs.

Temperature ranges are typically –50°C to 150°C but may be as wide as –265°C to +310°C.

Some manufacturers give plots of the effect of change of viscosity on the performance of their meters (Baker 1991b). The bearing, as well as the blade design, will have an influence on this. Acceptable viscosity limits for smaller sizes (less than 80 mm) may be up to 15mPas (15cP) and for sizes above 100 mm up to 50mPas (50cP). These viscosity limits prevail as a result of increasing nonlinearity with reducing Reynolds number.

10.2.7 Other Mechanical Design Features

Internals may be made of stainless steel, but other materials such as Teflon and polythene have also been used.

In order to increase the pulse rate from impellers with typically six to eight blades, a shroud ring with holes or a high permeability soft iron alloy is used. This increases the number of pulses due to the holes changing the magnetic reluctance and also allows viscosity compensation to improve low speed performance. The shroud can also give greater strength and reduced blade vibration.

In some small designs of turbine meter the blades have a T shape so that at the tip they are longer (longer chord) than at the hub. One reason given for this is to improve insensitivity to viscosity change which may result from a greater driving torque achieved by increasing the blade lift at the maximum radius. It may also flatten the characteristic hump at low speed. The author is not aware of published data to confirm these tendencies.

Some designs can be used bidirectionally, but of course require calibration in each direction.

Response time constants measured with water or with a liquid of similar density range from 0.005 to 0.05 seconds for a 50% flow rate change or up to 0.17 seconds to reach 63% of a step change final value.

After use with corrosive fluids, the meter should be cleaned with solvents. However, maximum rotational speeds should be carefully observed.

10.2.8 Cavitation

Cavitation is caused when the local pressure drops below the vapour pressure of the liquid and results in a two-phase flow of increased volume and therefore causes overreading.

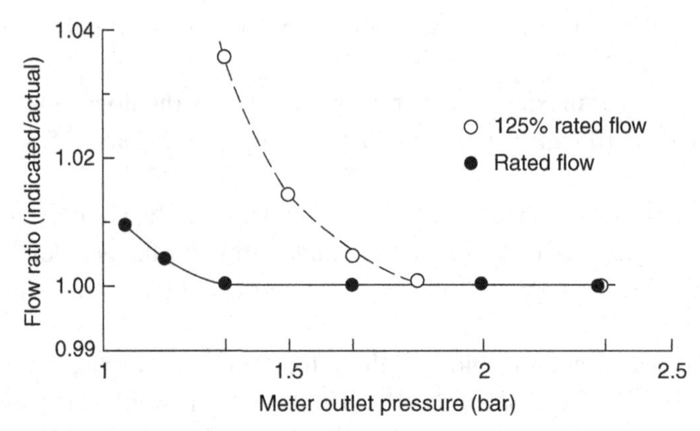

Figure 10.12. A typical example of cavitation effects on calibration factor (Shafer 1962; reproduced with permission from ASME).

Figure 10.12 demonstrates this effect where the vapour pressure of the hydrocarbon was less than 0.007 bar. It may also cause damage to the internals. A simple rule is that the pressure downstream of the meter should be at least one bar above atmospheric pressure. More generally the rule given is that the back pressure should be at least twice the pressure drop across the meter plus the absolute vapour pressure times 1.25. This recognises that as well as the pressure loss through the meter at rated flow of between 0.3 and 0.7 bar and the pressure needed to accelerate the flow past the turbine wheel of 0.3–1.0 bar there is also the minimum pressure level to prevent cavitation.

Bucknell (1963), from his work on liquid oxygen and hydrogen, suggested that a minimum safe margin for most meters was a back pressure of four times the meter permanent pressure loss above the vapour pressure.

10.2.9 Sensor Design and Performance

A direct mechanical drive to register may be used in large meters. However most commonly, to obtain rotor speed, blade passing is sensed by the change in the magnetic field around the sensor. The terms given to the various methods suffer from some variation and the following are suggested for consistency (Olsen 1974):

a) **Inductive:** magnets are embedded in the hub or blades and a pick-up coil with a soft iron pole piece senses their passage;
b) **Variable reluctance:** a permanent magnet with pick-up coil is positioned in the body of the flowmeter near the propeller/rotor which senses the variation of the flux due to the passage of each blade of highly permeable magnetic rotor material;
c) **Radiofrequency (RF):** an oscillator applies a high-frequency carrier signal to the coil in the pick-up assembly and the passing of the rotor blades modulates the carrier. At very high frequencies of signal transmission the skin effect is such that the electromagnetic field is essentially reflected by the passing blades with negligible drag effect on the wheel.

d) **Photoelectric:** a light beam is interrupted by the passage of the blades. Optical fibre methods which are intrinsically safe may also be used where the light reflects off the blade tips. The problem with this method is that the windows tend to become fouled and so light transmission through the liquid is not recommended in many applications. Place and Maurer (1986) proposed an optical system which overcame this problem where the blade passage rocked a magnetic element and the rocking was sensed optically;
e) **Magnetic reed switch:** contacts are opened and closed by magnets in the rotor or some rotating part of the meter.

The resulting signal must be amplified with care and shielded from extraneous noise due to voltage sources or magnetic fields as spurious pulses can introduce significant errors.

Vibration may also cause microphonic effects. Screened cable should be used and the distance from the flowmeter to the pre-amplifier should be as short as possible and typically not greater than 2m.

Typically the pulse is amplified from about 15 mV root-mean-square (r.m.s) to 8 V amplitude (increases with flow rate) and is of 0.5 to 20 millisecond duration. Pulse rate ranges from 50Hz to 3kHz, and impedance from 300 Ohms to 1,500 Ohms. A maximum transmission cable length has been given as 1,000 metres by one manufacturer. The signal is then applied to a signal converter, an electronic gearbox, to convert pulse rate into m^3/h and total into m^3. A high pulse density is desirable where a meter is to be calibrated against a volumetric standard such as a prover loop where generally it is recommended that 10,000 pulses or more should be collected per unit reference volume. Eide (1991) described a 12 inch Brooks Compact Prover and a 3 inch Brooks turbine meter modified to give 27,000 pulses/m^3 which were used to calibrate pipe provers. The uncertainty of the water draw was better than ±0.003% and the master meter's uncertainty was 0.02%.

Ball (1977) demonstrated that below about 1/3 fsd the drag due to a magnetic pick-up could cause differences compared with an RF pick-up, and at 1/6 fsd indicated an error of 4% rising further with decreasing speed.

10.2.10 Characteristics

The typical characteristic of the precision turbine meter is shown in Figure 10.13. There are four regions to note:

i) the highest accuracy region;
ii) an extended range which should be used only briefly and with care to avoid excess bearing wear;
iii) the normal range of operation that includes both the high-accuracy region and also a lower-flow region which is often shown as having a "hump"; and
iv) a low-flow region where drag forces are dominant and where the performance of the meter is less reliable.

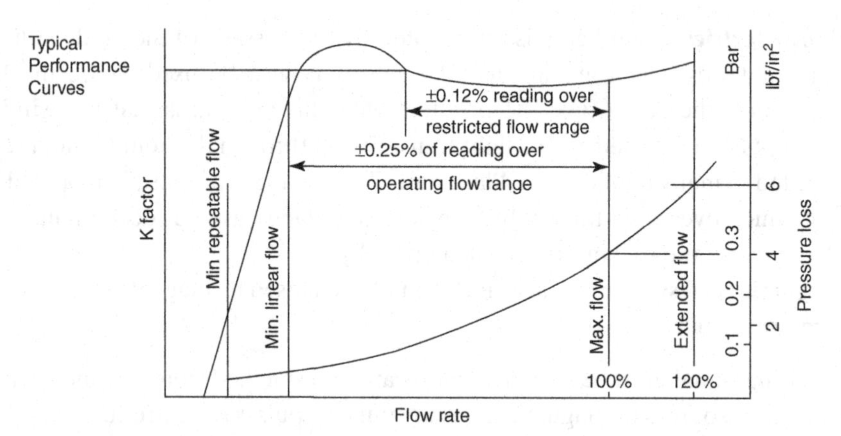

Figure 10.13. Characteristic for a precision liquid meter (KDG Instruments – reproduced with permission from Emerson Process Management).

Most of these regions and their causes are obvious. The one which is often puzzling is the hump region in (iii). Tsukamoto and Hutton (1985) have confirmed theoretically that a hump in the characteristic can be caused by movement of the point of transition of the blade boundary layer near the leading edge of the blades from turbulent in the high-accuracy region to laminar. Griffiths and Silverwood (1986), as indicated earlier, have suggested that changing the profile of the blade trailing edge by filing off the corner leads to a removal of the hump and an increase of speed of rotation of more than 1% due, presumably, to changes in the point of separation and reduction of drag (Figure 10.8). See also Appendix 10.A.3 for further discussion.

Pressure loss arises largely from the pressure reduction at the rotor and bearing supports and the failure fully to recover the total pressure downstream for an area ratio $m = A_1/A_2$ which will give a maximum pressure loss of:

$$\Delta p_{loss} = \tfrac{1}{2}\rho V_1^2 K$$

For pressure change due to area change only (neglecting friction), the loss factor, $K = (m^2 - 1)$, and for values of $m = \frac{4}{3}$, $\rho = 10^3$ kg/m^3 and $V_1 = 10$ m/s this gives a pressure change of 0.39 bar (3.9×10^4 N/m^2). Experimental values of K range from 0.4 to 0.9 depending on the geometric details of the meter. For $K = 0.5$, $\rho = 10^3$ kg/m^3 and $V_1 = 10$ m/s the pressure loss is 0.25 bar.

10.2.11 Accuracy

Typical values of measurement uncertainty given by manufacturers are ±0.12% to ±0.25% for the high accuracy range and ±0.25% to ±0.5% for the lower part of the normal range. This is sometimes given as a linearity. The meters are capable of calibration to achieve these values and possibly slightly better than ±0.1 to ±0.5% over a turndown ratio of 6:1 at least and up to 18:1 for thrust-compensated designs. With achievable uncertainty of ±0.15% they can meet the requirements of oil pipeline and custody transfer applications.

Repeatability is typically given as ±0.05% for meters of less than or equal to 50mm and ±0.02% for meters of greater than or equal to 75mm. Such precision values do not take into account bias errors often present in calibration equipment. Mattingly et al. (1977) obtained day-to-day laboratory repeatability of ±0.07% to ±0.12%. Shafer (1962) made the important point that before calibration the meter needs to be soaked in the calibration fluid and run for some time beforehand. He obtained impressive performances of 0.1 to 0.2% agreement as typical of year-to-year repeatability.

These meters have been widely used as transfer standards, and for best performance the meter is calibrated with upstream and downstream pipework and an upstream straightener. Twin meter packages of turbine and other meter types have been used for transfer standards, while Minkin et al. (1966) used a twin turbine meter arrangement to give extra confidence.

Withers, Inkley and Chesters (1971) tested 5 × 75 mm meters on water, kerosene, gas oil and spindle oil and obtained repeatability to better than ±0.05% (95% confidence) on all. However the characteristics were very varied when used with different liquids and temperatures. The temperature effect was shown to be due more to viscosity than to expansion. Despite the apparent variation the authors claimed to be able to determine a universal curve for each meter in terms of Reynolds number.

Minkin et al. (1966) reported extensive tests on two groups of meters ranging from about 20mm to 50mm diameter in which they showed the following results.

a) The calibration factor for full scale had a maximum deviation from the mean over two years of 0.6%;
b) The calibration factor for water and liquid hydrogen for a good meter (horizontal and vertical orientation) varied by less than 1% over three years apart from one occasion with 2% change;
c) The orientation of meters affected lower-range values more than full scale;
d) The effect of thermal expansion resulted in a calibration factor ratio for $C_{water}/C_{liquid\ hydrogen}$ of 0.991 to 0.994 theoretically with wider experimental variation (cf. Shamp 1971 who reckoned registration for liquid nitrogen at −190°C was 1.5 to 1.9% less than for water at about 15°C).

It is also necessary to calibrate in the same orientation as installation.

10.2.12 Installation

Installation is one of the major causes of meter error. The ideal long straight length of upstream pipe required to achieve a fully developed (usually turbulent) profile is seldom possible even in calibration facilities. It may be worth considering the calibration of the flowmeter together with surrounding pipework and straightener, to eliminate, as far as possible, the effect of upstream flow effects. Because of the susceptibility of turbine meters to asymmetry and swirl, generated by upstream

Table 10.2. *An example of upstream spacings from manufacturer*

Fitting	Upstream Distance
Reducer	15D
Swept elbow	20D
Two swept bends in the same plane	25D
Two swept bends in perpendicular planes	40D*
Valve	50D

* Consider installing a straightener

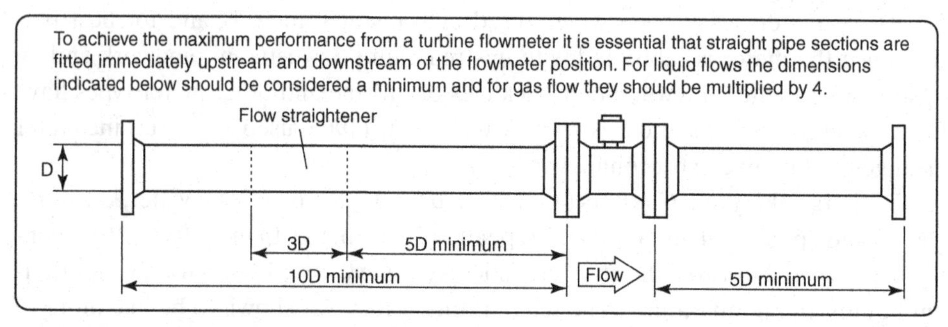

Figure 10.14. Installation for a precision liquid meter as suggested by Quadrina Ltd. (reproduced with permission).

fittings, it is common for manufacturers to provide recommendations for upstream pipe configurations, including the use of flow conditioners. In most cases these will, presumably, be to avoid additional error. These are sometimes accompanied by a diagram. A typical example is shown in Figure 10.14. This suggests that an upstream straight length of 10 diameters is required, with the downstream end of a flow straightener, three diameters long, positioned at least five diameters upstream of the meter. It is always slightly surprising that some other manufacturers, who provide a useful diagram like this one, suggest that without a flow straightener the user should still allow only, say, 10 diameters of straight pipe upstream of the flowmeter.

ANSI/API have suggested from experience that an upstream length of 20 diameters, and a downstream length of five diameters, provided effective straightening in many installations assuming there is no swirl. In addition maximum allowable misalignment of the meter bore and the preceding pipework has been given as 5%. Values are given in Table 10.2 for upstream distances for particular fittings for a commercial instrument. However where swirl is suspected a suitable flow straightener should be installed.

Wafer construction has been available, allowing installation between flanges. Presumably great care is needed in the positioning of this meter to obtain satisfactory results. Published data suggest that some of these figures for upstream spacing may be over optimistic, and that the uncertain effects of an upstream valve indicate the need to avoid this configuration and if unavoidable to use a straightener.

Table 10.3. *Some values from tests by Mattingly and Yeh (1991) on installation effects*

Fitting	Upstream Distance (Diameters)	Change in Meter Factor (%)
Without tube bundle flow conditioner		
Single elbow	30	Less than 0.1
Two swept bends in perpendicular planes	25	+1
With tube bundle flow conditioner		
Single elbow	15	–0.1
Two swept bends in perpendicular planes	10	+0.2

Larger hub-to-tip ratios appear to be less sensitive to profile and the larger the tip clearance the less the sensitivity. Salami (1984b) confirmed that a commercial meter with a hub-to-tip ratio of about 0.5 and a tip clearance of 10% was almost insensitive to changes from uniform to turbulent profile (cf. Tan 1976 and Jepson and Bean 1969).

Millington, Adams and King (1986) reported data from tests on 24 meters, the dimensions of which ranged from about 11 mm to about 34 mm, with hub-to-tip ratios between about 0.4 and 0.5, tip clearance between 1% and 4%; flat, twisted and T-shaped blades; and integral straighteners of about 1D to about 1.6D. Their results suggested that a single bend (R/D = 1.5) caused an error of about 0.3% at 5D upstream, but that a double bend causing swirl could cause serious errors at less than 10D (cf. Brennan et al. 1989 and Mattingly et al. 1987 on the effects of swirl).

Mattingly and Yeh (1991) reported tests on installation effects for turbine meters at a Reynolds number between 10^4 and 10^5 in a 50 mm water flow facility. In addition they showed, for swirl caused by the double elbow out of plane configuration, that the swirl angle dropped from about 18° to about 6° after 40D and to about 3° after 80D (cf. Hutton 1974 and Salami 1985). Some of the values from their plotted data are given in Table 10.3.

Mattingly and Yeh concluded that the tube bundle flow conditioner may introduce errors. As an example, they gave the case of a meter 10D downstream of a single elbow. In this case, the conditioner caused the meter factor to shift from –0.2% to –0.8%. This suggests that the conditioner/flowmeter combination should be tested together (cf. Baker 1993 and Mottram and Hutton 1987 for other published work on installation).

10.2.13 Maintenance

Problems can arise from: defective bearings; jammed bearing or rotor; damaged blades; defective pick-up; break in the transmission path or amplifier failure; extraneous signal interference.

Servicing should take place at least annually. If possible lines should be drained and the flowmeter removed for inspection for foreign matter and to check for free rotation of bearing and proper tip clearance. If necessary the meter should be washed with clean fluid or an appropriate cleaner. If the impeller touches the casing, it may indicate a damaged or worn bearing. It will then probably be necessary to return it to the manufacturer for replacement. Electrical continuity should be checked and the correct behaviour of the sensor should be confirmed.

The nature of the output pick-up signal waveform will often reveal the existence of vibration and non-uniform rotation (worn bearings) when examined with an oscilloscope.

10.2.14 Viscosity, Temperature and Pressure

Temperature differences between calibration and operation cause dimensional changes, viscosity change, density change and velocity pattern shifts (Gadshiev et al. 1988). Manufacturers may provide correction factors (cf. Hutton 1986 concerning a universal Reynolds number curve). Flowmeters are also affected by changes in pressure. See also Sections 9.A.1.2 and 3 for PD meters. Dr Mattingly, in a private communication, has reinforced the need to allow for such changes, noting also that the meter may consist of various different materials.

Temperature increase will cause an increase in the dimensions of the flow tube and its components (Equation 9.A.11)

$$1+\frac{\Delta l}{l}=(1+\alpha_{\mathrm{m}}\Delta T_{\mathrm{m}}) \tag{10.7}$$

where l is a length, α_{m} is the temperature coefficient of expansion and ΔT_{m} is the increase in temperature of the material of the flowmeter. The increase in length is Δl due to this temperature increase. For the change in cross-sectional area the equation becomes

$$1+\frac{\Delta A}{A}=(1+\alpha_{\mathrm{m}}\Delta T_{\mathrm{m}})^2 \tag{10.8}$$

which can be approximated, since the linear expansion is small, by

$$1+\frac{\Delta A}{A}=(1+2\alpha_{\mathrm{m}}\Delta T_{\mathrm{m}}) \tag{10.9}$$

The flow causes the turbine wheel to rotate. The increase in temperature will cause the wheel to expand. Due to the expansion of the hub and flow tube, the inlet flow to the wheel will impact on the wheel at a larger radius,

$$1+\frac{\Delta r}{r}=(1+\alpha_{\mathrm{m}}\Delta T_{\mathrm{m}}) \tag{10.10}$$

For a constant inlet velocity the wheel tangential blade velocity will be approximately constant, while the angular velocity of the turbine wheel will be smaller by a factor of $(1-\alpha_{\mathrm{m}}\Delta T_{\mathrm{m}})$ The combination of increased cross-sectional area in the duct and the reduction of angular velocity of the turbine wheel may result in an under-reading of the meter and a change in the K factor:

$$K' = K(1 - 3\alpha_m \Delta T_m) \tag{10.11}$$

However, if different materials are used for the turbine meter tube, hub and blades, the calculation may need to allow for this. (But note that careful selection of materials may be advantageous to the meter's temperature insensitivity.)

As well as possible small changes to the liquid due to pressure, pressure difference between the inside and outside of the flow tube will cause a change in the cross sectional area of the tube. [Equation (9.A.17)] gives the change in radius

$$1 + \frac{\Delta r_o}{r_o} = 1 + \frac{r_o \Delta p}{E t_o} \tag{10.12}$$

where r_o is the radius of the flow tube, t_o is the tube wall thickness, Δp is the pressure difference between the inside and outside of the flow tube and E is Young's modulus. The area increase will again require a factor of two. However, the change in the flow area and any change in the internals will, presumably, depend on the details of the meter design. It is likely that normal pressure changes will have a negligible effect on the dimensions. A fuller derivation of the effects of temperature and pressure change using a non-dimensional approach was given by Mattingly (2009).

Linear range decreases progressively above 1 cSt and virtually disappears between 50 and 100 cSt. Helical blades are affected much less by viscosity change than constant angle blades. Lower flow rates, smaller meter sizes and high viscosity all lead to decrease in the range of linear operation. Other causes are gum, varnish and other deposits on the bearings. A running-in period will usually overcome this problem. Care is needed in specifying liquid type when purchasing a turbine meter for a particular application because the viscosity of, for instance, JP-5 fuel can vary from 0.8 to 10 or more centistokes (cf. Ball 1977, who gave a polynomial series for the effect of viscosity).

To compensate for viscosity change a universal viscosity curve (UVC) has been used. This gives a plot of the K factor against frequency/kinematic viscosity and the data for various viscosities may then approximately collapse onto a single curve, Figure 10.15. If the viscosity changes are due to temperature changes then, in principle, it would be possible to have a look-up table within the system software, and a temperature sensor, so that the viscosity could be obtained from the temperature and the K factor adjusted to allow for these changes. An important development of the UVC is reviewed in Section 10.A.3.

10.2.15 Unsteady Flow

The effect of pulsation in liquid flow is much less than in gas flow because the density of the rotor is closer to that of the fluid. There may, however, be a tendency to overestimate the flow rate. Grey (1956) gave values of time constant for a range of meter sizes and showed that, in the range 12 to 150 mm sensing element diameter, the value was in the range 1 to 5 milliseconds at full flow rate, and 2 to 9 milliseconds at half

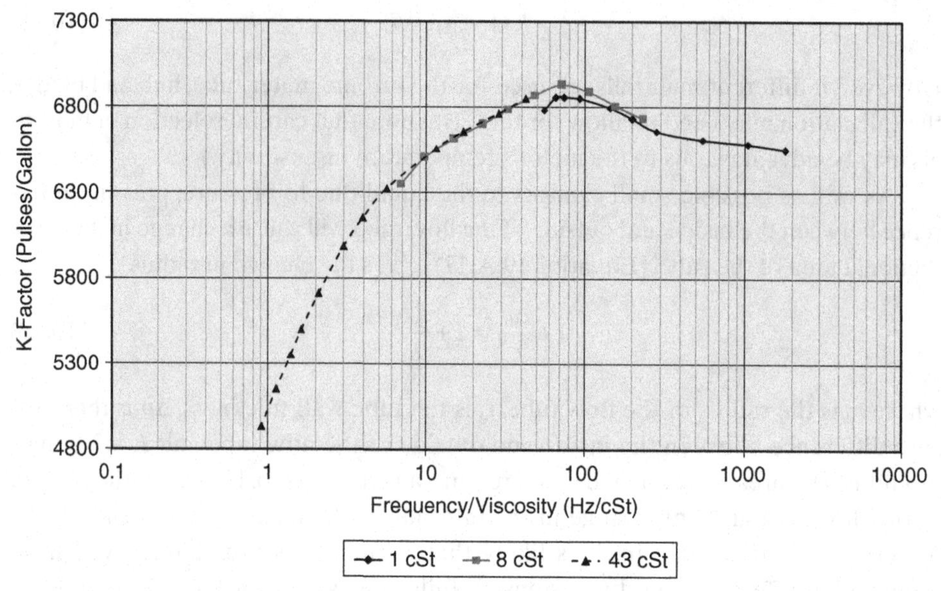

Figure 10.15. An example of a universal viscosity curve (UVC) for three viscosities (reproduced with permission of Badger Meter Europa GmbH).

flow rate. Increased blade angle was also noted to increase the time constant. Higson (1964) suggested that the equation:

$$\frac{q_2 - q_t}{q_2 - q_1} = exp^{-t/t_R} \tag{10.13}$$

where q_1 is the initial flow rate, q_2 is the final flow rate, q_t is the flow rate at time t and t_R is the relaxation time, gives the transient behaviour of a meter. He showed that, for the 20mm meter used, change to 63% of the step change took about 0.6 revolutions and 6 to 110 ms. To return to equilibrium takes about $4t_R$.

One assumes, although to the author's knowledge there is no experimental flow data, that when the wheel slows down as the flow in the pipeline stops, the liquid or gas between the blades will be carried round in the wheel with little interchange with the upstream or downstream fluid.

10.2.16 Multiphase Flow

Baker (1991a) contains a list of references relating to the use of the meter in multiphase flows. The turbine meter has been used in experiments to model nuclear plant operating a long way from normal conditions (e.g. Ohlmer and Schulze 1985). Under the resulting flows, high precision is not necessarily required. Much of the published data for water/air flows do more to warn of the unpredictability than to offer hope of high precision and some data appeared to exhibit a hysteresis effect in vertical upward water/air flows. A small increase in air content from about 4% to about 5%

caused an error increase of about 15%, and this was not immediately removed by reducing the air content back to 4%. A possible explanation was that a vortex structure within the meter had trapped air and caused it to be drawn into the bearing. Upstream jet mixing may reduce errors. Meter speed has also been found to increase with increased solids concentration for the same volumetric flow rate. I consider that the safe rule is to avoid using the meter in multiphase flows.

Mark, Sproston and Johnson (1990a) and Mark et al. (1990b) sought to interpret the data from the pulse spacing in a turbine meter for void fraction measurement in a two-phase flow. They claimed that fluctuation in the time between pulses was caused by the presence of the second phase. They termed this variation "signal turbulence" and claimed that at any flow rate it was directly proportional to void fraction.

Johnson and Farroll (1995) claim that while errors in turbine water and air flows can be as high as 12.5% for a void fraction of 25%, measurement of the fluctuations in the turbine rotor speed can be used to measure void fraction and water flow rate to an improved precision. The reader is referred to the original article for more details, but should be cautious when making general deductions from this approach.

Shim, Dougherty and Cheh (1996) suggested a method to measure the flow rates of each phase in a two-phase flow. They used two-inch (51mm) flowmeters in vertical upward and downward flows. They derived an empirical equation from previous work, but fuller justification for the changes to the equation would be useful. Essentially the plots show the meter reading correctly for water flow, but not rising as fast as the increase in volume flow due to the added air volumes (cf. Minemura et al. 1996).

10.2.17 Signal Processing

Electrical noise may be the most troublesome element in turbine meter systems. Two sensors are used where increased reliability and self-checking is required. However, the signal-to-noise ratio should be high so that the counter is not affected by the noise.

Footprinting of turbine meters can be used as a means of real-time measurement monitoring and control. This has been used as a means to extend calibration intervals (Gwaspari 1990). The use of the full frequency spectrum, for so long discarded as noise, opens up the possibility of condition monitoring particularly where the theoretical understanding is sufficient to link the spectrum distribution with mechanical and electrical causes (Higham, Fell and Ajaya 1986; Turner, Wynne and Hurren 1989; however, cf. Cheesewright, Bisset and Clark 1998).

10.2.18 Applications

Typical applications are:

in the process industries – providing precision, corrosion resistance, intrinsic safety, good temperature and pressure rating and ease of installation;

in oil (including crude) pipelines – where shroud rings may be used with nickel rivets to increase the count rate by 4:1; duty payable on hydrocarbons; an ability to cope with abrasive sand-laden crudes;
in a range of mechanical engineering test rigs;
in the drinks and dairy industry;
in cryogenics – with ball and journal bearings in liquid hydrogen, oxygen, nitrogen, argon, carbon dioxide, normal and superfluid helium; rapid changes in density which occur due to temperature changes; cryogenic fluids since they are clean and of low viscosity. The most appropriate bearing is one with stainless steel balls, non-metallic cage and preferably self-lubricating properties. Rotors for cryogenic service are usually of a nickel alloy to ensure compatibility with liquid oxygen.
in sanitary products – the exacting requirements for precision measurement are appropriate for turbine application;
as a secondary standard – to provide a transfer standard for the calibration of other meters;
as a reference system – consisting of two meters in series so that the reference system is self-checking. Of course the meters need to be isolated from each other by flow conditioning in the line between them;
for high pressure – with very high pressure capabilities; and
for high and low temperatures.

Typical applications for dual rotor meters may be:

subsea where the ratio of rotor movement may give indication of viscosity change and meter operational problems;
aircraft, automotive and defence equipment.

(cf. Baker 1993 for application references).

10.2.19 Advantages and Disadvantages

The compactness of the turbine meter gives it a considerable advantage in large metering systems over the positive displacement meter. The pressure drop may be twice as great for a PD meter. Short-term repeatability is excellent, being better than 0.02%, while long-term repeatability may be only 0.2% over a period of six months, so that, for custody transfer, regular proving under operating conditions is necessary. Although viscosity effects for large meters used with crude oil are less than for smaller meters, nevertheless a 200 mm meter will be affected by as much as 1/2% for changes from 1–30 cSt and will need to be reproved for large changes due to fluid change or temperature. Above about 50 cSt it will probably be necessary to use positive displacement meters. It is best for light products with viscosities which change little.

Other **advantages** are: pulsed output; reliability over extended periods; rapid response; and **disadvantages** are: particles affecting bearing and wheel; sensitive to installation and swirl which may affect calibration.

10.3 Precision Gas Meters

10.3.1 Principal Design Components

Figure 10.16 shows a typical gas turbine meter. While many of the previous comments concerning liquid meters are relevant, it is clear that there are significant design differences. The most obvious is the large hub and comparatively small flow passage. The reason for this is to impart as large a torque as possible on the rotor by moving the flow to the maximum radius and increasing the flow velocity. The second difference is the frequent use of a worm and gear output drive, resulting from the requirement by some national authorities for a mechanical display. It is common, however, to include an electrical output as well as the mechanical register. Bonner and Lee (1992) recorded some of the important innovations introduced into the design in the early 1960s, such as flat helically twisted blades with overlap, and blade tips which extended into a recess in the outer wall of the flow passage.

Lee, Blakeslee and White (1982) described a design in which the main rotor was followed by a second rotor to sense the condition of the installed flowmeter. The speed of the second wheel as a ratio of the first would change if the flow leaving the first wheel was deflected more by rotational constraints on the first wheel. It was possible for changes in the upstream flow also to affect the ratio. Self-correction was claimed for this device.

10.3.2 Bearing Design

Shielded ball bearings are sometimes used. External lubrication is necessary in some designs using instrument oil. Special lubricants are available for use with oxygen. Under normal use lubrication should be sufficient two or three times per year. Some designs, such as sealed ball bearings, do not need external lubrication. Sealed ball bearings are suitable for gases contaminated with solid particles.

10.3.3 Materials

Rotor material is typically Delrin or aluminium, usually the latter for sizes greater than 150mm. Sometimes stainless steel is used.

10.3.4 Size Range

For 25 mm diameter the range is from about 0.8 m^3/h to 10 m^3/h, for 50mm diameter the range is from about 5 m^3/h to 100 m^3/h (with a minimum flow response of 1.2 m^3/h) and for 600mm diameter the range is from about 1,000 m^3/h to 25,000 m^3/h. One type offers a 30:1 turndown ratio. Numbers of blades are typically 12 to 24, and maximum pulse frequencies can be 3 kHz. Maximum pressure rating is up to 100 bar. These figures can vary significantly between manufacturers.

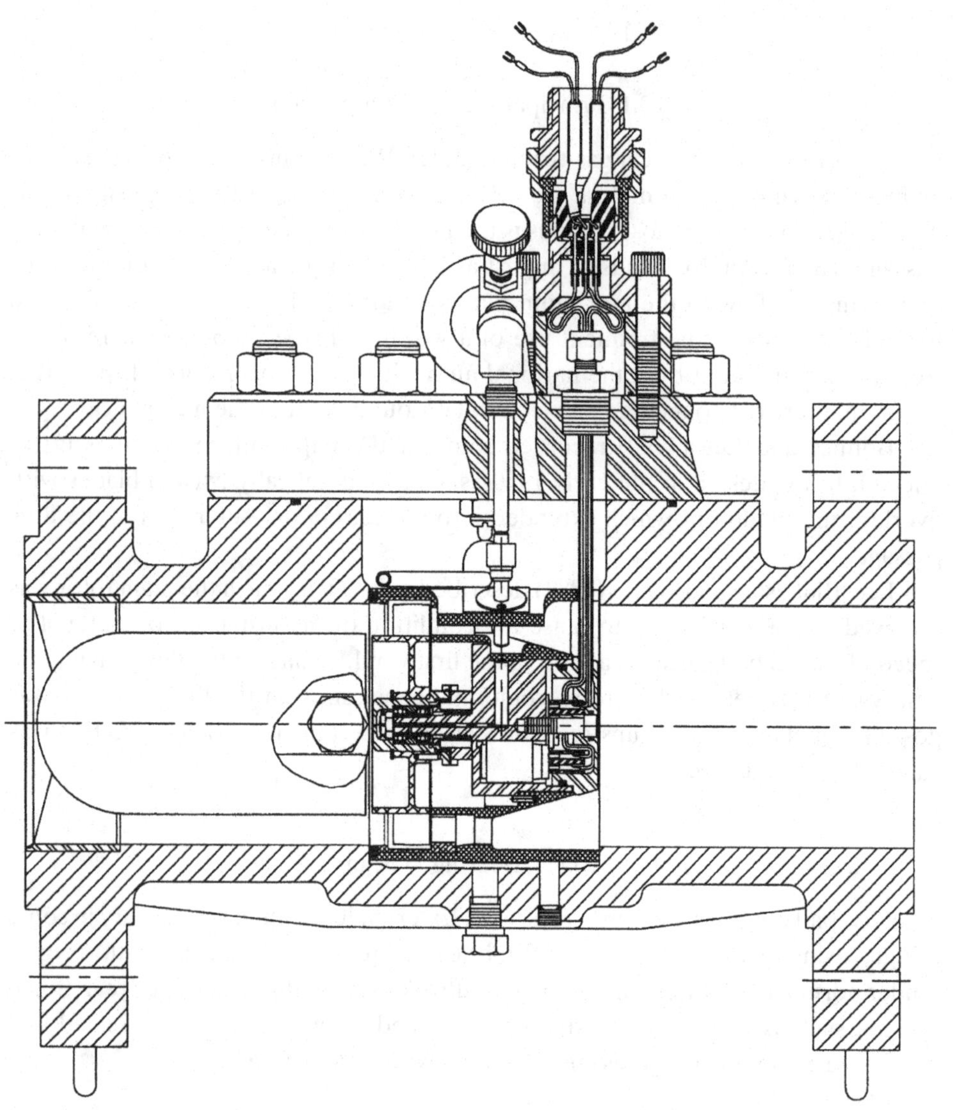

Figure 10.16. Diagram of 100 mm gas turbine meter (reproduced with permission of Emerson Process Management Ltd.).

The pressure loss in the meter at maximum flow rate is 5.5 mbar for the 50 mm diameter meter and 14 mbar for the 600 mm diameter meter. Pressure loss is of course directly related to the density, and therefore to the pressure, of the flowing gas. Reference should be made to manufacturers' data for given conditions.

10.3.5 Accuracy

A typical specification for uncertainty may be ±2% from the minimum flow rate, Q_{min}, to 20% of the maximum flow rate, Q_{max}, and ±1% from 20% Q_{max} to Q_{max}. Some linearity claims of ±0.5% are made.

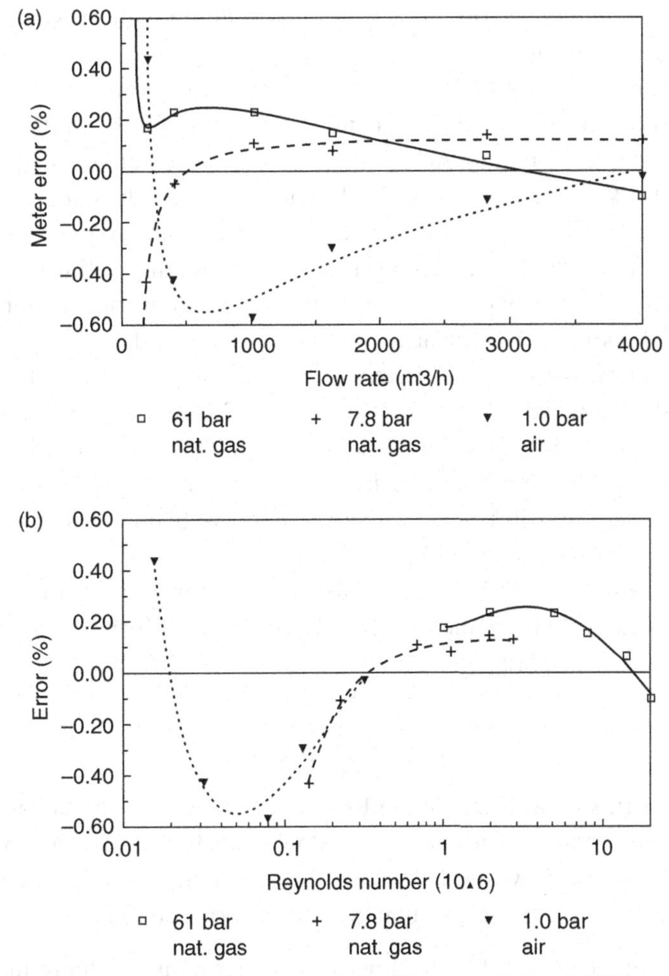

Figure 10.17. Meter error of a turbine gas meter (Van der Grinten 1990; reproduced with permission from Nederlands Meetinstituut): **(a)** Versus the flow rate at working conditions **(b)** Versus the Reynolds number based on the inner meter diameter.

The linear performance may be as good as ±0.5% on about 20:1 turndown with repeatability of ±0.02%. Maximum flow velocities can be up to 30 m/s. Other data has shown remarkable calibration stability with shifts of order +0.2% in nine years with about 10^8 m^3 of natural gas having been transported through the turbine meters at 8 bar line pressure.

Van der Grinten (1990) gave an error curve for gas turbine meters which allowed for drag of gas between blades, annulus boundary layer and friction in the bearing. Figure 10.17 demonstrates this curve fit and also shows the behaviour of the meter with changing pressure and gas.

The turndown of the simple turbine meter increases proportionally with the square root of the gas density ratio. At a pressure of 20 bar, the turn down can be as high as 100:1 compared to 15:1 at a few millibars gauge working pressure (Griffiths and Newcombe 1970). Watson and Furness (1977) claimed that for low-pressure nitrogen a flow range of 5:1 would be achieved, while on high-pressure natural gas the range might be as much as 30:1.

Van der Kam and Dam (1993) found that turbines can operate within an envelope of ±0.5% down to about 25% maximum flow, and of ±1% in the lower range. They also found that a pressure range of 1 to 10 bar caused errors well within 0.5% compared with 1% in older turbines. The Reynolds number dependence of turbines may, in some cases, allow the change in density to be related to curves of Reynolds number dependence. Their data demonstrates a repeatability within 0.1%. Another report (Erdal and Cabrol 1991) on calibration of 6 × 6 inch turbine meters suggested repeatability of about 0.24%, linearity of about 0.42% and day-to-day repeatability of about 0.05% or less, for periods in excess of four years. de Jong and van der Kam (1993) suggested shifts in calibration of +0.2% to 0.3% and Koning, Van Essen and Smid (1989) claimed shifts of order 0.1% in 10 years. Gasunie's experience suggests that the shift is of order a few hundredths of a per cent per year. Van der Kam and de Jong (1994) claimed that 50:1 range with an error curve band of less than 0.5% over the whole range was no longer exceptional.

Van der Grinten (2005) gave a step-by-step procedure for a Reynolds number–based interpolation method for calibrations of turbine gas meters and application to intercomparisons.

10.3.6 Installation

Work that the British Gas Engineering Research Station carried out confirmed that these meters are extremely insensitive to flow disturbance and in most practical cases no upstream or downstream straight pipe is required (Fenwick and Jepson 1975; cf. Harriger 1966). This is presumably due to various effects:

a) the reduction of any swirl in the flow at the larger radius annulus due to conservation of angular momentum and also due to the flow straighteners;
b) the flow development in the small annulus which follows a large contraction; and
c) the integrating effect due to the linear relationship between lift coefficient and incidence for small incidence angles.

They suggested that flow straighteners should only be used upstream if there is a possibility of persistent swirling flow at inlet (cf. Mottram and Hutton 1987).

For inlet flow disturbance van der Kam and Dam (1993) suggested that the inlet flow conditioner could be quite effective in removing swirl, so that swirling flow from a double elbow out-of-plane (with a 40° swirl angle) may not cause errors of more than 0.3%. Also changes in pipe size before the meter are insignificant. In extreme cases a tube bundle may deal with the problem. Roughness does not affect performance. Temperature effects are small for a 20°C variation, but have been difficult to examine because of the lack of a suitable rig. The turbine meter is not suitable for wet or dirty flows. Gases should be clean and free of liquids and dust, or else a filter with five micron filtration quality or better should be used. The upstream pipework should be cleaned before installing the meter (cf. Bonner 1993 and ISO 9951).

According to Harriger (1966) a short-coupled installation with as little as four diameters upstream (2D straightener and 2D straight pipe) is allowable. However swirl and pulsation can cause significant errors. Meters have a built-in flow straightener which removes some of the swirl. Straightening vanes may be necessary if pipe fittings are at five diameters or less upstream. The meter should be carefully centred and protuberances should be avoided for at least five diameters upstream. Downstream the pipework should be full bore, but no other restrictions are placed on it. Van der Kam and van Dellen (1991) for a 300 mm gas turbine meter suggested that 10D upstream length was sufficient to keep the meter within the limits allowed and 15 D if swirl was present.

Mickan et al. (1996a, 1996b) and Wendt et al. (1996) reported experimental investigation of flow profiles in pipes and their effects on turbine gas meter behaviour. The experiments covered laser Doppler anemometry profile measurements and installation effects on a turbine meter with flow straighteners, a single bend and two bends out-of-plane and also with the possible addition of a plate closing half the pipe between bends. The interested reader is advised to read this paper for the very useful data. The errors they tabulated appeared, all but one, to be less than 1%.

George (2002) discussed research undertaken on turbine flowmeters in support of the revision of AGA Report No. 7. The paper identified two developments in turbine technology since the last issue in 1996 of this report, dual rotor and extended range meters. His conclusions in summary were:

- for short coupled, close coupled, swirl and combined close coupled and swirl, results were within ±1% for the four meters calibrated;
- proper integral flow conditioners at meter inlet reduced bias to less than ±0.25%;
- no significant difference in bias appeared to result from single or dual rotor use;
- performance shifts due to pressure should be further explored.

Islam et al. (2003) reported tests of a turbine flowmeter in air with distorted inlet profiles and with integral straighteners.

Drift in gas meters has been reported (Balla and Takáras 2003) during the first year or so, possibly due to bedding in or to deposits of liquid and/or solid left from building the pipeline. Checks on pipe roughness, condition of straighteners, centring of installed meter, visual inspection etc. (Fosse, Ullebust and Ekerhovd 2008) were recommended.

Excessive speeds can damage the meter, but 20% excess may be allowable for short periods.

Temperature should be measured downstream within two diameters. One manufacturer gives a temperature range of −10°C to 50°C.

If, as a result of changes in the process conditions, liquid condensate is produced in the pipework, means of drainage should be provided.

10.3.7 Sensing and Monitoring

The alternative means of measuring the rotation of the wheel are the mechanical system involving a gear train which can result in retarding forces due to gear losses, losses in the magnetic coupling and loads from the meter register and calibration adjustment, or the electromagnetic design in which the loads are much reduced. For a high-frequency signal, a proximity switch operating off the aluminium blades, metal strips contained in the hub or a follower disc on the main shaft, with magnetic probe or proximity switch, allow up to 3 kHz. For low frequencies of 1 to 10 pulses per revolution a reed switch or slot sensor are used.

Reeb and Joachim (2002) reported on a tool for on-line monitoring of gas turbine meters, the AccuLERT G-II (FMC Measurement Solutions), which appeared to claim to be capable of detecting and analysing both mechanical and flow-related metering errors. A description of the basis of this is given at (http://info.smithmeter.com/literature/docs/mn02008.pdf). The AccuLERT claimed to monitor the pulse rise period ratio, the fall period ratio and the standard deviation. In addition, it appeared to monitor other key velocities, times, variations etc. in operation to assess the condition and operation of the meter.

10.3.8 Unsteady Flow

The gas turbine meter is affected by varying or pulsating flows. The increasing flow creates higher incidence angles on the turbine blades and the turbine wheel accelerates fast. However, when the flow decreases the blades presumably stall with low retarding force and hence low deceleration. The effect is to give an overestimation of the total flow. The turbine bearings may be damaged if subject to fluctuating flows for extended periods. Head (1956) gave a pulsation factor which for turbine meters was:

$$\frac{q_{\mathrm{i}}}{q_{\mathrm{v}}} = \left(1 + \alpha b \Gamma^2\right) \tag{10.14}$$

where q_{i} is the indicated flow rate, q_{v} is the actual flow rate, $\alpha = 1/8$ for sinusoidal variation of flow, b is taken as unity for non-following meters and Γ is full flow amplitude relative to average flow. Head gave $\Gamma = 0.1$ as the practical threshold for significant error.

The transient analysis is given in Appendix 10.A.2. From this analysis a speed decay curve for no flow is obtainable, Figure 10.18. We can obtain from this curve the rotor coast time to standstill and the final slope, S, of the curve, which is related to the ratio of the non-fluid drag and the inertia and hence to the condition of the bearings. However, de Jong and van der Kam (1993) were doubtful as to its value for high pressures. The reader is also referred to the paper by Lee and Evans (1970), who describe how they obtained speed of decay curves using an externally applied mechanical friction loading method and also gave typical values of inertia, for example

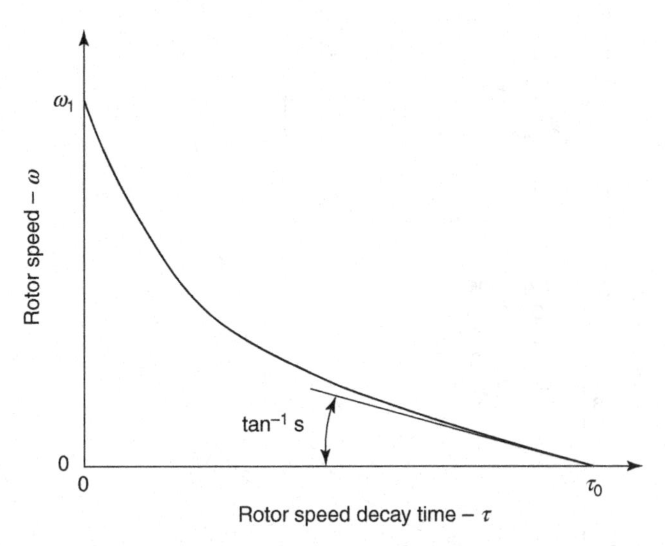

Figure 10.18. Rotor speed decay curve for no flow [Equation (10.A.20)] of a given meter during coasting process of the spin test (Lee and Evans 1970; reproduced with permission from ASME).

for a 150 mm low-pressure meter with plastic rotor $I = 0.242 \times 10^{-3}$ kg m^2 and for a high-pressure meter with aluminium rotor $I = 0.486 \times 10^{-3}$ kg m^2. They also gave $\eta = 0.2$ as used in Appendix 10.A.2.

Lee, Kirik and Bonner (1975) obtained the effect of sinusoidal pulsation on errors. Assuming the worst situation where rotor inertia is too great to follow pulsation they obtained for a pulsation index of 0.1, about 0.5% error and for 0.2 about 2% error where the pulsation index is given by:

$$\Gamma = \frac{V_{max} - V_{min}}{V_{max} + V_{min}} \tag{10.15}$$

Figure 10.19 is adapted from Fenwick and Jepson (1975) and shows the effect on a turbine meter of a square wave pulsation. McKee (1992) found errors rising from zero for a 2% variation to 1.5% for 6% and above (cf. Atkinson 1992, who developed a computational analysis of the meter error for near-sinusoidal flow pulsations and Cheesewright et al. 1996, who queried the lack of reports of flows with pulsating waveforms).

Fenwick and Jepson (1975) gave as an example a 100 mm meter subjected to on-off pulsations with a period of 60 seconds resulting in a 40% over-registration (cf. Grenier 1991).

Jungowski and Weiss (1996) described the performance of a 4 inch (100mm) turbine in pulsating (5 to 185 Hz) air flow, and that this caused an over-reading (as is to be expected) of about 1% for r.m.s to mean velocity ratio of 0.1 and 4% for a ratio of 0.2. In an interesting paper the possibility that flow oscillations in gas due to acoustical effects could cause spurious readings of a turbine flowmeter was considered by

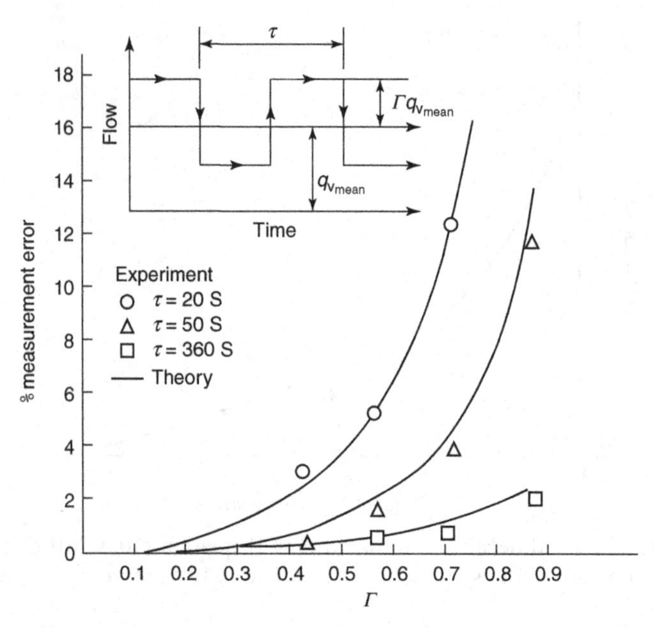

Figure 10.19. Effect of modulating flow on the performance of a 4-in. (100-mm) turbine meter (Fenwick and Jepson 1975; reproduced with permission of British Gas).

Stoltenkamp et al. (2003), who developed and tested a theoretical model to explain the cause of false readings

I worked with some data from tests where a gas conversion process required variations in natural gas flows from high to low flow and back with sudden change and significant errors which the methods of Jepson and others predicted.

10.3.9 Applications

One manufacturer claims that its meters can be used with all non-aggressive gases and fuel gases: town gas, natural gas, refinery gas, coke-oven gas, propane, butane, liquid-gas/air mixtures, acetylene, ethane, nitrogen, carbon dioxide (dry), air, all inert gases.

It is unusual for the turbine meter to be applied to flows of oxygen for the following reasons.

a) Any lubricant used must be compatible with oxygen;
b) A maximum flow rate of order 10 m/s may be specified for oxygen flows in pipework that may oxidise and this tends to be rather low for a gas turbine meter.

Pfrehm (1981) described a system adapted from long accepted and accurate liquid measurement techniques to the mass measurement of ethylene gas. It consisted of a meter, density measurement, flow computer and a piston bi-directional pipe prover. The meter is claimed to provide accuracies of ±0.2% with linearity from 20% or less to 100% of maximum flow.

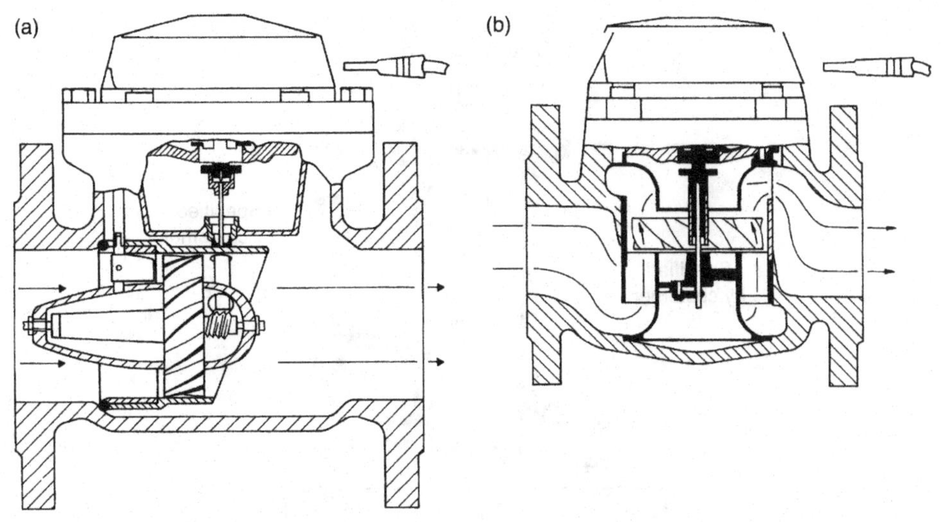

Figure 10.20. Diagrams of water meters (reproduced with permission of Meinecke): **(a)** Gas and Water Meter Manufacturers meter; **(b)** Gas and Water Meter Manufacturers meter with shaft perpendicular to the pipe.

10.3.10 Advantages and Disadvantages

Newcombe and Griffiths (1973) made the following suggestions.

a) Changes in friction and blade form due to mechanical deterioration or damage reduce the meter's rangeability and give a slow registration. Filtering limits the rate of deterioration, but regular proving tests are required. Spin-down tests can indicate bearing deterioration.
b) Rapid flow variations cause the meter to read fast. For a 10 minute off/10 minute on cycle one meter read 3% fast.
c) Swirl will affect registration unless a straightener is fitted.
d) Registration shifts between low- and high-pressure operation and due to high bearing friction can be as large as 2%.
e) Security of gas flow is not affected if the meter fails.

To these van der Kam, Dam and van Dellen (1990) add reliability, high accuracy, dual measurement and back-up system in a device with proven technology.

10.4 Water Meters

10.4.1 Principal Design Components

Figures 10.20(a) and (b) show diagrams of water meters.

A common design of this meter allowed for a complete mechanical insert to fit into a flanged iron section of pipe with a specially made central containment section. The insert consisted of: flow straighteners, which also incorporated the upstream bearing; the rotor, which was usually of helical design and may have had a substantial

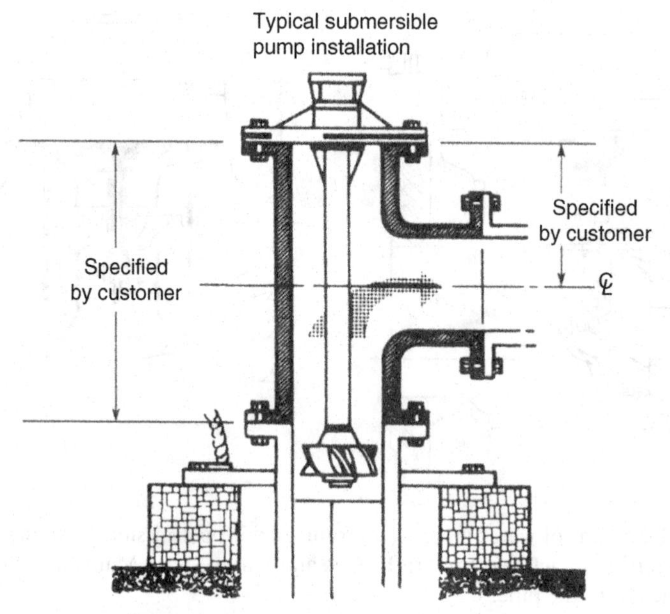

Figure 10.21. Hydrant meter (KDG Instruments – reproduced with permission from Emerson Process Management).

clearance between the blade tips and the casing; the rotor shaft, the motion of which was transmitted via a right angle worm gear assembly into a watertight gear box; and a rudder (or trimming vane) upstream of the rotor, which provided a mechanical calibration adjustment by altering the angle at which the inlet flow stream hit the rotor blades. Water may have entered the gear box via a filter, and the calibration rudder adjuster within the gear box was accessed through the calibration port provided in the top cover. When the cover was fitted over the gear box to form a watertight joint, the output from the gear box was transmitted via a magnetic coupling to the totalising register of the meter.

Figure 10.20(b) shows a meter in which the shaft is perpendicular to the pipe providing a simpler transmission, but a more tortuous flow path and, presumably, resulting in a higher pressure loss. Special designs have been available and Figure 10.21 shows one installed on a hydrant. Various more recent designs may be available.

10.4.2 Bearing Design

Manufacturers should be asked for information on this. Stainless steel shafts and in some cases stainless steel bearings are used.

10.4.3 Materials

For temperatures up to 40°C polythene propellers are suitable, but for meters for hot water the internals are of hot-water-resistant plastic up to 150°C and for temperatures up to 200°C the measuring element is red bronze and the wheel is stainless steel. Other materials used are brass and stainless steel.

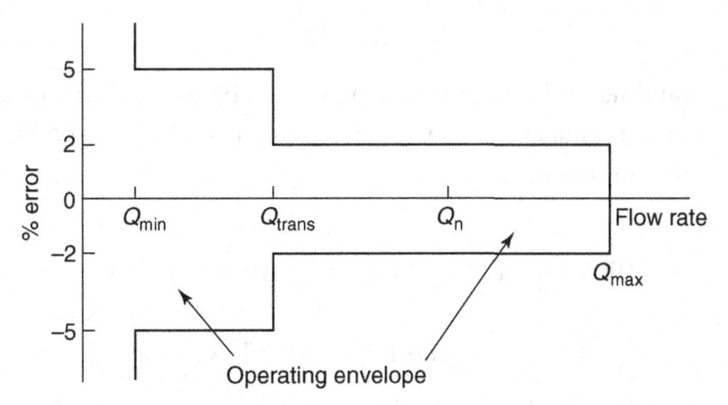

Figure 10.22. Operating envelope for a water meter based on ISO/BS specifications.

10.4.4 Size Range

Meters range from 15mm to 500 mm. One manufacturer has introduced a dual system which combines meters of very different capacity and uses a valve to control the flow through the larger meter.

10.4.5 Sensing

This is usually via a gear train, but may also make use of optoelectrical methods, reed switches etc.

10.4.6 Characteristics and Accuracy

Figure 10.22 shows a typical operating envelope for a water meter. The identified flow rates Q_{min}, Q_{trans} or Q_t, Q_n and Q_{max} appear to be labelled as Q_1, Q_2, Q_3 and Q_4 in the standard (ISO 4064-1) and in the MID*, where Q_{max} is the maximum flow rate for short periods, Q_n is the normally recommended continuous flow rate, Q_{trans} is the flow rate between the two zones where the maximum allowed error specification changes and Q_{min} is the lowest flow rate at which the meter operates within the permitted error bounds.

The allowable errors etc. will depend on the Class identified from the standard as appropriate for the application. The values of ±2% and ±5% in Figure 10.22 may relate to Class 2 (ISO 4064-1) within limited temperature ranges.

However, the reader involved in specification of meters for water flow measurement is strongly advised to obtain the relevant and current ISO and/or MID* documents to ensure that their installation meets the standards required.

10.4.7 Installation

A typical requirement for installation may be that fittings should be at least 5D upstream. In addition air and solid matter should be avoided.

* Measurement Instrument Directive DIRECTIVE 2004/22/EC OF THE EUROPEAN PARLIAMENT AND OF THE COUNCIL of 31 March 2004.

10.4.8 Special Designs

Irrigation and mainline designs are inserted through openings in the pipe, manholes, saddles or by a bolted flange in a T piece. Accuracy is unlikely to be high. Sizes can be as great as 1,000 mm diameter.

10.5 Other Propeller and Turbine Meters

10.5.1 Quantum Dynamics Flowmeter

One design used a patented twin-turbine configuration whereby the downstream slave turbine drove the shaft on which the indicator turbine bearings rotate freely. Under normal conditions the rotational speeds of the indicator and slave turbines are closely matched. At high flow rates the rotational speeds of the indicator turbine and the slave turbine/shaft assembly will begin to diverge. An integral flow-straightening device upstream of the indicator turbine reduces swirl and hence the effect of upstream flow distortion. With a special design of RF pickup, claims of ±0.01% linearity for a turndown of between 350 and 500:1 have been made. In gaseous flow these rangeabilities were claimed to correspond to a mass turndown in excess of 1000:1.

10.5.2 Pelton Wheel Flowmeters

Pelton wheel flowmeters operate like a hydraulic Pelton wheel by using the impulse due to the fluid jet momentum. To achieve this, the flow is constricted to a small outlet and hence forms a high speed jet. A typical design is shown in Figure 10.23. It was designed for use with liquids. An upstream filter with 60 mesh insert may be recommended. A typical lowest flow range may be 0.18 to 1.8 l/min with a repeatability claimed as ±0.25%. The passing of the rotor tips is sensed by a reluctance pick up.

Other designs on the Pelton wheel principle have also been used to produce meters capable of flow rates as low as 0.06 l/h to 0.75 m^3/h with repeatability of ±0.3% of fsd. Alternatively they have been used as a bypass flowmeter which operates across an orifice to allow higher ranges.

10.5.3 Bearingless Flowmeter

This meter had a floating rotor. After the inlet the liquid divided and flowed upwards and downwards past the twin rotors which were on a common shaft. The flow lifted the rotor off its seating due to the lower static pressure in the flow than under the rotor. The flow at both ends of the shaft passed the blades on the rotor wheels, before leaving through two channels which joined at the exit. Sensing magnets were sealed into the rim of the lower rotor and their passage was sensed by an inductive coil. The basic instrument had a size range of about 15 mm to 80 mm diameter with flow range from about 0.2 m^3/h to 5.2 m^3/h.

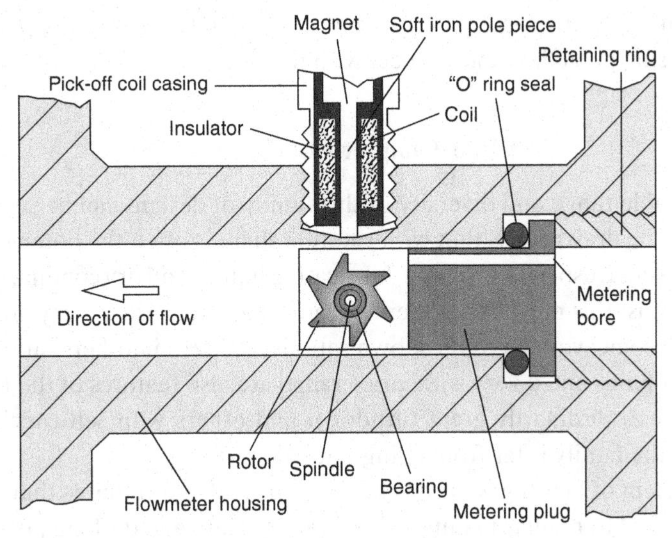

Figure 10.23. Pelton-wheel-type flowmeter (reproduced with permission of Nixon Instruments).

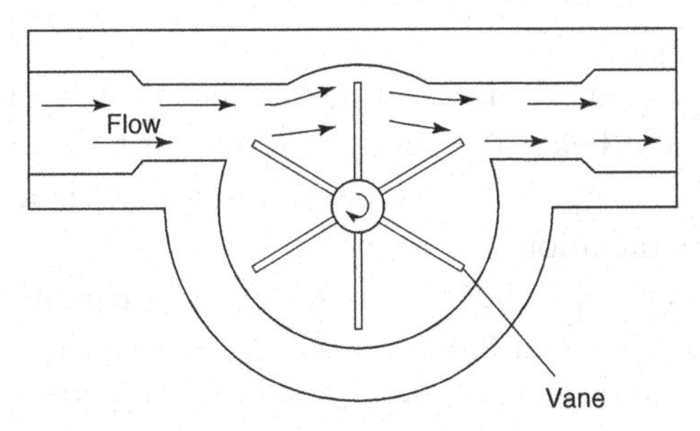

Figure 10.24. Vane flowmeter (reproduced with permission of Emerson Process Management Ltd.).

10.5.4 Vane Type Flowmeters

These cannot be claimed as high-precision instruments. The description "inferential", sometimes applied to turbine flowmeters, is probably appropriate for these meters, as the flow of fluid through the meter is inferred from the effect on the rotor of its interaction with a proportion of the fluid flow.

Figure 10.24 shows a diagram of the principle of operation. The liquid hits the vanes and imparts momentum to them. The clearance between the vanes and the casing distinguishes this meter from the positive displacement meter and allows the fluid to slip past the vanes on the return.

The flow ranges for this meter covered 0.45 to about 12 m³/h with a maximum pressure drop of about 1.4 bar and a maximum allowable kinematic viscosity of about 48 cSt.

Other devices of a similar type are the angled vane flowmeter with three to four blades on a shaft angled at about 45°, designed with a removable body, and a further design operating on the same principle but in the converse way to a paddle wheel. The fluid passing pulls the ridged wheel with it.

10.6 Chapter Conclusions

The considerable range and diversity of this family of designs can be seen from this chapter. The one drawback is that with rotation there is often the potential problem of wear to consider, especially for non-lubricating liquids and dry contaminated gases. However, against this must be set their extraordinary versatility. Very high accuracy is achieved in some; very large turndown ratio is achieved in others; suitability for a wide range of fluids and a very wide price range are also features of the family. New designs with extraordinarily great turndown and others with self-checking ability indicate that the family is far from dying out.

The problem of bearing wear has been controlled, but requires that all types of meter are subject to regular maintenance. Van der Kam and de Jong (1994) claimed that the gas turbine meter, calibrated at the operational pressure, was one of the most accurate gas meters then, and could be used more for custody transfer. Gasunie had suffered very few mechanical failures (blade failure or damaged bearings) [see also van der Kam and Dam (1993)].

Great care must be exercised in ensuring that the installation does not invalidate the calibration. Transient behaviour in gases results in an over-registration of the flow. New sensor technologies may lead to lower drag and higher integrity of signal, particularly in adverse radiation environments.

Intelligent designs which build in a self-checking or condition monitoring facility are likely to be of particular value for metering high value fluids in fiscal and custody transfer applications. Dijstelbergen and Richards (2012) proposed incorporating two meters in one body in combination with a powerful flow computer for gas flow measurement, resulting in good diagnostic capabilities for custody transfer.

New designs with:

- new sensing methods (e.g. Yu et al. 2000 who developed an optical fibre magneto optic sensor);
- great turndown ratios;
- self-checking or condition monitoring (e.g. Schieber 1998 on an accuracy checking module with second rotor, Anabtawi and Howlett 2000 on a neural network method for detection of blade contamination);

are likely to be of particular value for metering high-value fluids in fiscal and custody transfer applications.

Merritt (2001) reported on the use of ceramic ball bearings in metal races to cope with hydrogen sulphide.

Svedin, Stemme and Stemme (2001, 2003) described a meter with a constrained turbine wheel (static turbine flowmeter) which converted volume flow into a torque. The device used a micromachined silicon torque sensor. The optimum size was found to be a blade length of 2.7 mm with an angle of 30°.

Yinping (2007) described a turbine flowmeter based on magnetic suspension bearings and claimed to have better linearity and repeatability than with ordinary bearings. Tests appeared to suggest ranges up to 1:50, and possibly able to cope with corrosive media.

Theoretical analysis of flow through the meter should take account of the very small deflection compared with the substantial deflection which occurs in a power turbine. Ferreira (1988) attempted some computational work on the flows within some turbine meters and compared these with LDA measurements. The results were interesting and original, but the nature of the flow and the very small deflections which result in the case of a well-designed rotor suggest that future computations should take account of these features in a perturbation-type analysis.

Appendix 10.A Turbine Flowmeter Theoretical and Experimental Research

10.A.1 Derivation of Turbine Flowmeter Torque Equations

Blade Aerodynamics

There are two distinct approaches to the blade forces in the literature. Using cascade theory, Tsukamoto and Hutton (1985) give the driving force as

$$T_{\mathrm{d}} = \rho N \int_{r_{\mathrm{h}}}^{r_{\mathrm{t}}} r V_{\mathrm{z}}^2 s (\tan\beta_2 - \tan\beta_1)\, dr \qquad (10.A.1)$$

which equates the torque to the change in angular momentum of the fluid passing through the blade row. Replacing β_1 by $\beta - i$ and β_2 by $\beta - \delta$, and noting that both i and δ will be very small, we may rewrite Equation (10.A.1) as

$$T_{\mathrm{d}} = \rho N \int_{r_{\mathrm{h}}}^{r_{\mathrm{t}}} r V_{\mathrm{z}}^2 s [\tan(\beta - \delta) - \tan(\beta - i)]\, dr \qquad (10.A.2)$$

$$= \rho N \int_{r_{\mathrm{h}}}^{r_{\mathrm{t}}} r V_{\mathrm{z}}^2 s \left(\frac{\tan\beta - \delta}{1 + \delta\tan\beta} - \frac{\tan\beta - i}{1 + i\tan\beta} \right) dr \qquad (10.A.3)$$

Ignoring terms in $i\delta$ etc. ($1° = 0.0175^{c}$),

$$T_{\mathrm{d}} = \rho N \int_{r_{\mathrm{h}}}^{r_{\mathrm{t}}} r V_{\mathrm{z}}^2 s \frac{(i - \delta)(1 + \tan^2\beta)}{1 + (i + \delta)\tan\beta}\, dr \qquad (10.A.4)$$

or if we ignore $(i + \delta)$ compared with unity,

$$T_{\mathrm{d}} = \rho N \int_{r_{\mathrm{h}}}^{r_{\mathrm{t}}} r V_{\mathrm{z}}^2 s (i - \delta)(1 + \tan^2\beta)\, dr \qquad (10.A.5)$$

The alternative aerofoil theory approach is to use lift and drag coefficients. These are given per unit length of the blade by

$$L = \frac{1}{2} \rho c K C_{\mathrm{L}} \left(\frac{V_{\mathrm{Z}}}{\cos\beta_{\mathrm{m}}} \right)^2 \qquad (10.A.6)$$

$$D = \frac{1}{2}\rho c C_D \left(\frac{V_Z}{\cos\beta_m}\right)^2 \tag{10.A.7}$$

where K is the factor that allows for the change in lift coefficient between an isolated aerofoil and a cascade. The torque on the rotor is now given by

$$T_d = \frac{1}{2}\rho N \int_{r_h}^{r_t} \frac{rV_z^2}{\cos\beta_m} c(KC_L - \tan\beta_m C_D)dr \tag{10.A.8}$$

This is essentially the term Blows (1981) gives for the aerodynamic lift torque minus the aerodynamic drag torque. Next, Equations (10.A.2) and (10.A.8) are shown to be equivalent.

Equating the axial force to the force due to change in static pressure:

$$\begin{aligned} Z &= (p_1 - p_2)s \\ &= -\frac{1}{2}\rho(V_1^2 - V_2^2)s + (p_{01} - p_{02})s \end{aligned}$$

Equating the tangential force to the rate of change of angular momentum,

$$Y = \rho V_z^2 s[\tan(\beta - \delta) - \tan(\beta - i)]$$

Substituting for V_1 and V_2, we obtain

$$\begin{aligned} Z &= -\frac{1}{2}\rho V_z^2 s[\tan^2(\beta - i) - \tan^2(\beta - \delta)] + s(p_{01} - p_{02}) \\ &= -\rho V_z^2 s[\tan(\beta - i) - \tan(\beta - \delta)]\tan\beta_m + s(p_{01} - p_{02}) \end{aligned}$$

We may write the lift in terms of Y and Z as

$$L = Y\cos\beta_m + Z\sin\beta_m$$

and, substituting for Y and Z,

$$\begin{aligned} L = {} & \rho V_z^2 s[\tan(\beta - \delta) - \tan(\beta - i)]\cos\beta_m - \rho V_z^2 s[\tan(\beta - i) - \tan(\beta - \delta)]\frac{\sin^2\beta_m}{\cos\beta_m} \\ & + s(p_{01} - p_{02})\sin\beta_m \end{aligned}$$

So

$$L = \rho V_z^2 s[\tan(\beta - \delta) - \tan(\beta - i)]\sec\beta_m + s(p_{01} - p_{02})\sin\beta_m$$

and for the drag

$$D = Z\cos\beta_m - Y\sin\beta_m$$

and substituting for Y and Z

$$\begin{aligned} D = {} & \rho V_z^2 s[\tan(\beta - \delta) - \tan(\beta - i)]\sin\beta_m + s(p_{01} - p_{02})\cos\beta_m \\ & - \rho V_z^2 s[\tan(\beta - \delta) - \tan(\beta - i)]\sin\beta_m \end{aligned}$$

So

$$D = s(p_{01} - p_{02})\cos\beta_{\text{m}}$$

Substituting into Equations (10.A.6) and (10.A.7), we obtain

$$\begin{aligned} KC_{\text{L}} &= 2\frac{s}{c}[\tan(\beta-\delta) - \tan(\beta-i)]\sec\beta_{\text{m}}\cos^2\beta_{\text{m}} + \frac{s}{c}\left(\frac{p_{01}-p_{02}}{\frac{1}{2}\rho V_{\text{z}}^2}\right)\sin\beta_{\text{m}}\cos^2\beta_{\text{m}} \\ &= \frac{2s}{c}[\tan(\beta-\delta) - \tan(\beta-i)]\cos\beta_{\text{m}} + C_{\text{D}}\tan\beta_{\text{m}} \end{aligned} \quad (10.\text{A}.9)$$

where

$$C_{\text{D}}\tan\beta_{\text{m}} = \frac{2s}{\rho c}\left(\frac{p_{01}-p_{02}}{V_{\text{z}}^2}\right)\sin\beta_{\text{m}}\cos^2\beta_{\text{m}}$$

which shows that Equations (10.A.2) and (10.A.8) are equivalent.

The theoretical expression given by various authors for the lift coefficient is

$$C_{\text{L}} = 2\pi\sin\alpha$$

or in fuller form by Batchelor (1967)

$$C_{\text{L}} = 2\pi\sin\alpha(1 + 0.77\,t_{\text{B}}/c)$$

Lift curves are reproduced in Figure 10.9, and drag curves can be found in Wallis (1961). The angle α is that between the far field and the line of zero lift of the blade, which for a flat plate is the angle with the plate. It appears to be conventional to take the direction of the far flow field for a cascade as the mean direction between inlet and outlet and α for a cascade as the angle between this mean direction and the blade angle (cf. Thompson and Grey 1970). Thus for a cascade of closely spaced flat plates, the exit angle relative to the blades will be zero, whereas the inlet angle will be equal to the far field. Thus the value of α will be half the incidence angle *i*. So following convention,

$$C_{\text{L}} = 2\pi\sin i/2 \quad (10.\text{A}.10)$$

for the closely spaced flat plates, and, in this limit, according to Weinig (1932):

$$K = \frac{s}{c}\frac{2}{\pi\cos\beta} \quad (10.\text{A}.11)$$

Thus

$$KC_{\text{L}} = \frac{s}{c}\frac{4\sin i/2}{\cos\beta} \simeq \frac{s}{c}\frac{2i}{\cos\beta} \quad (10.\text{A}.12)$$

and Equations (10.A.5) and (10.A.8) are identical provided $\delta = 0$ and $C_{\text{D}} = 0$.

For values of s/c not tending to zero, we shall find it useful to use the values of K obtained by Weinig (1932) and given by Wislicenus (1947), and it will be necessary to write the lift coefficient in terms of the true incidence angle i. From Figure 10.3,

$$\frac{\tan(\beta - i) + \tan(\beta - \delta)}{2} = \tan\beta_{\mathrm{m}}$$

so that

$$\beta_{\mathrm{m}} \simeq \beta - \frac{i + \delta}{2}$$

but

$$\beta - \alpha = \beta_{\mathrm{m}}$$

so

$$\alpha = \frac{i + \delta}{2}$$

and

$$C_{\mathrm{L}} = 2\pi\left(\frac{i + \delta}{2}\right) = \pi(i + \delta) \tag{10.A.13}$$

and hence

$$T_{\mathrm{d}} = \frac{1}{2}\rho N \int_{r_{\mathrm{h}}}^{r_{\mathrm{t}}} \frac{rV_{\mathrm{z}}^2}{\cos\beta_{\mathrm{m}}} c\left[K\pi(i + \delta) - C_{\mathrm{D}}\tan\beta_{\mathrm{m}}\right] dr \tag{10.A.14}$$

If we make the assumption that both δ and C_{D} may be neglected in the case of a well-designed turbine wheel, we have

$$T_{\mathrm{d}} = \frac{\pi}{2}\rho N \int_{r_{\mathrm{h}}}^{r_{\mathrm{t}}} \frac{rV_{\mathrm{z}}^2 cK(r) i(\mathrm{r})}{\cos\beta_{\mathrm{m}}} dr \tag{10.A.15}$$

Finally, it is useful to consider the actual values of C_{D} and δ. We can first of all obtain from Equations (10.A.5), (10.A.8) and (10.A.10)

$$\frac{1}{2}\frac{c}{\cos\beta_{\mathrm{m}}} K\pi i = \frac{s}{\cos^2\beta} i$$

$$K \simeq \frac{2}{\pi}\frac{s}{c}\frac{1}{\cos\beta_{\mathrm{m}}}$$

Then applying this to C_{D} and δ making use of Equations (10.A.5) and (10.A.14)

$$\frac{1}{2}\frac{cK\pi\delta}{\cos\beta_{\mathrm{m}}} - \frac{1}{2}\frac{c}{\cos\beta_{\mathrm{m}}} C_{\mathrm{D}}\tan\beta_{\mathrm{m}} = -\frac{s\delta}{\cos^2\beta_{\mathrm{m}}}$$

Hence,

$$C_D = 4\frac{s}{c}\frac{\delta}{\sin\beta_m}$$

From Wallis (1961), as i tends to zero, C_D tends to 0.017. For $\beta = 45°$ and $s/c \simeq 1$,

$$C_D \simeq 0.017 \simeq 4\sqrt{2}\delta$$

so that

$$\delta \simeq 0.003^c \simeq (1/6)°$$

So if we neglect the drag terms [Equation (10.A.14)],

$$K\pi\delta - C_D\tan\beta_m \simeq 2\frac{s}{c}\frac{1}{\cos\beta_m}\delta - \frac{4s}{c}\frac{\delta}{\sin\beta_m}\tan\beta_m$$

we are actually neglecting

$$-2\frac{s}{c}\frac{\delta}{\cos\beta_m} \simeq -0.0085$$

We can compare this with the size of the drive torque

$$K\pi i = 2\frac{s}{c}\frac{i}{\cos\beta_m}$$

Thus the neglected term is δ/i of the drive term.

Drag Terms

Tsukamoto and Hutton (1985) identified four drag torques:

- T_B, bearing drag;
- T_w, hub disc friction drag;
- T_t, tip clearance drag;
- T_h, hub fluid drag.

In the first three, the torque is obtained from an equation of the form

$$T = \text{Shear stress} \times \text{Area} \times \text{Radius} \times \text{Reynolds number} \times \text{Function of Reynolds number}$$

For instance, for the bearing drag, we obtain

$$T_B = \left(\frac{\mu r_j\omega}{t}\right)(2\pi r_j B)(r)\left[\frac{\rho(r_j\omega)t}{\mu}\right]f(\text{Re}) = \pi\rho(r_j\omega)^2 r_j^2 B(\text{Re})^n$$

where

$$f(\mathrm{Re}) = \begin{cases} \dfrac{2}{\mathrm{Re}} & \text{for Re} < 1{,}000 \\ \dfrac{0.016}{\mathrm{Re}^{0.25}} & \text{for Re} > 1{,}000 \end{cases}$$

where Re is the Reynolds number based on the radial gap and circumferential speed. Blows (1981) obtained the same expression if Re < 1,000. From the work of Tsukamoto and Hutton (1985), the dependence on ω ranges from $T \propto \omega$ to $T \propto \omega^{1.8}$. At low speeds, the dependence is in the range ω to $\omega^{1.5}$, whereas at high velocities the dependence tends to $\omega^{1.75}$ to $\omega^{1.8}$.

The hub fluid drag is

$$T_\mathrm{h} \propto \frac{1}{2}\rho W_\mathrm{h}^2 C_\mathrm{h}$$

where W_h, is the relative velocity at the hub and C_h is a constant for a particular design. It will therefore behave like an adjustment to the main aerodynamic drag term coefficient.

Flowmeter Equation

We may now equate

$$T_\mathrm{d} = T_\mathrm{B} + T_\mathrm{w} + T_\mathrm{t} + T_\mathrm{h}$$

and we may note that, with increasing speed, the terms T_B, T_w and T_t will become less important compared with T_d and T_h. If we combine the T_h drag with the aerodynamic drag, we obtain from Equation (10.A.8)

$$\int_{r_\mathrm{h}}^{r_\mathrm{t}} \frac{rV_\mathrm{z}^2 c}{\cos\beta_\mathrm{m}}(KC_\mathrm{L} - C_\mathrm{D}' \tan\beta_\mathrm{m})dr = 0$$

where C_D' has incorporated T_h. If we then substitute for $C_\mathrm{L} = \pi i$ and approximate C_D' as constant,

$$\int_{r_\mathrm{h}}^{r_\mathrm{t}} \frac{rV_\mathrm{z}^2 c}{\cos\beta_\mathrm{m}} K\pi i dr = \int_{r_\mathrm{h}}^{r_\mathrm{t}} \frac{rV_\mathrm{z}^2 c}{\cos\beta_\mathrm{m}} C_\mathrm{D}' \tan\beta_\mathrm{m} dr \tag{10.A.16}$$

By iteration, this equation enables the wheel speed to be obtained for a given flow profile in the annulus (cf. Newcombe, Archbold and Jepson 1972 for an expression for gas meters).

10.A.2 Transient Analysis of Gas Turbine Flowmeter

Lee and Evans (1970) used a simple analysis to derive the behaviour of a meter by observing the spin-down as an indicator of changed-bearing friction. The transient equation (Lee et al. 1975; cf. Bonner 1977) is

$$\frac{Id\omega}{dt} = T_{\mathrm{d}} - T_{\mathrm{r}} \tag{10.A.17}$$

where the driving torque T_{d} may be obtained from Equation (10.A.1) for one radius and unit blade length as

$$T_{\mathrm{d}} = \rho N r V_{\mathrm{z}}^2 s(\tan\beta_2 - \tan\beta_1) \tag{10.A.18}$$

or, since $V_{\mathrm{z}} \tan\beta_1 = r\,\omega$ and putting

$$T_{\mathrm{d}} = \rho N r\, s V_{\mathrm{z}} \frac{(V_{\mathrm{z}} \tan\beta - r\,\omega)}{(1+\eta)}$$

where η allows for the blade deviation factor. Also putting $T_{\mathrm{r}} = \rho A r^3 C_{\mathrm{f}}\, \omega^2/2 + T_{\mathrm{n}}$, where the first term is fluid drag and the second term T_{n} is nonfluid drag assumed constant, Lee et al. (1975) then obtained

$$D_1 \frac{dW}{d\tau} + VW + D_2 W^2 = V^2 - D_3 \tag{10.A.19}$$

where

$$D_1 = \frac{(1+\eta)I}{r^2 \rho A V_0 T}$$

$$D_2 = \frac{(1+\eta)\tan\beta\, C_{\mathrm{f}}}{2}$$

$$D_3 = \frac{(1+\eta) A T_{\mathrm{n}}}{r \tan\beta \rho A^2 V_0^2}$$

and $V = V_{\mathrm{z}} / V_0$, $W = r\omega/V_0 \tan\beta$ and $\tau = t/T$, the fundamental period of the pulsating flow. This equation is soluble on a computer. For certain cases, an exact solution is possible. Lee and Evans (1970) give the speed decay curve for no flow ($V = 0$) as

$$\tau = \frac{D_1}{D_2}\left(\frac{1}{(D_3/D_2)^{1/2}}\right)\left(\tan^{-1}\frac{W_1}{(D_3/D_2)^{1/2}} - \tan^{-1}\frac{W}{(D_3/D_2)^{1/2}}\right) \tag{10.A.20}$$

This curve is shown in Figure 10.16. We can obtain from this curve, as shown by Lee and Evans (1970), that the rotor coast time to standstill is

$$\tau_0 = \frac{D_1}{D_2}\left(\frac{D_2}{D_3}\right)^{1/2} \tan^{-1}\left[W_1\left(\frac{D_2}{D_3}\right)^{1/2}\right] \tag{10.A.21}$$

and just before this point

$$D_1 \frac{dW}{d\tau} = -D_3 \tag{10.A.22}$$

so that, from the slope S, we have

$$S = -\frac{D_3}{D_1} \tag{10.A.23}$$

10.A.3 Recent Developments

Meters for Liquids

Sun and Zhang (2004) appear to have developed the theoretical approach outlined in Section 10.A.1 based on aerofoil theory. They gave some comparisons between prediction of performance curves and experimental data which appear encouraging. Mickan et al. (2010) used a meter model and, on the basis of their data, concluded that a dynamic model allowed a correction for variation of flow rate e.g. with piston provers. This work may be important since new primary standards for high-pressure gas volume may be based on piston prover technology (Mickan et al. 2010).

Modelling Meter Performance

Mattingly (2009) discussed the limitations of the universal viscosity curve for liquids (UVC), and argued the case for introducing additional dimensionless parameters. He showed that

$$\mathrm{Re}\times\mathrm{St} = \left(\frac{DV\rho}{\mu}\right)\left(\frac{fD}{V}\right) = \frac{fD^2\rho}{\mu} = \mathrm{Ro} \tag{10.A.24}$$

where the Roshko number, Ro, becomes the independent variable and where the Strouhal number is given by St, and where D is the diameter of the meter bore, V is the fluid velocity, ρ is the density, μ is the dynamic viscosity and f is the turbine frequency. When this approach is used the "viscous breakaway" phenomena, which caused problems for the UVC characterisations, were still observed for different viscosities. These viscous breakaways severely limit the turndowns for the respective meters. To address this situation and to improve meter performance characterisation that includes the viscous breakaway, he then derives a new pressure-velocity-frequency parameter

$$= C_{\mathrm{p}}\frac{\mathrm{Re}}{\mathrm{St}} = \frac{p}{\rho V^2}\frac{DV\rho}{\mu}\frac{V}{fD} = \frac{p}{\mu f} \tag{10.A.25}$$

where C_{p} is the pressure coefficient and p is the absolute static pressure of the liquid. It should be noted here that the dimensional quantities selected for the non-dimensional parameters should be the respective "characteristic" quantities, i.e., the most influential quantities affecting the measurement. Since the only pressure available was the static pressure, this was selected. It could be that the differential pressure across the meter might be a more characteristic pressure, but data might be needed to decide this and, if so, it would require another pressure measurement.

The result of this important approach is to create a three-dimensional surface characteristic for both liquid and gas turbine meters, which may allow a greatly increased turndown range for the meters, and thereby an improved flow rate measurement characterisation for liquid and gas turbine meters.

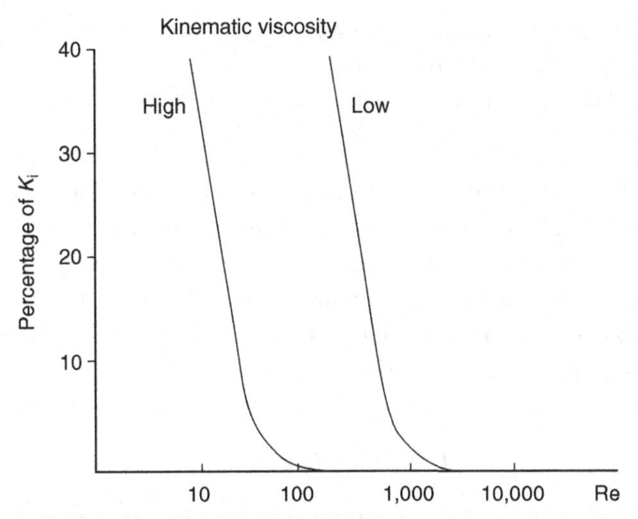

Figure 10.A.1. Bearing static drag term appears to be the main cause of the fanning of the characteristic with kinematic viscosity (after Pope et al. 2012).

It should also be noted that non-dimensional characterisations of meter performance cannot assess the respective influences of the different components of drag or lift. The intrinsic value of the non-dimensional approach is the significant increases in meter turndowns that can be achieved.

Pope et al. (2012) presented a model which they called the "extended Lee model", with reference to the work of Lee [Lee and Evans (1965), Lee and Karlby (1960)]. They identified "fanning phenomena" i.e. viscous breakaways, most strongly because of the bearing static drag, and which causes a fan of the characteristics of the meter at low Reynolds number, rather than a single curve, in the region which the authors call the "bearing dependent range", Figure 10.A.1 cf Figure 10.5. In addition to the model development, the authors have also demonstrated the use of mixtures of propylene glycol and water (PG+W) as a less hazardous test liquid than Stoddard solvent for calibration of the meters.

The object of the authors was, rather than modelling the details of the meter blading etc., to set up a model capable of being fitted to calibration data.

The model equation they propose is:

$$\frac{\omega}{q_v} = K_i - C'_D(\text{Re}) - \frac{C'_{B0}}{\rho q_v^2} - \frac{C'_{B1}\nu\omega}{q_v^2} - \frac{C'_{B2}\omega^2}{\rho q_v^2} \tag{10.A.26}$$

where ω is angular frequency, q_v is the volumetric flow rate, ρ is liquid density and ν is kinematic viscosity. The other terms are:

K_i the ideal meter factor (rad/m^3).

$C'_D(\text{Re})$ the dimensionless drag coefficient with turbine geometric constants included, which depends on whether the blade boundary layer is laminar or turbulent.

C'_{B0} the static drag of the ball bearings.
C'_{B1} the viscous drag of the ball bearings.
C'_{B2} the axial thrust and dynamic imbalance of the ball bearings.

The authors' experimental work made use of a dual rotor turbine meter similar to that in Figure 10.11. To obtain the various terms in the equation, an extensive calibration would be necessary over a range of kinematic viscosities. This suggests that in most cases the user would be advised to work within the viscosity independent range. The authors suggest that the fanning is due to the static drag term.

The authors claim that their model greatly improves the uncertainty when measuring low flows in the viscosity-sensitive region.

Unsteady Liquid Flow

Lee, Cheesewright and Clark (2004) discussed the effect of pulsating flows on the reading of small liquid turbine meters. They developed the theory and compared their results to new data which they obtained from a rig which provided a pulsating flow. They were able to obtain, for small liquid turbine meters in a sinusoidal pulsating flow, the amplitude attenuation, which could be significant, and the over-registration, which appeared to be small, although possibly greater than the claimed uncertainty for the meter. They proposed approximate analytical expressions for calculating the likely values of these quantities for a given meter in a given liquid.

Meters for Gases

von Lavante et al. (2004c) presented numerical predictions using FLUENT. The three-dimensional unsteady flow field predicted appeared to explain some observed behaviour. They computed flows for maximum rate, for 25% of maximum, 10% of maximum and 10% of maximum at higher pressure. They computed the flows in the low range for both turbulent and laminar and appeared to obtain significant differences. Some secondary flows in the stator for laminar conditions were thought to be a possible source of irregularity of output signals.

Woltman and Other Meters

Palau et al. (2004) described CFD calculations and experiments on the installation effects for Woltman meters. They used the CFD to obtain the flow profiles entering the meter, and then applied the data to obtain the torques applied to the blades of the meter (e.g. as in Section 10.A.1).

Zhen and Tao (2008) reported on a computational study and experimental tests of a meter referred to by them as a tangential type turbine flowmeter, but which appeared to be similar to a vane type flowmeter (Section 10.5.4).

11

Vortex Shedding, Swirl and Fluidic Flowmeters

11.1 Introduction

This chapter deals with measurement techniques based on instabilities in fluid behaviour. In preparing this chapter for the first edition, I benefitted from an unpublished note in 1988 by Dr M. V. Morris of Cranfield University, who drew my attention to some early references on this type of meter. New material has been added for this edition.

11.2 Vortex Shedding

The development of flows past a cylindrical bluff body, in relation to the size of the Reynolds number based on the diameter of the cylinder, is given in Figure 11.1 (Cousins 1977); see Roshko (1954) and Fage and Johansen (1928) for further details.

At a Reynolds number of about 40, the familiar shedding phenomenon starts. However, small changes of detail in the pattern of the shedding are illustrated by the diagram for certain Reynolds numbers. Frequency changes occur at these Reynolds numbers. The vortex-shedding meter depends on the relationship between the shedding frequency and the flow rate, so that these changes could result in nonlinearity and poor repeatability in a meter. (Figure 11.1 is discussed in a little more detail in Appendix 11.A.1, and an order of magnitude calculation is given in Appendix 11.A.2.)

Figure 11.2 is a simple diagram to suggest how, as a vortex rolls up downstream of a noncircular bluff body, it first forms, fills with rotating fluid, and then draws in fluid with opposite rotation, which eventually cuts off the vortex and allows it to move downstream. In Appendix 11.A.2 a simple argument is used to suggest that the frequency of shedding f is about

$$f = V/2\pi w \tag{11.1}$$

where V is the velocity past the body of width w. An important parameter, the Strouhal number, is given by

$$\mathrm{St} = f\,w/V \tag{11.2}$$

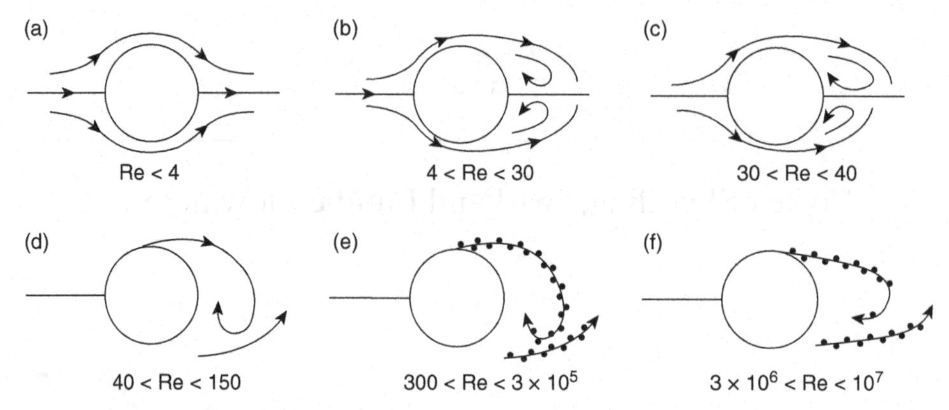

Figure 11.1. Flow patterns around a cylindrical body: **(a)** For Re < 4; **(b)** initiation of separation for 4 < Re and up to about Re = 30; **(c)** lack of separation symmetry behind cylinder for about 30 < Re < 40; **(d)** stable and viscous shedding for 40 < Re < 150; **(e)** stable shedding with turbulent shear layer for 300 < Re < 3×10^5; **(f)** stable shedding with turbulent boundary and shear layers for 3×10^6 < Re < 10^7.

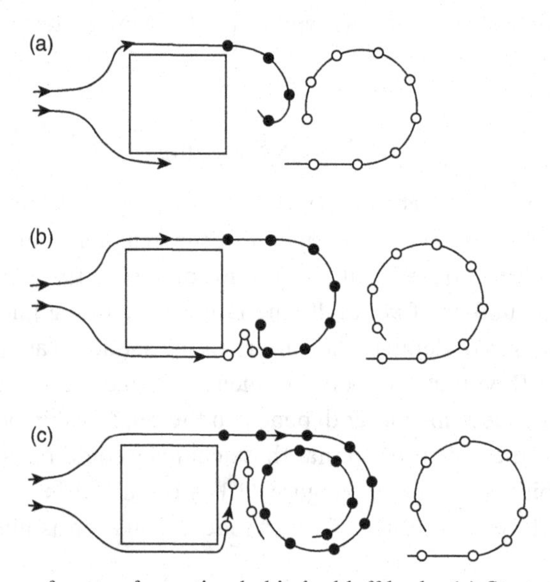

Figure 11.2. Diagram of vortex formation behind a bluff body: **(a)** Start of formation of new vortex; **(b)** opposite vorticity starting to be entrained; **(c)** opposite vorticity breaks and separates vortex.

This relationship has been applied to long, slender cylinders. Our application is to the vortex-shedding flowmeter where the flow is in a pipe and the width of the shedder (bluff body) is between a quarter and a third of the pipe diameter, and thus is not a long, slender cylinder. However, as we shall see, this length/width ratio is advantageous in terms of the strength of vortex shedding and leads to coherent shedding, whereas longer and more slender shedding bodies result in weaker and less coherent shedding.

11.3 Industrial Developments of Vortex-Shedding Flowmeters

The use of this instability for velocity measurement was recognised at an early stage (cf. Roshko 1954) and has led to its use in flowmeters and probes. Medlock (1976, 1986) gave some of the background to its development, and states that the first successful vortex-shedding meter was the invention of Alan E. Rodely, who with D. F. White formed a company, Eastech, in 1968 to develop, manufacture and market the new meter. Ogawa (2006) may also be consulted on the background history.

Most modern meters use bluff bodies that are not circular in cross-section and that have a sharp shedding edge to remove the shifts resulting from changes in the boundary layer. Coherent shedding takes place along the length of the bluff body if the length (in this case the diameter of the pipe) is in the range of about 2.5 to 4 bluff body widths (Zanker and Cousins 1975). By *coherence* we mean that the vortex is shed from end to end of the bluff body on the same side at the same time.

For a cylindrical bluff body, it is found that the strength of the force caused by the vortex shedding is greatest when D/w is about 2.8 because the shedding becomes more coherent and enhanced. By this means, the shedding becomes approximately two-dimensional. For a flowmeter where the shedder spans the pipe, it is advantageous if its width is about a quarter of the pipe diameter. The wall of the tube then acts as end plates for the shedder and encourages coherent and strong vortex shedding. Coherence and stability of the shedding range has been achieved down to about $\text{Re} = 10^4$, where the Reynolds number is based on bluff body width (cf. Takamoto and Komiya 1981, who described a positive feedback system to extend the range below $\text{Re} = 2{,}000$).

In flow measurement, strong pressure pulsations caused by the shedding will enhance the signal, resulting in easier detection of shedding.

Three-dimensionality in the flow stream (e.g. turbulence, velocity gradients and swirl) has detrimental effects on the quality of the vortex shedding (Zanker and Cousins 1975; cf. El-Wahed and Sproston 1991 and Robinson and Saffman 1982). Except in carefully controlled conditions, a slender cylinder in the flow will create vortex shedding, which varies in strength and phase along its length (cf. Saito, Hashimoto and Wada 1993).

11.3.1 Experimental Evidence of Performance

The vortex flowmeter, therefore, consists of a bluff body, usually of noncircular cross-section, spanning the pipe and having a width of a quarter to a third of the pipe diameter (Figure 11.3). Zanker and Cousins (1975) provided a useful review of early work on vortex flowmeters (cf. Tsuchiya et al. 1970 and Cousins et al. 1973) and the various effects that take place.

As the body becomes thicker (w increases) and the velocity past the body, therefore, increases for a fixed volumetric flow rate, Equation (11.1) suggests that the frequency and hence the pulses per unit volume of fluid may change. We shall refer to this ratio as the K factor (see Section 1.4). Figure 11.4(a) for a cylindrical bluff body

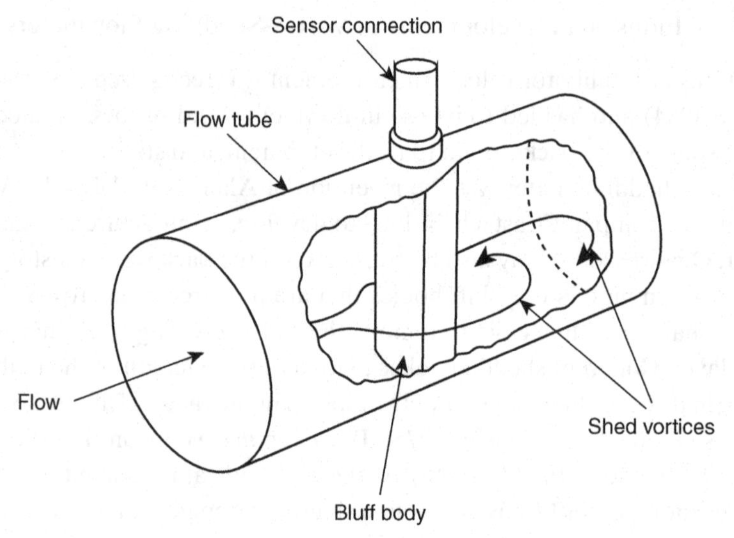

Figure 11.3. Diagram of vortex flowmeter.

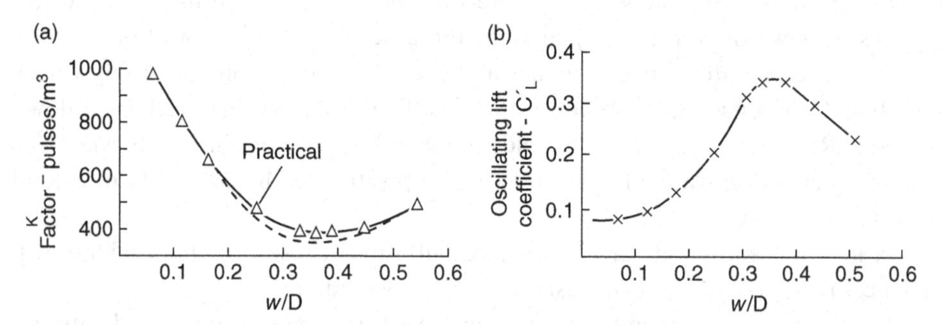

Figure 11.4. Pulses per unit volume and lift force for cylindrical bodies (from Zanker and Cousins 1975; reproduced with permission from NEL): **(a)** Pulses per unit volume against blockage ratio; **(b)** lift coefficient against blockage ratio.

shows this factor reducing to a minimum at a w/D of about 0.35. When w is of order 0.1D, the strength of the lift coefficient is low [Figure 11.4(b)], but when it has the value 0.35D, it is at a peak. Zanker and Cousins suggested that this strength was due to two effects that fortunately occur together:

- coherent shedding due to the shorter length of the bluff body between the end plates of the wall of the tube;
- the accelerating flow through the smaller space between the body and the tube wall, leading to a more uniform profile and less dependence on the upstream flow profile.

An important reason for the choice of the minimum is that small variations in w/D have a minimal effect on the frequency.

The constancy of the meter factor is shown in Figure 11.5 for a Reynolds number range of nearly 1,000:1 with air and water where the scatter was usually

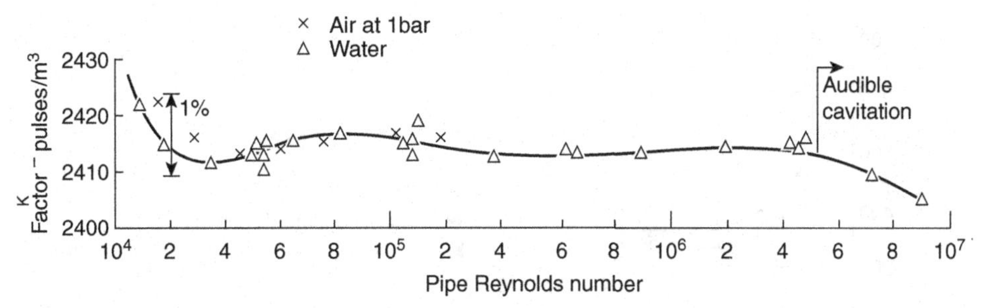

Figure 11.5. Typical calibration curve for a 3 in (75 mm) vortex meter (from Zanker and Cousins 1975; reproduced with permission from NEL).

within ±0.5% (Zanker and Cousins 1975). White, Rodely and McMurtie (1974) gave a calibration curve for a 3 in (75 mm) meter in air over a 360:1 turndown (with a lower Reynolds number of 10^4) and showed that scatter was of order ±1% over the range. A major problem in achieving this size of turndown is how to sense at the varying level of signal. If velocity changes by 100:1, then pressure head variation due to the velocity will change by 10^4:1. Goujon-Durand (1995) confirmed that a minimum Reynolds number of Re = 30,000 is probably necessary to avoid serious increase in nonlinearity. Takahashi and Itoh (1993) showed a slight decrease in Strouhal number (about 2%) from Re = 20,000 to Re = 50,000 for heavy oil, water and air.

Inkley, Walden and Scott (1980) obtained data to show the small effect of temperature change through viscosity and density on the meter. Change of liquid and liquid viscosity appears to have caused calibration shifts of order 0.5%. If the meter is designed close to the minimum in Figure 11.4(a), any variation due to temperature expansion will be minimised. The major change was due to cavitation. The authors agreed with White et al. (1974) that a universal curve should be obtainable and foreshadowed the work of Takamoto and Terao (1994) and, no doubt, others.

11.3.2 Bluff Body Shape

Figure 11.6 shows the effect of body shape on the optimum value of w/D and also shows that the K factor decreases from the value for circular bluff bodies. The truncated triangular or trapezium shape appears to be best if length to width is in the range 1.2 to 1.5. The rectangular body with length to width about 0.6 may have strong shedding.

Lucas and Turner (1985) confirmed that a splitter plate attached to the bluff body improved signal quality, which, presumably, is the reason for the T and truncated triangular body shapes.

El-Wahed and Sproston (1991) experimented with various bluff body shapes in air. The sensing used a modulated signal from naturally occurring flow-induced streaming currents. The shedders had end plates to create two-dimensional conditions.

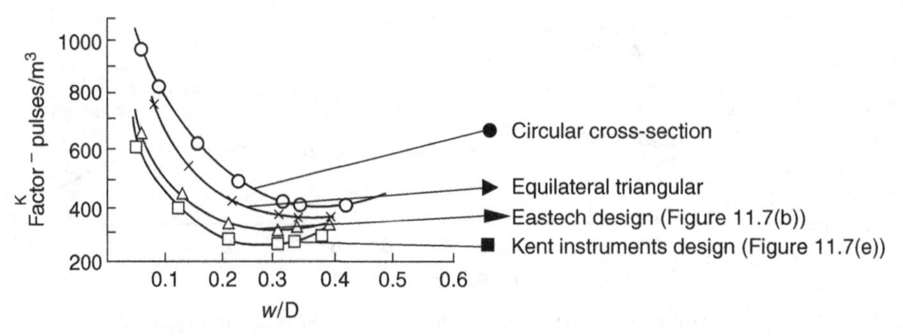

Figure 11.6. Optimum *w/D* for various body shapes (from Zanker and Cousins 1975; reproduced with permission from NEL).

The main conclusions were that the best position for the sensor was attached to the front face of the bluff body and that:

- the cylinder with a transverse slit was slightly better than the solid cylinder (cf. Igarashi 1986);
- the rectangular shedder produced a high signal-to-noise ratio, suggesting that the body produced a large drag, which in turn encouraged the vortex to sweep across the rear face, reinforcing and stabilising the shedding; and
- the T-shaped shedder was the best, suggesting that the tail helps control the shedding and produces a high signal-to-noise ratio, which is approximately constant throughout the entire range.

Figure 11.7 shows the cross-section of various bluff body shapes that have been used in flowmeters. Bentley and Nichols (1990) suggested that, for a thin plate, the vortex field is much farther downstream than for a square. The shape in Figure 11.7(g) may result in a considerable amplification of the vortex strength and hence of the pressure variation.

There appear to be mixed views on the use of composite shapes (cf. Majumdar and Gulek 1981 and Herzl 1982). The conditions for optimum shedding from dual bodies (Bentley and Benson 1993) relate to

- the thickness-to-width ratio of the two bodies;
- the spacing of the bodies; and
- the positions of maximum vorticity.

For dual bluff bodies, the movement of the boundary layer through the gap is essential for vortex enhancement. There are two options (Bentley, Benson and Shanks 1996) for position of a second bluff body: either the second body is far enough away for there to be shedding from both, or the second is less than a critical spacing, which leads to shedding taking place downstream of both, with greater shedding stability. Bentley et al. identified certain spacing, which may be preferable, but the small size of the spacing between bodies may present a problem in small meter sizes.

It is interesting to note that the dual bluff body is essentially providing a feedback path between the two bodies of fixed dimensions and so, presumably, defining the volume needed to fill the feedback path and the time in a more stable way, hence

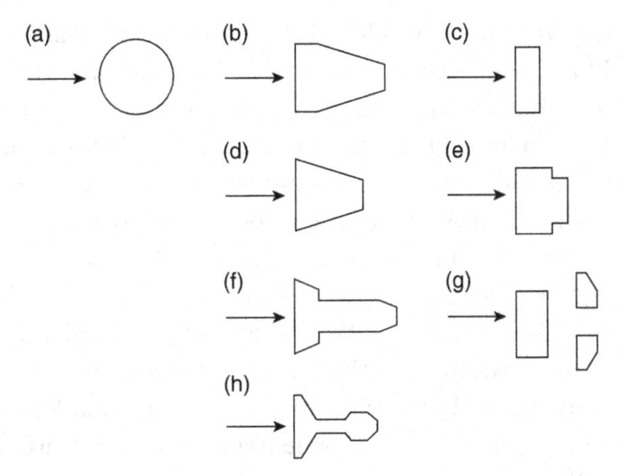

Figure 11.7. Diagrams of cross-section of bluff bodies (some after Yamasaki 1993 and some from manufacturers' brochures).

the greater stability. Tests by Bentley and co-workers suggested that this is still further enhanced by introducing a throat in the gap, by making the downstream body a double wedge with the vertex pointing upstream.

Some manufacturers' designs allow the replacement of the sensor without closing down the line. One that uses a dual bluff body (essentially a triangular shape followed by a piezoelectric sensor mounted outside the flow line on a symmetrical wing aerofoil that senses the downstream pressure variations) provides for the piezo crystal's accessibility during operation.

Turner, Popiel and Robinson (1993) proposed a new body shape (cf. Zanker and Cousins 1975) for vortex-shedding flowmeters, consisting of a cylinder split across its lateral diameter, with a concave rear surface. They claim a high signal-to-noise ratio and an almost constant Strouhal number over a wide Reynolds number range. Miau et al. (1993) described a T-shaped vortex shedder consisting of a trapezoidal bar of fixed shape, with a rear plate whose length can be varied. The optimum length of plate was found to be in the range 1.56–2.0 times the width of the vortex shedder. They obtained a turndown ratio of about 17:1.

One manufacturer (at least) may have offered a triple body. The outer bodies (cylinders) create the vortex street, and the centre body senses the vortices. It should give similar performance for flow in both directions.

Fu and Yang (2001) studied the hydrodynamic vibration in a dual bluff body vortex flowmeter. Numerical data made use of a large eddy simulation model. The experiments were on a 40 mm diameter meter. Triangular-section bluff bodies appear to have been used, and the dual arrangement was claimed to strengthen the hydrodynamic vibration. Differential sensors may have been used across the lateral faces of the bluff body.

Peng, Fu and Chen (2004) also undertook experiments on a dual triangular bluff body [apparently recommended by Bentley et al. (1996) and following Fu and Yang's numerical analysis] in a 50 mm circular pipe. Their results appear to indicate that their dual bluff bodies were less linear than single bluff bodies. Peng, Fu and Chen (2008b) reported further on their tests on dual triangular bluff bodies. The paper may have identified improved low flow sensitivity and that the optimum spacing for the dual bluff bodies may be the distance between shed vortices.

Ozgoren (2006) obtained experimental vortex wake structures downstream of square and circular cylinders obtained in a free surface water channel.

A paper by Peng et al. (2010, 2012a) concerned the effect on shedding from a circular cylinder, of a slit across the diameter of the cylinder and perpendicular to the flow direction (see El-Wahed and Sproston 1991 mentioned earlier). The slit appears to improve vortex shedding quality. Tests in water and air of slit widths ranging from 0 to $0.3d$, where d denoted the diameter of the circular cylinder, resulted in an optimal range of slit width, 0.1-$0.15d$. This gave the best quality shedding signals and the most linear relationship between Strouhal number and Reynolds number.

von Lavante et al. (2007) used experimental and (two-dimensional) numerical methods to assess the effect of shape change due to wear e.g. rounding of sharp edges, which may occur if the meter were used with wet-gas and the consequences on meter accuracy.

11.3.3 Standardisation of Bluff Body Shape

Takamoto and Terao (1994) have taken a basic design of bluff body of the truncated triangle type (Figure 11.7b). Figure 11.8 shows the essential dimensions, and the paper sets out the tolerance on these and the tolerance with which the body should be set. Thus, a selection of the dimensions and tolerances (rounded to reduce significant figures in some cases) are given in Table 11.1. The alignment refers to the orientation of the bluff body relative to the axis of the pipe. The perpendicularity is relative to the pipe diameter. This is a very interesting approach. It is interesting to note that the flow area is close to that for an orifice of $\beta = 0.65$. In their tests, the value of H/D ranged from 0.20 to 0.40 for $w/D = 0.28$, and the Strouhal number varied by about 8% for this range. Their results also suggest that $H/D = 0.35$ results in a Strouhal number, which only varies by about 1% for a range of Reynolds number from 2×10^4 to 10^6.

11.3.4 Sensing Options

A major limitation of the meter is the transducer, and as a consequence various sensing methods have been tried and all have advantages and disadvantages. These methods include:

- Ports in the sides of bluff bodies, such as those in Figures 11.7(a) and (b), leading to a transverse duct which allows movement of fluid to be sensed by an internal hot wire, or a thermistor;
- Temperature variation to sense changes by attaching thermistors to the upstream face of a bluff body such as that in Figure 11.7(b) or (d);
- Pressure on lateral (electrical capacitance) diaphragms which deflect in a bluff body such as in Figure 11.7e;

Table 11.1. *Selection of dimensions and approximate permitted error for a standard flowmeter*[a]

Dimension	w/D	H/D	h/D	L/D	θ	Alignment	Perpendicularity
Value	0.28	0.35	0.03	0.912	19°	0°	0°
Maximum error	0.1%	0.7%	6.6%	0.15%	0.4°	0.5°	0.3°
Effect on K factor (%)	0.13	0.09	0.13	0.08	0.05	0.05	0.06

[a] Refer to the original paper by Takamoto and Terao (1994) for precise values.

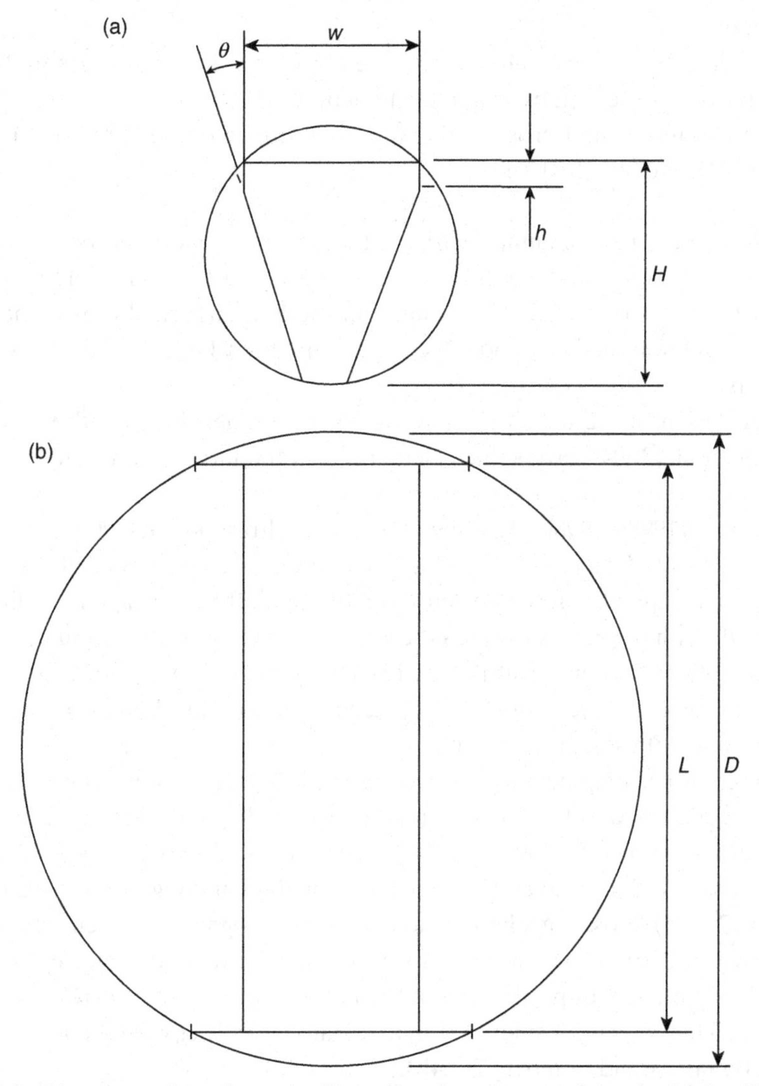

Figure 11.8. Diagram to show the main dimensions for a proposed standard (after Takamoto and Terao 1994): **(a)** Cross-section of bluff body; **(b)** cross-section of pipe.

- Pressure on a small diaphragm in the side of a bluff body such as Figure 11.7f;
- Pressure sensed through ports in the sides of a bluff body such as those in Figures 11.7(b) and (d), or downstream of the bluff body in the fluid or in a symmetrical wing aerofoil. The pressure changes create very small bending of a sensitive element (at low flows the tip may move by as little as 3×10^{-8} m). This movement can either be measured by strain gauge or by capacitive methods.
- Flexure of whole or part of the bluff body (Figure 11.7h) measured with strain gauges;
- Flexure of the whole bluff body sensed by strain gauges in the end supports of the body;
- A shuttle ball in an internal cavity in the bluff body linked by ports to the fluid, with movement sensed by magnetic induction;
- A beam of ultrasound, crossing the flow downstream of the bluff body, is modulated by the vortex shedding.

Pressure is the most common variable to be sensed, evidenced by: ports in the bluff body, a pressure sensor built into the bluff body surface, a flexible diaphragm, forces on the body, movement of a shuttle in the body. Thermal sensors have also been used, built into the body and affected by fluid flows. Fluids should be clean and noncorrosive.

Meter designs using ultrasonic sensing have been developed with varying success. Music et al. (2004) gave some calculations of the modulation of ultrasound in the vortex shedding.

Hans et al. (1998a, 1998b) described tests using ultrasound detection. Two papers by Windorfer and Hans (2000a, 2000b) looked at the correlation of ultrasonic and pressure signals, and also at the optimisation of the bluff body shape including a triangular body with the vertex towards the approaching flow. They also suggested that frequency obtained from the ultrasound signals was at twice the frequency of the pressure signals, since the beam is perpendicular to the bluff body axis, and senses fluctuations on both sides of the body.

Ultrasound has long been considered as a good method of measuring the shed vortices in vortex flowmeters by their modulation of the beam, and it has been suggested that its use might allow a more slender shedding body (Hans 2002a, 2002b and Hans and Windorfer 2003). In addition to digital handling of the shedding signal (Hans 2003b), cross-correlating the shed vortex signals between two stations (Hans and Lin 2005) could provide a self-monitoring function (cf. Lin and Hans 2006, 2007). The effect of positioning of the bluff body, orientation and wear on the output signal have been considered (Lin and Hans 2004; Ricken and Hans 2004; von Lavante, Banaszak and Lefebvre 2004b).

Lynnworth et al. (2006) used a strut, about half the width of those normally used, to shed vortices. The Reynolds number range, partly with air and partly with water, was 1,000 to 200,000. One path of an ultrasonic clamp-on flowmeter system was used to obtain the shedding frequency. The authors also investigated the optimum shape to combine flow rate, with torsional investigation of density etc.

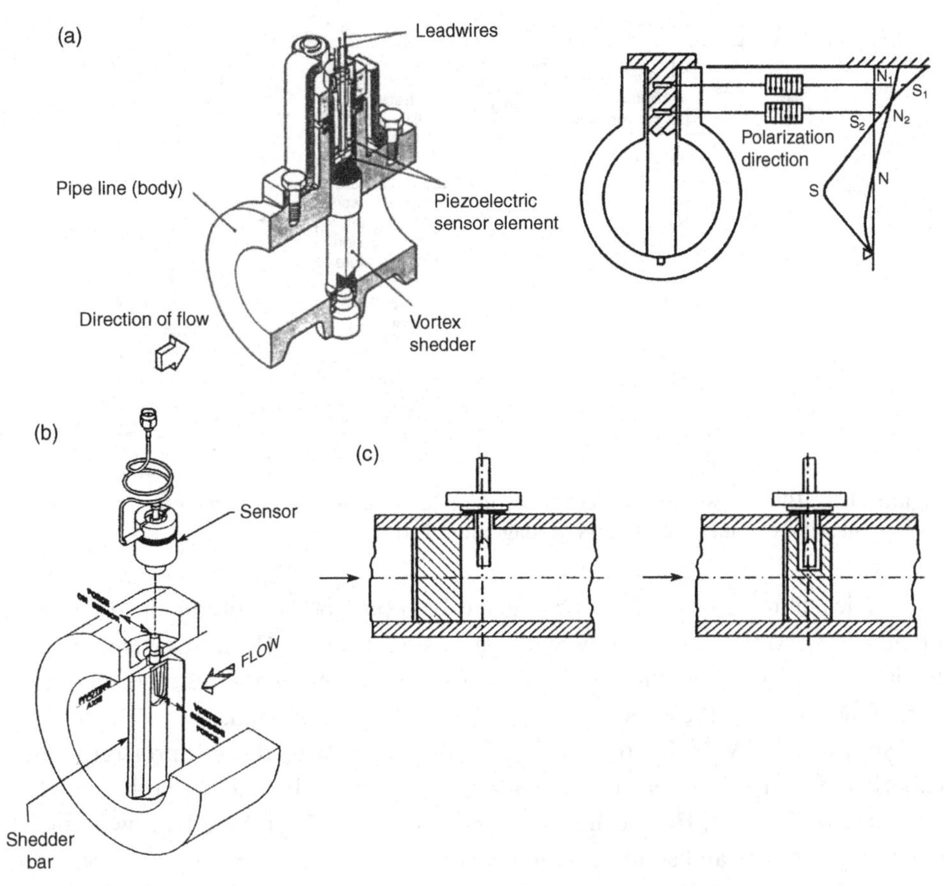

Figure 11.9. Commercial designs of flowmeter sensing systems: **(a)** Basic construction and vortex shedder bending moment diagram showing signal components *S* and noise components *N* (reproduced with permission of Yokogawa Europe B.V.). **(b)** Shedder bar/sensor operation (reproduced with permission of Emerson Process Management Ltd.). **(c)** Sensor positions (Hogrefe et al. (1995) reproduced with permission of the publisher).

Takahashi and Itoh (1993) described a design that used the alternating lift on the shedder, which was sensed through two piezoelectric elements [Figure 11.9(a)]. By using two sensors, effects due to vibration of the pipeline were reduced. A stable zero is created by using the strength of the vortex-shedding signal (proportional to ρV^2) and creating a low flow cut-off when this does not tally with the value of V from the vortex-shedding frequency. Another arrangement is shown in Figure 11.9(b); here the vortices lead to a small flexural movement of a section of the shedder bar, and a piezoelectric element inside the sensor senses this minute movement. Figure 11.9(c) shows the position of a sensor bar, which can be inserted into or placed downstream of the shedder body.

Herzog (1992) used optical fibres within the bluff body of a vortex-shedding flowmeter to sense the frequency. One fibre appears to have been subject to the transverse forces through ports in the body, whereas the other was totally enclosed. The bending of the first fibre caused changes to the quality and intensity of the light, and both speckle pattern and intensity were used to obtain the frequency (cf. Marshad and Irvine-Halliday 1994 and also Wen Dong-xu 1990).

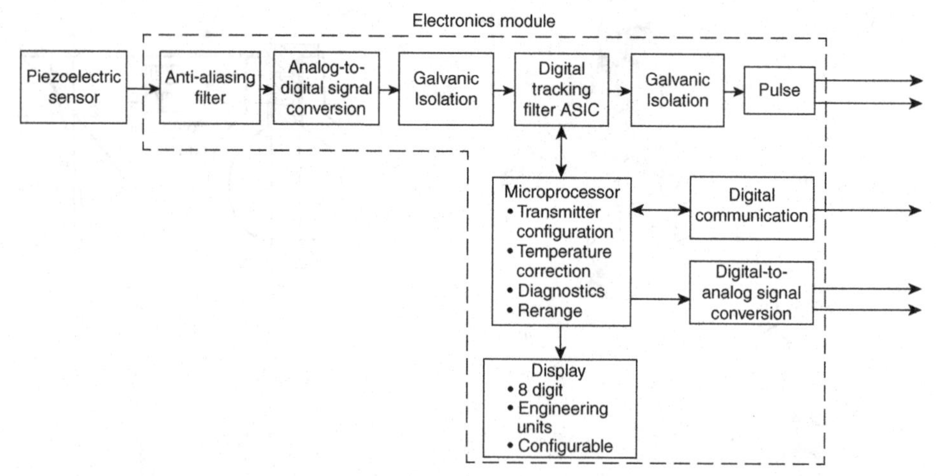

Figure 11.10. Block diagram of electronics module for commercial flowmeter (reproduced with permission of Emerson Process Management Ltd.).

Optical fibres have been attached to a rubber bluff body to obtain the drag force on the bluff body (Philip-Chandy, Scully and Morgan 2000). The fibres were grooved to cause an intensity modulation as the body bent. Using two fibres on each of two perpendicular faces, the direction of flow could also be obtained.

Sproston, El-Wahed and Johnson (1990) and El-Wahed and Sproston (1991) experimented in air with electrostatic sensors in the walls, behind the body or on the front face of the body. The mechanism on which this sensing is based is flow-induced streaming currents, and so its variation can be used to measure flow variation. (See Bera et al. 2004 induction sensing method).

Piezoelectric elements may be sensitive to pipe vibration and degrade the signal-to-noise ratio. Stiffening the meter body, using filtering and a low cut-off (Kawano et al. 1992) may partially overcome these problems.

Von Lavante et al. (2013) described the numerical simulation of a pressure detection system using a pressure chamber which can cope with temperatures up to 500°C or more.

Figure 11.10 shows a block diagram for a signal converter. The use of digital signal processing, now a standard in virtually all flowmeters, was reported on by Ostling and Oki (2001) to improve the signal-to-noise ratio, and has now become capable of providing a warning when conditions create sensing problems. Ghaoud and Clarke (2002) developed a model to evaluate the signal which contains elements from the installation as well as the primary shedding frequency (see Amadi-Echendu *et* al 1993). They evaluated various frequency tracking methods and were able to demonstrate experimentally an improved turndown ratio. Clarke and Ghaoud (2002, 2003) also described and tested a dual-phase-locked-loop system. A multi-channel sensing system for vortex flowmeters based on an array of 12 sensors was proposed by Mudd and Bentley (2002) as a means of improving repeatability

Zhang, Sun and Wu (2004a) suggested that a technique they term "de-noising" allowed them to extend the range of a vortex flowmeter at the lower end. Peng,

Wang and Fang (2012b) applied a de-noising approach to the signal from a vortex flowmeter in an oscillatory flow and appear to have had some success. Zheng and Zhang (2008) proposed a method for the vortex signal processing based on "double-window relaxing notch periodogram", to improve the flowmeter and solve the problem that the useful vortex frequency is hard to measure at low flow rates because of background and harmonic noise. However, the problem at low flow rates may be more of a fluid flow problem than a sensing problem.

Akresh, Reindl and Vasic (2010) suggested that the investigation of the vortex signal in the high turbulent range is rarely mentioned in the literature, but they used a typical vortex flowmeter and extended its measuring range up to 55%, by using various filtering techniques to extract the flow signal.

11.3.5 Cross-Correlation and Signal Interrogation Methods

Coulthard and Yan (1993a) suggested the use of the vortex wake as a tracer for use with cross-correlation techniques. The bluff bodies that were used presented a low blockage to flow and appeared to give similar accuracy to those of vortex-shedding frequency measurement. This presumably indicates that the cross-correlation technique is capable of the same discrimination as the frequency measurement electronics. Coulthard and Yan (1993b) investigated the effect of various bluff body shapes but found little difference between them (however, cf. Terao et al. 1993).

Amadi-Echendu, Zhu and Higham (1993) described a further piece of work to analyse the raw signal from a flowmeter, in this case a vortex flowmeter, to ascertain information beyond pure flow information. It is not clear from this paper the extent to which this will be possible. Clearly condition monitoring from this source would be very valuable. However, there is a need to link the frequency data as closely as possible to the physical effects so that the changes can be intelligently used. Some recent developments are mentioned in Section 11.3.11.

11.3.6 Other Aspects Relating to Design and Manufacture

Itoh and Ohki (1993) claimed (as does one manufacturer) a vortex flowmeter design which obtains mass flow rate by using the lift on the central body divided by the frequency of pulsation. This depends on the assumption that the lift coefficient and the Strouhal number are both constant so that the lift is proportional to ρV^2, and the frequency is proportional to V. As a result, the ratio is proportional to ρV (cf. Yamasaki 1993).

Zhang, Huang and Sun (2006) suggested the use of the pressure drop across the bluff body to obtain ρV^2 where ρ is density and V is velocity, and by using the vortex shedding frequency to obtain velocity, to derive the mass flow rate.

Miller, DeCarlo and Cullen (1977) gave some impressively consistent K factors from the production process showing random uncertainty in K of order ± 0.8% for 50-mm meters, ±0.4% for 75-mm meters and ±0.3% for 100-mm meters.

11.3.7 Accuracy

Typical performances of meters have been given as: repeatability, about ±0.2% rate; linearity, better than ±1% for a turndown ratio of up to 40:1 for liquids and up to 30:1 for gases. For liquids at higher flow rates, the uncertainty appears to be in the range 0.5–1% of rate. At low flow rates, it is the same range of values but as a percentage of full-scale deflection. The change will be at about Re = 20,000, with a minimum value of Re of about 4,000 and an upper limit of about 7×10^6. The type of transducer may affect the low flow limit.

For gases and steam, the uncertainty has been in the range 1–1.5% of rate, but with Re < 20,000 likely to increase uncertainty.

Temperature variation is likely to affect precision, and compensation may be included (e.g. 0.3%/50°C). BS ISO TR 12764:1997 makes the point that the meter geometry can be affected by temperature and pressure change. In the case of an increase in temperature there would appear to be two corrections which may need to be made resulting from:

1. the increase in the cross-section of the meter: this will cause volumetric flow, q_v, to increase for constant mean velocity, $\bar{V}$;
2. the increase in the width of the vortex shedder: because the Strouhal number, St, is approximately constant for a given geometry of meter, and $\mathrm{St} = fw/\bar{V}$ [Equation (11.2)] where f is the shedding frequency, w is the width of the bluff body and $\bar{V}$ is the velocity, this will cause w to increase and $f/\bar{V}$ to decrease.

Temperature increase will cause an increase in the dimensions of the flow tube and its components as discussed in Chapter 10. Length change was given in Equation (10.7) as

$$1+\frac{\Delta l}{l}=(1+\alpha_m \Delta T_m)$$

where l is a length, α_m is the temperature coefficient of expansion and ΔT_m is the increase in temperature of the material of the flowmeter. The increase in length is Δl due to this temperature increase. The change in cross-sectional area was approximated, since the linear expansion is small, by Equation (10.9)

$$1+\frac{\Delta A}{A}=(1+2\alpha_m \Delta T_m)$$

If meters with different materials and hence various values of α_m are used, this will need to be allowed for in calculating the area change, and the shedder frequency. If the mean velocity, $\bar{V}$, is constant, the volume flowrate, q_{v0}, at the initial temperature will increase due to the increase in A, since

$$q_v = A\bar{V}$$

and will become

$$q_v = q_{v0}(1+2\alpha_m \Delta T_m) \tag{11.3}$$

The shedding frequency is given by

$$\mathrm{St} = f\,w/\bar{V}$$

and since St is approximately constant and we are considering $\bar{V}$ as constant, increase in w, the width of the shedder, due to temperature increase, will result in a decrease in f

$$f = f_0(1 - \alpha_\mathrm{m}\Delta T_\mathrm{m}) \tag{11.4}$$

Since q_v increases at constant $\bar{V}$ and f decreases due to enlargement of the shedding body, the total effect on the K factor will be

$$K = K_0(1 - 3\alpha_\mathrm{m}\Delta T_\mathrm{m}) \tag{11.5}$$

For the effect of pressure difference see Section 10.2.14. It is likely that normal pressure changes will have a negligible effect on the dimensions. For a non-dimensional derivation of temperature and pressure change on turbine meters, relevant to vortex meters, see Mattingly (2009).

Cousins has commented that it is necessary to collect a large number of pulses to achieve good repeatability due to variation of 10–20% in the shedding period. As a consequence, long integration times should be expected. This variation in the shedding period may be known as jitter.

Miau et al. (2005) undertook a series of tests in both gas and water facilities to obtain uncertainty and lack of linearity. The authors indicated that the uncertainty for gas flow was better than ±0.8% (cf. Miau et al. 2004).

Storer, Schroeder and Steven (2009) discussed the possibility of the vortex meter combined with DP technology.

11.3.8 Installation Effects

Table 11.2 shows the change in error for increasing installation lengths (Takamoto et al. 1993b). Some of the results suggest that the error does not always decrease with spacing. The worst case has been used, from the results of four meters.

Mottram (1991) commented that "The difficulties of deciding what should be the minimum straight pipe length between a given type of flowmeter and a particular upstream pipe fitting are notorious. Even when the geometry is standardized, as is the case for the sharp edged orifice, the issue is still controversial. This is due to the typical shape of the envelope enclosing the experimental results for the calibration shift plotted against the straight pipe length separating the meter from the upstream fitting. Minimum straight lengths may differ by a factor of two." Mottram's tests were for a vortex meter with an area ratio approximately equivalent to a $\beta = 0.6$ orifice area ratio. We compare in Table 11.3 Mottram's and Takamoto et al.'s (1993b) values for the vortex installation lengths and those from ISO 5167 for orifice meters with $\beta = 0.6$ and $\beta = 0.75$.

Mottram gave a general rule that vortex sensitivity was similar to that for a high beta ratio orifice plate. This should be used with caution even though reasonable

Table 11.2. *Results from Takamoto et al. (1993b) that appear to give the worst case uncertainty (rounded up to the first decimal place) at various spacings*

	Bend 90° (%)	2x Bend (%)		Reducer (%)	Expander (%)	Gate valve open (%)	
		Same Plane	90° Plane			100%	50%
5D	1.9	2.1	2.1	0.9	4.2	0.3	5.1
10D	0.6	0.7	1.1	0.7	1.4	0.4	1.5
20D	0.4	0.4	0.8	0.3	0.3	0.4	0.5
30D	0.3	0.2	0.6	0.3	0.3	0.3	0.4

from his results and is in close agreement with Ginesi and Annarummo (1994), who suggested using the 0.7 beta ratio lengths. However, Mottram's comment that minimum lengths may differ by a factor of two is borne out by Takamoto et al.'s (1993b) results.

Further results from Mottram and Rawat (1988) (using a rectangular bluff body and hot-wire anemometer probe to detect the vortices) suggested that pipe wall roughness, presumably resulting in profile changes, can cause 3% variation in calibration (cf. Witlin 1979).

If we compare manufacturers' requirements, they have been fairly consistent in suggesting 5D downstream although at least one suggested 10D, but for upstream spacing, they give values as in Table 11.4. These are, again, compared with orifice spacings, for $\beta = 0.6$ ratio orifice (zero additional uncertainty) and $\beta = 0.75$ ratio orifice (±0.5% additional uncertainty).

The values in Table 11.4 suggest that as far as bends are concerned, a rule of thumb might be to take the mean of the $\beta = 0.6$ orifice with zero additional uncertainty and the $\beta = 0.75$ orifice with ±0.5% additional uncertainty as a guide for vortex meter spacing based on some manufacturers' literature. However, there appears to be a more severe effect due to reducers, expanders and valves about which the manufacturers are clearly cautious. Further, although the vortex meter values are more in line with Takamoto et al. (1993b), they do not fully cover the more pessimistic results of Mottram (1991) and we need to exercise considerable caution.

On a more optimistic note, Section 11.3.11 gives some information on the trend in industrial instruments to the use of signal analysis to adjust the meter's response to disturbed profiles and, possibly, to allow shorter upstream lengths after pipe fittings.

Manufacturers have warned against the protrusion of gaskets into the flow upstream and adjacent pipe bores of different size to the meter. The inlet pipe should be free of weld beads and thermometer pockets and should be smooth; the pipework should be aligned with the meter. Takahashi and Itoh (1993) suggested $\frac{1}{4}$% error for installation in a pipe of different bore to the meter, and one manufacturer gave the effect as less than 0.6% for a step of 4–10% greater than the flowmeter ID resulting from the pipe schedule.

Table 11.3. *Installation lengths (diameters) between upstream fitting and meter for changes within ±0.5% compared with spacings for orifice plate meters*

	Bend 90°	2 × bend		Reducer	Expander	Gate valve open	
		Same plane	90° planes			100%	50%
Orifice zero additional uncertainty							
$\beta = 0.6$	42	42[c]	65[d]	9[e]	26[f]	14	
Orifice ±0.5% additional uncertainty							
$\beta = 0.6$	13	18[c]	25[d]	5[e]	11[f]	7	
vortex[a]	55		60				30
vortex[b]	13	12	43	14	16	5	16
Orifice zero additional uncertainty							
$\beta = 0.75$	44	44[c]	75[d]	13[e]	36[f]	24	
Orifice ±0.5% additional uncertainty							
$\beta = 0.75$	20	22[c]	18[d]	8[e]	18[f]	12	

[a] Mottram's (1991) meter with blockage equivalent to $\beta = 0.6$
[b] Takamoto et al. (1993b)
[c] worst-case spacing with bends closer than 10D
[d] worst-case spacing with bends closer than 5D
[e] reducer 2D to D over a length of 1.5D to 3D
[f] expander 0.5D to D over a length of D to 2D

Table 11.4. *Installation requirements (in diameters) as indicated in manufacturer's literature compared with orifice plate requirements*

	Orifice $\beta = 0.6$ (±0%)	Vortex	Orifice $\beta = 0.75$ (±0.5%)
90° bend	42	10–20	20
T	29	10–20	18
2 bends			
in same plane	42[b]	20–25	22[b]
in perpendicular planes	65[c]	30–40	18[c]
Reducer	9[d]	10–20	8[d]
Expander	26[e]	10–40	18[e]
Valve[a]	14	20–50	12
Downstream spacing	7	5–10	4

[a] Globe valve fully open
[b] worst-case spacing with bends closer than 10D
[c] worst-case spacing with bends closer than 5D
[d] reducer 2D to D over a length of 1.5D to 3D
[e] expander 0.5D to D over a length of D to 2D

Note that swirl, which can be caused by successive bends in perpendicular planes, can be very long lasting and 40D may be inadequate to deal with it.

Cousins (1977) commented on the severe effect of swirl in changing the calibration and even in suppressing the shedding, and suggested at least 50D upstream spacing from the source of the swirl for a swirl angle of 10° and more spacing for larger angles. However, Laneville et al. (1993) showed that, for swirl indices (Ω = injected swirling flow rate/total flow rate) less than 0.3, the vortex-shedding signal quality from a vortex flowmeter with trapezoidal body is equivalent to the zero swirl case, and the frequency is still within about 2% of the swirl-free value for a spacing 18D downstream of the swirl generator. These apparently differing results will call for caution.

Miau et al. (1997) showed that the effect of swirl of 10° and 20° on the performance of the meter was very severe and could lead to errors of 15%. Miau et al. (2002) also looked at the effect of elbows on the signal. The disturbance caused by a butterfly valve installed upstream of the meter was studied numerically (for DN25) and experimentally (for DN50) by von Lavante et al. (2010). The authors suggested that 8D pipe length from the valve to the meter was sufficient to minimise the influence in the numerical prediction and to make its effect marginal in the experimental tests. This spacing appears to be very much less than that in Table 11.4.

Pressure taps downstream should be 2D to 7D, and temperature taps should be 1D to 2D downstream of the pressure ones. A flow straightener of 2D length separated by 2D from the disturbance should have 5D to the meter. Mottram (1991) obtained good results using a Mitsubishi conditioner with a different spacing. For a bend, 90° offset bends, orifice plate with β = 0.645 and fully open globe valve, his results were within a band of ±1% with 5D between the fitting and the conditioner, 2.5D between the conditioner and bluff body and 2.5D downstream of the bluff body.

Other mounting constraints are not usual, but pipework must ensure that the meter runs full. The pipework should also be free of vibration and clear of debris before running a new meter.

11.3.9 Effect of Pulsation and Pipeline Vibration

It is known that vortex meters may be affected by pulsatile flow, particularly where the pulsating frequency or its harmonics are close to the shedding frequency. The meter may lock on to the pulsation frequency and give an error (cf. Amadi-Echendu and Zhu 1992). Some sensing techniques may be more susceptible to error due to pulsation than others. Some may be designed to select the shedding frequency by eliminating pulsation or vibration using a push-pull arrangement.

Hebrard, Malard and Strzelecki (1992) confirmed that pulsations can cause large metering errors. Pulsations of an amplitude of only a few percent may be sufficient to produce metering errors, particularly when the frequency is near the fundamental or first harmonic of the shedding frequency. The lock-on frequency is most likely to occur when the pulsation frequency is twice the vortex-shedding frequency, and lock-on can occur (Al-Asmi and Castro 1992) even down to pulsation amplitudes of about 2%. Outside the lock-on range, there can still be significant differences between the shedding frequency in steady and pulsating flows, and free stream turbulence may increase the problem.

Manufacturers may suggest that disturbances upstream of an unsteady nature should be given a wide berth. Malard et al. (1991) found that there was an increase of about 2% and a decrease in the signal-to-noise ratio when the turbulence intensity increased from 3.3 to 10.3%. Their work also confirmed the errors due to pulsatile flows.

As indicated earlier, sensors are designed to minimise the effect of pipeline vibration, and the manufacturer should be consulted for information. However, process noise may cause zero flow signals but may be diagnosed by looking at the raw signal (Ginesi and Annarummo 1994).

Miau et al. (2000) reported on impulsive forces applied to the pipe and the effect on the signals from a vortex flowmeter. Improving the design of the piezoelectric sensor reduced effects due to structural vibration. Repetitive impulsive forces at a frequency greater than the vortex shedding frequency degraded the signals. They gave advice on the design to minimise any effects, but also confirmed the problem of the signal being degraded because of continuing vibration.

Peters et al. (2000) confirmed the errors of two sorts due to pulsations and pipe vibrations. In the former there is the known lock-on effect, but also the distorted signal can cause an irregular pulse from the Schmitt trigger. In the latter the meter may not be able to distinguish between the shedding frequency and the vibration of the pipe.

11.3.10 Two-Phase Flows

Hussein and Owen (1991) found that the correction factor for the vortex meter for wet steam of high dryness factor is closest to the value $x^{-1/2}$ (Figure 11.11). BS 3812:1964 is quite close. Pressure was shown to have little effect on correction factors. Hussein and Owen (1991) gave the wise advice that an upstream separator is advisable, and that there are several types giving a value of $x > 95\%$ with reasonably low pressure loss.

The possibility that the meter might be developed to sense the droplets of wet steam is an interesting challenge for the signal processing experts.

The meter has virtually the same calibration for liquid and gas flows and might therefore be expected to behave well in two-component flows. However the published work highlights some severe problems.

The first is that the components are separated by the vortex motion. Hulin, Fierfort and Condol (1982) found that shedding in vertical water-air flow was stable up to about 10% void fraction. They also found that the vortices trap bubbles [Figure 2.9(b)] with an associated decrease in the strength of the pressure fluctuations. Their later work (Hulin and Foussat 1983) on water-oil flows shows some contrasts. The separation is much more complete for water-air due presumably to the

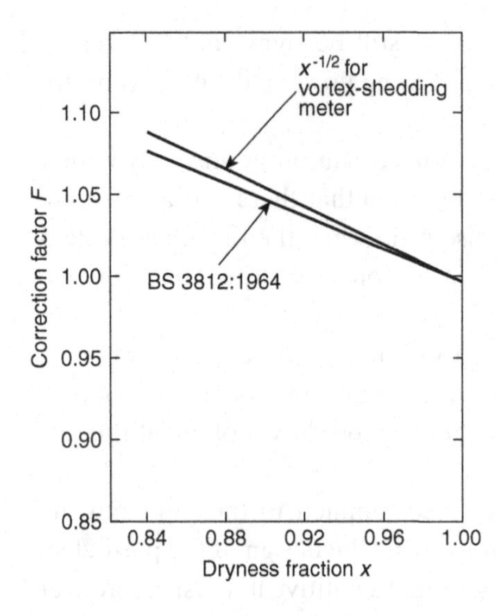

Figure 11.11. Comparison between wet steam correction factors for vortex meters (after Hussein and Owen 1991).

density difference, and this leads to a much narrower vortex emission bandwidth. The stability at oil concentrations greater than 30% was notable, and the cores of the vortices were continuous oil.

The second is that the bubbles may trigger certain sensors. Baker and Deacon (1983) obtained results suggesting that, even though vortex shedding continued with increase in void fraction in a vertical upward water-air flow, bubble impact may cause additional pulses. Above air volume concentrations of 1%, the meter error increased by more than 1% for each additional 1% air concentration.

A third problem, to which Washington (1989) attributed poor results in wet-gas, may be slip between the components. A homogeneous second component in the flow may be measured but will usually result in a loss of accuracy due, in part, to the fact that the instrument measures the volumetric flow.

However, the meter may have found an important niche in the measurement of wet-gas flows. Signal sensing and analysis has been used by Endress+Hauser for wet-steam measurement (A. Capper, Endress+Hauser September 2013, "Prowirl 200 new E+H vortex meter"). This would suggest that a value of temperature is obtained, and experimental data in the form of a lookup table for wet steam are stored on the meter. It would then, presumably, be possible to obtain a measure of the mass flow rate from the characteristics of the signal, or from the drag on the shedder, and hence the wetness might be obtained. There is a suggestion that condensate moves with its own wave motion, and it may be possible to correlate this with the actual dryness fraction and, using signal analysis, to measure the wetness of the steam and, if necessary, to provide an alarm.

For superheated steam the addition of a pressure reading should allow the state of the steam to be identified. With this software the meter may be capable of acting as a heat meter, possibly with an additional temperature input.

Another important area of two-component flows was discussed by Xing and Zhang (2009), who tested a vortex meter in oil-water two-phase flow and claimed to have found errors within 4% with oil fractions varying from 5% up to 40% by volume.

11.3.11 Size and Performance Ranges and Materials in Industrial Designs

As indicated earlier, the low end is limited by Reynolds number behaviour, and the effect of Equation (11.1) is that, as the size increases, the frequency becomes very low and makes the meter unusable.

Diameter ranges from about 15 to 300 mm or more are quoted by various manufacturers. Flow rates through the meters are up to about 9 m/s for liquids and 75 m/s for steam and gas. Flow ranges for liquids may be from a minimum of less than 1 l/s to a maximum of about 600 l/s and for gases 15 l/s or less to 3,000 l/s or more.

Range is of order 20:1 on gas and steam and 10:1 for liquids (up to 80:1 is sometimes given). This creates a constraint with the minimum Re, which may inhibit its application where the precise range is unknown or spans two meter sizes. In others, a smaller-sized meter may be necessary to cope with the range. Over-ranging up to 20% is generally allowable, but the low flow end may be more due to the strength of the signal.

Minimum Reynolds number is variously given as 3,800–5,000, although meters may be of lower accuracy below 10,000 and upper limits of 500,000 and up to or greater than 1,000,000 may be given.

Viscosity should be below 8 cP (the viscosity of cooking oils), although it may be possible to measure with reduced range up to 30 cP (Ginesi and Annarummo 1994). A low-flow cut-off is usually set somewhat below Re = 10,000, and this may be a disadvantage for some applications.

The temperature range may be from less than –200° to over 400°C. Pressure may be up to 150 bar or more depending on the transducer. Pressure drop on a meter of nominal internal diameter equal to the pipe is normally less than about 40 kPa (6 psi) on water flow.

Pulse rates will vary enormously depending on the size of the meter. For a small meter of about 20 mm diameter, the rate will be of order 200 pulses/l, whereas for a large meter of 300 mm diameter the rate will be of order 0.02 pulses/l. Response time may be of order 0.2 s or three shedding cycles for a change to 63% of input.

Material of the bluff body is most commonly stainless steel but may be titanium or another material.

Vortex meters, like turbine meters, are limited at the upper end of their range by possible cavitation. Casperson (1975) suggested the possibility of hysteresis effects due to cavitation and the need to avoid operating in this regime. To avoid cavitation in liquid flows, some manufacturers give a formula of the following type for the minimum back pressure 5D downstream:

$$p_{g\,\mathrm{min}} = A\Delta p + Bp_{\mathrm{v}} - p_{\mathrm{atmos}}$$

where Δp is the pressure drop across the meter, p_v is the saturated liquid vapour pressure, and p_{atmos} is the atmospheric pressure. A may have a value of about 3.0, and B has a value of about 1.3. Δp is of order 1 bar at 10 m/s.

These meters may also be subject to compressibility effects.

As an example of industrial developments, and the use of signal sensing and analysis, to extend the performance of the meters, Endress+Hauser have introduced some novel hardware and software into its designs (A. Capper, Endress+Hauser September 2013, "Prowirl 200 new E+H vortex meter"). The company suggests that typical meter usage is about 50% for steam, 30% for liquids and 20% for gases. The manufacturer uses capacitance sensors with primary vibration compensation. The sensors are positioned with care relative to the centre of gravity of the unit, may be capable of coping with water hammer and temperature shock, and may have a lifetime calibration. An uncertainty is claimed for liquids of 0.65% and for gas/steam 0.9%.

It may, also, be possible to use signal analysis to allow shorter upstream inlet lengths and a wider turndown, by sensing the turbulence pattern and deducing the nature and proximity of upstream fittings. An adjustment may be possible within the software for this. The company claims, following reducers and elbows, that the inlet length can be reduced possibly to 10D (with an additional 0.5% uncertainty) and that a turndown as much as 80:1 may be achievable.

These meters are probably not suitable for hygienic applications.

11.3.12 Computation of Flow Around Bluff Bodies

Some early work obtained good agreement between numerical and experimental results for Re up to 1,000 for confined two-dimensional flows around rectangular cylinders (Davis, Moore and Purtell 1984). Matsunaga, Takahashi and Kuromori (1990) showed some impressive computer-generated flow patterns around a trapezoidal body in a two-dimensional flow. Comparison of Strouhal number obtained computationally and experimentally agreed quite well at Re > 1,000 but diverged for lower Re and were less good for Re of order 10,000. Figure 11.12 shows the vortex-shedding mechanism as interpreted by the authors.

Johnson (1990) computed the unsteady laminar flow around a cylinder and obtained agreement within 10% for the Strouhal number in comparison with experimental values for part of the range. Computer simulations by El-Wahed, Johnson and Sproston (1993) again obtained good agreement with the experimentally acquired Strouhal number and indicated that the T-shaped shedder produced the largest vortex.

Some of the computational fluid dynamics (CFD) work has concerned 2-D solutions for vortex meters (Thinh and Evangelisti 1997). The straight vortex body in a circular pipe may call for more of an understanding of how upstream distortion and turbulence distribution feed into the creation of the shed vortices. The paper by von Lavante et al. (2010), mentioned earlier, modelled the inlet flow resulting from a

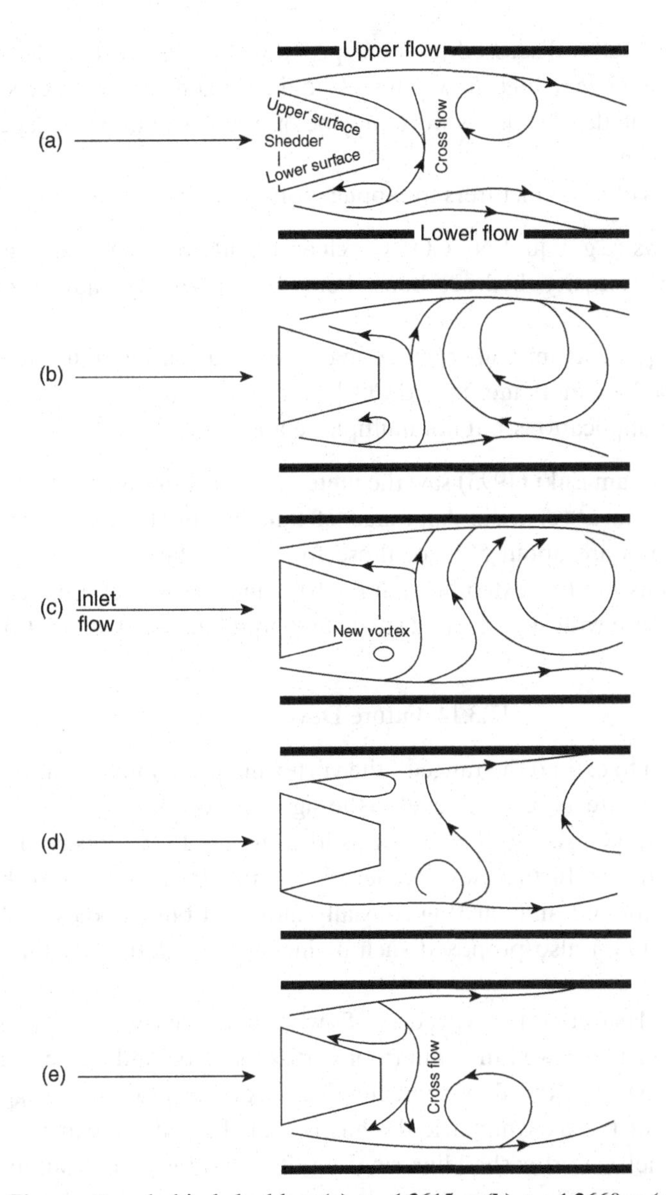

Figure 11.12. Flow pattern behind shedder: **(a)** t = 1.2615 s; **(b)** t = 1.2660 s; **(c)** t = 1.2690 s; **(d)** t = 1.2735 s; **(e)** t = 1.2765 s (reproduced with the agreement of Elsevier from Matsunaga, Takahashi and Kuromori 1990).

butterfly valve. Chen et al. (2010) used CFD to simulate the flow field for a vortex meter with different inflow conditions.

11.3.13 Applications, Advantages, and Disadvantages

Furness and Jelffs (1991) sought to make a case for using vortex meters to measure flows in and out of refineries rather than depending on tank transfer. Proving, however, may be problematic due to the poor period repeatability.

Siegwarth (1989) discussed tests on specially designed vortex-shedding meters for application to very high flow rates of liquid oxygen in the space shuttle main engine ducts. The details of this work may be of interest to those engaged in vortex flowmeter design.

The claims of manufacturers for applications include

- liquid flows (e.g. liquid N_2, CO_2, O_2, clean liquids, distilled water, glycol, some acids, low-viscosity hydrocarbons, benzene, diesel, hydraulic oils, creosote and tar),
- gas flows [e.g., steam (superheated and saturated) and various gases including compressed air, methane, N_2 and CO_2], and
- cryogenic applications, but not multiphase applications.

In general Yamasaki (1993) saw the meter as an alternative to the orifice meter, provided it could cover a similar range of fluid parameters. (One manufacturer claims that costs are about 50% of those for orifice plates.) The sensing method is the main constraint to extending parametric range. However, temperature range now allows its use with superheated steam and liquid natural gas (LNG).

11.3.14 Future Developments

There is a need to extend the range of the vortex meter, to improve application data (e.g. installation effects) and to develop the signal processing.

Zanker and Cousins in their work as long ago as 1975 foresaw the dual bluff body, and, no doubt, further development is possible in the design of these bodies. They also mentioned annular ring/coaxially mounted bluff bodies, and Takamoto and Komiya (1981) also proposed such a ring-shaped bluff body that sheds ring vortices.

Miau and Hsu (1992) surveyed the flow around axisymmetric discs and rings as vortex shedders, showed the pattern of vortex loops behind a disc, and were particularly interested in the flow characteristics near the pipe wall, suggesting that an advantage of the axisymmetric designs is that the sensing can be on the pipe wall. Axisymmetric vortex-shedding rings intuitively offer an elegant axisymmetric version of the meter. Tai et al. (1993) discussed optimum size and positioning and signal-to-noise ratio. Cousins and Hayward (1993; cf. Cousins, Hayward and Scott 1989) described developments of the vortex ring flowmeter and discussed sensing options. They made the point that ring vortices are far more stable than those from straight bluff bodies. They suggested that a ring/pipe diameter ratio of 0.4 is about optimum. They found that the signal-to-noise ratio varies markedly across the pipe and down the centreline, peaking at 3 to 4 ring widths downstream. They found that the inner ring vortices give a clearer signal. The provisional design which appeared to be optimal was a double ring with a tail and gave a repeatability of ±0.1% or better with only 0.3 velocity head pressure loss.

The combination of electrostatic sensing as described by El Wahed and Sproston (1991) with ring vortex shedders could be very attractive. The sensor could be in the wall of the tube or even at various points to give greater intelligence to the meter and forms of self-monitoring.

11.4 Swirl Meter – Industrial Design

This section is concerned with the industrial design of the swirl meter, which has some fluid behavioural similarities to the vortex whistle (Chanaud 1965). In the whistle, initial swirl is created by tangential flow into an upstream plenum from which the helical vortex emerges.

11.4.1 Design and Operation

A diagram of this type of flowmeter is shown in Figure 11.13. As the flow enters the meter, inlet guide vanes cause it to swirl. As the flow moves through the contraction, the angular momentum created by the guide vanes is largely conserved, and so the angular velocity will increase. The vortex filament develops into a helical vortex, as shown in the diagram, which moves to the outside of the tube. The frequency with which the helix passes the sensor provides a measure of the flow rate. The flowmeter section then expands, and the swirling jet at the wall surrounds a region of reverse flow on the axis. This motion is removed before the fluid leaves the meter by means of a de-swirler. Dijstelbergen (1970) explained the precession as being due to the increasing pressure in the diffusing section causing a force on the exiting swirl, which is converted to a lateral momentum change. Ricken (1989) gave further details of a model to describe the behaviour of the meter.

Heinrichs (1991) described a sensing system for a swirl meter in which the rotation is sensed by using two diametrically opposed pressure ports. In the measuring tube of the swirl meter, a helix-shaped vortex appears, rotating with a frequency proportional to the volumetric flow rate. In this way, the differential pressure eliminates common mode signals and hence the problem of a high line pressure and flow fluctuations. The push-pull arrangement suggests that the range may be extended. Capacitive sensors seem to be less suitable, whereas the results for piezoelectric ones are more promising.

11.4.2 Accuracy and Ranges

An industrial design has been available for liquids, gases or steam with vortex rotation frequency obtained using piezo sensors. Measurement uncertainty of better than 1% of rate was claimed by the manufacturer over the upper 80% of the range for the smaller sizes, with better than 2% for the 10–20% part of the range. For sizes above 32 mm, uncertainty of better than 1% of rate was possible over the upper 90% of the range.

Flow ranges were, for liquids, from 0.2 to 2 m^3/h up to 180 to 1,800 m^3/h and, for gases, from 5 to 25 m^3/h up to 1,000 to 20,000 m^3/h. Values for steam should be obtained from the manufacturer. Turndown for some sizes for liquids was up to

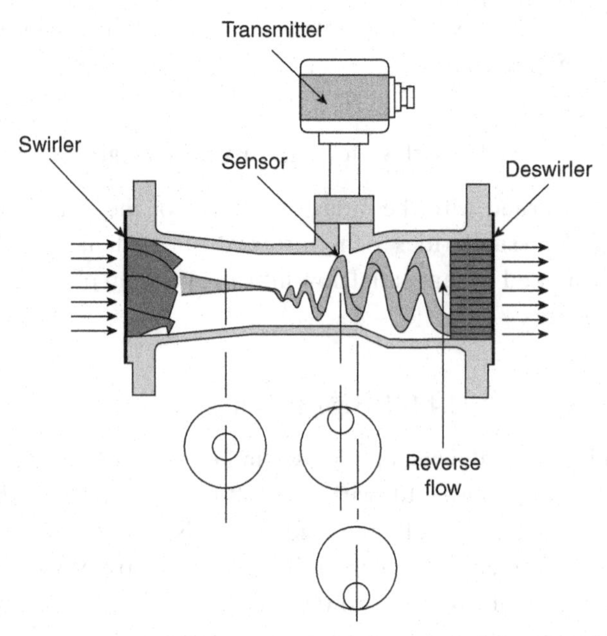

Figure 11.13. Swirl flowmeter (with the permission of ABB).

about 30:1, and for gases it was between 5:1 and 20:1. Maximum frequencies ranged for liquids from 100 Hz for a 20 mm diameter meter to 13 Hz for a 400 mm meter and for gases from 1,200 Hz for meters of 20 mm down to 150 Hz for a 400 mm diameter meter. A 10% over-ranging was allowed in gases, but in liquids cavitation had to be avoided. Maximum velocity is 6 m/s for liquids and 50 m/s for gases.

For liquids at the top of the flow range, pressure loss ranged from about 200 to 400 mbar or more, and for gases the range is from 25 to 70 mbar or more.

11.4.3 Installation Effects

For bends, contractions and valves upstream, 3D straight pipe before the inlet flange is recommended; and 1D for bends and expansions downstream. A contraction downstream is not permitted. See Peng et al. (2008a) on range of flow with oscillation.

11.4.4 Applications, Advantages and Disadvantages

The meter was recommended by the manufacturer for applications where absolute reliability with minimum maintenance was essential and was claimed to be suitable for wet or dirty gases and liquids.

11.5 Fluidic Flowmeter

The third instability we consider is that for the fluidic flowmeter. A diagram of a typical design is shown in Figure 11.14. The fluidic flowmeter uses the Coanda effect. If

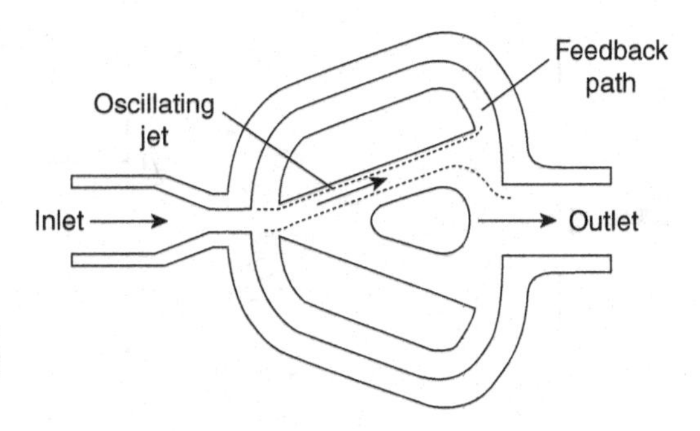

Figure 11.14. Diagram of a fluidic flowmeter (reproduced with permission of Professional Engineering Publishing).

a two-dimensional diffuser has too wide an angle, the flow will separate and attach to one wall of the diffuser. It may attach to either wall and can be displaced from one to the other by a suitable disturbance. The fluidic flowmeter introduces such a disturbance, which causes the flow to oscillate between the diffuser walls at a frequency proportional to the velocity of flow. There are several parallels with the vortex meter. Types of sensors used in vortex flowmeters could be applied to this device. There is likely to be room for design optimisation. The flow range may be wide, and certainly the device looks promising for low flows.

The mechanism by which the flowmeter operates is that the flow, having attached to one wall, passes into a feedback channel. The fluid when it exits from this channel knocks the main jet to the other side of the diffuser, and the process starts again. Thus the period is related to the time for the flow to move from the entry of the flowmeter to the entry of the feedback channel and communicate, via that channel, to the entry. This time will be inversely proportional to the velocity of flow, and so the oscillating frequency will be proportional to the flow rate.

11.5.1 Design

Boucher and Mazharoglu (1988) developed one design of fluidic target meter, which operated down to Re of about 70 but with considerable change in Strouhal number at low Re. Hysteresis may have a small effect in starting and stopping (cf. Boucher 1995).

Sanderson (1994) described a new fluidic water meter using electromagnetic sensing. This is a particularly appropriate method with a wide dynamic range and a linear response to flow rate. It is also usable in a constant magnetic field design because of the fluctuating nature of the flow. The sensing electrodes were made of stainless steel, and earthing electrodes were also provided. The magnetic materials considered for this device included sintered ferrite and bonded and sintered neodymium. The general configuration is shown in Figure 11.15. Sanderson's preferred arrangement of magnetic sensor was for the measurement to take place in the main jet with magnetic material in the diffuser walls giving a field strength of about 0.3 T. The two pairs of electrodes were in the side walls acting 180° out of phase in a push-pull arrangement. Over most of its range, the meter was linear to within ±2% (from about 5 to 500 ml/s). Further linearization was possible within

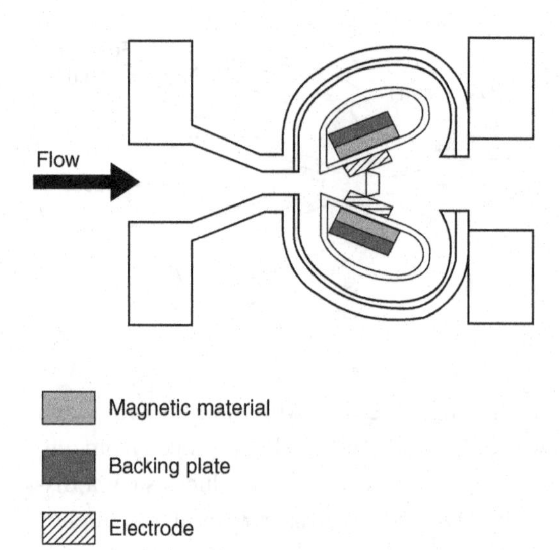

Figure 11.15. Fluidic flowmeter with magnetic sensing (reproduced with the agreement of Elsevier from Sanderson 1994).

the electronic circuitry. The meter operated from a 3.6-V lithium battery with a 10-year lifetime.

Sakai, Okabayasi and Yasuda (1989) described a dual sensing system gas flowmeter for domestic use in Japan. It was designed to operate from 3,000 l/h down to 150 l/h with a piezoelectric differential pressure sensor and seems to have achieved ±2.5% for 150–600 l/h and ±1.5% above 600 l/h. For the range 3–150 l/h, an integrated circuit thermal sensor was used. This had a heated area with temperature sensor areas on each side of the integrated circuit chip. The characteristic of this sensor appears to have a varying coefficient, and it is not clear what final precision was claimed for low flows. Sakai et al. saw the prospect of using one lithium battery of 5,000 mAh, which would last 10 years in continuous service by lowering the consumption of the circuits.

Measurement uncertainty of a preproduction gas meter on field tests in Japan is claimed (Yamasaki 1993) to be within ±1.5% of reading with a 30:1 turndown ratio. Upper flow limits are defined by the allowable loss in the meter. Computer simulation was used as an aid in the design.

Nishigaki et al. (1995) used LDA methods to obtain the flow patterns in a fluidic gas flowmeter and concluded that the patterns were almost the same across a Reynolds number range from 291 to 2,160.

The development of a meter for various Japanese gas companies (Aoki et al. 1996; Sato et al. 1996) has included numerical analysis. It has a more complicated array of internal bodies with a U-shaped initial target. A 60:1 turndown appeared to be achieved.

Martinelli and Viktorov (2011) discussed a mini fluidic oscillating flowmeter which used three micro fluidic bistable amplifiers. The air flow rate was up to about 34 cm^3/s to keep within the linear laminar range. For water the equivalent maximum appears to be about 8.7 cm^3/s.

Simões, Furlan and Pereira (2002) simulated microfluidic oscillators using the ANSYS modelling program for steady state and transient behaviour. In a further paper (2004 and 2005) they discussed the design and operation of microfluidic oscillators (outlet diameter of the supply nozzle of order 280μm), which they tested using nitrogen, argon and carbon dioxide. Fang, Xie and Liang (2008) used CFD to model a fluidic flowmeter.

11.5.2 Accuracy

Okabayashi and Yamasaki (1991) found that accuracy could be improved by

i. optimising the shapes of the side walls and the position of the target and
ii. incorporating a flow straightener to make the jet two-dimensional.

They claimed a linearity of ±0.8% of reading within the range 0.15–5.0 m^3/h with a repeatability of ±0.2% of reading. The frequency appeared to be about 20 $Hz/m^3/h$. Sensing was by a piezoelectric sensor (using a polymer piezoelectric film) that does not consume power and an amplifier with very low current consumption.

11.5.3 Installation Effects

It seems likely that these devices will be virtually insensitive to upstream disturbance because the inlet duct within the meter will have changed section and will be equivalent to several diameters. There is also likely to be flow straightening.

11.5.4 Applications, Advantages and Disadvantages

The prime application of these devices is for utility flows where they meet requirements of

- small size,
- large turndown,
- low power and
- long life without maintenance.

With no moving parts and compact size, these devices should show advantages over existing meters for utility (water and gas) flows. However, the sensing range will require capability for large turndown if the fluid turndown range is to be realised.

11.6 Other Proposed Designs

A wake oscillation was observed by Mair (1965). With an axially symmetric and streamlined body followed by a disc, the wake gathers in the cavity and puffs out regularly. With a torpedo-shaped body of diameter D, a coaxial disc of diameter w and spacing x downstream of the flat rear of the body, Mair found that with $x/D \sim 0.5$ and $w/D \sim 0.8$ the vortex was trapped between the body and the disc and resulted in

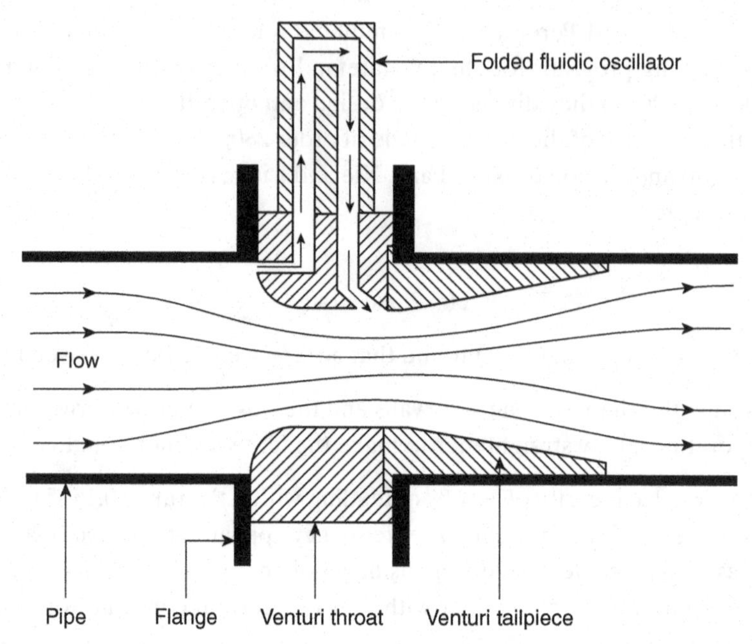

Figure 11.16. Combined Venturi nozzle with a folded fluidic flowmeter (from Parkinson 1991; reproduced with permission of the Institute of Measurement and Control).

a low drag, whereas for $x/D \sim 0.3$ and $w/D \sim 0.8$, a high drag and oscillation occurred, which he attributed to breathing of the cavity.

Parkinson (1991) described a combination of a Venturi nozzle with a folded fluidic flowmeter (Figure 11.16). Boucher et al. (1991) also designed a folded version of the fluidic flowmeter to create a bypass for a Venturi meter, In this way, the fluidic principle can be applied to pipes in the range 40–150 mm with turndown ranges of 24:1 to 40:1. The sensing in the fluidic flowmeter was by means of a miniature air-flow sensor on a silicon chip based on thermal convection. Wang et al. (1998) developed a similar idea: a Venturi with fluidic bypass which generated strong enough pulses inside a well so that they could be sensed at the surface (see also Wang et al. 1996).

Shakouchi (1989) demonstrated an alternative geometry of fluidic meter where a rectangular jet issues into a larger space. The oscillating vane meter was discussed in Section 8.8.

11.7 Chapter Conclusions

The vortex meter offers a very attractive alternative to the orifice plate for some applications and gains over the orifice plate by its frequency output and larger turndown. Its application to dry saturated steam may prove to be one of the most important contributions of this device. It would be interesting to explore means of sensing, by impact or by entrainment in the shed vortices, the amount of water in the flow and thus to obtain a dryness factor. The developments by industry may suggest research has taken place.

The prediction of performance, the effect of inlet flow and turbulence level and the means of sensing are areas for development. These will require consideration of optimum geometry and, in the case of the vortex flowmeter, the axisymmetric bluff body.

Much of the computational fluid dynamics (CFD) work has concerned two-dimensional solutions. The axisymmetric vortex body should lend itself to this, but the straight vortex body in a circular pipe may call for more of an understanding of how upstream distortion and turbulence distribution feed into the creation of the shed vortices. There may be some useful analytical work that would shed light on this and might lead to a weight function for the meter.

The sensing systems are somewhat limited. Pressure, force etc. have problems on turndown. Temperature sensing may offer scope for some applications. However, the electromagnetic solution is very attractive for liquids, and as well as application to fluidic flowmeter, I have, for some time, felt that the vortex flowmeter could be combined with a magnetic flow tube to give self-checking possibilities.

Further development of gas meters and suitable sensors is likely to continue.

At a more pedestrian level, but of great practical value for the vortex meter, would be further comparative installation effects of commercial meters, together with tests on the effect of pipe size difference, nonalignment, weld beads etc. and also vibration. This might be combined with the development of a standard vortex-shedding meter as foreshadowed by Takamoto and Terao (1994).

The fluidic meter appears to have found a particular niche in the utilities, and several are being, or have been, developed. They appear to offer a large turndown, provided a suitable sensing system is available. One such system is the magnetic system developed for water flows (Sanderson 1994).

The CFD analysis of the fluidic flowmeter is also, mainly, two-dimensional, but the effect of end walls and the turbulence level of the inlet flow may need further assessment, and it seems unlikely that the internal geometry has been finally optimised.

At the opposite extreme would be the investigation of intrinsic flow noise due to vortex shedding in existing pipework, which could be sensed and calibrated as a flowmeter system. If such noise exists, an intelligent sensor could learn its calibration and then provide for a utility measurement.

Some other oscillatory devices have been suggested. Sato and Watanabe (2000) undertook an experimental study on a vortex whistle as a flowmeter.

A predecessor of the vortex-shedding flowmeter had a bluff body which oscillated with the flow. Bird patented this device with a pivoted bluff body in 1959 (Medlock 1986; cf. Baker 1998). Because of the problems of moving elements in flowmeters, the industry has moved to the vortex meter with various sensing elements. Turkowski (2003), however, reported on the optimisation of a mechanical oscillator flowmeter. He further (2004) reported on the influence of some fluid properties on the instrument's performance. It would be interesting to know whether there is commercial interest in such a meter.

Lua and Zheng (2003) undertook numerical computations and experimental water tests on a device which they called a target fluidic flowmeter, but which appears, also, to have features of a vortex shedding flowmeter. A hot-film sensor was used to detect the shedding.

Mahulikar and Sane (2005) reported a piece of theoretical research, making use of experimental data, on a meter which depended on the rotation of a ball in a vortex chamber.

These ideas, whether or not they find a commercial outlet, suggest interest in the continuing development of these meters. Some interesting developments have been reported in this chapter, and further development is likely in terms of range, reliability, signal processing, low maintenance, accuracy and price, which should ensure that they enhance their position as a respected option for flow measurement.

Appendix 11.A Vortex Shedding Frequency

11.A.1 Vortex Shedding from Cylinders

Flow around a cylindrical body develops as in Figure 11.1. At a low Reynolds number [Figure 11.1(a)], the flow pattern on each side of the cylinder is almost symmetrical (for Re $\ll 1$ there is symmetry cf. Tritton 1988). The symmetry between the flow approaching the cylinder from upstream and that leaving it downstream is removed with increasing Re. When Re exceeds about 4 (Tritton 1988), eddies, which are more or less symmetrical [Figure 11.1(b)], are formed behind the cylinder and, as shown in Figure 11.1(c), start to become unsymmetrical before a Reynolds number of about 40 is reached. It is from this point that we obtain the remarkable phenomenon of vortex shedding.

The lack of symmetry in Figure 11.1(c) breaks down into a periodic flow as illustrated in Figure 11.1(d). At this stage, the flow is stable and viscous (Roshko 1954). The boundary layer on the cylinder and the shear layer in the wake are presumably laminar. The shedding is very regular in this region. In the range omitted from Figure 11.1, namely $150 < Re < 300$ (or $200 < Re < 400$ according to Tritton), transition from laminar to turbulent takes place in the shear layer, and the transition is accompanied by less regular shedding. Once the regime in Figure 11.1(e) is reached, the shedding becomes regular once more. A further transition in the region $3 \times 10^5 < Re < 3 \times 10^6$ occurs as the boundary layer changes from laminar to turbulent, causing irregularity (Tritton 1988). In this case, Tritton illustrates the laminar separation followed by turbulent reattachment and then, again, the occurrence of separation. The regime in Figure 11.1(f) has regular shedding and a later separation point. Tritton comments that markedly periodic shedding remains a characteristic of the flow up to the highest values of Reynolds number at which observations have been made, about 10^7.

11.A.2 Order of Magnitude Calculation of Shedding Frequency

Using a numerical solution, Abernathy and Kronauer (1962) showed that two shear layers result in the build-up of large vortex regions. Their diagrams also showed that one shear layer gets caught in the other shear layer and eventually results in a region

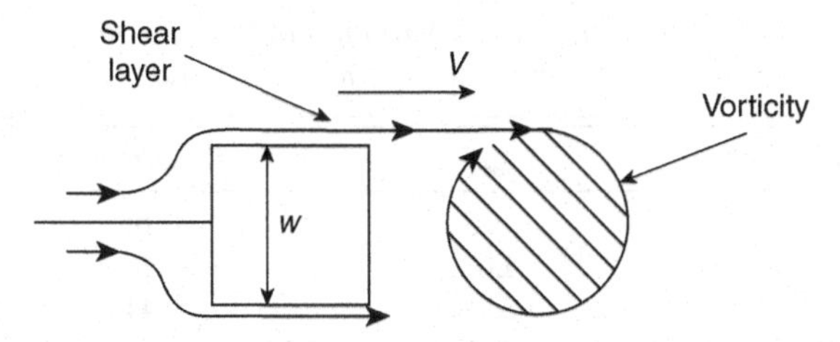

Figure 11.A.1. Simple model of vortex shedding.

where vortices of opposite sense cancel each other and break the shear layer (cf. Robinson and Saffman 1982 on the stability of Karman vortex arrays).

Figure 11.2 is a simple diagram of this mechanism (cf. Figure 11.12). In Figure 11.2(a), one vortex has just separated and the next is beginning to roll up. The roll-up is caused by the influence of each new vortex on those that have gone before. In Figure 11.2(b), the large vortex roll has curled back on itself and has started to entrain the vortex sheet of opposite sense on the other side of the body. In Figure 11.2(c), this entrainment has advanced to the point where the upper vortex sheet is nullified and the roll-up of the vortex is stopped as the whole vortex roll is set free (cf. Gerrard 1966).

A simple model to obtain a value for the roll-up time of the large vortex is given by Figure 11.A.1. The vorticity shed will be of order V/δ, where δ is the shear layer thickness. The rate of shedding will be approximately the mean velocity in the shear layer $V/2$ multiplied by the layer thickness δ. Thus the vorticity shedding rate is

$$\frac{V}{\delta}\times\frac{\delta V}{2}=\frac{V^2}{2}$$

The vorticity has to fuel a vortex with a diameter approximately equal to the size of the bluff body. The total vorticity within this roll will be

$$\begin{aligned}\int \omega\cdot \mathbf{da} &= \int \nabla\times V\cdot \mathbf{da}\\ &= \oint V\cdot \mathbf{ds}\\ &= \pi w V\end{aligned}$$

The time to fill a vortex of diameter w with $\pi\, w\, V$ is approximately

$$\tau=\frac{\pi w V}{V^2/2}=2\pi w/V$$

Thus an order of magnitude value for the frequency again is

$$f=\frac{V}{2\pi w} \tag{11.A.1}$$

A useful parameter is the Strouhal number given by

$$\mathrm{St}=f\,w/V \tag{11.A.2}$$

Table 11.A.1. *Approximate variation of Strouhal number based on mean pipe velocity for various body sizes*

w/D	V_{max}/V	St corrected
0.1	1.15	0.18
0.3	1.62	0.26
0.5	2.75	0.44

and using the approximation of Equation (11.A.1) we obtain a value of St = $1/(2\pi) = 0.16$. Goldstein (1965) gave 0.18 for a circular cylinder at a Reynolds number in the range 300–100,000.

Because of the reduced cross-section of the flowmeter due to the presence of a bluff body, the actual velocity past the body V_{max} will be greater than that in the upstream pipe V. For an incompressible fluid, area × velocity is constant. The area ratio will be given approximately by

$$\frac{A_{min}}{A} = \frac{\pi D^2/4 - wD}{\pi D^2/4} \tag{11.A.3}$$

Thus for an incompressible fluid

$$\frac{V_{max}}{V} = \frac{1}{1 - 4w/\pi D} \tag{11.A.4}$$

and so combining Equations (11.A.2) and (11.A.4) and assuming the appropriate velocity for Equation (11.A.2) is V_{max}, we obtain

$$\frac{f}{q_v} = \frac{4\mathrm{St}V_{max}/V}{\pi w D^2} \tag{11.A.5}$$

$$= \frac{4\mathrm{St}}{\pi D^3}\frac{1}{(w/D)(1 - 4w/\pi D)} \tag{11.A.6}$$

and so the approximate Strouhal number of 0.16, which we obtained based on the flow past the body, will apparently be increased if based on the velocity in the main pipe according to Table 11.A.1. For comparison, Takamoto and Terao (1994), for a *w/D* of 0.28, give St in the range 0.245–0.265 for Re from 10^5 to 10^6. The limitations of this simple approximation are shown by the effect of change in length of the bluff body from 0.4D to 0.2D found by Takamoto and Terao.

Zanker and Cousins (1975) assumed that the roll vortex was slightly larger than the bluff body diameter by between about 10 and 50%. This created a *vena contracta* confining the flow downstream of a sharp-edged bluff body to a smaller area near the wall in which the maximum velocity increases more. Adding their factor K to Equation (11.A.6), we obtain the same equation as they did:

$$\frac{f}{q_v} = \frac{4\mathrm{St}}{\pi D^3}\frac{1}{(w/D)(1 - 4K\,w/\pi D)} \tag{11.A.7}$$

It is arguable that a further K should appear as a multiplier in the denominator to allow for the larger vortex, which has to be filled. However, this is probably stretching a simple model too far, but interestingly it appears that for a w/D of about 0.3, this extended form of the denominator is less affected by the value of K than for, say, 0.1 or 0.5.

One can see how the change in the upstream conditions could influence the shedding frequency. Upstream disturbances will alter the vorticity in the flow and so reduce or increase the frequency of shedding and possibly the regularity. These changes are illustrated by Mottram (1991) for the effect of a more peaky profile due to pipe roughness and an increased Strouhal number. In this case, both the increase in local velocity at the body and the increase in vorticity will have effects. Other results discussed in this chapter show, surprisingly, that this type of meter is quite susceptible to the effect of a reducer upstream, presumably again because of the effect on the turbulence spectrum.

Terao et al. (1993) made two interesting observations.

- Vortices moved faster downstream than the mean velocity.
- Areas of intense vorticity existed at the wall.

The first of these two must affect the use of vortex shedding with cross-correlation techniques. The second suggests that three-dimensional computational solutions must include this observation (cf. Majumdar and Gulek 1981 for an early paper on shedding from various prisms).

12

Electromagnetic Flowmeters

12.1 Introduction

The possibility of inducing voltages in liquids moving through magnetic fields was known by Faraday in 1832, but the first flowmeter-like device was reported by Williams in 1930. The first real advance in the subject came from the medical field where Kolin (1936, 1941) introduced many ideas which are now standard practice. In 1941 Thürlemann gave the first general proof, essentially of Equation (12.2) as set out later in this chapter (see also Thürlemann (1955) and Shercliff (1962) for more background).

The industrial interest arose in the 1950s with:

- the Tobiflux meter (Tobi 1953) in Holland for rayon viscose, sand-and-water and acid slurries;
- Foxboro, to whom the patent was assigned in 1952;
- the first commercial instruments in 1954 (Balls and Brown 1959).
- nuclear reactor applications;
- the work which resulted in an essential book by J. A. Shercliff (1962).

In this chapter we shall concentrate on the application of the flowmeter to fluids that are of low conductivity, such as water-based liquids (Baker 1982; see also 1985). The flowmeter has also been proposed for use with non-conducting dielectric liquids and some designs have been built for this purpose. This is briefly discussed in the appendix section 12.A.9. It has also been used with liquid metals and this is briefly discussed in the appendix section 12.A.10.

12.2 Operating Principle

We start with the simple induction, which occurs when a conductor moves through a magnetic field. Figure 12.1 shows a copper wire cutting the flux of a permanent magnet. The wire is moving in a direction perpendicular to its length and perpendicular to the magnetic field with velocity V, and the result is a voltage generated between its ends of value BlV where l is its length and B is the magnetic flux density.

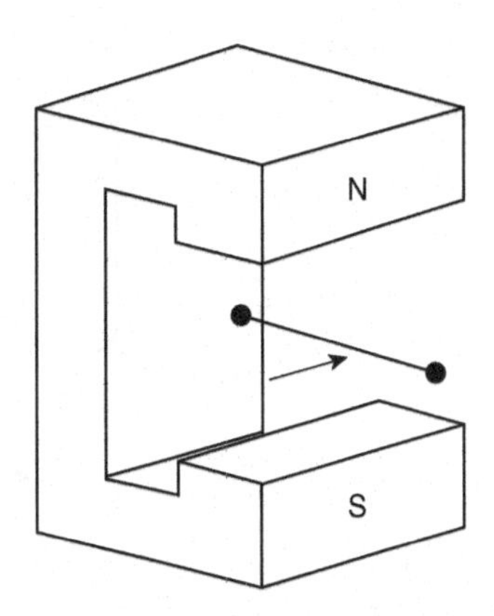

Figure 12.1. A wire moving through a magnetic field generates a potential difference between its ends.

Figure 12.2 shows a diagram with the essential features of an electromagnetic flowmeter. The liquid flows in a circular cross-sectional tube. A magnetic field is created across the pipe, usually by coils excited by an alternating current. The tube itself must be made from nonmagnetic material so that the magnetic field can penetrate the tube. In this diagram, we may imagine filaments of liquid spanning the tube from one electrode to the other and moving through the magnetic field at different speeds generating voltages between their ends as was the case with the copper wire. To avoid these being shorted out, the tube is lined with an insulating material. The voltages in the liquid are measured between electrodes, which are set in the wall of the tube.

Referring now to Figure 12.3, we have depicted the case where more than one wire is moving through the magnetic field. Wire P is in a region of strong magnetic field B and has a velocity V. Wire Q is in a region where the field is about the same size, but it only has a velocity $V/2$. Wire R, although moving with velocity V, is in a region of weak magnetic field, say, $B/4$. Thus the magnetic induction in each wire of length l will be different:

$$\begin{aligned} \text{In P} \quad \Delta U_\text{P} &= BlV \\ \text{In Q} \quad \Delta U_\text{Q} &= BlV/2 \\ \text{In R} \quad \Delta U_\text{R} &= BlV/4 \end{aligned} \tag{12.1}$$

Now if the ends of the wires are connected, currents will flow because the potentials differ, and the result will be that ΔU_P will be reduced by the ohmic loss. Carrying this same argument back to Figure 12.2 results in a complicated picture of voltages and circulating currents. Despite this apparent complexity, the actual operating equation of the flowmeter is quite simple for a range of conditions. The voltage between the electrodes is given by

$$\Delta U_\text{EE} = BDV_\text{m} \tag{12.2}$$

where B is the magnetic flux density in tesla, D is the diameter of the tube in meters, and V_m is the mean velocity in the tube in meters per second.

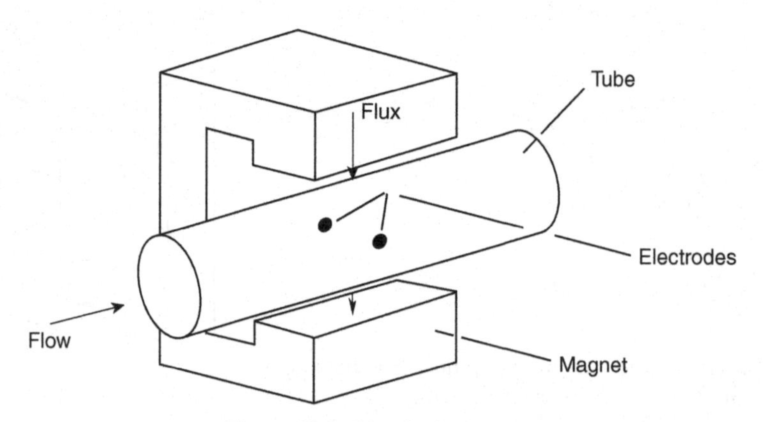

Figure 12.2. Simple flowmeter.

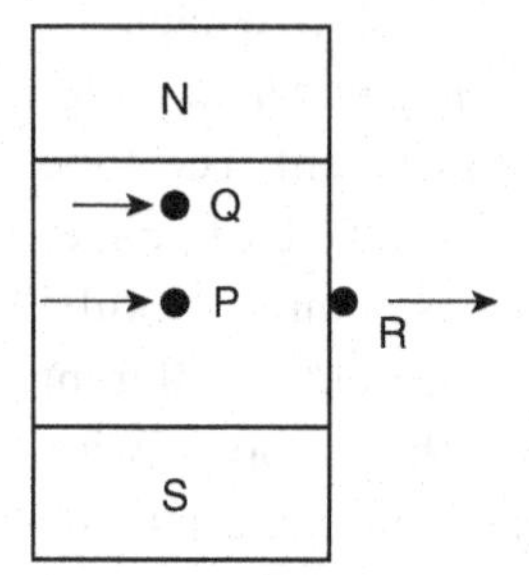

Figure 12.3. Three wires moving through a magnetic field of varying spatial strength at different speeds.

This is the basic equation for the flowmeter, and it can be shown to be valid only when

- the magnetic field is uniform and
- the velocity profile is axisymmetric.

In modern designs, these requirements are seldom satisfied, and so designs have been developed to give an output signal that is as little affected by the flow profile as possible, in compact designs with magnetic fields far from uniform.

12.3 Limitations of the Theory

Uniform Field – It can be shown that in theory the only field that is uniform is one from a magnet of infinite extent.

Axisymmetric Profile – All fully developed pipe flow profiles are axisymmetric. But it is not always convenient to allow the necessary upstream pipe length to ensure a fully developed profile.

Shercliff (1962) suggested a means of predicting the effect of distorted profiles using what he called a weight function (Appendix 12.A.3). He calculated the shape of this function for a uniform magnetic field meter [Figure 12.4(a)]. In its simplest form, this indicates how important the flow is in any part of the pipe cross-section

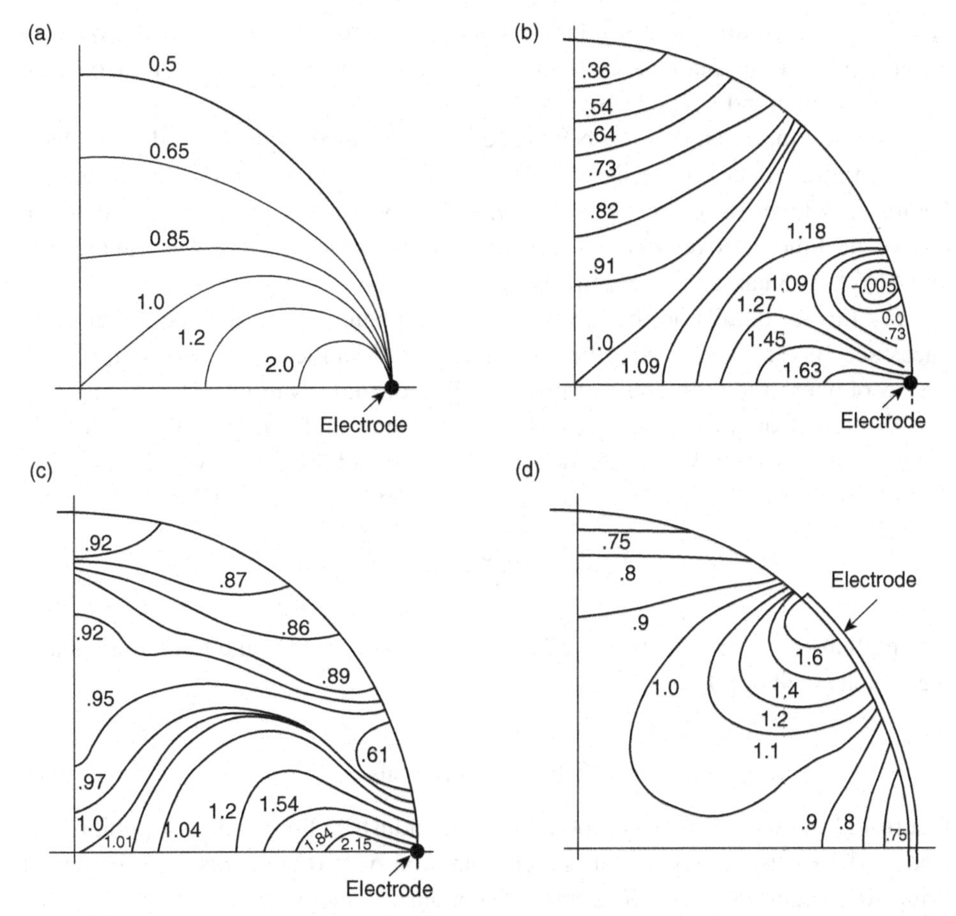

Figure 12.4. Weight functions: **(a)** For a uniform field point-electrode flowmeter (Shercliff 1962). **(b)** For a point-electrode flowmeter with rectangular-shaped coils and circular yoke (Al-Rabeh and Baker 1979). **(c)** For a point-electrode flowmeter with diamond-shaped coils and circular yoke (Al-Rabeh and Baker 1979). **(d)** For a large electrode flowmeter with rectangular-shaped coils and rectangular yoke (Al-Khazraji and Baker 1979) (reproduced from Baker 1985 with permission from R. Oldenbourg Verlag GmbH).

and how great its influence will be on the flow signal. The velocity at each point in the cross-section of the flowmeter is multiplied by the weight function. Thus the weighting function of 2 near the electrodes will cause the flow in that part of the pipe to have twice the influence on the signal. If all the flow went past that point, and the flow everywhere else was stationary, the signal would be twice as big as it should be.

This important concept has been extended by Bevir (1970) and Hemp (1975) and has proved a very useful method of assessing the performance of this and other types of flowmeter.

However, Figure 12.4 compares the distribution due to Shercliff with others. Figure 12.4(b) shows the distribution for flowmeters with rectangular-shaped coils, a common design. Figure 12.4(c) shows the distribution for flowmeters with diamond-shaped coils with their corners along the axis of the pipe and towards the electrodes. These are quite similar to some commercial designs, and the plots of weighting functions can be seen to give a much more uniform distribution, particularly in the diamond-shaped case.

Another approach that has been used is to introduce large electrodes that have an integrating effect on the signal and again improve the theoretical performance. Figure 12.4(d) shows an example of a large electrode meter, and it can be seen that the weight function over most of the cross-section varies only between about 0.75 and 1.6 (cf. Al-Khazraji and Hemp 1980).

For further details on the development of this meter and for additional references, see Baker (1982, 1983, 1985) and Hemp and Sanderson (1981). A thorough review of operating problems was provided by Cox and Wyatt (1984).

The effect of short magnetic fields causing non-uniformity of field results in the kind of shorting taking place between P and R in Figure 12.3, which was called end-shorting by Shercliff. If we define a value S, the flowmeter sensitivity, by

$$S = \frac{\text{Voltage generated}}{BDV_{\mathrm{m}}} \tag{12.3}$$

we find that $S = 1$ for Equation (12.2), but for a field length equal to the pipe diameter, it falls to about 0.8.

12.4 Design Details

Figure 12.5 shows the main components of an industrial flowmeter with point electrodes. The industrial flowmeter is a combination of two elements – the sensor or primary element and the transmitter or secondary element.

The sensor or primary element consists of the metering tube with insulating liner, flanged ends, coils to produce a magnetic field and electrodes. In addition, some means to produce a reference signal proportional to the magnetic field is usually provided. On AC-powered systems, this has, typically, been a current transformer, or search coil. On DC systems, it may be appropriate to measure the excitation current while constant.

The transmitter or secondary element, which is sometimes known as the converter,

a. amplifies and processes the flow signal;
b. eliminates spurious electromotive forces;
c. ensures that the transmitter is insensitive to electricity supply variation, radio interference etc.;
d. must have the required level of safety; and
e. communicates with the user's control computer in the most efficient way.

12.4.1 Sensor or Primary Element

The metering tube (Figure 12.6) will usually be nonmagnetic to allow field penetration. It may be available in a range of diameters from about 2 mm to about 3,000 mm

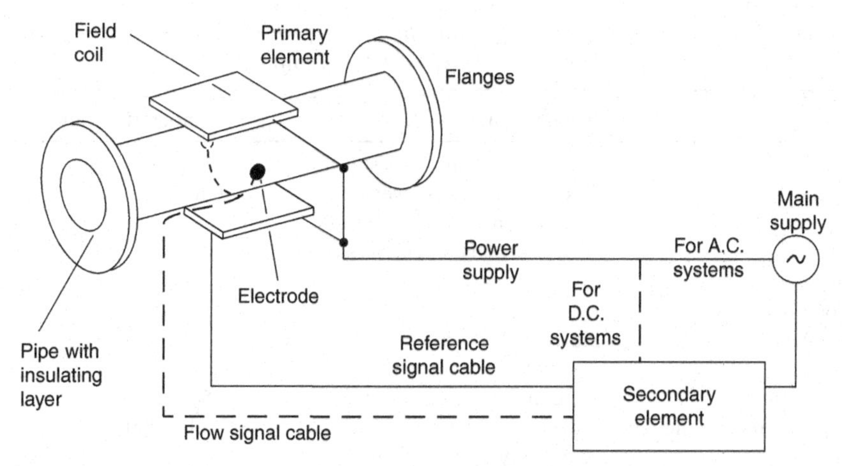

Figure 12.5. Main details of an industrial meter.

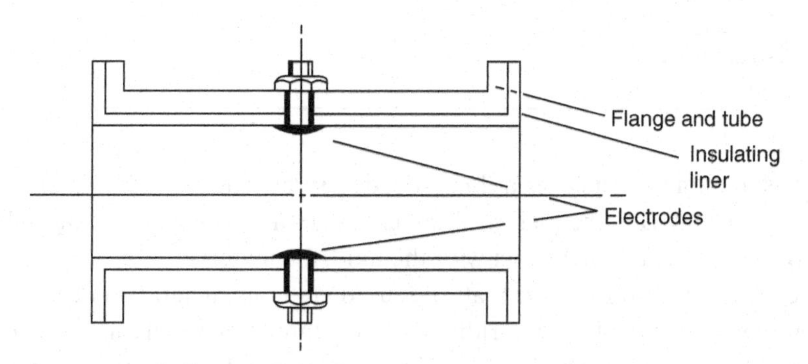

Figure 12.6. Metering tube.

(10 ft). Flow ranges may be from 0 to 28,500 m^3/h or more. The inside surface is insulated to prevent signal shorting. Electrodes are positioned at opposite ends of a diameter perpendicular to, and central in, the magnetic field. They are typically small (5–20 mm diameter) and are sometimes referred to as point or button electrodes. Large electrodes have been used in a few designs, and these may subtend about 90° and approach one diameter in length.

Typical materials for insulation are given in Table 12.1. The material is often moulded around the tube end, forming a composite with the flanges. Too high velocities [i.e. greater than 4 m/s (15 ft/s)] may lead to liner wear, which can be reduced with liner protection (Ginesi and Annarummo 1994). Grounding/earthing rings have been used in the past, where the pipes were lined or non-conducting, to provide a reference point for the measuring electrodes. In some designs, reference plates or electrodes, which are not necessarily earthed, have been provided. An earth link could, where the liquid is carrying an electric current, provide a track to earth and cause galvanic damage. In this case, the meter may be allowed to float electrically and would be made safe via an isolating transformer.

The electrodes normally have to make contact (cf. non-contacting electrodes, Section 12.6) with the fluid and therefore break through the tube lining (Figure 12.6).

Table 12.1. *Liner materials with an indication of approximate temperature limits that should be checked with the manufacturer*

Material	Application	Temperature Limits (°C)
Natural rubber	Wear and chemical resistance	–20 to 70
Neoprene	Good chemical and wear resistance in presence of oil and grease	0 to 100
Teflon	Abrasion and chemical resistance	0 to 90
Ebonite		
Fluorocarbon, Polyurethane, Elastomer	Slurry	
Polyurethane	Wear and impact resistance	–50 to 70
Polytetrafluoroethylene (PTFE)	Wear resistance chemically inert good for foodstuffs	–50 to 200
Ceramic 99.9% alumina (Al_2O_3) with cermet ($Pt–Al_2O_3$) electrodes moulded in before firing		

They are sometimes formed as a dome-headed screw and bolted in, pulling down onto the liner material, the electrical tag connection to the leads being bolted on finally. Because they are in contact with the fluid, the material of the electrode must be chosen with care. Some of the materials used are nonmagnetic stainless steel (for nonaggressive liquids), platinum-iridium, Monel, tantalum, titanium, zirconium (for aggressive liquids) and Hastelloy-C. Stainless steel may be recommended for slurries. Note also the ceramic combination of liner and electrode.

In paper pulp and possibly other applications, the impact of paper or other material on the electrodes can cause noise. According to one manufacturer, a porous ceramic coating over the electrode may reduce this problem.

Because the electrode must make contact with the liquid, various means have been tried for electrode cleaning including

- scraping (a scraper or brush may pass through the centre of the electrode and be turned to scrape the surface) (cf. Rose and Vass 1995),
- burn-off (the passing of a large enough current to remove surface deposits while the rest of the electronics is disconnected),
- ultrasonic cleaning (vibrating the electrode with ultrasound to cause local cavitation and hence cleaning),
- electrodes removable in service, and
- bullet-nosed electrodes.

The method should be chosen in relation to the likely nature of the deposit. In many cases, electrodes appear to be self-cleaning: the flow past them keeps deposits at bay. Coatings on meter internals can have more, the same or, most commonly, less conductivity than the bulk fluid. In modern DC systems, the input impedance may be

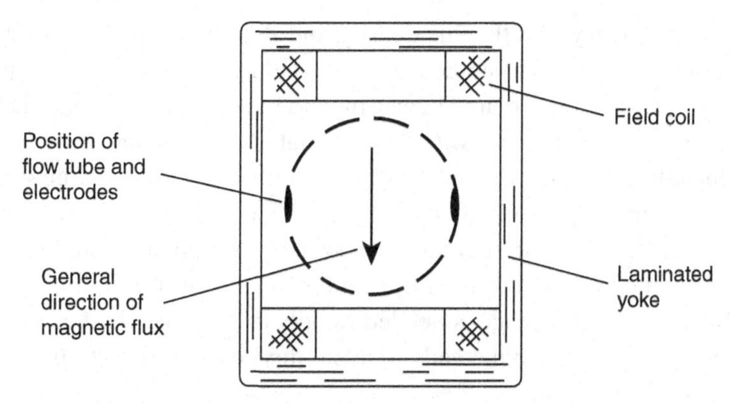

Figure 12.7. Magnetic field coils and yoke.

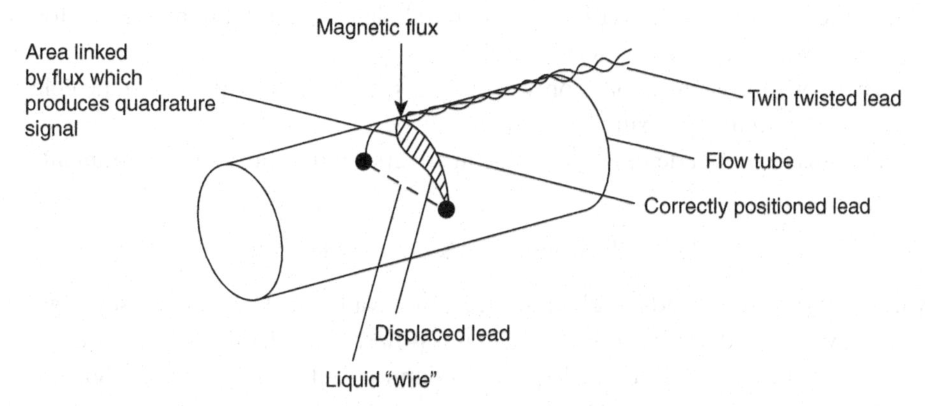

Figure 12.8. To show flux linking of signal leads.

high enough to make the effect of deposits negligible. However, high source impedance leads to thermal (Johnson) noise on the electrode signal, so although the high input impedance means there is no systematic error, the repeatability of the system will be lower.

The magnetic field is usually produced by a pair of coils and a laminated yoke (Figure 12.7). Typical power consumption used to be 10–100 W, but the lowest powers now can be as little as 0.5 W and, presumably, on average a lot less than this since long life on battery power is now possible.

As a result of using AC excitation, there was a danger of inducing a transformer-type signal, by the changing magnetic flux linking a loop of the electrode leads and fluid combination. Figure 12.8 indicates a poorly laid lead and the resulting area, which is linked by the changing flux. This area does not need to be large to result in a signal of a size comparable with the flow signal. It will be a quadrature signal (the phase will be 90° out of phase with the induced flow signal) given approximately by

$$\text{Quadrature voltage} \sim 2\pi fBA$$

where f is the frequency, B is the flux density, and A is the area of the resulting loop projected onto a surface perpendicular to the field direction. For example, if f is 50 Hz, B is 0.02 T, and A is 10^{-4} m^2 (1 cm^2), then the quadrature voltage will be about 0.6 mV. This should be compared with a flow signal for 0.1 m diameter tube and 5 m/s of 10 mV. In addition the phase angle is shifted from 90° by losses in the steels of the magnetic circuit making its elimination by mechanical design or electronics more difficult. The use of DC excitation has been seen as the way to overcome this problem, by sensing the flow signal when the field is briefly constant in time. This raises other problems. A large voltage is needed rapidly to ramp up the field, with its big induction coil, and then hold it steady to avoid flux-coupling effects, before making the flow measurement.

The primary element installation should not cause undue stresses in the tube and should ensure that the tube is always full. It is usual to mount the metering tube so that the electrode diameter is in a horizontal plane to avoid bubbles at the top of the tube open circuiting the electrodes.

The flow tube is most commonly of stainless steel, to allow the magnetic field to penetrate. Maximum pressure for sensors may be as much as 1,000 bar.

Designs may include options for use in adverse and hazardous environments.

12.4.2 Transmitter or Secondary Element

Various types of secondary element are now available. The longest-serving AC types have used 50 or 60 Hz. Because mains power is used, the magnetic field and, hence, also, the flow signal are high. The common but more recent designs use a low-frequency square wave of various patterns, which thereby allows quadrature signals to decay before the flow signal is sampled. This type is referred to here as the square wave excitation (or DC) system, but it comes under a variety of names depending on the manufacturer (cf. Brobeil et al. 1993). The description DC is used with caution because early meters did, indeed, attempt to operate on true DC without success. Field strengths may be lower in DC designs, but electrical noise from slurries and electromechanical effects will be the same as for AC. Most recent designs of DC meter may, therefore, have special high-power supply modules to address this.

Bonfig et al. (1975) described one of the first successful designs, which is referred to as keyed DC field. Hafner (1985) described another system, switched DC, which had features for noise reduction (e.g. both passive and active shielding), electrochemical effects, periodical zeroing of the amplifier, multiple sampling of the signal, increased field frequency (up to 125 Hz), digital filtering and the use of flow noise analogue circuitry. With low power [down to 1.5 W (as well as size and weight)], the design achieved intrinsic safety requirements and operated off a battery. Microprocessor control also provided self-testing, temperature compensation, interchangeable primaries and secondaries and communication. In addition, electrodes provided earthing and checking for empty tube. Herzog et al. (1993) dealt with switched DC design, which included a periodic integration of

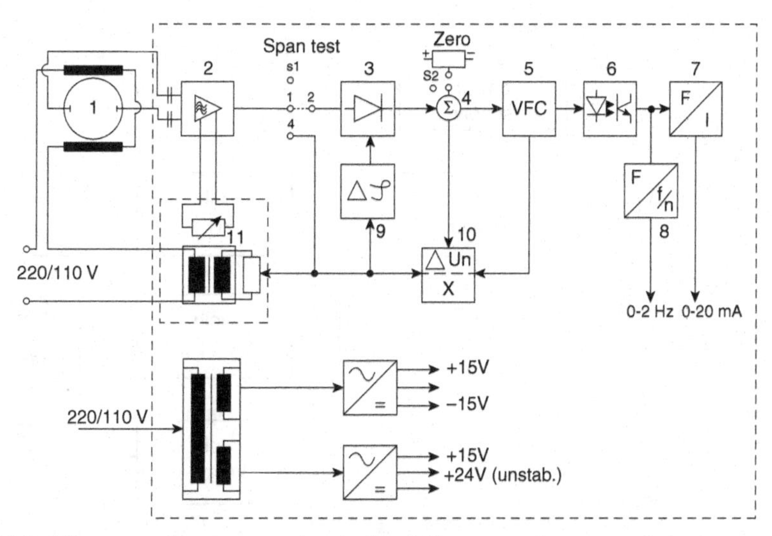

Figure 12.9. Diagram of converter circuit for AC system: 1, sensor; 2, input amplifier; 3, demodulator (phase-selective); 4, summation point; 5, voltage/frequency converter; 6, optocoupler (galvanic separation); 7, frequency/current converter; 8, pulse output (option); 9, reference; 10, division selection (coil current compensation); 11, current converter, reference source, measuring range module (reproduced with permission from Endress+Hauser). (It appears that this design is no longer available.)

the electronic reference point, and discussed a third electrode circuit for partially filled tubes.

The output is usually 0–10 or 4–20 mA, and two or three range switches are provided to allow a full-scale output reading to be achieved with flows from 1 to 10 m/s. However, this is being superseded by the power of microprocessor technology, which allows a move towards smart/intelligent instrumentation with autoranging, digital transmission and a much greater range of possibilities.

Figure 12.9 gives a block diagram of a typical AC circuit. The demodulator used the reference signal to remove quadrature voltages, and the circuit obtained the ratio of flow signal to reference signal. (This type of circuit has been largely superseded by the DC type.)

The block circuit of Figure 12.10(a) indicates the approach used in the DC systems. The sampling at times τ_n, τ_{n+1} and τ_{n+2} is illustrated in Figure 12.10(b), which exaggerates the drift of the baseline of the square wave signal due to electrochemical and other effects, and the reason for three samples is then apparent. Manufacturers have developed various square wave patterns often incorporating a dwell period with zero applied magnetic field, for instance.

The manufacturer's limits on length and specification of signal cable should be carefully followed. The cable is usually twin-twisted-shielded for AC excitation. However, with low-power operation possibly using a battery, two-wire operation may be offered.

Zero drift may be found in some instruments, but is usually small and is probably due to incomplete suppression of unwanted (particularly quadrature) voltages.

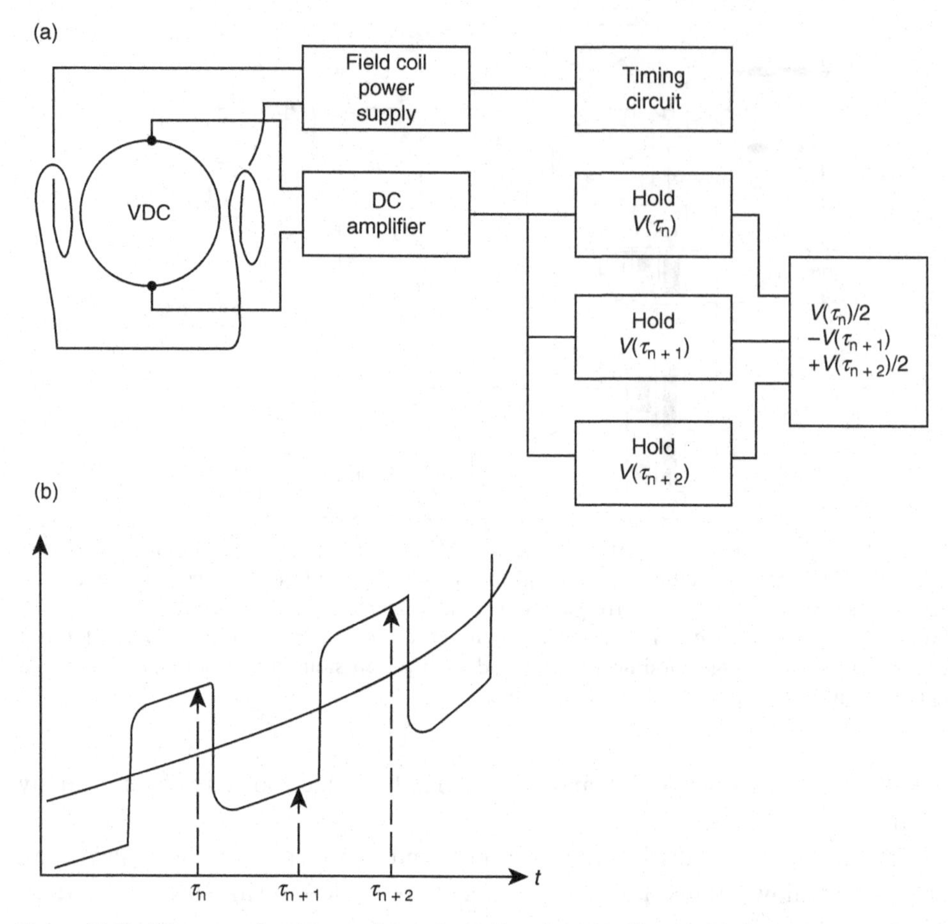

Figure 12.10. Diagram of converter circuit for DC system: **(a)** Circuit; **(b)** signal from flow tube.

The DC systems usually claim to eliminate this, although with a low flow cut-off, this may be hard to confirm. A low flow cut-off is usually set at 1% of upper range limit (Ginesi and Annarummo 1994) or possibly lower.

The overall uncertainty of the converter may be claimed to be about 0.2% for quite wide tolerance of mains voltage, quadrature signal, temperature fluctuation etc. Very low flows can be measured, but with decreasing precision.

Commercially available transmitters will increasingly offer:

- response – of order 0.1 s,
- turn down – up to 1,000:1,
- flow ranges – part of 0.005–113,000 m^3/h,
- pulse rates – from 0.01 to 10 l/pulse.

Manufacturers list some of the following features:

- sensor power supply and transmission carried on intrinsically safe two-wire cable for both power and signal;
- digital signal transmission through AC modulation of analogue signal;

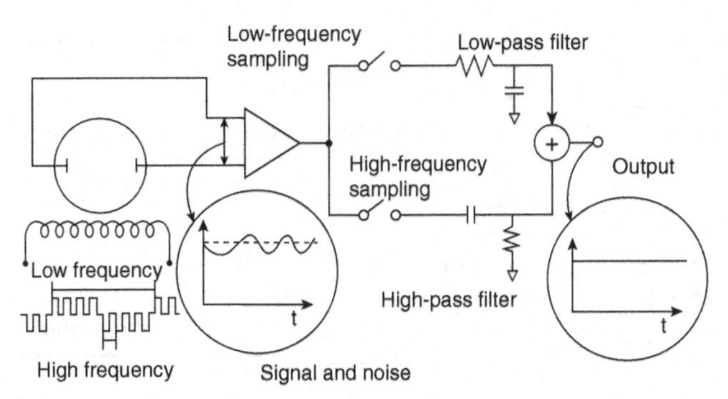

Figure 12.11. Dual-frequency operation (reproduced with permission of Yokogawa Europe B.V.).

- Interelement Protection, IP65 for transmitter;
- dual frequency (Figure 12.11) giving benefits of both high and low frequencies: the frequencies are processed through separate channels before being brought together to give both low-flow stability and low noise (cf. Matsunaga et al. 1988);
- interference immune transmission;
- self-checking or multiply checked data;
- empty pipe detector and alarm using an electrode to sense tube empty conditions (Ginesi and Annarummo 1994);
- grounding electrode;
- check for fouling of main electrodes;
- flow in both directions measurable with suitable electronics;
- auto ranging.

Application-specific integrated circuits (ASICs) offer advantages such as automatic system monitoring with diagnostics for reverse flow and other errors and alarms, dual ranging and serial interface communication (Vass 1996).

12.5 Calibration and Operation

Because of manufacturing variation between meters, they require calibration, and this is usually done by the manufacturer. One manufacturer, for instance, offered a 13-point calibration for a master meter. This is known as *wet calibration*. The term *dry calibration* is used in connection with electromagnetic flowmeters to imply calibration by measurement of the magnetic field and deduction of flow signal from this. The relationship between the field at a specific point and the overall performance of the meter is not as straightforward as Equation (12.2) might imply, and any dry calibration should be treated with caution.

The operation of the meter should not, generally, be affected by variation of liquid conductivity, provided that the conductivity is uniform over the region of the flowmeter. This also presupposes that the conductivity is sufficient to ensure a

Table 12.2. *Output resistances of meter tube based on electrode diameter of 0.01 m*

	Liquid Conductivity		Resistance
	(S/m)	(μS/cm)	(ohms)
Best electrolyte	~10^2	~10^6	1
Sea water	~4	~4×10^4	25
Tap water	~10^{-2}	~10^2	10,000
Pure water	4×10^{-6}	4×10^{-2}	25,000,000

primary device output resistance of two or more orders of magnitude less than the input resistance of the secondary device. However, severe variations in conductivity may cause zero errors in AC-type magnetic flowmeters. Even though some suggest that DC pulse types should be unaffected by such changes above a minimum threshold (Ginesi and Annarummo 1994), one manufacturer appeared to take an opposite view, recommending AC type for two-phase, slurries, low conductivity flows or inhomogeneous flows with rapidly varying conductivity. However, continuous development of the DC type may ensure that they are equally suitable.

The output resistance of the primary device is approximately given by

$$R \approx \frac{1}{d\sigma} \text{ohms} \tag{12.4}$$

where d is the electrode diameter, and σ is the conductivity.

From Equation (12.4), we can obtain typical resistances of a meter with electrodes of diameter 0.01 m, as shown in Table 12.2.

A secondary device with a typical input resistance of 20×10^6 ohms could cope with the first three liquid conductivities in Table 12.2, but not with the last one. Manufacturers may limit conductivity minima for certain sizes of meter. For instance, for 25–100 mm, conductivity may be acceptable only down to 20 μS/cm. However, at least one manufacturer offers a meter capable of use with down to 0.05 μS/cm conductivity.

Entrained gases cause errors due to breaking of electrical continuity, non-uniformity of conductivity and uncertainty as to what is being measured. The flowmeter should be positioned where these are negligible.

12.6 Industrial and Other Designs

It is not my intention to detail all the features of industrial designs nor the common operational problems (see, for instance, Baker 1982) but rather to give the main types that reflect the theoretical developments. Figure 12.12 shows a typical industrial design. After early designs, which were long in an attempt to create a uniform field, all are now more compact and Figures 12.4(b,c) provide a means of performance comparison with Figure 12.4(a) for the uniform field design.

Figure 12.12. Industrial design (reproduced with permission from Endress+Hauser): showing main components.

Internal coils recessed into an enlarged cross-section of the flow tube have been used in the past to avoid the need for nonmagnetic pipes. If used today, they are rare.

Manufacturers have developed electrodes with shapes and sizes optimised for individual applications. Large electrodes have been tried (Figure 12.13) and should result in a more uniform weight function [Figure 12.4(d)]. Yoshida, Amata and Frugawa (1993) suggested using short axial length arc-shaped electrodes subtending about 45° for partially filled pipes.

However, noncontacting electrodes offer additional advantages by permitting a continuous lining to the flow tube and avoiding crevices around the electrodes. They may also be less sensitive to fouling (cf. Rose and Vass 1995). Noncontact designs may allow operation with fluids of conductivity down to 0.05 μS/cm compared with normal minimum values of 5 μS/cm. Other design advantages indicated earlier may not have been fully exploited in commercially available instruments (Figure 12.14; cf. Hussain and Baker 1985; Brockhaus et al. 1996).

An electromagnetic meter has been proposed (Rose and Vass 1995) with three pairs of electrodes – one on the horizontal diameter and the other two pairs at the end of horizontal chords below the diameter. This arrangement is claimed to allow the meter to run partially empty. The field coils are excited in series for a flow signal. However, when used in opposition, it is possible to obtain the level of liquid in the pipe. With this arrangement, when the pipe is full, the mean voltage on the diametral electrodes is zero, but imbalance in the signal increases as the pipe empties. Installation requirements are more stringent with 10D straight pipe upstream and

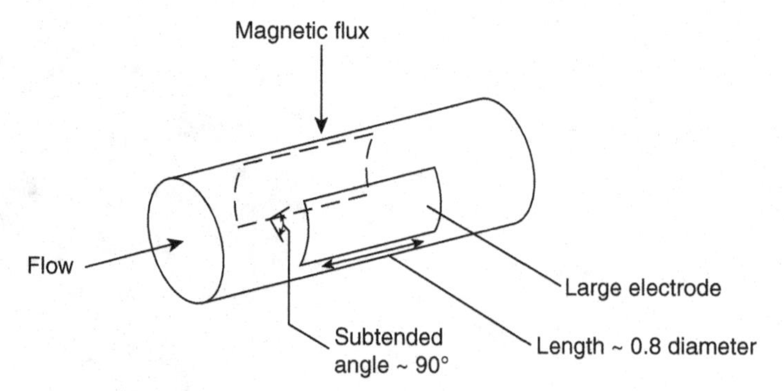

Figure 12.13. Large electrode flowmeter.

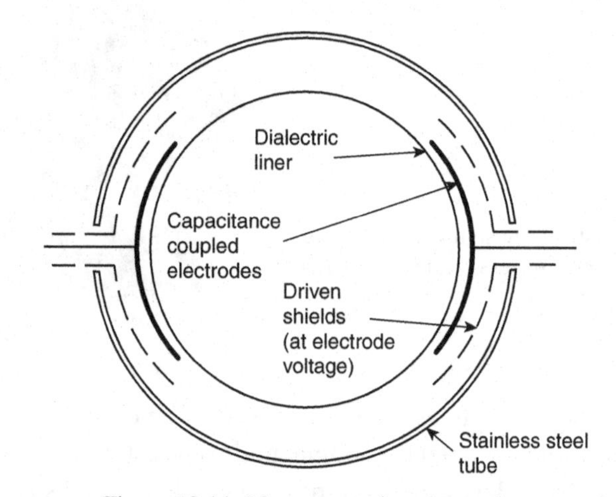

Figure 12.14. Non-contacting electrodes.

5D downstream. The pipeline should also have only a small slope to ensure tranquil flow. Minimum conductivity for the commercial version is claimed as 50 μS/cm, and uncertainty is claimed as 1.5% span down to 10% fill level.

Doney (1999a, 1999b) of ABB also discussed, from a commercial viewpoint, use of the meter in partially filled pipes, with three pairs of electrodes, and able to sense the level of fill and hence the volume flow to within about 5%. A high-frequency signal is injected through a bottom pair of electrodes and sensed at three pairs of measurement electrodes which are also used for the measurement of flow rate. The amplitude gives a measure of the fill level. A turndown of 1,000:1 is mentioned in the article (cf. Yao, Wang and Shi 2011).

Incontri (2005) reported on advances in electromagnetic flow measurement technology. Downward pressures on instrument cost have resulted in the lower price range of many manufacturers. Reliability appears to be continually improving, and instruments are being offered with additional sensing features, such as conductivity change and self-diagnostics (see Section 12.A.8). Bates and Turner (2003) undertook fluid flow studies associated with a commercial flowmeter with a reduced internal diameter "throat". They identified the required spacing between meters when mounted in line. This in turn allowed the development of a calibration facility with multiple meters in series (see also Thomas et al. 2001).

Where meters with such a throat are used as a pair, for instance as a transfer standard, it may be important for the pressure in the line to be sufficient to prevent cavitation from one throat affecting the next meter downstream.

Power ratings will cover a wide range (from 0.5 W for new designs up to 300 W for a noncontact flowmeter).

Uncertainty claims for most designs are of order 0.5% of flow rate, but manufacturers of noncontacting designs may initially be more cautious.

Hofman (1993) proposed the use of ceramic materials to increase stability particularly with temperature change. PTFE and the electrode gaskets can result in changes due to use, temperature change and pressure over periods of, say, two years. With ceramic, by incorporating sintered electrodes, creep is reduced from order 1% to order 0.1% or better. The problem with ceramic is temperature shock, but new ceramics may allow rapid changes of up to 100°C. Hofman also suggested a conical inlet to improve profile at the cost of a small increase in pressure loss. With special materials to ensure magnetic field stability, he claimed, on a 10 mm tube, ±0.1% on a 10:1 turndown (4.2 m/s upper range value with a possible maximum of 12 m/s) for 20–85°C compared with ±0.4% for PTFE over the same range.

Polo, Pallas-Areny and Martin-Vide (2001, 2002) described analogue signal processing in an AC electromagnetic flowmeter by a novel circuit design to amplify and demodulate the voltage picked up by the electrodes and to minimise interference. The amplifier included a fully differential front end.

Industrial developments were reported in 2003 (Anon 2003) relating to the two-wire electromagnetic flowmeter which was intrinsically safe.

12.7 Installation Constraints – Environmental

Some manufacturers indicate that they prefer vertical installation, with upward flow. This ensures that entrained gases are carried out and do not affect electrode contact.

The electrode to detect if the pipe is not running full should then be positioned at the top of the meter as installed.

12.7.1 Surrounding Pipe

Wafer construction flowmeters may be affected by surrounding pipe. Bevir (1972) calculated the effect of length of magnetic field and liner on sensitivity. Using a magnetic field, which represented typical industrial practice, he obtained values for uniform flow profile, of which a sample are given in Table 12.3.

Non-conducting pipes abutting the flowmeter will be equivalent to an infinite liner length. Thus a flowmeter with field length of 1D and liner length of 1D will lose 9% in sensitivity when inserted in conducting pipework.

Table 12.3. *Results obtained from Bevir (1972) for the effect on sensitivity of field and liner length*

Field Length	Liner Length				
	0.2D	0.5D	1.0D	2.0D	∞D
0.2D	0.211	0.335	0.380	0.388	0.389
0.5D	0.211	0.493	0.638	0.661	0.662
1.0D	0.211	0.497	0.790	0.865	0.867
2.0D	0.211	0.497	0.790	0.966	0.979
∞D	0.211	0.497	0.790	0.966	1.000

Baker (1973) calculated sensitivity reductions of about 5% for liner length reduction from 2D to 1D for a rectangular coil circular yoke flowmeter of 0.8D coil length.

Although the values in Table 12.3 are not directly related to the effect of steel pipework on the magnetic field distribution, it is reasonable to assume that the effect of such pipework will be less than lengthening the field. Thus for a 1D liner length, we note that increasing field length from 1D has no tabulated effect. If, however, the flowmeter is connected in non-conducting pipe, then field length change from 1D to 2D will cause a sensitivity increase of 13%. Assuming, for example, that the edge field increase due to connection into steel pipe will be of order 5% of the effect of increasing the length of the field, a 13% change (in the 5% component of the field) will be less than 1% on the sensitivity.

Christensen and Willatzen (2010) investigated the effect of magnetic pipes connected to electromagnetic flowmeters. Their computations appear to suggest changes of less than 1% decrease in signal when the meter is connected into magnetic pipework.

Manufacturers may require earthing straps, particularly in industrial environments, to connect up- and downstream metal pipes, to prevent significant current flows through the fluid interfering with the small flow signals.

12.7.2 Temperature and Pressure

Fluid density changes must be allowed for in calculating mass flow, but here we are concerned with accuracy as a volumetric flowmeter. Pressure will cause hoop stresses in the pipe resulting in strain, but the effect is likely to be negligible (Baker 1985).

Temperature increase will cause expansion of the flow tube and of the windings. The effect of the latter will depend on the method of compensation for field strength variation. If a search coil is used, this may also be affected by temperature change. Using a coefficient of expansion α of 10^{-5}/°C, Baker (1985) obtained a maximum error due to temperature of about −0.3%/100°C. However the effect of temperature and pressure on the liner, which may well dominate, has not been considered.

12.8 Installation Constraints – Flow Profile Caused by Upstream Pipework

12.8.1 Introduction

In this chapter, four examples of rectilinear weight function distributions have been given [a fifth is given in Figure 12.A.1(b)]. From these, it should be possible to deduce the performance of different types of flowmeters for various installations.

Rather than use actual fittings, some experiments have used localised jets and orifice plates to distort the flow profile in the meter. Shercliff (1955) used a localised wall jet to demonstrate the validity of the weight function distribution in Figure 12.4(a). Cox and Wyatt (1984) used this method to show the insensitivity of a contactless design to severe profile changes.

It is necessary to remember that in real installations the designer seeks a meter that is affected as little as possible by changes from the calibration profile. These changes, in most cases, represent small perturbations of the datum profile. The electromagnetic flowmeter is insensitive (Baker 1973) to a number of profiles that have reverse symmetry about diametral planes. Thus if such a profile is added to the datum profile, there should be no change of signal.

12.8.2 Theoretical Comparison of Meter Performance Due to Upstream Flow Distortion

Al-Khazraji et al. (1978) used an upstream orifice to distort the flow profile (Table 12.4). The diameter of the orifice was half that of the pipe, and the centre of the eccentric orifice was at half the pipe radius.

Where experiments were done, they appeared to confirm these trends. These computed values suggest that a disturbance upstream of the flowmeter will have minimum effect if its plane of symmetry is perpendicular to the electrode plane.

Approximate calculations based on eccentric orifice profile and Shercliff's (1962) weight function distribution [Figure 12.4(a)] suggest that errors for a uniform field flowmeter would be ±1% depending on the jet orientation.

Halttunen (1990) used data on flow profiles and combined them with magnetic field data obtained using a Hall probe and a computerised traversing and measurement system. He also gave experimental data from installation tests. The tests appear to have been confined to a single elbow test for one configuration of electromagnetic

Table 12.4. *Calculated signal change (%) from uniform profile (Al-Khazraji et al. 1978)*

Orientation of Eccentric Orifice Peak Velocity	Small Electrode		
	Rectangular Coil [Figure 12.4(b)]	Diamond Coil [Figure 12.4(c)]	Large Electrode [Figure 12.4(d)]
At electrodes	3.3	1.5	0.03
45° displaced	2.4	0.9	−0.03
90° displaced	1.4	0.0	−0.02

Table 12.5. *Conclusions that can be drawn from the plots Halttunen (1990) obtained on signal change due to upstream disturbance*

	2D(%)	5D(%)	10D(%)	15D(%)	20D(%)
Single elbow	−1.5 to 0.5	±0.5	±0.5	small	small
Double elbow 90°	±2.5	±1	±1	±1	±1

Note: at about 40D error becomes small.

meter. The agreement is quite good and suggests that this type of combined approach to installation testing may be worth pursuing. The conclusions that can be drawn from the plots are given in Table 12.5. The flowmeter's (electrodes and magnetic field) orientation relative to upstream orientation of fittings may cause an additional ±0.5% at 5D or less spacing. I was pleased to see a meter in these tests named The Shercliff after someone whose contribution to the development of this meter has been so important!

Luntta and Halttunen (1989) confirmed the effect of a bend but noted that four different meters (400 mm ID) gave very different results when downstream of an eccentric orifice plate and of a reducer.

12.8.3 Experimental Comparison of Meter Performance Due to Upstream Flow Distortion

Scott (1975b) was the first to publish data on electromagnetic flowmeter operation downstream of pipe fittings. As Mr D. Halmi commented in the discussion of this paper, it is not possible to generalise from such data. This is mainly because of the variation in meter design.

Much of the useful data is inaccessible to the public. For instance, de Jong (1978) has published a few results from his tests which omit full flowmeter design detail. His most useful conclusions appear to be that, for a gate valve 5D upstream and at least 50% open, errors were within 0.5% of specification.

Tsuchida, Terashima and Machiyama (1982) gave the requirements in Table 12.6 for minimum upstream straight pipe to ensure uncertainty within ±0.5% of full scale.

Deacon (1983), after tests with a gate valve upstream (at least 50% open), one swept bend (radius/D = 1.5), two swept bends in perpendicular planes and a reducer, was able to conclude that the worst error with 5D spacing was less than 2% from calibration (Table 12.7). For gate valve 50% open and for one bend, the orientation of the flowmeter electrodes relative to the pipe fittings did not affect the magnitude of the error.

These data indicate considerable variation for a particular installation. One bend with 5D spacing caused about 0.3% error for a large electrode flowmeter and up to 2.0% error for a small electrode flowmeter. The small electrode flowmeters exhibited errors that could be, in some cases, a factor of two different. For one small electrode flowmeter, opening the gate valve from 50 to 75% *increased* errors by a factor of more than four when the orientation caused a jet past the electrodes.

Bates (1999; cf. 2000a) dealt with small misalignment of the electromagnetic flowmeters, and suggested that with a smaller upstream pipe of 45 mm compared

Table 12.6. *Tsuchida et al.'s (1982) values for minimum upstream straight pipe to ensure uncertainty within ±0.5% of full scale*

Pipe Fitting	Minimum Upstream Straight Pipe
Reducer	3D (direct connection)
Expander Ball valve	5D
Single elbow Two elbows in perpendicular planes Gate valve	10D
Butterfly valve	15D

Table 12.7. *Deacon's (1983) values of maximum error over all tests for upstream components*

	Spacing	Max. Error (%)
Reducer	2.5D	0.8
Gate valve (at least 50% open)	5D	1.2
One bend Two bends in perpendicular planes	5D	2

with 50 mm for the flowmeter bore, and with misalignment of 3 mm the errors were of order 1%. He showed that when the centre line of the inlet pipe is misaligned relative to the larger bore of the flowmeter, the orientation of the smaller pipe relative to the electrode position had an effect on the error size, as would be expected from weight-function theory. England (2014)* confirmed from Sentec's experience that misalignment was one of the most significant errors with smaller flanged meters for which there is no means to centre the flange. Bates (2000b) has also suggested that the error for a flowmeter downstream of a mitre-bend/reducer combination, when the electrode plane is about 7.7D downstream of the reducer outlet, can range upward from 1% for higher flow rates to 2–4% for lower flow rates.

12.8.4 Conclusions on Installation Requirements

Tentative conclusions on the likely influence of installation profiles follow.

a. The error due to a reducer installed next to the flowmeter is likely to be less than 1%.
b. An error of up to 2% should be assumed for other fittings separated from the meter with 5D of straight pipe.
c. An error of 1% may occur for other fittings separated from the meter with 10D of straight pipe.

* Personal communication from Dr Mark England.

d. The orientation of the flowmeter at 5D spacing or greater does not simply correlate with the size of the error.
e. At least 3D spacing downstream should be allowed.
f. Misalignment and other upstream combinations may increase error.

Manufacturers may be less conservative, but spacing such as 3D to 5D upstream, with 3D downstream, should be treated with caution. Despite data such as de Jong's (1978), valves should not be installed close upstream of flowmeters and, even if fully open, are likely to have an effect of ±0.5% even if at least 15D upstream. In all cases, it should be recognised that, because of the design differences between magnetic flowmeters, it is not possible to give absolute guidelines that are equally valid for all designs.

12.9 Installation Constraints – Fluid Effects

12.9.1 Slurries

The electromagnetic flowmeter has found a special niche (almost a universal solution) for slurry flow measurement and also for some complex non-Newtonian fluids. A slurry will affect the flowmeter because of flow distortion, conductivity variation and magnetic permeability. The work quoted in Section 12.9.3 enables us to make an informed guess as to the errors, if we have some idea of the size of the conductivity variation due to the presence of slurry. In addition, the distribution in Figure 12.4(b) suggests that a flattened profile, due to a non-Newtonian fluid, say, could cause negative errors. The likely performance on an untried fluid should, therefore, be undertaken with caution.

In a vertical line, which is a common installation position in slurry service, it will also be necessary to allow for any slip that may occur between the conducting fluid and the non-conducting solid. For instance, a settling slurry in a vertical pipe with zero net flow will record the movement upward of water displaced by the settlement downward of the solid matter (as a colleague noted when working for a flowmeter manufacturer).

A magnetic slurry (Baker and Tarabad 1978) will result in a change of size and distribution of magnetic field. If the flow signal is compensated for change in field current size, then the increase in fluid permeability will cause an uncompensated increase in flux density and flowmeter signal.

Where the flow signal compensation is achieved by a search coil monitoring magnetic flux density, then careful design should ensure that magnetic fluids have a negligible effect on flow signal.

A magnetic slurry may also be deposited in the flowmeter magnetic field. The cure for this may be to run the flowmeter with a lower field strength.

Eren (1995) provided an example of the use of frequency analysis of the output signal generated by DC-type electromagnetic flowmeters. The frequency and amplitude components are dependent on slurry characteristics of the flow. Eren implied that these signals can be related to the density and other characteristics of the slurries.

12.9.2 Change of Fluid

Although conductivity of the fluid should not affect the flowmeter signal, there has been some suggestion that electrochemical effects of the fluid may do so, and that this may affect the performance of the modern square wave field excitation flowmeters. However, as noted earlier, Ginesi and Annarummo (1994) commented on the effect of severe variations in conductivity in AC-type magnetic flowmeters but suggested that DC types should be unaffected above a minimum threshold. Few data are available, but Baker et al. (1985) compared the performance of an AC-excited and a DC-field flowmeter for a change in fluid from pH = 7 to pH = 4 and found no significant effects.

12.9.3 Non-Uniform Conductivity

Non-uniformity of conductivity in the flowmeter can cause changes in flowmeter signal due to the changes in the size of the shorting currents which result. Thus a higher conductivity layer near the flowmeter wall will cause higher shorting currents and a reduced flow signal. For certain conditions of turbulent flow, field shape and conductivity profile, a signal change of up to 3% may occur (Baker 1970a), and it will be higher for laminar profiles. This is probably most significant in multiphase flow applications with a second non-conducting phase (cf. Ginesi and Annarummo 1994) because the continuous conducting component will generate the same signal as for a uniform fluid provided the fluid is homogeneous.

12.10 Multiphase Flow

Baker and Deacon (1983) have reported experiments on the behaviour of an electromagnetic flowmeter in a vertically upward air-water flow and have shown that volumetric flow of the mixture was measured with errors of less than –1% up to void fractions of 8%. For void fractions greater than this, the negative error increased.

Bernier and Brennen (1983) have analysed the flowmeter's performance with uniform field and obtained in addition to the following equation for a dispersed second phase without slip:

$$\Delta U_{EE} = BDV_m = \frac{4B}{\pi D}\frac{q_L}{(1-\alpha)} \tag{12.5}$$

where q_L is the liquid (or conducting phase) volumetric flow rate, and α is the void fraction, the same equation for annular flow. They also showed that, for small values of α, the expression is the same for "cylindrical voids" parallel to the flowmeter axis. Their data for bubbly and churn flows fall within ±2% of the predictions up to void fractions of 18%. Murakami, Maruo and Yoshiki (1990) developed an electromagnetic flowmeter with two pairs of electrodes for studying gas-liquid two-phase flow.

Cha, Ahn and Kim (2001) also obtained the signals and noise from a flowmeter to verify Equation (12.5), and noted that the signal amplitude could provide evidence of the passage of slugs of gas. They made some observations regarding the signal for various flow conditions and simulated some using acrylic rods down the flowmeter tube. Cha et al. (2003a, 2003b) also undertook experiments with a liquid metal two-phase flow of sodium and nitrogen and concluded that the meter might be able to identify flow regimes.

Further work on two-phase flows was reported by Ahn, Do Oh and Kim (2003), who used the less common current-sensing electromagnetic flowmeter for two-phase flow which they noted was sensitive to liquid conductivity. They studied annular flow (cf. O'Sullivan and Wyatt 1983; Wyatt 1986). They used a finite difference method to obtain the three-dimensional virtual current distribution.

The use of the electromagnetic flowmeter for slurry flows was mentioned earlier in the context of its early development. Peters and Schook (1981) have developed the idea of using the signal fluctuation as a measure of slurry concentration. It is also possible that special design of the magnetic field could be used to obtain concentration information.

Krafft, Hemp and Sanderson (1996) looked at the passing of gas bubbles through the flowmeter in terms of the electric dipoles that result.

12.11 Accuracy Under Normal Operation

Manufacturers' claims for uncertainty (often referred to as accuracy) may lie somewhere near the following:

Uncertainty	Flow range as % of FSD
±0.3% rate	100–50%
±0.5% rate	50–15%
±1.0% rate	15–5%
±1.5% rate	5–2.5%

In other cases, the value may be given as a percentage of rate or a velocity (e.g. ±0.015 m/s), whichever is greater.

Repeatability is likely to be of order ±0.2% rate. The analogue output may be less precise than this (e.g. ±0.1% FSD).

One manufacturer, at least, has offered meters in the range 2.5–6 mm with uncertainty of order 0.8% rate at the upper end of the flow range.

There has been some discussion in the technical press relating to the accuracy of this instrument over large turndown ranges. Research may, therefore, be needed to consider the fundamental limits to accuracy related to

a. flow steadiness and turbulence, and the sensitivity of the meter to profile variation over the Reynolds number range;
b. the field stability and the usually neglected size of currents flowing in the liquid;
c. the effect of charge distribution, current size and conductivity variation; and
d. the dimensional stability of the flowmeter's components, and compensating systems with variation in ambient conditions.

12.12 New Industrial Developments

It is interesting to note that recent developments have made use of earlier ideas. Li, Yao and Li (2003) discussed the theory of the rectangular electromagnetic flowmeter and gave an example of its application in micro-flow measurement. Rectangular section meters were, of course, discussed by Shercliff in his book (1962). One very elegant device should be mentioned, which could well offer other applications than that for which it was developed. Gray and Sanderson (1970) developed a differential flowmeter using two adjacent rectangular flow channels so that the same magnetic field traversed both and the electrodes were so connected as to give the difference of the two signals due to the two flows and hence the differential flow. They claimed to be able to measure differences of 1 to 10 cm^3/min in a flow of 500 cm^3/min to ±10% and possibly even to measure down to 0.5 cm^3/min.

At least two manufacturers have developed rectangular tube meters which can allow a small gap between magnetic pole pieces, possibly using larger electrodes, and increased flow velocity through the meter. Better coil design with lower power and a meter less affected by upstream disturbance should result (see Shercliff 1962).

Other innovations for one of the manufacturers (KROHNE) are: a virtual reference to improve isolation of components within the meter; non-contacting electrodes using capacitive pick-up (cf. Figure 12.14; Hussain and Baker 1985) sinter-fused onto the outside of the ceramic flow tube may avoid compatibility constraints, allow operation in depositing liquids, be less profile sensitive and can operate in low liquid conductivity; partially filled pipelines (cf. Rose and Vass 1995); diagnostic features including reversal of segments of magnetic field polarity, to check for poor installation, and also to check for entrained gas, low conductivity, electrode corrosion, deposits on electrodes, electrode short circuit, liner damage, partial filling and external magnetic fields (cf. Section 12.A.8).

Another interesting development by Sentec has resulted in an innovative design for residential water metering with a good performance, Figure 12.15. It has a lofted rectangular section (throat), Figure 12.15(c), which accelerates the linear velocity (and hence increases the signal level) by two to three times, and has the benefits of rectangular section meters mentioned earlier, including tolerance to

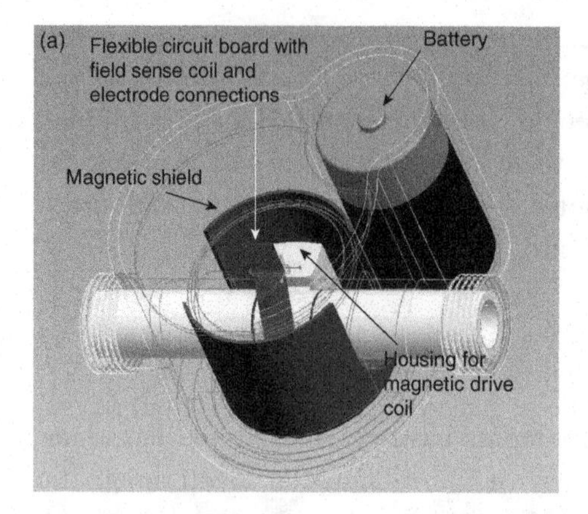

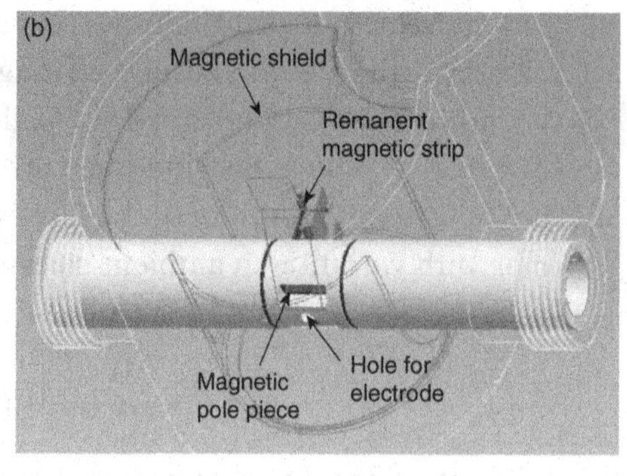

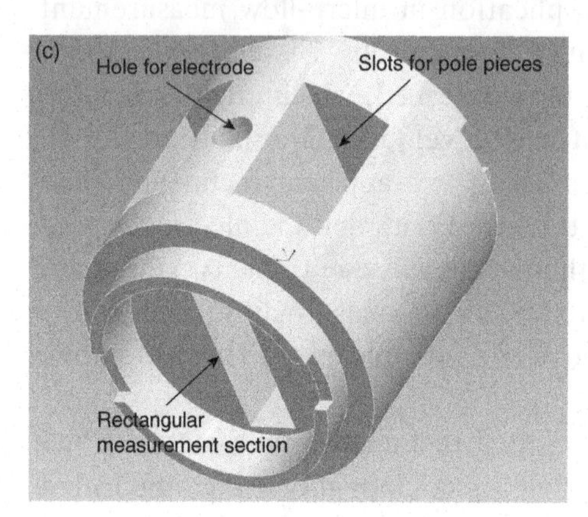

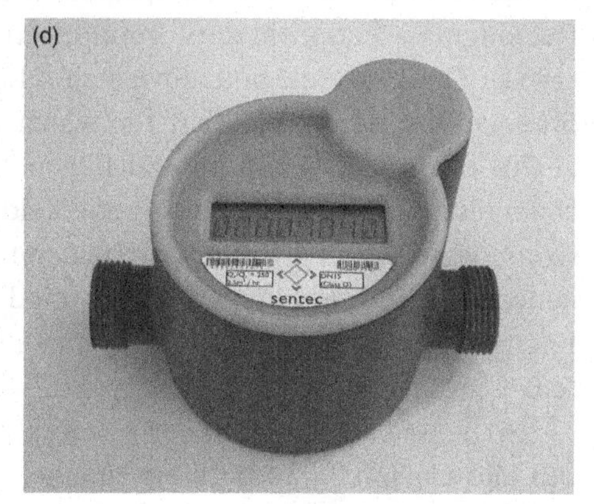

Figure 12.15. Sentec's Sterling flowmeter (reproduced with the permission of Sentec Ltd) **(a)** magnetic and electrical components. **(b)** component positions. **(c)** central component of sensor showing rectangular flow duct and positions of pole pieces and electrodes. **(d)** photograph of the flowmeter.

poorly conditioned flow. To achieve low power consumption, the magnetic field is generated by magnetising a semi-hard remanent material, Figure 12.15(b), using a very low duty cycle electromagnet, achieving a square-wave magnetic field at about 1Hz. The electrodes are electrochemically active, based on ruggedized silver/silver chloride chemistry, to achieve a very low noise at the operating frequency, Figure 12.15(b,c). The flow tube material is injection-moulded glass-reinforced PPS, which has good mechanical and surface electrical properties. The meter is shown in Figure 12.15(d).

A "dry calibration" device was designed based on a search coil which responds to the change of field when its polarity changes and feeds a voltage to the electrodes. In order to check for an empty pipe the meter measures impedance between the electrodes using an orthogonal frequency to avoid interference with normal operation. It is claimed to have a battery life of up to 20 years, and has been successfully commercialised (cf. Amata 1986; Salmasi et al. 2001).

12.13 Applications, Advantages and Disadvantages

12.13.1 Applications

The applications of this meter for liquid flow measurement are extensive. It is suitable for essentially any conducting liquid, and to my knowledge, it has been unsuccessful on hardly any to which it has been applied. One industrial expert once claimed to me that his only major problem had been icing sugar! Failure is more likely to be due to flow problems or incompatibility. If it is applied to a two- or multicomponent flow, then the continuous component must be conducting, and the signal will, essentially, be due to the velocity of this component. If it is applied to a liquid metal, the physics becomes more complex. A brief description of this application is given in 12.A.10.

Applications include viscous fluids, corrosive chemicals, erosive slurries and start-up and shut-down operations, but the flow tube should run full (some manufacturers offer versions capable of part-filled tubes), and the electrodes should not be open circuited by air bubbles (Ginesi and Annarummo 1994). The tube should, if possible, be vertical with upward flow. If horizontal, the electrodes should be in a horizontal diameter. If the meter is at a low point in the line, then the danger of slurries or other fluids coating the electrodes should be monitored. Coatings can have a conductivity that differs from the fluid and will then behave like a partially conducting wall, as well as changing the internal diameter of the meter. Likelihood of deposition is reduced if the velocity is kept above about 2–3 m/s (6–10 ft./s) through the meter. Electrodes of conical shape may reduce deposits, and electrode-cleaning systems can be used. Non-Newtonian fluids may alter the response. Wear of the liner can also result from abrasive slurries from close elbows etc. and can be reduced by liner protection. The cleaning fluid must also be compatible. Additives can cause non-uniform conductivity.

The AC technology was preferable for applications with large amounts of entrained air, with slurries that are nonhomogeneous or have non-uniform particle size, high solids content or solids with a tendency to clump, and with pulsating flows. This included about 15% of the industry including pulp and paper flows. However DC-pulsed technology appears to be replacing AC technology as an important option.

Effects of radiofrequency interference (RFI) should be virtually eliminated in new meters. However, signal cables must be screened and earthed according to the manufacturer's instructions.

Rose and Vass (1995) discussed adaptations of electromagnetic flow technology to difficult processes:

Chemical	acids, bases, polymers, emulsions, latex solutions
Pharmaceutical	spray coatings, flavouring, hygienic
Mining and minerals	slurries of iron ore, taconite, magnetite, pyrite, copper, alumina
Food and beverage	beer, soda, toothpaste, milk, ice cream, sugar, juices
Water and waste	water, wastewater, raw sewage, primary sludge, digester flows
Pulp and paper	white and black liquor, brown stock, bleaching chemicals, additives
Nuclear fuel processing plant	on both radioactive and nonradioactive liquids (Finlayson 1992)

Other more recent published applications relate to:

- possible problems with liquid lead-bismuth flow (Kondo and Takahashi 2005);
- monitoring pump performance (Anon 2002);
- the measurement of stock flows using capacitance electrodes (Okada, Nishimura and Tanabe 2003);
- monitoring wastewater (Kwietniewski and Miszta-Kruk 2005);
- continuous digesters: extraction lines, blow lines and circulation lines (Okada and Nishimura 2000);
- boreholes (Arnold and Molz 2000);
- production of alkylate, and the precise measurement of concentration of sulphuric acid (Dunn et al. 2003).

To this list may be added slag, cement, slurries (abrasive), reagent charging and special applications like ultra-low rates, custody transfer, liquids with steam tracing, blast furnace flows, batching and erosive liquids.

At high (120 measurements per second) rates, AC metering allows measurement of pulsating flows from pumps etc.

One manufacturer offered a range of sizes from 2 to 25 mm for milk, and others will offer their own range and specification for hygienic and sanitary applications. The meters are suitable for high-speed batch processing of these and other applications and can achieve 0.2% repeatability.

12.13.2 Advantages

a. The theory shows that the response is linear (apart from effects due to profile change), and the only reason the meter may not register satisfactorily down to zero flow is zero drift. It is one of the few meters capable of this performance and has been unfairly criticised as a result, for the zero drift that is observable. Modern designs often have a low flow cut-off that avoids this problem.

b. Clear bore is of most value in fluids that contain solids or in fluids that can be damaged by passing through constricted flow passages.
c. There are no moving parts.
d. Sensitivity to upstream fittings is comparable to other meters and only seriously bettered by positive displacement and Coriolis meters or by ultrasonic meters with more than two beams.

12.13.3 Disadvantages

The major disadvantage is its restriction to conducting liquids only, and although laboratory designs have operated on non-conducting liquids (e.g. transformer oil), to my knowledge only one or two commercial designs have attempted to operate in this regime. However, this application is discussed in Section 12.A.9. It is unlikely that such a meter could operate with normal gases.

Its sensitivity to upstream distortion has sometimes been suggested as a weakness. In my view, it is one of its strengths. Very few meters operate with less detrimental effect from upstream disturbance than this one. The other feature that was also suggested as a disadvantage from time to time was zero drift, the fact that early designs were prone to large errors at very low flows. Again, there are virtually no flowmeters capable of operating remotely linearly over the range of this one or down to the flow rates of which this is capable. Indeed, at least one commercial meter is claimed to be capable of 1,000:1 turndown.

12.14 Chapter Conclusions

The theoretical analysis of the electromagnetic flowmeter is well understood and has been developed by Shercliff, Bevir, Hemp and others with some contributions from the present author. This analysis allows the response of the flowmeter to various flow profiles to be deduced but also, through the weight function analysis, allows a flowmeter to be designed with a predetermined level of insensitivity to profile change.

The weight function analysis is, itself, very important as it offers an approach that Hemp has exploited for ultrasonic, thermal and Coriolis meters. A similar approach may be possible in other types of meter.

This meter must be a prime contender in any flow measurement application with a conducting liquid. I have always taken the view that its successful commercial development for non-conducting liquids, although possible, was highly unlikely because the voltages needed to drive the field coils would be incompatible with the intrinsic safety requirements of many non-conducting (oil-based) liquids. Further work may overcome some of the expected limitations for this meter.

What other developments are likely? The idea of dry calibration could still be developed to provide condition monitoring of the field. For some time I have wondered if the inclusion of a vortex-shedding bar in the meter, to provide another approach to condition monitoring, would be worthwhile, even though it would remove the clear bore. It would provide two independent velocity measures, the electromagnetic steady signal, with a fluctuation superimposed related to the vortex shedding frequency. It might also be possible to include additional electrodes in the shedder to provide a guide to upstream flow distortion.

Another line of development, exemplified by Horner, Mesch and Trachtler (1996), is the use of rotating fields or other multiple coil arrangements. In their case, two pairs of coils and 16 electrodes appear to have been used to improve performance on non-axisymmetric flows.

With the developments in design, materials technology, manufacture and signal processing, this meter has become of increasingly competitive price, usually lower maintenance than those with moving parts, and steadily improving performance. This has included

- improved accuracy,
- extended turndown,
- reduced size and weight,
- reduced power, and
- microprocessor-based intelligence.

This has led to the possibility of battery operation with time periods up to 10 years or more and turndown claims of 1,000: 1. The potential of the meter has also encouraged the use of new materials, especially ceramics, to increase stability and, by careful design, to reduce noise from radiofrequency and electrochemical sources.

I have not reviewed flowmeters which make use of ionisation techniques such as Morgan and Aliyu (1993), Nakano and Tanaka (1990), Brain et al. (1975).

Appendix 12.A Brief Review of Theory, Other Applications and Recent Research

12.A.1 Introduction

We make use of a simplified form of Maxwell's equations

$$\nabla \times \mathbf{B} = \mu \mathbf{j} \tag{12.A.1}$$

$$\nabla \cdot \mathbf{B} = 0 \tag{12.A.2}$$

$$\nabla \times \mathbf{E} = -\dot{\mathbf{B}} \tag{12.A.3}$$

and Ohm's law with velocity included

$$\mathbf{j} = \sigma(\mathbf{E} + \mathbf{V} \times \mathbf{B}) \tag{12.A.4}$$

where $\mathbf{B}$ is magnetic flux density vector, μ is permeability, $\mathbf{j}$ is current density vector, $\mathbf{E}$ is electric field vector, σ is conductivity, and $\mathbf{V}$ is velocity vector.

We may ignore $\dot{\mathbf{B}}$ because the effect of this term may be eliminated by careful mechanical and electrical design. This allows the introduction of the scalar electric potential U:

$$\mathbf{j} = \sigma(-\nabla U + \mathbf{V} \times \mathbf{B}) \tag{12.A.5}$$

Taking the divergence of this equation and making use of the fact that the magnetic field produced by external coils will be virtually unaltered by the very small currents in the fluid so that

$$\nabla \times \mathbf{B} = 0 \tag{12.A.6}$$

we obtain the flowmeter equation

$$\nabla^2 U = \mathbf{B} \cdot \nabla \times \mathbf{V} \tag{12.A.7}$$

and boundary conditions for a non-conducting wall of

$$\frac{\partial U}{\partial n} = (\mathbf{V} \times \mathbf{B})_{\perp} \tag{12.A.8}$$

when n is the coordinate perpendicular to the wall. Since the velocity at the wall is normally parallel to the wall, this will become, for cylindrical coordinates,

$$\frac{\partial U}{\partial r} = -V_z B_\theta$$

This condition is assumed to exist everywhere on the wall of a conventional flowmeter with small electrodes. For a nonslip condition on velocity, the right-hand side is zero. However, if the electrodes are large, then the simplest assumption is that the fluid at the electrode surface is at the same potential as the electrode. The electrode is assumed to have the same potential at every point on its surface, and the total current flow into the electrode is assumed to be zero. If a contact resistance exists at the electrode surface, this will have to be allowed for in relating the value of the voltage on the electrode and in the fluid.

One important observation from Equations (12.A.2) and (12.A.6) is that $\mathbf{B}$ in the fluid may be expressed as the gradient of a scalar potential Φ; hence

$$\nabla^2 \Phi = 0 \tag{12.A.9}$$

and we observe that the magnetic field distribution obeys Laplace's equation.

The boundary conditions on the magnetic field will usually be those existing on the pole pieces and at the surface of the excitation coils or the slots into which they fit. Thus, Φ will have a constant positive value on the surface of one pole and an equal but negative value on the other.

So far we have assumed steady conditions. Often, however, the excitation is sinusoidal, and the voltages induced by transformer effect may not be negligible. We can demonstrate that the analysis so far is valid by noting that if we associate a sinusoidal excitation of known phase with the applied magnetic field, all the equations relating to flow-induced signal will have the same phase as the applied field, and the electric field derived from Equation (12.A.3) will have a $\pi/2$ phase shift and may be eliminated by suitable electronic design. Additionally, the effect of this field may be reduced by improving the symmetry of the flowmeter.

12.A.2 Electric Potential Theory

It may be shown (Baker 1968b) that the solution of Equation (12.A.7) for a circular pipe with axisymmetric flow profile is given by

$$U = \frac{1}{2}\int_0^r \left[V\left(\sqrt{\rho r}\right) + V\left(a\sqrt{\rho / r}\right) \right] B_\theta(\rho, \theta)\, d\rho \tag{12.A.10}$$

where r is the radial coordinate, $V(r)$ is the axisymmetric flow profile, θ is the azimuthal coordinate, and $B_\theta(r, \theta)$ is the azimuthal component of magnetic field, which varies with r and θ and is constant with axial position z. ρ is a dummy variable, and a is the pipe radius.

The potential difference between diametrically opposed electrodes ΔU_{EE} is then given by

$$\Delta U_{EE} = 2\int_0^a V\left(\sqrt{ra}\right) B_\theta(r, \theta)\, dr \tag{12.A.11}$$

For a uniform magnetic field B, this becomes

$$\begin{aligned} \Delta U_{EE} &= \frac{4B}{a}\int_0^a rV(r)\, dr \\ &= BDV_m \end{aligned} \tag{12.A.12}$$

where D is the pipe diameter, and V_m is the mean velocity. This equation resulted in the high expectations for the device that offered true bulk flow measurement. However, the assumptions – uniform field and axisymmetric flow profile – were too constraining in practice.

12.A.3 Development of the Weight Vector Theory

Shercliff (1962) showed that the response to a point-electrode uniform-field electromagnetic flowmeter, when subjected to an arbitrary rectilinear flow profile, could be represented by a weighting function W given by

$$W = \frac{a^4 + a^2 r^2 \cos 2\theta}{a^4 + 2a^2 r^2 \cos 2\theta + r^4}$$

where θ is zero in the direction of the magnetic flux. This is the function shown in Figure 12.4(a).

Bevir (1970) developed the powerful concept of the weight vector **W**. Bevir introduced the concept of the virtual current $\mathbf{j}_v$, which is related to a potential function U_v by

$$\mathbf{j}_v = -\sigma \nabla U_v \tag{12.A.13}$$

and is the current density that would exist in the flow tube in the absence of magnetic field and flow if unit current entered by one electrode and left by the other.

He showed that (cf. Baker 1982 for a simplified derivation)

$$\Delta U_{EE} = \iiint_{\substack{\text{Flow tube}\\ \text{volume}}} \mathbf{V} \cdot \mathbf{B} \times \mathbf{j}_v \, dv \tag{12.A.14}$$

where the integration is taken over the whole volume of the flowmeter tube. Bevir introduced **W** as

$$\mathbf{W} = \mathbf{B} \times \mathbf{j}_v \tag{12.A.15}$$

and hence

$$\Delta U_{EE} = \iiint_{\substack{\text{Flow tube}\\ \text{volume}}} \mathbf{V} \cdot \mathbf{W} \, dv \tag{12.A.16}$$

Bevir then showed that the necessary and sufficient condition for an ideal flowmeter, one which measured the mean flow regardless of flow profile, was

$$\nabla \times \mathbf{W} = 0 \tag{12.A.17}$$

An important point to note is that Equations (12.A.16) and (12.A.17) possess a generality beyond electromagnetic flow measurement. Bevir (1970) proposed a rectangular section flowmeter that satisfied Equation (12.A.17), and he carried out extreme tests to demonstrate its performance. Using non-contacting electrodes and a special magnetic field distribution, Hemp (with Al-Khazraji 1980 and with Sanderson 1981) has designed a flowmeter which in theory should be virtually insensitive to flow profile effects.

12.A.4 Rectilinear Weight Function

However, most designs have assumed a rectilinear flow and have used an appropriate weight function W'

$$W'(r,\theta) = \int_{-\infty}^{\infty} W_z \, dz \tag{12.A.18}$$

where W_z is the axial component of **W**. This results in

$$\Delta U_{EE} = \iint_{\substack{\text{Flow tube}\\ \text{cross-section}}} r V_z(r,\theta) W'(r,\theta) \, d\theta \, dr \tag{12.A.19}$$

This is essentially Shercliff's (1962) idea, which resulted in the well-known distribution for a uniform magnetic field [Figure 12.4(a)].

Many weight function distributions have appeared in the literature since Bevir's work, and we have noted three of these:

- For a point-electrode flowmeter with rectangular coils and circular yoke (Al-Rabeh and Baker 1979) [Figure 12.4(b)]. This type of flowmeter is of interest in that it approximates to some industrial flowmeters.
- For a point-electrode flowmeter with diamond coils [Figure 12.4(c)] and circular yoke, which approximates to one introduced in the 1960s and claimed to have a reduced sensitivity to flow profile distortion.
- For a large electrode flowmeter with rectangular coils and yoke (Al-Khazraji and Baker 1979), shown diagrammatically in Figure 12.13, for which the weight function is reproduced in Figure 12.4(d).

The improvement obtained in the uniformity of W' by use of diamond coils instead of rectangular coils is very noticeable, and there is even more improvement by using large electrodes. However, a major disadvantage of these is the change in sensitivity resulting from electrode fouling (Al-Khazraji and Baker 1979).

Hemp (1975) attempted to minimise the variation of weight function with the copper strip pattern in Figure 12.A.1(a). Figure 12.A.1(b) shows the resulting weight function distribution, which should be compared with those in Figure 12.4. The axisymmetric weight function was approximately constant for about 87% of the pipe falling to zero at the wall (Figure 12.A.2).

12.A.5 Axisymmetric Weight Function

In concluding this section on the theory, it is useful to refer briefly to the axisymmetric weight function. This is given by

$$W''(r)=\frac{1}{2\pi}\int_0^{2\pi}\int_{-\infty}^{\infty} W_z\,dz\,d\theta \qquad (12.A.20)$$

Its value for a uniform field is constant and unity. However, for most practical fields, it falls in value at the outside of the tube. This trend is illustrated in Figure 12.A.2, which gives plots for Shercliff's uniform field weight function [Figure 12.4(a)] and by interpolation for the others in Figure 12.4. The signal is given by

$$\Delta U_{\mathrm{EE}}=2\pi\int_0^a rW''(r)V_z(r)\,dr \qquad (12.A.21)$$

Thus as flow profiles change with increasing Reynolds number from being more peaked in the centre of the pipe to more uniform across the pipe, the signal will fall below direct proportionality.

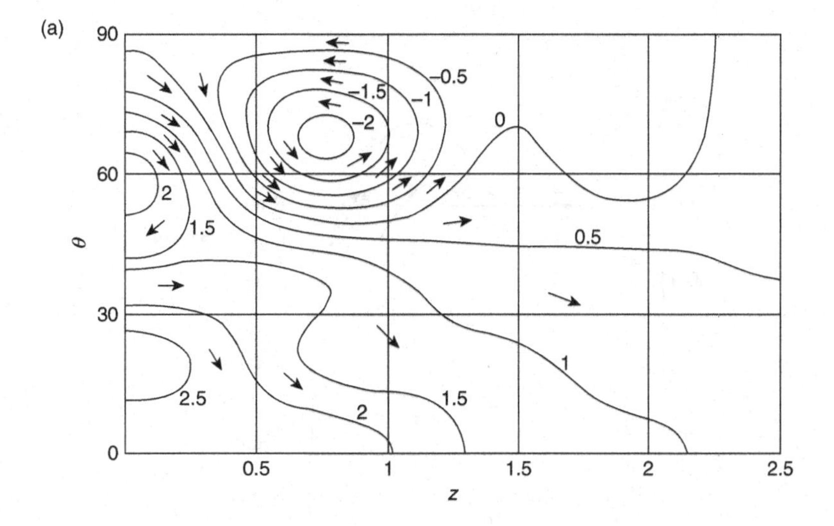

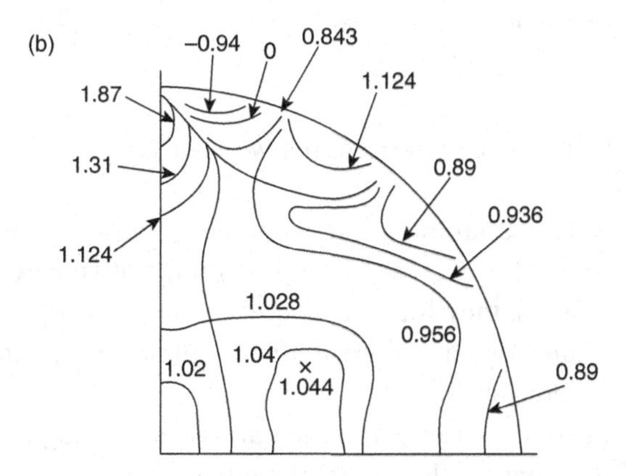

Figure 12.A.1. Hemp's (1975) improved magnetic field design: **(a)** Copper strip pattern associated with the improved magnetic field. The number associated with each line is the value of the magnetic field potential function, and the arrows, which are equal in size in each strip, show the direction of the current. **(b)** Weight function diagram for the improved magnetic field (refer to the original paper for further details; reproduced with permission from the author and IoP publishing).

12.A.6 Performance Prediction

Apart from the design problem, but related to it, has been the accuracy with which performance could be predicted from theory. This is the problem of dry calibration. Equation (12.A.12) suggested a very simple prediction for a uniform field flowmeter. However, the precision of such prediction has always been uncertain. Bevir, O'Sullivan and Wyatt (1981) have described a very elegant piece of work in which magnetic field measurements around the tube wall were used to compute the weight function, and very good agreement was obtained between prediction based on pipe flow profiles and actual meter response. They concluded that sensitivities would

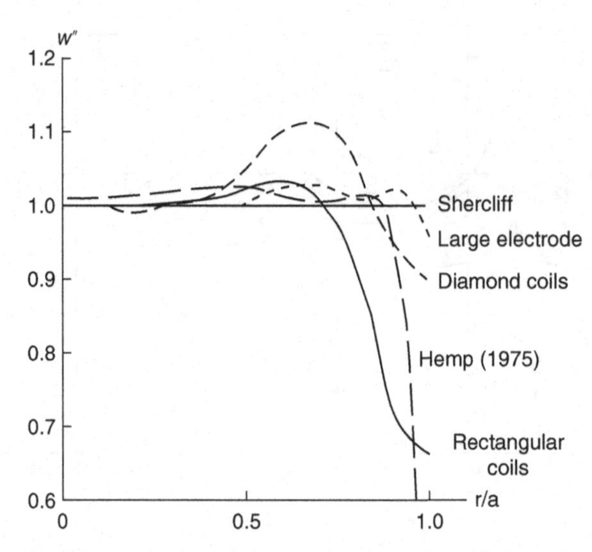

Figure 12.A.2. Axisymmetric weight functions based on Figures 12.4 and 12.A.1.

be predicted for most flowmeters by their procedures with an error of not more than 0.5%.

Al-Rabeh and Baker (1986) have attempted to predict the performance of a flowmeter from its magnet design. However, although good qualitative agreement on distribution of field was obtained, they found a discrepancy between predicted and actual field size. Thus with care, high accuracy prediction of flow signal may be achieved from actual magnetic field distribution.

Others have developed this approach by measuring and/or predicting the field at the pipe wall and using this to calculate the weight function.

12.A.7 Further Research

Further Theoretical Developments by Hemp and Others

Hemp and Wyatt (1981) explored the use of a "worst flow" to compare electromagnetic flowmeter performance.

Hemp and Versteeg (1986) developed the analysis of the flowmeter performance based on the magnetic field on the flowmeter surface. This approach would allow "dry calibration" of a meter based on magnetic field measurements on a tube next to the inner surface.

Krafft et al. (1996) discussed the use of the transformer effect signal due to bubbles to obtain both bubble velocity and velocity of the continuous phase of a bubbly flow.

Hemp (1991) developed the theory of eddy currents in electromagnetic flowmeters and suggested a means of self-calibration. Hemp also (1994c) looked at errors in potential measurement due to non-uniform contact impedance at the electrodes.

Clarke and Hemp (2009) examined the effects of eddy current in an electromagnetic flowmeter. They modelled a 2-D field model which allowed for eddy currents in the meter tube and magnet core material (all cylindrical and coaxial). It predicted with reasonable accuracy the gain and phase of the magnetic field and the drive coil impedance.

Research at Tsinghua University, China

Zhang (1998) revisited the weight function distribution (cf. Al-Khazraji and Baker 1979) and calculated the distribution for a uniform magnetic field with partially filled pipes with electrodes in the horizontal diameter and at 30° below the horizontal. In two further papers (1997, 1999) Zhang used a similar approach to find the virtual current when "two-dimensional" bubbles were present.

Zhang developed numerical prediction methods for the magnetic field and virtual current. Zhang (2001) described an idealised model used to calculate the 3D magnetic field distribution in an electromagnetic flowmeter with diameter of 500 mm and with five pairs of coils. The numerical results were compared with experimental data. The agreement of the calculation with the experiment confirmed the approach. The difference between theory and experiment and the possible causes of errors were discussed. In a subsequent paper Zhang (2002) discussed his work and the basic physical concepts and equations used. Examples were given of using the theory to analyse the effect of conducting pipe connections, errors caused by a pipe wall of different electromagnetic properties and gas-liquid flow. The model (Zhang 2001) was applied further (Zhang 2003). It was shown that the idealised model is practical for the design and the analysis of the flowmeter. Zhang and Li (2004) solved the Laplace equation in a complicated three-dimensional (3-D) domain by an alternating method. Virtual current potentials were obtained for an electromagnetic flowmeter with one spherical bubble inside. The solutions were used to investigate the effects of bubble size and bubble position on the virtual current. Comparisons were done among the cases of 2-D and 3-D models, and of point electrode and large electrode. The results showed that the 2-D model overestimates the effect, while large electrodes were least sensitive to the bubble. Zhang and Wang (2004) demonstrated a means of obtaining flow patterns from the signals of a meter with multiple electrodes and coils with impressive results. Zhang (2007) discussed measurement errors caused by asymmetry in a meter. Zhang (2010) has written a book (in Chinese) on electromagnetic flow measurement.

Research at Cranfield University, United Kingdom

Several papers have emerged from Cranfield University on electromagnetic flowmeters. Sanderson (2003) discussed factors affecting the dynamic performance of electromagnetic flowmeters. The dynamic performance of electromagnetic flowmeters is limited by the fundamental frequency of the magnetic field which is used to excite the flowmeter. Early medical developments used non-sinusoidal excitation up to

200 Hz, and initial industrial designs used sub-multiples of the line frequency e.g. $^1/_{16}$, with increased frequencies for batching. The reasons these fields cannot be increased in frequency without causing significant problems are identified including: time constant of coils, eddy currents induced in metals and liquid, drive circuit requirements and possibly higher voltages. Sanderson identified two potential solutions to overcome the limited dynamic performance whilst at the same time achieving good long-term stability. The first was to operate the meter at dual frequencies: the low frequency to obtain the base line performance and the high frequency to obtain the dynamic performance (Kuromori et al. 1988; Kuromori, Goro and Matsunaga 1989). The second related to applying Hemp's reciprocal theorem to the operation of the meter (Hemp 1988). In addition papers have discussed work on dielectric fluids. Rosales and Sanderson (2003) discussed a model for streaming current noise generation in electromagnetic flowmeters used with conducting fluids.

Other Research

The authors of a group of papers [Wang and Gong (2006), Wang and Lu (2006), Wang, Tian and Lucas (2007a), Wang, Lucas and Tian (2007b), Wang (2009), Wang et al. (2009b)] used a numerical method to solve the electromagnetic flowmeter equation and the weight function. The authors analysed the flowmeter geometry with electrodes at the ends of a non-diametral chord, with the possibility of deducing the profile. Wang et al. (2007a) used simulations which gave an indication of the behaviour of the meter in two-phase flows. They also stated that "induced potential differences, measured using an array of boundary electrodes, could conceivably be used to infer the axial velocity distribution of the flow." Leeungculsatien and Lucas (2013) appear to have deduced such a profile in one plane. (See also Xu et al. 2001, 2003.)

Guan et al. (2002) also undertook experiments designed to compare the performance between a multi-electrode flowmeter and a two-electrode electromagnetic flowmeter. They concluded that multi-electrode flowmeters outperformed two electrode flowmeters especially under the condition of non-axisymmetric flow or low-velocity flow. They suggested that this feature could extend the use of such flowmeters. It should be noted that this improvement has been recognised for a long time (Shercliff 1962).

12.A.8 Verification

An area of considerable interest at the present time is that of verification of electromagnetic flowmeters. The simplicity of Equation (12.2) probably led to the idea of simulation of the flow signal, in order to check the integrity of the amplifier and circuitry in the flowmeter. In addition the magnetic field could be checked and in many, if not all designs, a search coil was used to provide a compensation for any changes in the magnetic field caused, for instance, by mains supply voltage changes. Over the past 20 years or so the manufacturers have developed more sophisticated methods, partly embedded in the flowmeter, and partly in a special computer-based measuring system, although it appears that these systems are all becoming part of the meter hardware and software. The three primary objectives of the verification devices are:

- to simulate the flow signals and apply them to the meter electronics to check that the system is working correctly and is within specification;
- to check that the magnetic field is operating within specification by measuring its size or some related evidence such as the decay pattern of the field or exciting current;
- to check insulation and electrode cable integrity.

Walker (2001) discussed advanced diagnostics for these flowmeters which could indicate hardware and software problems. Brockhaus (2005) discussed the self-test functions of KROHNE's signal converter. Baker, Moore and Thomas (2004a, 2006a) reviewed the validity of claims made for currently available verification devices. Theoretical requirements to achieve the verification measurements were suggested and the validity of the approach assessed accordingly. Successful verification strategies will depend on the quality of the manufactured product and the administration of the verification procedure by manufacturers and users. Even if, on occasion, the value of the method is overstated, the approach offers a useful check on meter performance and has received strong support (e.g. Johnson 2002).

Baker, Moore and Wang (2004b, 2005) also discussed the use of dry calibration for modern flowmeter technologies, and considered that it might offer a way to reduce the frequency and cost of wet calibrations. The acceptability of dry calibration for the orifice plate was contrasted with the use of dry calibration for modern flowmeters.

The area where the current strategies may fail relates to the integrity of the flow tube itself. This could reflect the nature of the fluid and any build-up of deposits in the tube. It could also reflect problems with the integrity of the liner, and it could relate to the electrodes themselves. While all of these problems may be of lower probability than those tested, their possible existence reduces the strength of the claim to be able to check that the uncertainty is retained.

The University of Sussex, UK, has done much work in seeking to identify useful information from the "noise" which has often been discarded in flowmeter signals. Perovic, Higham and Unsworth (2001) discussed this approach for the electromagnetic flowmeter, and suggested that it might be possible to use signal analysis to identify flow regimes such as increased turbulence, swirl, cavitation and two-phase conditions, and to develop this into a fuzzy logic system. Perovic and Higham (2002) reported that the electromagnetic flowmeter's process measurement signals showed "noise" components which carried potentially useful diagnostic information not only regarding the operating conditions at the point of measurement but also regarding the status of the measurement system. If, instead of removing all the "noise", this useful data could be extracted, it could, possibly, provide a valuable additional input to the existing electronic verification devices which are likely to be important in the near future.

Hemp (2001a) also considered a dry-calibration technique in the manufacture of electromagnetic flowmeters, particularly of larger diameters, which made use of the axial component of the eddy current electric field in the water over each electrode, and might form the basis of a verification method. It requires a calibrated probe to measure the axial component of the eddy current electric field at the position of the flowmeter electrodes, and provided the meter design makes it essentially insensitive to flow profile, can be used as a check on its calibration.

The weight vector theory and Equation (12.A.15) suggested another indicator of the constancy of the signal source. Constancy of the weight vector should ensure constancy of the meter signal, and thus constancy of the virtual current and of the magnetic flux density should ensure that the signal is unchanged. Baker (2011) discussed this theory and experimental methods for extending the verification methods to take account of changes within the bore of the flowmeter. The method was to measure the value of a normalised virtual voltage, applied separately from the flow signal, preferably measured between the electrodes or between specially installed/adapted ones. The method looked promising, and would lend itself to optimisation in combination with the magnetic flux distribution. Thus the meter could be programmed to check the virtual voltage periodically to assess whether this indicated any changes which might affect the calibration.

A more major approach to condition monitoring, and one which would put up the manufacturing costs, would be the inclusion of a vortex shedding bar in the electromagnetic flowmeter, to provide independent flow data. The vortices should superimpose a variation on the flow signal due to the oscillating flow, which could be extracted from the electrode signal, although it would remove the clear bore. Yet another approach would be to provide attachment points for ultrasonic transducers. It would provide two independent velocity measures.

12.A.9 Application to Non-Conducting Dielectric Fluids

Work has continued on the challenge of an electromagnetic flowmeter for non-conducting liquids. Whether such a device would be incompatible with the intrinsic safety requirements of many non-conducting (oil based) liquids is not yet clear.

Cushing (1958, 1965, 1971) has pioneered the idea of using the electromagnetic flowmeter over a long period as indicated by these references. Two recent papers by him (Cushing 2001, 2002) further discussed the electromagnetic flowmeter for all fluids and their applicability to insulating liquids. In his description of the meter design he identified important features:

- the meter bore appeared to be a circular tube with a dielectric liner;
- the electrodes were large area ones made from photo-etched Mylar sheets so that the electrodes were divided into narrow strips to reduce eddy currents, and were positioned behind the dielectric liner. There was a sensing electrode on one side and a common electrode diametrically opposite;

- a "guard manifold" was placed behind and around the periphery of the sensing electrode;
- the electrode lead from the common electrode appeared to go to earth while that from the sensing electrode went radially outward;
- The magnetic flux at the centre of the meter appeared to be perpendicular to the tube axis and to the diameter through the electrode centres, with a 960Hz frequency square-wave excitation.

Cushing (2002) went through the details of the (multiple-sampling) signal handling and indicated the problems involved. From his wording I assume that he was successful in operating the meter, although there were, clearly, major spurious noise sources and quadrature signals to be controlled. However, he states that "zero-point drift was no more than about ± five percent", presumably of full flow signal.

A paper by Al-Rabeh, Baker and Hemp (1978) set out the theoretical basis (cf. Baker 1982 for a brief summary which corrects some typographical errors in the original paper). Thus if the flow is parallel to the pipe axis, there is normal symmetry and the excitation $\omega \gg V/l$ where V is the flow velocity and l is a characteristic length, then the equations are simplified and we may replace Equation (12.A.7) by

$$\nabla^2 U = Z\nabla \cdot (\boldsymbol{V} \times \boldsymbol{B}) \tag{12.A.22}$$

Equation (12.A.16) is unchanged, but Equation (12.A.15) becomes

$$\boldsymbol{W} = Z\boldsymbol{B} \times \boldsymbol{j}_V \tag{12.A.23}$$

where

$$Z = \frac{\sigma + (K/\mu)i\omega}{\sigma + \varepsilon i\omega} \tag{12.A.24}$$

and

$$K = \varepsilon\mu - \varepsilon_0\mu_0 \tag{12.A.25}$$

and ε is the permittivity of the medium, ε_0 is the permittivity of free space, μ is the permeability of the medium and μ_0 is the permeability of free space. (The reader should consult the original paper for the full equations. This simplified version eliminates the terms containing electric charge which introduced a nonlinearity).

Cushing (1965, 1971) suggested that a mass flow measurement might be obtained by measuring the permittivity of the medium (from the inter-electrode capacitance) and employing the Clausius-Mossotti relation

$$\rho = \frac{1}{M}\frac{\varepsilon - \varepsilon_0}{\varepsilon + 2\varepsilon_0} \tag{12.A.26}$$

where ρ is the fluid density, ε and ε_0 are, respectively, the permittivity of the fluid and of free space, and M is a constant related to the polarizability of a single molecule of the medium.

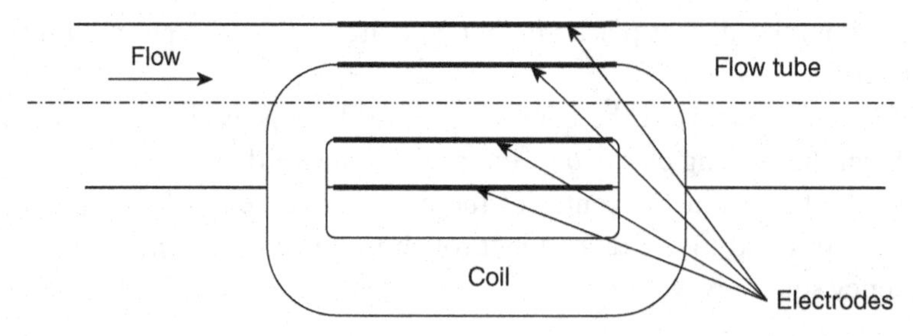

Figure 12.A.3. Outline sketch of the coil and electrodes of the dielectric flowmeter of Al-Rabeh (1981).

Al-Rabeh (1981) did other interesting work at Imperial College, London, as my research student developing experimental investigations and demonstrations of the validity of the induction with dielectric materials. Wilson (1901, 1904) and Wilson and Wilson (1913) devised an experiment to demonstrate the induction of a voltage across a dielectric insulator moving in a magnetic field. Their apparatus consisted of a substantial solenoid along the axis of which a tube of ebonite, with brass tubes on its inner and outer surfaces, rotated at high speed. The voltage generated between the inner and outer brass tubes was measured and shown to agree with the expected voltage generation induced by the rotation. Al-Rabeh (1981) repeated the Wilson experiment, and achieved satisfactory agreement with predicted values of voltage induction.

Al-Rabeh (1981) then proceeded to design and build a flowmeter using an axial coil to produce an azimuthal magnetic flux in the flow. A sketch of the design is shown in Figure 12.A.3.

Major problems identified by Al-Rabeh were the sources of noise and the problem of extracting the measured voltage from a source with such a high impedance. Although satisfactory signals were reported, the experiment identified major areas needing attention.

Cranfield University undertook further investigations, and two of theses were reported. Amare (1995) wrote one entitled "Electromagnetic flowmeter for dielectric liquids", and in a subsequent paper (1999) commented that "The choice of magnetic field frequency and the design of the electrostatic shield is critical for the function of the flowmeter." The design appears to have similarities to that of Cushing. The magnetic field excitation frequency was 1.5 kHz. Using Equations (12.A.24) and (12.A.25) Amare obtained

$$Z = \frac{\varepsilon' - 1}{\varepsilon'} \tag{12.A.27}$$

where $\varepsilon' = \varepsilon / \varepsilon_0, \mu / \mu_0 \approx 1$ and $\sigma = 5.7 \times 10^{-11}$. Amare suggested that a flowmeter output voltage with water would be about twice that with oil (BP180). Using his value of $\varepsilon' = 2$ results in a value of Z in Equation (12.A.27) which would tally with this.

The second thesis, that of Durcan, was published as a book in 1998, entitled "Development of baseline stability in an electromagnetic flowmeter for dielectric liquids".

Much of the research at Cranfield University seems to have been focussed on the problems of noise in the system. Researchers included Al-Rabeh, and some of the research was published by Hemp et al. (2002). They described experiments to understand the cause of noise from charges in the fluid. Using a pumped circuit with a large electrode flow tube installed in it, the noise sources were investigated. Apart from Johnson noise due to the high input impedance of the electronics, it was found that, with the lowest conductivity oil tested, the noise was significantly reduced by filtering out small particles in the flow. When an oil of greater viscosity was used the Reynolds number could be reduced to about 2,300 at which point the flow was probably on the borderline of laminar, and the noise signals were again reduced, only significantly to increase with flow in the low-turbulent region. The low noise may also have been partly due to the higher conductivity of this second liquid compared with the first. Further, charge was injected into the flow by using sharpened brass probes excited at high voltage. The result was a large increase in the static charge noise. There appeared to be an effect from increased temperature. The authors appear to have concluded that the electrodes of the flowmeter should be as large as possible within the region of the magnetic field to increase the signal/noise ratio. No flow signals, with insulating liquid, were measured in this research (private communication from Dr J. Hemp 2014).

Rosales, Sanderson and Hemp (2002a, 2002b) presented an analytical model of the generation of noise by turbulence modulation of the background charge distribution in electromagnetic flowmeters working with dielectric liquids. The model was tested for highly insulating oil at different flow speeds and electrode lengths, and the authors claimed that the results showed good agreement with the available experimental data. Hemp and Youngs (2003) discussed problems in the theory and design of electromagnetic flowmeters for dielectric liquids and investigated theoretically the magnitudes of zero offsets and zero drifts originating from magnetic flux linkage between the coils of the electromagnet and the loop formed by the electrode cables in an electromagnetic flowmeter for dielectric liquids. The dependence of such zero offsets on liquid properties, frequency of operation etc. was explained. Youngs carried out very careful and useful measurements of the conductivity and permittivity of the various liquids (private communication from Dr J. Hemp 2014).

The work described has identified many of the problems relating to this device, and it presents a challenge to anyone who can identify a market opportunity for this type of flowmeter. I have reservations as to whether the flowmeter could be made intrinsically safe for the likely liquids it would need to meter. I should be happy to have my doubts allayed and to witness a commercially successful design which followed up Cushing's pioneering efforts.

For a different approach to the measurement of insulating fluids see Al-Rabeh and Hemp (1981).

12.A.10 Electromagnetic Flowmeters Applied to Liquid Metals

Sharma et al. (2010) reviewed some past work in relation to fast breeder reactors and the use of electromagnetic flowmeters for liquid sodium, particularly in relation to the Indian nuclear power programme. They claimed that flow measurement took place mainly at pump outlet in the primary circuit, in the secondary sodium circuits and at the outlet of sub-assemblies to detect blockage. They mentioned different types of flowmeters. Further references to research in this area were provided by Priede, Buchenau and Gerbeth (2011a). A review of the international situation with regard to fast breeder reactors (FBR) by Cochran et al. (2010) appears to suggest that only two countries were working with FBRs.

The electromagnetic flowmeter has a long history of use with liquid metals and Shercliff (1962), who appears to have been the first person to set out the various types of electromagnetic flowmeter for liquid metals, gave an important review of the literature and the designs which had been proposed. Apart from induced voltage flowmeters, and in particular, the saddle coil flowmeter (Figure 12.A.4), this included the effect of liquid metal flows on the magnetic field, causing distortion and field sweeping. This resulted from the currents generated in the flowing liquid metal which caused secondary magnetic fields, and hence an apparent field distortion or sweeping. He noted that the performance of the induced voltage flowmeter would depend in some cases on their axial length and on the magnetic Reynolds number $Re_m = \mu\sigma Vl$ where μ is the permeability of walls and fluid assumed to be equal to that of free space, $4\pi.10^{-7}$, σ is the conductivity, V is the velocity and l is the typical length scale. The field interaction also allows forces, created by the induced fields, to be used as a means of measuring the flow rate. The reader is encouraged to refer to Shercliff's book (1962) in which he sets out the equation expressing the conflict between the convection and diffusion of the magnetic field and the equation for incompressible viscous flow with the added electromagnetic force term.

I undertook theoretical and experimental work on some of the flowmeter designs, and reviewed the literature then (Baker 1977). The reader is referred to the reviews of Shercliff, Baker, Sharma and Priede, to which I have added some of my research references not included in any of these.

The following types have been discussed as indicated and may be suitable both as a flowmeter and possibly as a probe:

- permanent magnet flowmeter (Figure 12.2) (Shercliff 1962);
- saddle coil flowmeter (Figure 12.A.4) (Thatcher, Bentley and McGonigal 1970; cf. Tarabad and Baker 1982);
- axial current flowmeter (Shercliff 1962);
- motion-induced magnetic field (Cowley 1965);
- eddy current (Lehde-Lang) flowmeter (Figure 12.A.5) (Lehde and Lang 1948; Shercliff 1962; Baker 1977; Baker and Saunders 1977; Sharma et al. 2010; Priede et al. 2011a);
- sensing field sweeping, caused by fluid motion in a duct or around a probe, at the point of symmetry between two opposing applied axisymmetric magnetic fields around the duct (Baker 1969; Weigand 1972), in a similar way to Figure 12.A.5 but possibly using direct current coils and a Hall effect probe to sense field sweeping;

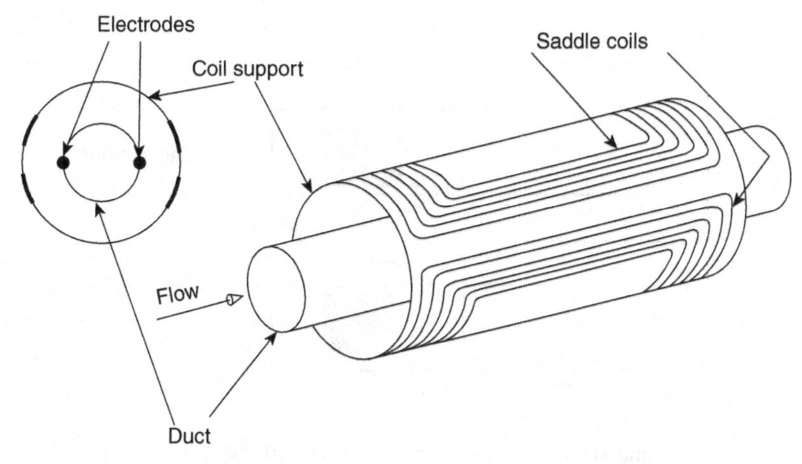

Figure 12.A.4. Saddle coil flowmeter due to Thatcher et al. (1970) (diagram from Baker 1977 reproduced with the agreement of Elsevier).

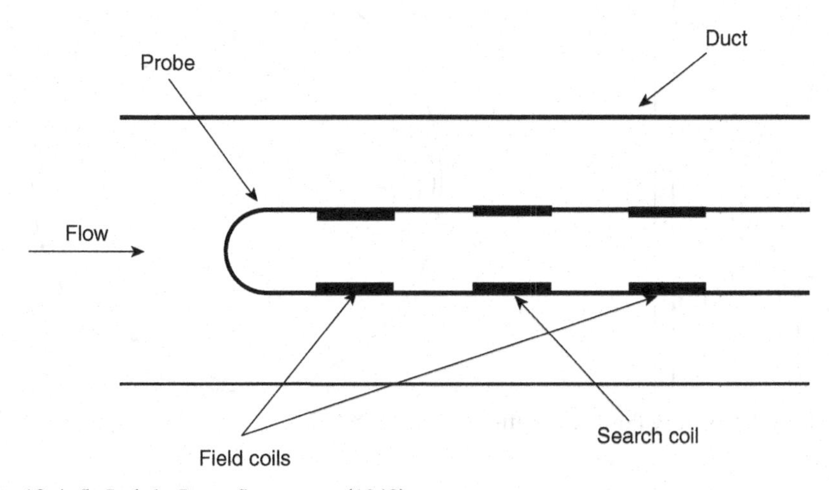

Figure 12.A.5. Lehde-Lang flowmeter (1948).

- force flowmeter (Figure 12.A.6) (Shercliff 1962; cf. Priede et al. 2011a for references to recent work on this type of flowmeter),
- rotating magnet flowmeter (Figure 12.A.7) (Shercliff 1962; cf. Priede et al. 2011b for references to recent analysis of a similar type of flowmeter with a magnet rotating above a free surface liquid metal flow),
- pulsed field flowmeter (Tarabad and Baker 1983).

Shimizu, Takeshima and Jimbo (2000) undertook a numerical study on an electromagnetic flowmeter in a liquid metal system based on magnetic field and electric potential field. Shimizu and Takeshima (2001) appear to have undertaken further studies also aimed at the application of the flowmeter in a liquid metal fast breeder reactor main coolant loop, and Sharma et al. (2012), with the same application in mind, modelled and tested an induced voltage electromagnetic flowmeter in sodium

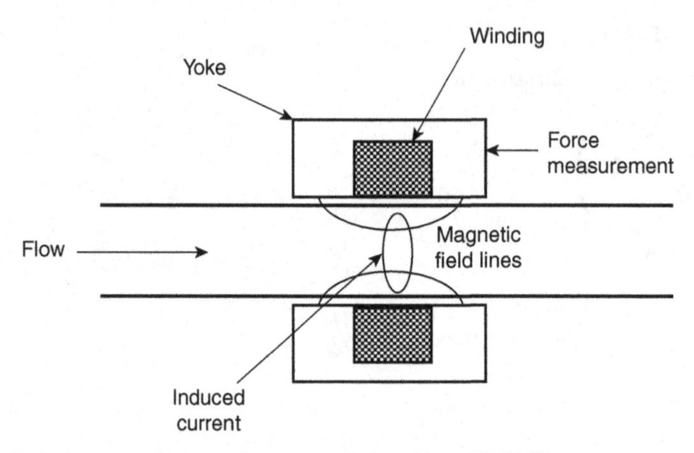

Figure 12.A.6. Axisymmetric force flowmeter (after Shercliff 1962).

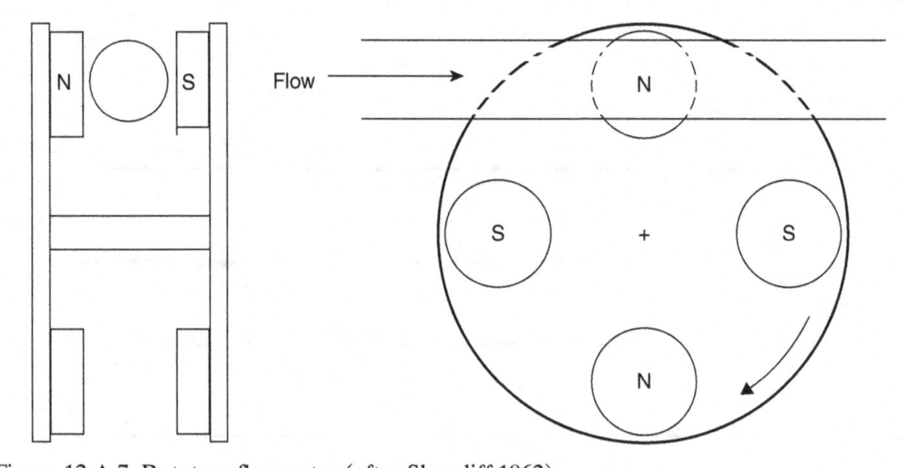

Figure 12.A.7. Rotatory flowmeter (after Shercliff 1962).

flow, using a permanent magnet. They appear to have obtained agreement within about 2%.

The use of electrodes may be a problem with some liquid metals (Priede et al. 2011a), and an alternative contactless method is the eddy-current flowmeter, which measures the flow-induced perturbation of an externally applied magnetic field (Lehde and Lang 1948; cf. Cowley 1965).

The design attributed to Lehde and Lang (1948) consisted of three coils, situated in a submerged probe, Figure 12.A.5 or, as Shercliff observed, the coils could be around the flow tube. In Figure 12.A.5 the outer coils are alternating-current (ac) energised in opposition, creating two opposing magnetic fields which cancel each other on the mid-plane. The interaction of flow with these fields generates circular currents which, in turn, generate a secondary magnetic field, and the centre coil senses this induced secondary field. Thus fluid motion around the probe or through the tube is sensed. An alternative to this was proposed (Baker 1969, 1970b) making use of direct-current (dc) excited coils causing axisymmetric fields around the duct or probe and using a probe, such as the Hall effect probe, on the axis midway between the coils, to sense the induced magnetic field due to motion of the fluid past the coils. In addition the effect on the flow pattern was considered (Baker 1970b). An alternative, as Shercliff suggested, would be to have the centre coil creating the magnetic field and the outer coils sense the field distortion.

This type of meter was considered for use in fast nuclear reactors with sodium core flow, and our research was reported (Baker 1977; Baker and Saunders 1977), which was funded by the United Kingdom Atomic Energy Authority. Due to the axisymmetry and the observation that, as a result, all induced currents would flow in circular paths, the meter was tested using a simulation which replaced the sodium flow with brass tubes moving at different speeds to give an approximate model of a flow profile. In the event, the use of this meter did not appear to be pursued by the sponsors. However, Sharma et al. (2010) appear to suggest that the eddy current flowmeter is again being considered for use in a prototype fast breeder reactor (PFBR) and they reported a numerical simulation of the flowmeter. Priede et al. (2011a) have developed a phase shift–sensing method for such a flowmeter. They also noted effects such as thermal expansion which can cause some asymmetry between the coils. In their paper they noted that the flow can disturb not only the amplitude of an ac magnetic field but also its phase distribution and they analyse the basic characteristics of such a phase-shift flowmeter. They referenced some of their earlier publications on this approach. They also discuss its technical implementation and test results. Poornapushpakala et al. (2014) discussed the development of fast response electronics for an eddy current flowmeter for the measurement of sodium flow in fuel sub-assemblies and primary pump bypass line in a prototype fast breeder reactor.

Another approach was to set two coils outside the tube (Tarabad and Baker 1979, 1982) and the upstream one was pulsed to create a magnetic field in the flow which would be swept downstream with the flow and would be sensed by the second coil giving a time of transit.

A solid was also used to simulate the liquid metal flows (Minchenya et al. 2011) in an open channel electromagnetic flowmeter which used the Lorentz force caused by the liquid flow through a permanent magnetic field. [See also Stelian (2013) on calibration of a Lorentz force flowmeter using numerical methods.] Jian and Karcher (2012)used time-of-flight Lorentz force velocimetry to avoid the need to know the conductivity of the liquid.

This is an interesting area of work which would benefit from further theoretical research. It appears to be of commercial interest in one or two countries at the present time, but the future is clearly uncertain and is likely to depend on the wider uptake of nuclear power generation.

13

Magnetic Resonance Flowmeters

13.1 Introduction and Some Early References

In the first edition of this handbook, I mentioned that flowmeters depending on magnetic resonance (MR) had been suggested, but that I was not aware of an instrument on the market. Nuclear magnetic resonance (NMR now, apparently preferably referred to as MR) essentially marks the fluid and measures transit times (Genthe 1974; King and Rollwitz 1983). Gol'dgammer, Terent'ev and Zalaliev (1990) showed that the magnetic resonance signal amplitude of a gas-liquid mixture in diamagnetic fluids depended on the liquid content regardless of the physicochemical properties of the medium and conditions of flow. This linearity was destroyed if paramagnetic centres were added to the fluid. De Jager, Hemminga and Sonneveld (1978) described their experimental data from water flow velocity measurements in the range of 0.5 to 5 mm/s by means of nuclear magnetic resonance using a sequence of inhomogeneous 180° pulses and a gradient in the stationary magnetic field. MR had been used in liquid-gas flows to measure the mean velocity of the liquid averaged over the liquid volume and also the average of the liquid fraction (Kruger, Birke and Weiss 1996). The combination provided the mass flow of the liquid. In tests using water as the liquid, the authors claimed to determine the mass flow to within ±5%.

Scott (1982), in "Developments in Flow Measurement–1", comments that "Sensing of the flow relies on the application initially of a magnetic field to the nuclei of hydrogen or fluorine in the flow and the subsequent (downstream) removal of this influence. The nuclei enter the magnetic section with their individual magnetic orientations completely random and leave with their nuclear orientations aligned in the direction of the magnetic field. In the detector section which follows they approach a tagger coil and then a receiver coil. The former injects a short high intensity burst of radio frequency waves into the meter. This demagnetises a 'window' in the magnetised fluid and the window is discerned by the detector coil as a momentary drop in output signal. A knowledge of the distance between the tagger and the receiver, coupled with a time-of-flight measurement between them leads to the mean flow velocity." The reader is referred to Scott's book for early references to this technique.

Calamante et al. (1999) reviewed the use of magnetic resonance imaging techniques to measure cerebral blood flow and discussed the potential of the method.

Since the early methods of NMR already mentioned, more recent methods use a different approach.[1] The signal detected in MR can be regarded as a complex number with both phase and magnitude. Therefore it is possible to encode information in the phase of the signal, and it proves to be very natural to encode the velocity in the phase of the signal. Most velocity measurements in MR use this approach. However, flow metering using MR has tended not to use this so-called phase encoding approach. Instead, methods relying on the increase or decrease in signal intensity have dominated. This is likely because such an approach simplifies the design of the instrument, although it may also be advantageous if there is a difference between the velocity of the oil and water phases. A more recent review has been written by Elkins and Alley (2007) on applications to fluid motion.

13.2 Developments in the Oil and Gas Industry

Ong et al. (2004) gave a useful review of instrument development linked to the oil and gas industry and identified some earlier instrument developments.

At the end of the 20th century they indicated that the technique had been mainly applied to well logging. Early designs of flowmeter apparently had problems with turbulent flows. The authors concluded that in 2004 there was no meter available with this technology either for downhole or for subsurface use in the oil and gas industry.

They developed an in-well nuclear magnetic resonance (NMR) multiphase flowmeter. They identified one of its advantages as the ability to distinguish different phases by means of molecular properties such as relaxation time and resonant frequencies.

In *Fundamentals of Multiphase Metering* (Schlumberger 2010), the authors saw advantages such as insensitivity to phase distribution, and stated that the development technology had shown some good downhole results. One limitation they saw was "the slow response time because the relaxation of the fluid after excitation is measured in seconds".

In a recent paper, Fridjonsson, Stanwix and Johns (2014) demonstrated the use of earth field NMR for flow metering. Their intention is to develop an instrument suitable for multiphase flow measurement.

13.3 A Brief Introduction to the Physics

Before progressing to recent designs of flowmeter, it is useful to review the physics. I am not knowledgeable in this area of physics and have found Hornak's (1997–1999) notes very useful in preparing an explanation in this introduction, as well as benefitting from the description of Hogendoorn et al. (2013).

Magnetic resonance (MR) may occur when an applied magnetic field fills the region of the fluid in a flowmeter and when a second electromagnetic field, for

[1] This paragraph is based on a private communication from Dr D. Holland, 2014.

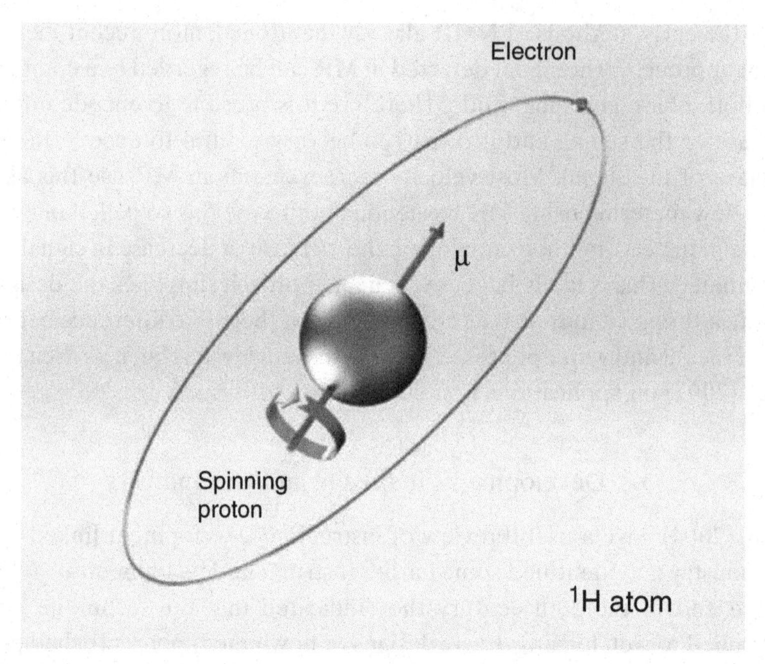

Figure 13.1. ^{1}H atom consisting of a proton and an electron. The spinning proton has a magnetic moment, μ (Hogendoorn et al. 2013) (reproduced with permission from KROHNE).

instance a radio frequency (RF) signal, is applied to the fluid in the magnetic field. The applied magnetic field causes the nuclei spins to behave as though they were magnets and align themselves with the applied vector field, **B.** When, in addition to the applied magnetic field, an RF field of a specific frequency is applied, a nucleon with spin can absorb a photon of a particular frequency, and this will affect the distribution of nuclei across different energy levels as well as the entropy of the ensemble of nuclei.

Spin is a property, like charge or mass, of the protons and neutrons (Figure 13.1), but may not be evident in some materials (Hornak 1997–1999). Under certain circumstances, the particle spins in a macroscopic quantity of liquid can cancel out and so no spin is observable; for example, ^{12}C and ^{16}O are invisible to MR (Hogendoorn et al. 2013). Nuclei having an odd number of protons or neutrons, such as ^{13}C or ^{1}H, will experience a force perpendicular to both the direction of the applied magnetic and RF fields, which, in turn, results in a precession motion around the direction of the background magnetic field vector, **B** (Figure 13.2). This precession motion is referred to as Larmor precession. Its frequency is directly proportional to the flux density of the magnetic field:

$$2\pi f = \gamma|\mathbf{B}| \tag{13.1}$$

where f is the frequency of the absorbed or emitted radiation, γ is the gyromagnetic ratio and $|\mathbf{B}|$ is the scalar magnitude of the magnetic flux density vector, **B**, of the imposed field. The energy of the absorbed photon must exactly match the energy

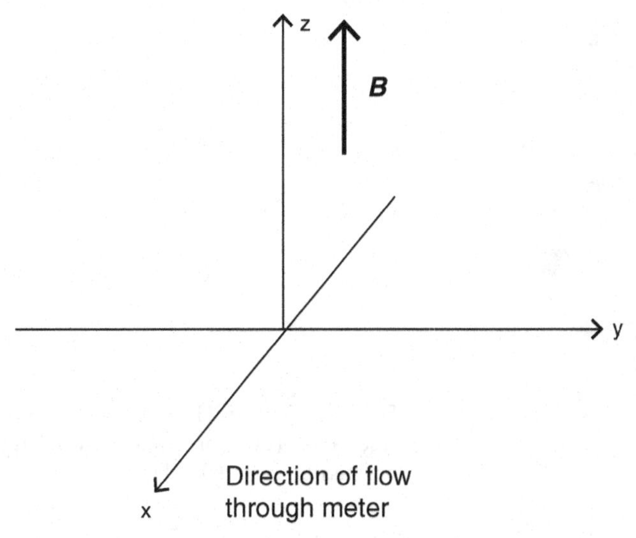

Figure 13.2. Diagram to show the directions of magnetisation.

difference between two spin energy states for the energy state of the proton to be changed. The energy of a photon is given by

$$E = 2\pi hf \tag{13.2}$$

where h is Planck's constant. The energy level of the photon will need to be

$$E = h\gamma|\mathbf{B}| \tag{13.3}$$

The conventional way of writing the symbol for an element is ${}^{A}_{Z}\mathrm{E}$ where E is the symbol used for the element e.g. H for hydrogen, A is the mass number or the total number of protons and neutrons, and Z is the atomic number or the number of protons (ref: http://en.wikipedia.org/wiki/Atomic_number). Thus ${}^{1}_{1}\mathrm{H}$ is hydrogen with one nucleon, a proton, and ${}^{2}_{1}\mathrm{H}$ is hydrogen with two nucleons, a proton and a neutron and so is an isotope of hydrogen. Since the lower number is synonymous with the element's symbol in our discussion, it will be omitted.

Hogendoorn et al. (2013) state that for most atoms, such as ${}^{12}C$ and ${}^{16}O$, the individual magnetic moments of protons and neutrons offset each other, and the effective magnetic moment of such nuclei vanishes. Those atoms are invisible to MR. Other nuclei possess either an odd number of protons (e.g. ${}^{1}H$), or neutrons (e.g. ${}^{13}C$) or both protons and neutrons (e.g. ${}^{2}H$). If these atoms or isotopes have a sufficiently high natural abundance, they can be used in the MR technique. Hydrogen protons (${}^{1}H$) (Figure 13.1) provide the strongest MR response and are targeted in most oil-field applications of MR (Hogendoorn et al. 2013). Two distinct alignments of the nuclei relative to the direction of an external magnetic field (also called spin states) are possible at the atomic level, either with (Figure 13.3) or against the external magnetic field. That is because direction and magnitude of the angular momentum of a nucleus are quantized, so that they can only vary by integer multiples of Planck's constant. There is approximately the same number of proton nuclei aligned

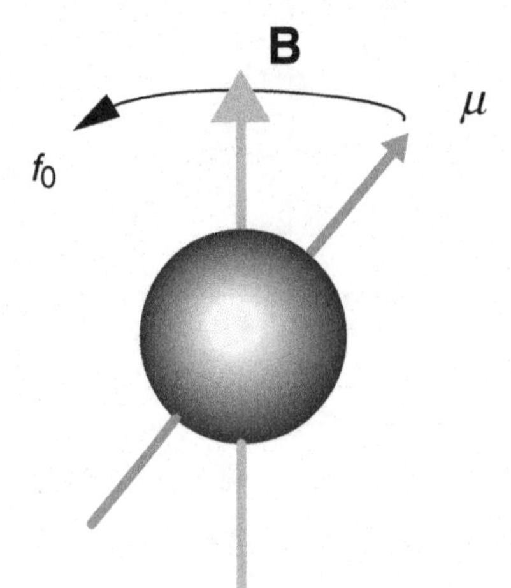

Figure 13.3. The ^{1}H proton starts to precess with the resonance (Larmor) frequency f_0 (reproduced with permission from KROHNE).

with the main magnetic field as against it. Those aligned, due to the nucleus being at a lower energy, are slightly favoured. This creates a macroscopic magnetisation.

An important value for the flow measurement application is the response time of the fluid when exposed to changes in the external magnetic field. This time constant is referred to as a spin lattice relaxation time:

$$M = M_0(1 - e^{-t/T_1}) \tag{13.4}$$

where M is the magnetisation with an equilibrium state of M_0. This equation can only be true if $M = 0$ when $t = 0$. Thus the change in magnetisation at T_1 is defined by the equation

$$M = M_0(1 - e^{-1}) \tag{13.5}$$

It is a common practice in NMR technology to describe the motion of the spin ensemble using a coordinate system that rotates with the Larmor frequency. In this way, the precession motion of the spins can be ignored and only the amplitude changes of the net magnetisation projected onto the direction of the applied magnetic field need to be considered. This amplitude variation follows Equation (13.4) and can be described as a magnetisation build-up in the z direction.

Thus the magnetisation follows an exponential as it returns to its equilibrium value of M_0, as indicated in Equation (13.4) (Figure 13.4). T_1 is the longitudinal relaxation time or the spin lattice relaxation time. For a multiphase fluid, the pattern is complicated by an array of time constants related to each component of the fluid.

The Hahn echo effect is used in the flowmeter and illustrated in Figure 13.5. With the applied magnetic field in the z direction so that the nucleons have orientated themselves in this direction, a 90° radio frequency pulse is applied in the x direction, causing the nucleons to be directed along the y direction. After this pulse the magnetisation vector rotates with the Larmor frequency in the x-y plane. Because we observe the average due to a large number of nucleons covering a wide region with varying magnetic field and fluid make-up, the Larmor precession occurs

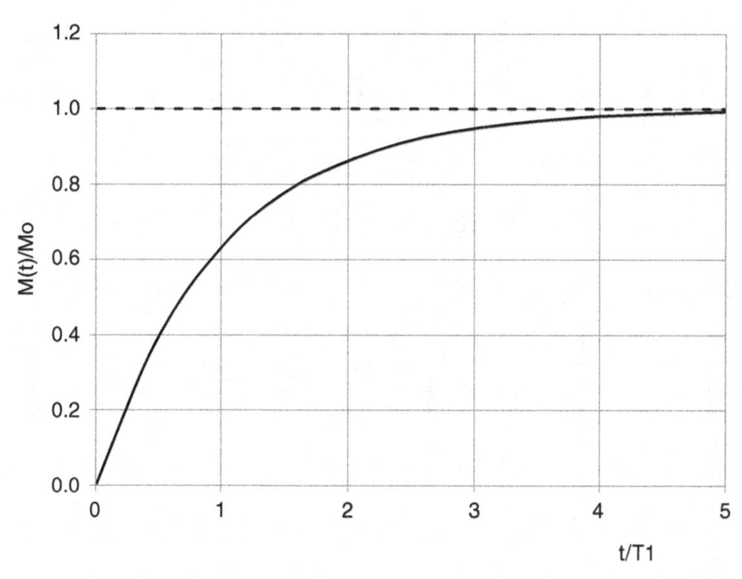

Figure 13.4. Build-up of magnetisation for a single-phase fluid (after Hogendoorn et al. 2013).

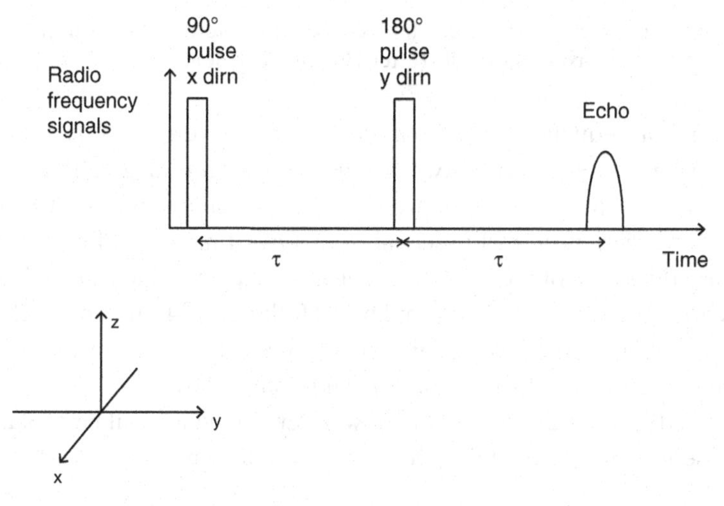

Figure 13.5. Hahn spin echo (after Hogendoorn et al. 2013).

at various rotational speeds, and the phase coherence between precessing spins disappears with slower and faster nucleons. After a time, τ, a 180° pulse is applied to the nucleons and they are flipped along the y-axis, so that the slower nucleons are ahead of the faster nucleons. The result of this change is that the faster nucleons catch up the slower ones, reuniting the phases, and after time, τ, an echo can be sensed. Hahn echoes can be created multiple times by repeating the 180° pulse. Figure 13.6 illustrates this. After the 90° pulse, the first 180° pulse follows at a time of τ later and the first echo is a further τ after that. The figure then indicates that the succession of 180° pulses followed by echoes can be continued. This experimental sequence is known as the CPMG pulse sequence where the initials relate to the originators of

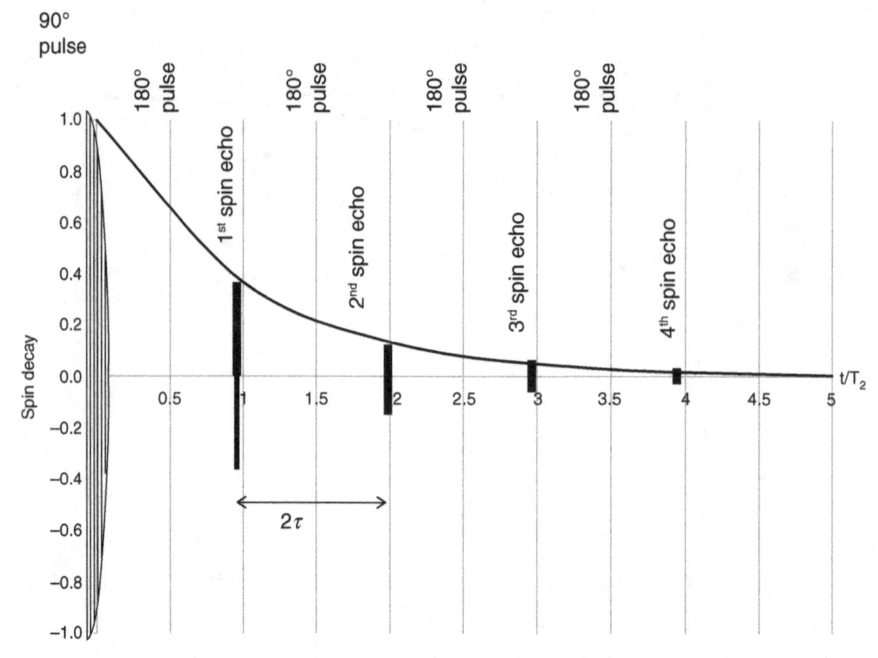

Figure 13.6. Decay of repeated Hahn echoes characterised by the transverse relaxation time, T_2 (after Hogendoorn et al. 2013).

this experimental sequence The amplitudes of successive echoes decrease because of molecular-level interactions between spins. One might imagine that each spin is a little bar magnet and when this approaches another spin, it will slightly distort the magnetic field experienced by the second spin. Over time these distortions will add up and cause the phase of the spins to become incoherent in an irreversible manner (based on private communication from Dr D. Holland, 2014). It appears that various parameters and times may be varied in order to obtain further information about the fluid in the meter (http://en.wikipedia.org/wiki/Spin_echo).

It is this effect which is used in the flowmeter and which will be explained. First of all it is useful to understand the geometry of the flowmeter.

13.4 Outline of a Flowmeter Design

As indicated in this chapter, there have been earlier designs of magnetic resonance flowmeter. The meter described in this section is, to this author's knowledge, the most recent industrial development, and we shall consider the essentials of this meter shown in Figure 13.7.

The flow tube is made of non-magnetic material to allow the magnetic fields to pass through it, and also to allow the instruments to sense and measure the magnetic fields' behaviour within the tube from outside the tube. The main magnetic field, applied externally, is across the pipe in the z direction, and the pre-magnetisation section of the flowmeter has the field in three sections, so that the field length can be varied for reasons which will be explained later. For the initial two magnets, each

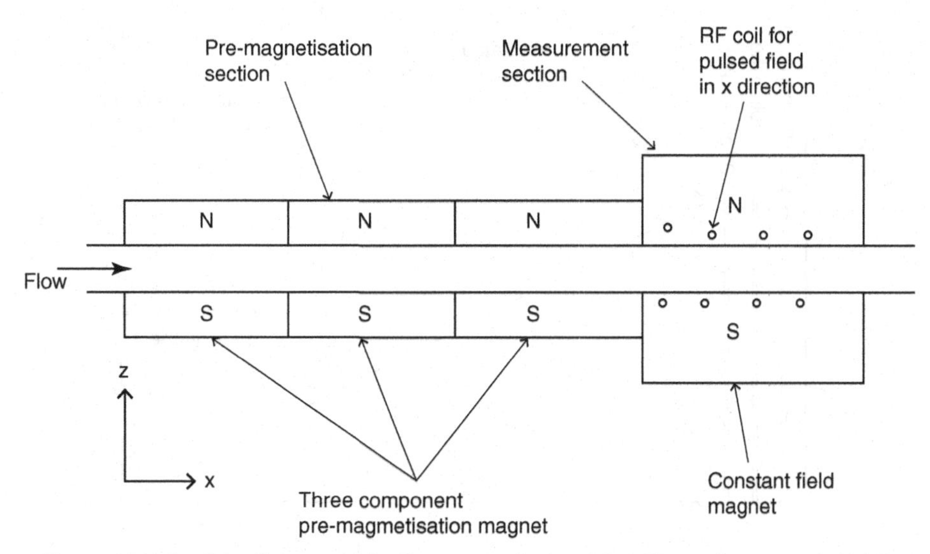

Figure 13.7. Outline diagram of the flowmeter design (after Hogendoorn et al. 2013).

consists of two magnets which are constructed on concentric mountings so that they can either enhance each other or cancel each other, and are rotated by a motor to the required position. Within the measurement section is the RF coil wound around the tube and creating a field axially along the tube. The main (external) magnetic field in the measurement section is of the same flux density as that in the pre-measurement section, but the homogeneity is an order of magnitude greater. The magnetic field appears to be of order 0.01 to 0.1 tesla.

The flow enters the pre-magnetisation section and the nucleons experience a torque turning them into line with the main (externally applied) magnetic field in the z direction. When the fluid enters the measurement section, the RF coil is energised with the Larmor frequency, the fluid nucleons are reoriented by the 90° pulse and magnetisation is moved into the x-y plane. As the fluid flows through the measurement section, that fluid fraction which has been affected by the 90° pulse will progressively flow out of the measurement section. Thus each time the 180° pulse is activated the echo intensity will be proportional to the amount of fluid remaining within the measurement section which has been affected by the RF 90° pulse. The signals from the Hahn echoes will, therefore, follow a linear curve (the linearity may depend on flow profile), Figure 13.8, from which the flow rate can be obtained by taking the slope of the line (although allowance should be made for the decay as in Figure 13.6). The 180° pulses are sent from the same coil as the 90° pulse, but the 180° pulses have a phase shift relative to the 90° pulse, which depends on the pulse sequence, and can be varied (private communication from Dr Jankees Hogendoorn).

However, various calibrations and corrections will be required. We have noted the reduction of the echo in Figure 13.6. For low flow rates and fast relaxation, an allowance will be required for this. The strength of the signal will depend on the number of nucleons affecting it, and thus a calibration will be required which may be obtained from a water flow rig.

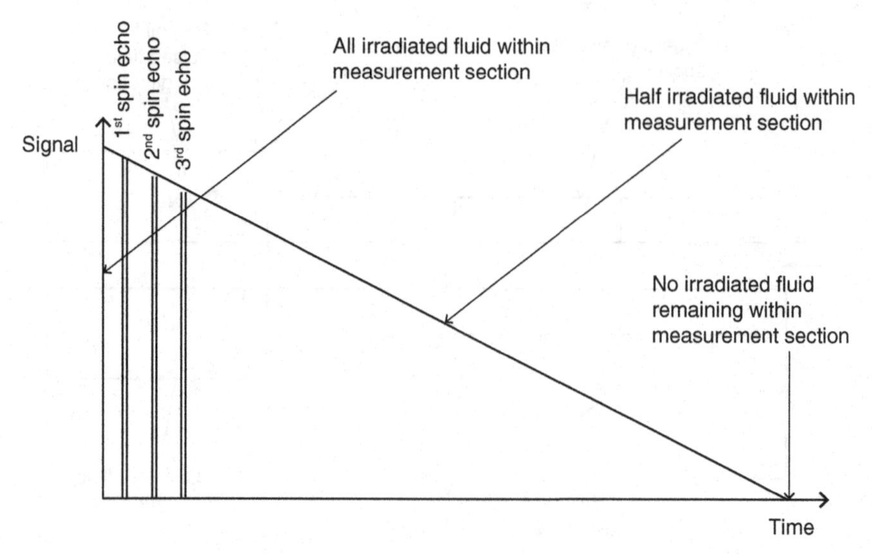

Figure 13.8. Convective decay method to obtain velocity (after Hogendoorn et al. 2013).

The presence of gas in the multiphase flow is sensed by the instrument and measurements are made of the gas volume fraction, Hogendoorn et al. (2015).

If α is the gas fraction (or gas hold-up) in the flow, and if the signal amplitudes are known for all liquid, S_l, and for all gas, S_g, then the measured signal amplitude, S_{meas}, will be given by

$$S_{meas} = \alpha S_g + (1-\alpha)S_l$$

from which α can be obtained as

$$\alpha = \frac{S_{meas} - S_l}{S_g - S_l} \tag{13.6}$$

This approach can be extended to the case where the liquid phase is made up of oil and water. However it may be necessary to do a small amount of iteration to obtain the fractions of the three components, after the ratio of oil to water, as described below, has been obtained.

The reason for the multiple magnet segments in the pre-magnetisation section is to allow the separate flow rates of oil and water to be obtained. The magnetisation of oil is quicker than that of water; i.e., oil typically possesses shorter T_1 relaxation times than water. Figure 13.9 shows, approximately, the magnetisation of these liquids after they have passed through the total externally applied magnetic field of the flowmeter (1.8 metres). The oil magnetisation has reached within about 5% of complete magnetisation by $x = 1.2$, whereas the water magnetisation is still changing over the region from $x = 1.2$ to 1.8. If the magnetisation starts at the entry to the measurement section ($x = 1.5$), neither oil nor water will have reached their full magnetisation by $x = 1.8$. The magnetisation level will then be as for $x = 0.3$ in Figure 13.9. These values, with appropriate mathematical analysis, should lead to a value for the ratio of oil to water in the flow. This ratio should, with the additional variation of

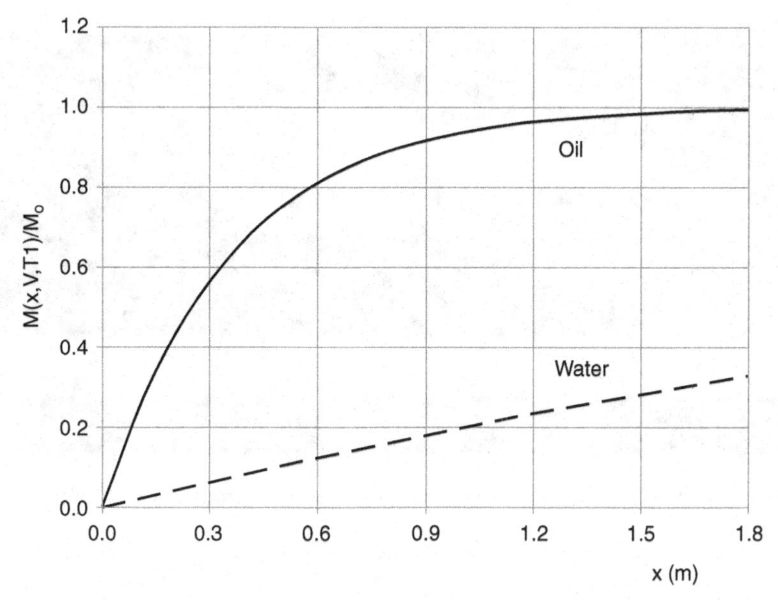

Figure 13.9. Approximate build-up of magnetisation in oil and water and hence build-up of signal level against distance through the magnetic field (Figure 13.7) for a flow velocity of about 2 m/s (after Hogendoorn et al. 2013).

the length of the magnetisation in the pre-magnetisation section, allow the relative velocities of the oil and water phases to be assessed and hence a value for the slip velocity to be obtained.

In addition the instrument has pressure and temperature sensors. The instrument is also claimed to have various diagnostic capabilities based on the data extractable from the various measurements. mentioned earlier.

Figures 13.10 and 13.11 show the commercial flowmeter and a cutaway view of it. The meter has a 4 inch-diameter, full-bore tube (100 mm), weighs about 1,000 kg and has an overall flange-to-flange length of about 3 metres. It appears to have been extensively tested and resulting modifications made. Hogendoorn et al. (2013) indicated that they were expecting the first industrial version to be available for testing in various applications in the near future. Testing has been under way since January 2014 on various multiphase flow loops and since October 2014 a meter has been installed in the field.

13.5 Chapter Conclusions

It is interesting to compare the KROHNE meter with that described by Scott (1982). It appears in the early designs described by Scott that the radio frequency pulse created a marker in the fluid which was then sensed at a downstream point a known distance away and hence the velocity was obtained. This contrasts with the Hahn echo method used by KROHNE which is presumably more reliable. (See also last paragraph of Section 13.1.)

This development has introduced a new industrial design into the flow measurement field and taken a significant step forward in the application of NMR to

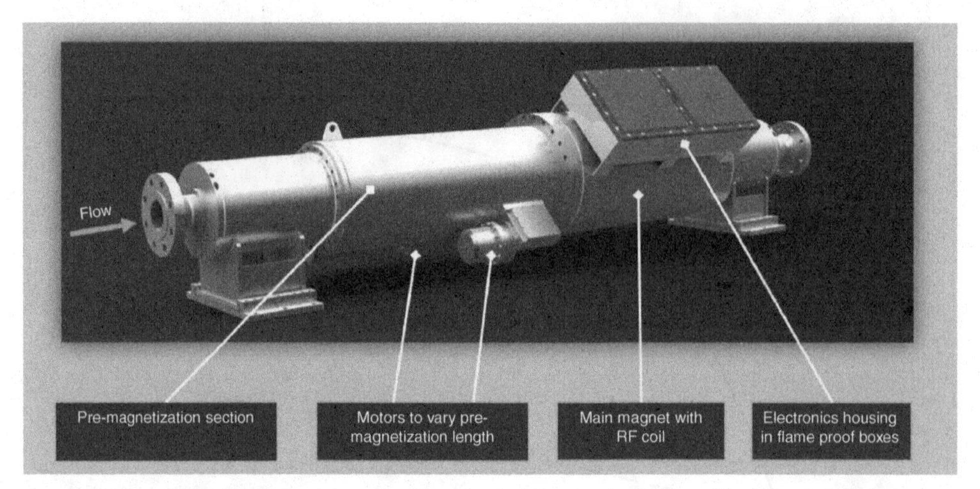

Figure 13.10. KROHNE magnetic resonance flowmeter (reproduced with permission from KROHNE).

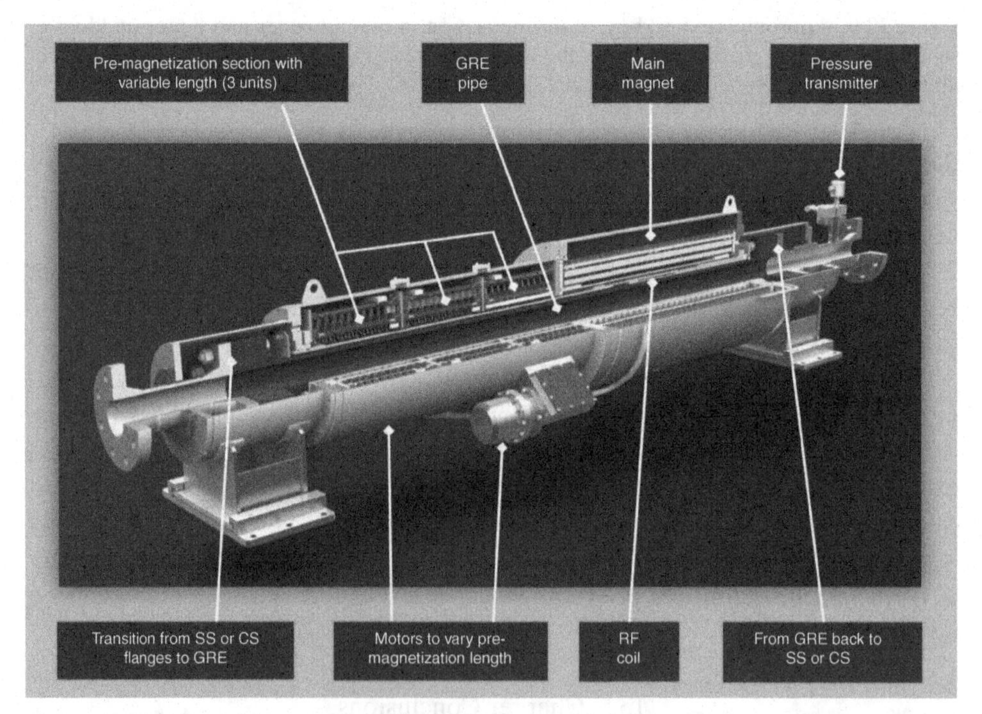

Figure 13.11. Internal layout of the KROHNE magnetic resonance flowmeter (reproduced with permission from KROHNE).

flow measurement. The physics on which other devices in this book are based might be termed pre-Maxwellian and pre-quantum theory. This meter requires physical concepts which may depend on effects bordering on quantum theory. Its introduction and commercialisation is likely to open a very interesting approach which could have wide implications.

14

Ultrasonic Flowmeters

14.1 Introduction

The first proposal for the use of ultrasound for flow measurement, according to Thompson (1978), seems to have been in a German patent of 1928. It was not until after 1945 that the idea became more widely proposed, when the development of piezoelectric transducers made ultrasonic applications really attractive. Fischbacker (1959) provided an early review of ultrasonic flowmeters. Sanderson and Hemp's (1981) review provided a further source of information.

The main types of meter are:

Transit-time, also known as a time-of-flight flowmeter, the most accurate of the family, is available as a spool piece or clamp-on meter for liquids and gases. Ultrasonic meters can also be retrofitted into a pipe. Measurement uncertainty will be from a fraction of a percent for spool piece meters to of order 5% for clamp-on. A transducer in direct contact with the fluid is said to be "wetted", as compared with one which is fixed to the outside of the pipe, or has a protective layer between it and the fluid. Transducers for gases are usually expected to be in contact with the gas i.e. wetted, a rather inappropriate word for this context! Clearly, for the clamp-on versions this is not the case.

The **Doppler** meter is suitable for predominantly liquid flows. It has an important niche in some applications where there is an adequate second component in the fluid to provide reflection of the beam, but the meter is unlikely to operate satisfactorily if the flow does not have a continuous liquid component. It is sensitive to installation. Measurement uncertainty is unlikely to be better than about ±2% of full scale and at worst it may be indeterminate. With gating of the returning pulse a more powerful instrument should be achieved. It was initially oversold to such an extent that many people assumed that, when an ultrasonic flowmeter was mentioned, it was a Doppler meter.

The cross-correlation flowmeter should be capable of reasonable precision (between the previous two devices), requires a disturbed liquid or multiphase flow to operate satisfactorily, and has mainly been of interest for measurement in multiphase flows.

However, recently there has been a new clamp-on design which is claimed to be for gases.

It is possible that other meters which have ultrasonic sensing, such as the vortex-shedding meter, may be referred to as ultrasonic meters.

Beyond these there is the possibility of using the deflection of the ultrasonic beam to measure flow rate, but to my knowledge this has not been done commercially. There is also the possibility of measuring density and so turning the meter into a mass meter.

It is important not to confuse the transit-time meter with the Doppler meter and to appreciate that the transit-time ultrasonic meter is a far more accurate instrument than the Doppler meter, and should be considered seriously in a wide range of applications, especially involving non-conducting liquids or gases.

We shall treat each type of ultrasonic flowmeter in turn. All are based on the fact that ultrasound is made up of acoustic waves at frequencies above the audible range. As a form of sound it travels with the speed of sound relative to the medium, and consists of a compression wave in gases and liquids. However, where the wave is transmitted through a solid it can also move as a shear or transverse wave, due to the elasticity of the solid.

Sanderson (2004) edited an issue of the *Journal of Flow Measurement and Instrumentation* on ultrasonic flow metering. It includes papers on non-Newtonian fluids, installation and pulsation effects. This chapter should demonstrate that transit-time technology is of high accuracy (Sanderson 1999).

14.2 Essential Background to Ultrasonics

In Figure 14.1(a) we note that the ultrasound is carried by the fluid, so that its speed is the sum or difference of its own, c, and that of the fluid, V. This is the basis of the transit-time meter, which uses the difference in time of transit in an upstream and a downstream direction. If the sound is crossing the flow then the apparent speed is obtained from the hypotenuse of the triangle in the third example of Figure 14.1(a).

In flowmeters, the continuous wave [Figure 14.1(b)] is sometimes used. The diagram shows the main features of the wave pattern, with the wavelength and the high- and low-pressure regions. The frequency, f, is related to the sound speed, c, and wavelength, λ, by $f = c/\lambda$. The intensity of the transmission varies according to the angle of the perpendicular axis through the transducer. Three patterns are shown in Figure 14.1(c) for three different ratios of the transducer radius, a, to the wavelength. The higher the frequency, and the shorter the wavelength, the more pencil-like the transmitted beam. A narrow beam is preferable to a wide one, as it is less likely to cause spurious reflections, and its path will be better defined. For a speed of sound in water of 1414 m/s a frequency of 400 kHz will result in a wavelength of about 3.5 mm, while 1 MHz will result in 1.4 mm. For a speed of sound in air of 343 m/s a frequency of 40 kHz will result in a wavelength of about 8 mm, while 100 kHz will result in about 3.4 mm. Thus for air with $\lambda = 2a$ the diameter of the piezoelectric crystal to obtain the middle distribution in Figure 14.1(c) at 40 kHz will be about 8 mm,

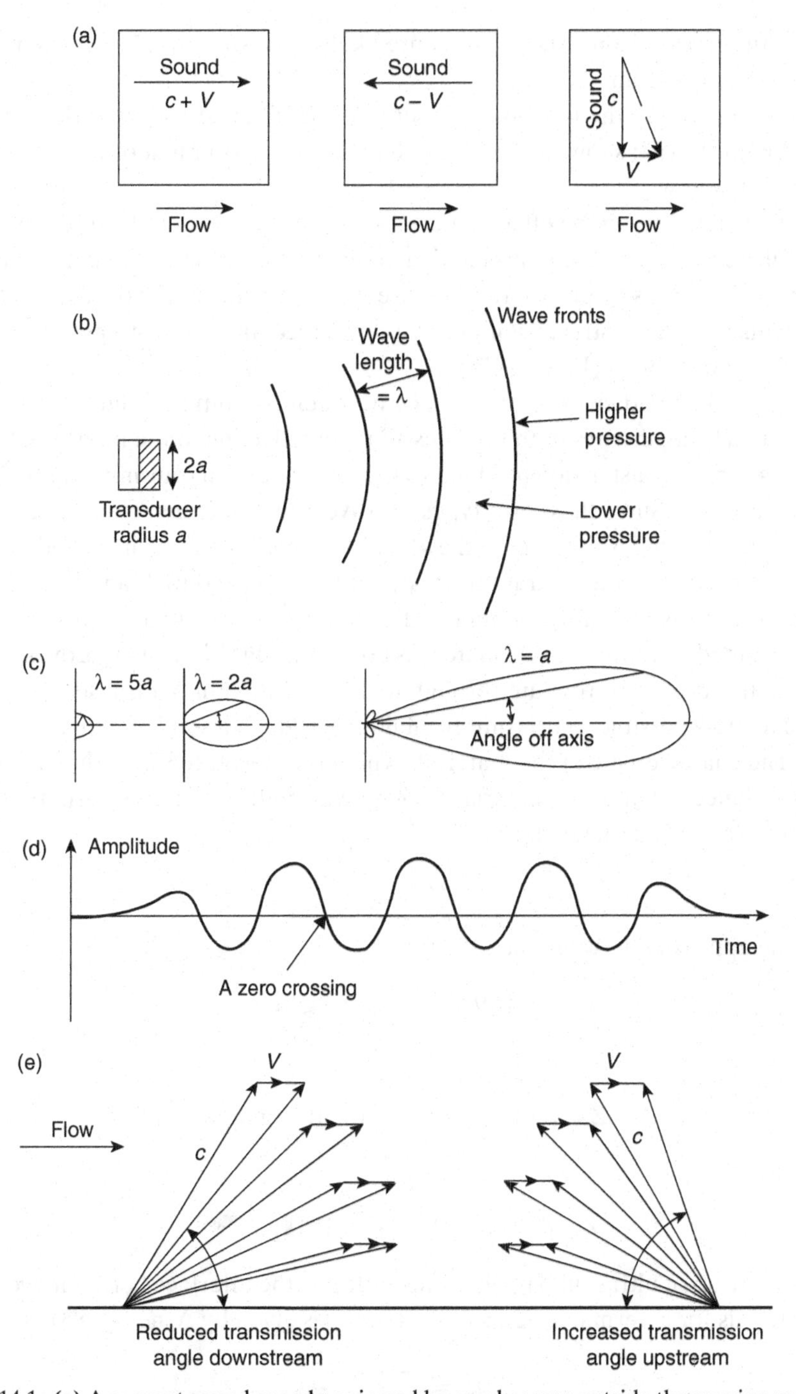

Figure 14.1. **(a)** Apparent sound speed as viewed by an observer outside the moving medium. **(b)** Ultrasonic waves move forward with the speed of sound and with a frequency λ from a source. **(c)** Radiated intensity variation with angle from a piston transducer set in a plane wall (after Morse and Ingard 1968). **(d)** Wave packet. **(e)** Ultrasound waves (like sound waves) can be bent by the flow and reinforced or dispersed.

whereas the crystal diameters for water are likely to be such that $\lambda \ll a$, and result in a narrow intense beam.

More commonly the ultrasound is sent in a small burst (or pulse or packet) of waves. Figure 14.1(d) shows an idealised diagram of a short burst of waves transmitted across the flow.

However, the sweeping effect, Figure 14.1(e), will mean that the direction of the beam will be a compromise between high and low flow rates to ensure that it arrives at the receiving transducer. Another feature of the bending of the beam is that there is a tendency for downstream sound to be reinforced and for upstream sound to be diffused and weakened (Lamb 1925).

When ultrasound crosses an interface between two different fluids or between a fluid and a solid, there is both transmission and reflection at the interface. Liquids and gases can only sustain longitudinal waves, but solids can transmit longitudinal or transverse waves. Since these two types of wave have different sound speeds, there may be multiple angles of waves in the solid. This change is known as mode conversion, see Krautkramer and Krautkramer (1990) referenced by Mahadeva (2009).

The transmission will be greater if the product of density and wave speed (or acoustic impedance) for the two materials are similar in size. Thus, it proves difficult to transmit ultrasound from air, through metal, and then into a transducer. On the other hand it is possible to transmit from liquid through a solid and then to a transducer. The characteristic of the materials which is relevant to this is the impedance.

Impedance = density of material through which ultrasound is transmitted × velocity of ultrasound in the material

$$Z = \rho \times c \tag{14.1}$$

and using approximate values at 20°C for air

$$Z = 1.19 \times 343 = 408 \text{ kg m}^{-2}\text{s}^{-1}$$

for water

$$Z = 1000 \times 1414 = 1.41 \times 10^6 \text{ kg m}^{-2}\text{s}^{-1}$$

and for steel

$$Z = 8000 \times 5625 = 45 \times 10^6 \text{ kg m}^{-2}\text{s}^{-1}$$

The proportion of ultrasound power transmitted at the interface between two different materials for a normal incidence plane wave is given by Asher (1983) as

$$P_T = \frac{4Z_1Z_2}{(Z_1 + Z_2)^2} \tag{14.2}$$

and that reflected as

$$P_R = \frac{(Z_1 - Z_2)^2}{(Z_1 + Z_2)^2} \tag{14.3}$$

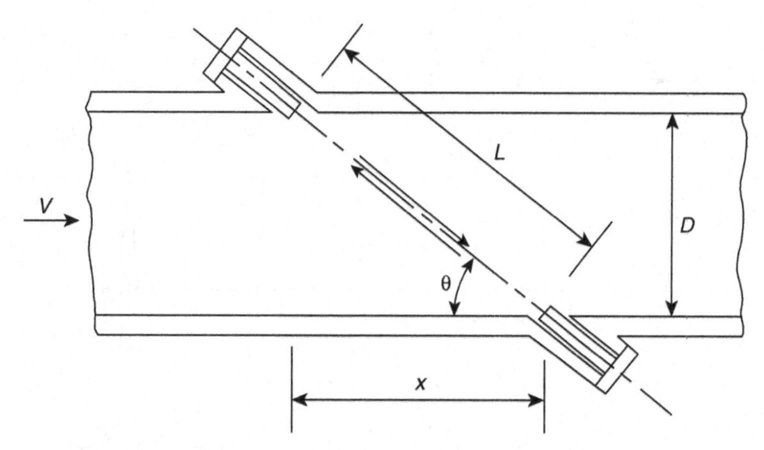

Figure 14.2. Geometry of the transit-time flowmeter.

Thus while transmission from steel to water is about 12%, that from air to steel is very small. In discussion at a symposium on flow measurement (Trans. Soc. Instrum. Eng. June 1959) Bertele of ICI raised the problem of getting enough energy into the gas. However, as mentioned, some manufacturers now offer clamp-on ultrasonic gas meters in which the ultrasound necessarily passes from solid to gas and gas to solid, a result not considered realisable a few years ago.

The final point to make is that attenuation increases with increasing frequency, and tends to be greater in gas than in liquid.

14.3 Transit-Time Flowmeters

14.3.1 Transit-Time Flowmeters – Flowmeter Equation and the Measurement of Sound Speed

In this section the operation is described, and the key flowmeter equations are given. The simple theory of these devices is given in Appendix 14.A with a reference to some fuller solutions.

The transit-time flowmeter depends on the slight difference in time taken for an ultrasound wave to travel upstream rather than downstream. Thus waves are launched each way, their time of transit is measured, and the difference can be related to the speed of the flow. Apart from the smallest-bore flowmeters, it is usual to send the beam across the flow but not at right angles to the flow, so that there is a component of the fluid velocity along the path of the acoustic beam. Figure 14.2 is a simple diagram of an ultrasonic flowmeter geometry.

Transit-time Flowmeter

We can show that for a uniform velocity in the pipe (Appendix 14.A) we can obtain an expression in terms of the transit times of upstream, t_u, and downstream, t_d, wave pulses and the time difference, Δt.

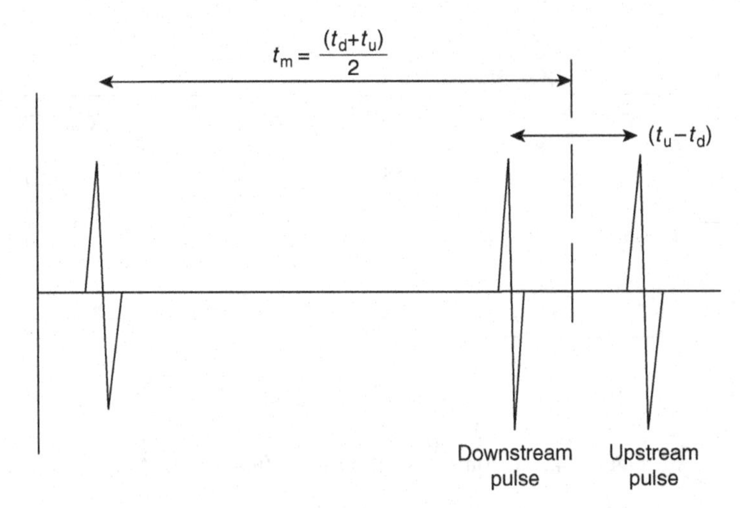

Figure 14.3. Diagram to explain transit-time measurement.

$$V = \frac{L^2 \Delta t}{2Xt_u t_d} \quad (14.4)$$

where L is the path length, and X is the axial spacing between transducers. This is one form of the equation, and others are set out in Appendix 14.A.

The reason for the product $t_u t_d$ in Equation (14.4) is to eliminate the speed of sound, which otherwise appears in the equation. By measuring the mean time of transit, and $\sqrt{t_u t_d}$ approximates to this, and knowing the transit distance we can obtain the speed of sound.

In the transit-time method pulses of sound of a few cycles in length are transmitted in each direction (Figure 14.3), and the time taken for them to reach the receiving transducer is measured. Possible orders of magnitude for pulse sizes may be, for liquid: pulse packets of about 6 cycles with 0.5 to 5 MHz carrier, and for gas: pulse packets of about 13 cycles with 100 to 500 kHz carrier.

We obtain from Equation (14.4) assuming V is constant across the pipe,

$$q_v = \frac{\pi D^2 V}{4}$$

$$= \frac{\pi D^2 L^2 \Delta t}{8Xt_u t_d} \quad (14.5)$$

It is important to appreciate the value of the times involved in these measurements. If the flowmeter is to discriminate to better than 1% we may define $E(\Delta t)$ as $\Delta t/100$ and for two meters we obtain the values in Table 14.1 (Baker 2002b/2003).

Table 14.1 highlights the problem of the very small time differences which we need to measure in the transit-time flowmeter. For the smaller tube this is to better than one nanosecond. However, one advantage of this system is that any echoes or standing waves have time to die down, another is that the same path can be used for each direction of the pulse, and finally, the individual measurements can be used as a tolerance check to ensure that spurious measurements are rejected (Scott 1984).

Table 14.1. *Transit-time meter times*

Diameter (mm)	$(t_u + t_d)/2$ (s)	Δt (s)	$E(\Delta t)$ (s)
100	10^{-4}	10^{-7}	10^{-9}
300	3×10^{-4}	3×10^{-7}	3×10^{-9}

$\theta = 45°$, $V = 1$ m/s, and c is taken as 1430 m/s the value for water at 4°C.

The correction which must be made to adjust the measured times of transit for a clamp-on flowmeter for use in Equation (14.5) is explained in Section 14.A.1 (Transit time). The allowances for delay times in the transducer wedges and in the wall are identified.

Sing-Around System

An alternative way to eliminate sound speed is the sing-around method (Figure 14.4) discussed by Suzuki, Nakabori and Kitajima (1975; cf. 1972). In this method each of the paths, upstream and downstream, is operated in such a way that when the receiving transducer receives the pulse, it triggers a new pulse from the transmitting transducer. Thus the frequency of pulses is dependent on the velocity of sound and of the flow. We find that the frequency of pulses is given by

$$\Delta f = 2XV / L^2 \tag{14.6}$$

which has the virtue of relating V to the frequency difference without requiring the value of c, the sound speed. Again we can obtain typical values for the frequencies from Table 14.2 (Baker 2002b, 2003). Making the approximation that V is constant across the pipe we obtain

$$q_v = \frac{\pi D^2 L^2 \Delta f}{8X} \tag{14.7}$$

The limitation of the sing-around method relates more to the very low frequencies and the long time periods needed to achieve sufficient accuracy. To achieve 1% uncertainty we shall need to have a measuring period of greater than one second for the most favourable of the cases in Table 14.2. Carlander and Delsing (2000) reported tests of a 10 mm sing-around design with upstream distortion and pulsation.

Timing can also be achieved in some cases by obtaining the phase shift between the two arriving waves to obtain the time difference (Baker and Thompson 1975). Other problems such as standing waves and spurious transmission paths which cannot be gated out may need to be considered in relation to any of the methods chosen. Unwanted reflections within the pipe and their effect on the measurement uncertainty may be reduced by the use of higher frequencies. These attenuate reflections, but also attenuate the signal. Cunningham and Astami (1993) produced curves of upper and lower frequency limits against pipe diameter for various reflection settling times. The upper limit assures signal attenuation is not too great while the lower limit is to ensure that attenuation of reflections is sufficient. They reckoned that the received signal amplitude should be at least 30% of that transmitted.

Table 14.2. *Sing-around meter frequencies*

Diameter (mm)	$V = 1$m/s		V = 10m/s	
	f_d (Hz)	Δf (Hz)	f_u (Hz)	Δf (Hz)
100	10,117	10	10,162	100
300	3,372	3.3	3,387	33

$\theta = 45°$, $V = 1$ m/s, and c is taken as 1,430 m/s, the value for water at 4°C.

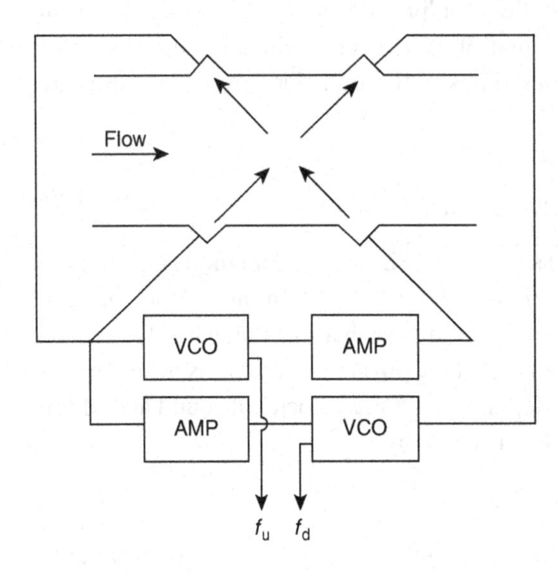

Figure 14.4. Diagram to explain sing-around system.

Mass Flow Rate

An alternative approach, rather than measuring the value of c, is to use the fact that $c^2 = 1/(k_s\rho)$ for a particular liquid and $c^2 = \gamma p / \rho$ for a particular gas [Equation (14.A.3)]. For a liquid

$$\Delta t = 2k_s X \rho V$$

$$= \frac{8k_s X q_m}{\pi D^2} \tag{14.8}$$

where the adiabatic compressibility would need to be known, and since it varies with temperature, one would need to include some temperature compensation. For a gas

$$\Delta t = \frac{2\rho XV}{\gamma p}$$

$$= \frac{8Xq_m}{\pi D^2 \gamma p} \tag{14.9}$$

thus relating the mass flow rate of a gas to the time difference, the pressure of the gas and the geometry of the flow tube. Table 14.3 gives values of γ, and a flowmeter using this method would need, as well as the pressure measurement facility, a selection switch for the type of gas.

Table 14.3. *γ for various gases from Haywood (1968)*

Type of Gas (number of atoms)	Examples of Gases	Value of γ	Mean Value of γ^a
Monatomic	Argon	1.67	1.67
Diatomic	Air	1.40	1.40
	Nitrogen	1.40	
	Oxygen	1.40	
3–5 atoms			1.29
3 atoms	Carbon dioxide	1.31	
5 atoms	Methane	1.31	
8–9 atoms			1.17
8 atoms	Ethane	1.19	
9 atoms	Propylene	1.15	

[a] Based on number of atoms.

An alternative method to obtain the mass flow of gases is to obtain the density from the value of the impedance discussed earlier. Guilbert and Sanderson (1996a) have described such an ultrasonic mass flowmeter which combines a fairly standard configuration of ultrasonic flowmeter with an impedance density cell, in which the impedance of the liquid is obtained through the interface between the liquid and a material which divides the cell into compartments. The material selected was polyamide/imide. They also used the time between reflected signals to obtain the sound speed in the divider and hence its impedance, see also Hirnschrodt et al. (2000) [cf. Chapman and Etheridge (1993) who described the evolution of ultrasonic flowmeters for gas].

Van Deventer (2005) described a new design of meter capable of measuring mass flow rate using transducers operating as densitometers. Rychagov et al. (2002) considered methods of obtaining density and mass flow measurements based on a meter with quadrature integration of the flow profile [cf. Matson et al. (2002), who discussed using density and achieving mass flow, and Nygaard, Mylvaganam and Engan (2000) on impedance measurements to obtain gas density and so to measure mass flow].

The very important development of the clamp-on flowmeter will be discussed later. At this stage it is appropriate to make the point that the simple theory set out earlier is valid for the beam transmission so long as it is in the fluid. As mentioned earlier, the transmission between different materials is more complicated (see Figure 14.A.2).

14.3.2 Effect of Flow Profile and Use of Multiple Paths

Single Path

The simple analysis outlined earlier has assumed that the flow profile is uniform. In fact, if we allow for varying velocity along the path of the ultrasonic beam Equation (14.4) may be rewritten in the form

$$V_{\mathrm{m}} = \frac{L^2 \Delta t}{2 X t_{\mathrm{u}} t_{\mathrm{d}}} \tag{14.10}$$

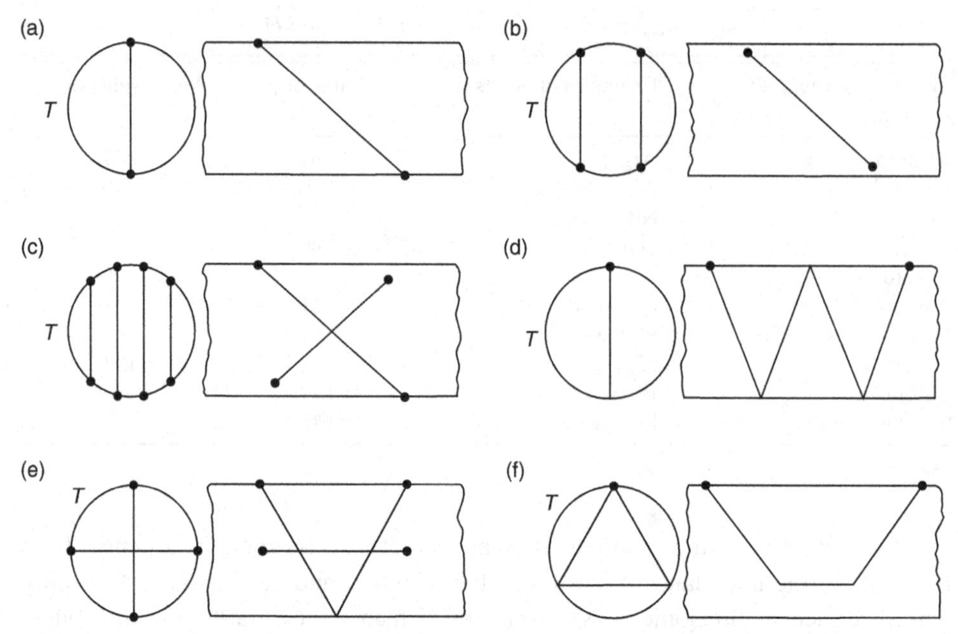

Figure 14.5. Acoustic beam positions [T (top) indicates possible preferred orientation in liquid flow].

where V_m is the mean velocity along the path of the beam because the flowmeter averages velocity along the path of the beam. In general this average does not correspond to the area average of the velocity required for an ideal flowmeter. Unfortunately for circular pipes with paths crossing the diameter, Figure 14.5(a), the mean velocity along the path gives too heavy an emphasis to the flow at the centre of the pipe. We, therefore, need to introduce a factor F to allow for the difference between the mean velocity in the pipe cross-section and the mean velocity along the diametral path of the ultrasound, V_m,

$$F = \frac{\text{mean velocity in the pipe}}{\text{mean velocity along the diametral path of the ultrasound}}$$

This implies that change of profile from laminar to turbulent and through the turbulent range will affect the calibration even though the flow profile is fully developed and not distorted by upstream fittings etc. Equations are given in Chapter 2 which simulate the profile shape for circular pipes and which could be integrated along the path of the ultrasound and across the pipe section to obtain the value of F. Figure 14.6 shows the variation of F, based on Table 2.2 and Equation (2.6), against Reynolds number. It is easy to show (Kritz 1955) that the effect of integrating a laminar profile across a diameter of the pipe causes a signal about 33% high due to the overweighting which the velocities near the axis cause, compared with a uniform profile. Through the turbulent range, using the $1/n$ power law (see Chapter 2; see also Section 2.A.1 for other profiles) for a Reynolds number variation between 10^3 and 10^6, a variation of about 3% occurs. Apart from Kritz, others have provided calibration factors

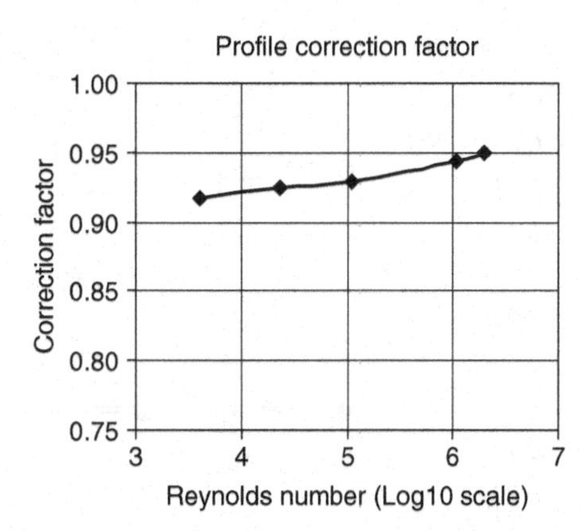

Figure 14.6. Plot of profile correction factor, F, based on relation between n and the Reynolds number.

(Del Grosso and Spurlock 1957; Kivilis and Reshetnikov 1965). Thus for a meter which is constructed this way there is an inherent variation in calibration factor, which cannot be wholly allowed for in the transmitter circuitry because of Reynolds number dependence on factors other than velocity only. Scott (1984) made the point that a change in viscosity from 1 cP to 10 cP causes a change in profile which will account for a −1% error. Manufacturers may build in software to obtain the value of F allowing, as far as possible, for parameters such as temperature which will affect the Reynolds number, the pipe profile and the consequent value of F.

Mid Radius Paths

One way to improve this is to offset the paths from the axis [Figure 14.5(b)], in which case a position at about the mid radius is found to be a good compromise (Appendix 14.A). Baker and Thompson (1975, 1978) showed the effect of offset paths, Figure 14.7, for varying Reynolds number on the signal with the paths offset at 0.5R, 0.505R and 0.523R from the axis. This shows that a position of 0.523R is ideal for the turbulent range, whereas 0.505R is probably a reasonable compromise for the whole Reynolds number range. They also suggested that the meter should have low sensitivity to swirl, and to flows which have symmetry of size, but opposite sense about the axis. This concept has been developed by others such as Lynnworth (1978), Drenthen (1989) and Ao et al. (2011) using variants of the path in Figure 14.5(f) with half radius or near half radius and internal reflections off the pipe wall. A diametral path has also been added in some cases.

Multipath

Estrada, Cousins and Augenstein (2004) discussed the background of the multipath meter conceived in 1966 which made use of four paths [e.g. Figure 14.5(c)] and selected their positions so that a Gaussian quadrature integration (a numerical method for integrating general curves, which is similar but more complicated than

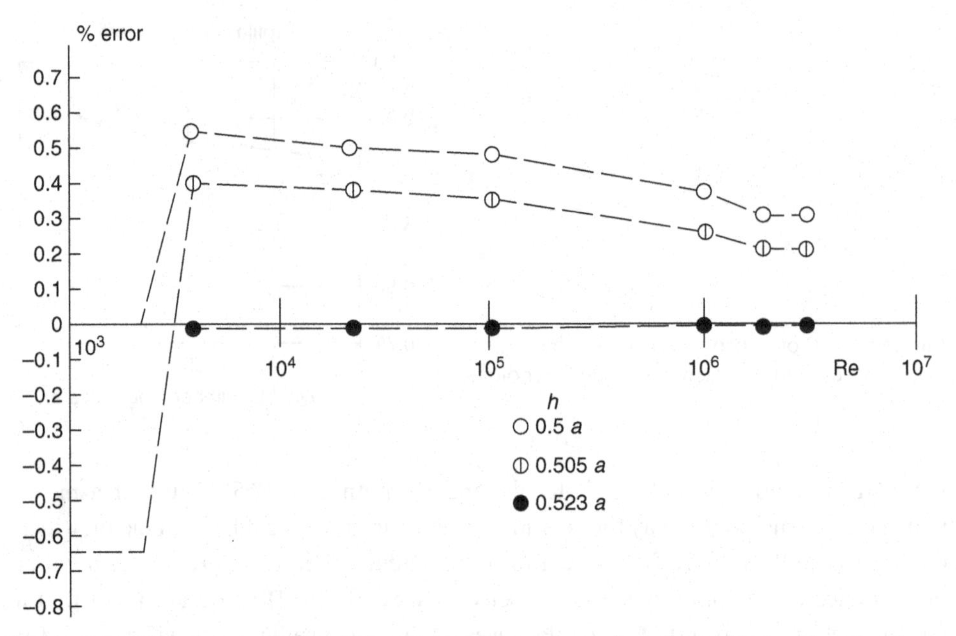

Figure 14.7 Calculated values of the percentage error against Reynolds number for three positions of the acoustic paths (after Baker and Thompson 1975).

Simpson's rule – see Appendix 14.A) could be used to give a correct mean velocity signal for profiles formed from polynomials of 7th power or less (Hastings 1968, 1970). Again the experimental evidence since these early papers confirms the quality of the integration achieved, and the potential accuracy of the design.

Using the velocity measured on multiple paths Vaterlaus (1995) gave the following positions for multiple paths and the weight to be given to each:

Planes	Distance from centre plane	Weight
2	±0.520	0.5
3	0	0.444
	±0.774	0.277
4	±0.339	0.326
	±0.861	0.174

The value given for two paths agrees well with Baker and Thompson (1975). Taylor and Cassidy (1994) quoted weightings for four planes which to three significant figures agreed with these (cf. Pannell, Evans and Jackson 1990 on a generalised approach to positioning chordal paths).

A variation on the positioning of the paths, as an alternative to the Z-geometry arrangement in Figure 14.5(a), uses internal reflection of the beams to allow transmitter and receiver to be on the same side of the pipe [Figures 14.5(d,e)]. This is known as reflex mode. This can be either V with one reflection or W with three reflections. It can also be used to create subtle reflections which essentially give the path position the benefits of off-axis integration [Figure 14.5(f)], and multiple paths around the pipe to give even better integration of the profile. An example of this results in the paths lying on an equilateral triangle if viewed axially down the pipe, thus using the mid radius point, but also covering a large area of the tube (Drenthen and Huijsmans 1993). Vontz and Magori (1996) described a meter with a helical sound path. It had a tube of square cross-section in which two transducers and two special reflectors were mounted.

Jackson, Gibson and Holmes (1989) developed an ultrasonic flowmeter using one transmitting transducer and a lens system made of perspex which produced a diverging ultrasound beam detected by three widely spaced receiving transducers. The chords of the three paths are all designed to be of different lengths and angles. In the test meter a further path was used across the flow to obtain the sound speed. It is not obvious that this design contributes significantly to the normal design with three complete paths, or indeed, the designs using reflecting beams to obtain off-diametral paths. The advantage of reciprocating paths is that the sound speed is obtained using the mean.

Jackson, Gibson and Holmes (1991; cf. 1989) described a three-path meter where the path angles were chosen to obtain information about the flow profile. The paths were in a diametral plane and with chordal paths on each side at angles of 44° and 52° to the diametral plane. For axisymmetric profiles performance was of order 2% (assumed to be of rate). Asymmetry was detected unless there was a symmetry about the diametral plane but precise flow measurement with asymmetry was not obtained.

Van Dellen (1991) demonstrated compensation by using multiple paths. Such designs allow self-checking routines where speed of sound is compared, and successive measurements can be used.

Multipath meters are an option for hydroelectric turbine efficiency measurement. Lowell and Walsh (1991) reported on the use of eight-path, crossed-plane meters and concluded that ±0.5% of flow rate could be achieved with appropriate relative positioning of acoustic plane and upstream fittings.

Drenthen (1996) proposed an important multipath design. He proposed two triangular paths with double reflections from the wall and three diagonal paths with single reflections, thus achieving a total of 12 traverses of the pipe and an ability to sense swirl and asymmetry in the profile. The diagram from his patent is shown in Figure 14.8. An 18-beam design by Guizot (2003) has the 18 paths in nine different planes. There should be adequate redundancy to ensure high reliability and capability to deal with and possibly analyse flow velocity profile perturbations and swirl.

(See also Smith and Morfey (1997) on the effect of beam bending due to velocity variation across the pipe).

Other Suggestions

Bragg and Lynnworth (1994; cf. Lynnworth 1989, 1994) proposed the use of a single port with two axially spaced transducers. As an example they suggested that a spacing of 60 mm axially for a velocity of 1 m/s and a sound speed of 343 m/s would

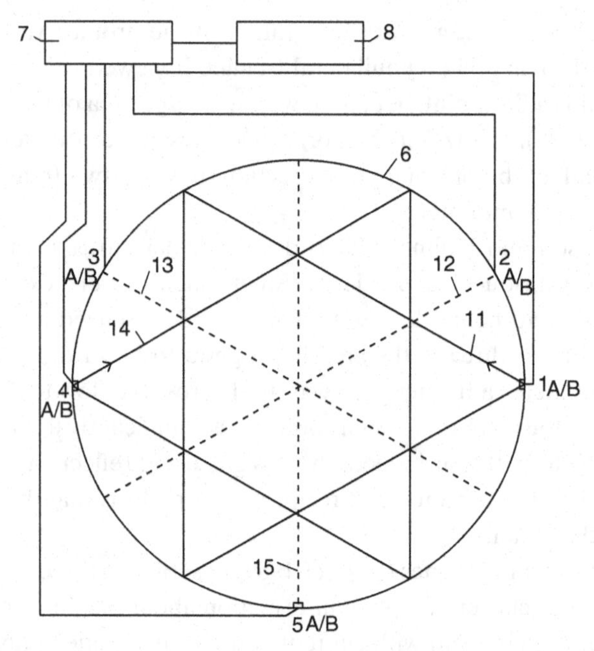

Figure 14.8. Path layout in a multibeam flowmeter after Drenthen (1996).

result in a 1 μs transit time difference or a 10 ns resolution i.e. 1%. One application might be non-combustible gases. They also suggested that two angled transducers in the same port can transmit and receive a signal which forms a triangular path in the pipe cross-section and thus senses swirl. Lynnworth, Hallewell and Bragg (1994) developed a profile-measuring method using the one port with either a traversing reflector or a series of fixed reflectors.

A very small meter has been reported by Ishikawa et al. (2000) with a pipe diameter of less than 1 mm for very low flow rates of less than 1 ml/min.

14.3.3 Transducers

One of the most important features of the mechanical design of the meter is the transducer and its mounting. This must achieve:

- efficient transmission and reception of acoustic signals through the interface;
- negligible acoustic transmission through the body of the flowmeter;
- accurate and permanent positioning;
- no adverse effects due to operating fluid;
- trouble-free performance.

Transducers fall into a number of categories and these will be worth bearing in mind in the following discussion:

Factory installed: wetted transducers – open cavity
non-wetted transducers – open cavity
non-wetted transducers – filled cavity

Note: For any of these the spool piece may be factory wet calibrated. Only the first is normally used for gases.

Retrofit: wetted transducers – open cavity
non-wetted transducers – open cavity

Note: Only in situ calibration possible. Dry calibration by very careful measurement may be acceptable.

Baumann and Vaterlaus (2002) described a method based on the "drill and tap" hot-tapping process which they claimed allowed the installation of wetted transducers in ducts with as much as 35 bar pressure.

Clamp-on: non-wetted transducers – transmission through pipe wall

Figure 14.9 shows some typical mounts for transducers. Figure 14.9(a) shows a wetted transducer in an angled mounting block. The crystal emits a pencil beam of sound through perpendicular interfaces and it is not therefore refracted. Figure 14.9(b) shows an alternative arrangement with mounts perpendicular to the tube and transducer with an angled end in which the crystal is mounted. If the liquid is aggressive or particle laden, some form of window may be required. In this case the transducer may be mounted as shown in Figure 14.9(c) where a metal window is used but where all interfaces remain perpendicular to the beam. The cavity is a source of flow disturbance and solid deposit. Both may cause signal failure or flow measurement errors. Figure 14.9(d) has the cavity smoothed off with a filler of some sort. Ideally this filler should have an acoustic refractive index similar to the liquid, thus keeping refraction to a very low value, and the bending of the acoustic ray on crossing the interface will be small. (I have tended to use ray rather than beam where the geometry is concerned). However, changes of fluid or temperature may introduce errors. An important advantage of the mounts shown in Figures 14.9(a) and (b) is that alternative transmission around the tube wall and outside the fluid is reduced and in practice the transducer mount can be made to absorb it. With the mounting systems in Figures 14.9(c) and (d) the signal processing must discriminate between the fluid-borne signal and the spurious signals around the wall of the tube. Sanderson and Hemp (1981) gave some details of ways to increase the bandwidth of the piezoelectric plate.

Figure 14.9(e) shows a clamp-on system in which the transducer mount is entirely external to the tube and may be mounted on existing tube. Here the problems of the type found in the mounts shown in Figures 14.9(c) and (d) are greatly increased.

For very small pipe sizes the limitations of pipe diameter and consequent short pulse times can be overcome by a flow tube and transducer mounting system of the type shown in Figure 14.9(f). The flow enters a section of tube in which the acoustic beam is axial and hence there is no limitation on the separation of transducers apart from those resulting from the application. Temperley, Behnia and Collings (2004)

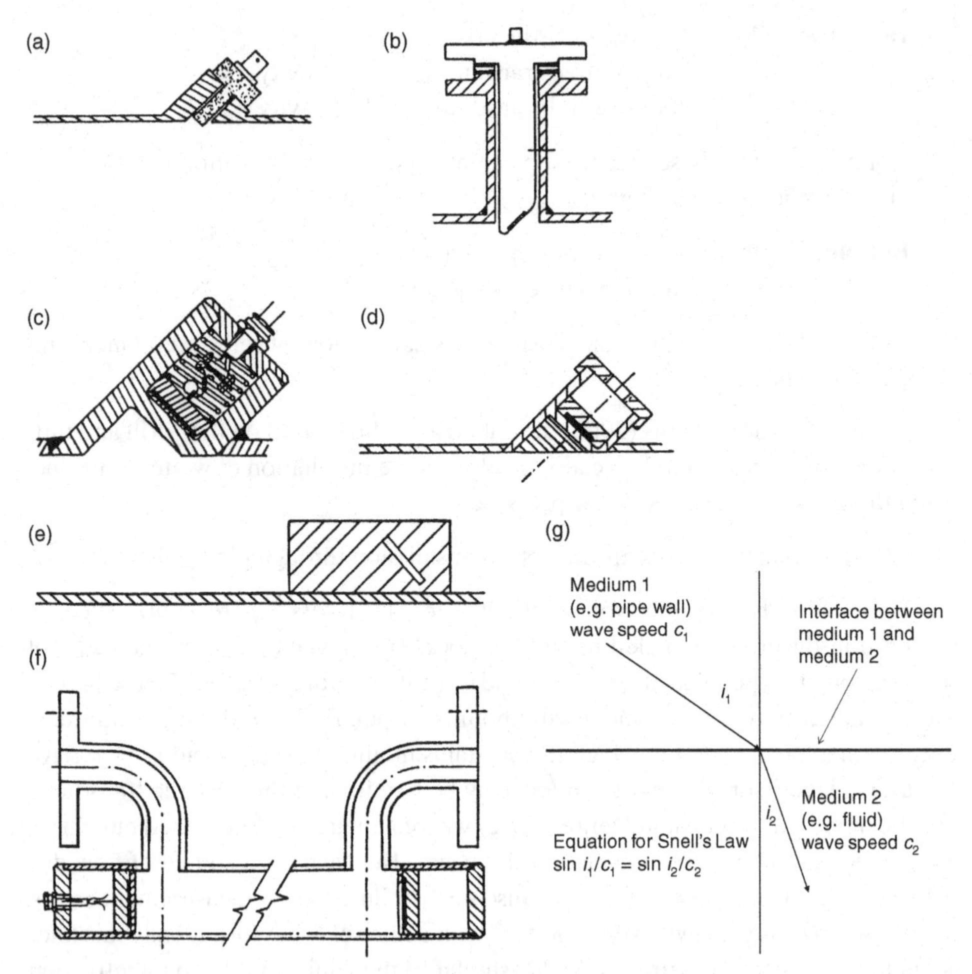

Figure 14.9. Transducer mounts; **(a)** in line – wetted; **(b)** angled – wetted; **(c)** in line – window protection; **(d)** in line but with cavity filled with epoxy or other filler; **(e)** clamp on transducer with crystal embedded in the mount; **(f)** in line transmission for small tubes with either wetted or protected transducers; **(g)** Snell's Law.

used CFD and laser Doppler anemometer on such a design, using a layout similar to Figure 14.13(b), but with larger transducer housing to give a smooth flow into the pipe. Sanderson and Al-Rabeh (2005) described a novel ultrasonic flowmeter for low flow rates in small tubes, using axial propagation. They used two piezo rings of transducers spaced axially along the tube to transmit and receive signals using Hemp's weight theory and reciprocal theorem. A long wavelength was used, comparable with or greater than the pipe diameter, and with a frequency of 200 kHz.

For the clamp-on meter, in addition to the problem of wave transmission through the meter body from one transducer to the other and the possible impedance mismatch, another problem is that of refraction through the pipe wall. The problem can be explained by reference to Snell's Law. Figure 14.12 shows the path of a shear wave through the interface from the transducer block to the pipe wall, and then from the pipe wall to the fluid. Snell's Law defines the relationship between the velocity of waves in each medium and the angle between the ray and the perpendicular to the surface at the point where the ray passes through the surface interface, Figure 14.9(g). If c_1 is the speed of the sound wave in medium 1 and c_2 is the speed of the sound wave in medium 2, and if the angle between the incident ray and the perpendicular to the surface in medium 1 is i_1 and if the angle between the refracted ray and the perpendicular to the surface in medium 2 is i_2, then $\sin i_1 / c_1 = \sin i_2 / c_2$, Figure 14.9(g). Since the speed of waves in the pipe wall is very much greater than that in a liquid the ray becomes close to the perpendicular after refraction into the liquid. The wave speed in the solid depends on the type of wave, but if we take the speeds of sound waves used with Equation (14.1) as an example (in steel 5625m/s, in water 1414 m/s, in air 343 m/s) we obtain for a range of incidence angles the following refraction angles:

Incidence angle in pipe wall i_1	Refracted angle in fluid i_2	
	water	air
45	10.2	2.5
60	12.6	3.0
75	14.1	3.4
90	14.6	3.5

From this it can be seen that the refracted angle in water is small, but that in air is very small. The smallness of this angle, which for gas means that the ray is almost perpendicular to the pipe axis, results in very small components of the gas velocity in the ray direction, on which the measurement depends. The application of Lamb waves has been one approach to address this.

For these reasons, until recently in gas applications it had been considered essential that the transducers were in contact with the fluid and so were mounted as shown in Figures 14.9(a) and (b) or possibly Figure 14.9(f). However, the appearance of clamp-on ultrasonic meters for gas has changed this perception.

To overcome the impedance mismatch, a low-impedance transducer was reported (Collings et al. 1993) which, when used in a meter, allowed natural gas flows of 0.013 to 8 m^3/h with a temperature range of −13°C to 47°C.

It should also be remembered that for high-pressure service the transducers must be flanged to the correct line pressure (rather than screwed). Transducers which project into the tube may avoid attenuation and other problems due to air collection within the pockets. To reduce turbulence caused by the transducer cavities on a 20 mm ID ultrasonic flowmeter for gas, it appeared that nets were placed to create a wall which transmitted the ultrasound (Hakansson and Delsing 1992). Raišutis (2006) investigated the flow profile in the transducer recess. See also Løland et al. (1998) on transducer cavity flows.

Lynnworth et al. (1997) described the use of relatively high acoustic impedance transducers using solid piezoelectric materials for air and gas flows. They

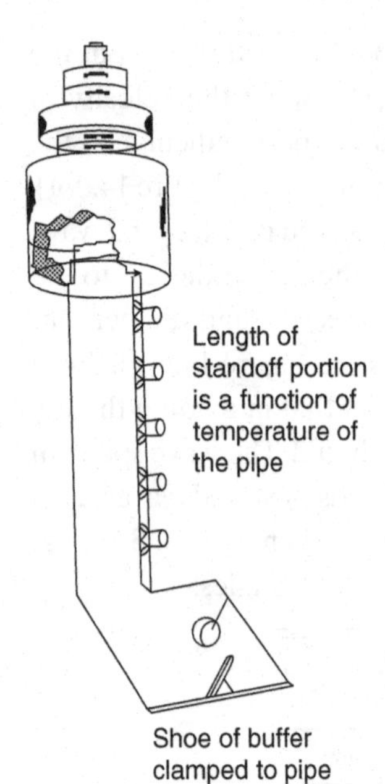

Figure 14.10. Ultrasonic shear wave clamp-on buffer waveguides for hot liquids. Lynnworth, US Patent 6,047,602, issued April 11, 2000.

were capable of being subjected to temperature extremes, pressures from vacuum to 10 bar, vibration and shock, and the end faces could be contoured to create different distributions of sound wave. See also Ao et al. (2014) "Ultrasonic coupler assembly". Lynnworth (1988) also suggested the use of clamp-on buffer rods to allow for extreme temperature (e.g. ±200°C): for instance, in wetted use in cryogenics or in the flow of quench oil in a carbon steel pipe at 260°C. Liu, Lynnworth and Zimmerman (1998) and Lynnworth (1999) described buffer waveguides for use with hot liquids and appeared to suggest that they can operate continuously with liquids up to 260°C (see for example Figure 14.10). They recommended their use to transmit the ultrasound signal between the hot fluid or pipe to the transducer.

Important research has been developed on high temperature-resistant capacitive transducers for high-temperature gas (up to 450°C) suitable for engine exhaust measurement applications, and with fast response capable of following the flow variations (Kupnik et al. 2003, 2004, 2006b; Kupnik, Schröder and Gröschl 2006a). They have also used directional arrays for transmission in air capable of producing narrow beams of sound at e.g. 5kHz.

O'Sullivan and Wright (2002) found electrostatic transducers suitable for ultrasonic measurement of gas flow in a pipe.

Wright and Brini (2005) described investigations into applying high-frequency capacitive transducers for gas metering. The impedance match between the thin membrane and the gas appeared advantageous.

A gap discharge sound transmitter has been developed (Martinsson and Delsing 2009, 2010; Karlsson and Delsing 2013) which appears suitable for use in ultrasonic meters.

14.3.4 Size Ranges and Limitations

Single- and twin-path designs may be available in 80 to 2,000 mm diameter or greater although for very large sizes retrofit may be recommended. The arrangement for small tubes is likely to be for 10 to 80 mm.

When retro-fitting, there is always a tolerance on the pipe dimensions, both diameter and wall thickness, which will affect the deduction of mean flow from ultrasound path velocity and the measurement needs to take account of the actual separation of transducers (Scott 1984). For instance, in a carbon steel pipe of 100 to 1,000 mm diameter the tolerance may be ±1% on diameter and on wall thickness in the range 3 to 5 mm ±10%. In addition, erosion, corrosion and painting of the pipe make it improbable that the error can be reduced below 0.5% of rate. Manufacturers may offer retrofit for one, two or more path, which with precise measurement should be capable of high accuracy.

14.3.5 Clamp-on Meters

Baldwin and Rivera (2012) reviewed the development of clamp-on ultrasonic flowmeters from the 1950s with early Doppler devices, with the date of the first prototype time-of-flight meter given as 1972 and the first gas clamp-on meter in the early 2000s. The authors commented that: "The developments that led to clamp-on measurement of gas were improvements in signal processing (Digital Signal Processors), Lamb Wave transducers (signal transmission harmonized with pipe resonance), and exterior sound dampening materials to absorb the pipe signal." (See also Panicke and Huebel 2009) Figures may be as low as 5 bar for application to steel pipes (plastic pipes even lower than atmospheric pressure) as the minimum gas pressure for a clamp-on meter depending, of course, also on factors such as wall thickness and outside pipe diameter. There may also be the possibility of operating as low as −170°C and up to about 600°C using extension devices between the hot pipe and the piezo, presumably with a similar concept to that in Figure 14.10. A commercial design of clamp-on meter is shown in Figure 14.11. Figure 14.11(a) provides a diagram of the shear wave transit-time flowmeter with internal reflection, and indicating the time of transit of the up- and downstream waves. Figure 14.11(b) is a photograph of a meter designed for coping with high-temperature flows. Figure 14.12 gives an indication of the region where shear and Lamb wave transducers operate in a clamp-on flowmeter. See also Herremans et al. (1989), who described a Lamb wave design of clamp-on flowmeter in which two segmental transducers on opposite walls of the pipe were energised so as to create a travelling wave, within the pipe wall and in the axial direction, of a frequency and wavelength to create the Lamb waves which travelled within the pipe wall and caused longitudinal waves of ultrasound to be transmitted into the fluid in the pipe. The transducers acted as both transmitters and receivers.

Abe et al. (2013) undertook a feasibility study on a new design of clamp-on ultrasonic flowmeter with a carbon fibre-reinforced plastic tube as a meter body because of its excellent characteristics, particularly the high attenuation of ultrasound signals

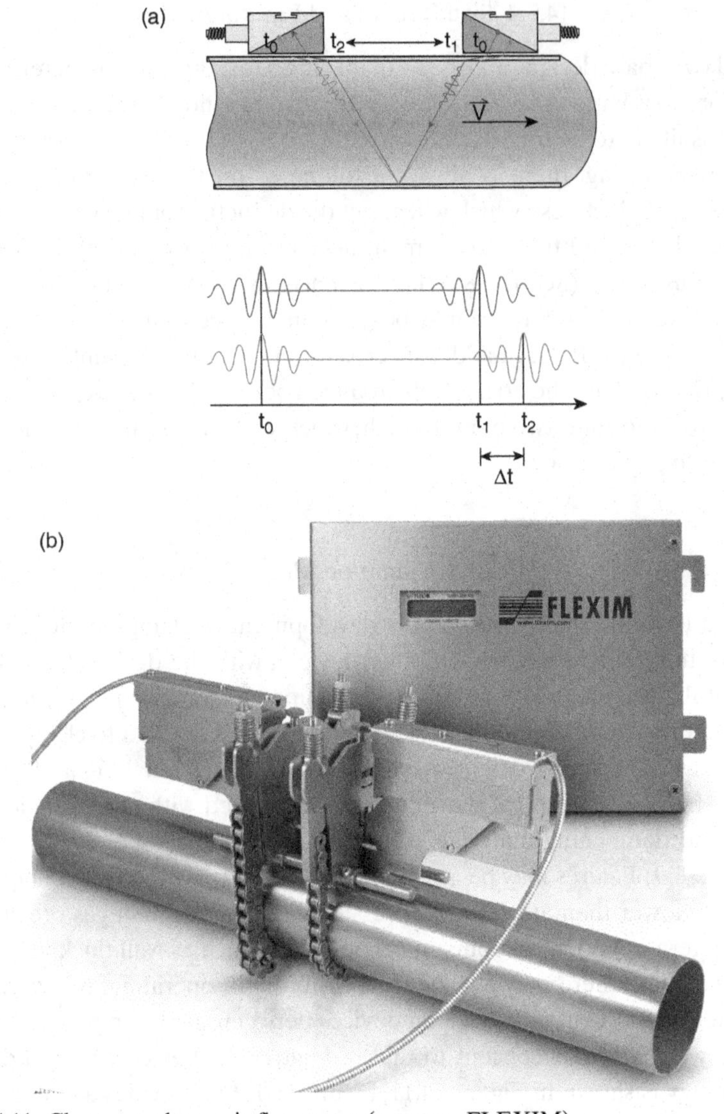

Figure 14.11. Clamp-on ultrasonic flowmeter (courtesy FLEXIM):
(a) Diagram of shear wave transit-time flowmeter with internal reflection indicating the time of transit of the up- and downstream waves
(b) FLEXIM clamp-on flowmeter designed for use on extremely hot or cold pipes by using the WaveInjector mounting fixture.

and the consequent reduction of unwanted acoustic signal transmission from the transmitter to the receiver, leading to a very high signal/noise ratio. They quote promising results with deviations within ±0.2% of rate for the higher flow rates.

Chun et al. (2013) appear to suggest that a clamp-on flowmeter with both Z-paths and V-path is capable of a better level of uncertainty at low flow rates. Their results suggest a marginal improvement at the bottom end of their flow range. They "conjectured" that this could be due to "the cancelling effect of flow unsteadiness".

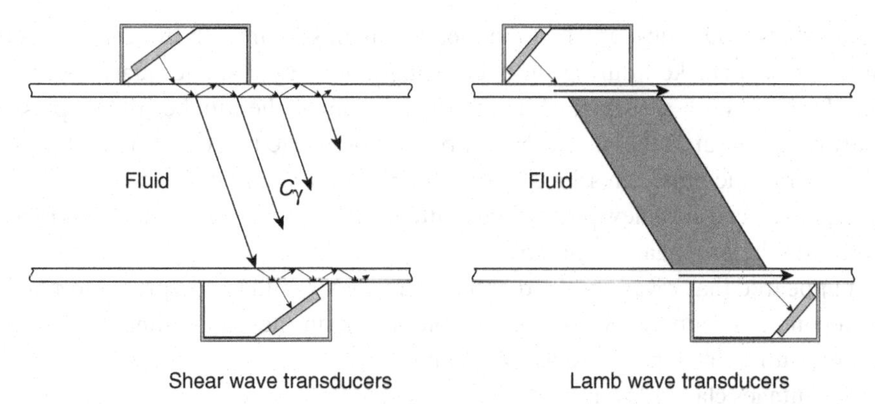

Figure 14.12. Simple images to indicate the difference in the pipe wall regions carrying the Shear and Lamb waves (courtesy FLEXIM).

14.3.6 Signal Processing and Transmission Timing

The measurement system essentially requires the ability to measure very small time periods, or periods to very high precision. This increment could be of order 10^{-9} s (a nanosecond). Clearly the technology in this area has moved rapidly, and the problems encountered in small time measurement are finding new solutions. Part of this requirement is how to identify the precise point in the transmitted wave at which the timing starts and stops. This may be identified by a zero crossing of the received wave, Figure 14.1(d), or correlation techniques may be developed.

In some recent research we had the use of a commercial meter which measured the time of arrival of the up- and downstream flow signals using a zero crossing method. A trigger level was set to intercept the acoustic wave at a suitable fraction of the amplitude of the wave. When a wave arrival triggered the system the following zero crossing was used to measure the acoustic wave arrival time. The trigger level was chosen to obtain the zero-crossing following the second positive-going peak of the wave, Figure 14.1(d). Under certain circumstances, if the trigger level is set too near a peak, the measurement might be made at the wrong zero-crossing. Cross correlation might provide an alternative to obtain transit time measurements (Mahadeva 2009).

Tables 14.1 and 14.2 have identified the design requirements for the signal processing, for example that transit-time meters should be capable of measuring timed differences to better that 10^{-9}s (one nanosecond).

To achieve this, various clock circuits have been used. In an early commercial meter a ramped voltage system provided a time measure. Digital methods have introduced a new era. In addition there are problems in ensuring that the zero-pass, or whatever part of the wave is used, is consistent between transmitters and receivers. Szebeszcyk (1994) discussed some aspects of the design of the ultrasonic flowmeter and transducers, use of first zero crossing of the signal, optimum thickness of matching layers etc.

Vaterlaus (1995) described a time measurement system with two counters. One counter N_t was clocked during the measuring period by a stable quartz oscillator with a frequency, $f = 64$ MHz, while the other counted the number of samples, N_n. A start pulse initiated the measurement by activating the two counters and triggering the ultrasonic pulse emission. During the measurement period, $\tau = 30$ms, every pulse received caused a new pulse to be emitted. The two counts N_t and N_n were used to calculate the propagation time from N_t/N_n.

It appeared that τ was obtained precisely from N_t/f and hence, provided τ ended after a complete value of N_n, it was possible to obtain the transit time as $N_t/(N_n f)$ and a resolution of 60 ps in 116 μs was claimed.

Advantages claimed were:

- velocity resolution to 0.8 mm/s;
- precision depended on the stability of the oscillator;
- jitter in digital logic was averaged.

Another solution to the timing problem (Pavlovic et al. 1997) used:

a) a transmitted pulse in two halves, the first at one amplitude and the second at twice the amplitude, presumably allowing more precise timing edges;

b) the downstream and upstream times were then given by

$$t_d = N_d T + \Delta\tau_d$$

$$t_u = N_u T + \Delta\tau_u$$

where N_d and N_u were the whole number of counter periods of a 2.2 MHz square-wave train obtained from a 17.7 MHz oscillator through a divider by 8. N_d and N_u were obtained by standard counter methods;

c) $\Delta\tau_d$ and $\Delta\tau_u$ were measured through the phase difference between transmission and received signals, and a two-stage measurement system using a coarse and fine approach. An integrator was used to obtain $\Delta\tau_d$ and $\Delta\tau_u$.

The effects of error due to temperature on pipe diameter, recesses, delays in electronics and transducers were also addressed.

Since such short times are being measured, system designs need to address delays in the cables, the transducers and the converter (Scott 1984). Time delays of microseconds could lead to errors if the head is calibrated on water for use on oil (±1.5%), or if the temperature of the water changes by 10°C (±0.1%). Further delays are caused in the window, or in the transmission through the wall, ranging from about 3μsec to about 15μsec. These delays will also depend on the choice of materials.

Hemp (1988) proposed a reciprocity approach to the elimination of zero instability. This requires that the transducers are driven, say, by a voltage pulse, and the received signal is then sensed by its current pulse. The path is then reversed. Van Deventer and Delsing (2002) suggested that Hemp's reciprocal principle was only valid for identical transducers, so that small manufacturing variation appeared to cause measurement errors in the times of flight (see also van Deventer (2005) on possible contradictions in the reciprocity theory for ultrasonic flowmeters). See also Yang, Cao and Luo (2011) on a proposed forced oscillation method which they claimed would reduce zero errors and temperature effects and did not use the reciprocal principle. Lunde et al. (2005) discussed the electroacoustic reciprocity principle as a means of simplifying "dry calibration". They extended earlier work by taking into account finite-valued electrical impedances of the electronics and the transducers employed in the meter, deriving specific design criteria for "sufficient reciprocal operation" in terms of requirements for the electrical impedances of the electronics and transducers, and giving criteria for transducer manufacturing reproducibility, in terms of bounds for variations of the phase of the transducer impedances. They also considered the advantages and disadvantages of reciprocal operation.

See also Roosnek (2000) on pulse transit times measurement and peak labelling, Zhang and Li (2015) on improving the accuracy of time-difference measurement (including references to related research) and Brassier and Hosten (2000) and Brassier, Hosten and Vulovic (2001) concerning a single-path meter for gases with frequency of 500 kHz.

Additional Sensing and Processing

The ultrasonic signals offer more data than is necessarily exploited. This is particularly the case for multiple beam methods. The possibility of using comparative methods to check the validity of the measurements on each path, the information on upstream disturbances and other information should be extractable from the data. See Section 14.3.7, which mentions some ideas on this due to Zanker and Freund, Jr. (2004). One aspect is using density to obtain mass flow measurement. Lansing (2000) discussed smart monitoring and diagnosis for ultrasonic gas meters, noting that the four common diagnostic features are speed of sound by path, path gain levels, percentage of accepted pulses and signal-to-noise ratio. Lansing, Herrmann and Dietz (2007) used a four-path meter with the addition of a single diametral path, which when compared to the four-path output was much more sensitive to profile change.

It appears that with additional sensors, pressure, temperature etc., and the molar fractions of CO_2 or N_2, it may be possible to obtain the composition of natural gas (Morrow and Behring 1999), and also mass and energy flows (Lunde and Frøysa 2002). Frøysa and Lunde (2004) developed this idea, noting problems in obtaining density, while calorific value may also be obtained (Frøysa and Lunde 2005). Frøysa et al. (2006a) reported field data and Frøysa et al. (2006b) gave a brief summary of these developments.

A meter was developed for flare gas with means to obtain molecular weight of the gas from the transit time of acoustic waves, and hence mass flow (Smalling et al. 1984; Smalling, Braswell and Lynnworth 1986); cf. Matson et al. (2002). Remaining technical challenges in flare gas flowmetering were extremely high flows (> 80 m/s) in emergency situations, high flow noise and beam drift (Sui et al. 2010).

McDonald and Sui (2013) discussed the use of temperature compensation to allow for dimensional change etc. and proposed a method using known material properties which result from temperature change to obtain the temperature.

14.3.7 Reported Accuracy

14.3.7.1 Reported Accuracy – Spool Piece Meters

Van Dellen (1991) suggested the meter might meet custody transfer accuracies, and this now appears to be the case (see Selvikvag 1997), as indicated by Zanker and Mooney (2010) who gave an important review of the development of the gas meter and particularly noted that 25 years had elapsed since their first use for custody transfer. ISO 17089-1:2010 gave various maximum allowable errors depending on class of gas meter, range and size with ±0.7% of rate appearing to be the smallest of these allowable errors. The standard should also be consulted on diagnostics and type testing: testing a type of ultrasonic flowmeter by simulation to check that it can cope with "the real world".

In the UK sector, by consensus, the total combined uncertainty for custody transfer as given by Nesse and Bratten (2013) is ±0.25% (dry mass) for liquid.

Reports indicate that good accuracy can be achieved by dry calibration – the use of flowmeter dimensions to obtain calibration (de Boer and Lansing 1997; see also Drenthen and de Boer 2001), but Kegel and Cousins (2012) appeared more cautious, making the point that measurement of transducer spacing and angle may not yet be precise enough to predict the calibration within required uncertainty limits. Zanker and Freund (1996) suggested that the difficulties of calibrating a large (30 inch or 750 mm) diameter meter may make dry calibration preferable. This would, in part, reflect the limitations in the flow calibration facilities at large volume flow rates. Zanker and Freund, Jr. (2004) noted the data available to evaluate meter performance, and to determine requirements for maintenance and recalibration, and presented diagnostics and indicators to check meter operation, also capable of identifying any incorrect chordal measurement (see also Zanker 2006). Peterson et al. (2008), as well as suggesting a possible rule for the periods between calibrations, noted the inherent diagnostics in the ultrasound beams which could provide verification. Trostel, Clancy and Kegel (2010), on the basis of data for several years, saw the possibility of quantifying recalibration time intervals based on operational change. Hall *et* al (2010) seemed to conclude that the need for recalibration is best based on diagnostic logs rather than the number of years for which the meter has been in service. Kegel and English (2011) introduced a mathematical model which could assist in determining a recalibration interval. See also Kneisley, Lansing and Dietz (2009) on the effectiveness of condition monitoring.

The performance of these meters is continually improving. Percentages given are assumed to be of rate (reading) unless indicated otherwise. For fiscal applications with oil, repeatability appeared to be of the order of 0.01% (Dahlström 2000). Yeh and Mattingly (2000) undertook further tests on commercial meters and suggested that the zero flow and remove-replace performance had improved. They confirmed that the very high "dry calibration" expectations for multipath meters may be achievable with uncertainties of ±0.2% or better.

Delsing (1991) tested the zero-flow performance of a sing-around ultrasonic flowmeter with multiple paths. The meter was in a 25 mm line. The author estimated uncertainty of ±1% rate for flows as low as 6 cm/s, and suggested that this would open up the meter's range to 100:1.

Holden and Peters (1991) found that for undisturbed flow the British Gas 300 mm meter readings lay within 0.4 and −0.8% of the reference turbine meters. For a 150 mm meter the variation was between 1.4% and −0.35%, and a 500 mm meter appeared to be within about 0.3%. (Percentages of rate appear to be implied.)

A six-inch (150 mm) ultrasonic meter calibration against sonic nozzles (with repeatability of 0.04%, Erdal and Cabrol 1991) gave a day-to-day repeatability of 0.1%.

Grimley (1996) tested commercial meters and concluded that they were capable of accuracies within 1% tolerance and with repeatability better than 0.25%. Pressure change, for instance 4.5 MPa caused shifts of 0.4%. It was also suggested that for dry calibration there will be a need to understand the effect of parameter changes.

Lunde, Frøysa and Vestrheim (2000) also noted that ultrasonic flow metering was recognised as an alternative for fiscal metering of gas. However they also noted some challenges for improved accuracy and traceability including: transit time corrections, reduced sensitivity to operational factors and installation conditions, increased use and confidence in dry calibration. Niazi and Gaskell (2000) commented that the uncertainty of the multipath ultrasonic meters they tested was better than ±0.4% when tested under ideal conditions, but confirmed the problems of flow-generated noise on ultrasonic meters.

Johnson, Harman and Boyd (2013) reported the use of a blow-down rig to calibrate an eight-path meter of approximately 0.9 m ID, obtaining a discrepancy between the meter and the calibration facility of about 2% which they considered might have been due to the rather flat profile which emerged from the bell mouth at the entrance of the working section where the meter was installed. Kuo et al. (2012) also appear to have observed temperature variation and profile effects in a blow-down rig when testing a 100 mm ID meter with eight paths.

14.3.7.2 A Manufacturer's Accuracy Claims

Liquids. Performance claims for liquid meters based on one manufacturer appear to be of order:

- ±1% of rate (2 beam);
- ±0.5% of rate or better (3 beam);
- ±0.15 to 0.2% rate (5 beam)

(from Product Overview Flow Measurement, KROHNE Messtechnik GmbH (2014) in http://krohne.com/en/dlc/brochures-flyers/) with possible ability to cope with small amounts of solids or gas; process temperature ranges possibly as much as −200°C to +250°C; pipe sizes may range from less than 25 mm to 3 m; fluid velocity in the pipe may range from 0.03 m/s to 10 m/s or higher.

Gases. Performance claims for spool piece gas meters based on one manufacturer appear to be of order

- ±1% to 1.5% of rate (2 beam);
- ±0.1 to 0.2% rate (12 beam);

(from Product Overview Flow Measurement, KROHNE Messtechnik GmbH (2014) in http://krohne.com/en/dlc/brochures-flyers/) for process gases. Process temperature ranges possibly as much as –40°C to +180°C. Pipe sizes may range from less than 50 mm to 0.6 m.

However, the technology in the hands of ambitious manufacturers is resulting in claims of very high order, possibly vying with the uncertainty of the flow rigs available. Drenthen et al. (2009) presented a design of gas meter with 12 chords which they claimed was capable of 0.2% uncertainty provided there was at least 5D of upstream straight pipe, which was insensitive to swirl, providing information on the velocity profile and had an ultrasonic beam to check for contamination on the bottom of the pipe, and other features.

14.3.7.3 Clamp-on Accuracy

Liquids Cascetta (1994) compared a clamp-on transit-time ultrasonic meter with an electromagnetic meter in situ in a water distribution network with 400 mm ID. The ultrasonic meter generally overestimated the flow rate, and the difference between the ultrasonic and electromagnetic flowmeter was mainly in the range ±1% to ±5% of rate.

Sanderson and Torley (1985) described an intelligent clamp-on flowmeter and claimed an uncertainty within ±2% of reading. They concluded that the wedge angle of the transducers (estimated uncertainty of ±1°) could account for much of the error. They gave plots of baseline drift equivalent to about 30 mm/s. They demonstrated the value of Hemp's (1979) reciprocal drive system on baseline stability (for instance where there was temperature change), which reduced drift to about one eighth of that for a conventional system. They found that an error of ±2.5% in transducer separation resulted in errors within 0.5%. The meter used additional transducers to locate optimal axial separation of transducers, and for self-calibration. The transducers allowed pipe wall thickness and internal dimensions of the pipe to be measured. They also gave a very useful analysis of the likely errors in the system.

Cairney (1991) tested a clamp-on flowmeter which had dedicated transducers with a specific pair for each different size of pipe, with a scale card within the electronics. The manufacturer claimed uncertainty of ±3% of reading without calibration and Cairney suggested that for most of the tests the uncertainty was within ±5%. He also found that removal and replacement of transducers led to errors, typically in the range 5% to 10%, but with care this might be only about ±2% and, if necessary, the computed speed of sound should be checked. It was also found that: pipe material had no noticeable effect; performance was similar in direct and reflex (one or more internal reflections off the pipe wall) modes; rust on the inner surface is more likely to affect the cross-sectional area of the pipe than the signal transmission. Nominal pipe dimensions should not be used, since, for example, 1 mm error in ID in a pipe of 50 mm diameter results in about 4% error in cross-sectional area; incorrect axial spacing of the mounting tracks was not significant, but incorrect angular spacing caused errors up to ±10%, confirming Sanderson and Torley's (1985) observations. However see Figure 14.A.3. Although Cairney's data is more than 20 years old, it indicates some of the problems which, if lessened, are still possible, and require careful application of these devices.

The results of tests to compare a portable transit-time flowmeter (manufactured by Tokyo Keiki in Sept 1988) which had two sets of clamp-on transducers with a two-path profiling system were described by Lynch & Horciza (1995). The transducers were fixed magnetically to the carbon steel pipe. The outside diameter was measured and the wall thickness obtained with a sonic gauge. The performance appears to have been remarkably good within 0.6% of a calibrated Venturi. Four portable meters (three from Tokyo Keiki 1 MHz and one single-path from Panametric 0.5 MHz) were tested in V and Z modes and appear to have agreed with an eight-path chordal meter to within 1%, suggesting a convenient and cost-effective means of testing small hydro-plant and turbine performance.

NEL (1997b) noted that temperature change could lead to changes in the refraction angle (cf. Svensson and Delsing (1998) on deviations in operation).

Baumoel (1994) described a meter which appeared to include velocity up to 30 m/s (100 ft/s), and uncertainty of up to ±0.5% of flow rate.

David (2012) undertook an experimental survey of clamp-on flowmeters varying water temperature, pipe material and upstream fittings, allowing for relative installed angular position of the ultrasonic beam and spacing from fitting to meter, but also looking at other effects due to incorrect setting up etc. The results include the meter beam in the same plane as the inlet and outlet pipes to and from a 90° bend showing a predominant bias due to the bend of 10% at 5D and 3% at 15D. The angle between the beam and the plane of the bend appeared to have a small effect for spacings of 10D or more. The results for two bends in perpendicular planes appear to be somewhat similar in bias size at their minimum but showed considerable variation with rotation. There appeared to be considerable variation between the seven meters tested, but this may be due to the problem of mounting variation. Temperature caused a change of about 2% for increase in temperature from 20° to 70°C. The author emphasised the need for care in setting up etc.

Commercial devices allow a range of materials for the pipe including steel, stainless steel, cast iron, vinyl chloride, FRP and asbestos, in sizes from 15 mm or less, provided the beam length is sufficient, to 4m or more. The presence of a lining of tar, epoxy, mortar or Teflon appeared surmountable. They may be available in both reflex or direct mode. The meter should be capable of measuring flows in continuous liquids with low attenuation of the beam and no air bubbles etc.

However, recent manufacturer uncertainty claims appear to be of order ±1% to 2% of rate, but from my experience it seems likely that this is the optimum achieved with great care and very precise set-up, and that values approaching ±5% are more realistic, allowing for pipe and other uncertainties which can introduce substantial errors in the calculated volumetric flow rate.

Earlier manufacturers' claims may still be valid, allowing for installation uncertainties:

- repeatability if nothing changed ±1%
- uncertainty D < 25 mm ±10%
- uncertainty 25 < D < 50 mm ±5%
- uncertainty D > 50 mm ±3%

reflecting the importance of precise pipe dimensions. Some manufacturers offer pipe wall ultrasonic thickness gauges, and the uncertainty will then include the uncertainty of such a device.

Gases. In a private communication, Lynnworth (dated 2014) made the point that a "reader should be aware of the importance of the acoustic impedance of the pipe material. Transmission of ultrasound from a clamped-on transducer into a given gas is much affected by the pipe material's characteristic acoustic impedance as well as pipe thickness, frequency, gas composition, pressure, and other factors."

Lynnworth (2001) introduced a clamp-on transit-time ultrasonic flowmeter (Panametrics, now GE) intended for a range of gases including steam, methane, difficult gases and at various conditions, and Frail (2005) described the complementary clamp-on gas correlation sensing meter (cf. Jacobson, Lynnworth and Korba 1988; Scelzo 2001; Ao et al. 2002).

Ting and Ao (2002) reported an evaluation of the clamp-on gas transit time flowmeter. Natural gas pressure range was about 14 to 76 bar and velocities 3 to 18 m/s. Claimed performance for dry gas was within ±2% uncertainty. Ao and Freeke (2003) outlined general application requirements. The papers suggested applications including wet, sour natural gas, compressed air, hydrogen and petrochemical process gases in thick-walled lined pipes, copper tubing and stainless steel pipes. See also Cloy (2002) and Sims and Rabalais (2002).

Espina and Baumoel (2003) tested the Controlotron WideBeam™ system which used sound that passed diametrically through the fluid. They claimed that the instrument not only met its specified accuracy and repeatability range, but showed potential to achieve custody transfer accuracy (cf. Baumoel 2002).

This is a remarkable technology which has achieved what many of us thought out of reach – a gas clamp-on flowmeter. Other companies, such as FLEXIM, have developed such an instrument.

The uncertainty quoted may be of order ±3% of rate or better depending on the actual flow rate for pipe diameters from 10 mm or less to one or more metres, with flow rates up to 25m/s or more. However, the setting up and operation of these instruments may require considerable understanding, care and perseverance.

Van Essen (2010) reported tests on a gas clamp-on meter carried out at Groningen (500 kHz Lamb wave mode), and van Luijk and Riezebos (2014) also reported some field applications of the clamp-on meter. The results appeared to have been satisfactory.

14.3.8 Installation Effects

In addition to the effects due to upstream pipework, it should be remembered that with these precision instruments in which the accuracy of mounting transducers and general dimensional stability is important, temperature gradient effects within the measuring region of the flow tube (Willatzen 2001), stresses in pipe fixtures, unsuitable mounting, effects of wall roughness (Calogirou, Boekhoven and Henkes 2001) and even Joe Bloggs who regularly taps the pipe with his spanner as he walks by or climbs on it as a step, may have detrimental effects on precision. Whitson (2008) gave a paper on a "general methodology for geometry related pressure and temperature corrections" [cf. Morrison and Brar (2005) on temperature difference between meter pipe and gas at low flow].

Brown *et* al (2010) commented that the primary mechanism by which thermal gradients affect the performance of ultrasonic meters is not velocity profile, it is refraction, resulting from the fact that the ultrasound paths must cross a sound velocity gradient.

The impact of fouling and corrosion should be observable from its effect on the ultrasound beams, and their reflection off the pipe wall allows detection of changes to the conditions of the pipe wall (Drenthen et al. 2010 see also 2011). The paper considered: design matters, detection of fouling, condition of upstream pipes etc.

Noise sources in the ultrasonic range, such as control valves, regulators, compressors and flow conditioners (Vermeulen et al. 2004) can interfere with the ultrasound. Signal processing techniques may be available in such cases. Noise suppression algorithms have been tested to overcome the problem (Kristensen, Lofsei and Frøysa 1997). See also Krajcin et al. (2007) on regulator noise effects on gas meters.

14.3.8.1 Effects of Distorted Profile by Upstream Fittings

Some early reported installation effects (Al-Khazraji et al. 1978) obtained by combining measured flow data for an eccentric orifice discharge at 5.5D with theoretical calculations suggested that a single-path meter could have errors of 5% to 16%, a twin-path meter ±1.6%, a four-path meter less than 1% errors.

Unless stated otherwise, the effects mentioned in the following paragraphs were obtained without flow conditioners. If flow is bidirectional, adequate straight lengths are presumably necessary in upstream *and* downstream directions to achieve a stated accuracy.

Heritage (1989) reported an important series of tests on the performance of transit-time ultrasonic flowmeters. The first part of her work indicated the surprisingly large failure rate of new flowmeters.

Table 14.4. *Installation tests on water for single-path meter (after Heritage 1989)*

Disturbance	Spacing between downstream flange of disturbance and upstream flange of meter (error range in percentages)		
	5D	10D	15D
Gate valve 50% closed by movement	Not obtained	−1 to −2.5	+1.5 to −3
Swept bend (R/D= 1.5) parallel	Not obtained	+4.5 to −6.5	−2 to −3.5
Swept bend (R/D=1.5) perpendicular	Not obtained	−3.5 to −4.5	+3.5 to −3
Two swept bends	Not obtained	+7.5 to −11	−0.5 to −7
Reducer	Not obtained	+2 to −4.5	+2 to −1

Table 14.5. *Installation tests on water for single-path clamp-on meter (after Heritage 1989)*

Disturbance	Spacing between downstream flange of disturbance and upstream flange of meter (error range in percentages)		
	5D	10D	15D
Gate valve 50% closed by movement	Not obtained	+3 to −2.5	+1.5 to −0.5
Swept bend (R/D=1.5) parallel	Not obtained	+3.5 to −3	−2.5 to −3.5
Swept bend (R/D=1.5) perpendicular	Not obtained	−2 to −2.5	+2.5 to −3.5
Two swept bends	Not obtained	−5 to −10	+0.5 to −6
Reducer	Not obtained	−2.5 to −4.5	0 to −0.5

Tables 14.4 to 14.6 summarize the installation tests of Heritage (1989). In the original paper the results of two single-path meters are reported separately, but are combined in Table 14.4. The effect of the gate valve is taken regardless of orientation. This is partly because the differences did not appear substantial and the safe assumption is the range given. The same is true for the two bends which were in perpendicular planes. However, in this case the variation between the meters is so great that it appears to outweigh any effect of orientation.

Halttunen (1990) used data on flow profiles and applied them to the ultrasonic flowmeter analyses to obtain the effect of flow distortion on the performance. The paper also gave experimental data from installation tests. Each method this author used appeared to give a slightly different datum, a salutary reminder that all such

Table 14.6. *Installation tests on water for dual-path meter (after Heritage 1989)*

Disturbance	Spacing between downstream flange of disturbance and upstream flange of meter (error range in percentages)		
	5D	10D	15D
Gate valve 50% closed by movement	0 to +1.5	+0.75	Not obtained
Swept bend (R/ D=1.5) parallel	0 to +1	+0.25	Not obtained
Swept bend (R/ D=1.5) perpendicular	+0.75	−0.75	Not obtained
Two swept bends	−1.5 to −3	+0.5 to −1.5	Not obtained
Reducer	+1 to +0.5	+1 to +0.5	Not obtained

Table 14.7. *Halttunen's (1990) installation data*

	5D	10D	20D	40D	80D
	Ultrasonic flowmeter – single path				
Single elbow	−5%	−2.5%	−2%		
Double elbow 90°	−7%	−5%	−3%	−1%	−1%
	Ultrasonic flowmeter – dual path				
Single elbow	−1%	−0.5%	small		
Double elbow 90°	±1.5%	±1%	±1%	+0.5%	+0.5%

measurements and predictions are subject to some uncertainty. The conclusions which can be drawn from the plots are given in Table 14.7.

The orientation for single-path ultrasonic flowmeters may cause an additional ±1% at 10D or less spacing and for dual-path ultrasonic flowmeters may cause an additional ±0.5% at 5D or less spacing.

The following claims of Vaterlaus (1995) appear, possibly, optimistic, but probably reflect the (surprising) variation found by Heritage (1989). For a single-path meter, errors within 1% if at least:

- 10D is allowed between a reducer and the meter
- 20D for an elbow or a T
- 25D for two elbows in one plane
- 40D for two elbows in perpendicular planes
- 50D for a partially open valve or for a pump.

Hakansson and Delsing (1992) tested the effect of upstream disturbance on a 20 mm ID single-beam sing-around ultrasonic flowmeter for gas. The results indicate that the laminar/turbulent flow change at Re in the range 2,500 to 4,000 caused a calibration shift of about 11% and that for the Re range 4,000 to 11,000 calibration drops by about 1%. The authors reckon that, within the turbulent region:

- for a single bend in the plane of the beam, about 40D upstream straight pipe are required to contain the errors to within ±1%;
- for two bends in perpendicular planes, 80D upstream straight pipe does not contain the errors to within +1%.

Multipath (apparently two paths, and not necessarily off axis) flowmeters were tested on water to obtain profile effects due to a bend. 10D upstream length was considered to cause a calibration shift of 1% (Johannessen 1993).

Holden and Peters (1990, 1991) concluded from tests on a four-path ultrasonic flowmeter operating on high-pressure gas, in both fully developed flows and also with upstream disturbances, that 10D upstream and 3D downstream is sufficient for the data to fall within ±1%. The meter can also deduce swirl and turbulence in a disturbed flow, and may operate satisfactorily with less than 10D upstream at the lower end of the flow range.

The effect of bends, step changes in diameter and pressure reduction were investigated by a joint industry project on multipath meters. Van Bloemendaal and van der Kam (1994) concluded that in well-developed conditions uncertainty of 0.6% should be achievable without calibration but with careful determination of meter dimensions and zero setting. They reckoned that: an additional ±0.5% should be allowed for 10D minimum spacing from a bend; small changes in pipe diameter had negligible effect; but swirl and noise from pressure reductions could have severe effects resulting in 2% to 2.5% additional uncertainty.

Lygre et al. (1992) described a five-path gas ultrasonic flowmeter. The paths were positioned so that three were on one diagonal plane, while the other two lie between the three but on a reverse diagonal plane. It was designed to operate with 10D upstream and 3D downstream. They claimed that for a bend upstream the installation length may be reduced to 5D. Their results suggested an uncertainty within ±0.5% over a 4:1 turndown.

Grimley's (1997) results suggested that thermowells at 5D or less may cause errors of order 0.6%.

Yeh and Mattingly (1997) predicted the performance in high Reynolds number flows and downstream of an elbow. They concluded that multipath meters were desirable for high accuracy and concluded that, if the effect of ray bending were ignored, it was not likely to introduce errors in low-velocity flows. Rychagov and Tereshchenko (2002) reviewed the mathematical approaches to quadrature integration of the flow profile to obtain high-precision estimations of flow rate. They also suggested the possibility of reconstructing the velocity profile for both symmetric and arbitrary profiles using transform techniques. See Hamidullin, Malakhanov and Khamidoullina (2001) on the basics necessary for the design of an ultrasonic flowmeter capable of coping with a range of flow profiles.

The power of computational methods has caused an increasing move towards computational fluid dynamics (CFD) and other numerical simulation of flows and installation effects (cf. Barton and Boam 2002). Moore, Brown and Stimpson (2000; see also Moore and Brown 2000), following Salami's (e.g. 1984a) methods for turbine meters, developed equations to describe flow patterns in circular pipes. From these they identified one which had a resemblance to the profile downstream of a single bend, and others which had resemblance to flows downstream of double out-of-plane bends. The ultrasonic beam paths used in this paper appear to have all been diametral beams on multiple angles. For a parameter

$$H = \frac{v_{\text{actual}}}{v_{\text{measured}}}$$

where v_{actual} is the true mean actual cross-sectional velocity in the pipe section, and v_{measured} is the average of the velocities measured on each ultrasonic beam. H, for most real flows, will be less than unity, and the authors showed how H varied with angular position relative to the "single bend" profile. The plot, of a parabolic-like shape, ranged approximately from H = 0.93 to H = 1.03. It appeared that, for the same profile with two orthogonal diametral beams, the variation of H with angle was of order 2% or better. Moore et al. (2000) demonstrated the possible sensitivity to flow profile of meters which, essentially, use diametral paths rather than half-radius paths. The effect of flow profile appears to cause changes in measured flow compared with actual flow of up to 20% for one path, and up to 3% even for the four-path version. It would be interesting to see the equivalent results for the half-radius types.

Brown, Augenstein and Cousins (2006) gave an interesting comparison between the integrating effect of various multipath designs from two-path to eight-path in axisymmetric flows at various Reynolds numbers, and also when subjected to a distorted and non-axisymmetric flow profile. Their results for axisymmetric profiles appear to show errors rather larger for the two mid-radius path design than those in Figure 14.7. Their results for four or more paths were within 0.1 to 0.15% of the correct average velocity in the pipe. Their results for the three-path design were similar to those for the two path given in Figure 14.7. They plotted results for a very wide range of profile shapes beyond that which appears usually to be applied to turbulent profile simulation (index n = 6 to 10 in Equation 2.6).

For the non-axisymmetric profiles, Brown et al. (2006) gave results for angles of the meter beams relative to the disturbance. In order to give a single value that indicated the performance of each design for each profile they took the root-mean-squared (r.m.s.) value of error over all orientations. For one profile which Moore et al. (2000) indicated might resemble the flow downstream of a double out-of-plane bend, the r.m.s. errors were: 2 beam 0.52%, 3 beam 0.23%, 4 beam 0.12%, 5 beam 0.08%. A second velocity profile used in their research gave r.m.s. errors: 2 beam 1.25%, 3 beam 0.43%, 4 beam 0.2%, 5 beam 0.13%. It may not be straightforward, with profiles such as they used, to relate the distribution of velocity to that which would occur downstream of an actual upstream disturbance.

Frøysa, Lunde and Vestrheim (2001) investigated the effect of various installation conditions using a ray model for the flowmeter and CFD predictions of the profiles. This was a very thorough piece of theoretical research which considered the effects of high flow rates on the beams. They considered a single bend and two bends in perpendicular planes both at 10D upstream of the meter.

Coull and Barton (2002) presented both laboratory tests and CFD modelling results of investigations of installation effects on the performance of meters.

O'Sullivan and Wright (2002) used finite element simulations to obtain pipe flow patterns. Hu, Wang and Meng (2010b) have used a numerical model to estimate the likely error in a multipath ultrasonic flowmeter on the flow field in the penstock of the 3-Gorge hydropower station [cf. Tresch, Gruber and Staubli (2006), Grego and Muciaccia (2008), Moore, Johnson and Espina (2002), Lüscher et al. (2007) and Staubli et al. (2008)].

Zheng et al. (2013; see also 2010) modelled the effect on the reading of a meter for installation downstream of a 90° single bend by CFD and a "flow pattern model". Their results appear to suggest that a single-beam flowmeter should be positioned at least 20D downstream of an elbow to achieve a signal shift of order not more than 2%, and preferably with the beam in the plane of the inlet and outlet pipe axes.

See also Salami (1984a), AGA (2007), Barton and Brown (1999), Coull and Boam (2002), Walsh (2004), Zhang, Hu, Meng and Wang (2013) and Hu, Zhang, Meng and Wang (2015).

Luntta and Halttunen (1999) used a neural network to study velocity profile dependence.

Démolis et al. (1998) used ultrasonic tomography to characterise internal flow in pipes. Kurnadi and Trisnobudi (2006) appeared to use eight pairs of transducers on the pipe circumference to obtain a tomographic flow profile pattern. Yeh, Espina and Osella (2001) suggested that a four-path meter with software based on CFD and experimental data may offer a means for identifying flow profiles.

Gibson (2009) appeared to claim CFD predictions for a 12-inch (300 mm) gas flowmeter (diametrical and mid-radius paths) were within 2–6% of measurements. Frøysa, Hallanger and Paulsen (2008) obtained values for the flow profiles at the outlet of a gas distribution manifold. The pipework appeared to include components known to cause swirl and profile distortion. They also used CFD to obtain the resultant profiles in the entry to the ultrasonic flowmeters and this confirmed that there was flow profile distortion. Swirl was also present. This emphasises the importance of suitable upstream pipework and a meter able to cope with profile distortion.

Drenthen et al. (2009) found a new design of multipath meter had a baseline meter performance of ± 0.1%, and that this was retained provided disturbances were at least 10D upstream.

Carlander and Delsing (2000) tested a small meter downstream of pipe fittings and in a pulsating flow. They suggested that an increased noise level due to turbulence from pipe fittings might be useable for diagnosing installation problems. See also Berrebi et al. (2002, 2004a&b) on installation and pulsation errors..

A pipe of 0.2 m diameter with air flowing through it was set up with 16 transmitters and 16 receivers in a transit-time mode using 40 kHz and with wide-spreading beams so that each receiver could receive from each transmitter, offering a tomographic approach to profile and the mean velocity measurement (Kurnadi and Trisnobudi 2006).

NEL's (1997b) rule of thumb that single, two and multipath meters need, respectively, 20, 10 and 5D upstream between the meter and a fitting perhaps needs a bit more detail.

The overall conclusions from these sources of data do not appear to have been changed significantly by new data since the first edition of this handbook. I assume that, so long as the beams intercept the flow profile in the same positions, the uncertainty should not have changed appreciably. The two cautions on this would appear to be:

i) the increasing value and use of numerical simulation and CFD which may not include a full model of the beam interactions or the developing flow profile;
ii) the spread and consequent subsidiary paths of the primary beam.

My suggestions are:

a) A single-path meter will have a calibration shift of up to 33% for changes from laminar to turbulent, and up to 3% for changes in the turbulent range.
b) For a single-path meter to remain within 2% of calibration the following spacings should be allowed:
   15D for a reducer
   20D for a bend or a T
   40D for two elbows in perpendicular planes.

   For distances less than these the following may be conservative:

Fitting	Spacing	Error
Reducer	10D	±5%
Bend	15D	±4%
Two bends	15D	±7%

c) A two-path meter has calibration shift of 0.7% (NEL 1997) or less for changing Reynolds number and should have the following spacings to remain within 1% of calibration:
   10D for a reducer
   10D for gate valve (50% closed) or bend
   20D for two swept bends
d) A conservative estimate is that a four-path meter should retain its performance to within ±1% in liquid or gas with 10D upstream and 3D downstream. Individual manufacturers, and with more paths, may reduce this uncertainty to as little as 0.1%. The user for whom this figure is critical is advised to check with great care the basis for it. For a recent and wide ranging review of installation effects the reader should refer to the paper by Miller and Hanks (2015).

14.3.8.2 Pipe Roughness and Deposits

Dane and Wilsack (1999) investigated the influence of upstream pipe wall roughness on ultrasonic flow measurement. They found, over the range of conditions investigated, an increase in meter reading of 0.1 to 0.2% for an increase in arithmetical mean roughness of about 5μm to about 20μm.

A theoretical study of wall roughness changes on single path meters was reported by Calogirou et al. (2001), and another of calibration change due to internal fouling which reduced ID, deposited on transducers and changed velocity profile (Lansing 2002a, 2002b).

Karnik and Geerligs (2002) examined the effect of steps between the meter and the adjacent spool piece and roughness on a multipath ultrasonic meter.

It appears that an ultrasonic flowmeter subjected to dirty natural gas may have a reduced internal diameter (ID), experience a build-up on the transducer faces and have a changed flow profile due to pipe roughness changes (Lansing and Mooney 2004).

14.3.8.3 Unsteady and Pulsating Flows

Mottram's (1992) key comment "if you can't measure it, damp it!" is balanced by his advice that ultrasonic transit-time meters are probably not affected. However, Hakansson and Delsing (1994) noted possible problems, including: aliasing, where the sampling frequency picks samples which do not give a true average, resulting in an error which should be avoidable; flattening of the flow profile, an effect of pulsating flow, resulting in the incorrect averaging of the flow based on the diametral path flowmeter. This error may be avoidable if recognised.

Pulsation sources in pipe systems also appear to have an impact on multipath flowmeters (van Borkhorst and Peters 2000; see also 2002). Sources range from fractions of a Hertz for process dynamics, through reciprocating machinery, flow-induced pulsations, rotating machinery to valve noise of more than 10 kHz.

Berrebi et al. (2004a, 2004b) examined the metering errors due to pulsating flow. They considered the prediction error (resulting from the delay time between up- and downstream pulses being of the same order of magnitude as the period of the flow pulsation) and the zero crossing error. They appeared to find that the former was the more serious. They also noted the effect of pulsation on flow profile. In addition they were able to estimate the period of the pulsations and the period appears, then, to have been used for the integration time, resulting in a reduction of error due to the pulsations. Lansing (2012) reported on pulsation tests on SICK 4-path and single-path meters, and the author's final comment was that "flow during typical levels of pulsation (could) be measured accurately with a four-path ultrasonic meter sampling at 10 samples per second". However, the reader should refer to the paper for details.

See also McBrien and Geerlings (2005) on standing waves in pulsating flow.

14.3.8.4 Multiphase Flows

I have observed, in some field data from an ultrasonic flowmeter, a behaviour which could result from entrapment of air. A meter in a low head flow, where air was entrained with the water, periodically failed. The possibility that in such a flow the transducer cavities could cause small local vortices which would entrap the air and block the ultrasonic beam offered a possible explanation.

Lenn and Oddie (1990) have looked at how the scattering of ultrasound from secondary components in the flow, both liquid and solid, could be detected, and they considered that it was possible to measure some concentrations and velocity using a variety of signal processing techniques, particularly at low concentrations.

There has been a commercial transit-time device claimed to be capable of operating in flows of drilling mud, although the author's experience is that it is usually recommended that commercial devices should not be applied to other than single-phase flows.

Brown (1997) suggested that ultrasonic meters may be able to cope with oil/gas and oil/water flows, but Johannessen (1993) put a limit of 10% oil-in-water and reckoned that they were unlikely to cope with more than 0.5% by volume of air in liquid, in agreement with NEL's (1997b) summary.

Eren, Lowe and Basharan (2002) showed that, although particles and bubbles in the flow can weaken the acoustic signal, the signal actually carries useful information which may be used to identify the nature of the flow components.

Wada, Kikura and Aritomi (2006) discussed the possibility of using echo signal for pattern recognition in two-phase flow.

14.3.8.5 Flow Straighteners and Conditioners

It should be noted that, in some cases, the flow profile may need to be modified by flow conditioning. Hogendoorn et al. (2005) discussed flow conditioners and straighteners and referred to standards in relation to fiscal metering such as American Petroleum Institute (API 2011) standard: "Measurement of Liquid Hydrocarbons by Ultrasonic Flow Meters". It should also be noted that, in some cases of gas flow, the presence of a flow straightener may introduce noise which may affect the meter sensing. The problem of acoustic interference may cause errors in meters. Riezebos et al. (2000) discussed whistling flow straighteners. It appears that high ultrasonic background and resulting vibration causes more frequent errors in measurement of path times.

Delenne et al. (2004) assessed the performances of ultrasonic meters downstream of piping configurations with and without flow conditioners. They confirmed that ultrasonic meters were still sensitive to flow perturbations, despite sophisticated techniques. A CFD study (Hallanger 2002) also suggested errors could be related to the tube bundle flow conditioner upstream of the meter, which generated local jets after the tube bundle without sufficient downstream length for adequate mixing. See also Griffith, Cousins and Augenstein (2005) on the effect on the meter which may be detrimental.

Brown and Griffith (2012), in a paper with a useful summary of the development of multipath meters, reviewed benefits and problems of flow conditioners on the performance of multipath meters and on the added pressure loss and possible fouling resulting from their installation. They also demonstrated that the benefits may be achieved by suitable spacing or, in one design, by an upstream reducer before the meter. As they indicated in a subsequent paper (Brown and Griffith 2013), "there are benefits, limitations and trade-offs to be taken into account when considering using a flow conditioner with an ultrasonic meter." They also described a new design of conditioner/straightener which in their diagram consisted of three circular "ligaments" concentric with the pipe, each of which was divided up by radial "ligaments" so that the innermost division was undivided, the annulus outside this was divided into four passages, the annulus outside this was divided into six passages, and the final annulus into eight passages. The thickness/axial length of the plate appears to have been recommended as 0.2D. The fabrication of this device will need some care to ensure that none of the "ligaments" introduce non-axial flows. This design was intended for use with four-path ultrasonic flowmeters.

14.3.9 Other Experience of Transit-Time Meters

Vieth et al. (2001) described a new design development of an 8 inch (200 mm) meter for gas with 200kHz operating frequency, reduced bore, two double reflection paths forming a triangle across the cross-section to deal with swirl and a single reflection path through the diametral plane. Stobie et al. (2001) noted, as pointed out by Zanker, that Reynolds number correction for fully developed profiles may not deal with distorted profiles. For stratified flow (Zanker 2001) a vertical beam reflected from the liquid surface has been used. Panneman (2001) compared sound speed from gas parameters to check meter operational changes.

Errors due to temperature differences between top and bottom of the pipe, and between reference and meter under test, may arise. Temperature in the meter may be overestimated and gas density in the line if equilibrium has not been achieved could result in errors. Lagging and careful path layout may improve the situation [van den Heuvel and Kemmoun (2005); cf. van den Heuvel et al. (2009) and Morrow (2005)].

14.3.10 Experience with Liquid Meters

Boer and Volmer (1997) described a five-path liquid meter claimed to be for use in maintenance-free custody transfer applications. It had two paths to measure swirl and inlet and outlet cones to condition the profile.

Transit-time meters were used in the electricity supply industry (Cairney 1991). Transducers were grouted into the wall of an octagonal section culvert. It was also possible to use radioactive-isotope dilution methods for in situ calibration with an uncertainty of ±1% or better with a 95% confidence level. The ultrasonic flowmeter was reckoned to have an overall error of ±1.5%.

In the Ataturk Hydro Power Plant in Turkey two systems were used to compare flows for loss due to bursting (Vaterlaus 1995). The pipes were up to and more than 6 m diameter. Taylor and Cassidy (1994) discussed the use of ultrasonic meters at BC Hydro in Canada.

Yeh and Mattingly (2000) reviewed improvements in zero flow and remove-replace performance of commercial meters. They confirmed that the "dry calibration" of multipath meters may be achievable with uncertainties of ±0.2% or better.

Critical factors for high-viscosity flow measurement (Hogendoorn et al. 2009) were: signal attenuation, temperature dependence of the viscosity, unsteady flow profile in laminar/turbulent transition.

Nesse (2007) described experience with five-path 6inch and 12inch ultrasonic flowmeters on high-viscosity oil. The export metering station had a 30-inch bidirectional ball prover with 10 m^3 volumes. The meters did not fulfil the linearity requirements of the Norwegian Petroleum Directorate. However, following a test programme, new K-factor models and new weighting factors for the ultrasonic paths for high viscosities appear to have been successfully implemented in the meters' flow computers. The paper provided a very useful guide to flowmeter control and development. Hwang, Lee and Kim (2004) also discussed weighting factors and appeared to obtain good results.

Sluše and Geršl (2013) modelled cavitation of LNG in the transducer cavities of a flowmeter. The paper is illustrated with interesting flow patterns and meter simulations, and figures indicate the possible pressure drop which could result from flow past the cavity for a fluid velocity of 7.5 m/s of about 27 kPa. [See also Figure 2.9(c).]

Temperley, Behnia and Collings (2000) tested a design (similar to a dog-bone or dumb-bell design, Figure 14.13(b)) for liquid with the ultrasonic beam axially along the duct. This design had a more restricted and faired entry and exit than that in Figure 14.13(b) which has slight similarity but was for gas flow measurement. The authors had reservations about the meter's ability to obtain an average flow rate in the flow tube.

Miller and Belshaw (2008) investigated the performance of two (four-inch multipath and chordal) ultrasonic meters at liquid viscosities up to 300 cSt. One of the meters appeared to have an abrupt signal change at turbulent/laminar Reynolds (Re) number and both showed some Re dependence. Gas entrainment (~1%) had a severe effect.

In order to meter flows of heavy oils with high and temperature-sensitive viscosities in the transition region, Brown et al. (2009b) used a nozzle with a diameter ratio of less than 0.64 to stabilise and flatten the velocity profile at entry to a flowmeter in the laminar/turbulent transition region. This appeared to lead to a more gradual transition and to improved repeatability and reproducibility.

[See also Falvey (1983), who, working under very different conditions in a river delta, experienced refraction due to salinity and temperature gradients.]

14.3.11 Gas Meter Developments

Zanker and Mooney (2003) gave a useful reassessment of ultrasonic gas metering. In about 1988 when the first fiscal meters came to market typical claims were:

1. No moving parts.
2. No obstruction to the flow.

3. No pressure loss.
4. No calibration required.
5. Large turndown ratio.
6. Bidirectional flow measurement.
7. Velocity measurement independent of gas properties.
8. Installation requires minimal pipework.
9. No periodic maintenance required.
10. Long life and stability.

The authors assessed how these claims stood in 2003. Points 1, 5, 6 and 9 were unchallenged. Points 2, 3 and 8 should be valid, but insistence on flow conditioning had negated them. Point 4 was valid when 1% uncertainty was adequate, but higher demands had changed this. Point 7 was basically true. Point 10 was an open question with too little experience. They saw transducers as the limiting factor in terms of extremes of ambient and flow conditions.

British Gas's development of a multipath ultrasonic flowmeter was discussed in several papers (cf. Nolan, Gaskell and Cheung 1985) and site tests were reported. Four-path meters were tested ranging from 150 mm (6 inch) diameter to 1050 mm (42 inch). In particular, tests of a 300 mm (12 inch) meter were reported. The signal was used to derive, without any empirical constants, a flow rate which was initially checked on a traceable calibration stand and found to be well within ±1% of rate over nearly 12:1 turndown. On site it was calibrated within the uncertainty of that method, and appeared to give a performance of order ±0.5% or better. The meter was developed and sold by Daniel Industries.

An uncertainty of 0.5% was attainable for a custody transfer application with a multipath ultrasonic meter (Beeson 1995). NorAm installed the first custody transfer meter in North America in 1994. One problem experienced was that a small oscillating flow caused a small flow indication, dealt with by increasing the low flow cut-off. The installation is the most critical step under line pressure. Having removed the outer paint layer and used ultrasonic gauging for the pipe wall thickness, the OD was measured and the ID calculated. Two collars were welded in the correct positions and valves installed on the collars, and the line was hot-tapped. Probes were then inserted as well as pressure and temperature sensors. Beeson also recommended on-board isolation to avoid ground loop problems.

Dry calibration is possible, but the speed of sound needs to be verified by measurement of the gas constituents from a gas chromatograph, and the gas in each case must be thoroughly mixed. A standard, which consisted of two multipath meters in series with a bank of sonic nozzles calibrated against a primary standard weigh tank, was reported. It appears that the meters ranged from 0.4 m to 0.75 m (16 in to 30 in).

The development of the Siemens gas meter was described by Sheppard (1994) (cf. Chapman and Etheridge 1993), and also aspects of communication, automatic meter reading and temperature compensation. The meter used a W path configuration [Figure 14.13(a)] where the beam was reflected within the tube three times. It was claimed that this gave good integration across the flow. The timing resolution at

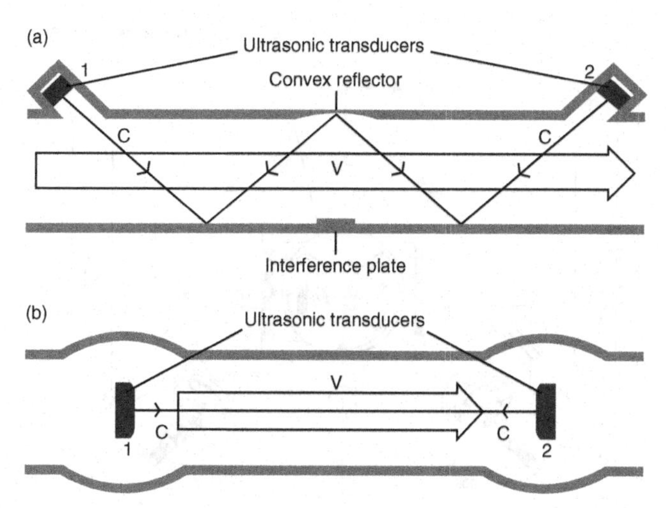

Figure 14.13. Ultrasonic gas meter
(a) W beam type.
(b) Dog bone type.
(Sheppard 1994 reproduced with the agreement of Elsevier).

the low flow rates defined by British Gas (40 l/h) was of order a few nanoseconds. Low power required a two-stage timing design, and resulted in a 10-year life on one D cell battery. Hemp's reciprocity requirement was observed. The flow was sampled randomly but with a basic 0.5Hz frequency. The specification was ±1.5% from 6,000 l/h down to 80 l/h, and ±3% from 80 l/h down to 40 l/h. Koechner, Melling and Baumgartner (1996) described the tube as of rectangular cross-section, 30 mm by 6.3 mm, with the transducers recessed. The small width and W path increased velocity and improved averaging. The reflector supplied some focussing to the beam and an interference plate suppressed single-reflection (V) path signals. Bignell et al. (1993a; cf. 1993b) also described an ultrasonic domestic gas flowmeter. The operation was for −10°C to 50°C within about ±1.5%. Bignell (1994) also described a meter with axial transmission through a "dog-bone" shape [(cf. Sheppard 1994 and Figure 14.13(b)] with the transducers in the enlarged "ends" around which the inlet and outlet flow passed to and from the "bone" between the "ends". The tube length was 0.56 m with diameter of 11.95 mm. The transducers were PVDF with 125 kHz, low enough to avoid absorption and high enough to avoid ambient frequencies and to give a rapid zero crossing. The author claimed that he had achieved a 300:1 turndown and saw the possibility of achieving 1,000:1 in the future.

A meter with a single path is shown in Figure 14.14(a), with dual paths in Figure 14.14(b), and a multipath gas meter is shown in Figure 14.14(c). It was claimed that for a flow range of 0.3 m/s (1 ft/s) to about 20 m/s (70 ft/s) the four-path design was capable of an uncertainty with calibration of ±0.5% rate plus an additional factor depending on size of meter etc. of 0.2% to 0.1% full scale for sizes 100–600 mm and temperature range of −20°C to 40°C. The whole operated on a clock resolution of 10 ns with nearly 1,000 measurements per minute and for a range of temperature of −20°C to 40°C. Van der Kam and Dam (1993) described the replacement of orifice

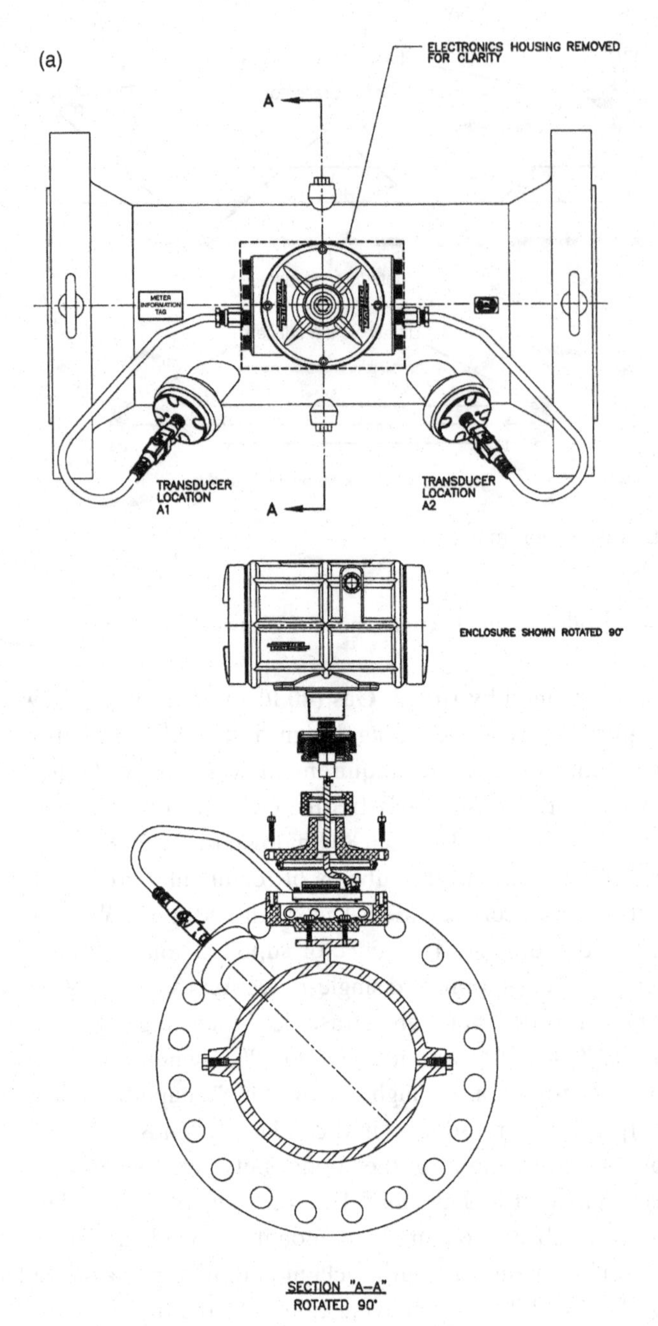

Figure 14.14. Ultrasonic gas meters
(a) drawing of single path;
(b) drawing of dual path;
(c) drawing of multipath.
(reproduced with permission of Emerson Process Management Ltd.).

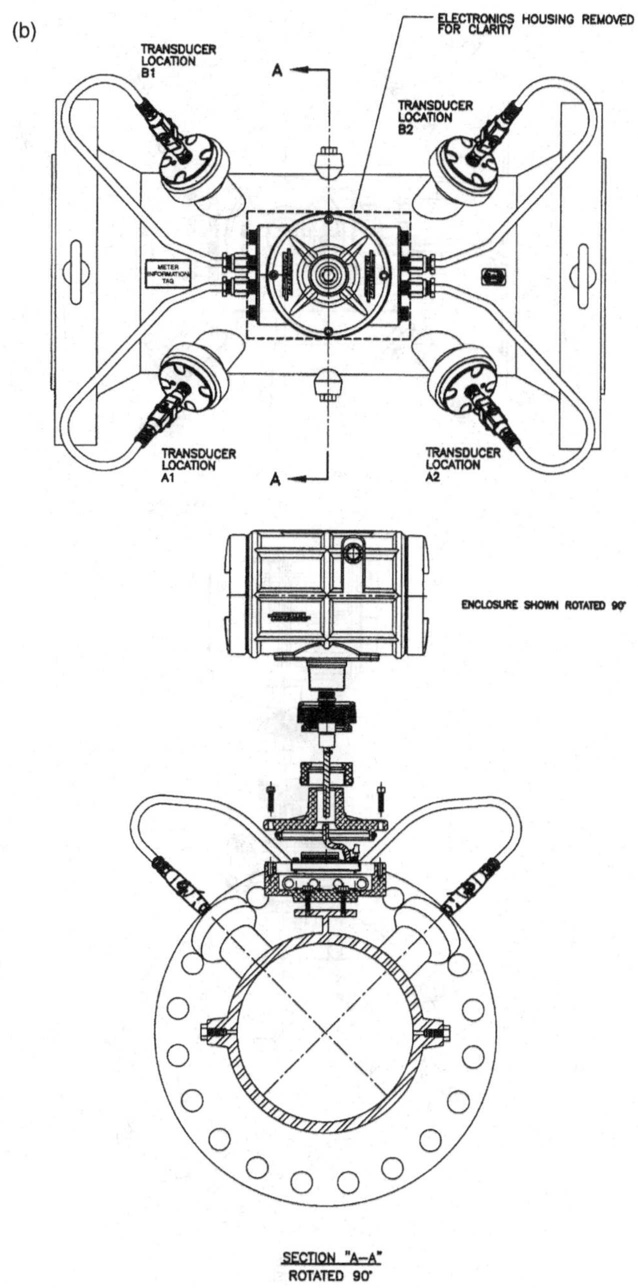

Figure 14.14 (*continued*)

meters in export stations of Nederlandse Gasunie and suggested that for wet or dirty flows the four-path ultrasonic would possibly be best of all.

A five-path meter was reported (McCarthy 1996) which was developed in cooperation between Gasunie and Instromet. Trials on an export line resulted in a performance within 0.2% of installed turbine meters.

To overcome the beam bending, Mylvaganam (1989) used a ray-rescue offset angle which appeared to be taken as the half-width of the transmission lobes of the

(c)

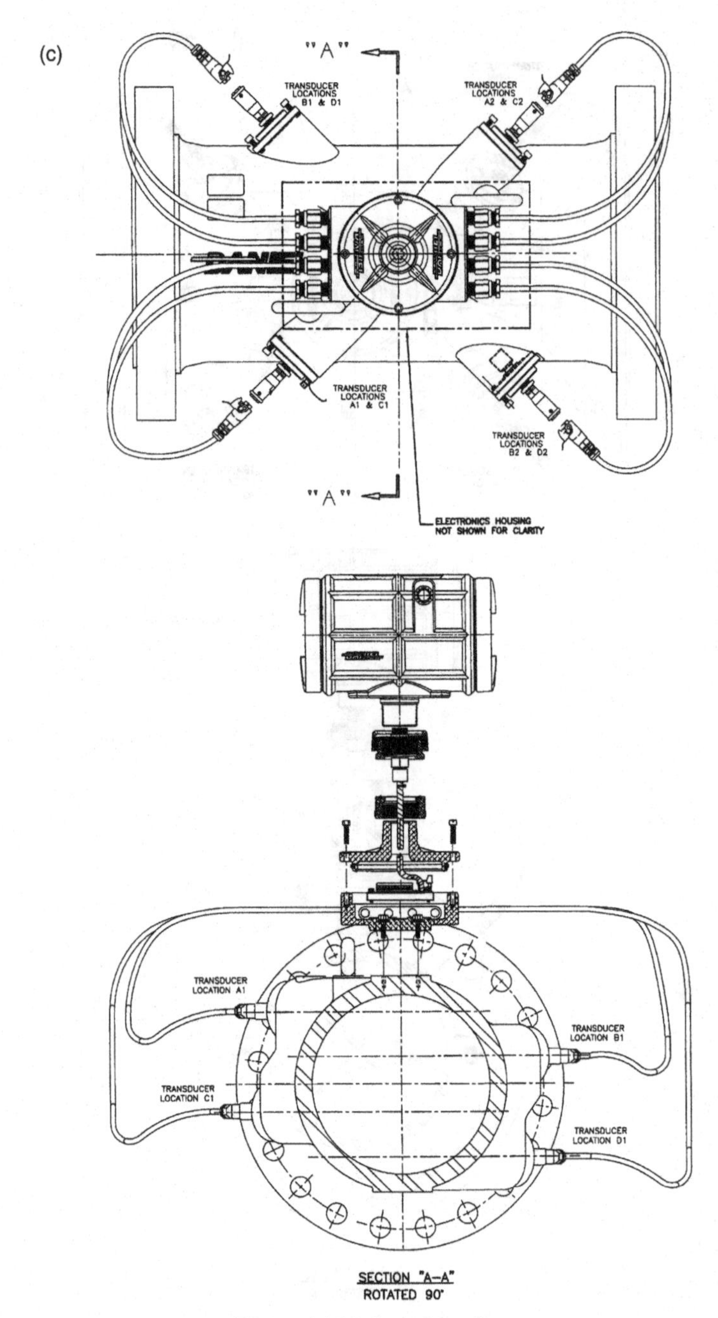

Figure 14.14 (*continued*)

transducer sound energy pattern. He pointed out that this would also prevent standing waves in small tubes. He also briefly described the transmission of ultrasound which was achieved in a chirp pulse where the frequency was ramped up in the time of the pulse, and resulted in pulse compression at the receiver through correlation (Cook and Bernfeld 1967). At low flows a continuous wave was used. The author appeared to claim 3% to 5% uncertainty with 95% confidence level. Other solutions have been discussed in this chapter.

For very high flow rates (Mach number of greater than 0.2) Guilbert and Sanderson (1996b) have calculated the shape of wall for a reflex meter to focus the beam onto the receiving transducer. The result was a slight hump about 1 cm maximum in a duct of 10 cm with transducer spacing about 10 cm.

Gorny, Gillis and Moldover (2012) discussed a long wavelength acoustic meter which was part of NIST's programme on the measurement of greenhouse gas emissions, in this case aimed at CO_2 etc. from coal-burning power plants. Their aim appeared to be the reduction of uncertainty from about 5% down to about 1%. A plane wave is generated in the duct and moves along axially. Hemp (1982) showed that such an axial wave can provide a correct average of the flow in the duct.

The 12-chord gas meter, already mentioned (Drenthen et al. 2009), claimed to be capable of 0.2% uncertainty provided there was at least 5D of upstream straight pipe. This indicates the continual development of these meters to today's ones with more paths, higher precision, less sensitivity to upstream fittings and increasingly powerful software, and self-monitoring capabilities.

14.3.12 Applications, Advantages and Disadvantages of the Transit-Time and Related Designs

Applications The meters appear to be suitable for any homogeneous liquid which admits an ultrasonic wave, and may be able to cope with small amounts of solids (possibly up to 5%) and/or gas (possibly up to 2%) entrained in the flow.

With its introduction into some national regulations for fiscal measurement of oil and gas this meter is becoming increasingly widely used in gas export station (Sloet and de Nobel 1997) flow line measurement (Agricola 1997) with the added benefits of self-monitoring (Sakariassen 1997).

Gallagher and Saunders (2002) considered meter intelligence such as self-checking: master clock, sound speed, path lengths, fluid parameters, distortion/strength of received signals, deposits, path angles, software and compare paths.

Sanderson and Yeung (2002) provided guidelines for the application of clamp-on transit-time flowmeters.

Particular applications which have been tried or which look promising, and some which should be treated with caution, are:

- liquid hydrocarbon custody transfer requiring uncertainty better than 0.15% of reading (Mandard et al. 2008);
- water flow in steam power plants, and hydroelectric plants (Erickson and Graber 1983);
- heavy crude oil (with a caution regarding Reynolds number compensation implemented (Folkestad 2001);
- fiscal measurements on stabilised crude (Decker, Leenhoven and Danen 2001);

- for oil with a second component in the liquid flow: gas in liquid with gas volume fractions (GVF) up to 70%; water in oil flows across the water-cut range, the reader should be cautious and may find Brown and Coull's (2001) paper useful. Note the transit-time meter attenuation at greater than 1% GVF and possible severe effect on signals, while a Doppler meter may give results [cf. problems in a flow of water in oil (Cousins, Augenstein and Eagle 2005)];
- caution on difficulties of use with high-viscosity products rather than on lighter hydrocarbon products. May provide accurate measurement over a wide range of crude oil applications if properly applied, proven and operated (Kalivoda and Lunde 2005);
- fiscal gas metering (Hannisdal 1991);
- natural gas meter technology (Buonanno 2000);
- general purpose gas meter capable of use with air flows, nitrogen, argon, chlorine;
- steam flow measurement in chemical plant (Kuchler 1999);
- gas consumed in a process (Womack 2008);
- gas industry (Stidger 2003);
- gas production measurement application (Fish 2007);
- liquid oxygen and hydrogen (Buchanan 2003);
- CO_2-rich applications, CCS (Carbon Capture and Storage) (Lansing, Ehrlich and Dietz 2009);
- flare-gas flow measurement (Raustein and Fosse 1991); cf. Mylvaganam (1989) and see Johannessen (2001) on legislative changes relating to zero flare operation and environmental sensitivities;
- possible flowmeter for wet-gas pipelines capable of measuring film thicknesses and mist content (Gopal and Jepson 1996);
- pipeline management combined with leak detection (Baumoel 1996);
- nuclear fuel–processing plant (Finlayson 1992), for both non-radioactive and radioactive liquid flows;
- air flow rate into an IC engine (Park et al. 1997);
- potential for domestic use if manufacturing cost is acceptable (Bignell 2000); cf. Buonanno (2000) on field tests;
- heat flow measurement (all metal construction) for low flow range (Baumgartner 2001);
- pulverised fuel mass flow (Millen et al. 2000);
- flowmeter for irrigation flows (Ziani, Bennouna and Boissier 2004);
- large-bore sewage systems (Harmuth and Erbe 2002) and water, treated wastewater and chemical processes (Marshall 2004);
- reservations on clamp-on meters to check performance of billing meters (Svensson and Delsing 1998);
- transfer package [two eight-path meters, with straight runs and conditioner (Brown et al. 2009a)]. Norsk Hydro recommended calibrating (pairs of) meters together with the complete metering run, having achieved results within ± 0.175% between meters (Nesse et al. 2003);

- monitoring calibration facilities – explored by NIST Fluid Flow Group – using an advanced ultrasonic flowmeter (Yeh et al. 2001);
- weights and measures authorities in various countries have approved the flowmeters for fiscal applications (cf. API (2011) standard "Measurement of liquid hydrocarbons by ultrasonic flow meters using transit time technology" (Hogendoorn et al. 2004; Kalivoda and Lunde 2005) [cf. Hofstede et al. (2004) on a meter for liquids with three beams for custody transfer applications with light crudes and refined liquid hydrocarbons]. Relevant standards appear to be (Smørgrav and Abrahamsen 2009) OIML R 137-1 and AGA Report No 9 and ISO 17089;
- possible custody transfer measurement of LNG (Hogendoorn, Boer and Danen 2007). Authors noted challenges resulting from cryogenic conditions (cf. Brown et al. 2007).

Sui et al. (2013) investigated and used CFD models to understand the operation of an ultrasonic meter with coal seam gas and how temperature, pressure and gas composition affect the signal. Many of the wellheads are in environmentally sensitive areas and on various counts the authors suggest that ultrasonic metering is a suitable solution to monitor the flows.

CO_2 or mixtures containing CO_2 have problems of attenuation and changes of density and sound speed which need careful evaluation (Harper, Lansing and Dietz 2009). They discussed this further (2012), giving their view that the meter is capable of metering gas with high levels of CO_2 and even pure CO_2 provided careful attention was given to operating frequencies and that, to avoid too great attenuation, the path should not be a reflecting one. At the other extreme, Hoshikawa et al. (2005) discussed the development of a meter for hydrogen.

A particular application which has been important concerns the measurement of feed water flows in nuclear power plants, which is critical in calculating reactor thermal power. Gribok et al. (2001) discussed inferential measurement of feed water flow rate, and Roverso, Ruan and Fantoni (2002) gave a brief but clear explanation of the flow measurement problem in nuclear reactors. Venturi meters, frequently used, may suffer from a deposit in the tube leading to an overestimate of flow, hence the options of transit-time or correlation flowmeters. Ruan et al. (2003) discussed combining neural networks with cross-correlation and other options for nuclear feed water flow measurement. See also Roverso and Ruan (2004), Estrada et al. (2004), Fathimani et al. (2007), Mori, Tezuka and Takeda (2006), Estrada (2002), Peyvan, Gurevich and French (2002), Hauser, Estrada and Regan (2004), Tominaga, Yudate and Cormier(2005), Jung and Seong (2005), Jae et al. (2006) and Yun and Park (2008).

Laan (2012) set out developments in the technology, presumably from the manufacturer's viewpoint. The development is in transducers for demanding applications: pressure and temperature, the acoustic development, the digital signal processing, and then there is the estimate of uncertainty. The paper gives examples of extreme temperature requirements for certain applications such as solar power plant monitoring, with resultant problems of rapid sound speed changes and density and viscosity changes leading to Reynolds number changes, and at the other extreme cryogenic liquefaction of natural gas. At one extreme is the need to keep the transducer below 200°C with a liquid temperature of 600°C and at the other cryogenic extreme LNG at −160°C. The paper also indicates the accurate custody transfer resulting from the technology. There are also the high-temperature steam and gas flows which need to be monitored. The paper also makes the point about long-term stability.

Applications for the clamp-on were given by Baumoel (1994) as: fuel mass metering, hydraulic fluid flow metering, leak detection, engine lubricant, ground support of all aircraft fluid systems, rocket fuel and oxidiser metering and space vehicle coolant flow metering. The sound speed for hydrocarbon fluids was shown to drop over a range from about 0°C to 40°C (30°F to 100°F) by about 12% and differs for different hydrocarbons. With the relationship between temperature and density the meter can deduce the type of fuel in the line, whether there is water in the line and the mass flow of the fluid.

Advantages (Beeson 1995; Mylvaganam 1989): non-intrusive, no moving parts, compact, high rangeability, ease of installation, cost savings and improvements in maintenance with self-diagnostics, accuracy may not be affected by build-up of contaminants, fully pigable, bidirectional flow capability, no lined pipe restriction. Hot tapping can be used under pressure. Self-checking using signal strength and quality alerts the instrument technician to problems.

De Vries et al. (1989) claimed that transducers could be installed in underground gas pipes from ground level, so that using reflection mode a measurement uncertainty of 2% could be achieved with a repeatability of 0.2%.

Disadvantages: Problems when applied to gases include: low efficiency of launching waves due to impedance differences, absorption increases with frequency, beam bending in large flow lines with high speed flow.

Aliasing effects may be present with pulsating flow and Beeson (1995) suggested using asynchronous sampling techniques.

14.4 Doppler Flowmeter

14.4.1 Simple Explanation of Operation

This type of meter depends on the Doppler frequency shift which occurs when sound bounces off a moving object as shown in Figure 14.15. In the Doppler meter the waves need to reflect or scatter off something moving with the flow. If they reflect off a stationary object, then they retain their wavelength and frequency. If, however, they reflect off a moving object, the wave fronts will hit the moving object with a time interval which is not the same as their period in a stationary medium. As a result, the reflected wave will have a new period, frequency and wavelength. It is almost invariably a clamp-on design for liquids. Figure 14.15 is a diagram of the effect.

The arrangement for typical commercial Doppler meters is also shown in Figure 14.15. The transmitting and receiving transducers may be in the same block and held on the outside of the tube. On the other hand, they may be in separate

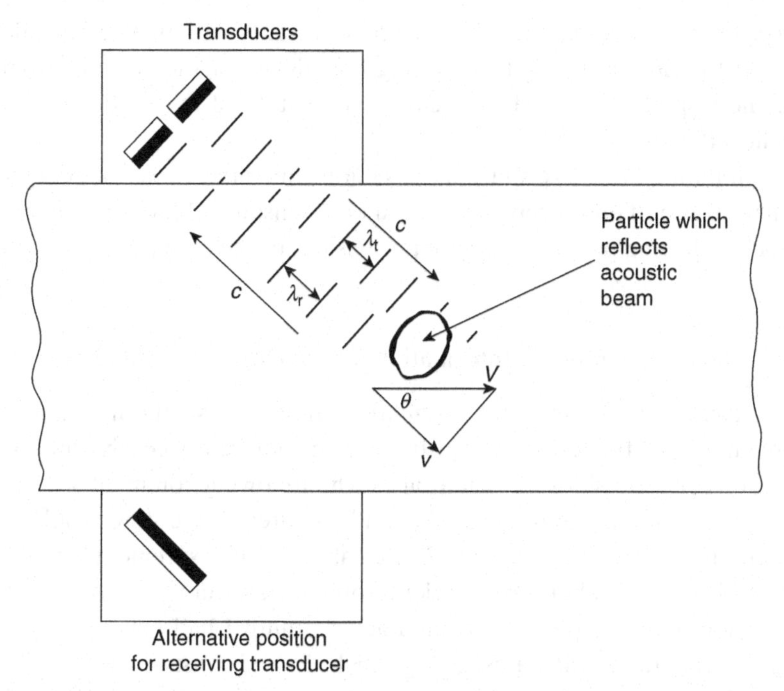

Figure 14.15. Diagram of Doppler flowmeter showing the Doppler effect and typical arrangements of the transducers for a commercial device.

blocks (Figure 14.15) and can then be positioned on the same side or on opposite sides of the tube.

We now face one of the major uncertainties in the operation of these devices. What is the velocity of the reflecting object compared with the axial mean velocity of the flow in the pipe? This depends (Figure 14.15) on:

- what the reflecting surface is;
- where the reflecting surface is;
- the velocity of the object and its relative velocity to the flow in size and angle.

We can start by taking the simplest assumption that the particle is moving with the flow and that the ultrasound path makes an angle with the axis of θ. So in terms of V the velocity in the pipe is:

$$\Delta f = 2 f_t \frac{v}{c} \cos\theta \tag{14.11}$$

We can obtain a typical value of Δf assuming a transmission frequency of, say, 5 MHz, and a flow rate of water of 10m/s. Δf will then have the value of about 50kHz (assuming 1,414 m/s for sound speed). Clearly this will not be one sole frequency coming back from the fluid, but a range of frequencies, and some method will be built into the converter to identify the favoured frequency.

At least one manufacturer may offer a profile measurement capability. This is usually achieved by range gating the returning signal so that the distance of penetration into the pipe is known and hence the velocity at that point.

Uncertainty claims may be ±2% of full scale, although great care and understanding of the flow is needed in gaining confidence in the reading from these devices. The response is very dependent on the nature of the particles or bubbles which reflect the ultrasound.

Poor mounting can cause spurious reflections, and pipe vibration may give false flow signals. Transmit and receive may be in one transducer block or may be in separate blocks, allowing positioning on opposite sides of the pipe (Figure 14.15).

14.4.2 Operational Information for the Doppler Flowmeter

One manufacturer suggested that installation should allow 6D upstream and 4D downstream. It may be best to mount transducers adjacent to each other for large pipes, but on opposite sides for small pipes. The following comment was noted by this author: "the pulses are reflected from a large area of the flow profile, giving a good representation of mean velocity. Repeatability of measurement is to specification, and with 'on-site' calibration checked, volumetric accuracy is assured." I would be less certain than this particular manufacturer, until I had experience of a prolonged set of calibrations and a usage log which showed consistency.

Ranges may be from, say, 0.3 m/s to 6 m/s, with a temperature range of –20°C to 80°C.

14.4.3 Applications, Advantages and Disadvantages for the Doppler Flowmeter

Cairney's (1991) comment on Doppler flowmeters was that "these were the first type of ultrasonic flowmeter produced commercially. Experience has shown that they were oversold as an all-purpose flowmeter. When there were particles in the fluid measured they have been of some use, but most fluids flowing through power station pipelines are clean, which makes them far less effective."

Manufacturers suggest that these meters can be used for: mining slurries, coal slurries, sewage, sludge, raw water, sea water, pulp (paper), acids, emulsion paint, fruit juice, yoghurt, vegetable oil, citric acid, glucose, contaminated oil, cement slurry, lime slurry, industrial effluent.

Fischer, Rebattet and Dufour (2013) described their development of a pulsed Doppler system which allowed them to obtain flow profiles very rapidly. Their objective was to obtain flow profiles through hydraulic pipes and turbo machines. Pulsed Doppler gates the reflected pulse, so that the reflection point can be identified, and greatly enhanced precision achieved. Some research is reported briefly on this type of meter in Section 14.A.5, including a calibration of a pulsed Doppler meter using air bubbles as reflectors which appeared to achieve an expanded uncertainty of ±0.26% or better (Tezuka et al. 2008b). See also Murakawa *et al* (2014) on the effects of the number of pulse repetitions and noise on the velocity data. Pulsed Doppler should be distinguished from the more basic type.

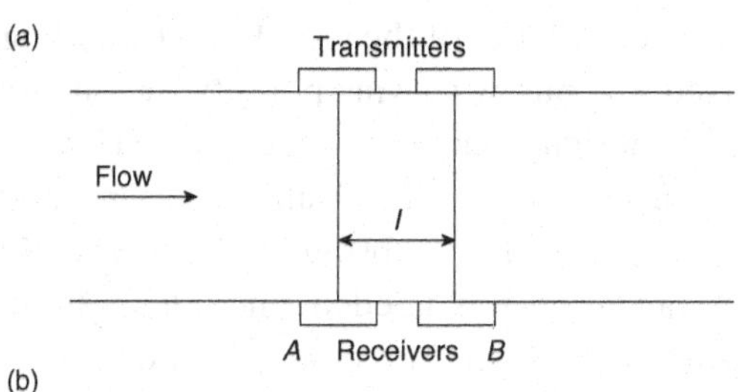

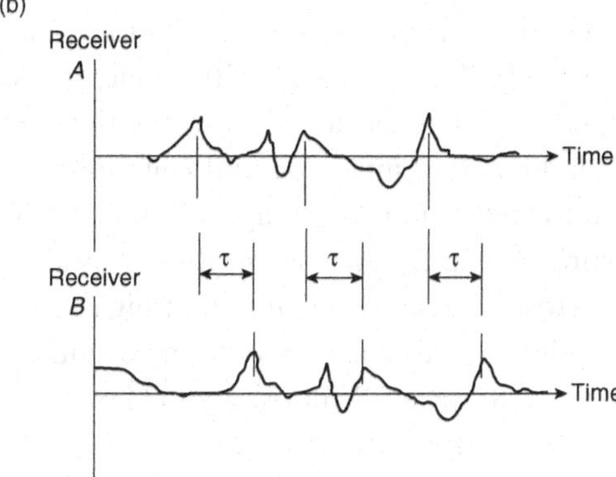

Figure 14.16. Correlation flowmeter **(a)** diagram of the geometry; **(b)** traces from channels A and B.

14.5 Correlation Flowmeter

14.5.1 Operation of the Correlation Flowmeter

If two beams cross the flow at a known distance apart, L, as in Figure 14.16(a), and the received signals are compared as in Figure 14.16(b) to find a similar pattern, the pattern for channel B will be found to be displaced a time τ_m from that for channel A. From this it is simple to deduce that the flow has taken time τ_m to move a distance L or

$$V = \frac{L}{\tau_m} \tag{14.12}$$

Correlating the two signals requires some complex electronics. The fluctuation in the signal must be adequate, and for this reason it will probably need to be created artificially by an upstream disturbance in the flow if heavy turbulence, bubbles, second phase etc. are absent naturally. The mathematical concept of cross-correlation is defined by the equation (Keech 1982)

$$R_{yx}(\tau) = \lim_{T\to\infty} \frac{1}{T}\int_0^T y(t)x(t+\tau)\,dt \tag{14.13}$$

where $x(t)$ and $y(t)$ are the upstream and downstream signals, respectively. The value of time delay τ_m corresponding to the maximum of $R_{yx}(\tau)$ provides a measure of the flow transit time between the two beams spaced L apart. The calculation of the integral in Equation (14.13) is achievable with modern computation. The problem is to do it cheaply and fast. For instance, sufficient precision may be achievable by using only the polarity of the signal, and by cutting the signal into sufficiently small bits of fixed size, but by varying polarity, the integration may be simplified and speeded up without significant loss of precision. (However, see Yang and Beck 1997.)

It has found a particular niche in multiphase flows where the disturbance to the signal will be substantial. However, Keech (1982) suggested that some form of validation was needed since the system could give incorrect estimates of velocity: spurious noise could give a correlation, decay of disturbances at low flows, other movement such as cavitation-generated shockwaves or intense pressure waves, oscillatory flows. Multichannel devices may overcome some of these effects. Worch (1998a) discussed a correlation flowmeter which needed no bluff bodies with 40 and 50 mm pipe and claiming a measurement error of less than 2 to 3% for Reynolds number range 25,000–250,000 and down to lower values. (See also Worch 1998b for two-phase flows.) Skwarek and Hans (2000) also reported work on cross-correlation meters and highlighted the different origin and treatment of amplitude- and phase-modulating events and the possible extension of applications to gases. Skwarek, Windorfer and Hans (2001) discussed the measurement of pulsating flow with ultrasound. Battye (2001) discussed acoustic considerations including standing wave patterns which might affect the design of demodulators for the ultrasonic correlation flowmeter. With transducers in contact with the water to minimise unwanted resonances, the standing wave patterns were, nevertheless, thought possibly to contribute up to ±2.5% uncertainty in the flow reading without in situ calibration. With non-wetting transducers, the uncertainty might be larger. The author continued to see the meter as having a particular application to multiphase flows.

Hans (2002a) suggested turbulence in gas could be used for cross-correlation, and suggested reasons for discrepancy between measured and actual flow rate (Hans 2003a). A theoretical model (Schneider, Peters and Merzkirch 2003) appeared to show that the measured velocity was profile dependent. The research appeared to relate to a clamp-on meter. Jenkins et al. (2006) identified pressure pulsations in power plant as important on sound speed, and on path length changes due to vibration.

Lysak et al. (2008a, 2008b) developed an analytical model of the cross-correlation flowmeter sensing based on the turbulence model and making use of CFD.

Gurevich et al. (2002) reported a performance evaluation and field application of the CROSSFLOW™ clamp-on ultrasonic cross-correlation flowmeter.

See also Frail's (2005) paper mentioned in Section 14.3.7.3 under **Gas clamp-on.**

14.5.2 Installation Effects for the Correlation Flowmeter

Paik, Mim and Lee (1994) used clamp-on transducers with 1 MHz crystals, 35° wedges and 2D or 3D spacing in some tests of installation effects. Their baseline tests were with 120D upstream straight pipe. The errors they found are shown in Table 14.8.

There may be other experimental studies of installation effects of which I am not aware, but since this is the only data provided in this book it should probably be used with caution.

Table 14.8. *Error in signal due to upstream fittings (after Paik et al. 1994)*

Fitting	Diameters Upstream			
	5	10	20	30
Single elbow	−7%	−4%	(−2%)	(−2%)
Double elbow out-of-plane	−2 to −4%	+5%	+2 to 6%	0 to 4%
Double elbow in-plane	−6%	−5%	−2%	−2%

( ) possible scatter.

14.5.3 Other Published Work on the Correlation Flowmeter

King (1988; cf. King, Sidney and Coulthard 1988 and Sidney, King and Coulthard 1988a, 1988b) mentioned the development of an ultrasonic cross-correlation meter for multiphase flow at the National Engineering Laboratory, Scotland, in collaboration with Moore, Barrett and Redwood. This development undoubtedly exploits one of the most important applications of this technique, and increasingly sophisticated electronics should allow more information about the flow components and their velocities to be deduced.

Coulthard and Yan (1993c) reviewed correlation flowmeter results for two-phase mixtures:

Solids/liquids	satisfactory operation
Oil/gas	at void fractions up to 25% and velocities up to 7 m/s scatter of results was ±4.5% without mixing and ±2% with mixing.

Single-phase gas and liquid are also usable, but may need a bluff body to enhance the signal.

Xu et al. (1994) described a clamp-on ultrasonic cross-correlation flowmeter for liquid/solid two-phase flows, with brief indications of how the various components operate. Calibrations of these meters against an electromagnetic meter gave differences of about ±5% for a 0.4% by weight paper pulp flow, and about ±3% for a crude oil flow.

Battye (1993) suggested measurement uncertainty over 10:1 of 2% with ±0.5% scatter but that temperature variation in the acoustic path, if not allowed for, might reduce linearity. Other work by Kim et al. (1993a) suggested ±2.2% or for clamp-on ±2.8%. Yang and Beck (1997) claimed that, with a highly intelligent cross-correlator, a measurement uncertainty of 1% could be achieved.

Lemon (1995) discussed a technique which is referred to as acoustic scintillation flow measurement and appeared to refer to, or be related to, cross-correlation flow measurement using the disturbances caused to the ultrasonic beams in traversing the flow.

It should be remembered that while cross-correlation has been used with ultrasonics and is included here for that reason, it has much wider application. Chen et al. (1993) reported work using two pairs of radiation sensors. Their application was to medium-consistency pulp in the paper industry. Flows approach Newtonian at high rates but at low flows behave more like a plug with a laminar annulus of

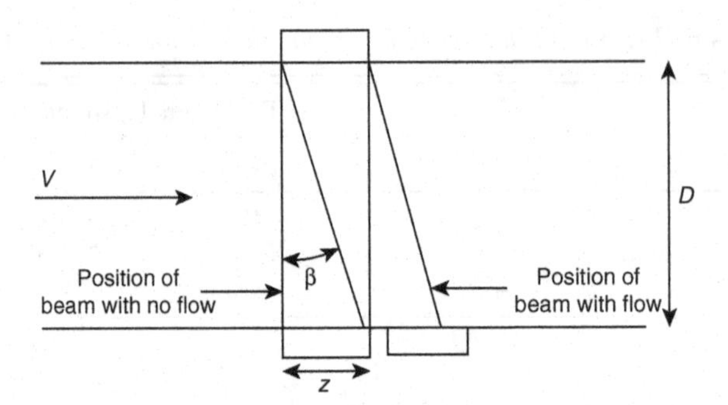

Figure 14.17 Flowmeter concept using beam sweeping.

continuous pure liquid between plug and wall. As flow increases the annulus flow becomes turbulent with lower consistency suspension. Eventually complete plug disruption takes place. Flow rates may be 0.3 to 3 m/s. The correlation can be based on inhomogeneities, flocs in the annulus or temperature changes, and repeatability of 2.5% may be possible. Cf. Ohira et al. (2003) capacitance sensing.

Readers are referred to Beck and Plaskowski's (1987) book on this type of flowmeter. See also Chapter 17 on multiphase flowmeters.

14.5.4 Applications, Advantages and Disadvantages for the Correlation Flowmeter

The cross-correlation ultrasonic flowmeter appears to require a significant disturbance in the flowing fluid. Where the disturbance is due to multiphase flow, the meter comes into its own and provides a meter which can satisfactorily cope in a non-intrusive way with such flows.

14.6 Other Ultrasonic Applications

Flowmeters could be designed to make use of the beam sweeping caused by the flow. These have been proposed, and some research which appears to use this method has recently been published by Kawaguchi et al. (2015). Figure 14.17 shows a diagram of a beam crossing the flow approximately at right angles, but being swept sideways because of the fluid motion. On the receiving side of the flow passage are two or more sensors which measure the strength of the received signal at each sensor and so allow the electronics to deduce the deflection due to the flow. The angular deflection, β, is given approximately by

$$\tan\beta = \frac{V}{c} \tag{14.14}$$

and the deflection on the receiving wall will be

$$z = \frac{VD}{c} \tag{14.15}$$

Ultrasonic sensing is also used in open channel flowmeters, flumes and weirs to sense the level of the upstream surface from which is deduced the flow rate (Herschy 1995).

In vortex meters one method of sensing the oscillation of the shedding is ultrasound, which is disturbed by the vortices and provides a signal which can be used to extract the frequency of shedding.

Coulthard and co-workers have also suggested using the vortex shedding combined with a correlation-sensing system to obtain flow measurement. There may be problems with using such a regular marker in this technique and one where vortices may move faster than the fluid.

Teufel et al. (1992) described new techniques for measuring the profile in pipes, and applied it to oscillating flows. The object of the paper was to compare LDV and the new ultrasonic velocity profile sensor (UVP). The authors saw the techniques as complementary and it is possible that the second may offer an industrial approach where profile determination is important.

Olsen (1991) described tests of what appeared to be an anemometer using sonic and ultrasonic signals and allowing velocities in a wind tunnel to be measured in three directions simultaneously.

Joshi (1991) described a surface-acoustic-wave (SAW) flow sensor. The oscillation frequency is temperature dependent, and so if the element is heated above ambient, and placed in the flowing gas, the frequency will be flow-rate dependent.

Guilbert, Law and Sanderson (1996) used a pulse of heat through the pipe wall to mark the liquid, and the passing of the marked liquid was sensed by an ultrasonic beam using the difference in transit time for cold and hot liquid.

14.7 Conclusions on Ultrasonic Flowmeters

This has been an area of great growth and is one of continuing potential. It is experiencing a great deal of research and development with increasingly impressive results in performance and application.

One feature of ultrasonic flowmeters which various authors have made is the large amount of information available from the system, but not completely used, which will give indication of operating changes or problems. In my earlier books (1988/1989, 2002b, 2003) I suggested that ultrasonic technology might develop a clamp-on meter (possibly a master meter) which sensed:

- wall thickness;
- the quality of the inside of the tube;
- the turbulence level;
- profile from a range-gated Doppler system: cf. Yamamoto, Yao and Kshiro (2004) on combined transit-time and Doppler meter;
- flow measurement from a transit-time system;
- correlation to provide information about a second phase;
- density from the impedance and sound speed;
- condition (self-) monitoring.

Such a device could measure diameter and other pipe details and could programme a much simpler device for permanent installation at the site.

Many of these features now appear together. Sanderson and Torley (1985) combined wall thickness and some dimensional and condition monitoring with intelligent control. Range-gated Doppler has been exploited in medical applications. (See also Mahadeva et al. 2010.) Birch and Lemon (1995) discussed the use of Doppler on multiple paths with range gating to obtain profiles of flows in open channels and discharges. A commercial device may allow operation in both transit-time and Doppler modes. Correlation has been used for multiphase oil and gas flow measurement. A clamp-on Doppler and a clamp-on correlation appear to have been developed for gas flow measurement. Guilbert et al. (1996) have used the device for mass flow measurement. Lynnworth (1990) has found that the phase velocity of flexural waves in the wall of small-diameter pipes was dependent on the density of the liquid inside. As the technology becomes more widely accepted the microprocessor control is likely to become more powerful and to introduce multiple facilities.

A technology which appears to be developing is the use of transducer arrays which can produce a beam which is steerable by electronic means and offers possibilities of varying angle and position of the beam within the pipe while in operation.

The development of instruments for domestic gas measurement, which are within the tight budget for the utilities, operate off a battery with 10 years or more of life and have sophisticated external communication ability, does not appear to have resulted in a design in widespread use. Further innovation could be very profitable in this area.

Perhaps the ultrasonic flowmeter is the realistic rival to the Coriolis meter in gas mass flow measurement, and has great versatility.

This powerful technology appears capable of meeting fiscal demands for metering liquid hydrocarbons and gas and, therefore, limits of its precision need to be well understood.

My last point in the list at the beginning of this section was condition monitoring, and it appears that various people and organisations have been developing protocols to maximise the available data from the ultrasonic beams. The potential for verification approaching in situ calibration is very promising, and should be given high priority in research plans. An important industrial development is described by Vermeulen, Drenthen and Den Hollander (2012). This is an expert system which makes use of the information contained in the meter's sensing system continuously to give information about the operation and state of the meter.

Appendix 14.A Mathematical Methods and Further Research Relating to Ultrasonic Flowmeters

14.A.1 Simple Path Theory

Transit-Time

Figure 14.2 provides a diagram on which to base the maths. We can work in terms of trigonometric functions or dimensions.

We start with the downstream-going wave, and note that its speed will be greater than the sound speed by the component of the flow in the direction of the path. This component is $V \cos \theta$. The distance to be travelled is D cosec θ, where D is the pipe diameter so that the time taken for the downstream wave to travel from the transmitter to the receiver is

$$t_d = \frac{D \operatorname{cosec} \theta}{c + V \cos \theta} \tag{14.A.1}$$

A similar expression for the upstream wave time, t_u, is

$$t_u = \frac{D \operatorname{cosec} \theta}{c - V \cos \theta} \tag{14.A.2}$$

We can now obtain the difference between these two wave transit times (Figure 14.3)

$$\begin{aligned}
\Delta t &= t_u - t_d \\
&= \frac{D \operatorname{cosec} \theta}{c - V \cos \theta} - \frac{D \operatorname{cosec} \theta}{c + V \cos \theta} \\
&= \frac{D \operatorname{cosec} \theta / c}{1 - \left(V/c\right) \cos \theta} - \frac{D \operatorname{cosec} \theta / c}{1 + \left(V/c\right) \cos \theta}
\end{aligned}$$

Combining the two fractions on the RHS over a common denominator and ignoring terms in $(V/c)^2$ we obtain

$$\Delta t = \frac{2VD \cot \theta}{c^2} \tag{14.A.3}$$

Thus we see that, provided we know the speed of sound, we can find the value of V from the time difference and the geometry of the meter. The requirement to know the value of c is not difficult if we measure, as well as the time difference, the actual transit times each way and take the mean, t_m. We can make the approximation

$$\begin{aligned}
t_m^2 &= (t_u - \Delta t/2)(t_d + \Delta t/2) \\
&= t_u t_d + \Delta t(t_u - t_d)/2 - (\Delta t)^2/4 \\
&= t_u t_d + \Delta t(\Delta t)/2 - (\Delta t)^2/4 \\
&= t_u t_d + (\Delta t)^2/4 \\
&\approx t_u t_d
\end{aligned} \tag{14.A.4}$$

where, using values in Table 14.1, we have neglected a term of about 10^{-6}. We can now replace c^2 by $L^2/t_u t_d$ and $\cot \theta$ by X/D. We can then rewrite Equation (14.A.3) by

$$\Delta t = \frac{2VXt_u t_d}{L^2} \tag{14.A.5}$$

From this we can obtain Equation (14.5).

However, when we are concerned with a clamp-on flowmeter we have a further consideration (see e.g. Mahadeva 2009). The time measured by the meter timing circuit, t_{umeas}, will be from the pulse leaving the transmitting transducer, passing through the wedge, t_{wedge}, and pipe wall, t_{pipe}, through the fluid, t_u, and finally through the pipe wall and wedge to the receiver. Thus

$$t_u = t_{umeas} - 2(t_{wedge} + t_{pipe})$$

and

$$t_d = t_{dmeas} - 2(t_{wedge} + t_{pipe})$$

and to obtain t_u and t_d will require, not only the measured time up- and downstream, but values, calculated or measured, of the two terms t_{wedge} and t_{pipe}. The values will need to take account of environmental changes as well as the spacing of the transducer wedges along the pipe which may cause a change to the angle of incidence. This in turn may cause changes to the angle of the rays within wedge and wall and hence to the time taken for the pulse to pass through the wedge and the pipe.

Sing-Around

An alternative way to eliminate sound speed is the sing-around method (Figure 14.4) discussed by Suzuki et al. (1975; cf. 1972). In this method each of the paths, upstream and downstream, is operated in such a way that when the pulse is received by the receiving transducer, it triggers a new pulse from the transmitting transducer. Thus the frequency of pulses depends on the velocity of sound and of the flow. We can obtain the period between pulses from Equations (14.A.1) and (14.A.2), and the pulse frequency will be the inverse of the period, or for the downstream-going pulses

$$f_d = \frac{c + V\cos\theta}{D\operatorname{cosec}\theta} \tag{14.A.6}$$

A similar expression for the upstream pulse train is

$$f_u = \frac{c - V\cos\theta}{D\operatorname{cosec}\theta} \tag{14.A.7}$$

If we now take the difference between these frequencies we obtain

$$\begin{aligned}\Delta f &= \frac{2V\cos\theta\sin\theta}{D} \\ &= V\sin 2\theta / D\end{aligned} \tag{14.A.8}$$

We can write this in terms of dimensions, since $\sin 2\theta = 2DX / L^2$ as

$$\Delta f = 2VX / L^2 \tag{14.A.9}$$

which has the virtue of relating V to the frequency difference without requiring the value of c, the sound speed. Again we can obtain typical values for the frequencies from Table 14.2 (Baker 2002b, 2003). Making the approximation that V is constant across the pipe we can obtain Equation (14.7).

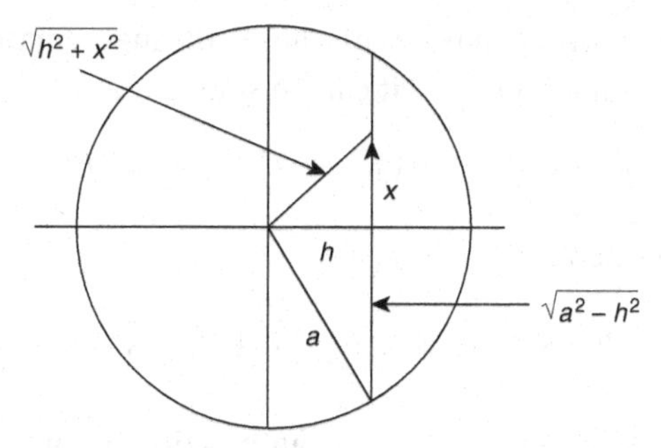

Figure 14.A.1. Geometry of the pipe cross-section for chordal integration.

14.A.2 Use of Multiple Paths to Integrate Flow Profile

This simple analysis has assumed that the flow profile is uniform. In fact if we allow for varying velocity along the path of the ultrasonic beam Equation (14.A.3) may be rewritten in the form of an integral

$$\Delta t = \frac{2\cot\theta}{c^2} \int_{\text{along chord}} V(y)dy$$

$$\Delta t = \frac{2V_{\mathrm{m}} Y\cot\theta}{c^2} \quad (14.A.10)$$

The flowmeter averages flow rate along the path of the beam.

For integration along the chord shown in Figure 14.A.1

$$\Delta t = \frac{2\cot\theta}{c^2} \int_{-\sqrt{a^2-h^2}}^{+\sqrt{a^2-h^2}} V\left(\sqrt{h^2+y^2}\right)dy \quad (14.A.11)$$

If the path is on the diameter two extremes of laminar profile and uniform profile yield a Δt ratio for equal volumetric flows of 4:3. Thus an ultrasonic flowmeter with diametral sensing has an intrinsic calibration shift across the laminar-turbulent profile change approaching 33%.

Baker and Thompson (1975) showed the improvement possible by using approximately mid-radius transducers and recalculating Equation (14.A.11) for values of $h \approx a/2$. By using the approximate expansion for the turbulence velocity profile of:

$$V(r) = \frac{(1+2n)(1+n)}{2n^2} V_{\mathrm{m}} (1-r/a)^{1/n} \quad (14.A.12)$$

where n varies with Reynolds number as:

Re	4×10^3	2.3×10^4	1.1×10^5	1.1×10^6	2.0×10^6 to 3.2×10^6
n	6.0	6.6	7.0	8.8	10

the signal variation can be obtained as in Figure 14.7.

The basis for using more than two paths lies in the mathematics of numerical integration. Linked to the name of Gauss it can be shown that a polynomial of the form

$$f(t) = c_0 + c_1 t + c_2 t^2 + c_3 t^3 + \cdots + c_{2n+1} t^{2n+1}$$

can be precisely integrated as

$$\int_{-1}^{+1} f(t)\,dt = A_0 f(t_0) + A_1 f(t_1) + A_2 f(t_2) + \cdots + A_n f(t_n) \tag{14.A.13}$$

Or a polynomial of degree $2n + 1$ or less can be correctly integrated by an expansion such as in Equation (14.A.13) provided the constants A_n and the positions t_n are correctly chosen. It is also possible to show that an integral can be rearranged to the interval −1 to +1. Thus with four positions and with suitable weighting, the integral of a polynomial of up to order seven can be correctly evaluated.

It is as a result of considerations such as these, and possible refinements based on likely profiles in the flow, that weightings for multiple paths (cf. Vaterlaus 1995) have been obtained. Zheng, Zhao and Mei (2015) claimed to have improved the numerical integration based on Gauss quadrature over the Gauss-Jacobi method. It appears that this was tested on two two-dimensional profiles due to Salami (1984a) and some experimental data..

14.A.3 Weight Vector Analysis

Hemp (1982) has developed the ideas from electromagnetic flowmeter theory for ultrasonic flowmeters. As in the case of electromagnetic flowmeters, the condition for an ideal flowmeter, one where the signal is independent of flow profile, is

$$\nabla \times \mathbf{W} = 0 \tag{14.A.14}$$

and Hemp obtained a flow signal expression using his reciprocal theory of

$$\pm\left(\mathrm{U_I}^{(2)} - \mathrm{U_{II}}^{(1)}\right) I = 2\int \mathbf{V}_s \cdot \mathbf{W}\mathrm{dv} \tag{14.A.15}$$

where the negative sign is for electromagnetic and/or magnetostrictive transducers and the positive one for electrostatic and/or piezoelectric, I is the driving current, $\mathrm{U_I}^{(2)}$ and $\mathrm{U_{II}}^{(1)}$ are the received voltages and $\mathbf{V}_s$ is the undisturbed flow. The weight vector is then given in terms of ρ_m the mean density in the meter and $\mathbf{v}_o$ the acoustic field for two distinct cases (1) and (2) when opposite transducers transmit, as

$$\mathbf{W} = \rho_m[(\mathbf{v}_o^{(2.)}.\nabla)\mathbf{v}_o^{(1)} - (\mathbf{v}_o^{(1)}.\nabla)\mathbf{v}_o^{(2)}] \tag{14.A.16}$$

Hemp suggested that a flowmeter in a rigid walled duct with plane wave modes, a meter with radial transmission and a meter with a large area transmission, one example of which is the industrial design for small pipes, are approaching ideal configurations.

As well as the fundamental weight vector [Equation (14.A.16)] it is possible to define weight vectors for phase shift and amplitude shift (in sine wave operation) and for pulse operation (Hemp 1998).

14.A.4 Development of Modelling of the Flowmeter

Transit-time Flowmeter Models

Lygre et al. (1987) discussed numerical simulation of ultrasonic flowmeters, and it appears that this has been a research strand at CMI, Bergen, Norway, and at the University of Bergen [Frøysa et al. (2001); cf. Kocbach 2000 on piezo electric ceramic discs].

Zheng, Zhang and Xu (2011) studied acoustic transducer protrusion and recess effects on the performance of ultrasonic flowmeters. This paper should provide some initial insight for calculating specific design effects which could affect precision of the flowmeter. It may be more difficult to generalise the likely effects on meter accuracy. There may be some suggestion that protruding transducers are to be preferred to those recessed.

Clamp-on Flowmeter Models

There has been a lot of activity in the modelling of clamp-on ultrasonic flowmeters, and some of the publications are referred to later, and they in turn carry other references.

Panicke and Huebel (2009) discussed the increasing potential capabilities of clamp-on ultrasonic flowmeters including: measurement of gas and LNG flows, hot and viscous liquids. They suggested these resulted from design developments such as special transducer types, more sophisticated signal processing, pipe mounting and pipe coupling methods.

Funck and Mitzkus (1996) derived an acoustic transfer function for a clamp-on ultrasonic flowmeter and their results, both calculated and measured, appeared to be in good agreement, and suggest the curve of velocity error against transducer position which has been called the S-curve.

Koechner and Melling (2000) used ray tracing to model the behaviour of an ultrasonic transit-time flowmeter and the deviation due to flow, particularly where the velocity was higher in gas flows. The finite size of the transducer was allowed for and a Monte Carlo method was used to reduce computing time by reducing the number of rays computed. They referenced early work on ray-tracing applied to room acoustics (Krockstad, Strøm and Sørsdal 1968). Iooss, Lhuillier and Jeanneau (2002 see also Jeanneau and Piguet 2000) also described a numerical ray-tracing method to obtain the uncertainty due to velocity profile. They reported that they found flow rate to be overestimated by 0.35% due to the effects of the mean velocity profile. Temperature and turbulent velocity fluctuations were allowed for and the deviation of acoustic paths from straight lines was also considered.

Bezděk et al. (2003, 2004, 2007) have developed a very elegant method (Helmholtz integral/ray tracing method) for handling the mathematics of acoustic rays through the wall of the flowmeter in a clamp-on flowmeter.

Mahadeva et al. (2008, 2009) and Mahadeva (2009) developed the ray theory. Rays were followed from transmitting transducer through the wedge, pipe wall,

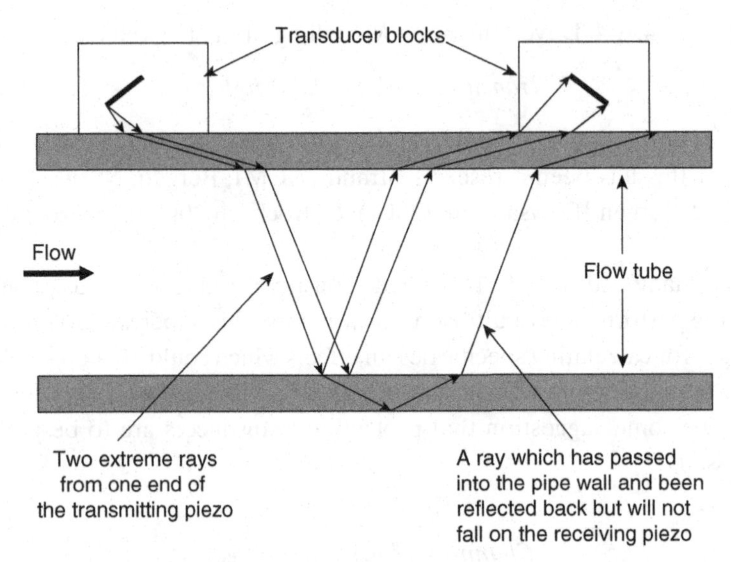

Figure 14.A.2. Diagram to show ray tracing (after Mahadeva 2009).

liquid and on to the receiving transducer. Each element of the transmitting piezo was considered a source of a wave (which could be thought of as based on Huygens' principle) and all the resulting rays that would arrive at the receiving piezo were combined, allowing for time delays, signal strength, pulse shape etc. to give the received pulse.

Figure 14.A.2 indicates some of these rays. Two rays are shown which leave one element of the transmitter and arrive at the two extreme points of the receiver. The diagram also shows a ray which passes through the liquid/solid boundary into the pipe wall and is reflected and passes back through the liquid. The ray shown would miss the transducer, but all the possible rays that can emanate from some point on the transmitter and reach some point on the receiver were included in Mahadeva's (2009) calculation by tracing a large number of rays from each element of the transmitter. For example, multiple reflections can take place when transiting the upper wall. The agreement with experiment (Mahadeva et al. 2008, 2009) appeared encouraging.

Mahadeva (2009) suggested that the reason for the typical S-curve (Figure 14.A.3) rather than the expected direct curve may be, at least in part, due to the reflected waves in the pipe wall. She suggested that the effect of wall thickness on the S-curve might allow information on the pipe wall to be extracted.

Koechner and Melling's (2000) use of Monte Carlo methods to reduce the ray computation would seem to be a useful development of Mahadeva's method.

Plane Wave Flowmeter

Willatzen (2001) calculated the effects of radial temperature gradients on the performance of a plane wave flowmeter. Willatzen (2003) further discussed flow measurement errors in the laminar flow regime and concluded that the transducer

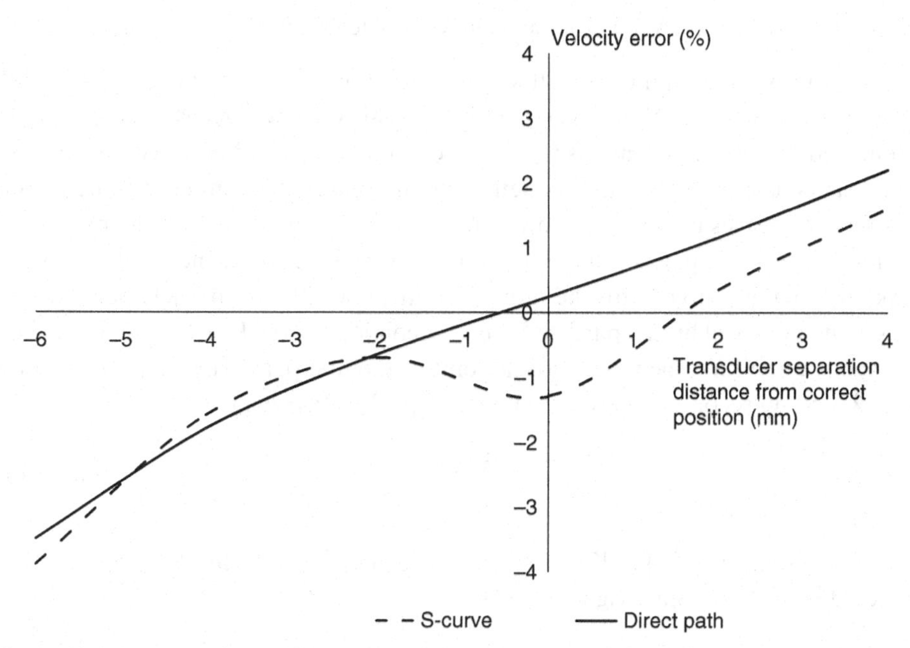

Figure 14.A.3. Approximate diagram of the "S" curve compared with the direct assumption [after Mahadeva, Baker and Woodhouse (2008, 2009) and Mahadeva (2009)].

diameter/pipe diameter ratio should be unity to obtain a correct measurement of the flow in the pipe. This appears to relate to Hemp's (1982) result. Wendoloski (2001) solved the case for a plane wave moving axially in circular and rectangular ducts and allowed the wave to be perturbed by the velocity profile. Willatzen explored (2004a) the importance of wall acoustic impedance in the prediction of meter performance, where the wall can act as an absorber as well as a reflector of acoustic waves. The acoustic impedance and the flow velocity both affect the level of error. Willatzen also reported (2004b) on different methods of calculating the performance for various flow profiles, which he indicated had a strong influence on the level of error.

Transducers

Lunde et al. (2003) investigated the effects of transducer diffraction on transit times, in relation to "dry calibration" methods and consequences for the ultrasonic flowmeter measurement accuracy in flow calibration and field operation. For the speed of sound measurement used in mass and energy flow metering, flow calibration may not reduce the error due to diffraction (cf. Vestrheim and Vervik 1996; Lunde et al. 2000; Vervik 2000). The time delay may depend on pressure, temperature, gas composition and flow velocity due to beam drift. Lunde, Frøysa and Folkestad (2007) and Lunde et al. (2008) in this and earlier work covered changes in: cross-section, ultrasound path, length of transducer ports, length of transducers and Reynolds number (cf. Hallanger, Frøysa and Lunde 2002).

14.A.5 Doppler Theory and Developments

Figure 14.15 is a diagram of the effect. A wave approaches a particle moving with velocity component v in the direction of the acoustic wave which has a velocity c, and is reflected. The velocity remains at c relative to the medium after reflection because sound speed is not affected by frequency (within reasonable limits). But the period between two waves being reflected will not be λ_t/c. The wave will hit the particle at a velocity $c\text{-}v$ and so peaks will hit the particle every $\lambda_t/(c\text{-}v)$ seconds. But successive peaks will make contact with the moving particle at different particle positions, so the second peak will hit the particle $\lambda_t/(c\text{-}v)$ seconds after the first, but will then need to travel $\lambda_t v/(c\text{-}v)$ to reach the starting point of the first peak. This additional travel takes $\lambda_t v/c(c\text{-}v)$ seconds. So the time between peaks is τ given by

$$\tau = \frac{\lambda_t}{c}\frac{1+v/c}{1-v/c} \tag{14.A.17}$$

Expanding Equation (14.A.17) using the binomial theorem, the frequency of the reflected wave, f_r, becomes (ignoring v^2/c^2)

$$f_r = \frac{c}{\lambda_t}\left(1-\frac{2v}{c}\right) \tag{14.A.18}$$

Hence the frequency shift, Δf, is given by

$$\begin{aligned}\Delta f &= f_t - f_r \\ &= 2f_t\frac{v}{c}\end{aligned} \tag{14.A.19}$$

Luo, Liu and Feng (2002) developed new signal processing for the Doppler signals using adaptive filtering and spectral analysis. Simulations using this system were encouraging. Luo, Liu and Feng (2003) discussed this system further. Rubio et al. (2006) suggested the use of a "warped discrete Wagner-Ville" distribution to obtain the instantaneous frequency for the Doppler meter. The work is related to blood flow measurement but should be of wider interest.

Kikura, Yamanaka and Aritomi (2004) looked at the effect of the size of the area sensed in a Doppler ultrasonic flowmeter. For multiphase flows Murakawa, Kikura and Aritomi (2005) suggested the use of differing sizes of ultrasonic transducers leading to a change in the measurement volume and sensing different sizes of particles. A special transducer was developed for the purpose.

Takeda and Shaik (2008) edited a special journal issue on Doppler methods which included the following papers. Tezuka et al. (2008a) described an ultrasonic pulse-Doppler flowmeter applied to hydraulic power plant penstocks. A bubble injection system was used to get adequate ultrasonic reflectors in the flow, and the system was capable of measuring the flow profile. Birkhofer et al. (2008) also discuss a pulsed Doppler application to cocoa butter crystallisation processes. Other papers covered calibration tests in which an expanded uncertainty of better than

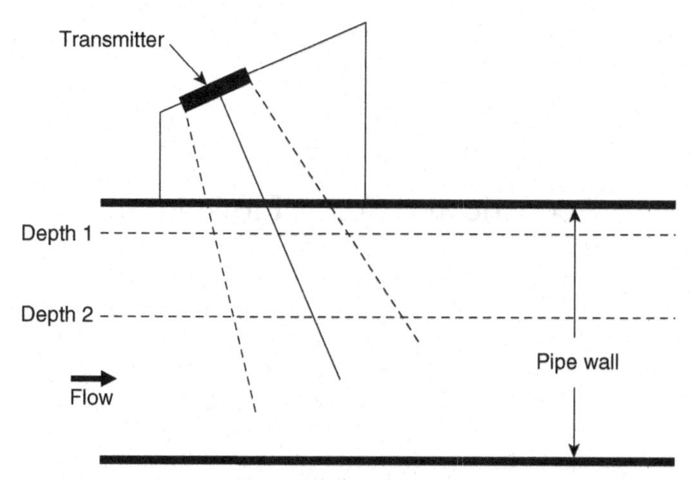

Figure 14.A.4. Approximate diagram of the Doppler flowmeter [after Mahadeva et al. (2010)].

±0.26% appeared to be achieved (Tezuka et al. 2008b), velocity vector profile measurements using multiple transducers (Obayashi et al. 2008), treatment of Doppler spectral information (Fischer et al. 2008), application to two-phase bubbly and slug flows using coaxial piezos with frequencies of 2 and 8 MHz (Murakawa, Kikura and Aritomi 2008) and other topics in ultrasonics related to flow measurement.

Messer and Aidun (2009) investigated the main effects on the accuracy of pulsed-ultrasound Doppler velocimetry in the presence of rigid impermeable walls (the implication being impermeable to ultrasound). They particularly considered system control factors, beam propagation, the shape of the beam and the intensity of the echo.

Franke et al. (2010) discussed the use of an ultrasound Doppler system for two-dimensional flow mapping in liquid metals. They appear to have used arrays of transducers to allow electronic traversing of the flow and applied it to the vortex flow in a vessel (cf. Cramer, Zhang and Eckert 2004).

Willatzen and Kamath (2008) considered the nonlinearities in ultrasonic flow measurement due to the fact that the sound speed in gases is smaller than in liquids and the flow velocity is a more significant fraction of the sound speed.

Mahadeva et al. (2010) undertook some interesting computation and experiments to study the effect of beam spreading on Doppler flow measurement errors. The model traced rays to obtain their interaction with the fluid flow at various depths and enabling an overall Doppler shift to be calculated (Figure 14.A.4). This allows the apparent flow profile to be deduced, in the same way as that for the experimental method. The effect of an annular profile was expected to be significant because of the difference in time at which the edges of the beam reached the annulus.

15

Acoustic and Sonar Flowmeters

15.1 Introduction

I have introduced this new chapter to cover two new flowmeters using sonar principles, those produced by CiDRA and those produced by Expro. In addition I have briefly covered related ideas and, in particular, acoustic chemometrics, and other applications of noise in flow systems.

Since preparing the 1st Edition of this Handbook, CiDRA's SONARtrac® flowmeter has become a significant additional meter making use of methods not previously developed to this author's knowledge, and finding an important application in minerals processing and other industrial flow measurement needs. I became aware of the instrument from an acquaintance who had experienced tests of the meter at SP, Sweden. This chapter provides a summary of the information available, virtually all of which comes from publications provided by CiDRA and from Expro Meters Inc. While SONARtrac® is described as passive sonar, the meter developed by Expro is described as active sonar. We shall discuss this further in Section 15.3.

My knowledge of this meter is, therefore, almost entirely dependent on the manufacturers, as I am unaware of any published data which is entirely independent of them, except for two Test Reports: T 1917 X 10 November 2010 (SP Technical Research Institute, Sweden) and T 1883 X 08 February 2008 (Alden Research Laboratory Inc) published by Evaluation International.

15.2 SONARtrac® Flowmeter

15.2.1 Basic Explanation of How the Passive Sonar Flowmeter Works

Sonar array-based flowmeters operate by using an array of sensors and passive sonar processing algorithms to detect, track and measure the mean velocities of coherent disturbances travelling in the axial direction of a pipe (O'Keefe, Maron and Rothman 2009a). These meters are primarily "clamp-on" meters.

These disturbances are grouped into three major categories:

1) disturbances conveyed by the flow;
2) acoustic waves in the fluid;
3) vibrations transmitted by the pipe walls.

Each disturbance class travels at a given velocity. For example, the flow will convey turbulent eddies, density variations or other fluid characteristics at the rate of the fluid flow. Liquid-based flows in pipes tend to be less than 10 m/s. The speed of sound (acoustic wave velocity) in water without gas bubbles is about 1,500 m/s. The third group, pipe vibrations, travel at velocities greater than the acoustic waves. Thus each class of disturbance may be identified and measured based on the differences in velocities. The meter uses sonar-based array processing techniques to "listen" to and interpret stress fields generated by turbulence and acoustic waves passing down the pipe.

15.2.2 A Note on Turbulent Eddies and Transition to Laminar Flow in the Pipe

A schematic of the relevant structures within a turbulent process flow is shown in Figure 15.1. As shown, the time-averaged axial velocity is a function of radial position, from zero at the wall to a maximum at the centreline of the pipe. The flow near the wall is characterised by steep velocity gradients and transitions to relatively uniform core flow near the centre of the pipe. The turbulent eddies are superimposed over time-averaged velocity profiles. These coherent structures contain fluctuations with magnitudes typically less than 10% of the mean flow velocity and are carried along with the mean flow. Experimental investigations have established that eddies generated within turbulence remain coherent for several pipe diameters. The sensor relies on turbulent flow but measurements on non-turbulent paste flows may be possible depending on process conditions (O'Keefe et al. 2008b, c). The expression for the Reynolds number is

$$\mathrm{Re} = \frac{\rho V D}{\mu}$$

where ρ is fluid density, V is flow velocity, D is pipe diameter and μ is dynamic viscosity. Transition from laminar to turbulent flow in pipes, under normal circumstances, is taken to occur at a Reynolds number of about 2,000. Most flows in the chemical and petrochemical industry have Reynolds number within the turbulent regime (Gysling and Loose 2003).

15.2.3 Flow Velocity Measurement

Flow velocity may be determined by focussing on the disturbances that are conveyed by the flow. These disturbances can be density variations, temperature variations, turbulent eddies or others. Within most industrial processes, the most common flow disturbance is turbulence (Figure 15.1). The overall mean velocity of the disturbances is equal to the flow velocity. The turbulent eddies exist throughout the pipe flow and have scales ranging from those defined by the pipe diameter down to small eddies which result from break-up of the larger ones. When eddies become small enough they are dissipated as heat through viscosity. However, they retain some coherence for several diameters before breaking up.

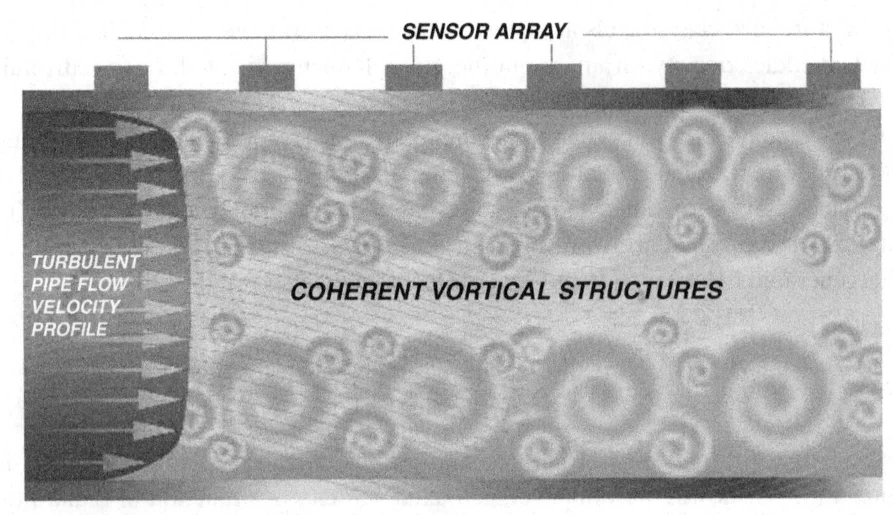

Figure 15.1. Approximate diagram of the pipe with sensor bands around to pick up acoustic signals (reproduced with permission of CiDRA).

Thus the average velocity of the fluid can be determined by tracking the average axial velocities of the entire array of turbulent eddies. This is achieved through an array of passive sensors and the signal processing algorithms, Figure 15.1. The small stress variations created by the passage of turbulences and other disturbances cause strain in the pipe wall which is sensed by the passive sensors wrapped around the pipe and which respond to the strains in the pipe without requiring interface gel etc. The electrical signals from each element of the array of sensors, which are equally spaced along the pipe, are interpreted as a characteristic signature of the frequency and phase components of the disturbance under the sensor. An array-processing algorithm combines the phase and frequency information of the characteristic signature from the group of sensor array elements to calculate the velocity of the characteristic signature as it propagates under the array of sensors. Volumetric flow rate is determined using a Reynolds number-based calibration procedure, which links the speed of the coherent turbulent structures to the volumetric flow rate (cf. O'Keefe et al. 2008a, 2008b; Gysling and Mueller 2004).

15.2.4 Speed of Sound and Gas Void Fraction (Entrained Air Bubbles) Measurement

In most installations there are acoustic waves propagating within the process pipes, which are generated naturally from a variety of sources including: pumps, flow through pipe fittings and bubbles within the fluid that generate acoustic waves through their natural oscillation. These acoustic waves are low frequency, and travel through the pipe with wavelengths much longer than the entrained gas bubbles. They can propagate in either direction down the pipe, Figure 15.2. The average velocity of these waves is obtained using the same sensors. The resulting measure of sound speed in the fluid passing through the meter provides a means of obtaining parameters of

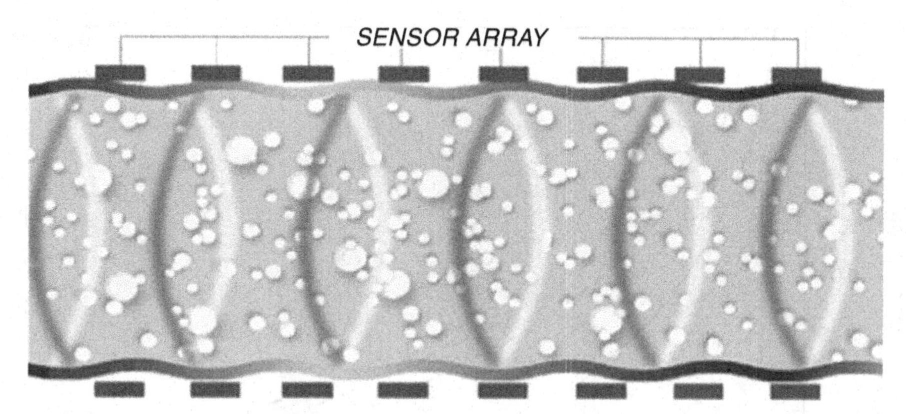

Figure 15.2. Illustration of naturally occurring acoustic waves propagating in pipe under the array of sensors (reproduced with permission of CiDRA).

the fluid, if it is known or, in the case of multicomponent fluids where the nature of each component is known, the average value of the mixture. For instance, the ratio of known gas and known liquid could be deduced (see Section 15.2.3).

In multicomponent fluids that consist of a gas mixed with a liquid or slurry, the acoustic velocity can be used to determine the amount of entrained gas (gas void fraction, Figure 15.3), when the gas is in the form of bubbles that are well mixed within the liquid or slurry. Because the wavelengths of the acoustic waves are much larger than the bubble size, a complex interaction takes place that sets the acoustic velocity to be a function of the gas void fraction. The particular values outlined by the curves in Figure 15.3 could be influenced by other factors, as well as pressure. Thus pressure at the location of the array-based instrument must be measured or calculated. The method has been used for entrained air applications ranging from 0.01% to 20% gas void fractions (cf. Gysling et al. 2007; Gysling and Loose 2005).

15.2.5 Localised Velocity Measurements

O'Keefe et al. (2008b) suggested that localised measurements can result in a measure of the profile of the flow in the pipe, but this may depend on setting an additional array of acoustic sensors around the pipe to sense flow around the cross-section of the pipe and in the vicinity of the wall (O'Keefe et al. 2009b, 2011).

15.2.6 The Convective Ridge

Operating in the convective mode, sonar-based flowmeters use the convection velocity of coherent structures (eddies) inherent within turbulent pipe flows to determine the volumetric flow rate. The sonar-based algorithms determine the speed of the turbulent eddies by estimating

- the frequency: that is the frequency of eddies passing an element of the sonar array

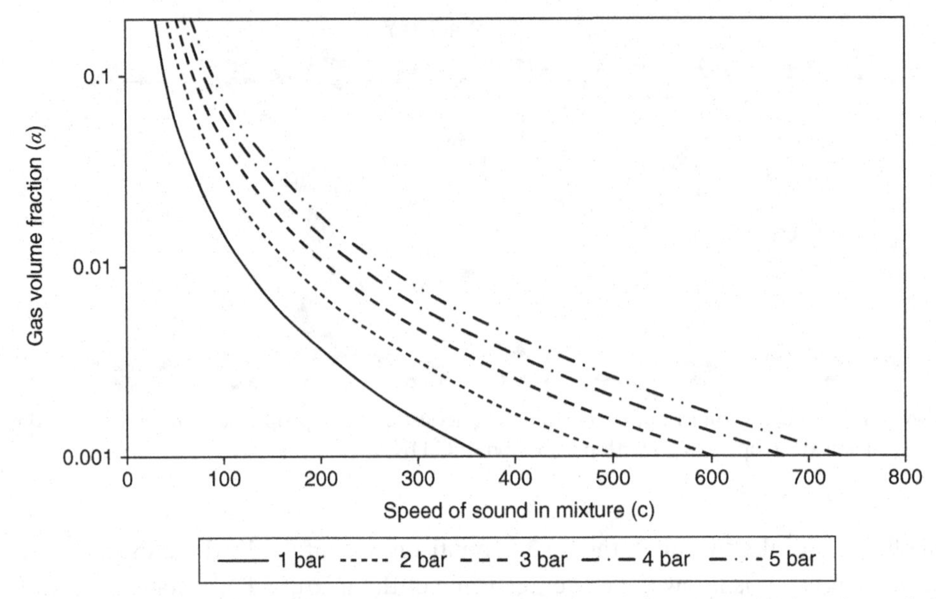

Figure 15.3. Relationship between gas void fraction, α (entrained air bubbles) and speed of sound, c, for various pressures (after Gysling et al. 2005).

- and "wavelength": that is the reconstruction of a "wave" formed by successive eddies and sensed by the sensor array as a variation in signal strength or wave "amplitude/phase", two characteristics of the flow field.

For a series of coherent eddies convecting past a fixed array of sensors so that the pressure fluctuations are registered by the sensor array, the frequency, ω, and wavelength, λ, are related through the expression:

$$\omega = kU_{\text{convect}}$$

where ω is in radians/second and is related to the frequency, f, by $\omega = 2\pi f$, k is the wave number, defined as $k = 2\pi / \lambda$ in units of 1/length, λ is the wavelength and U_{convect} is the convection velocity or phase speed of the disturbance. The manufacturers call this equation the dispersion relationship. If an array of eddies of similar size flow down the meter tube, the sensors may be able to identify a frequency as they pass, and may also be able to identify a wavelength for this size of eddy.

Figure 15.4 plots data extracted from the signal relating to an array of wavelengths and frequencies. From the slope of the dotted line, which the manufacturers refer to as the "convective ridge", the dominant velocity can be obtained. The triangle indicates how the velocity is obtained from this plot by dividing ω by k. The plot presumably includes all data coming from the sensor array, some of which will be for lower velocities and some will appear to indicate higher velocities due, presumably, to eddies caused to flow faster by the turbulence.

The implication, not obvious from the plot, is that moving away from the "ridge" leads to contours with fewer data than on the ridge. It appears that the peaked graph

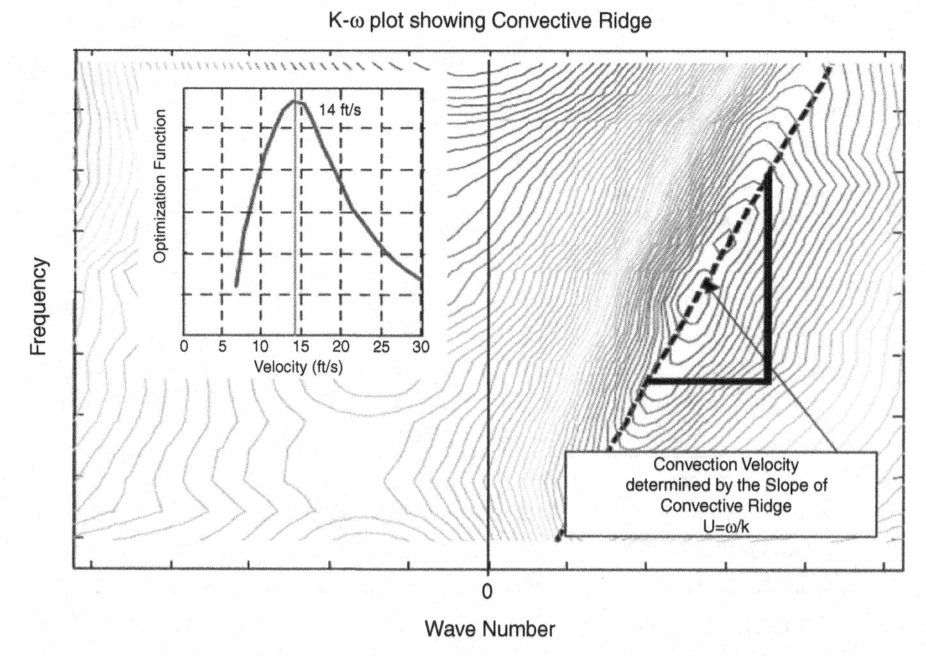

Figure 15.4. SONARtrac® convective ridge (reproduced with permission of CiDRA).

of "optimisation function" against velocity identifies the strongest signal and hence the dominant velocity of convection.

Plots, such as the one in Figure 15.4, are stated by the manufacturers to be "three-dimensional" power spectra where the power axis would be vertically out of the plane of the paper. The power of a sound field is decomposed into "bins" corresponding to specific wave numbers and frequencies. Thus, identifying the slope of the convective ridge provides a means to determine the convection speed of the turbulent eddies, and this may be adjusted, with calibration, to give the volumetric flow rate within a pipe.

A simpler way to look at the extraction of flow rate would be to take the time difference between eddies passing each sensor and with the sensor spacing to obtain velocity

$$U_{\text{convect}} = L / \tau$$

where L is the spacing between sensors in the array and τ is the time of transit between sensors of eddies. The method selected will, clearly, depend on the most convenient software interpretation of the signals, and the most stable and dependable result.

15.2.7 Calibration

The fundamental principle behind sonar-based convective flow measurements is that axial array sensors can be used in conjunction with sonar processing techniques to determine the speed at which naturally occurring turbulent eddies convect within

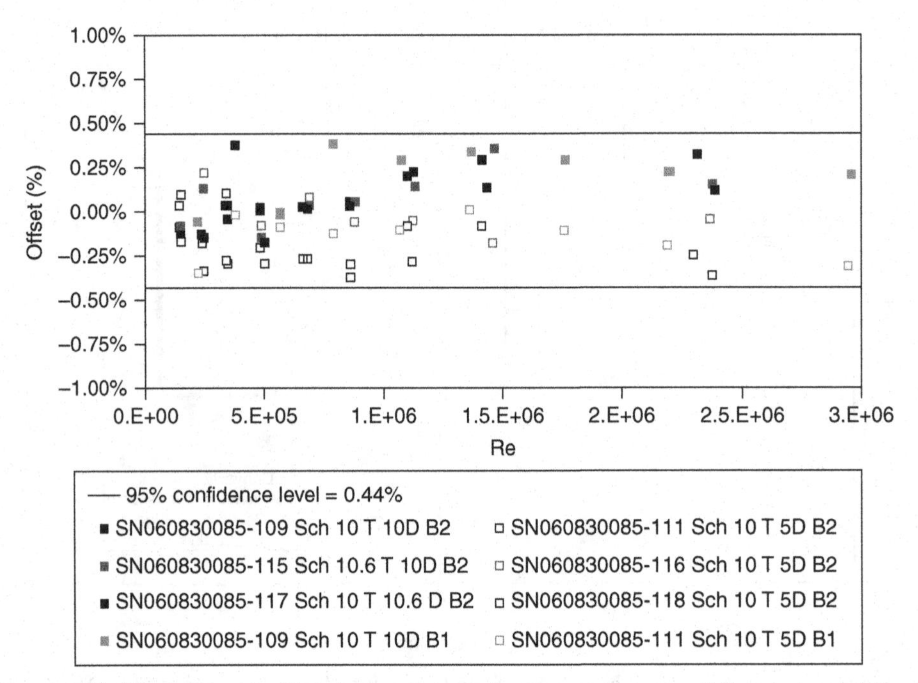

Figure 15.5. Illustration of calibration consistency from meter to meter. All meters are claimed to have the same calibration coefficients. Serial numbers cover specific sensors, test locations, pipes, etc. (Rothman, O'Keefe and Thomas 2009) (reproduced with permission of CiDRA).

a pipe. The slope of the convective ridge can be calibrated to the volumetrically averaged flow rate as a function of Reynolds number. Figure 15.5 shows the volumetric flow rate measured by the calibrated sonar-based flowmeters plotted versus reference flow. As shown, the sonar-based convective flowmeter reported the flow rate to within ±0.5% uncertainty.

15.2.8 Sound Speed Used to Obtain Fluid Parameters

The speed of sound may be used to obtain the void fraction in the liquid, and for this purpose Figure 15.3 is used. The basic equations for this are as follows. If α is the gas fraction, ρ_l is the density of the liquid in the flow, ρ_g is the density of the entrained gas, c_l is the speed of sound in the liquid and c_g is the speed of sound in the gas, then the density of the mixture is given by (Gysling et al. 2005)

$$\rho = \alpha\rho_g + (1-\alpha)\rho_l$$

and the speed of sound of the mixture, c, can be obtained from the equation

$$\frac{1}{\rho c^2} = \frac{\alpha}{\rho_g c_g^2} + \frac{1-\alpha}{\rho_l c_l^2}$$

Thus, if the speed of sound can be measured in a mixture and provided the values of the densities, and the sound speeds are known for the components of the mixture, the equations can be solved for the gas fraction, α, and hence for the mixture density. Figure 15.3 shows an example of a plot of the relationship between void fraction and speed of sound.

15.2.9 Additional Sensors

In a patent application O'Keefe et al. (2011) proposed the use of a ring of conformable ultrasonic (piezoelectric film based) transducers at various positions around the pipe which would provide data on pipe wall thickness. The ultrasonic transducers emit a pulse perpendicularly into the pipe wall, which reflects off the inner surface of the pipe and is received by the transducer. The time elapsed will then provide a measure of the wall thickness provided the velocity of the ultrasound wave is known. If the sound speed is about 6,000 m/s in steel, and the pipe thickness is 0.01 m, the elapsed time will be of order 3 μs, requiring very precise time measurement. This system appears to be implemented to allow:

- identification and measurement of deposition on the pipe wall;
- sensing of pipe wear due to abrasive solids content in the flow;
- compensation for the inner cross-section area of the pipe in the presence of either scale build-up or wall thinning, thus providing for an increased accuracy in the conversion of the flow velocity to a volumetric flow rate.

Currently the wall thickness measurement is not part of the flowmeter, but is performed by another instrument that CiDRA has which is designed for pipe wear monitoring.

15.2.10 Clamp-on System

Figure 15.6 shows the meter being installed on a pipe. In the commercial version of this measurement principle, a flexible band of passive sensors is wrapped around and tightened onto the pipe. In the case of the basic array, this is a dry fit that does not require gels or couplants. The sensor band (wrap around sensor) is normally 19.5 inches (50 cm) long in the axial direction of the pipe axis and equal to the circumference of the pipe in the orthogonal dimension.

These photographs show the sensor array which is fixed around the pipe, so that disturbances in the flow are sensed by each of the sensors and enable the speed of the fluid flow and the speed of sound to be deduced. Since the turbulent eddies are distributed throughout the pipe, the software is able to obtain an approximate average of the time taken for them to pass the sensors. When there is a longer period between sensing eddies this would, presumably, be due to the eddies being near the wall.

The typical installation procedure is outlined in Figure 15.6. First the pipe is wiped down and any high points are sanded or filed away. Second, the flexible sensor band is wrapped around the pipe and a series of captive screws on the sensor band are used to tighten the band onto the pipe. Each screw uses a stack of spring washers

Figure 15.6. Installation procedure from (top-left) pipe preparation through cleaning and light sanding of pipe to (top-middle and top-right) mounting of the flexible, lightweight sensor band to (bottom-left) installation of the sensor cover and to (bottom-middle and bottom-right) connection of sensor cover to transmitter via watertight cable (reproduced with permission of CiDRA).

to allow for pipe expansion and contraction, as well as to ensure a set clamping force. Third, a protective cover with signal conditioning and diagnostics electronics is installed over the sensor band, and the sensor band is connected to the electronics in the cover. Fourth, the cable from the sensor head to the transmitter is installed and wired to the transmitter. Fifth, the front panel menu on the transmitter is used to configure the system (Markoja 2011).

15.2.11 Liquid, Gas and Multicomponent Operation

The ability of the meter to measure entrained gas in liquid flow also provides a means to correct density for measurements obtained from vibrational or nuclear density meters. It appears to have been successfully used for entrained air applications ranging from 0.01% to 20% gas void fractions with an uncertainty of 5% of the reading (Markoja 2011).

An additional facility which can be added to the instrument appears to allow the flow profile in the meter to be measured, so that, if the flow is laden with solid matter, the change in profile can be monitored (O'Keefe et al. 2009b). This consists of further passive strain sensors, but rather than a single array in which each element of the array extends across most of the circumference of the pipe, a series of shorter sensor elements are used to create multiple arrays with each array located at a different position. The circumferential positions usually include one at the bottom of the pipe, one between the bottom and the side, one at the side of the pipe, one between the side and the top of the pipe, and one at the top of the pipe.

This adaptation adds to the measurement of averaged velocity within the pipe, and allows the profile to be measured, particularly where solids are being carried by the flow and are tending to settle out. The meter can measure local velocities by using specially shaped sensors located at various heights around the circumference of the horizontal pipe. This enables it to function as a stratification monitor (Maron and O'Keefe 2007). The software measures the rate of movement of the solid/liquid mixture and will indicate the reduction in speed from top to bottom of the pipe.

Gysling et al. (2007) used the flowmeter to obtain the wet-gas mixture flow rate using the convection time of coherent flow patterns past sonar arrays. The piping pressure loss gradients were also used as a basis for obtaining liquid content.

Tests performed on non-Newtonian fluids (such as Bingham fluids) suggest that the meter can operate on these; it is not the case for all non-Newtonian fluids.

15.2.12 Size Range and Flow Range

From the manufacturer's literature the following was obtained:

- Pipe diameters 2″ to 60″ (51 mm to 1.52 m)
- The length of the clamp-mounted lightweight environmental enclosure, also called the sensor head, was 0.91 m for sizes up to about 0.9 m diameter and 1.3 m long for pipes greater than 0.91 m diameter.
- Flow rates: liquid 3 ft/s to 30 ft/s (0.91 m/s to 9.1 m/s) on stiff pipes such as those composed of steel. Lower flow rates are typically achieved on plastic pipes.

  gas > 20 ft/s (> 6 m/s)

- One version can cope with 0–20% GVF, but the accuracy of measurement of the entrained gas is less predictable. A variant of the meter can also measure gas holdup. A further option for slurry pipes provides a measure of pipe wear.
- Operating temperature ranges are:
  - transmitter −40°C to +60°C (but LCD only down to −20°C)
  - sensor head process temperature −40°C to +125°C (for one model)
  - sensor head ambient temperature −40°C to +60°C
- Power requirement is 25 W.

15.2.13 Signal Handling

As indicated, the acoustic waves in the meter are sensed for passage of turbulent structures past the sensor array, and for passage of acoustic waves as they pass through the meter. The software segregates the sensor waves and, thereby, eliminates those due to unwanted noise and vibration (O'Keefe et al. 2012).

The signal output and operation is programmable by keypad or PC, and the system has a self-diagnostic capability. Remote data logging and retrieval is possible. Output signals are 4–20 mA, pulse/frequency and various bus protocols.

15.2.14 Accuracy Claims

Uncertainty is claimed as ±1% of reading. This calibration was performed at Alden Labs in Massachusetts. In addition, although not traceable, slurry-based field verification of the calibration was performed at several mining sites using a tank draw-down test.

Repeatability appears to be claimed as ±0.3% of reading.

Figure 15.5 shows traceable calibrations based on a NIST facility where several meters were compared, and the clamp-on equipment appeared to be within ±0.5% from manufacturing setup. The implication of this, if further test results confirm this stability, is that calibration requirements may be stable with time, but do need some adjustment or initial calibration. A typical uncertainty of ± 1% (O'Keefe et al. 2008b) appeared to be claimed in the field and ± 0.5% under reference conditions or after in-field supplemental calibration. This, presumably, assumes that for clamp-on designs the pipe cross-section is known precisely and/or that the clamp-on meter has been independently calibrated. The adjustment indicated is by a few percent and depends on the Reynolds number (O'Keefe et al. 2008b). McQuien et al (2011) commented that calibration accuracy from meter to meter as well as from temperature effects and aging is dependent on maintaining the spacing between the sensor elements and maintaining the stability of the clock used in the digitiser. The spacing between the sensors is fixed when they are bonded to a stainless steel sheet in the manufacturing process, and cannot be adjusted by the user. The clock stability is claimed to be better than 0.01% and thus is 50 times better than needed to maintain the flowmeter's claimed uncertainty. It was suggested, as a result, the impact of clock stability can be neglected (Markoja 2011).

To achieve these uncertainties will require careful control and compensation for changes in process and ambient conditions. For instance, temperature change will alter density and viscosity and, consequently, Reynolds number with resultant profile changes, which may affect the output signal of the meter and the uncertainty values. In most situations the impact on the final calibration adjustment is claimed to be small, and as the Reynolds number increases the calibration adjustment becomes less sensitive so that, for larger pipe diameters and high Reynolds numbers, the calibration adjustment loses its Reynolds number dependency and is left with an offset of a few percent. If temperature dependency adjustment becomes important, the information from the temperature sensor built into the sensor band can be used.

15.2.15 Installation Effects

Pipework

An example of the meter is shown in Figure 15.7. It is recommended that the flowmeter is installed a minimum of 15D downstream of an elbow and 5D upstream of an elbow. If these straight pipe runs are not available, the meter will continue to operate, but there may be a slight flow rate offset. For those applications where better accuracy is required, calibration on a process specific pipe configuration can be

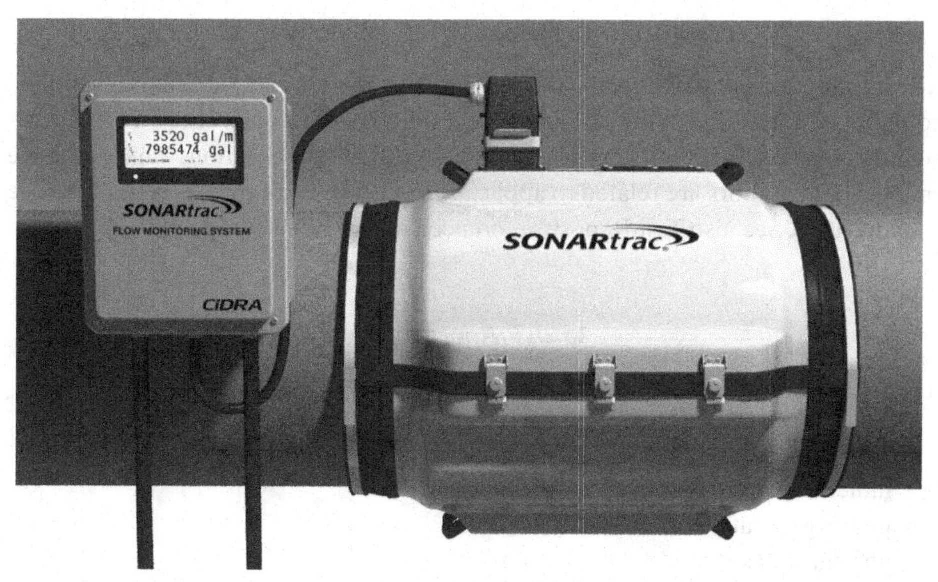

Figure 15.7. Photograph of the SONARtrac® flowmeter (reproduced with permission of CiDRA).

performed in the laboratory, and a calibration coefficient developed (Markoja 2011). In situ calibrations are discussed in Chapter 4 with their limitations on precision which may be inadequate for this meter.

Pipe and Liner Material, wall Thickness and Scale Build-up

Pipe materials, for which it is claimed the meter operates satisfactorily, are: steel, PVC, HDPE and fibreglass.

Pipe linings through which it can operate are claimed to include: rubber, urethane, cement and Teflon-lined pipes.

The meter is claimed to operate through scale build-up on the inside of the pipe (Markoja 2011). In this case where there would be consequent reduction in cross-section for the flow, it is implied that an estimate of the resulting cross-section should be entered into the meter's transmitter (O'Keefe et al. (2008c). Maron and O'Keefe (2007) claimed to have operated the meter on a pipe of diameter 30″ (0.75m) high-density polyethylene pipe with a wall thickness of 2.25″ (5.6 cm).

Noise and Vibration Effects

Problems of performing this measurement in an industrial environment include resolving the relatively low-level eddy disturbances from the relatively high background noise levels. This noise includes acoustics and vibrations generated from pumps and valves. The strength of the array-processing algorithm is its ability to isolate and measure the velocities of the low-level effects of eddies within the flow.

15.2.16 Published Information

The information which I have gleaned comes from papers and publications supplied to me by CiDRA and some obtained from the Web. Although these are predominantly authored by CiDRA personnel, some of those publications which do have independent authors are related to application experience of the meter, for instance on mine tests. See also Test Reports referenced at end of Section 15.1.

15.2.17 Applications

The manufacturers claim that it is applicable to:

- oil sands processing
- minerals processing
- power generation
- nuclear industry
- chemical industries
- pulp and paper
- consumer product industries
- water and waste water
- fluid treatment
- food and beverages

The technology is able to measure flow rates of clean liquids, high-solids-content slurries, pastes, liquids and slurries with entrained air, froth lines and flotation feed lines with entrained air, slurry lines with magnetite and other magnetic ore, slurry lines with abrasive or corrosive materials, high-pressure lines and lines exhibiting scale build-up.

The consistency of the signal from different clamp-on attachments, Figure 15.5, might suggest the possibility of using the meter as a "transfer standard" to check calibration of installed meters on site. However, more experience and data would be needed before there is adequate confidence for this.

15.3 ActiveSONAR™ Flowmeter

The ActiveSONAR™ meter, Figure 15.8, uses an array of transducer pairs, 180° opposed which are mounted perpendicular to the pipe wall. One transducer transmits; the other receives. Some of the technology appears to have been developed in collaboration with CiDRA.

Expro refers to this configuration as a pulse-array configuration, where pulses of acoustic energy are sent through the pipe wall, through the flowing medium and through the opposite pipe wall. The signal is modulated by the flowing medium, and the receiving transducer picks up the modulated signal and sends it to the data-processing system, where sonar processing is performed on the sensor array signals to determine flow velocity.

Figure 15.8. Expro's ActiveSONAR™ clamp-on flowmeter for single and multiphase volumetric flow measurement (reproduced with permission of Expro Meters, Inc.).

15.3.1 Single and Multiphase Flows

For single-phase flows, the cross-sectional area of the pipe is used, and also the pressure and temperature (for gas flows), to determine flow rate. The cross-sectional area is derived from pipe OD and the wall thickness using a commercial ultrasonic thickness gauge. For multiphase flows multiphase processing algorithms are applied to determine phase flow rates (gas, oil, water) based on pressure, temperature, BS&W (basically sediment and water-cut) and a customer-supplied hydrocarbon definition, Figure 15.9. The algorithms used for multiphase processing are proprietary information, but Figure 15.9 suggests a general approach with fluid data, a phase chart, pressure, temperature, water-cut and flow rate obtained from the sonar meter, possibly with some identification of the individual phases.

The main difference between this technology and ultrasonics appears to be in the use of sonar processing algorithms rather than cross-correlation to determine flow velocity (description based on private communications from Michael Sapack, Expro Meters Inc).

15.3.2 Brief Summary of Meter Range, Size etc.

- **Ranges**
  - Flow ranges – 1.5 ft/s to 150 ft/s (0.5 to 50 m/s)
- **Meter dimensions**
  - Pipe Sizes: ANSI 2″ to 32″ diameter (50 mm to 800 mm);
  - Sensor head length – approximately 1 foot (300 mm);
- **Materials**
  - Exposed components are constructed from 316SS and power-coated aluminium;

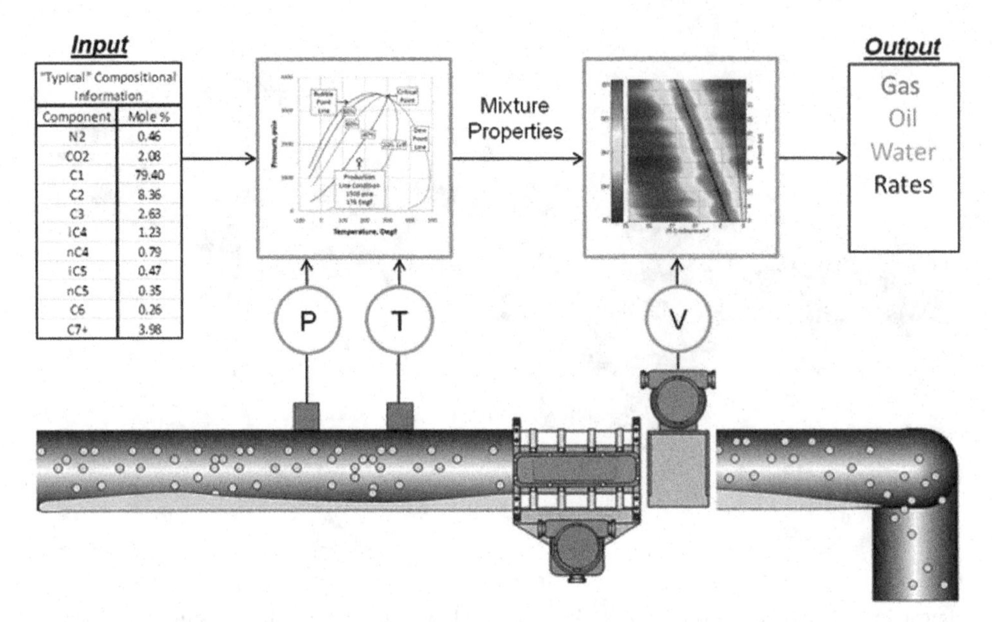

Figure 15.9. Diagram to illustrate the approach of Expro to multiphase metering (reproduced with permission of Expro Meters, Inc.).

- **Clamp-on mechanism**
  - Sensor head uses 316SS mounting clamps, mounted to pipe. Transducers are installed in the mounting clamps;
- **Calibration**
  - ±2% of reading over the flow range specified earlier. This is for single-phase gas or liquids. The manufacturer considers that the ActiveSONAR™ meter can be "type" calibrated, and that each meter does not require individual calibration on a facility traceable to NIST;
- **Applications:** Main applications include:
  - Oil and gas wellhead gas/condensate production surveillance – Type 1 or 2 wet-gas
  - Oil and gas water and gas injection/gas lift
  - Single-phase gas and liquid flow rate measurement
  - Drilling mud.

Shields et al (2013) discussed the use of the ActiveSONAR™ meter to minimise losses associated with well testing on the East Brae platform.

15.4 Other Related Methods Using Noise Emissions

In the first edition of *Flow Measurement Handbook* I commented, in relation to the use of noise sources in pipework (Baker 2000, p. 278), that it might be worth investigating the intrinsic flow noise due to vortex shedding in existing pipework, which could, possibly, be sensed and calibrated as a flowmeter system by an intelligent sensor which could learn its calibration and then provide for a utility measurement.

The idea was to obtain the frequency spectrum and relate this to the flow rate. The sonar meter appears to have developed a very powerful technology making use of the passage of turbulent eddies and the speed of sound in the fluid to obtain its make-up, which is finding an important role in areas of process flow measurement.

The use of signal patterns appears to have some potential, as indicated also by work on acoustic chemometrics, but may need more development of the physical theory and mathematics to give substance to the ideas. The acoustic chemometric technique was described by Esbensen et al. (1998):

"In one particular sense, acoustic chemometrics is simple: obtaining problem-dependent 'acoustic signals' (by relevant technical means), which – followed by some form of pertinent signal analysis – are subjected to chemometric data analysis. In this context it is often the power of multivariate calibration that comes to the fore."

Acoustic chemometrics is a general process-monitoring technique. The concept is that acoustic spectra contain information on the process state. Arvoh et al. (2012a), together with other earlier references [Arvoh et al. (2012b), Esbensen et al. (1998, 1999), Evans and Blotter (2002), Evans, Blotter and Stephens (2004), Kim and Kim (1996) and Kupyna et al. (2008)], have applied this technique. See also Ibarz et al. (2008) for a domestic application.

The basis of the technique applied to flow measurement appears to be the use of the noise spectra emanating from the flow. The vibrations caused by the flow are measured in a process plant by means of accelerometers or other instruments installed in optimum sites, and the technique appears to be then to associate the signals from the accelerometers with the flow rate in various parts of the system, and with the additional effect of variation of key parameters such as temperature. In this paper (Arvoh et al. 2012a), it was applied to the estimation of reject gas and liquid flow rates in compact flotation units for produced water treatment, to obtain both liquid and gas flow rates.

A driver for this type of system may be in the application of non-intrusive measurement techniques. There may be scope for adapting this technique to flows in the oil industry, particularly multiphase flow.

However, as suggested earlier, there appears to be considerable scope for further fundamental mathematical and physical justification for this approach.

Another method would be to generate noise. Cheung et al. (2001) appeared to do this in their proposed method for gas flow measurement which averaged the flow both along the tube and across the section of the tube. It made use of the one-dimensional plane wave propagation. Their meter included a sound source and a series of equally spaced microphones along the pipe. Their experimental data indicated that the meter was linear up to 27m/s and gave encouragement to pursue the concept.

15.5 Chapter Conclusions

The sonar flowmeters represent a very interesting new meter development and would appear to have potential for further research, analysis and development. It appears already to have found some niche areas where it meets a need.

The way the meter operates has been indicated earlier, but there does not appear to be any more fundamental physical analysis of the meter's operation which has been published, presumably for reasons of commercial confidentiality. There would appear to be scope for some mathematical work to develop the understanding of the meter, for instance in the flow profile measurement and the behaviour at the wall.

The nature of the acoustic generation of noise from turbulence, a complex theory, does not appear to be essential to the basic concepts of this meter, although it may be important in the refining of the meter. Since one of the meters senses the noise using acoustic bands around the pipe, it appears likely that the eddies will be sensed equally around the pipe, but with a possible reduction in intensity for eddies nearer the axis of the pipe.

This sonar technology has introduced a new and powerful approach to flow metering which is to be welcomed.

As well as further research on this technology, there may be other approaches which use aspects of the acoustic techniques and which should be explored further. The topic of acoustic chemometrics has been the subject of publications, but to this author, appears to lack the theoretical examination of the physics which it requires if it is to be taken up more widely.

Other possibilities have been referenced or suggested which indicate that this is an interesting field with potential for further exploration.

16

Mass Flow Measurement Using Multiple Sensors for Single-Phase Flows

16.1 Introduction

The measurement of mass flow, a fundamental requirement for any fluid, has been an elusive goal due to the problems of developing a suitable flowmeter. Despite this, the availability of mass flowmeters has increased greatly over the past 25 years. This is partly due to the increasing value of products (Hall 1990), but it is also due to an increasing realisation that volumetric flow measurement is often inappropriate. In addition, the advent of the Coriolis flowmeter has stimulated engineers to find other mass flowmeters. General reviews of mass flowmeters were given by Sproston, Johnson and Pursley (1987), Betts (1990) and Medlock and Furness (1990).

Mass flow measurement is sometimes categorised as direct (true) or indirect (inferential). However, it may be useful to allow a few more than two categories.

a. True (direct) mass flow measurement by a single instrument is rare. It appears that to achieve it we need to use one of the fundamental acceleration laws. We can do this by creating:
   - the force (or torque) resulting in a linear (or angular) acceleration (Chapter 19 describes an example) or
   - the force that produces Coriolis acceleration (Chapter 20).
b. Fluid-dependent thermal mass flow measurement uses the temperature rise resulting from heat addition but is affected by other parameters such as the specific heat of the fluid (Chapter 18).
c. Multiple differential pressure flowmeters used in a dedicated system (Section 16.2) depend on the nonlinearity of the flowmeter equation.
d. Indirect (inferential) mass flow measurement combines volumetric flow rate or momentum flow rate with a density measurement (Section 16.3). (An alternative is to combine volumetric and momentum flow measurements.)
e. Multiphase flow measurement almost always requires multiple measurements (Chapter 17).

In this edition of the book, I have included a brief survey of mass flowmeters for multiphase flow in Chapter 17.

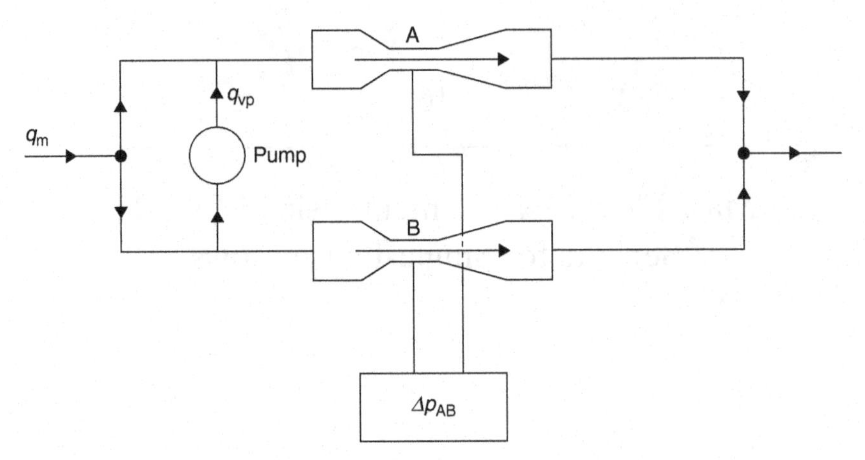

Figure 16.1. Twin Venturi system.

16.2 Multiple Differential Pressure Meters

Before turning to the one method that is commercially available, for completeness we shall look at two methods that have been suggested but, to my knowledge, have not been used commercially.

Twin Venturi System

The twin Venturi system depends on the use of two identical Venturi meters (or other differential pressure devices) that are arranged in parallel balanced paths as shown in Figure 16.1. By means of a metering pump, which transfers q_{vp} from one line to the other, the flows may be unbalanced. If the pump is off, so that the flow is split equally between the two paths and is, therefore, the same in each meter, the differential pressure across each meter is also the same. If the pump is started, then one Venturi meter has a lesser and the other a greater flow passing through it. The pressure difference between the throat pressures Δp_{AB} is then measured, and the mass flow rate q_m is deduced.

We first note that the pressure drop to the throat of each Venturi is given by

$$\Delta p_A = \Delta p_B = K(q_m/2)^2/\rho \tag{16.1}$$

where K is a constant, ρ is the density of the fluid, and q_m is the total mass flow through both Venturis, and so $q_m/2$ is the total mass flow through one of the Venturis when the flows are equal.

If the main flow, q_m, splits equally, then if the pump is started and transfers fluid as shown, and provided that the upstream pressures at inlet to the Venturis are not affected by the transfer, the equation for the pressure drop to the throat of A will be

$$\Delta p_A = K(q_m/2 + \rho q_{vp})^2/\rho \tag{16.2}$$

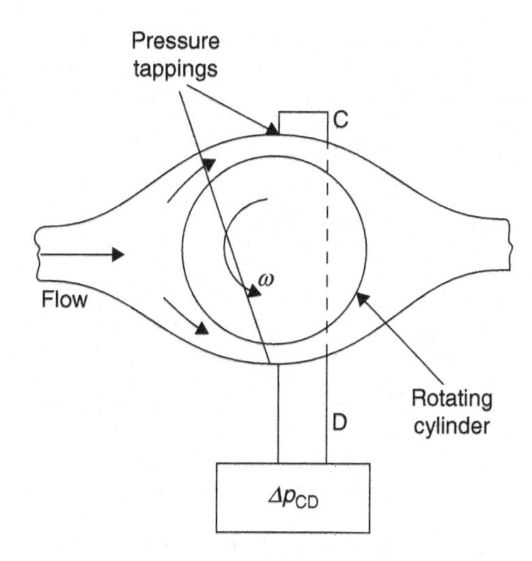

Figure 16.2. Brand and Ginsel's system.

and to the throat of B

$$\Delta p_{\mathrm{B}} = K\left(q_{\mathrm{m}}/2 - \rho q_{\mathrm{vp}}\right)^2/\rho \tag{16.3}$$

Thus the differential pressure will be

$$\begin{aligned}\Delta p_{\mathrm{AB}} &= K\left(q_{\mathrm{m}}/2 + \rho q_{\mathrm{vp}}\right)^2/\rho - K\left(q_{\mathrm{m}}/2 - \rho q_{\mathrm{vp}}\right)^2/\rho \\ &= K\left(q_{\mathrm{m}}^2/4 + \rho q_{\mathrm{m}} q_{\mathrm{vp}} + \rho^2 q_{\mathrm{vp}}^2\right)/\rho - K\left(q_{\mathrm{m}}^2/4 - \rho q_{\mathrm{m}} q_{\mathrm{vp}} + \rho^2 q_{\mathrm{vp}}^2\right)/\rho \\ &= 2K q_{\mathrm{m}} q_{\mathrm{vp}}\end{aligned} \tag{16.4}$$

Hence

$$q_{\mathrm{m}} = \frac{\Delta p_{\mathrm{AB}}}{2K q_{\mathrm{vp}}} \tag{16.5}$$

Thus the precision of q_{m} depends on Δp_{AB}, K the Venturi constant assumed to be the same for each meter, and q_{vp}, which is set by the pump. However, there are clearly various effects such as the incompressibility of the fluid and the disturbing effect of the transfer flow, which would need careful consideration.

Brand and Ginsel's System

This is illustrated in Figure 16.2. It has similarities to the twin Venturi method. The flow on each side of the rotating cylinder passes through a contraction. An imbalance in the flow is introduced by rotating the central cylinder and the pressures at the contraction throats are again compared.

Medlock (1989) refers to an MSM1 flowmeter made by Professor Dr W. J. D. van Dijck and associates at the Technical University of Delft. The MSM1 meter (Massa Stroon Meter) was $\frac{3}{4}$ in. and consisted of a solid cylindrical rotor spinning at a constant angular velocity within the meter casing. The rotor induced circulation, and so half of the annulus passed more fluid, and the other half passed less. A differential

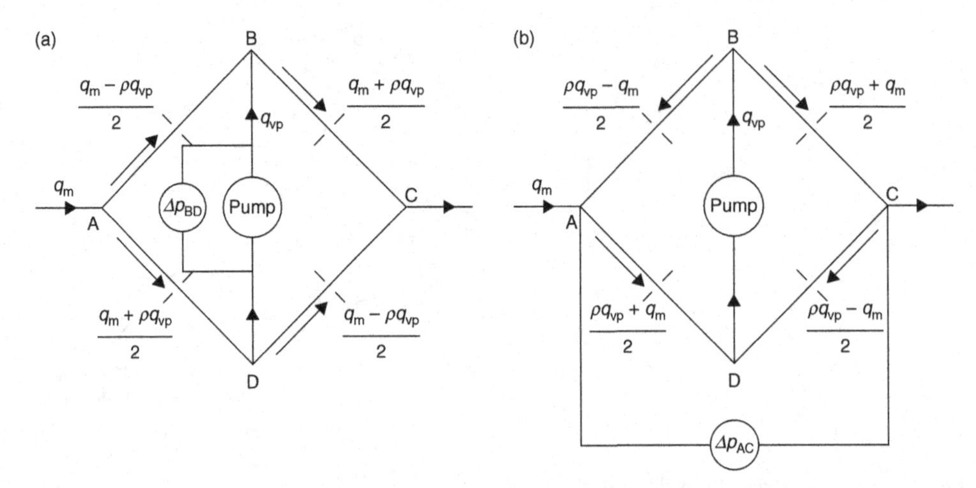

Figure 16.3. Wheatstone bridge flowmeter: (**a**) High flows; (**b**) low flows.

pressure was established between the two throats in the same way as in the differential Venturi system, and the mass flow rate was deduced from the differential pressure and the rotational speed.

16.2.1 Hydraulic Wheatstone Bridge Method

A Wheatstone bridge system has been successfully developed as a commercial instrument. See also Svete et al. (2009, 2013).

16.2.2 Theory of Operation

The theory of operation of this type of flowmeter is illustrated by Figure 16.3. Four matched orifices are arranged in a hydraulic Wheatstone bridge network with a constant volume pump producing a recirculating flow. The process mass flow is q_m, and the pump volumetric flow is q_{vp}. The pressure drop is given by Δp_{BD} in Figure 16.3(a) for the situation when $q_m > \rho q_{vp}$ (high flows). If the flow q_m were to divide equally between the upper and lower halves of the Wheatstone bridge, then the pressure drop through each would be [as for Equation (16.1)]

$$\Delta p = K\left(q_m/2\right)^2/\rho$$

Applying this relationship to the four orifices in the bridge, when the pump creates an unbalanced flow, as in Figure 16.3(a), we obtain the following expressions:

$$\Delta p_{BD} = \Delta p_{AD} - \Delta p_{AB} \tag{16.6}$$

$$= K\left(q_m/2+\rho q_{vp}/2\right)^2/\rho - K\left(q_m/2-\rho q_{vp}/2\right)^2/\rho \tag{16.7}$$

$$= K q_m q_{vp} \tag{16.8}$$

$$q_m = \frac{\Delta p_{BD}}{K\,q_{vp}} \tag{16.9}$$

Similarly, for the case of Figure 16.3(b), the pressure difference is given by Δp_{AC} when $q_m < \rho q_{vp}$ (low flows), and we obtain

$$\begin{aligned}\Delta p_{AC} &= \Delta p_{AB} + \Delta p_{BC} \\ &= -K\left(q_m/2 - \rho q_{vp}/2\right)^2/\rho + K\left(q_m/2 + \rho q_{vp}/2\right)^2/\rho \qquad (16.10)\\ &= K q_m q_{vp} \qquad (16.11)\end{aligned}$$

$$q_m = \frac{\Delta p_{AC}}{K\, q_{vp}} \qquad (16.12)$$

16.2.3 Industrial Experience

The meter consists of four precisely matched orifices forming the hydraulic Wheatstone bridge. A recirculating pump establishes reference flow in the bridge. At zero measured flow rate, pressure drop across the bridge is zero. As measured flow passes through the meter, it upsets the balance in the bridge such that a differential pressure signal, which is linear and proportional to true mass flow rate, is generated.

The resulting flowmeter is claimed to give true mass flow rate and may be unaffected by changes in fluid, temperature, density and viscosity. It can be of stainless steel construction and capable of operation at elevated temperatures and with low lubricating liquids.

Typical flow ranges may be

- 0–2.3 kg/h (0–5 lb/h) for microflows with a turndown of 50:1 and a minimum flow of 0.05 kg/h;
- 0–900 kg/h (0–2,000 lb/h) for low flow ranges and 0.5 kg/h minimum flow for low ranges; and
- 0–23,000 kg/h (0–50,000 lb/h) for high flow ranges with about 100:1 turndown.

A wide range version in the low flow range may be capable of 500:1 turndown. Best uncertainty may be of order 0.5% of rate plus 0.02% FSD, and repeatability 0.25% of rate, although in some designs an additional amount based on FSD may be added. Pressure drop may be in the range from about 0.1 to 2 bar. Depending on the design, fluid temperature may be in the range from about –18 to +150°C and pressure up to about 67 bar gauge.

16.2.4 Applications

This device is particularly suited for fuel flow measurement applications such as engine emission control and fuel economy testing. Another application may be fuel efficiency testing in boilers. It produces an output signal that is linearly proportional to mass flow rate.

16.3 Multiple Sensor Methods

If the fluid has unknown properties, we shall probably need to combine the flow measurement with a density measurement. However, if the fluid has known parameters, we may be able to use equations that describe the fluid's state. The following are examples of both.

- A volumetric meter (e.g. turbine) with density, or pressure and temperature measurement. This is a relatively common practice. A combination of electromagnetic flowmeter with a gamma density-sensing cell has been available commercially. The mass flow will be given by

$$q_m = \rho V A \tag{16.13}$$

where ρ is the fluid density, V is the mean velocity through the meter, and A is the cross-section of the meter. There is likely to be a correct position at which to measure the density. There will, therefore, be an overall uncertainty stemming from the various components

$$\frac{\delta q_m}{q_m} = \frac{\delta\rho}{\rho} + \frac{\delta V}{V} + \frac{\delta A}{A} \tag{16.14}$$

or the equivalent root-mean-square expression. Similar error calculations will be appropriate to the other methods mentioned here.

- A vortex meter that combines frequency measurement with measurement of lift on the shedding body. Itoh and Ohki (1993) claim a vortex flowmeter design that obtains mass flow rate by the use of the lift on the central body divided by the frequency of pulsation. This depends on the assumption that the lift coefficient and the Strouhal number are both constant, so that the lift is proportional to ρV^2, and the frequency is proportional to V. As a result, the ratio is proportional to ρV.
- Ultrasonic flowmeters. The speed of sound is related to the density

$$c^2 = \gamma p / \rho \tag{16.15}$$

where γ is the ratio of the specific heats, and ρ is the density of the fluid. A similar equation for liquids is

$$c^2 = 1/(k_s \rho) \tag{16.16}$$

where k_s is the adiabatic compressibility. Because c^2 appears in the equation for time-of-flight ultrasonic meters, it can be combined through these equations to give velocity times density and hence mass flow. This assumes that for gases γ is constant, which is a good approximation for air. The value of γ depends on the number of atoms in the gas molecule. The value of p will be needed. For a liquid, the value of k_s will be needed (Baker 1976).

- Alternatively for ultrasonic flowmeters the impedance of the fluid is related to the density and may be used to obtain its value (Guilbert and Sanderson 1996a).
- A differential pressure flowmeter with the density of the fluid obtained either directly or, say, for a gas via pressure and temperature measurements (cf. Figure 5.20).
- A combination of velocity measurement to obtain V, with momentum ρV^2, allows the value of ρV to be found.

Medlock (1989) used the term *hybrid* to describe meters combining momentum sensing and volume sensing, such as the Venturi tube with an electromagnetic flowmeter built into it or Venturi and turbine meters (Frank, Mazars and Rique 1977; Reimann, John and Muller 1982) or the turbine meter that also senses drag on the wheel or on a separate drag plate (Reimann et al. 1982) or gauze (Cole 1985). Wong, Rhodes and Scott (1981) suggested the use of the pressure drop in a swirl generator in combination with a Venturi.

16.4 Chapter Conclusions

These metering methods may not be widely used, but may well find niches in research or in manufacturing industry, in addition to those in Section 16.2.4, where they fulfil a need not otherwise met by the more common types. The exception is likely to be some of the concepts in Section 16.3 and particularly the use of the speed of sound in ultrasonic flowmeters to obtain mass flow rate.

I have included this chapter for completeness, in the hope that those who have an interest in such a device will be able to build on past experience.

17

Multiphase Flowmeters

17.1 Introduction

The measurement of mass flow of multiphase fluids is likely to range across wide industry areas as diverse as food processing and subsea hydrocarbon extraction, but at present the technology appears mainly to be driven by the petrochemical industry.

The measurement also needs to take account of flows where the nature of the second component may or may not be known, and the proportions or indeed the existence of the second or even third component may not be known.

I am very conscious that others are more knowledgeable on multiphase flowmeters than I, and it is not appropriate to attempt to duplicate their published work on the subject. This is, therefore, a brief view of this important industrial area and a review of some published literature of which I am aware including some early references, together with pointers to key specialist books on the subject.

Some of the earliest data appear to be related to nuclear plants under emergency conditions. Extensive experiments were undertaken, and special designs and combinations of meter were tried in order to obtain a system which would signal fault conditions in the nuclear reactor duct flows (e.g. Baker 1977; Baker and Saunders 1977 in relation to sodium flows in fast breeder reactors). Some applications of traditional flowmeters to two/multiphase flows may be found in the chapters of this book specific to those meters.

Dykesteen (1992) identified the development of small oil wells in the North Sea where a platform to support a separator would be too costly as one reason for needing a compact and subsea multiphase meter. Subsea reservoir management needs to take place without the use of a separator, and where different satellite wells are operated under different licences, individual flow measurement will be required. The multiphase meter should, if possible, provide component fractions, component velocities and component densities.

An attempt was made (Baker 1991a) to draw together the references and the most useful data on the likely response of bulk flowmeters when placed in such flows (cf. Baker 1988b, 1989). Subsequently Whitaker (1993) reviewed further developments. In this chapter I have attempted to draw together more recent published work of which I am aware, but would particularly recommend to readers that they obtain one of the recent books specifically on multiphase flow measurement, for instance Falcone, Hewitt and Alimonti (2009), Schlumberger (2010) or Corneliussen et al. (2005).

17.2 Multiphase and Multi-Component Flows

The term *multiphase* tends to be used both for multi-component flows and for multiphase flows, and the wrong term may be used in this book from time to time. The term has come to cover both flows.

These complex multi-component flows (discussed briefly in Chapter 2) have been extensively studied, and flow pattern maps have been developed to indicate the conditions under which the various flow regimes occur (e.g. Butterworth and Hewitt 1977; Hetsroni 1981). A brief but relevant introduction to the subject is also included in Schlumberger (2010).

However, these maps tend to be limited in their application to particular fluid combinations, pipe sizes and orientations. To measure the flow in any of these regimes requires a large number of variables to be interpreted from measurements. Mixing or conditioning of the flow may be possible in some applications, but may not greatly improve the situation. Separation of the components is also possible in some applications, and some of the designs mentioned later attempt this, but again this is not possible or convenient in all applications.

The use of mixing to create an almost homogeneous flow has drawbacks. The flow downstream of a mixer is extremely disturbed, and such a position would certainly not be recommended for precise conventional flow measurement. In many cases the flow also starts to separate as soon as it is mixed. The flowmeter would need to be calibrated in these flow conditions, and such a calibration would need to take account of the effect of varying component ratios. The possibility of using the pressure loss across the mixer as a flow measurement may be used in some devices.

17.3 Two-Phase/component Flow Measurements

Hall, Griffin and Steven (2007) suggested that *wet-gas* should be defined as a flow of a relatively small amount of liquid of any composition by volume in a flow that is predominantly of gas by volume, and took the upper boundary of *wet-gas flow* as having a *Lockhart-Martinelli parameter*

$$X = \frac{q_{m,l}}{q_{m,g}}\sqrt{\frac{\rho_g}{\rho_l}} \tag{17.1}$$

of approximately 0.30 (cf. Reader-Harris and Graham 2009). An alternative definition of wet-gas may be gas volume fraction (GVF) > 0.98 (see Table 17.1).

Four ranges in Table 17.1 were suggested (Schlumberger 2010) to define wet-gas.

The measurement of water-cut is a special case of multiphase flow measurement where the flow has two components, oil/water.

Table 17.1. *Gas volume fraction (GVF) based on wet-gas type: where GVF is the ratio of the gas volumetric flow rate to the total volumetric flow rate at line conditions (Schlumberger 2010)*

Wet-gas type	Type 1	Type 2	Type 3	Multiphase
GVF (%)	98–100	95–98	90–95	<90

Oddie and Pearson (2004) provided a useful review of the most important techniques for gas-liquid, gas-solid, liquid-solid and liquid-liquid flows (cf. Oddie, Stephenson and Fitzgerald 2005 on a new differential pressure concept of particular relevance to flows with more than one component).

17.3.1 Liquid/Liquid Flows and Water-Cut Measurement

Buhidma and Pal (1996) concluded that wedge and segmental orifices were feasible for emulsions of oil-in-water. Li et al. (2008a) experimented with Venturi, turbine and oval gear meters for oil-water flow with oil volume fraction of 15% to 85%. Xing and Zhang (2009) tested a vortex meter in oil-water flow and claimed errors within 4% with oil fractions from 5% to 40% by volume. Lucas and Jin (2001) used resistivity sensors for cross-correlation flow measurement for oil-in-water flows. Meribout et al. (2010) appeared to use acoustic measurements, impedance, ultrasonic sensors, capacitance and conductance sensors with a Venturi to obtain water-cut.

17.3.2 Entrained Solid in Fluid Flows

Flows of solids in liquid: slurries, paper pulp etc. are frequent and various meters have been used, in particular the electromagnetic flowmeter. One paper I have come across using a different approach is that by Zhou and Halttunen (2004), who used an electrical tomography approach for pulp flow measurement.

It is not my intention to discuss solid/gas flow measurement, but the reader is referred to: a review by Yan (1996), Yan (2000) as editor of one issue of the *Journal of Flow Measurement and Instrumentation* (Volume 11 No. 3 Sept. 2000), Soderholm (1999), Zhang et al. (2004c) on electrostatic measurement for pulverised fuel, see also Motta, Schmedt and Souza (2011) on pulverised coal mass flow measurement, Benes and Zehnula 2000 on the generation of an acoustic surface wave, Deloughry et al. (2001) on process tomography based on capacitance measurements, Ibrahim, Green and Dutton (2000) on optical fibre sensors for particles and droplets, Thomasson et al. (1999) on optics for flow measurement of cotton, Melick and Robinson (2006) on high-temperature cement particle flows using triboelectrical methods, Carter, Yan and Cameron (2005) on imaging and electrostatic sensors for particle size and flow rate, Yan, Xu and Lee (2006) on an electrostatic sensor using a neural network, Sun et al. (2008) on cross-correlation with capacitance sensors, Peng, Zhang and Yan (2008c) on a theoretical analysis of an electrostatic sensor.

17.3.3 Metering Wet-Gas

Wet-gas flows are one of the most important gas/liquid flows. Operating regimes may be complex. There appear to be three general approaches to this measurement:

- using differential pressure (DP) devices, such as the Venturi and the V-cone, and applying correlations such as those of Chisholm, de Leeuw and others to derive the mass flow rates of gas and liquid;
- using more than one flowmeter: both DP meters or including an ultrasonic or other modern flowmeter in combination with a DP or other device;
- using a meter similar to a multiphase meter designed for wet-gas and possibly incorporating a density-sensing meter or similar.

Differential pressure devices, such as the orifice, Venturi and V-cone, are widely used. See Evans, Ifft and Hodges (2007) for some tests. Cooley et al. (2003) had reservations about dual differential meters, but other meters such as ultrasonic and vortex, multiphase meters capable of measuring in the wet-gas region, correlation techniques, tracers which mix with one phase only and pattern recognition may be other possibilities (Jamieson 2001).

Differential Pressure (DP) Flowmeters

Steven and Hall (2009) suggested a new correlation developed from Chisholm's. Li et al. (2008b) and Li, Wang and Geng (2009) described a meter with dual slotted orifices. However, the predominance of Venturi meters in wet-gas flows is probably related to their robustness to erosion and the impact of slugs of liquid, as well as to their well-researched behaviour (Reader-Harris and Graham 2009) (see Steven 2000). Testing wells which produce 1–2% liquid in the presence of 99–98% gas by volume appeared to be very difficult since the mixture is far from homogeneous and the liquid tends to adhere to the walls, creating an annular flow, and is dragged along by the gas core. In addition the gas will carry droplets and, sometimes, slugs of liquid, and it seems likely that these will not necessarily be homogeneous or of the same composition as the liquid on the walls (Fairclough 2000). See de Leeuw and Dybdahl (2002) on wet-gas flow measurement, and the use of Venturi-based flowmeters. Cf. de Leeuw et al. (2005). Tests suggested that the de Leeuw gas flow correlation was suitable for hydrocarbon gas flow rate to within ±2–4%, and the liquid flow correlation gave total liquid rate to approximately ±10–15% (see de Leeuw 1997). The tests of Stewart et al. (2003) appeared to indicate parameters affecting the over-reading: gas pressure, gas velocity, the Lockhart-Martinelli parameter and the β value, and encouraged care when applying a correction factor that it cover the relevant parameter range (cf. Britton, Seidl and Kinney 2002). It appears that while Geach and Jamieson (2005) found the de Leeuw correlation was most appropriate, Cazin et al. (2005) had reservations about the accuracy of existing correlations when used with high-pressure flows. Salque et al. (2008) calculated the liquid and gas mass flow rates, allowing for fluid properties, direction and evolution of flow and the Venturi, film thickness, dispersion factors etc. Lupeau et al. (2007) discussed the influence of an upstream annular liquid film on wet-gas flow and developed a model to predict its influence. Reader-Harris and Graham (2009) gave a review of wet-gas correlations and derived an improved model for Venturi over-reading in wet-gas (see Section 6.A.3), and numerical modelling by He and Bai (2012) of wet-gas flows appeared to support the suggestion that a liquid film formed on the inlet wall of the Venturi accounting for the over-reading.

Other gas/liquid tests on the Venturi have been reported by: Strzelecki et al. (2000), who obtained test results for air-water which fell between Murdock and de Leeuw correlations (see Steven 2000); Boyer and Lemonnier (1996), who described sensitivity to slip between phases; Rosa and Morales (2004), who reported modelling and experiment on a vertical bubbly flow (air-water or air-glycerine) with void fraction of about 12%; Reis and Goldstein Jr. (2008), who used a Venturi with a capacitance device with helical electrodes to obtain horizontal two-phase mass flow rate; Steven (2008a), who reported work on dimensionless numbers applied to two-phase flows in Venturi meters. Huang et al. (2005) used capacitance tomography for void fraction and a Venturi for flow rate. Meng et al. (2010b) used a Venturi with an electrical resistance tomography sensor, and appeared to achieve better than 5% error for mass flow of bubbly and slug flow and 10% for annular and stratified flow. See van Mannen (1999) on tracer-Venturi combinations.

V-cone Stewart et al. (2002) provided information on V-cone flowmeters in wet-gas from NEL tests. Correlations were obtained for the meter and uncertainty was claimed to be generally of order 2%. Steven and Lawrence (2003) discussed repeatability. They also suggested that a disturbance 10 diameters upstream of the V-cone meter inlet had no noticeable effect on its performance. Steven, Kegel and Britton (2005) gave an update on V-cones metering wet-gas. Steven (2007) discussed the additional use in flow measurement that the expansion section of a DP meter could provide. He reviewed work on the V-cone for wet-gas. For further flow performance see Steven (2009).

Zhang et al. (2010) discussed high gas void fraction (GVF) and low-pressure gas-liquid two-phase flow measurement based on a dual-cone flowmeter.

Wedge Meters Steven et al. (2009) discussed the potential for the wedge meter with wet-(natural)-gas flows.

Ultrasonic Meter

Gopal and Jepson (1996) described an ultrasonic flowmeter for wet-gas capable of measuring liquid film and mist contents in gas flow. Up to eight flush-mounted ultrasonic transducer pairs were used to obtain transit time and attenuation of signals, but it was capable of handling stratified, annular and mist flows. See also Fish (2007).

Acoustic/sonar Meters

Gysling et al. (2007) used a sonar-based clamp-on flowmeter to obtain the wet-gas flow rate. The piping pressure loss gradients were also used as a basis for obtaining liquid content (cf. Gysling, Loose and van der Spek 2005 and Chapter 15).

Meters Specifically Designed for Wet-Gas

Johansen et al. (2007) described a prototype wet-gas and multiphase flowmeter. It consisted of a Venturi nozzle followed by a straight piece of tube of nozzle diameter with an array of pressure sensors which measured the pressure fluctuations from the convecting turbulent structures.

Bø et al. (2002) discussed the Roxar Flow Measurement wet-gas meter, implying that it used microwave water detection technology and differential pressure flow measurement, and may be able to detect changes in the water production to better than ±0.01% by volume with uncertainty of ±0.1% by volume when GVF (gas volume fraction) > 98.5%. For GVF 91–98.5%, the water detection uncertainty appeared to have been ±0.2% in cases with a water/liquid ratio (WLR) < 50% while the water was systematically overestimated and it appeared that the meter finally failed in the case of continuous water (WLR>50% – this may be system specific).

On one application the Dualstream II wet-gas flowmeter appeared to have been used to measure both gas and liquids to within ±5% (Wood, Daniel and Downing 2003). Jacobsen et al. (2004) focussed on issues relating to meter validation at a wet-gas test facility, commissioning and testing once the meter was installed subsea, and also solving problems of hydrate formation in impulse lines.

An interesting development from Elster-Instromet Ultrasonics appeared to combine a 6 in (152 mm) Venturi with ultrasonic meters at the inlet, throat and outlet (van Werven et al. 2006). The ultrasonic meter included triangular beam paths.

Other Flowmeter Combinations for Gas/Liquid

Zheng et al. (2008a) used turbine flowmeter and conductance sensor. See also Shim, Dougherty and Cheh (1996), mentioned in Chapter 10; Opara and Bajsiae (2001) reported on the use of an electromagnetic flowmeter in down flow; Mi, Ishii and Tsoukalas (2001a) used an electromagnetic flowmeter and impedance measurement of vertical air-water slug flow; Cha, Ahn and Kim (2001) obtained verification of Equation (12.5) and noted that signal amplitude provided evidence of the passage of slugs. Ahn, Do Oh and Kim (2003) used the less common current-sensing electromagnetic flowmeter for two-phase flow (cf. O'Sullivan and Wyatt 1983; Wyatt 1986). See also Zhang (2002) on gas-liquid flows and Wang, Tian and Lucas (2007a) on simulations of the flowmeter in two-phase flows. Al-lababidi and Sanderson (2004) used a clamp-on ultrasonic meter, a two-phase slug flow model and a conductivity probe.

Liu et al. (2001) described a neural network to reduce greatly errors caused by two-phase flow in a Coriolis meter. Rieder, Drahm and Zhu (2005) developed a theory for a Coriolis meter using a moving resonator model to supplement the "bubble effect" model of Hemp and Sultan (1989); see also Hemp and Yeung (2003). See also Henry et al. (2006) for specific applications, and Gysling (2007) on operating frequencies.

Stewart and Hodges (2003; see also Stewart et al. 2002) noted that, for wet-gas flow measurement, the liquid flow rate may not be important and could be determined by other means, allowing a standard single-phase gas meter, with known response to the presence of liquid to be used, with a correction applied. They noted that liquid on the pipe wall may slow a turbine wheel; their results for the Coriolis meter were not encouraging, but appeared to consider the vortex results promising.

Skea and Hall (1999a) tested Venturi, positive displacement, turbine with various blading and Coriolis flowmeters, in flows of oil with gas fractions up to 15% by volume. Their results appeared to be that:

- 100 mm PD and Venturi for lower (less than 6% by volume) gas fractions and higher flow rates were found to be more accurate than the other meters and obtained flow rate to within ±2%;

and (Skea and Hall 1999b) that:

- the effects of the second component (3-15% by volume) in flows of water-in-oil and oil-in-water on single-phase flowmeters were within ±0.4% for turbine and PD meters and the others tested were within ±1% generally, but on occasion were as high as 5–10%.
- in both papers the data appeared to be more spread than these figures would suggest.

Other work reported was by: Dong and Zong Hu (2002) concerning a rotational drum device which extracted a fixed proportion of the flow and then separated the liquid and gas; Shen et al. (2005) concerning a theorem which they applied to optical probes to sense air-water interfaces; Tan and Dong (2006 cf. Dong et al. 2005) used electrical resistance tomography with cross-correlation techniques in a vertical gas-liquid flow; Jin, Miao and Li (2006) applied a technique they referred to as symbolic sequence analysis to analyse two-phase flow measurement signals; Kim, Ahn and Kim (2009) measured void fraction and bubble speed in slug flow with a three-ring conductance probe. See also (Ito, Kikura and Antomi 2011) concerning a 25 mm by 3 mm channel, who used conductivity measurements to sense bubble velocity, volume and void fraction.

17.4 Multiphase Flowmeters

17.4.1 Categorisation of Multiphase Flowmeters

Dykesteen (1992) suggested that multiphase meters operated in one of three ways:

i) by total separation of components,
ii) by measurement of the multiphase flow rate, but separation of a sample for component measurement,
iii) by in line measurement of all parameters.

(i) and (ii), quite old techniques, have been standard practice in the oil industry, but the advent of increasingly precise in-line multiphase flowmeters has moved much of the development to (iii), with which this section will be primarily concerned.

For other applications than oil production, (iii) has usually been the aim. An example was the early work in the nuclear industry such as the combination of turbine and target meters for loss-of-coolant nuclear applications (cf. Goodrich 1979 and other papers at the same conference). Another example was the Venturi meter which was combined with both capacitance and gamma densitometer, Figure 17.1(a), results from which are shown in Figure 17.1(b) (Kratzer and Kefer 1988; Smorgrav 1990). One advantage of the Venturi is that it appears to homogenise the flow downstream of the throat. Other examples are Arnold and Pitts' (1981) proposal for a positive displacement meter with a gamma densitometer, temperature and pressure sensing (Figure 9.20) and Priddy's (1994) tests of a positive displacement design (Figure 9.21).

Xiaozhang (1995) suggested using magnetic and differential pressure flowmeters with capacitance/impedance transducer.

17.4.2 Multiphase Flowmeters (MPFMs) for Oil Production

Three books/reports of which I am aware are Falcone et al. (2009) and Schlumberger (2010), and, an earlier one, Corneliussen et al. (2005). I shall not attempt to duplicate what others have written with their authority and thoroughness, and I would encourage the reader, for whom this area is important, to obtain the publications. My contribution is to provide a very brief summary of published work of which I have become aware. Thorn, Johansen and Hammer (1997) reviewed developments in three-phase flow measurement and anticipated an instrument capable of measuring each phase with an uncertainty of about ±5%, to be non-intrusive, reliable, flow regime independent, and suitable for the full component fraction range. The authors identified the main means of component fraction measurement: gamma-ray attenuation and electrical impedance methods. For component velocity measurement they mentioned cross-correlation and the Venturi meter if the flows were well mixed. They saw future developments in the areas of performance specification, standardisation and in situ calibration. In an important recent paper Thorn, Johansen and Hjertaker (2013) provide an instructive review of current instruments and design options.

It seems likely that multiphase/component meters will increasingly be used in other industries. However, current papers are predominantly related to the oil business.

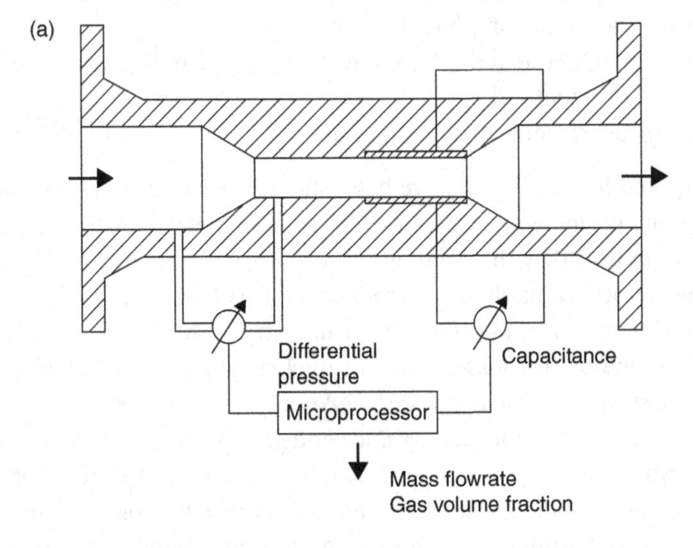

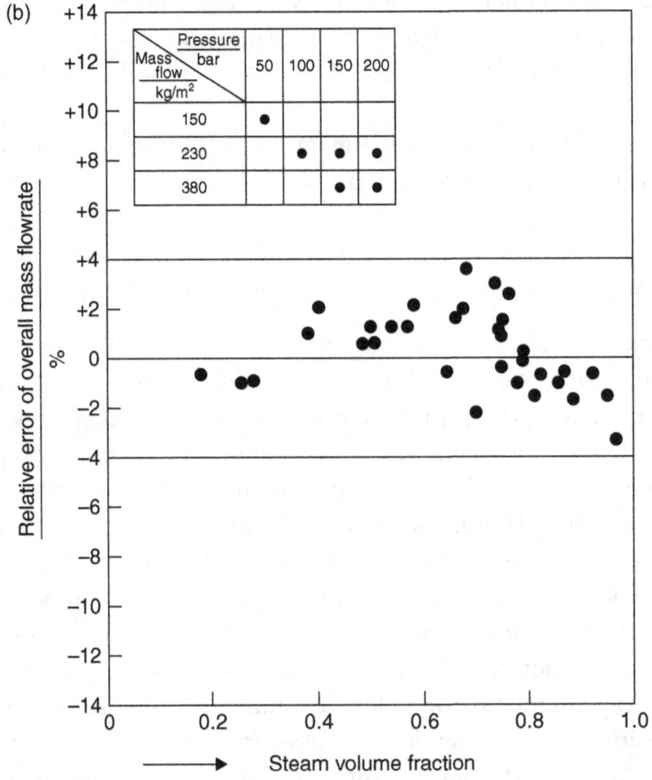

Figure 17.1. Mass flowmeter due to Kratzer and Kefer (1988) (reproduced with permission of Cranfield University): **(a)** Geometry; **(b)** error in steam-water flows.

Separation of Components

The Texaco SMS subsea metering system (Figure 17.2) reported by Dean et al. in 1990 combined an inclined gravity separator, a microwave sensor in a side stream sampling loop for water-cut, a differential pressure liquid flow rate meter, a vortex

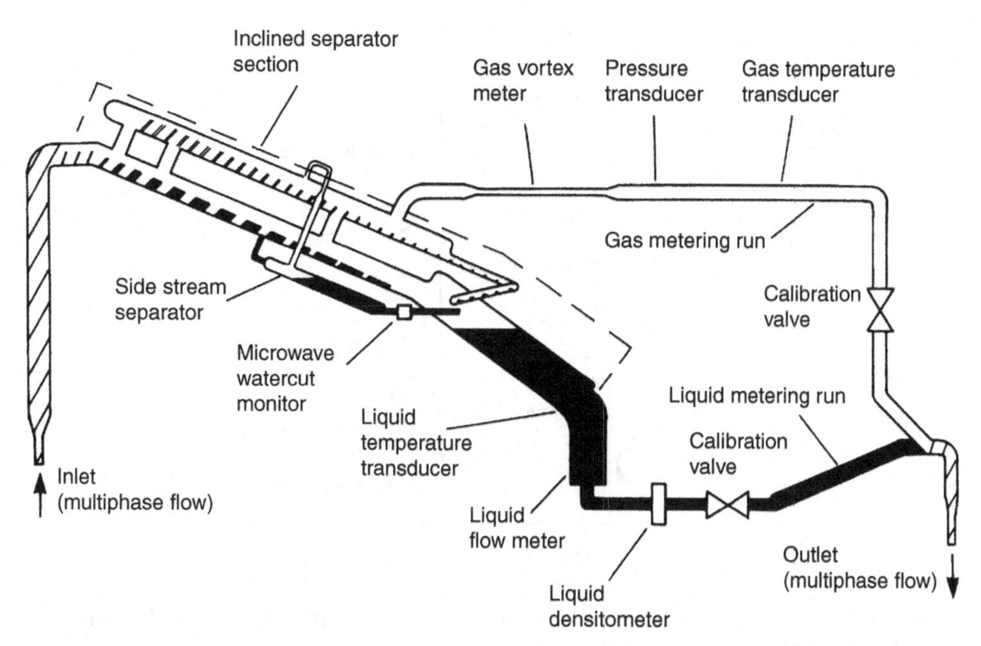

Figure 17.2. Subsea metering system (Dean, Dowty and Jiskoot 1990a; reproduced with permission from the NEL).

shedding meter on the gas outlet and temperature and pressure measurements (cf. Dean et al. 1990a, 1990b).

Thorn et al. (2013) listed current industrial flowmeter designs of this type which incorporate separation. Some of the more recent designs of this type appear to have introduced a cyclone for separation in a vertical entry tube with liquid gas interface sensing. Flows then go either upwards to the gas line with a flowmeter such as a vortex meter to measure the flow rate, or to the liquid line with water-cut and Coriolis or other meters. In addition there may be other measures of liquid and gas parameters. In Figure 17.3 I have provided a simple diagram, based on one design, in which I have attempted to encapsulate these features.

Separation of a Sample for Component Measurement

Early publications reported the use of: turbine meter, gamma densitometer, pressure, temperature and flow separation (Hall and Shaw 1988); correlation flowmeter and gamma densitometer after jet mixing, and sampling/separation to obtain liquid and gas components (Anon 1988; Kinghorn 1988; see also King 1988); jet mixer and turbine (King 1988; Millington and King 1988); static mixer, Venturi or pressure loss across the mixer, density from gamma meter, and also sampling the mixture and a Coriolis used to obtain the water-cut (Anon 1994); ultrasonic cross-correlation (King 1988) (see also Sidney, King and Coulthard 1988a, 1988b). Tuss (1996) described tests of an Agar multiphase meter with: PD meter for total flow, Venturi for gas/liquid flow rates, water-cut monitor and a gas bypass loop for gas volume flow.

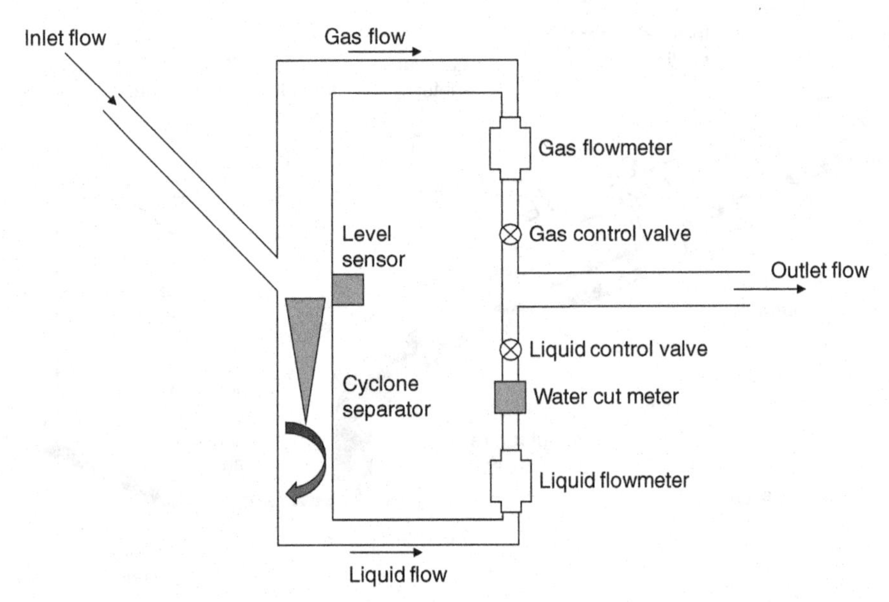

Figure 17.3. A diagram to indicate the features of a more recent flowmeter with separation (after Weatherford International Ltd).

In-line Measurement of All Parameters

Fluenta developed a meter using cross-correlation of capacitance signals for flow rate, and also used these sensors for water-cut and with a gamma densitometer to obtain make-up of the fluid (Millington, Frantzen and Marshall 1993; see also Dykesteen et al. 1985; Frantzen and Dykesteen 1990). Further description implied also an inductive sensor and a Venturi meter, pressure and temperature transmitters and pvT information was obtained from an equation of state (Leggett et al. 1996; see also Caetano et al. 2000 on further tests of a Fluenta meter). Torkildsen, Helmers and Kanstad (1997) mentioned possible effects of flow regime and salinity.

The Framo meter appeared to consist of: static mixer, Venturi with differential pressure, pressure and temperature transmitters and multi-energy gamma meter. There has been collaboration between Framo and Schlumberger involving Vx technology. The most recent development of this technology for a smaller meter and low rates appears to be the Schlumberger Vx Spectra.

The ISA device (Priddy 1994) (Figure 9.21) used a helical screw positive displacement meter with gamma densitometer. CSIRO, Australia, developed a meter-based on cross-correlation and dual energy gammas with, in addition, pressure and temperature measurements and with specific gravities etc. of the components known (Watt 1993). See also Hartley et al. (1995), who noted that it might have been better for water-cut to use microwaves. The Kongsberg/Shell meter used capacitance for interface in the channel, the oil/water fraction and by cross-correlation for the slug velocity and to estimate the individual flow rates. The Multi-Fluid International

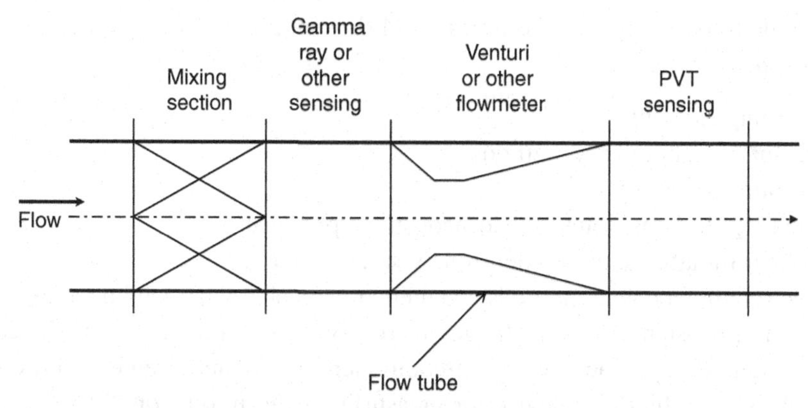

Figure 17.4. Diagram to show typical components of an in-line flowmeter. All components may not be used, and the order of components may differ from that shown.

(MFI LP meter) used microwave and gamma for composition and cross-correlation for velocity between two dielectric sensing sections (Okland and Berentsen 1994). See also pulsed neutron techniques (Anon 1993). Whitaker and Millington (1993) and Whitaker (1993) at that time concluded that phase flow rates should be measurable to approximately 10%, and Whitaker (1996) confirmed ±10% of reading for: Agar, Fluenta, Framo and MFI meters.

Fueki et al. (1998) described a meter without radioactive sources; Tokarczuk, Sanderson and High (1998) explored the application of nuclear magnetic resonance (NMR); Mohamed and Al-Saif (1998) described an MPFM using a fluidic-flow diverter to separate gas and liquid; see also Mohamed, Al-Saif and Mohamed (1999) on testing of two MPFMs. See also Gainsford (1990) on meters using microwave and gamma; van Santen, Kolar and Scheers (1995) on dual energy gamma- or X-rays.

See also Hall, Whitaker and Millington (1997) and reviews (Okland et al. 1997) on use by Statoil, of installation of a Fluenta meter (Slater, Paterson and Marshall 1997), and on cooperation between Petrobras and Fluenta (Caetano, Pinheiro and Moreira 1997).

17.4.3 Developments and References Since the Late 1990s

The trend has appeared to be towards a full-bore meter which allowed measurement of all components, with flow measurement using a Venturi, or correlation method, capacitance, microwave or gamma for component fractions, and possibly a static mixer to homogenise the flow and eliminate flow regime effects. Figure 17.4 is a simple diagram to identify these features and in addition the measurement of pressure and temperature (Torkildsen and Hanssen 1996). Other flowmeters have been mentioned in terms of separation and for full flow magnetic resonance, sonar and Coriolis may be options. The options on density measurement to combine with flow measurement to obtain the mass flow rate of the components in a multi-component stream appear to be:

- Weighing elements
- Pressure and buoyancy methods
- Vibrating elements
- Acoustic methods: sound speed, acoustic impedance
- Nuclear radiations: pulsed neutron, X-ray, gamma
- Electromagnetic field methods: magnetic resonance, electrical impedance (capacitance, resistance), static charge, microwave; measurement of resistance and capacitance might achieve multi-component ratio measurement (Dykesteen et al. 1985); a full-bore device for measuring water-in-oil from 0 to 80% using capacitance methods may have been available; microwaves have been used for the measurement of water-in-oil.

Vibratory methods may not be suitable for multi-component flows since bubbles and particles may not fully follow the vibration. Optical (transmission problems) and thermal (heat transfer across boundaries) methods have not been listed since they, also, are unlikely to be suitable for both measurements (cf. King 1990).

It appears that some countries will not allow nuclear sources to be imported or make it very difficult to do so. This suggests that methods which do not require nuclear radiation are likely to be more acceptable in the future.

In 2001 Dykesteen identified the dominant MPFMs as MFI (Roxar), Fluenta, Framo and Agar. He also commented on some of the early history of meters and companies:

- in 1997 Kvaerner Oilfield Products signed an exclusive licence with CSIRO (Commonwealth Scientific Industrial Research Organisation) for DUET,
- in 1998 Schlumberger and Framo Engineering merged their technologies and manufacturing expertise in the multiphase flow measurement area,
- that Roxar was a development between Smedvig Technology and MultiFluid ASA,
- that in 2001 Roxar acquired Fluenta and Roxar Flow Measurement AS was formed, wholly owned by Roxar ASA.

Falcone et al. (2001, 2002, 2005) gave useful reviews of multiphase. Abro, Kleppe and Vikshåland (2009) gave an estimate of current use: for production optimisation and management, in some fields with meters for fiscal allocation for oil and wet-gas as subsea production systems from where unprocessed flows could be taken to platforms by pipeline.

Schlumberger (2010) listed the multiphase meters at that date. The list in Table 17.2 is based on that of Thorn et al. (2013). I have interpreted the operating features of the meters from the literature. This is a best attempt at correctness of information, but I am conscious that misinterpretation and changes with time mean that all information should be checked before use. I am also conscious that company

Table 17.2. *Multiphase flowmeters (MPFM) according to Thorn et al. (2013). The table is reproduced with permission of the authors and © IOP Publishing. All rights reserved*

Company	Designation	*	Flow rate measurement	Composition measurement
Accuflow Inc.	AMMS	S1	gas: vortex liquid: Coriolis	 oil/water: microwave
Agar Corporation Inc	MPFM 50	D1	Coriolis	microwave
Aker Solutions ASA	DUET	D1	cross-correlation	dual energy gamma
Haimo Technology Inc.	MPFM	S2/D1	gas: Venturi	gas: single energy gamma oil/water: dual energy gamma
Jiskoot Quality Systems	Mixmeter	D1	pressure drop across in-line mixer	dual energy gamma
MB Petroleum Services	MB Flow Master	S1	gas: vortex oil/water: Coriolis	 oil/water : microwave
Multi Phase Meters AS	MPM	D1	Venturi	single energy gamma/ high frequency electromagnetic waves
MEDENG Ltd	MD 04	D2	flow modelling from primary sensor output signals	
Neftemer Ltd	Neftemer	D1/D2	analysis of primary sensor output signals	multi energy gamma
Petroleum Software Ltd	ESMER	D2	flow modelled from primary sensor output signals	
Phase Dynamics Inc	CCM	S1	gas: Coriolis oil/water: Coriolis	 oil/water : microwave
Pietro Fiorentini	Flowatch 3I	D1	cross-correlation/ Venturi	impedance
Pietro Fiorentini	Flowatch HS	D1	cross-correlation/ Venturi	single energy gamma/ impedance
Pietro Fiorentini	Flowatch High GVF	S1	gas: vortex oil/water: Venturi/ cross-correlation	oil/water: impedance
Roxar Flow Measurement	MPFM 1900VI	D1	cross-correlation/ Venturi	single energy gamma/ impedance
Roxar Flow Measurement	MPFM 1900VI Non-gamma	D1	cross-correlation/ Venturi	impedance
Roxar Flow Measurement	MPFM 2600	D1	cross-correlation/ Venturi	impedance
Roxar Flow Measurement	MPFM 2600 Gamma	D1	cross-correlation/ Venturi	single energy gamma/ impedance
Roxar Flow Measurement	Subsea MPFM	D1	cross-correlation/ Venturi	single energy gamma/ impedance
Schlumberger Ltd	PhaseWatcher	D1	Venturi	dual energy gamma

(*continued*)

Table 17.2. (*cont.*)

Company	Designation	*	Flow rate measurement	Composition measurement
Schlumberger Ltd	PhaseTester	D1	Venturi	dual energy gamma
TEA Sistemi SpA	LYRA	D1	Venturi	single energy gamma/ impedance
Weatherford International Inc.	Red Eye MMS	S1	gas/liquid: vortex oil/water: Coriolis	 infrared absorption water-cut:

*S = liquid/gas separation required as part of the meter system,
S1 = total separation of liquid and gas
S2 = partial separation which may require wet-gas and gas-in-liquid measurements
D1 = direct measurement of multiphase flow
D2 = advanced analysis of output signals to obtain flow of components
(Thorn et al. (2013) should be consulted for a fuller description of these categories and for the sources of their tables from which this information is taken.)

product and service names are likely to be the property of their respective owners and should be respected as such.

Additional References Related to MPFMs on Features of the Meters

Published information included:

- tests of the Agar MPFM 300 and 400 series (Murugesan 2002);
- development and operation of Complete Resource Neftemer (Kratirov et al. 2006) which has a gamma source and a gamma detection unit. These units are mounted diametrically opposite each other on a vertical pipe section containing a vertically upward multiphase flow. For flow rate it may require correlation of signals. The detection unit was specially designed for Neftemer. It uses a sodium iodide crystal with a photomultiplier for the gamma ray detection. The gamma source appears to be caesium-137.
- that the Framo/Daniel meter used an ejector tube to mix the high gas volume flows, and had been tested in the North Sea (Tuss 1997; see also Murugesan 2002). It used multiple-energy gamma ray absorption (MEGRA), and obtained flow rates of oil, gas and water, totals, water-cut, gas volume fraction, mixture density and process temperature and pressure.
- Way and Wood (2002) reported on the selection of Framo/Schlumberger PhaseWatcher (Vx) and Skaardalsmo and Moksnes (2003) described their experience with this meter.
- a multiphase meter for viscous flows aimed at replacing test separators for heavy oils, and made use of a Venturi and a dual-energy gamma ray meter (Atkinson, Berard and Segeral 2000).

- the Fluenta meter (MPFM 1900VI) used a capacitive sensor, an inductive sensor, a gamma densitometer, a Venturi and a system controller. It may have been able to obtain instantaneous composition of the flow (Dykesteen 2000).
- the MFI and CSIRO meters used cross-correlation (Tuss 1997).
- experience with a 3 in (75 mm) Roxar (MFI model) meter used for continuous production testing in a highly sour field environment (Shen, Vierkandt and Ogden 2003). (MFI, MultiFluid International; for associated changes see Dykesteen 2001).
- an observation that the MPFM-eliminated test separator provided continuous monitoring and reduced maintenance (Letton, Svaeren and Conort 1997). The dual-energy gamma provided fractions of oil, water and gas. There was also a means of salinity determination from the gamma spectrum.
- reduced costs and time for tests, SCADA compatibility (Murugesan 2002). Three flowmeters were tested: the Daniel MEGRA meter, the Roxar/Fluenta MPFM 1900VI and the Agar MPFM 300 and 400 series.
- data on the BP/ISA Controls multi-stream meter (de Carvalho and Antunes 2000).
- See also Moreau (2000) on the problem of finding the best meter for the application; Warren et al. (2003) on use for offshore installation; Clarijs et al. (2003) on X-ray measurement.
- Some of the designs in Schlumberger (2010) were not on the list of designs given by Thorn et al. (2013) see Table 17.2.
- Conference papers from about 2010 to 2014, suggest that publications are concerned with field testing, comparison with separators and installation effects, although some papers have included new design experience. There is also, as one would expect, research into development through CFD, etc. The interested reader is encouraged to access these conference proceedings for the last few years, for example the North Sea Flow Metering Workshops. See, for instance, the field trial of the Pietro Fiorentini MPFM Mobile Unit on flow with high gas void fractions, low pressures, variable water-cut, tight emulsions, severe slug flow, large turndowns and varying flow temperatures which appears to have been successful (Watkins et al. 2014).

Schlumberger (2010) indicated that the portable PhaseTester was of width 1.5 m, depth 1.6 m and height 1.7 m.

The most recent multiphase metering technology from Schlumberger is the Vx Spectra, which is shown in Figure 17.5. The largest specification is of width 0.565 m, length 0.795 m and height 0.7 m. For this specification the maximum flow capacity is given as: for liquid flow rate 8,745 m^3/day, and gas flow rate at 10 MPa as 3.68×10^6 m^3/day, with a claimed repeatability (total mass rate at line conditions) of better than ±1% (http://www.slb.com/~/media/Files/testing/product_sheets/multiphase/vx_spectra_surface_multiphase_flowmeter_ps.pdf).

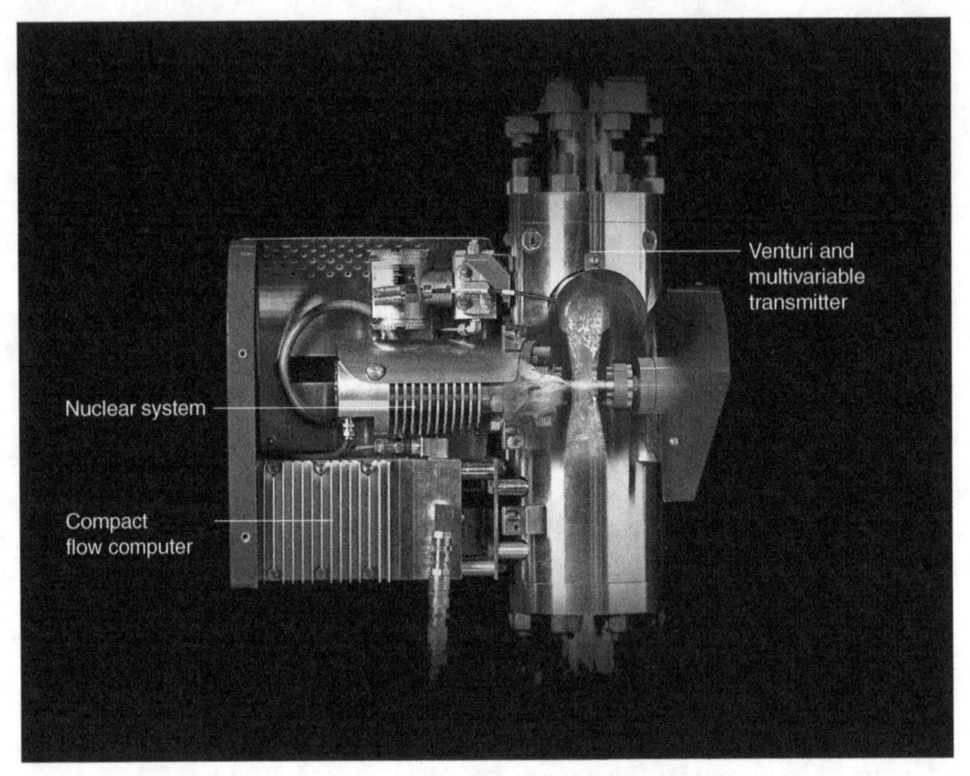

Figure 17.5. The Vx Spectra surface multiphase flowmeter comprises three sections: a Venturi and transmitter to measure total flow rate; a gamma source and detector to obtain holdup of oil, gas and water; and a flow computer to convert line flow measurements to standard conditions. (Vx Spectra is a mark of Schlumberger. This image is copyright Schlumberger, Ltd. Used with permission.)

Brief Notes on Other Recent Publications

Kettle, Ross and Deznan (2002) confirmed the good opinion in the industry for the dual-energy Venturi multiphase meter and the many benefits of the meter over previous separation technology. Theuveny et al. (2002a) reviewed the performance of mobile meters and Theuveny et al. (2002b) discussed the performance of the multiphase flowmeter using dual energy gamma/Venturi meters. Ali et al. (2003) used tracers injected for oil, gas and water into a multiphase flow to obtain individual component flow rates. Al-Bouri et al. (2005) mentioned measurement accuracy, the introduction of multiphase metering and the convenience of remote communication for the management of the wells. Falcone et al. (2001, 2002) gave a useful review of the situation then, and compared the various technologies on offer with future direction and research needs. Coull and Miller (2005) tested a conceptual design of multiphase flowmeter using ultrasonic technology.

Yan (2005a) edited an issue of the *Journal of Flow Measurement and Instrumentation* on tomographic techniques for multiphase flow measurement.

The reader is referred to the issue as some of the papers, although of considerable interest, have not been included in this review. Hampel et al. (2005) discussed ultra-fast X-ray tomography with an electron beam source. Prasser, Misawa and Tiseanu (2005) discussed methods capable of acquiring instantaneous void fraction distributions over the cross-section. Frøystein, Kvandal and Aakre (2005) discussed dual-energy gamma tomography for high-pressure multiphase flows. Wee et al. (2007) described a multiphase meter combining various measurements in a tomographic system, which used Venturi and a high-speed radio-frequency technique for water-cut, composition and liquid/gas distribution within the pipe cross-section. Zhao and Cao (2008) described an electrical impedance tomographic system for multiphase flow. Fosså, Stobie and Wee (2009) used a Venturi, gamma detector, advanced flow models, with a tomographic measurement system (cf. Wee and Scheers 2009) [cf. Tomoflow R100 a multiphase flow structure analysis based on electrical capacitance tomography (ECT) http://www.tomoflow.com/pdf/TFLR100flyer.pdf.]

Yan (2005b) also edited an issue of the *Journal of Flow Measurement and Instrumentation* on optical techniques for multiphase flow measurement. The reader is, again, referred to the issue as the papers have not been included here.

Beg and Toral (1993) reported on the use of pattern recognition techniques to obtain superficial gas-liquid flow rates. They reckoned that the techniques could be extended to oil-water-gas flows by using pressure sensing and orifice plate with capacitance and gamma ray. They introduced a new hydrodynamic system for scaling pipe diameter and fluid properties and used neural network pattern recognition techniques [cf. Toral, Beg and Archer (1990) computer package].

Wylie, Shaw and Al-Shamma'a (2006) used a radio frequency source, the spectrum of which could be interpreted to give the phase fractions of the flow. Johansen and Tjugum (2007) described a multiple-gamma-ray beam measurement technique. Basil, Stobie and Letton (2007) described an analytical approach to the performance assessment of a dual-gamma Venturi MPFM operating with heavy oil (Reynolds number less than 2,000), which calculated various parameters. Xue and Shen (2008) discussed a correlation technique for flow measurement using conductance. Hall, Letton and Webb (2008) suggested: to design in the means of verification, to consider a sampling port near the flowmeter and to consider including extra sensors which may be useful in future modelling. Meng et al. (2011) developed a multiphase meter which avoided separation and was used by passing the fluid down a sloped pipe (Dean et al. 1990a) to obtain stratified flow, and hence to obtain the flow rate of the components of the flow. It appeared to be for high water-cut fields and to achieve a promising performance. An alternative method using a neural network approach was mentioned by Cai et al. (2004). Wee *et al* (2013) described their MPM meter which measured multiphase flow rates with no separation or mixing device. A combination of Venturi flowmeter, gamma-ray densitometer, multi-dimensional, multi-frequency dielectric measurement system and flow models are used. A method for detection of scale build up in the MPM meter had been developed.

Methods Based on Modern Flowmeters

In Chapter 13 a new design of magnetic resonance flowmeter is claimed to be able to measure three-phase flows.

In Chapter 15 sonar flowmeters are discussed with capabilities in the measurement of some flows with more than one component.

In Chapter 20 on Coriolis meters the reader is referred to recent relevant work by Henry et al. (2013) which appears to be making good progress using a Coriolis meter with a water-cut meter in laboratory tests and field trials.

Multiphase Well Test, Downhole and Other Meters

This is possibly the earliest multiphase flow measurement area, but it is a specialist area for the petroleum industry and I have, therefore, only included recent references which I have come across in preparing this book. Theuveny and Walker (2001) confirmed that multiphase flowmeters improve well test data, and Mus et al. (2002) emphasised the added value of using a meter. Wangsa et al. (2005) reported on successful use of a portable multiphase flowmeter for high gas fractions and high water-cut. Araimi, Jha and Al-Zakwani (2005) compared exploratory well testing using a multiphase flowmeter (Venturi plus dual-energy gamma), with previous conventional well test equipment, and discussed the advantages of the new approach. Kragas et al. (2002, 2003a, 2003b) described the design, operation, installation, testing and data analysis of a downhole fibre optic flowmeter, which appeared to be able also to measure phase fractions, pressure and temperature. Ningde et al. (2005) discussed a design of turbine meter for measuring oil-gas-water three-phase flow in vertical upward pipes for production logging. The package also includes a radioactive source.

Other Possible Approaches to Multiphase Measurement – Check Valves

Falcone et al. (2009) in their book on multiphase flow metering have an interesting discussion of the possibilities of using check valves as flowmeters in combination with other measurements, possibly including water-cut, and pvT data. They also discussed the use of neural networks which might be combined with use of data from such a valve.

Other Possible Approaches to Multiphase Measurement – Virtual Flow Measurement

In this book I have identified one or two areas where suggestions have been made for deducing flow measurement from other data. The oil industry appears to be exploring the use of data, for instance pressure loss data which can be used to deduce a flow rate. If the pressure drop along a length of pipe is known, then Equation (2.12) can be used to obtain the velocity through the pipe provided the value of K, the loss coefficient, is known or deducible from e.g. Miller (1990), essentially using the pipe as a differential pressure flowmeter. The precise geometry of the pipe and the properties of the fluid will be needed for a flow estimation. The method can be extended to the whole of a system, giving means of cross-checking the calculated flows.

A note on Calibration

Multiphase or multi-component flow rigs are now available at calibration centres. However, it is worth making the point that, while single-component calibration rigs should have a long pipe length upstream of the meter to obtain fully developed flow, there are good reasons why such an approach is less appropriate for a multi-component rig:

i) The settling length may be impracticably long;
ii) The orientation of the pipe will affect the flow and a long settling length will prohibit other than horizontal orientations;
iii) It is virtually certain that actual installation positions will not have fully developed flow conditions;
iv) The enormous range of flow conditions requires that the flowmeter accuracy is not dependent on particular flow conditions.

A more appropriate design for such a rig is likely to be a test section preceded by a short upstream section and a manifold for component injection. By selection of the injection manifold options, flow regimes could be constructed upstream of the flowmeter. The whole unit including manifold, upstream section and test section would be short enough to rotate to any orientation, so that both horizontal and vertical as well as intermediate flows could be simulated. The outlet flow through a flexible pipe would enter a separator before recirculation. Such a rig may also have a small enough total volume to allow live crude to be used as one component for oil related flows.

17.5 Accuracy

The errors in the papers reviewed on multi-component/multiphase flows range from ±1% for an ultrasonic correlation flowmeter in paper pulp flow, for example, to as much as 25% for one of the offshore oil flow measurement systems (Fischer 1994). This range of error is symptomatic of the problem for the flowmeter designer. For control and distribution a measurement uncertainty of order ±10% may be acceptable. However, high accuracy is required for fiscal and custody transfer applications, and here the level which can be achieved by the designer is far from clear. Is even ±5% achievable in multi-component oil flow measurement within certain ranges and/or envelopes?

Dou, Guo and Gokulnath (2005) discussed the performance of a compact high GVF multiphase meter, which they appeared to claim had an uncertainty within ±2% absolute for water-cut and 10% relative for liquid and gas flow rates at 90% confidence level while optimising the size and cost. Moderate GVF (25–85%) were considered to be the range with the optimum performance, and where traditional single-phase meters are not a viable option.

17.6 Chapter Conclusions

In the first edition of this handbook I asked the question: separation or multi-component metering? The problems relating to the accurate measurement of multi-component flow are now better understood and being solved in a variety of innovative ways. Meters incorporating separation appear to have become much more compact and may continue to be an option for some applications, but it seems likely that they will have strong competition from the straight through multiphase meter. It seems virtually certain, however, that separation will continue to be required for calibration of multiphase flowmeters. Separation of the components and subsequent measurement of each component may offer the surest route to an accuracy for fiscal requirements, but increasingly the straight through multiphase flowmeter is offering a more compact and flexible solution.

One interesting observation, made to me by a colleague, and apparent from this chapter, is the convergence of the technology onto a few techniques. The Venturi has become a dominant component of the measuring systems, with correlation as an alternative. Gamma ray meters appear to be the most favoured of the options for phase proportions, although some countries may not allow their import. Alternatives which may be less accurate are capacitance/impedance devices.

Mixing or separation may be required, and for the latter the cyclone separator may be one option. Pressure and temperature differences are likely to be obtained, and an operational numerical model to allow for the fluids and their properties, the likely flow regime etc. is a part of some of these systems.

At least three flowmeters discussed elsewhere in this book are contenders: the magnetic resonance, the sonar and the Coriolis meters.

The virtual flowmeter is an interesting concept which may well have an important role, and reflects similar developments in other areas of the technology.

Dyakowski (1996) introduces the idea of applying process tomography to multiphase flow measurement, but it would seem that 20 years later the approach is still waiting to be more generally applied. Beg and Toral (1993) reported on the use of pattern recognition techniques to obtain superficial gas-liquid flow rates. Mi, Ishii and Tsoukalas (2001b) described a flow regime identification methodology with neural networks and two-phase flow models.

It is very likely that major developments will occur. These are likely to have improved component ratio measurement, and clever data handling to obtain the most information. We should also remember that other areas of industry beyond the oil industry will benefit from such developments.

The application of CFD is being used as an important tool in the continuing development of these meters. Barton and Parry (2012), in a paper entitled "Using CFD to understand multiphase and wet-gas measurement", gave some case studies of such flow modelling, the effects of heavy oil multiphase flow, the behaviour of multiphase swirling flow and modelling of wet-gas flows, with particular reference to the Venturi.

18

Thermal Flowmeters

18.1 Introduction

There are broadly two concepts of thermal flowmeter now available for gas mass flow measurement, one of which is also applicable to liquids. I shall follow the useful terminology in ISO 14511:2001 for the two types and encourage the reader to make use of the ISO document. The first is the capillary thermal mass flowmeter (CTMF), which has broad applications in the control of low flows of clean gases, but which can also be used with a bypass containing a laminar element to allow higher flow rates to be measured. The arrangement of heaters and coils between the various manufacturers differs, but the basic approach is the same, with heat added to the flowing stream and a temperature imbalance being used to obtain the flow rate.

The second is the full-bore thermal mass flowmeter (ITMF), which is available as both insertion probe and in-line type. It has a widely used counterpart in the hot-wire anemometer for measurement of local flow velocity, but, as the need for a gas mass flowmeter has become evident, it has been developed as a robust insertion probe for industrial usage and then as the sensing element in a spool piece flowmeter. It has been produced by an increasing number of manufacturers in recent years as a solution to the need for such a mass flowmeter.

However, the advent of the MEMS (micro-electro-mechanical sensors) or CMOS (complementary metal-oxide semiconductor) designs has opened a new set of flowmeters based on silicon chip technology. Their operation is similar to the CTMF, but usually on a smaller scale.

18.2 Capillary Thermal Mass Flowmeter – Gases

In various examples of the CTMF, the gas flows through a very small diameter tube that has heating and temperature-measuring sensors. For larger flows, this is used with a bypass laminar flow element in the main gas stream. In many cases, the mass flowmeter is in one unit with the control electronics and the gas control valve. However, in this chapter our interest is in the meter itself. These meters are designed for clean dry gases.

18.2.1 Description of Operation

In this design of thermal flowmeter, the gas flows through a very small tube (possibly with ID in the range of 0.2 to 0.9 mm), with a sufficient length to diameter ratio to ensure fully developed laminar flow, and on the outside of which there are heating and temperature-sensing windings. The heater winding transfers heat through the wall of the tube to the gas. If there is a gas flow, the heated gas is carried downstream. The downstream temperature sensor will then sense a higher temperature than the upstream sensor. This differential can then be used to deduce the mass flow. A diagram is shown in Figure 18.1(a). The equation for this is given in the steady state by

$$q_m = Q_h / (K c_p \Delta T) \tag{18.1}$$

where q_m is the mass flow rate to be measured in kilograms per second, Q_h is the heat input in Joules per second, K is a constant, c_p is the specific heat at constant pressure in Joules per kilogram Kelvin, and ΔT is the measured temperature difference in Kelvins. A knowledge of K, Q_h and c_p allows q_m to be deduced from ΔT. Meters may have a linearisation circuit built in to their electronics to correct the signal and to allow for the particular flow tube features.

In an alternative mode of operation [Figure 18.1(b)], the sensed temperature is used to adjust the heat into three heaters to keep them at a constant temperature. The energy required in this mode to restore equilibrium to the windings can be used to obtain the instantaneous rate of flow. This results in a sensor with a short response time.

Figure 18.1(c) shows a design that uses the temperature difference between two coils positioned along the tube axis, both of which are heated. The coils also act as resistance thermometers. When flow occurs, the downstream coil is cooled less than the upstream coil, and the differential temperature is used as a measure of the flow.

Figure 18.1(d) shows a design with two temperature-sensing coils and one heater coil. Of these the most common designs in industry are possibly those in Figures 18.1(a) and (c).

In order to accommodate a greater flow rate than is possible through the capillary, a laminar flow bypass is used [Figure 18.1(e)]. The total flow rate through capillary and bypass is then proportional to the flow through the capillary. In some designs, a control valve is incorporated. In others, an air-actuated valve may also be offered by the manufacturer.

Figure 18.2 shows an industrial CTMF with bypass through *n* identical elements, where *n* can be selected for the appropriate flow range. It also shows the valve, operated from a second amplifier, which controls the flow.

In another design [Figure 18.3(a)], the means of heating the tube is through a current passed along the tube wall, which, in turn, heats the fluid. Two heat sinks, one at each end, conduct away the heat that has passed up the tube. The fluid cools the tube, and the resultant temperature difference is measured in two places with the thermocouples TC_1 and TC_2; the flow rate can be deduced from the difference. As with the other designs of thermal flowmeter, this one can be used with a bypass.

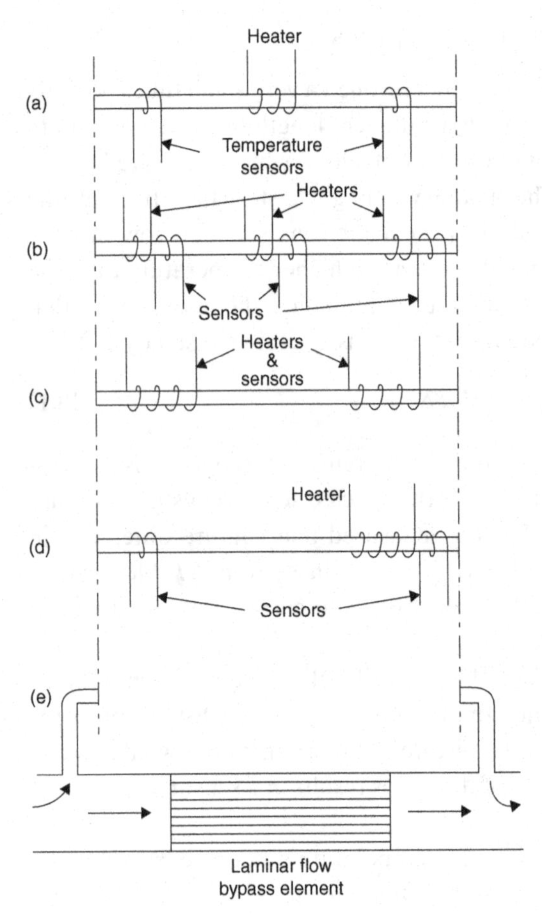

Figure 18.1. Diagram to show the various configurations of CTMFs: **(a)** Central heater with upstream and downstream temperature sensors. **(b)** Three heaters with sensors to ensure that they remain at constant temperature. **(c)** Two combined heaters and resistance thermometers. **(d)** One downstream heater with two sensors. **(e)** Bypass with laminar flow element.

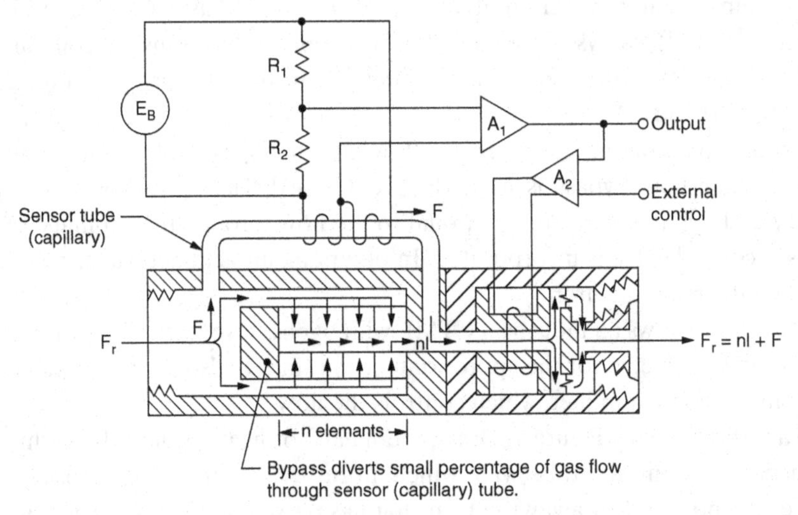

Figure 18.2. Industrial capillary thermal mass flowmeter showing schematic of flow paths and electrical diagram (reproduced with permission of BOC Edwards).

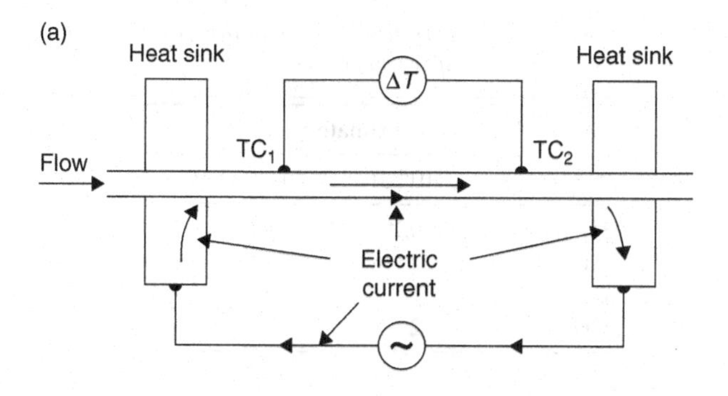

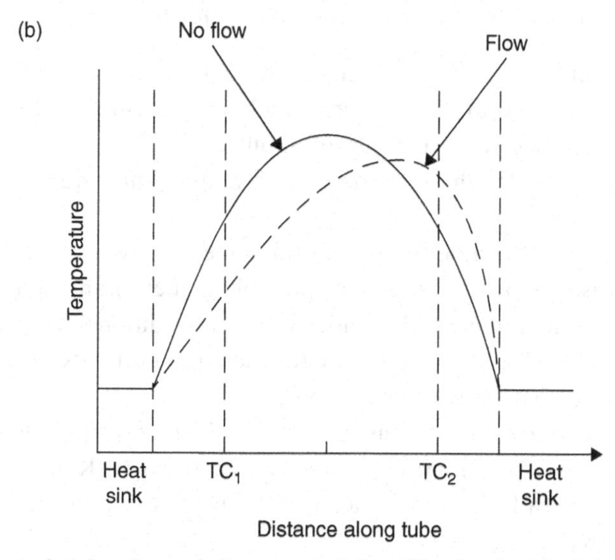

Figure 18.3. Heated tube thermal flowmeter (after Hastings): **(a)** Schematic diagram; **(b)** temperature distribution under static and flowing conditions.

The distributions of temperature for zero flow and for a small flow are shown in Figure 18.3(b). It can be seen that the temperature difference between the two thermocouples will be zero at zero flow but will increase with flow. The equation derived for this flowmeter is reproduced by Hemp (1995a) (cf. Appendix 18.A.4).

All these varieties depend on a knowledge of c_p, which depends on the gas and varies with pressure as shown in Table 18.1.

It is important for the calibration of the meter to be close to, or referenced to, operating conditions; otherwise, it will be necessary to look up the data for the gas and pressure at operation. Temperature variation has only a slight effect on the value of c_p. For example, for air it is 0.01%/°C, and for methane it is 0.11%/°C. Viscosity variation has a negligible effect unless the gas is near liquefaction. Calibration may be referenced to primary volumetric standards at atmospheric conditions and corrected normal temperature and pressure or standard temperature and pressure.

Table 18.1. *Variation of c_p with pressure for four gases*

Gas	Approximate % Variation	
	0–10 bar	40–100 bar
H_2	+0.1	+1.6
Air	+2	+16
O_2	+1	+18
CH_4	+2	+31

18.2.2 Operating Ranges and Materials for Industrial Designs

Rangeability is up to 50:1. Flow ranges are from 0 to 3 ml/min or lower to 0–100 l/min or even up to 1,000 m^3/h for special bypass designs. A 100-mm diameter meter would typically give a range up to about 250 m^3/h and a 200-mm diameter meter up to about 1,000 m^3/h. Note that, for air at ambient conditions, 1,000 m^3 has a mass of about 1.2 kg.

These meters may be able to operate for some designs up to about 200 bar or even, in some cases, 300 bar pressure or possibly higher. Manufacturers should be consulted on possible nonlinearities under vacuum conditions. Temperature operating range may be 0–65°C with a coefficient of less than 0.1%/°C. Pressure drop on the line is up to 0.02 bar for the maximum flow rate.

Materials in contact with the gas include 316, 316L and high-alloy ferritic stainless steel; anodised aluminium, brass and Viton; Buna-N; Kalrez; and Teflon or silicone O-rings. Ultra-high-purity seals may be used for applications in the semiconductor industry.

Power supply requirements usually relate to 15 and 24 V DC versions; when combined with a controller, these may draw up to about 180 mA, resulting in power consumption of order 3.5 W.

The manufacture of these devices entails careful winding and fixing of the sensing and heating wires onto the small-diameter capillary tube. The aim of the final testing and preparation of the meters is likely to be to reduce drift and ensure an accurate instrument. Manufacturers may recommend calibration checks after a period of, say, six months of use to verify there is no drift.

Thermal drift may be reduced against ambient temperature changes by insulating the unit. A solution may be to enclose the sensor in a temperature-controlled environment. Sensors may also suffer from RF interference, and may be incompatible with corrosive gases. Typically 30 minutes of warm-up time may be needed.

Flowmeters are likely to be factory-set to a preselected gas, pressure and flow range. Correction factors range from about 0.55 for ethane to about 1.40 for argon against air. These may also be referred to nitrogen.

Table 18.2. *Time to approach the final value after a flow change*

To approach within	Response time
5%	3τ
2%	4τ
1%	4.6τ
0.5%	5.3τ

18.2.3 Accuracy

Uncertainty is typically claimed as ±1% full-scale deflection or FSD (or for one design ±0.75% rate ±0.25% FSD) but may be ±1.5% for the top end of the flow range and ±4% FSD in some other cases, depending on design and the temperature and pressure range. Repeatability is typically 0.2–0.5% FSD, but in some cases it may be quoted as being of order ±0.25% of rate and in at least one case as 0.05% FSD. In addition, there may be pressure and temperature coefficients.

18.2.4 Response Time

An exponential curve is usually used to indicate the response of a flowmeter. The exponential curve is given by

$$S = S_0\left(1 - e^{-t/\tau}\right)$$

The time constant τ gives the time for the flow signal S to reach about 60% of its final value S_0. Table 18.2 gives the times to approach the final value.

τ is of order 250 ms to over 1 s. A standard instrument may take 1–5 s to settle to within 2% of final value for a 0–100% command change. However, manufacturers may claim a response of about 1 s for some designs.

18.2.5 Installation

There is unlikely to be any effect from upstream pipework for the basic device without bypass because the bore of the tube is so small that the profile will be readjusted before reaching the elements. For bypass designs, there may be an effect, and it may be necessary to include 20D upstream straight pipe (Gray, Benjamin and Chapman 1991). Designs should ensure that flow is truly laminar in the bypass, by suitable matching of size and pressure drop.

There may be significant effects on sensitivity that can seriously impair accuracy under certain conditions of installation (e.g. mounting angle), gas density and pressure when outside normal operating envelopes. Zero offset appears to be affected by meter orientation and may result from free convection when the meter is near vertical. Rarefied gas flows may also create problems.

There may be temperature and pressure sensitivities which the user should clarify with the manufacturer prior to purchase.

Gray et al. (1991) discussed environmental factors that affect mass flow controllers (MFC). The MFC was tested with a vacuum diagnostic system to which the MFC is connected. The test chamber was evacuated to below 50 mtorr, and the valve to the vacuum pump was then shut. The initial flow rate was set, and the pressure was measured as it rose. From this, the time to achieve controlled flow, actual flow rate, error and deviations from flow were obtained. They tested various effects and concluded that, to minimise effects, temperature extremes must be avoided and surroundings should not be allowed to raise the temperature of the gas boxes, pressure should be carefully regulated, and vibration should be avoided. Where possible, tubing distances should be minimised, and MFCs should be mounted in a normal upright orientation.

18.2.6 Applications

CTMFs are most commonly applied to low flows of clean dry gases above their dew-points (e.g. gas blending). Gases which manufacturers quote include air, acetylene, ammonia, argon, arsine, nitrogen, butane, carbon dioxide, carbon monoxide, chlorine, ethane, ethylene, fluorine, Freon 11, Freon 12, Freon 13, helium, hydrogen, hydrogen sulphide, krypton, methane, neon, nitrous oxide, oxygen, propane, propylene, silane and xenon.

The meter may be used with a wide range of gases for processes involving crystal growth, thermal oxide, diffusion, chemical vapour deposition, sputtering, ion implantation and plasma etching, and the gases involved will each require a correction factor against the calibration gas, possibly nitrogen, which may vary between as low as 0.17 for Freon-C318 (C_4F_8) or perfluoropropane, and as high as 1.543 for krypton. The manufacturer should be asked for the compatibility of any particular gas and the correction factor (Sullivan no date). These meters are widely used in the semiconductor industry for measuring flows of pure gases (Kim, Han and Kim 2003). In selecting a flowmeter, particular care should be taken where the inlet pressure is less than atmospheric or has a vapour pressure less than atmospheric.

18.3 Calibration of Very Low Flow Rates

One method of calibration of these flowmeters for low flow rates is by means of a special piston prover. A piston moves within a precision glass tube. The piston is connected to a counterweight by means of a tape that turns an encoder wheel. The temperature and pressure of the gas are measured while flow is taking place.

Positive displacement devices or piston meters may achieve calibration to an uncertainty of about 0.2% of reading over a range from 0.02 to 30 l/min (Sullivan,

Ewing and Jacobs no date). The volume flow at standard temperature and pressure (STP), subscript s, may be obtained from the actual, subscript a, using

$$q_{vs} = q_{va} \frac{T_s}{T_a} \frac{p_a}{p_s}$$

18.4 Thermal Mass Flowmeter – Liquids

18.4.1 Operation

The diagram in Figure 18.4 shows the layout of a meter with parallel tubes running between a heating block and a heat-sink block. At the midpoint of each tube was the temperature sensor. The heating caused an increase in temperature of 10 or 20°C. When the flow took place, heat was convected, and the difference between T_3 and T_4 varied with the mass flow rate. It appears that the heat was then conducted through the body of the meter. The liquid may have increased in temperature by as much as 15°C within the meter, but there was negligible change in liquid temperature outside the device.

Huijsing, van Dorp and Loos (1988) described a meter aimed at fuel consumption measurement in automobiles and light aircraft and also industrial processes. The external muffs (for heat transfer and temperature measurement) of other designs were replaced with internal copper blocks, perforated with multiple axial holes through which the liquid flowed, resulting in much greater thermal contact with the liquid.

Bruschi, Nizza and Piotto (2006) appear to describe a flowmeter composed of a probe and an external circuit based on a temperature control loop integrated on a 2 mm by 2 mm chip, which was bonded on top of a copper cylinder of 2 mm diameter and 2.5 mm length that made contact with the liquid and the temperature of which was cycled between two levels. They tested it on de-ionised water. Cf. Chung et al. (2003) and Koizumi and Serizawa (2008) micro meters.

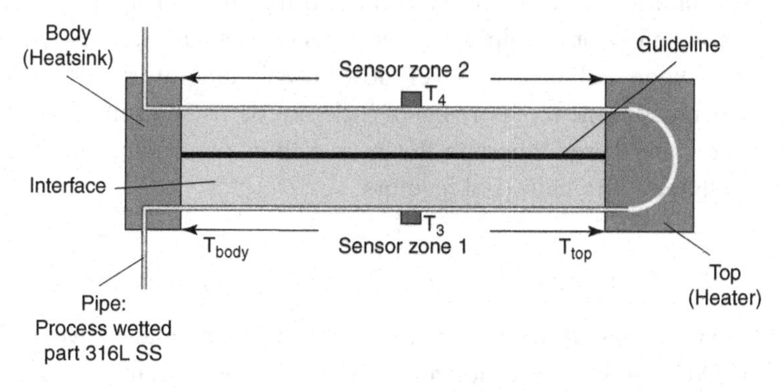

T_{top} - T_{body} = (20°C for Model 5881, 10°C for Model 5882).

Figure 18.4. Liquid mass flowmeter (reproduced with permission of Emerson Process Management Ltd.).

18.4.2 Typical Operating Ranges and Materials for Industrial Designs

For liquids, minimum measurable flow is about 0.002 kg/h with a maximum measurable flow of between 0.1 and 0.5 kg/h, depending on the design, and a turndown of up to about 15:1. Wider ranges may be obtainable. Uncertainty is about 0.5–1% FSD and repeatability 0.2% of rate for the upper end of the range and 0.1% FSD for the lower part of the range. Repeatability is of the order of ±0.5% FSD per month.

A bypass arrangement allows greatly increased flow range up to 100 kg/h, but the measurement uncertainty is likely to increase to about 2% FSD. However, if extreme range turndown ratios are claimed, it is likely that the manufacturer will give a performance envelope for the meter.

The ambient temperature range may be from 0 to 65°C, and pressure ratings can be as high as 400 bar. Maximum viscosity may be about 100 cP. Pressure loss at operating conditions is claimed as very small.

The response time for sensors is likely to range from 0.5 s up to about 10 s for the largest meters, but the response for controllers should be obtained from the manufacturer.

18.4.3 Installation

Pipe size connections range from $\frac{1}{16}$ to $\frac{1}{4}$ in. (about 1.5–6 mm) with materials of stainless steel and seals of Buna-N, Viton, PTFE or Kalrez. Process temperature change can cause +0.02%/°C, and ambient temperature effects can cause up to +0.03%/°C.

18.4.4 Applications

It may be suitable for measurement of very low flow rates of toxic, corrosive and volatile liquids (liquefied gases under pressure), catalysts in petrochemicals, pigments in paints, reagents in pharmaceuticals, flavorants, enzymes, antibiotics in food, odorants into natural gas, titanium chloride closing into furnaces for hardening metals, corrosion inhibitors in pipes and reagents in fermenters. It also may offer a means for controlling and measuring very low flows in laboratories concerned with microfiltration, measurement of porosity of rocks, fuel consumption, lubrication, gas/vapour mixtures and chromatograph flows, as well as controlling the injection of additives (catalysers) into industrial reactors.

18.5 Insertion and In-Line Thermal Mass Flowmeters

The in-line thermal mass flowmeters (ITMFs) represent an alternative design concept to the CTMF, and both insertion and in-line versions appear to be aimed at the same goal – mass flow measurement of gases in pipes and ducts.

In Figure 18.5 there is a diagram of an arrangement for an ITMF. The heaters and temperature sensors for either insertion or in-line designs may be contained

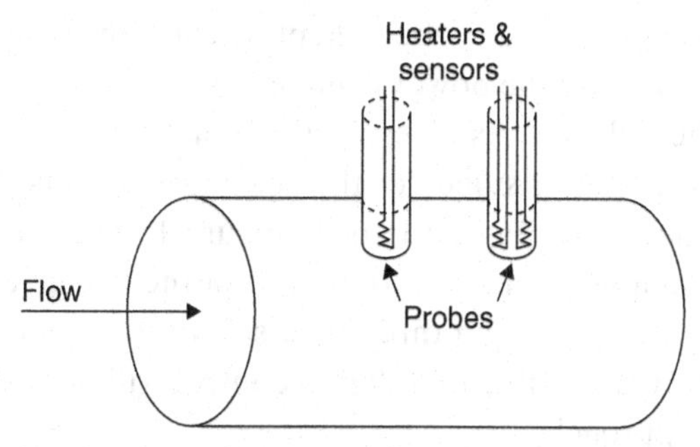

Figure 18.5. Diagram of an ITMF with either insertion probes or in-line design. In the latter, the probes are often in the same cross-section.

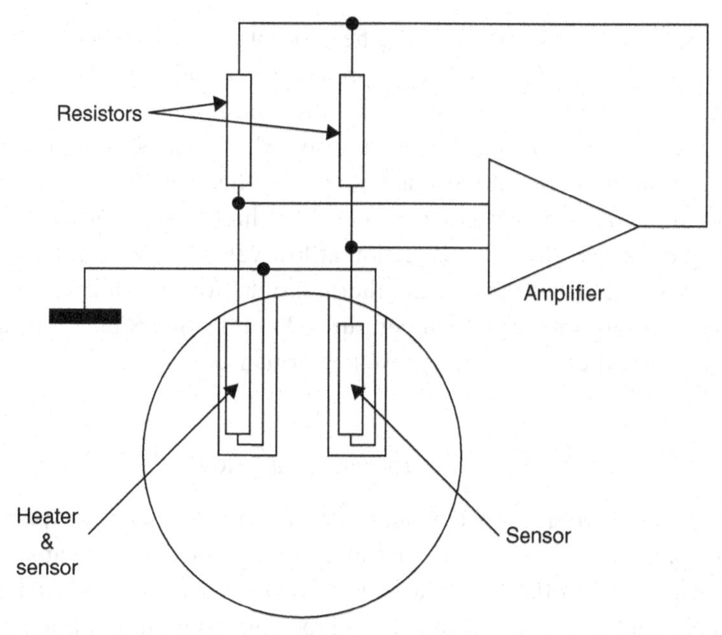

Figure 18.6. Schematic of in-line thermal mass flowmeter showing Wheatstone bridge arrangement.

in one or two probes, may use the same element for both functions, and, where two probes are used, will often, as in Figure 18.6, be in the same cross-section of the duct.

This flowmeter (Kurz 1992) uses a heated element in a flowing stream, the convective heat transfer of which depends on the temperature difference between the probe and the gas, on the flow rate and on characteristics of the gas. A second temperature probe is used to measure the temperature of the gas. Two types have been used: the constant power and the constant temperature anemometers. The temperature of the heated probe in the former varies with gas velocity and is used as the sensing parameter. The response is slow due to the thermal inertia of the probe, and the zero is unstable because natural convection becomes important at low flows. In the latter type, the probe is kept at constant temperature through a feedback control circuit, and the power needed to retain this constant temperature is used to obtain the velocity. As a result, the response is much faster, typically 1 s for a 67% change. The resistance of the sensing elements is connected into a modified Wheatstone bridge (Figure 18.6; cf. Figure 18.A.1), in which the voltage across the bridge is amplified and fed to the top of the bridge to retain a constant temperature difference. By using a three-wire sensor, the lead resistance can also be compensated for. It is possible that digital control will increasingly displace the Wheatstone bridge approach.

The relationship for these flowmeters, from Equation (18.A.8), is

$$q_h = k'(1 + Kq_m^n)\Delta T \tag{18.2}$$

where q_m is the mass flow rate, q_h is the heat supplied, and ΔT is the temperature difference between heated and unheated probes, k' is a constant that allows for heat transfer and temperature difference at zero flow, and K is a constant incorporating the area of the duct into which the probe is inserted, as well as gas and heat transfer constants. In Equation (18.A.8), n was equal to $\frac{1}{2}$. This may be approximately correct for low velocities but may be more like $\frac{1}{3}$ for higher velocities. Factors affecting the equation, as well as free convection at low velocities, are heat transfer from the probe to its base, usually small, and thermal radiation, which is small compared to forced convection for normal temperatures. Manufacturers may introduce additional sensors into the probes to improve performance.

18.5.1 Insertion Thermal Mass Flowmeter

An example of an ITMF is shown in Figure 18.7(a) and consists of two pairs of elements. In each pair, there is a resistance thermometer; coupled to these are, in one pair, a heated rod and, in the other, a similar unheated body as a mass equaliser. Flow past the heated element results in a heat loss, and the combination of this and the differential temperature allows the flow rate to be deduced.

Insertion probes may be inserted into and withdrawn, without process shutdown, from a pipeline of 30 to 500 mm diameter. They may also be available with multiple sensors for installation across a pipe [Figure 18.7(b)].

The uncertainty may be quoted, typically, as being of order ±1% FSD or ±3% rate, whichever is less, for turndown ratios of order 10:1. However, gas velocities up to about 60 m/s or more have been claimed with turndown up to 800:1. Manufacturers who claim wider turndowns will presumably provide a calibration that relates uncertainty to various points in the characteristic.

Various designs of these meters have been used for boiler applications (preheater and combustion air), combustion gas flows, hydrogen coolant flow, gas pipeline transmission flows, methane flows, nitrogen purges, flare gas flow, ventilation, heating and air conditioning and superheated and dry saturated steam.

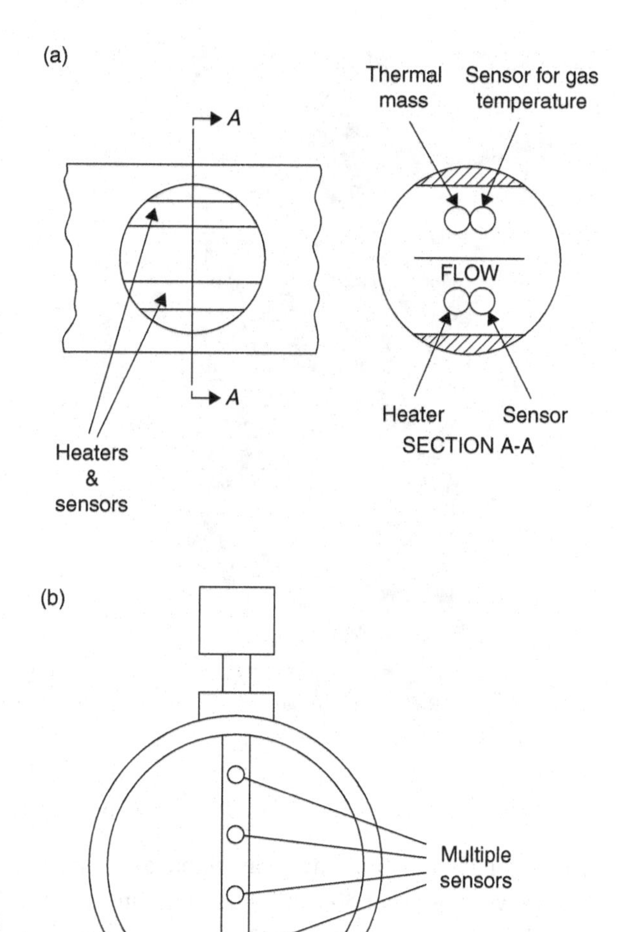

Figure 18.7. Diagrams of single- or multisensor probe: **(a)** Showing individual sensor; **(b)** showing multisensor probe in a pipe (after the design of FCI).

Lai et al. (1991) discussed the use of thermal mass air-flow sensors (essentially a hot-wire anemometer) for measuring the air flow into gasoline engines. A recommended arrangement is to have the mass air-flow sensors in a bypass of the intake air passage of an engine. Possible errors may arise due to variation in the profile at inlet caused by filter blockage.

18.5.2 In-Line Thermal Mass Flowmeter

Figure 18.8 shows a commercial ITMF with heated and unheated temperature-dependent resistors. Another approach that may be available uses two stainless steel wires in the flow stream in the throat of the transducer – one heated with a current, and a bridge circuit to adjust the current to keep the temperature between them, measured by their resistances, constant. There may also be designs that heat the fluid and sense its temperature from outside the tube, leaving an unobstructed bore to the tube.

Figure 18.8. Photograph of an in-line thermal mass flowmeter (reproduced with permission from Endress+Hauser).

18.5.3 Range and Accuracy

Meters of 25–200 mm may be available as flanged or, in some sizes, as wafer designs. Turndown is of order 50:1 or higher with uncertainty of 2% of rate and within a band of order 4% down to 1% FSD. Uncertainty may be quoted in the range ±1.5% to ±2.5% FSD. Repeatability may be given as 0.25–3% of rate or as 0.25–1% FSD, the lesser value being taken. The manufacturers give correction factors for various gases.

Flow ranges are from 2–125 kg/h to 300–8,000 kg/h or greater. Response time is about 0.2 s. Pressure drop is up to 10 mbar. Operating pressure is up to 40 bar. The temperature range is −50 to 300°C or greater, depending on the design. Hygienic versions are available. There may also be temperature effects of order ±0.3% rate or ±0.08% FSD/°C.

18.5.4 Materials

Materials of construction are typically stainless steel with seals of Viton rubber and encapsulated sensor elements, consisting of platinum resistance thermometers or thermistors, in stainless steel or glass.

18.5.5 Installation

Installation orientation may affect the performance, and manufacturers' instructions should be followed. Installation requirements recommended by one manufacturer

Table 18.3. *Installation requirements for a wafer or insertion thermal mass flowmeter compared with those for a $\beta = 0.5$ orifice plate (after Endress+Hauser)*

Upstream fitting	Thermal Flowmeter		Orifice plate ($\beta = 0.5$)	
	Upstream[a]	Downstream[b]	Upstream	Downstream
90° bend	20D	5D	22D	6D
2 × 90° bends same plane	25D	5D	22D[c]	6D
2 × 90° bends perpendicular planes	40D	5D	75D[c]	6D
Reducer	20D	5D	8D[d]	6D
Expander	20D	5D	20D[d]	6D
Control valve	50D	5D		

[a] Reduce by about 5D for flanged in-line flowmeter.
[b] Reduce by about 3D for flanged in-line flowmeter.
[c] Two 90° bends are separated by: ≤ 10D (same plane); < 5D (perpendicular planes)
[d] Reducer: 2D to D over 1.5D to 3D, expander: 0.5D to D over D to 2D

Note: upstream distances should be doubled for light gases such as helium and hydrogen.

are given in Table 18.3. For a flow conditioner of the perforated plate type based on the Mitsubishi design, this manufacturer recommended 8D upstream spacing with wafer and insertion-type meters, and with larger in-line meters, but allowed positioning of the flow conditioner at the inlet flange of the in-line design for 25- to 100-mm diameters. The manufacturer reckoned that, for a control valve downstream, there should be 10D separation (8D to flange of in-line flowmeter). Gaskets, flanges and pipes should be correctly sized and aligned. The table compares the ITMF with a $\beta = 0.5$ orifice plate and shows that the requirements are about the same apart from bends in perpendicular planes and the reducer. ISO 5167 for the orifice plate does not include or recommend partially open or variable valves upstream of the meter.

The influence of temperature variation can be as small as 0.05%/K rate and of pressure 0.2%/bar rate.

18.5.6 Applications

This may be an alternative meter for installations where orifice or vortex meters are commonly used for gas mass metering. Examples are chemical engineering and processing technology, semiconductor process gas measurement and control, combustion air measurement for large utility plant, mass flow rate for stack effluent, air sampling in nuclear plant and environmental facilities (Kurz 1992), flare gas systems, leak detection, food industry (e.g. CO_2 metering in breweries), internal works gas consumption measurement, dosing plants, nitrogen and oxygen flows in steelworks, burner control, combustion and preheated air, flue gas, methane, moist and corrosive gas from sludge digester, compressed air in hospitals and other medical applications, argon and helium flows in laboratory environments, waste air measurement and large ducts. These meters are unlikely to be suitable for steam, but the manufacturer should be consulted.

Haga, Sotono and Hanai (1995) used a thermal flowmeter for fuel vapour flow measurement and used electrical heaters to ensure that the piping was of a temperature to prevent liquefaction of the vapour. A liquid recovery container and a filter were provided at the inlet of the flowmeter. To prevent the remaining vapour in the piping from being liquefied, air purge was used. They gave expressions for specific heat and density as a function of pressure, temperature and vapour pressure. Haga et al. claimed that it had been possible to measure the vapour flow in real time for a moving vehicle and that it offered a practical meter.

18.6 Chapter Conclusions

This technology, which is of more modest price than the Coriolis, is likely to be exploited because it also (with ultrasonics) offers a means of mass flow measurement and possibly greater versatility for gases. Development could take place in various directions:

- improved theoretical understanding and design of the CTMF to minimise pressure, temperature and installation effects and to improve accuracy;
- further analysis of the ITMF designs to understand accuracy limitations, environmental and flow profile effects (it should be remembered that the accuracy of this instrument depends in part on the measurement of a small temperature difference within a gas of variable temperature);
- improved theoretical predictions of a meter which is insensitive to flow profile (Hemp 1994a/1995a);
- use of multiple probes;
- use of nonintrusive heating and temperature measurement; and
- self-checking with transit time to downstream probes.

Casperson (1993) described a device that used a thermocouple in a pulsed mode. Current pulses heat and cool the junctions, and the subsequent decay is monitored. With flow, the time constant will change. Being nonlinear, the device needed calibration.

An important development is microthermal sensors, which allow fabrication of air-flow sensors as well as signal-conditioning electronics on a single chip (van Dijk and Huijsing 1995) with a standardised digital output signal for a microprocessor. The one described is direction-sensitive.

Nguyen and Kiehnscherf (1995) described a flow sensor in which the channel with an area of 0.6 mm^2 and length 10 mm was etched, and the sensor contained polysilicon heaters and measurement resistances on the chip. The chip is bonded to Pyrex glass to complete the channel. Four modes of operation were built into the electronics to allow constant temperature and constant power operation. In this way, they tested with both liquid (0–500 ml/min or 0–10 ml/min) and gas (0–500 ml/min) (Figure 24.2).

Many papers have reported work on very small devices. For example:

- four different strategies for thermal flow sensing are: heat loss, heat addition, convection and time-of-flight (Ashauer et al. 1999). They presented numerical work on the propagation of heat pulses in the flow, and described the design and construction of a micrometer which combined two strategies: the thermotransfer (which appears to have a central heater with equally spaced temperature sensors on the up- and downstream sides as in a CTMF), and the time-of-flight, resulting in a possible flow range from 0.1 mm/s to 140 mm/s, where a range overlap may allow calibration checking;
- a CMOS microsensor thermal mass flowmeter in which the temperature sensors were in a sealed membrane with dielectric outer layers (Matter et al. 2003);
- a micro-machined hot film sensor for automotive applications of mass air flow (Strohrmann et al. 2004);
- a high-performance CMOS wall shear stress sensor of very small size with the sensor of 130μm x 130μm located in the centre of a silicon oxide membrane of 500μm by 500μm giving effective thermal isolation (Haneef et al. 2007): devices of this sort should lend themselves to incorporation in flowmeters;
- the micro-machining of a mass flow sensor which appeared to consist of a central thin-film heater and a pair of thin-film temperature sensing elements: the circuitry allowed two flow ranges (Wang et al. 2009);
- recent advances in MEMS mass flowmeter technology for natural gas may require improved calibration and verification (Deng et al. 2010): for a description of the meter design (turn-down ratio may be over 200:1) and a discussion of cost effectiveness and battery life see Huang et al. (2010) (cf. Wang et al. 2009);
- a thermal time of flight gas flowmeter, using MEMS thermal sensors which may be able to operate despite moisture/particles (Yao et al. 2010).

One further general comment: if calibration makes use of a gas which is not the working gas, the user should check very carefully with the manufacturer for possible problems. It may also be worth checking whether the ISO gives any guidance.

Appendix 18.A Mathematical Background to the Thermal Mass Flowmeters

18.A.1 Dimensional Analysis Applied to Heat Transfer

The heat flux density (Kay and Nedderman 1974) is given by

$$q_{\mathrm{h}} = f\left(\Delta T, k, c_{\mathrm{p}}, \rho, \mu, D, V\right) \tag{18.A.1}$$

where ΔT is the temperature difference between the heat source and the fluid, k is the thermal conductivity of the fluid, c_{p} is the specific heat at constant pressure, ρ is the fluid density, μ is the dynamic viscosity of the fluid, D is the pipe diameter, and V is the fluid velocity.

One simple form of Equation (18.A.1) for CTMFs is

$$Q_h = K\Delta T c_p q_m \tag{18.A.2}$$

where Q_h is the total heat input into the winding, and a knowledge of K, ΔT, and c_p allow q_m to be deduced.

We can identify dimensionless groups, which must have a functional relationship with each other. For this purpose, a number of heat transfer groups have been suggested

$$\text{Nusselt number}(\text{Nu}) = \frac{q_h d}{k\Delta T} \tag{18.A.3}$$

is the ratio of heat transfer to the rate at which heat would be conducted through the fluid under a temperature gradient $\Delta T/d$;

$$\text{Prandtl number}(\text{Pr}) = \frac{\mu c_p}{k} \tag{18.A.4}$$

is the ratio of kinematic viscosity to thermal diffusivity;

$$\text{Reynolds number}(\text{Re}) = \frac{\rho V d}{\mu} \tag{18.A.5}$$

is the ratio of inertia forces to viscous forces in the flow.

These give a general relationship for heat transfer:

$$\text{Nu} = f(\text{Pr}, \text{Re}) \tag{18.A.6}$$

A special case of this equation is that due to King for the hot-wire anemometer:

$$\text{Nu} = A + B\,\text{Pr}^{0.5}\,\text{Re}^{0.5} \tag{18.A.7}$$

where $A = 1/\pi$ and $B = (2/\pi)^{0.5}$. The hot-wire or hot-film anemometer provides a tool for the measurement of local fluid velocity.

18.A.2 Basic Theory of ITMFs

King (1914) investigated the relationship between heat transfer rate and flow velocity. Using Equation (18.A.7), the following expression gives heat loss per unit length of a cylinder of diameter d at a temperature of ΔT above the fluid temperature in which it is immersed:

$$q_h = k\Delta T + \left(2\pi k c_p \rho d V\right)^{0.5} \Delta T \tag{18.A.8}$$

where k is the thermal conductivity of the fluid, c_p is the specific heat of the fluid at constant pressure, ρ is the density, and V is the velocity assumed to be perpendicular to the cylinder (cf. Ower and Pankhurst 1966, who used c_v in place of c_p and also provided details and early references).

As the velocity increases, the second term becomes more important, and the first term, which is due to conduction and natural convection, becomes more important at very low flow rates. It is clear from this equation that it is necessary to measure q_h and ΔT and hence to obtain V assuming that the constants of the wire and the fluid are known. However, the relationship between q_h and ΔT is more usually obtained by calibration.

This, in turn, suggests that any device designed to measure velocity from this formula will need a means of measuring the temperature of the stream as well as the heat loss from a heated wire. This may be achieved by calibrating the probe for the proposed application.

The other important observation is that, when the density is not constant, the combination of velocity and density with the area of the meter will give mass flow rate.

King showed that in air the formula held down to a value of

$$Vd = 0.0187 \tag{18.A.9}$$

where V is in centimetres per second, and d is in centimetres. At very low flow rates, the natural convection in the region of the wire becomes significant in comparison with the forced flow. Collis and Williams (1959) gave a Reynolds number, below which convection becomes important, as

$$\text{Re} = \left(\frac{gd^3\Delta T}{T\nu^2}\right)^{1/2} \tag{18.A.10}$$

where ν is the kinematic viscosity, T is the ambient temperature of the air, and ΔT is the difference between the temperature of the wire and the ambient air.

Simplifying Equation (18.A.8) and introducing the heating current I through a resistance R, we can write

$$q_h = I^2R = C + DV^{0.5} \tag{18.A.11}$$

where C and D are functions of temperature. But if the resistance of the wire or cylinder is kept constant, then the temperature of the wire or cylinder will also be constant and so C and D will be constants of the meter for a particular fluid.

To maintain the constancy of the resistance, a bridge circuit such as that shown in Figure 18.A.1 is used where the null indicator ensures that the resistances are all equal, and the current, and hence the power, is measured by means of the resistor in series with the bridge. The advantages of using a constant resistance compared with a constant current are that it is possible to maintain a constant temperature and avoid thermal inertia, and that the equation is simplified as in Equation (18.A.11).

Using Equation (18.A.8) and allowing for temperature variation, we can rewrite it as

$$q_h = k\left(1 + Kq_m^{0.5}\right)\Delta T \tag{18.A.12}$$

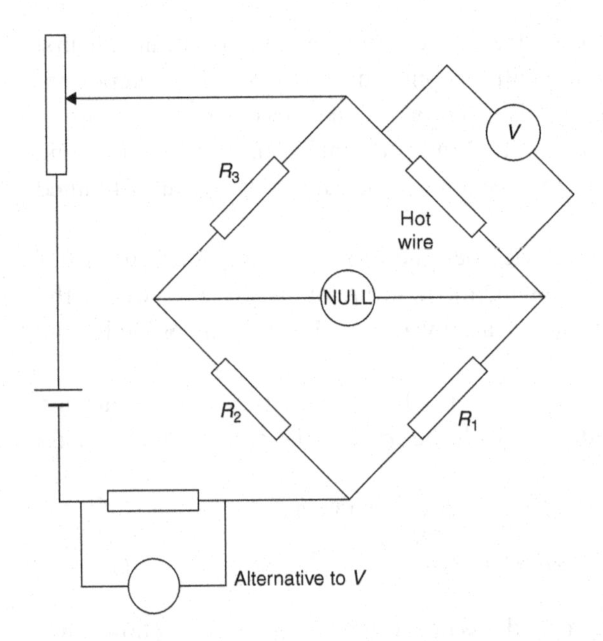

Figure 18.A.1. Constant temperature resistance circuit (after Ower and Pankhurst 1966).

where K is a constant incorporating the area of the duct into which the probe is inserted and the gas and heat transfer constants, q_m is mass flow rate, q_h is heat supplied, and ΔT is temperature difference between heated and unheated probes.

This provides the basic relationship for ITMFs with either q_h or ΔT constant.

18.A.3 General Vector Equation

The governing equation for flow with heat transfer under steady conditions is

$$\mathbf{V}\cdot\nabla T = \frac{k}{\rho c_p}\nabla^2 T + \frac{q_a}{\rho c_p} \tag{18.A.13}$$

where q_a is the added heat flux. This can be obtained from Hemp (1994a, Equations 2 and 5)

$$\mathbf{q}_h = -k\nabla T \tag{18.A.14}$$

and

$$\nabla\cdot\mathbf{q}_h = -\rho c_p \mathbf{V}\cdot\nabla T \tag{18.A.15}$$

with the addition of the final term, which relates to the local heat addition. If all heat is added across boundaries with the flow, then the last term disappears. Hemp uses Equations (18.A.14) and (18.A.15) to develop his weight function theory for thermal diffusion flowmeters. This is an elegant theory, and the interested reader is encouraged to refer to Hemp's papers. However, as he comments, the work may still be mainly of theoretical interest.

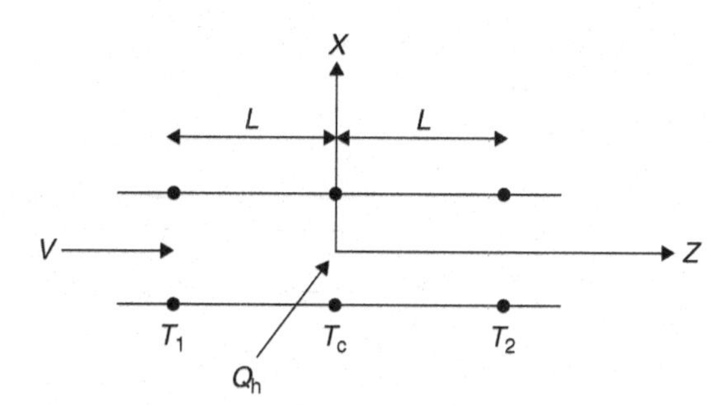

Figure 18.A.2. Diagram of tube to illustrate the use of finite differences.

We can write Equation (18.A.13) for flow in a two-dimensional channel as

$$V\frac{\partial T}{\partial z}=\frac{k}{\rho c_p}\left(\frac{\partial^2 T}{\partial x^2}+\frac{\partial^2 T}{\partial z^2}\right)+\frac{q_a}{\rho c_p} \tag{18.A.16}$$

We can simplify this for flow in a tube where we neglect temperature gradients in any but the axial direction. Referring to Figure 18.A.2 and rewriting the gradient and the Laplacian as finite difference expressions, we obtain

$$V\frac{(T_2-T_1)}{2L}=\frac{k}{\rho c_p}\frac{(T_1-2T_c+T_2)}{L^2}+\frac{q_a}{\rho c_p} \tag{18.A.17}$$

If we take the area of the duct to be A and the heat source to be at the centre, which has a temperature T_c, we can rewrite the equation in terms of mass flow:

$$\underbrace{q_m(T_2-T_1)}_{\text{convection term}} = \underbrace{\frac{2kA}{c_p}\frac{(T_1-2T_c+T_2)}{L}}_{\text{conduction term}} + \underbrace{\frac{2LA}{c_p}q_a}_{\text{heat input term}} \tag{18.A.18}$$

The first and last terms constitute the simplest approximation for the behaviour of the thermal gas flowmeter. Heat is convected away by the flow, and the difference of temperature between inlet and outlet is related to the heat input, the specific heat and the mass flow. This assumes that the first term on the right-hand side can be neglected. However, in the case of no flow, the heat added is conducted away, and the central term cannot be neglected because it gives the size of the conduction.

18.A.4 Hastings Flowmeter Theory

Figure 18.A.3 shows the heat flows in the Hastings flowmeter based on an approximate one-dimensional theory. From the heat flows in the diagrams, we can derive the simple equation of the meter. We consider first the wall, and specifically an element of length δz. Here we have a heat addition as a result of the electric current I, which flows down the tube wall of resistance R per unit length. Thus the heat input is $I^2R\delta z$.

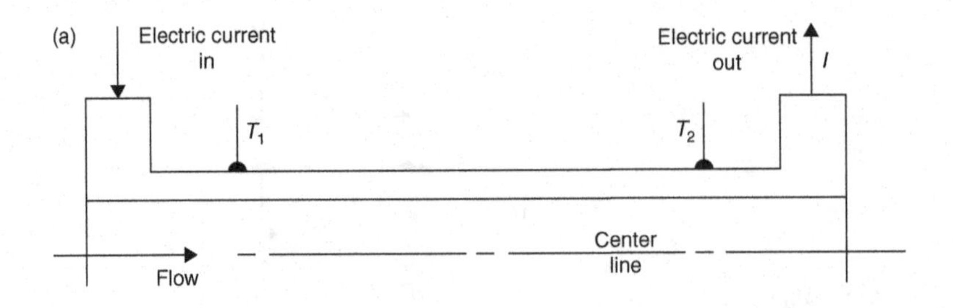

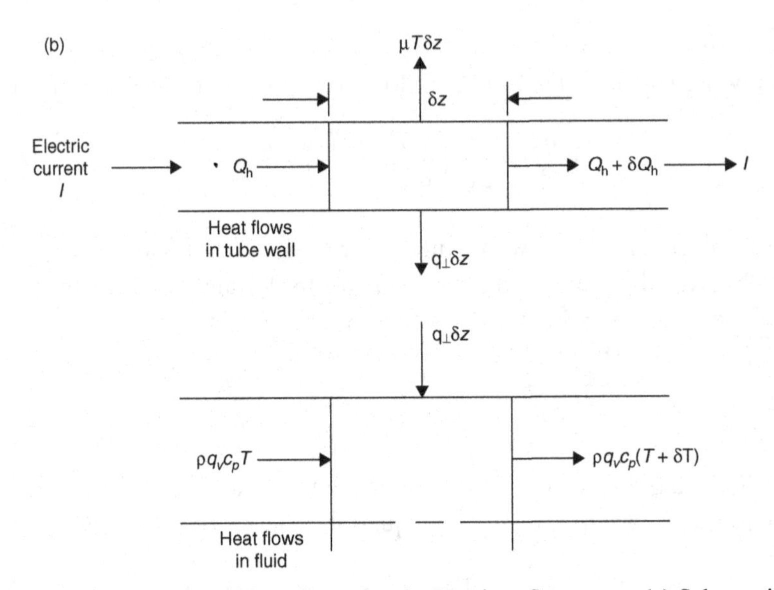

Figure 18.A.3. Geometry and heat flows for the Hastings flowmeter: **(a)** Schematic diagram; **(b)** heat flows within the meter.

From this element, heat flows along the wall, of amount δQ_h, which can be related to the temperature gradient in the wall

$$Q_h = -kA\frac{dT}{dz} \tag{18.A.19}$$

where A is the cross-section of the wall, and k is the thermal conductivity of the wall material. Heat also flows out as heat loss from the meter, μT, μ being a dimensional constant to give the heat loss based on the wall temperature, and it flows into the fluid, $q_\perp$. From this we have an equation

$$I^2 R\delta z = \mu T\,\delta z + q_\perp\,\delta z + \delta Q_h \tag{18.A.20}$$

$q_\perp \delta z$ then flows into the fluid, which is assumed to be at approximately the same temperature as the wall at each point along its length. This heat causes a rise in the temperature of the fluid. The rate of heat addition $q_\perp \delta z$ is equal to the product of the temperature rise of the fluid, the specific heat of the fluid at constant pressure and the mass flow rate. Thus we can write

$$q_{\perp}\delta z = \rho\, q_{v} c_{p} \delta T$$

or

$$q_{\perp} = \rho\, q_{v} c_{p} \frac{dT}{dz} \tag{18.A.21}$$

Finally, we can combine Equations (18.A.20) and (18.A.21) and use Equation (18.A.19) to eliminate Q_h

$$\frac{-dQ_h}{dz} - \rho q_v c_p \frac{dT}{dz} - \mu T + I^2 R = 0$$

or

$$kA\frac{d^2T}{dz^2} - \rho q_v c_p \frac{dT}{dz} - \mu T + I^2 R = 0 \tag{18.A.22}$$

This equation was quoted by Hemp (1995a), and he also quoted a solution due to Blackett and Henry (1930), which for low flow rates can be written in a linear form

$$T_2 - T_1 = Kq_v \tag{18.A.23}$$

Hemp suggested that the nonlinear complete solution was probably invalid outside the linear range due to the assumptions about uniform temperature across any cross-section of the flowmeter. For other solutions and more details, the reader is referred to Hemp (1994a, 1995a), Blackett and Henry (1930), Komiya, Higuchi and Ohtani (1988), Brown and Kronberger (1947) and Widmer, Fehlmann and Rehwald (1982).

18.A.5 Weight Vector Theory for Thermal Flowmeters

Hemp (1994a, 1995a) presented an extension of the weight vector theory to thermal flowmeters. Again the ideal weight vector must satisfy

$$\nabla \times \mathbf{W} = 0 \tag{18.A.24}$$

For a typical thermal diffusion flowmeter, the signal

$$T_B - T_C = \int \mathbf{V}\cdot\mathbf{W}dv \tag{18.A.25}$$

and the weight vector is given by

$$\mathbf{W} = \rho\, c_p T^{CB} \nabla T^{A} \tag{18.A.26}$$

where T^A is the temperature distribution resulting from unit heat flux injected into A (Figure 18.A.4), and T^{CB} is the temperature distribution when unit heat flux enters at C and leaves at B. It appears that the usefulness of Hemp's theory is limited at

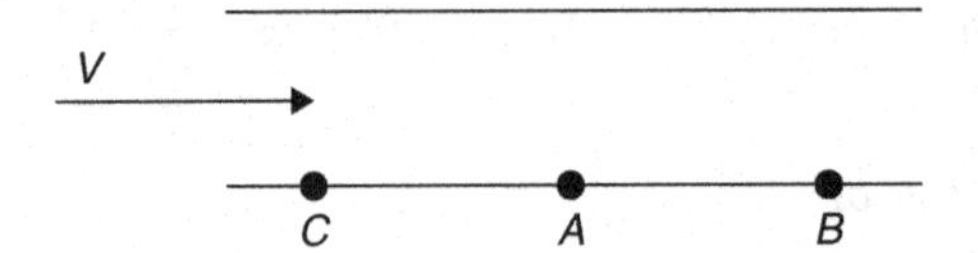

Figure 18.A.4. Diagram for Hemp (1994a) theory.

present by the nature of the cases for which it is valid and the flow rates that are low enough to ensure linear conditions.

18.A.6 Other Recently Published Work

Recent research reported in the literature and known to this author includes:

- a flowmeter which could be fixed to the outside of the pipe (Zhu and Min 1999), development of a thermal flow sensor (Kaltsas et al. 2007), multiple thermal sensors (Neda, Saito and Nukui 1997; Zhao and Cao 2008);
- meter models and analysis: novel design (Sazhin 2013), steady state and transient (Kim, Han and Kim 2003, 2007b; see also Han, Kim and Kim 2005), steady-periodic heating of a surface mounted film (Cole 2006);
- effects of line pressure (Olivier 1997);
- sensing method (Bera and Kumar Ray 2001);
- cross-correlation of heat pulses(Moriyama, Sukemura and Morishita 2001);
- manufacturing variation (Baker and Gimson 2001);
- mass flow measurement of gas-solids two-phase flow (Zheng et al. 2008b);
- bypass meter (Viswanathan et al. 2001; Viswanathan, Rajesh and Kandaswamy 2002), bypass for respiratory control (Kaltsas and Nassiopoulou 2004), capillary meter (CTMF) with bypass(Gralenski 2004), multiple bypass (Baker and Higham 1992).

19

Angular Momentum Devices

19.1 Introduction

I am indebted to Medlock (1989) for notes on three early devices that attempt to use change of angular momentum to obtain mass flow.

Katys (1964) used an electric synchronous motor with a special rotor supported internally in the pipe and a stator outside the pipe. Fluid passed through vanes in the rotor and acquired angular momentum. From the power used or the torque needed to drive the rotor, it was claimed that the mass flow could be deduced.

According to Medlock, the Bendix meter "measures the torque required to impart angular momentum to the liquid. One end of a calibrated spring is driven at a constant speed and the other end is connected to a freely rotatable turbine. The torsion developed in the spring is a measure of mass flow rate."

The twin rotor turbine meter (Potter 1959) attempted to measure mass flow by means of two in-line turbine rotors of different blade angles, which are joined by a torsion spring. Despite discussions with Medlock about this meter, I am not convinced that this is a true mass flow meter and suggest that the valid derivative of the Bendix meter is the fuel flow transmitter described later.

An early design was a device due to Orlando and Jennings (1954) shown in Figure 19.1. A constant speed motor drives an impeller imparting swirl to the liquid.

The angular momentum, which is removed from this liquid per second as it leaves the driven rotor and moves into the tethered rotor, is

$$X = \omega R^2 q_m \tag{19.1}$$

where ω is the angular velocity of the rotor, and R is the radius of the annulus in which the flow takes place. The force on the spring restraining the tethered rotor will be equal to the loss in angular momentum per second and will be indicated by the angular deflection of the tethered rotor. The mass flow will therefore be given by

$$q_m = s\theta / (\omega R^2) \tag{19.2}$$

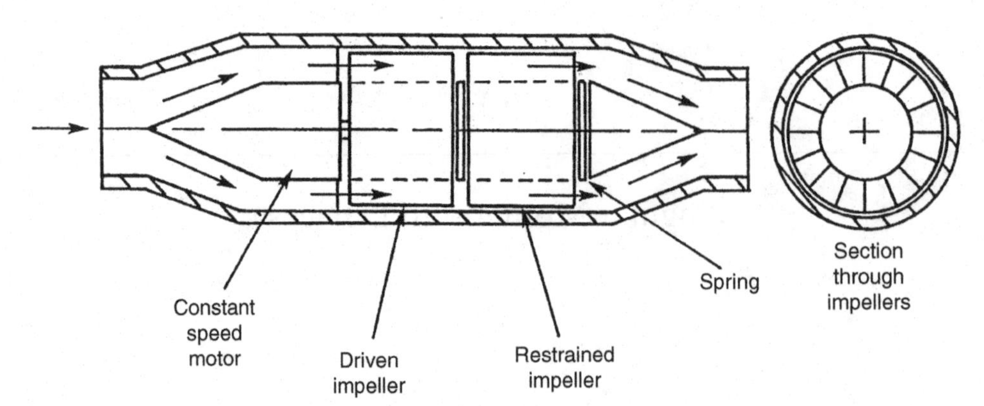

Figure 19.1. Orlando and Jennings' device.

Where θ is the angular deflection, and s is the spring constant. It is important to note the following.

i. The flow will not precisely follow the vanes unless sufficient length is allowed to force the flow to be axial relative to the blades.
ii. The flow profile in the annulus will modify the effective value of R.
iii. The motor speed must be known.

19.2 The Fuel Flow Transmitter

Although instruments of this type may not now be available commercially, I have retained this section, essentially unchanged, as I consider that it is an important example of how a rotating "turbine-type" meter can measure mass flow rate.

Much of the material in this section comes from a lecture by Rowland (1989). The main features of a particular commercial device are shown in Figure 19.2 and in exploded view in Figure 19.3. This design eliminates the electrical drive by extracting power from the liquid flow to drive the turbine and provides one example of commercial meters of this type.

19.2.1 Qualitative Description of Operation

As the flow enters the unit, a bypass assembly controls the amount of liquid passing through the power turbine in order to avoid the assembly's running too fast or too slow. The bypass mechanism consists of a spring-loaded bypass valve that contains the speed of the unit to within the range 70–400 rpm. At the lowest flow rates, all the liquid passes through the turbine, but as the flow rate increases, the build-up of pressure across the turbine progressively opens the valve and allows an increasing portion of the total liquid flow to bypass the turbine.

Depending on the opening of the bypass valve, a proportion of the flow passes through a stator with guide vanes that create swirl in the flow. The flow then enters

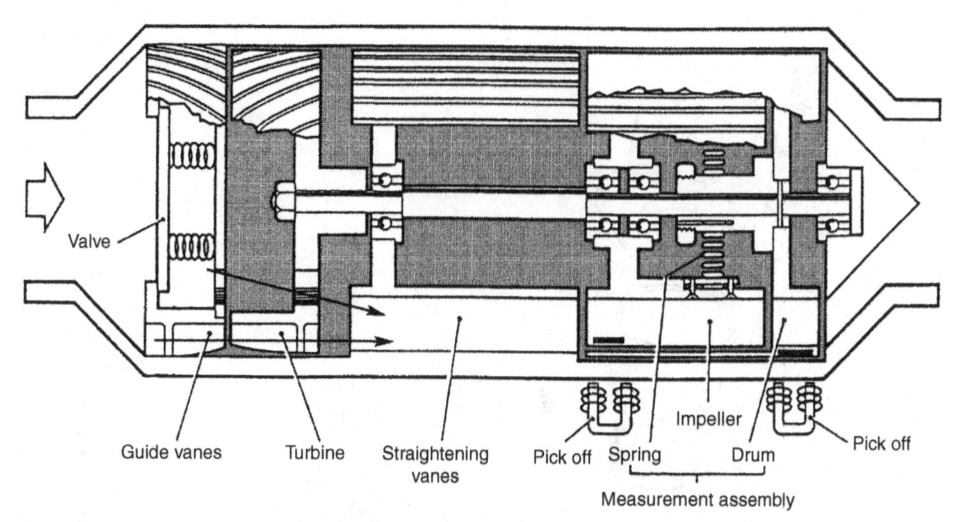

Figure 19.2. Diagram of the fuel flow transmitter (reproduced with permission of GEC-Marconi Avionics).

the power turbine, which drives a shaft with the measurement assembly on the other end. However, having extracted power from the flow in this way, it is essential that the remaining swirl after the power turbine is completely removed from the liquid.

After the turbine and bypass valve, therefore, the flows combine to pass through the straightening vanes, before passing into the driven measurement assembly. The meter proper then follows and in this particular design, unlike the example described earlier, measures the torque needed to impart the angular momentum to the liquid. Within the rotating measurement assembly, a tethered impeller imparts to the liquid the angular momentum. The ratio of blade length to mean space between the impeller vanes is 8:1. Clearances are kept to a minimum to reduce leakage, and the meter is insensitive to installation attitude.

In the process, the impeller experiences a torque that causes the restraining spring made of NiSpan-C902, a material whose change of stiffness with temperature can be adjusted by heat treatment, to tighten. To avoid overtightening the spring, the bypass control of speed also reduces the necessary deflection of the spring. This can be seen from the fact that, without the bypass, the angular velocity would be approximately proportional to the mass flow rate and so the torque would be proportional to the square of the angular velocity. By allowing some of the fluid to bypass the drive turbine, the angular velocity is reduced, and the speed need not rise in proportion to the mass flow rate. The relationship between flow rate and rotational speed is shown in Figure 19.4.

The bearings are precision ball races. Friction errors and vibration errors can result from spurious torques at the impeller bearings. The fuel acts as a lubricant and must therefore be clean (filtered).

The deflection is measured by means of a pair of electromagnetic pickup coils placed on the outside of the nonmagnetic meter body. A pair of small powerful

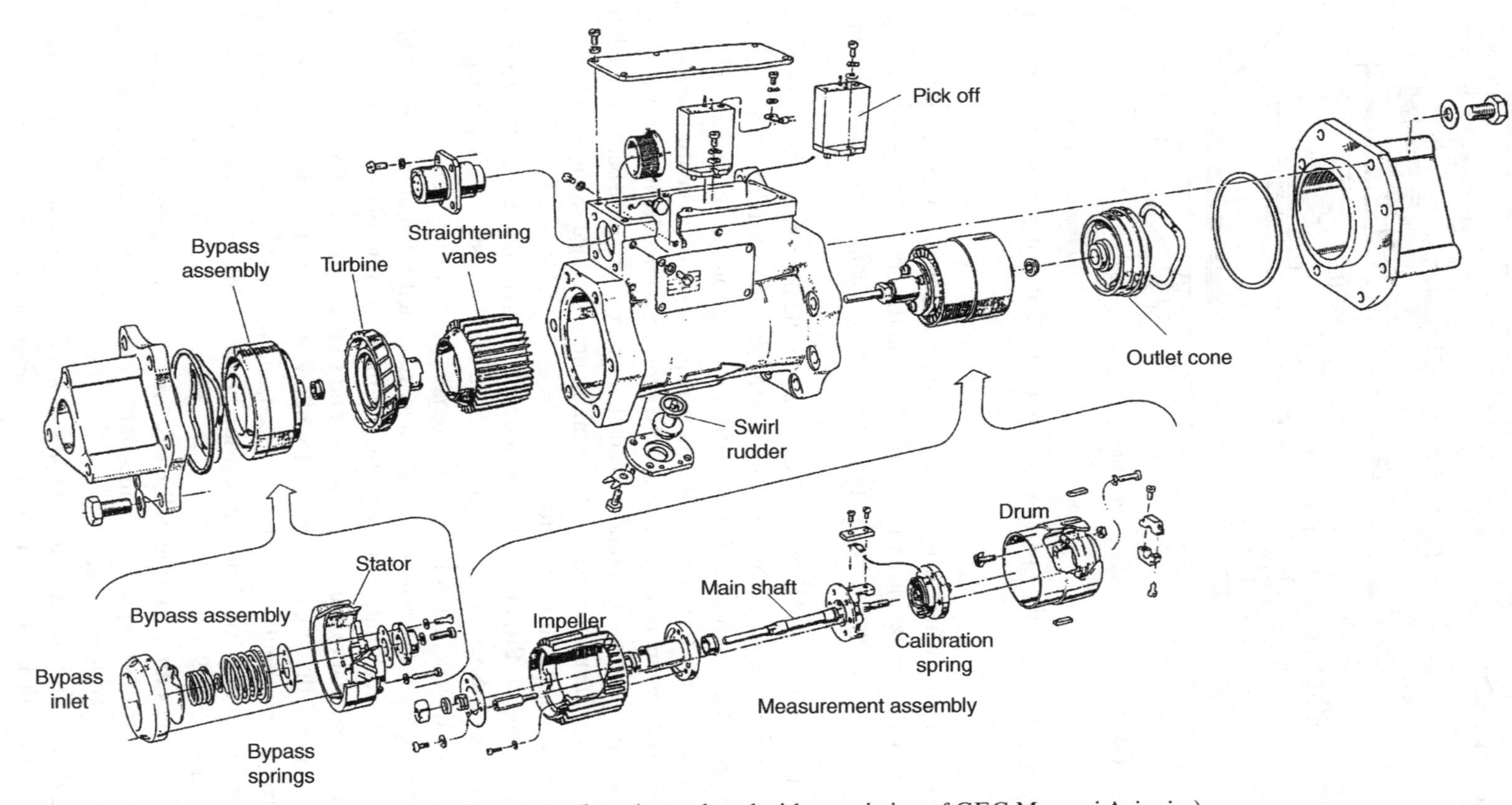

Figure 19.3. Exploded diagram of the fuel flow transmitter (reproduced with permission of GEC-Marconi Avionics).

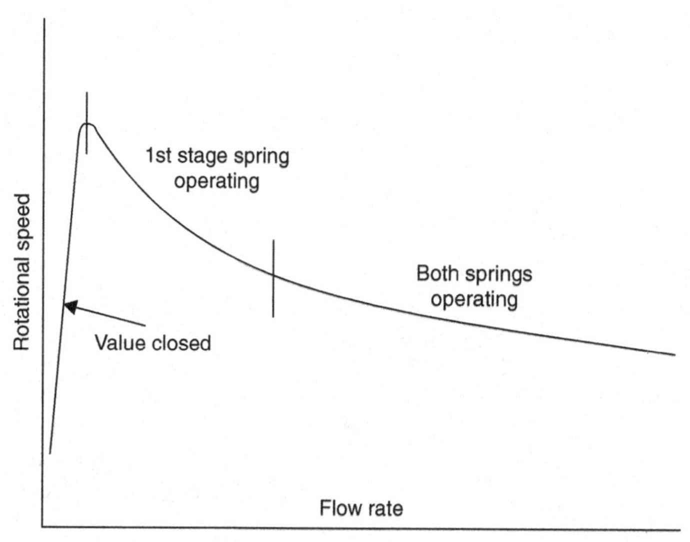

Figure 19.4. Relation between flow rate and rotational speed for the fuel flow transmitter (reproduced with permission of GEC-Marconi Avionics).

magnets displaced 180° apart is attached to the circumference of both drum and impeller. Signals are therefore generated for each 180° of rotation of the assembly.

19.2.2 Simple Theory

We can obtain an expression for the mass flow rate. We write the spring constant as s. The torque causes the impeller to be displaced by an angle θ from the rest of the measurement assembly. The displacement in the range 0–160° is measured by recording the time of transit of a marker on the assembly drum and a marker on the impeller. A knowledge of the time delay τ and the angular velocity of the assembly ω will then allow the angular displacement to be deduced. Maximum calibration spring hysteresis is approximately 0.05% of FSD. Thus

$$q_m = s\theta/(\omega R^2)$$

where the effective radius of the flow annulus is taken as R. This is affected by the velocity profile in the impeller and the shape of the channel between the vanes. Hub-to-tip ratio is generally greater than 0.7.

The time difference between the markers is

$$\tau = \theta/\omega \tag{19.3}$$

so that

$$q_m = s\tau/R^2 \tag{19.4}$$

and we, thus, note that the mass flow may be obtained as directly proportional to τ, provided s is constant, and our simple description of operation is adequate.

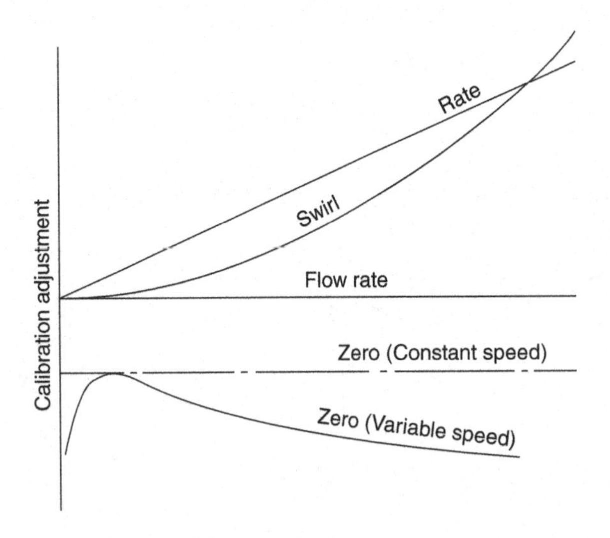

Figure 19.5. Calibration adjustments (reproduced with permission of GEC-Marconi Avionics).

19.2.3 Calibration Adjustment

The meter has three methods of calibration adjustment (Figure 19.5):

1. Zero adjustment achieved by moving the impeller pickup circumferentially out of alignment with the drum coil so as to introduce an apparent zero flow adjustment.
2. Rate adjustment achieved when the meter is partially disassembled by altering the clamping point of the spring and hence its effective length. This will alter the value of s in the preceding equations.
3. Swirl adjustment by means of a small rudder in the flow upstream of the measurement assembly creates a small local flow deflection and affects the performance throughout the characteristic of the meter.

19.2.4 Meter Performance and Range

The meter is adjusted to rotate at 150 rpm at maximum flow and between 150 and 300 rpm at other flow rates. Figure 19.4 indicates an increase in rotational speed at lower flows resulting from the action of the bypass valve to maintain spring wind-up. The assembly is fitted with a stop to prevent, at low flow rates, a reverse oscillation causing spurious readings. It has a flow range of about 115–4,500 kg/h. The temperature range is −40 to 150°C, and the maximum line pressure is about 130 bar with a maximum pressure drop of about 0.5 bar.

Accuracy appears to be claimed as ±1% of rate over a 7:1 flow range.

It is desirable to install the meter with a straight upstream length of pipework to avoid swirl and profile distortion effects. Because this is seldom possible on an aircraft, the calibration should be done with a representative piece of inlet pipework.

Referring to Equation (19.4), we can review the likely sources of uncertainty in this meter that will influence the claims for its accuracy. It is unlikely that the impeller completely imparts the assumed angular velocity to all the fluid. In addition, there will be some leakage and drag due to the gap between rotor, tube and vanes, although the shroud may reduce some of these. These may be approximately proportional to the mass flow rate, so error may be of second order. The spring may introduce nonlinearity or hysteresis and may give problems if not correctly installed. The time difference may be of order 100 ms, and so measurement will be to within about 100 μs. There are also the characteristic changes due to the means of zero adjustment. These may be incorporated into a calibration characteristic if a flow computer is available to adjust the flow signal accordingly.

The material of the body is high-strength aluminium alloy.

19.2.5 Application

The meter appears to have primarily one application to date. This is for the measurement of fuel flow in aircraft.

19.3 Chapter Conclusions

The fuel flow transmitter is an elegant solution but probably suffers, in an age when mechanical precision is being displaced by solid state, from being a mechanical rotating device. If the mechanical integrity can be assured and operating life is long enough, one would expect that sensing of the rotation, torque and other factors with intelligent secondary instrumentation would make it highly accurate.

It is surprising that it has not been exploited for more applications, and even for gas flow measurement. Presumably the main reason is the preference for a meter without moving parts.

The logical next step to overcome this would be to move to an oscillating or vibrating device to achieve the same result. Such devices have been suggested but may offer little advantage over the Coriolis, which we shall look at next. Any developments in this device will also be in competition with the rapidly developing ultrasonic methods that clearly meet many of today's requirements in terms of no moving parts and predominantly electrical sensors.

The future development of a device such as this, therefore, is unlikely to be extensive.

20

Coriolis Flowmeters

20.1 Introduction

This is one of the most important meter developments, and its application is impinging on many areas of industry. Its importance has been recognised by the national and international advisory bodies (see Mandrup-Jensen 1990 for an early description of work in Denmark on pattern approval and ISO developments).

20.1.1 Background

Plache (1977) makes the interesting point that mass cannot be measured without applying a force on the system and then measuring the resulting acceleration. This is a point which the present author has long considered a possible requirement and it is certainly supported in the Coriolis meter.

Possibly the first application of the Coriolis effect for mass flow measurement was proposed by Li and Lee (1953) [but see Wang and Baker (2014) for other possible contenders for the original concept]. The meter is shown in Figure 20.1. The T-piece flow tube rotated with the outer casing and was linked to it by a torque tube. As the flow increased, so the T-piece experienced a displacing torque due to the Coriolis acceleration and this was measured from the displacement of the T-piece relative to the main body.

In the Li and Lee meter the liquid was forced to move radially and therefore a force was applied to it through the tube. This force in turn was balanced by an equal and opposite one applied by the liquid to the tube. The force caused the tube to twist and the small relative rotation was sensed to obtain the mass flow rate.

It is interesting to note that Stoll (1978) made no mention of the technology which was to revolutionise mass flow measurement. Medlock (1989) referred to several designs which, although not identified as dependent on the Coriolis effect, have a very similar basis. Macdonald (1983) described a vibrating vane flowmeter. Vsesojuzny (1976) described a vibrating nozzle flowmeter (Figure 20.2) [cf. Hemp's (1994b) work on vibrating elements]. Finnof et al. (1976) described a vibrating tube flowmeter (Figure 20.3).

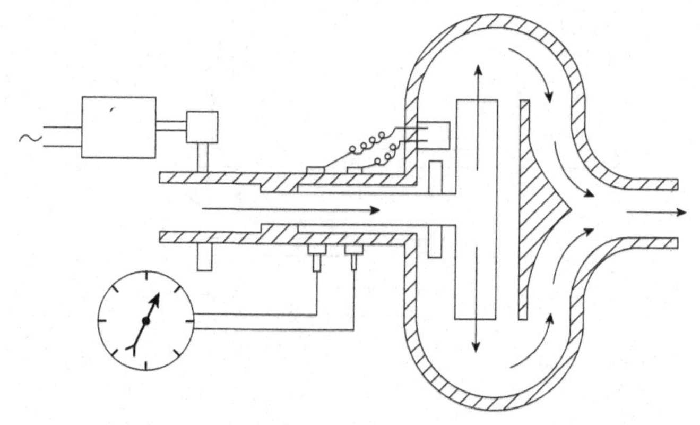

Figure 20.1. Li and Lee (1953) fast response true mass-rate flowmeter (reproduced with the permission of ASME).

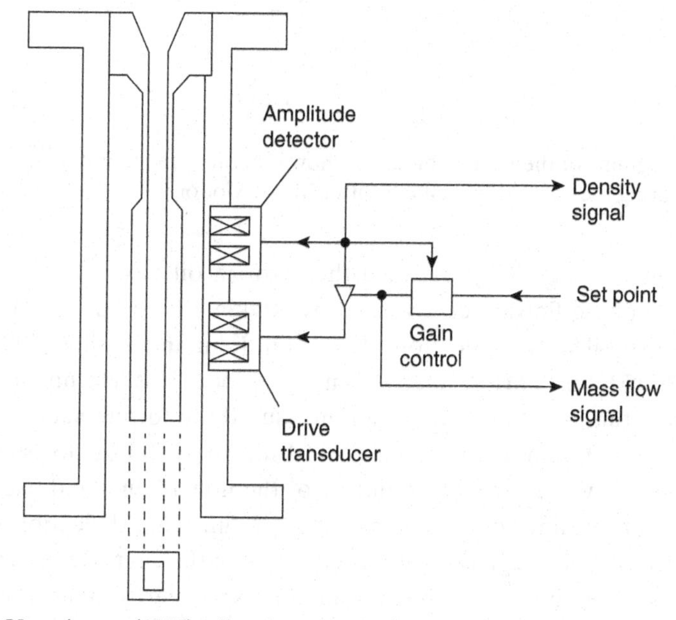

Figure 20.2. Vsesojuzny (1976) vibrating nozzle (reproduced with the permission of the Controller of HMSO).

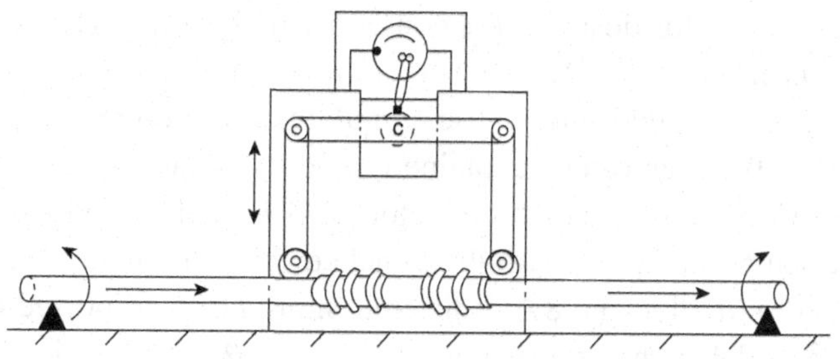

Figure 20.3. Finnof et al. (1976) vibrating tube (reproduced with the permission of the Controller of HMSO).

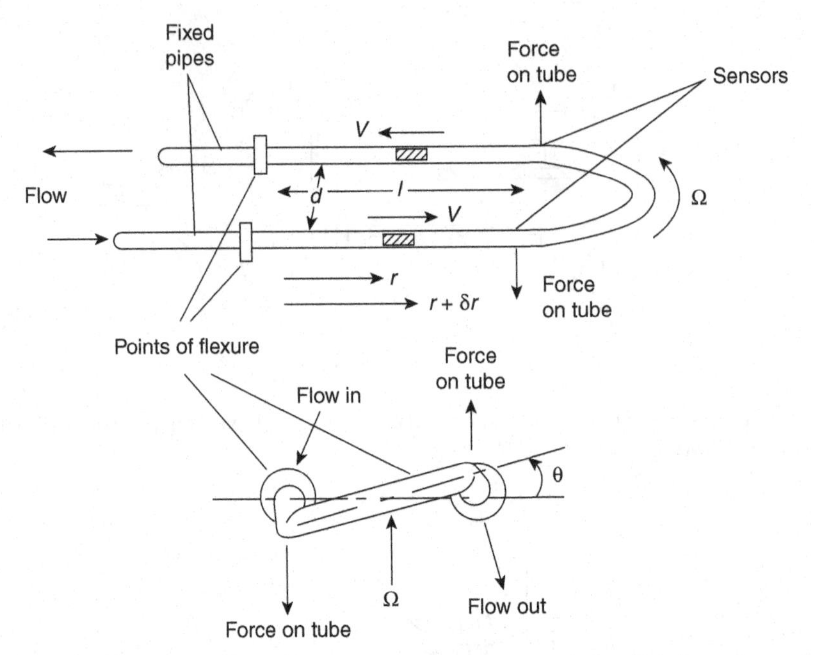

Figure 20.4. Diagrams of the U-tube meter to show motion of the vibrating tube and consequent forces during the upward movement (after Micro Motion).

Tucker and Hayes (1982) analysed the error in oil flow rate for a vibrating pendulum two-phase flowmeter invented by Rivkin (1978) and its suitability for North American oil well applications. Reimann, John and Muller (1982) tested a True Mass FlowMeter (TMFM) on a steam-water loop. This instrument created an angular momentum in the fluid and then measured the torque needed to remove it. John et al. (1982) built and tested a TMFM for up to 50 kg/s flows in air/water. They found that it was nearly independent of the flow regime and had an overall measurement uncertainty of less than ±1.5% of full scale deflection up to 20% mass fraction of air, although they observed a shift in the characteristic of 2.5% for mass fractions of air above 1%. This was an impressive performance. However the instrument is unlikely to be attractive for most applications because of the requirement to drive a rotor and the consequent design and maintenance problems.

Dimaczek et al. (1994) suggested the use of a radial turbine–type wheel to create a Coriolis effect meter for dosing flows consisting of gas/solid. They claimed that a measurement uncertainty of ±1% rate was achieved, and that it was used for flows of coal dust, quartz sand, feldspar, plastic granulates and foodstuffs.

Decker's (1960) design can be explained in terms of a modern Coriolis meter.

It appears to have been the meter developed by Finnof et al. (1976; cf. Smith and Cage 1985), which was first successfully launched onto the market in 1976 (Plache 1977), according to Medlock (1989). Tullis and Smith (1979) reported early tests on meters of ⅛, ¼ and 1 inch (approximately 3, 6 and 25 mm) which were promising, and led to optimism for wide application of the meters to liquids, gases and two-phase flows.

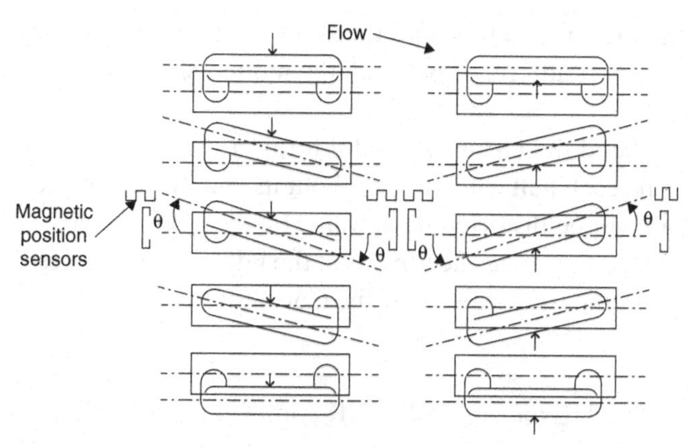

Figure 20.5. Variation of twist with vibration for U-tube meter (reproduced from Micro Motion's brochure with permission of Emerson Process Management Ltd.).

20.1.2 Qualitative Description of Operation

My explanation of the operation of these flowmeters is based on the U-tube design (Figure 20.4), rather than on the straight tube design. The theory is essentially the same for both, but I find it easier to explain for the U-tube flowmeter. We consider one of these vibrating U-tubes and we consider the period of vibration during which it rotates upwards. A simple theoretical treatment is given in Appendix 20.A.

Consider first a portion of the fluid moving outwards in one half of the U-tube. It will have a certain angular momentum because of the rotation of the tube upwards. As it moves outwards and the tube rotates upwards, the angular momentum will increase since the same portion of fluid will be at a greater radius from the centre of rotation and so will have to move at a greater velocity upwards. To achieve this it is necessary to apply a force on the portion of fluid in an upwards direction. As a result the portion of fluid will apply an equal downwards reactive force on the pipe. Now consider the other half of the U-tube and another portion of the fluid which is moving inwards. As this portion moves inwards its angular momentum will be reduced since the upwards velocity of the tube will be less. The tube will therefore have to apply a force downwards on the portion of fluid and this will, in turn, apply a force upwards on the tube. The result of all this is a torque twisting the tube so that the near side is forced down and the far side is forced up.

The principle of operation is based on the Coriolis acceleration, resulting from the flow of fluid through the flow tube. However, the motion is vibration rather than pure rotation. The forces are essentially the same, but are alternating. The vibration takes place in such a way that one end is essentially stationary and the other end is vibrating through an arc. This is illustrated in Figure 20.5.

As the tube rises the forces are at their maximum as it passes the mid-plane and at the top of the motion the forces become zero and the twist will cease. As the U-tube descends the forces reverse and the twist reverses so that again at the mid-plane the twist is at a maximum. It is then apparent that if the transit times of the two halves of the U-tube past the mid-plane are measured, the difference will be related to the twist in the tube and therefore to the mass flow through the tube. (There may be a small difference between the motion shown in Figure 20.5 and the actual motion.)

If we now unbend the U-tube so that it forms a straight tube with the vibrating drive at the centre, each half will "rotate" about its end. If we consider the centre rising, the inlet half will be distorted downwards by the Coriolis forces, and the outlet half will be distorted upwards. The sensors at the inlet and outlet positions will sense the lag, rather than the twist, caused by the Coriolis forces.

20.1.3 Experimental and Theoretical Investigations

Adiletta et al. (1993; cf. Cascetta et al. 1989a, 1989b) described a prototype which sought to overcome the limitations of commercial instruments' dependence on tube deformation by:

a) a system of rigid tubes;
b) a separate elastic suspension;
c) electromagnetic drive and capacitance transducers to infer the twist.

Their design appeared to be little affected by external vibrations.

Sultan (1992) described a single-tube meter of 28 mm OD built for laboratory experiments. It had a drive coil in the centre of the 1410 mm long straight pipe and two symmetrically positioned detector coils. For his rig, a change of water temperature of about 14°C caused a zero flow calibration shift equivalent to about 0.4 m/s (maximum flow rates tested were about 5.6 m/s). He accounts for this as due to differential thermal expansion and consequent differential damping characteristics. This will lead to frequency change. Tube temperature change will also cause a change in the modulus of elasticity and hence a frequency change. Sultan also showed that a pump in the flow circuit caused a zero fluctuation due to vibration equivalent to a flow of about ±0.1 m/s.

Kolahi, Gast and Rock (1994) described a prototype meter with similarities to the Micro Motion meter having two U-tubes. The design was very versatile in terms of component changes. Optical, capacitive and inductive sensors could be used. They recognised the difficulty of measuring gas flow using conventional Coriolis flowmeters because of the small Coriolis forces. The authors used a negative feedback method and they demonstrated a way to tune it so as to amplify the torsional amplitude greatly (see also Wang and Baker 2014).It was also possible to tune it so as to amplify the torsional amplitude by up to 100 times.

Hagenmeyer et al. (1994) reported a design of single-tube, compact Coriolis meter operating in a hoop mode, and claimed that experimental results were promising. (See Section 20.1.4.)

Important theoretical and experimental work has been reported by Clark & Cheesewright of Brunel University, UK. They undertook work on the response time of Coriolis meters and on pulsations. They identified two effects caused by flow pulsations that may excite additional tube motions (Cheesewright, Clark and Bisset 1999):

- internal vibrations of the tube by direct interaction;
- beat frequencies with the driven motion of the meter.

Cheesewright, Clark and Bisset (2000) also investigated the effects of external factors on the accuracy of Coriolis meter calibration:

- flow pulsations at Coriolis frequency
- flow pulsations at drive frequency
- mechanical vibrations at Coriolis frequency
- mechanical vibrations at drive frequency
- disturbances at other frequencies which indirectly excite the Coriolis frequency
- swirl in the inlet flow
- asymmetric inlet profile
- increased turbulence in the inlet flow
- two-phase flow (air/water)
- cavitation
- installation stresses.

They concluded that almost all Coriolis meters were affected by vibrations at the Coriolis frequency. Other frequencies could also affect the calibration. Inlet flow conditions had little or no effect, but air/water flows introduced errors. The authors concluded that while it was not generally possible to provide corrections for the error conditions, it would be possible to monitor the meter signals and generate warnings of most conditions apart from frequencies near the drive frequency. The authors defined Coriolis frequency as "the frequency corresponding to the mode shape which gives rise to the phase difference proportional to the mass flow".

While external frequencies at the drive frequency may cause errors (Cheesewright and Clark 2002), external vibrations not at the drive frequency could be isolated with correct signal analysis (Cheesewright, Belhadj and Clark 2003a). Clark and Cheesewright (2003b) tested the effect of external vibration and the dynamic response of commercial meters from five manufacturers. They found that all the meters showed errors when the vibration was at the drive frequency, but the extent of error for Coriolis and other frequencies depended on the design of the manufacturers' software. In further tests Clark and Cheesewright (2006) investigated the response to step changes in flow rate of representative commercial meters both of the flow tube and of the complete meter. They indicated that the dynamic response tended to be limited by the transmitter technology rather than the flow tube and, therefore, should be capable of considerable improvement.

Clark, Cheesewright and Hou (2003) used theory, finite element (FE) and experiment to examine the response of the meters to step changes of periods less than one cycle of the drive frequency and also looked at the response to ramp changes and low-frequency pulsations (cf. Henry, Clark and Cheesewright 2003a; Henry et al. 2003b). Further study of the response by Cheesewright et al. (2003b) using analytical,

FE and experimental data indicated that the time constant could not be less than one period of the drive frequency, and that longer response times may be due to additional damping in the signal processing to reduce the effect of unwanted oscillations (cf. Fan and Song 2003, in Chinese). Clark et al. (2006c) assembled a new meter with an adapted commercially available straight tube, using the Oxford digital transmitter. The meter was tested in the laboratory and in the field and had delays of order 4 ms between change of flow and change of output signal. Pope and Wright (2014) have obtained data on the performance of meters in transient nitrogen and helium flows.

20.1.4 Shell-Type Coriolis Flowmeter

So far we have considered the beam-type of Coriolis meter in which the pipe oscillates in beam-like vibration in one plane. In addition to this type of Coriolis meter, there is another version known as the shell-type Coriolis meter. Instead of the pipe oscillating in a beam-type mode (given the mathematical value n = 1), the pipe cross-section oscillates between a round shape and an oval or other shape, a "lobe-like" shape (n = 2, 3, …), possibly termed a hoop mode (Hagenmeyer et al. 1994). Kutin and Bajsić (1999) discussed the characteristics of the shell-type Coriolis flowmeter. They have done further work on this type and their paper gives references. The shell-type of Coriolis mass flowmeter has a thin-walled measuring tube which allows oscillations in various modes with more than one circumferential wave, resulting in the pipe circular cross-section deformation (Amabili and Garziera 2004). Exciters and sensors generate the modes and measure the distortion to obtain a measure of mass flow rate. The bending of the flow past the distorted wall causes the same forces as the flow in the beam-type bent tube (n = 1). This may sometimes be referred to as a straight-tube radial-mode meter. Elsewhere in this chapter there are a few other references to this type of meter, but the chapter is, predominantly, concerned with the beam type (n = 1) meter.

20.2 Industrial Designs

The Micro Motion meter was the first to appear in 1981 and resulted from patents filed in 1975, 1977 and 1978 (cf. Willer 1978 and Smith 1978). The initial device consisted of a single U-tube. A paper by Tsutsui and Yamikawa (1993) seemed to be using one tube plus a resonance vibrator as in the early Micro Motion meter.

Twin tubes were introduced in model D in 1983. The main features of this meter were: a single inlet splitting into the two tubes; a drive system to cause the tubes to vibrate; a pair of sensors to detect the movement of the tubes past the mid-plane; a strong point around which the tubes vibrated. In some meters the outer cover forms a pressure vessel so that in the event of the tubes failing, the fluid is contained. Failure should not be due to fatigue, as manufacturers should design so that tubes will not fail due to cyclic stressing. Tube failure may sometimes occur due to corrosive stressing.

Several designs followed Micro Motion and a selection of these from the 1990s, including those no longer available or where the design has been updated, are

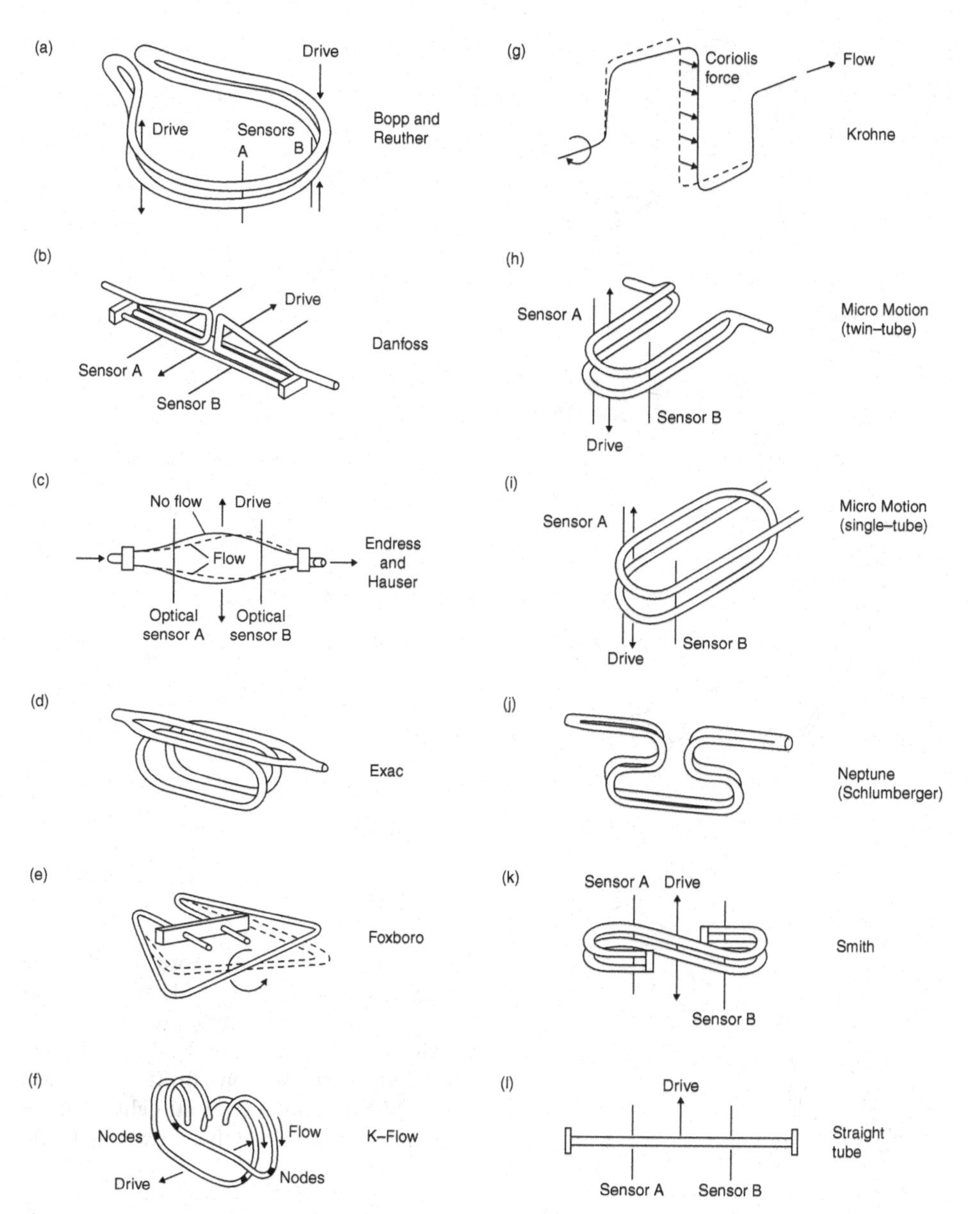

Figure 20.6. A selection of Coriolis-type mass flowmeters commercially in use now or in the past: **(a)** Bopp and Reuther; **(b)** Danfoss; **(c)** Endress+Hauser; **(d)** EXAC; **(e)** Foxboro; **(f)** K-Flow; **(g)** KROHNE; **(h)** Micro Motion (twin-tube); **(i)** Micro Motion (single-tube); **(j)** Neptune (Schlumberger); **(k)** Smith; **(l)** KROHNE, Endress+Hauser, et al.

shown diagrammatically in Figure 20.6 based on the author's understanding of the manufacturers' brochures. Apart from those shown: Bopp and Reuther, Danfoss, Endress+Hauser, EXAC, Foxboro, K-Flow, KROHNE, Micro Motion, Neptune (Schlumberger) and Smith, other manufacturers for which the author lacked details may have been Heinrichs and Yokogawa, and others may also have been available. It should be noted that some manufacturers offer various options.

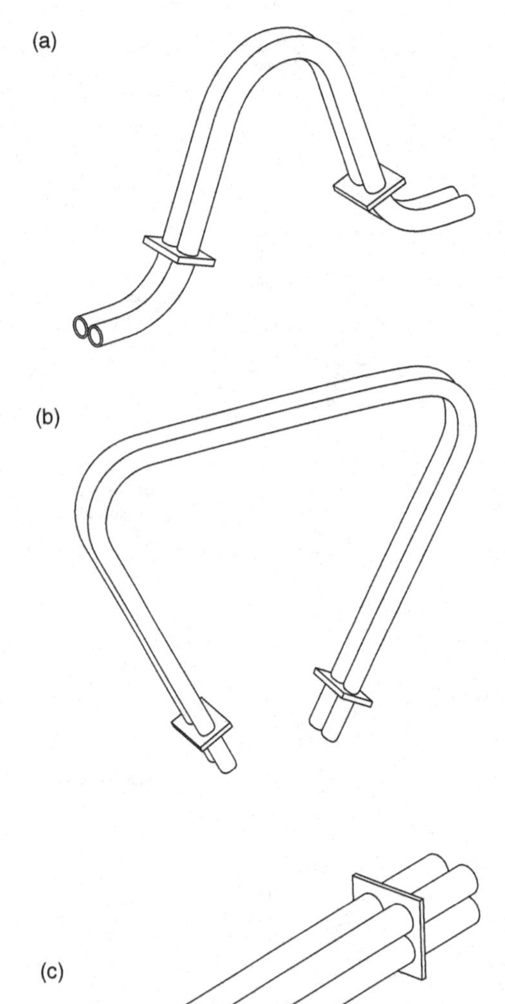

Figure 20.7. Sketches of more recent tube configurations of Coriolis flowmeters: **(a)** V-shaped with twin tubes also in double V with four tubes (e.g. Endress+Hauser, KROHNE, Micro Motion); **(b)** Ω-shaped twin tubes (Micro Motion); **(c)** straight multiple tubes after Hussain, Rolph and Wang (2010).

Most industrial designs in current operation use two parallel wetted tubes or two tubes in series. The deflection is measured from one tube to the other, so that external vibrations are largely eliminated and the measured shift between the tubes has the effect of adding the flows in the two tubes even if the flow split is not equal.

Single-tube versions (e.g. Hussain and Farrant 1994) are also available from several manufacturers and provide an attractive option (Reider and Drahm 1996; Yamashita 1996) (cf. Stansfeld, Atkinson and Washington 1988; Harrie 1991 for an earlier design). New designs of bent tube are being introduced which overcome some of the shortcomings of previous designs. Some of these designs are sketched in Figure 20.7, including a further development of multiple straight tubes.

20.2.1 Principal Design Components

Flow Tubes

The deflection of the tubes, in twin tube designs, is sensed from one tube to the other and this eliminates, to a large extent, external vibrations. The tubes are mounted to isolate them from external vibration and are designed to optimise phase shift between sensor signals, and to minimise pressure loss. Wagner (1988) discussed trade-offs for tube selection. For instance, by lengthening the meter tube, vibration of the tube is easier, stresses are reduced and Coriolis forces increase. However, this is at the cost of greater weight and pressure drop. In the case of single-tube meters, the structure is designed to provide the isolation needed from external vibration and to prevent the meter vibration penetrating outside the meter.

The problem of corrosion is a reason why some manufacturers use an outer containment vessel around the vibrating tubes in case of tube failure.

Foxboro introduced a technique called anti-phase excitation which used a double driver system to produce a rotating motion on the tube essentially at the flattened bottom of the U-tube. In this way the stresses were greatly reduced because the vibration was away from the pipe joints. This appeared to be similar to the Bopp and Reuther and the KROHNE designs. Watt (1990) described work on the computation of pipe stresses.

Drive Mechanisms

These are most commonly electromagnetically excited to oscillate at the chosen frequency. In some cases this is the natural (resonant) frequency, in others a harmonic of the natural frequency. Higher frequencies are claimed to reduce the effect of external disturbances by introducing free nodes of vibration and by working further above the most common surrounding frequencies. The control circuit then keeps the oscillation at the chosen frequency. For high viscosity or density additional power may be required in the drive. In addition, the use of digital control has ensured the possibility of controlling vibration when the meter is used with difficult fluids.

Sensor Types

In meters with twin tubes the sensor usually measures the relative motion between the two tubes, thus eliminating any common spurious vibrations. One method used in several designs has a coil on one tube and a magnet on the other. The relative velocity causes a voltage to be generated in the coil and in order to obtain position this signal is integrated. Most sensing appears to be by electromagnetic means, although an optical method was used for one earlier design with an arrangement of photodiodes and a modulating shutter (Vogtlin and Tschabold, Endress+Hauser).

In addition to mass flow rate the meter may provide: temperature, density, volumetric flow rate and viscosity. Temperature sensing is included in many, if not all, commercial instruments, for instance the use of a platinum resistance thermometer (PRT) with resistance set to 100Ω. Nicholson (1994) noted that the location of the temperature sensor within the meter housing had a significant effect on the precision with which fluid temperature was measured. His results suggested discrepancies ranging up to ±2°C. In KROHNE's single straight tube design, a strain gauge has been added together with a temperature sensor in order to obtain information from the meter for subsequent corrections by the electronics.

Kalotay (1994) used the pressure drop through the Micro Motion meter to obtain a value of viscosity. More recently Endress+Hauser developed a meter using the torsional vibration of the tube in a single straight tube meter, causing a shear stress in the fluid at the tube wall. This resulted in a torque due to the rotational motion of the fluid and a consequent measure of viscosity (Drahm and Bjønnes 2003). See also Drahm and Staudt (2004), Hussain (2003) and Standiford and Lee (2010).

Secondary Containment

Some manufacturers design the flow tube with an outer containment tube so that in the event of flow tube failure the process fluid is contained [cf. Hussain and Farrant (1994), who quoted a bursting pressure of 500 bar with in-house testing up to 300 bar].

Meter Secondary

As indicated earlier, the distortion in the tube due to Coriolis forces will cause the two sensing points to pass the mid-plane at different times. The phase shift at full-scale flow may be about 1°. At zero flow the phase difference should be zero although normally there is an offset and hence a zeroing routine.

As in many areas of signal processing, digital methods have become common in instrumentation. Henry (2003a; see also Harrold 2001) reviewed the improvements due to the digital approach and in particular the possibility of using the meter for short batching and batching to and from empty, due to its robustness to conditions when batching to and from empty. Henry (2003b) discussed a digital transmitter with a time constant equal to one cycle of the drive frequency and with a flow tube design to produce high first-mode resonant frequencies, improved meter drive and signal processing procedures with high dynamic response potentially available. Clark et al. (2006c) described a very high response rate Coriolis meter based on a commercial twin straight tube meter with a development of the Oxford digital transmitter. Their results suggested an order of magnitude, at least, increase in response speed.

Figure 20.8(a) shows the block diagram of a mass flow processing system. The drive coil vibrates the tube. Pickoffs transmit signals to the amplifiers which are then sent to the tube period detector and thence to the microprocessor. The microprocessor supplies a 4–20 mA output, a frequency output and the flow direction, together with a serial interface. The diagram shows the drive amplifier for the drive coil.

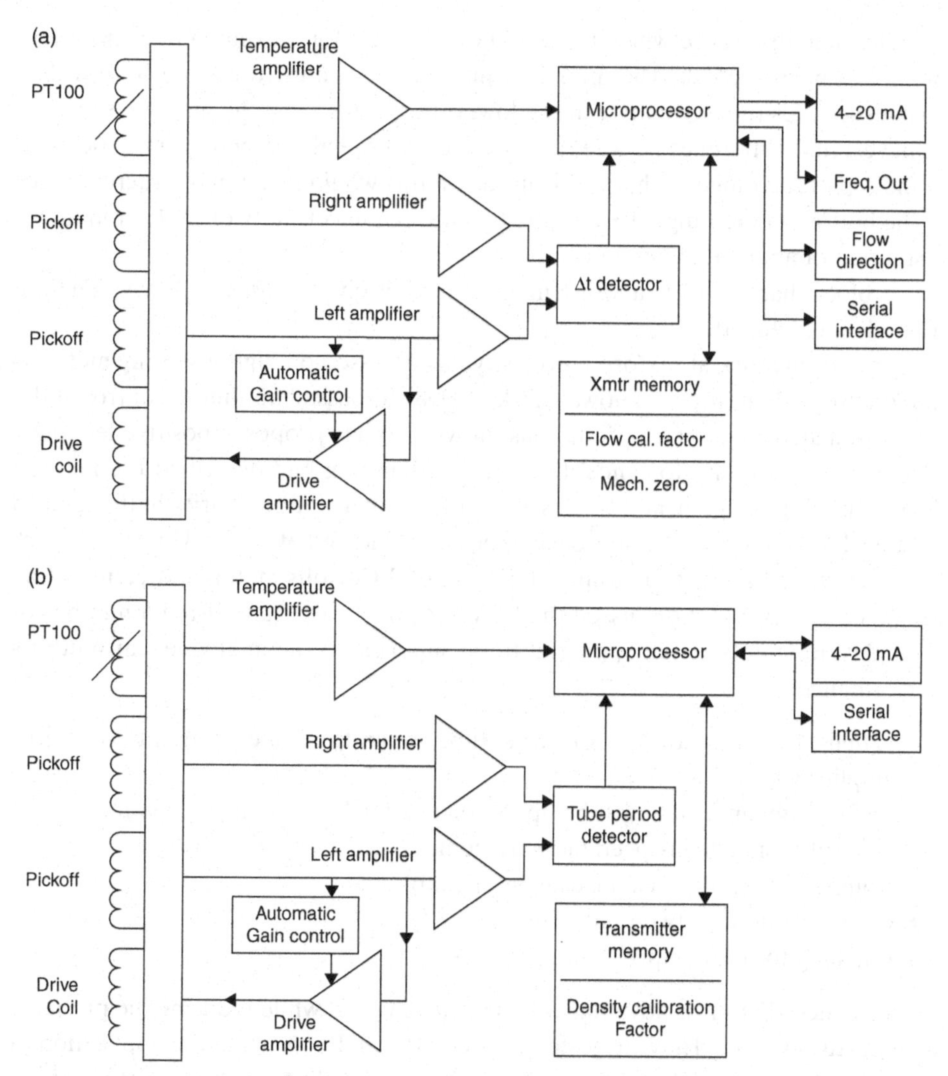

Figure 20.8. Micro Motion signal processing system (reproduced with permission of Emerson Process Management Ltd): (**a**) block diagram of mass flow processing; (**b**) block diagram of density processing.

The amplitude of the tubes' oscillation, obtained from the pickoffs, is maintained constant with the automatic gain control, below the fatigue limit of the tubes. The temperature measured with a PT100 element is fed to the microprocessor to allow compensation to be made for the effect of temperature on tube elasticity. The diagram also shows the transmitter memory, calibration factor and zero information retained in the meter.

Figure 20.8(b) shows the block diagram of the density-processing circuit with appropriate changes reflecting the requirements for density measurement and output data. Once the tube period of oscillation is determined, the microprocessor calculates the density and corrects it for tube temperature.

The time difference which needs to be measured for maximum flow may be of order 120 microseconds. The minimum measurement must therefore be less than 10 nanoseconds to discriminate at the lowest flow rates. Or if the meter designs can achieve a time difference between the two pick-off signals of about 60 microseconds at the top of the range, with a 100:1 turndown this would result in 0.6 microseconds at the bottom of the range. To measure this with an uncertainty of ±0.4% requires a resolution of about ±2.4 nanoseconds.

A block diagram from another manufacturer is given in Figure 20.9, with a little more detail included.

Important work at Oxford University has focussed on digital sensing methods and on use with multiphase flows. Clarke (1998) discussed nonlinear control of the oscillation amplitude of a Coriolis mass flowmeter. He proposed positive feedback of the output velocity to cancel the internal damping and described the method of determining the gain, and the insertion of an inverse nonlinearity in the loop to create a linear system. The approach appeared valid and effective. Henry (2000; cf. 2001) discussed a self-validating (SEVA) digital Coriolis meter, and Henry et al. (2000; cf. Henry 1995) discussed the SEVA meter further and listed a number of their developments which appeared to be superior to the analogue transmitter's performance:

- "High-precision control of flow tube operation, including at very low amplitudes.
- The maintenance of flow tube operation even in highly damped conditions.
- High-precision, high-speed measurement.
- Compensation for dynamic changes in amplitude.
- Compensation for two-phase flow.
- Batching to/from empty."

They claimed that these advantages had been achieved while reducing the probable manufacturing costs. They suggested future work, and noted that the application of the techniques to straight tubes was important, but might be a challenge to achieve. Morita and Yoshimura (1996) also described digital processing of the sensor signals and self-calibration. They claimed that the result was "100 times as accurate" as using a discrete Fourier transform.

20.2.2 Materials

Some materials of construction are: for wetted parts 316L stainless steel, NiSpan C, Hastelloy C-22, titanium, zirconium, tantalum and super duplex; for gaskets Viton and kalrez; and for non-wetted parts 304 or 316 stainless steel.

Manufacturers have had to address the problem of corrosion fatigue, which is a particular problem where materials are undergoing cyclic stress and an aggressive environment. Carpenter (1990) provided a useful introduction to material selection and gave information about alloy composition and examples of corrosion. It is necessary, therefore, in the Coriolis flowmeter, which depends on vibration and is applied

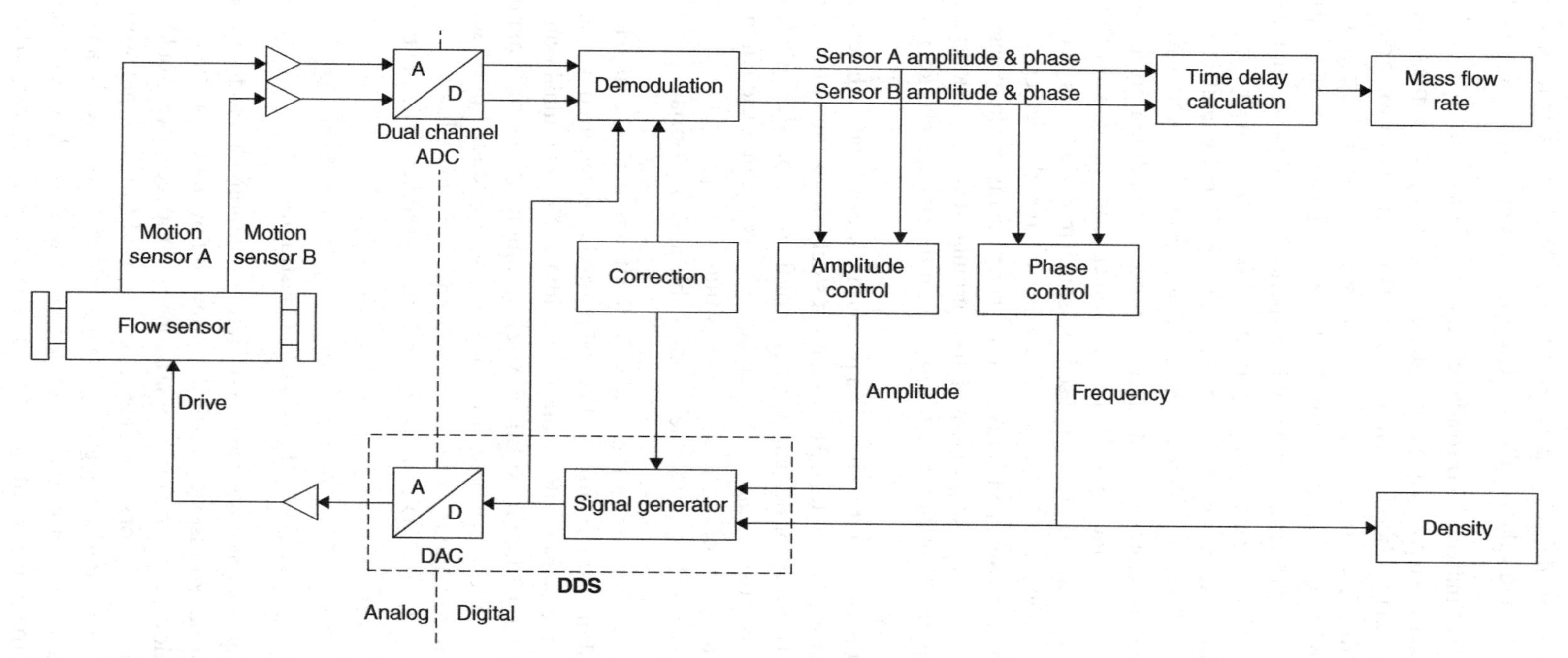

Figure 20.9. Digital signal processing of latest-generation Coriolis flowmeters (reproduced with the permission of KROHNE after Kunze, Storm and Wang 2014).

to some very aggressive fluids, to take particular care that corrosion fatigue does not occur and that if it occurs, the results are contained. This has been achieved by:

- correct choice of materials: corrosion resistance of 316L stainless steel is a result of its ability to form a protective oxide film due, primarily, to the presence of chromium, nickel and molybdenum. However this film can break down in certain fluids and cause pitting, inter-granular corrosion, stress corrosion cracking and corrosion fatigue (Micro Motion information). Hastelloy C-22 may provide resistance to both oxidising and reducing media and is superior to local attack due to chloride ions, because of the formation of a very stable oxide film which is little affected by chloride ions. However, there is a need for care in the manufacture of tubes from Hastelloy due to localised diffusion of carbon at grain boundaries resulting from problems in the drawing and annealing process. In addition, brazing with filler materials may cause inter-granular penetration cracking of the Hastelloy. (Micro Motion information). Titanium is used by some manufacturers because of its favourable elasticity, low density and compressive strength, coefficient of thermal expansion and corrosion and abrasion resistance to many different aggressive fluids (cf. Wagner 1988 on the problem of inter-granular stress corrosion, pitting etc.);
- control of amplitude of vibration to ensure that the stress cycles keep within the design envelope;
- containment of the vibrating tube in a pressure vessel so that in the event of failure, the process line is intact. 316L stainless steel may be an option for wetted parts (except with chlorides and halogens) of flow sensors. This grade of stainless provides corrosion resistance with the majority of fluids used in the process industries. However, other metals, such as titanium, appear to be increasingly used as a possible alternative where compatible and appropriate.

As described earlier, sensor tubes must be flexible to generate sufficient deflection and phase shift. Thus, relatively thin walled tubing is required. For example, a typical 25mm (1″) sensor may use 19mm (3/4″) tubing with a nominal wall thickness of 1.65mm (0.065″). Unlike most other system components, there is minimal corrosion and erosion allowance. Therefore, it is not recommended that users rely on standard compatibility guides, which often incorporate such allowances. (See Bell and MacLeod (2012) on CFD to assess the effects of particle erosion in two Coriolis meters).

20.2.3 Installation Constraints

Flow sensors may in some cases have preferred orientations due to the geometry of the meter pipework. For instance, vertical downward flow may sometimes result in a partially empty flow tube. For liquid applications any trapped gases should be able to escape, and for gas applications any liquid or condensate. In addition, pipework arrangements should be avoided which might lead to trapped air or gas, such as an inverted U.

Their sensitivity to upstream flow profile, in my understanding of the research to date, is unlikely to be significant except for meters which achieve uncertainty levels of better than 0.1%, although Cascetta et al. (1992) suggested that axial swirl and turbulence spectra might have an influence on the flowmeter performance of some designs (but see Hemp 1994b). Nicholson (1994) found some effects due to pipeline configuration but no effects due to compressive or tensile loadings on the meters. One possible application requiring caution, as discussed later, is for a turbulent/laminar transition.

Hemp (2001b) calculated the sensitivity of a straight meter with free ends and found a profile (Reynolds number) dependence. However, the question as to whether the Coriolis meter is affected by velocity profile has been re-examined by Bobovnik, Kutin and Bajsić (2005) for the shell-type meter (one which vibrates by distortion of the circularity of the pipe in modes $n>1$). They appeared to suggest that the signal may be reduced by as much as 8% when the Reynolds number drops to 3,000 and the flow profile (axisymmetric) is consequently changed. Kutin et al. (2005a), using a weight function approach, indicated that the weight function for axisymmetric velocity profile for a beam-type Coriolis meter (bending mode $n = 1$) was almost uniform for $L/R > 40$ where L is the length of the meter and R is the pipe radius. In contrast the effect of profile for the shell-type meter with $n = 2, 3$ etc. was, indeed, more significant. However, they also made the points that their results were for axisymmetric profiles and bends etc. had not been considered, and also that for the new generation of Coriolis meters aiming at uncertainties less than 0.1%, the effect of profile may need to be considered even for the beam-type meter. In a further paper (Kutin et al. 2005b) they considered further parameters on which velocity profile effects might depend. Kutin et al. (2006) discussed the effects of velocity profile further, and suggested that this might be responsible for small changes in signal, although there appeared to be a need for more evidence of its significance before this could be accepted with certainty. For low Reynolds numbers Kumar, Anklin and Schwenter (2010) suggested that a Coriolis meter may deviate under the influence of fluid-dynamic forces, and that the effect could be explained by a periodic shear mechanism interacting with oscillatory forces, with profile change from turbulent to laminar and a possible shift of order 0.7%. Bobovnik et al. (2013) have reported a numerical analysis of installation effects for a short straight beam-type tube which suggested possible effects of upstream flow distortion (bend, two bends out of plane and an orifice), but less than or equal to 0.1% except in the case of the orifice 5D upstream when the effect was about 0.2%. They made the point, however, that the size of these effects was likely to be smaller for commercial designs than for their short tube which also lacked some detailed design features.

The presence of two-phase flow introduces its own problems due to the splitting of the flow between the twin tubes, which takes place in many instruments, the possible unequal split of the phases between the twin tubes and the possible effect of the tubes on coalescence. Swanson (1988) described an in situ method that provided for recognition and compensation for coating or scale build-up within the meter. In all cases manufacturers should be consulted on best practice.

Pressure variation can cause the tube dimensions to change with possible effects on the sensitivity and zero stability of the flowmeter. Nicholson (1994) found K-factor shifts due to fluid temperature.

Levien and Dudiak (1995) gave a brief review of effects on Coriolis meters due to temperature, pressure, vibration and flow profile. Cascetta (1996) discussed the effect of fluid pressure on Coriolis flowmeter performance. He noted that since the meter operation is related to the elastic deformation of the tube, it will be necessary to compensate for changes of temperature and pressure. The stiffness of the system decreased with increase in temperature, while the rigidity of the tube increased with fluid pressure. The meter appeared to underestimate flow rate with increased pressure. Wang and Hussain (2010b) discussed the theory behind the effect of pressure change on the meter, and obtained general agreement with experimental tests while recognising that open questions remained.

Kiehl (1991) considered the adverse effect of an interference frequency close to the operating frequency of a Coriolis flowmeter, with particular reference to cross-talk between two similar meters working in close proximity in applications such as leak detection, difference measurements or parallel flow configurations. To avoid this, mechanical decoupling of the meters may be possible by using rubber or plastic hose between the meters, or by changing the cross-section with the insertion of a large cross-section pipe between the two meters. An alternative available from some manufacturers is to use two different working frequencies separated by a few Hertz: Kiehl instances 6.5Hz.

20.2.4 Vibration Sensitivity

The meter has been found to be affected by inadequate supporting structures which have a natural frequency of vibration within about 20% of the operating frequency and by external vibrations close to the natural vibrational modes of the meter. To avoid effects due to external vibration, some meters operate at higher frequencies than the resonant frequency, and at higher harmonics. Nicholson (1994) found only one of the eight meters which he tested to be affected by vibration, and in this one case was unable to explain the fact that it occurred at a frequency different from the meter's [cf. Adiletta et al.'s (1993) meter designed to reduce the effects of external vibrations].

Vetter and Notzon (1994) showed that small amplitude pulsation (±2% of the flow) disturbed the operation of the flowmeters when the frequency coincided with the torsional frequency of the measuring tube. The frequency range of concern is 50 to 1,000 Hz for most Coriolis meters. Reciprocating pumps are likely to have lower frequencies than these, but rotary pumps may cause problems. The authors suggested that below ±2% the effects may not be severe, but some of their data above this value indicated serious errors.

Early designs of flowmeter could be affected by change in meter supports, particularly for a single-tube meter (Storm, Kolahi and Rock 2001, 2002). The problem of support and of vibratory transfer between the meter and the neighbouring pipe has been recognised and addressed in some, if not all, commercial meters. Other effects, such as liquid properties and operating conditions, have been discussed elsewhere in this chapter where information is known to the author.

20.2.5 Size and Flow Ranges

Meters are available in nominal line sizes from about 1mm or possibly less to 350mm or more and for flow ranges from 0 to 0.1 kg/h up to 0 to 2,000 t/h (Wang and Baker 2014). The weight of the instruments may range from about 8 kg to more than 600 kg for the largest sizes. Power consumption is likely to depend on size, and for larger sizes may possibly be of order 10 watts.

Figure 20.10 shows the typical ranges for one manufacturer of straight-tube meters with pressure loss. It should be noted that more than one size is suitable for most flow rates, and the manufacturer may be able to fit flanges to suit the particular application. Turndown ratio is typically in the bracket 20:1 to 100:1. Typical temperature range is −240°C to +427°C and high-pressure versions in certain sizes are available up to about 390 bar or more. There may be a minimum static pressure of about 1.2 bar.

Manufacturers' data sheets may indicate any effects due to pressure variation, temperature variation, viscosity, line vibration, density or back pressure. Wang and Hussain (2010b) used a theoretical method to explore the pressure effects. However, they commented that most of the recent designs have a relatively small sensitivity to pressure changes. In the main the results presented from various manufacturers suggested that change of reading was of order 0.01%/bar in most cases.

Zero stability appears to be about 0.01% of fsd. My experience suggests that zero adjustment should take place as close to expected conditions as possible, and that static pressures too near atmospheric may not allow the highest precision to be achieved.

Single-tube meters may be available up to 75 mm diameter or more, with overall lengths of about 400 mm to 1.25 m or more, and flow rates from 15 kg/h to 1,200 kg/h up to 1,500 kg/h to 180,000 kg/h or more. Process temperature range is typically −25°C to 130°C and pressure to about 60 bar. Measurement uncertainty is from ±0.05% to ±0.3% or more depending on zero stability. The outer tube is likely to become a more important element in the fundamental design and may require weights positioned to give the correct natural frequency, as well as providing secondary protection etc. (cf. Hussain and Farrant 1994; Yamashita 1996; Rieder and Drahm 1996). For the straight-tube version, the structure must both sustain external pipework forces and, as far as possible, prevent vibration from entering or leaving the meter.

Koudal, Bitto and Wenger (1996) reported on a 3.5 mm ID design for flows up to 450 kg/h which had a special design of carrying plate with damping elements on the nodal line.

While indicating that the Coriolis meter approaches the ideal flowmeter, Reizner (2003) identified its size for larger pipe sizes and its cost as a shortcoming. Wang and Hussain (2010a) also discussed size ranges and noted that Coriolis flowmeters had

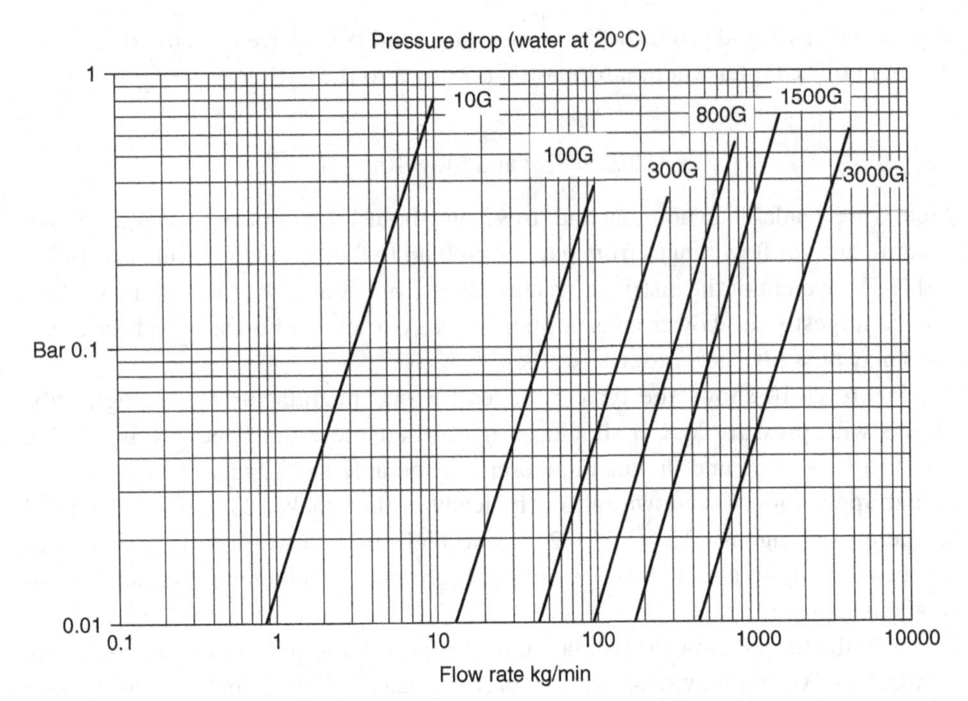

Figure 20.10. Typical flow ranges and pressure drops for straight tube meters (reproduced with permission of KROHNE Ltd).

been mainly developed and used in line sizes less than DN100 (or 4 in) because the size of the flow sensor itself could become too unwieldy and expensive for practical uses. This paper specifically reported the latest research and development using the straight-tube Coriolis technology to extend flow measurement capacity to a higher flow range.

20.2.6 Density Range and Accuracy

Most meters offer the option of density measurement for a typical range up to about 3,000 kg/m³. The uncertainty (over the calibration range) is from about ±0.2 kg/m³ to ±10 kg/m³. Repeatability is about ±0.1 kg/m³ to ±3 kg/m³.

In tests on water, naphtha and two different crude oils (Eide and Gwaspari 1996) all of the density outputs showed some dependence on either temperature or pressure. It seems likely that this has been reduced since then.

Density is a more problematic measurement in gas, but with careful calibration Pawlas and Patten (1995) claimed that it was possible.

20.2.7 Pressure Loss

To reduce pressure loss in the tube, a flow rate in the lower half of the normal range is recommended in some cases (Figure 20.10 is for straight tube design). Different manufacturers have different approaches to define maximum flow. Pressure loss

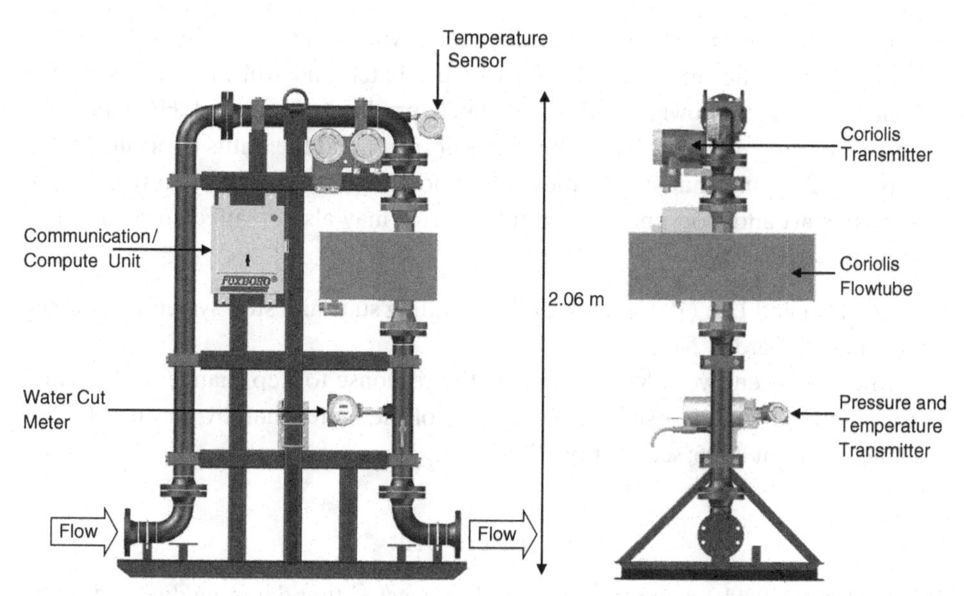

Figure 20.11. A Coriolis meter mounted on a skid, in combination with a water-cut meter, being trialled for multiphase flow field operation (reproduced with the permission of Schneider Electrico and with the agreement of Elsevier from Henry et al. 2013).

may be a criterion in some cases. With viscosities of 200 cSt or more the losses may be as high as 5 bar. However, for a single straight tube instrument the pressure loss is, essentially, the same as for the equivalent length of straight tube, thus introducing no additional pressure loss. Also for abrasive fluids a low measuring range with velocities below about one metre per second is recommended.

Cascetta et al. (1992) gave a plot of the pressure loss of a selection of commercial flowmeters, in a 25 mm test line with various pipe geometries, and obtained pressure drops in the range 1 to 1.3 bar for flow rates in the region of 3 to 4 kg/s for most of them. This should be compared with the straight tube results in Figure 20.10 which appear to be very much lower. Nicholson (1994) tested eight commercial 1 in (25 mm) meters and obtained pressure drops from 1.2 to 2 bar at 4 kg/s and noted that two of the meters gave audible signs of cavitation above 4 kg/s. I am not aware of more recent data on pressure losses, which, in any case, should be of similar order to these data.

20.2.8 Response Time

Trigas and Hope (1991) tested three meters from two manufacturers for step changes in the flow rate. A time constant was defined as the time to reach 63.2% (which is 1-1/e) of its steady-state value. The response time was then defined as six times this value. They found that repeatability improved as integration time increased, and as flow rate increased. The problems with long time constants are:

a) that the actual flow rate will not be followed precisely and the smoothing effect may cause errors to be introduced.

b) that for batch control, the integration time will essentially cause some of the batch flow to be missed at the start of the batch and will therefore stop the batch too late, allowing additional batch product (Wang and Hussain 2006 more recently explored batch flow measurement and the limits of accuracy and appeared to find that systematic and random errors were likely to result from empty start and stop, and some random error may also be affected by the flow update routines).

Paik, Lim and Lee (1990) found, for a standing start and stop system, a time lag for the meter of order 2s.

However, recent work has shown that the response to step changes in flow rate is more limited by the transmitter technology of the meter than by the tube (Clark and Cheesewright 2006; see Section 20.1.3).

20.2.9 Zero Drift

Keita's (1989a, 1989b) computational work suggested that damping due to dissipation, friction etc. and non-symmetry of the tubes were contributory causes of zero drift. He also appeared to demonstrate that lack of symmetry between the tube ends can lead to zero errors in water of up to 0.04%. He showed that fluid changes due to density and viscosity can cause zero shifts of more than 0.2%. Cascetta et al. (1992) also attributed sensitivity to zero drift to the unavoidable lack of symmetry between the two sensor tubes, but also to an unequal split of fluid which is not homogeneous.

A zero shift error causes the uncertainty to be given as of rate at the high ranges, and of span at the lower ranges (Ginesi and Annarummo 1994). Zero should be adjusted after installation and fitting, and may need adjustment for temperature shifts of 10°C or more. Downsizing may be used with caution to increase velocities and reduce coating, and also to reduce costs, but beware corrosion, erosion and pressure drop.

Eide and Gwaspari (1996) aimed to test six different makes of Coriolis meter on water, naphtha and two different crude oils. In each there was a five-point calibration. The zero error for most of the meters for most of the time was less than ±0.3% fsd and appeared not to be influenced by increasing temperature and pressure.

Wang et al. (2010) examined the zero drift in a Coriolis meter and some possible factors suggested, for this meter, related to the effects on the transducer's structure asymmetry, stress and non-uniform thicknesses, of a temperature changing environment. Enz, Thomsen and Neumeyer (2011; cf. 2010) examined the zero shift caused by imperfections in the flowmeter, introducing non-uniform pipe damping and mass, and the effect of temperature changes, confirming asymmetry in the axial distribution of damping as a cause of zero shifts.

Koschmieder and Röck (2010) also noted that the sensitivity and zero point of the Coriolis may change due to temperature gradients along the measuring pipe or due to mounting conditions. These changes have to be detected and corrected in order to assure high accuracy. They described a model-based approach to estimate the zero point during normal operation and one-phase flow. The approach exploited two characteristics of the measuring device: firstly the impact of mass flow on the oscillation in the second mode when the first mode is stimulated, which is the operation principle of nearly all Coriolis mass flowmeters, and secondly the impact of mass flow on the oscillation in the first mode when the second mode is stimulated. Both of these characteristics are realised by compensation of Coriolis forces. In this paper a model based control scheme was presented allowing the Coriolis flowmeter to be operated in the two compensation methods: (1) compensation of the second-mode oscillation when stimulating the meter in its first mode and (2) compensation of the first-mode oscillation when stimulating the meter in its second mode. As the characteristics of both of the compensation methods differ in their zero point, the lumped parameter model had to be modified by introducing sensor and actuator coupling that could be interpreted in terms of different actuator and sensor gains. If the actuator coupling was properly chosen, the zero point of the characteristics of both of the compensation methods and the characteristics of the deflection method appeared to be in good agreement. Exploiting the characteristics of both of the compensation methods, the zero point could be corrected either in a cyclic procedure or by separating the measurements in the frequency domain.

20.3 Accuracy Under Normal Operation

Typical performance for a 20:1 turndown for commercial instruments (with zero adjusted) is likely to be of order or better than (Wang and Baker 2014):

- Liquid uncertainty ±0.05 to 0.5% rate
- Gas uncertainty ±0.35 to 0.75% rate
- Zero stability possibly of order ± 0.002% to 0.1% for nominal flow rate (flow rate of water under 1bar pressure drop) or better
- Liquid and gas flow rate repeatability of order half uncertainty
- Density uncertainty for liquids may be in the range ±0.2 to ±10 kg/m^3

Some earlier values appear to be making claims for greater accuracy, but this may reflect a caution on the part of the manufacturers. Mencke (1996) reported errors for a large number of Coriolis designs ranging from 2 to 250 mm on various fluids including some non-Newtonian (liquid colour) and two component (liquid chalk; water with 76% solids) and obtained for:

- 100% to 50% flow range ± 0.01% to ± 0.35%
- 50% to 10% flow range ± 0.01% to ± 0.50%
- 10% to 1% flow range (three meters) ± 0.11% to ± 0.37%

Fyrippi, Owen and Escudier (2004) showed that the Coriolis flowmeter they tested operates within the manufacturer's specification with non-Newtonian liquids and appeared unaffected by the fluid rheology.

Uncertainty according to Harrie (1991) was 0.25% on mass and ±0.0002 g/cc on density. He gave typical results on zero offset due to pressure, temperature and density as:

- 0 to 40 bar 0.025% fsd
- 6.7°C to 30°C 0.05% fsd
- Air to water 0.05% fsd

See also Mencken (1989), Nicholson (1994) and Eide and Gwaspari (1996). Keita (1990) discussed the effect of gas compressibility which led to a loose coupling between fluid and vibrating pipe and hence to the possibility of errors.

Pawlas and Patten (1995), noting that increase of 100°C for a stainless steel tube causes a decrease of Young's modulus, a change in the stiffness of the tube, and caused about 5% change in signal, suggested that an RTD would allow compensation, and noted that elevated pressures caused tube stiffening due to radial stress, which could range from a negligible effect for small sensors to about 0.08% per 100 psi for larger sensors. Compensation is usually by a bias in the calibration constant for larger sensors. As an example they suggested the application to custody transfer for natural gas. A vehicle filling cycle starts at up to 50 lb/min (23 kg/min) with supply pressure of 3,500 psi (240 bar) and the vehicle pressure of zero gauge, and continues for about five or more minutes at decreasing rate until the vehicle is filled. The meters tested had errors in the range ±0.2 to 1.8%. At high mass flow rates the meter could not cope due to the high flow noise. Their plots suggested that water and compressed air gave similar results (air within about ±1% of the water), and that similar changes resulted from pressure change up to 1,450 psi (100 bar). They suggested that, using water calibration, the performance on air is better than ±2% and in much of the operation within ±0.5%, and over wide flow and pressure ranges.

For gases, there are various questions raised in the literature about the accuracy of a gas Coriolis meter. Effects due to compressibility and other features of the gas could mean that, for the highest accuracy, the meter becomes sensitive to type of gas. Hemp and Kutin (2006) also discussed errors due to compressibility of the fluid and showed that for gases and aerated liquids the error can be significant under certain conditions, particularly with the increased performance of Coriolis meters. Assuming that the flow velocity is not too large, they indicated that the fractional error, when small, was of order $\frac{1}{8}\left(\frac{\omega D}{c}\right)^2$ for mass flow rate and half this value for density, where ω is the operating angular frequency, D is the internal diameter of the pipe and c is the sound speed in the fluid.

My own experience reported in Baker et al. (2013) is that the precision of good-quality meters has greatly improved over the past couple of decades. With due attention to the factors which we set out, modern Coriolis meters should be able to achieve, for liquid flows, uncertainties of order 0.1% or better.

20.4 Published Information on Performance

The meter may be suitable for homogeneous two-phase flows and for heterogeneous flows provided the phases are of similar density so that when vibrated the two phases behave as if rigidly connected. Some other situations are as follows.

20.4.1 Early Industrial Experience

Two important papers were published at the North Sea Metering Workshop in 1988. These represent, possibly, the earliest available field data for operation of these meters in North Sea applications. Occidental (Lawson 1988) installed a Micro Motion D150 meter on the Claymore condensate system in a vertical downward flow. There appeared to be a discrepancy of about 0.35% between the meter and an orifice plate, and a final calibration of the meter showed an under-reading of about 0.5%. Chevron (Dean 1988a) installed an EXAC 2 inch class 1500 meter on the Ninian Central Platform on a liquefied petroleum gas (LPG) flow. The main problem appears to have arisen from zero errors and the resultant data were not very precise. Despite this Dean (1988b) saw advantages for petroleum liquids. Coriolis meters were used to improve crude oil measurement at individual wells throughout the Little Knife field (Liu, Canfield and Conley 1986). These early years were a good portent for the future (see Rezende and Apple 1997).

A problem then was communication (cross-talk) between two Coriolis meters, and proposed methods to overcome this were by using flexible pipework or by separating the operating frequencies (Kiehl and Gartner 1989). Manufacturers' data from then (Micro Motion: McKenzie 1989) suggested linearity of ±0.15% – ±0.25% with repeatability of ±0.05% or better, and seemed to suggest calibration changes of 0.3% or more as a result of temperature, pressure and fluid change (Endress+Hauser: Frankvoort 1989). Davis (1990) described tests of a Micro Motion 25 mm meter which operated for 15 months on the distillate export facility at St Fergus gas terminal which were promising with potential uncertainty of 0.1% on flow and density measurements within 0.1%. Grendstad, Eide and Salvesen (1991) described further tests with Micro Motion and Schlumberger meters. Some problems were encountered due to zero setting, stress caused by distortion between flanges and cross-talk. The meters were also tested on natural gas. Two of the meters were within the ±1% of rate which was the aim of the tests. Arasi (1989) found field performance of a meter on an 80% ethane 20% propane mixture resulted in errors ranging from +1% to −2% or more, and concluded that for field service regular proving would be necessary. Myhr (1991) reported tests on propane and propylene and suggested that the meters were sensitive to changes in pressure, density and viscosity. Further tests were reported by Erdal and Cabrol (1991) on a 1.5 inch (38 mm) Coriolis meter on natural gas, who found a repeatability over a year of 0.47%, and by Withers, Strang and Allnutt (1996) whose tests offshore resulted in performance within 0.4% of water calibration. These publications from the early developments of this meter should be compared with the very wide acceptance and use of the meter now.

20.4.2 Gas-Liquid

From early in the use of the meter it was observed that small amounts of gas in liquid could cause significant errors (Grumski and Bajura 1984; Nicholson 1994). Birker (1989) suggested that between about 0.5% air-in-water by volume up to about 4% the error in reading rose to about 10%. Nicholson put the figure much higher than this. Hemp and Sultan (1989) addressed this question theoretically and Rieder, Drahm and Zhu (2005) developed a theory for errors in two-phase conditions (Appendix 20.A) to supplement that of Hemp, and appeared to suggest errors of 2.4% for mass flow with gas concentration of 1.5% by volume at 5 bar. Gysling (2007) also described a theoretical model which gave good agreement for lower operating frequencies, but less good for higher frequencies.

Al-Khamis, Al-Nojaim and Al-Marhoun (2002) found the performance of U-tube meters deteriorated significantly when gas flowed with crude oil, while the straight tube was less sensitive. Orientation and tube geometry appeared to affect accuracy (see also Al-Taweel, Barlow and Aggour 1997; Al-Mubarak 1997). Reizner (2003) gave an example of void fractions of gas in liquid of as little as 2%, where the meter may not indicate any problem, but may have errors of as much as 20%.

20.4.3 Sand in Water (Dominick et al. 1987)

Measurement of flows of water with sand particles of predominantly 0.125mm (80%) and also 0.250mm (15%), 0.063mm (4%) and 0.355mm (1%) were made using a Micro Motion meter. It was found that for a homogeneous steady flow with loading from 3 to 22% of sand, the measurements were within 1% of the reference values. Mass flows were up to about 140 kg/min (cf. Agarwal and Turgeon 1984, who found that operating temperature and specific gravity affected the early versions which they were using).

20.4.4 Pulverised Coal in Nitrogen (Baucom 1979)

A Micro Motion meter with a U-tube of 12.7 mm (0.5 in) and a flow range of 0 to 200 kg/min (0 to 1.5 lb/s) was operated on nitrogen at 7.5 bar, and the outputs for mass flow rate and fluid density were recorded. The meter was initially calibrated using water, and it was then compared with load cell measurements of coal hopper weight loss. This latter measurement did not allow a high accuracy. The metered flow rates agreed well with the load cell measurements except where slugging of the coal and transport gas caused erratic meter behaviour. The meter behaved best for dense-phase coal flows, but for low coal-to-gas ratios the erratic behaviour occurred.

20.4.5 Water-in-Oil Measurement

De Kraker's work (1989) gave early suggestion that by using both mass flow and density the water-cut in oil production could be obtained. However, Liu and Revus (1988; cf. Liu et al. 1986) have given some impressive results of Chevron's water-cut measurement for a wide range of water-cuts. The claims appear to suggest results within about ±1 to 2% of actual water-cut. Young (1990) gave a plot of water content measured by an EXAC meter which suggested a low reading at 50% water-cut of up to about 10%. He also used it to obtain percentage of solids in liquid.

Anderson et al. (2004) tested two vertically installed Coriolis mass flowmeters (2″ Micro Motion CMF200) in oil/water mixtures of about 50% water-cut, and appeared to find differences in the density measurements, but less in the mass flow readings. Horizontal installations with various orientations appeared to result in significant under-reading in density in unconditioned flow when the oil and water phases became significantly separated at low velocities, and in some orientations water hold-up resulted as flow rate approached zero. Mixing immediately upstream of the meter appeared beneficial in all cases.

20.4.6 Two- and Three-Component flows

The possibility that digital control might allow Coriolis meters to be used in multiphase or multi-component flows was discussed by Adejuyigba et al (2004). Tombs et al. (2004) discussed their all-digital Coriolis mass flow transmitter, which was able to maintain flow tube operation throughout all conditions of two-phase flow. Key to this technology was the transmitter's ability to respond quickly to flow condition changes, and this was also reflected in the meter's ability to track changes in flow rates accurately with a fast dynamic response. They presented results from two- and three-phase flow experiments (cf. Yeung et al. 2004). Hemp and Yeung (2003) and Hemp, Yeung and Kassi (2003) reported initial tests of the digital meter on the Cranfield multiphase flow rig with air/oil and air/water in horizontal and vertical orientations, and the performance appeared satisfactory under some conditions, but with very low liquid velocities the meter could seriously over-read. Henry et al. (2004) reviewed the requirements for a Coriolis meter to obtain mass and density measurements in a two-phase flow, and identified a key aspect of this as the drive system for the vibrating tube (see also Henry et al. 2006).

Henry et al. (2006) provided a case study on the application of a Coriolis meter to two-phase flows and concluded that, while there is no generally applicable flowmeter for such flows, there may be the possibility of developing an approach for a specific meter, application and fluids. This continuing research has resulted in tests undertaken with a "skid", Figure 20.11, including a Coriolis meter and a water-cut meter to obtain the components of three component flow for oil exploration etc. (Henry et al. 2013). Using a three-phase neural net model to correct the results, they appear to have got close to achieving their target uncertainty envelopes of ±2.5% for total liquid mass flow against water-cut, ±5% for gas mass flow against gas volume fraction and ±6% for oil mass flow against water-cut up to 70% water-cut and ±15% for higher water-cuts up to 90%.

20.5 Calibration

Mencke (1989) described the process for obtaining limited approval in Germany in the light of the weights and measures regulations for new techniques of flow measurement. This involved test facility calibration and site tests following limited approval. Strawn (1991) reviewed proving methods and listed test sites approved by the American Petroleum Institute. Paik et al. (1990) described results from a gravimetric calibration system. Hayward and Furness (1989) reported on the development of a gravimetric prover. Rivetti et al. (1989) described another rig, partly for evaluation of Coriolis meters.

Grini, Maehlum and Brendeng (1994) described a calibration system for Coriolis gas flowmeters which depended on measuring the mass change in a gas bottle of nitrogen or some other gas. They claimed an uncertainty of better than 0.02%. Standard deviations for the two makes of Coriolis meter tested on nitrogen were 0.09% and 0.15%.

Stewart (2002) undertook experimental tests to ascertain the use of Coriolis meters in gas flow and the validity of a water calibration for such an application. The meters were within ±0.1% for the water calibration, and for one of the meters, mostly within about ±0.5% for air calibrations. They appear to have operated mostly within the meter specification. However, unexpectedly, one of the meter's performance did not appear to improve with increased pressure. Zero stability of both meters was very good throughout the test programme. the meters did not need re-zeroing. One of the meters was seriously affected by sonic nozzles upstream

Grimley (2002) presented some baseline and installation effects testing of Coriolis flowmeters with natural gas. This is a very useful paper which gives confidence in its data. Results suggest several important findings:

- water calibrations for gas meters showed promise;
- some meters are particularly sensitive to line pressure and manufacturers' pressure corrections should be used;
- baseline repeat tests indicated reproducibility of 0.15% to 0.25% for the 50mm meters tested at about 120–650 bar;
- bent tube designs showed insignificant changes due to upstream piping for all but one, unaccounted for, installation for one meter:
- the straight-tube radial-mode (shell-type) meter appeared more sensitive to installation tests with shifts in the range −1.5% to +4%.

Small volume provers (SVPs) have become increasingly used in industry for checking the calibration of custody transfer meters. Tombs et al. (2006) discussed the use of a small volume prover in conjunction with the Coriolis meter using the Oxford transmitter and highlighted some of the constraints on accuracy, for instance, associated with pulse rate and shortness of proving period.

Wang and Hussain (2010a) gave a description of a special calibration procedure used in the manufacturer's gravimetric water flow rig. An extensive test programme on a typical DN250 straight-tube flowmeter was reported. The test programme covered five different fluids and five different test references over four different locations. Test results were within the well-accepted custody transfer limit and confirmed the performance of straight-tube Coriolis flowmeters for flow measurement in the high flow range. They indicated that average values for each test point fell within the ±0.2% band.

Baker et al. (2009) proposed a method for in situ verification of Coriolis meters. Cunningham (2009) presented what was claimed as a new Coriolis verification technology which measured the stiffness of the flow tubes, and hence provided a relationship with the flow calibration factor uninfluenced by process conditions. See also Rensing and Cunningham (2010) on verification using embedded modal analysis.

In this section we have discussed published information on calibration, and touched on the matter of verification. With regard to the latter, it should be noted that one or more manufacturers have introduced methods to check various key operational aspects of the meter to ensure that, in relation to these aspects, the meter continues to operate as when it left the factory.

20.6 Applications, Advantages, Disadvantages, Cost Considerations

20.6.1 Applications

One manufacturer claimed that meters had been applied to flow measurement in polymer/monomer, ethylene oxide, milk, fruit juice and concentrates, liquid chocolate, processed egg, coffee extract, peanut butter, animal and vegetable fats, vegetable oil, titanium dioxide, asphalt, fuel oil and natural gas. Table 20.1 and Figure 20.12 give some of the constraints due to material compatibility. Agarwal and Turgeon (1984) described the use of early meters in coal liquefaction plants, including sour water, solvent, slurry and high-temperature solvent. The meter operated successfully on coal liquids with a high (39%) solids content. Blumenthal (1985) listed application of the Micro Motion meter to the pulp and paper industry: green liquor, black liquor, coating of filler slurries, dyes, fuel oils, tall oil and shear sensitive fluids.

Soderholm (1999) applied the meter to bulk solid flows. Other applications have been in the pulp and paper industry (Altendorf 1998), in adapted form as a possible non-intrusive pressure measurement (Smith and Kowalski 1998), and as part of a multiphase metering unit (Al-Taweel et al. 1997; Henry et al. 2013).

In some designs care may be needed in applications involving liquids with high vapour pressure (low boiling points) such as hydrocarbons like propane and butane, solvents, liquefied gases such as CO_2 and liquids which tend to produce gas. Such liquids might start to gas and form cavities which could lead to measurement errors, and so in applying the meter it will be important to check with the manufacturer that the operating conditions are suitable. This caution is even more important in the case of multiphase and multi-component flows.

Wang and Hussain (2009) analysed the effects of cryogenic temperatures on Coriolis mass flowmeters and addressed the problems of varying Young's modulus and of thermal dimensional effects. The calibration using a NIST cryogenic facility was said to be within "the claimed accuracy for reference conditions".

Other applications mentioned were: density measurement of cement slurry (Benabdelkarim and Galiana 1991); water-cut measurement; on-line well test data

Table 20.1. *Compatibility of 316L stainless steel*

Compatible Fluids	
Adhesives	Asphalt
Beer/beer foam	CNG
Ethylene oxide	Fatty acids
Fruit juices	Fuel oils
Isopropanol	Lime slurries
Liquefied gases N_2, O_2, LPG	Magnetic slurries
Milk/cream	Molten sulphur
Nitric acid	Olefins
Paint	Peanut butter
Phosphoric acid	Pie fillings
Polymers	Polypropylene
Potassium hydroxide	Sodium hydroxide
Sour crude	Tar sands
Urethane	
Incompatible Fluids	
Acetic acid (high concentrations and elevated temperatures)	
Ammonium chloride	
Bromine	
Calcium chloride	
Iodine (other than 100% dry)	
Mustard	
Seawater	
Dyes and inks in some cases	
Fluids containing halogen ions	

From Micro Motion user information with permission of Emerson Process Management Ltd.

(Taylor and Nuttall 1993); blending (Sims 1992); batching of binders in a mixing plant to provide mass delivery of the binder (Chateau 1991).

Miller and Belshaw (2008) investigated the performance of two (4 in (100mm) and 6 in (150mm)) Coriolis meters at liquid viscosities up to 300 cSt. They appeared to be within spec, but were more difficult to zero at higher viscosities. They may be more restricted by pressure loss at high viscosities. Gas entrainment (~1%) severely affected them.

Where density is known, or the liquid is of low value, or where repeatability is more important than accuracy, the Coriolis meter is less competitive (Ginesi and Annarummo 1994). However, the ability to deduce not only mass flow, temperature and density but, from these, percent solids and other data, makes the meter particularly useful. Two-phase fluids can cause problems, as discussed elsewhere in this chapter, and single-tube meters may be preferable. In either design, the largest particles must pass through the meter. Pitting may result and can cause problems due to thinner walls and cycling fatigue. The meter should remain full and preferably be in a vertical upward flow. All air must be bled or purged before zeroing and the piping should avoid air collection. The pump may also need to be running when zeroing is done. The manufacturer's instructions should be followed, particularly on how and where to support the meter, for instance by adjacent pipework and not vice versa. Upstream flow fittings are unlikely to affect meter performance significantly.

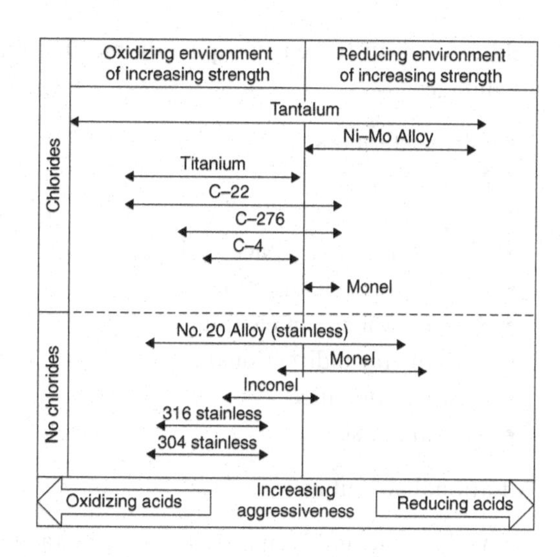

Figure 20.12. Compatibility of materials from Micro Motion (reproduced with permission of Emerson Process Management Ltd.).

Other recent reports and suggestions concerning applications of the meter are:

- **custody transfer** (Basrawi 2003a, 2003b; Mattar 2005);
- **production of alkylate** and precise measurement of **concentration of sulphuric acid** (Dunn et al. (2003);
- in the **gas industry**(Stidger 2003).
- unloading **railcars and tank trucks** (Mattar 2003, 2005); **filling rail wagons** (Riddle 2004); **batch metering, oil wellhead metering and metering difficult foodstuffs** (Mattar 2003, 2005); **ethylene oxide applications** where it is at or near the boiling point and there is a danger of nitrogen break-out (Mattar 2003).
- **particulate flows in air, such as cement, gypsum, fly ash, corn starch etc.,** a suggestion I find somewhat surprising (Fahlenbock 2005); cf. Boyle (2002) re gypsum; **two-phase flows and water-cut** (Mattar 2005) the Invensys/Oxford University collaboration on digital technology; cf. Henry et al. (2004, 2006).
- **cryogenic service**(Patten and Dunphy 2006; cf. Wang and Hussain 2009);
- **medicine (e.g. drug infusion)** and **micro-fluidic devices** (Clark, Wang and Cheesewright 2006a, 2006b) in relation to a micro machined Coriolis meter with range of about 0.1 to 10 mg/s;
- **paint and varnish, high-viscosity and high-pigmentation fluids** (Engelbert, Scheulen and Incontri 2007) used a straight single-tube meter;
- **transfer of fuel oil** (Gregory et al. 2008) if the latest meters can cope with single- and two-phase (oil-air) flow conditions.

20.6.2 Advantages

The advantages of this technology have been evident from the early days, and only needed the steady and significant improvement together with experience in the field to confirm the early expectations. One manufacturer suggested the following advantages:

- Obstructionless
- No bearings
- No probes
- Compact design (space saving)
- Low power consumption
- Suitable for steam cleaning
- Suitable for bidirectional flow
- Low maintenance
- Sanitary design

Additional advantages:

- wide range and high accuracy (Cascetta et al. 1992);
- true mass flow measurement of all liquids, slurries and foams (Cascetta and Vigo 1988);
- multi-parameter measurements (density, volume flow, temperature etc.) in addition to mass flow;
- very small effect due to velocity profiles;
- wide temperature range, from cryogenic up to +400°C.

Early expectations were: fiscal measurement applications as they offered significant capital and operating cost savings (Total Oil Marine – Davis 1990); fiscal duties and as transfer standards (Micro Motion 1993); but see Hannisdal (1991), who did not consider them an alternative for fiscal measurement.

20.6.3 Disadvantages

It is interesting to note early perceived disadvantages which are now largely overcome:

- temperature limitations, tube failure, cleaning due to configuration (Robinson 1986);
- difficult maintenance (cleaning and repairing) with sensor tubes clogging with fluids such as slurries, incomplete temperature compensation, stress effects, creep etc. (Cascetta et al. 1989a):
- high pressure loss; high initial cost; spring constant temperature sensitivity; suitable only for very high-pressure gases; calibration needed if density of liquid differs substantially from that of calibration fluid (Cascetta and Vigo 1988);
- in competition with ultrasonic multi-path meters for volumetric custody transfer for larger (greater than 100 mm) sizes;
- some hazardous environments might need secondary containment; might be limited by pressure drop and sensitivity to changes in temperature, pressure and pipe vibration; cost as a limitation (Walker 1992).

Walker (1992) was, rightly, optimistic about overcoming many of these limitations as has been shown in the past 20 years or more.

20.6.4 Cost Considerations

The initial cost of these meters appeared, in the early years, to be significantly greater than most other meter technologies. However, the introduction of lower-cost alternatives in the Coriolis range may have changed this perception. It is, however, useful for the prospective user to assess the appropriateness of the meter for a specific application in terms of:

a) Initial instrument cost compared with value of product being metered;
b) Cost of installation including any necessary pipework changes in orientation;
c) Cost of operation, maintenance and regular servicing;
d) Cost of recalibration including:

  * The adequacy of in situ methods;
  * The cost of removal and test stand calibration, and any increased uncertainty between test stand calibration and plant performance.

Even when they were writing, Liu et al. (1986), from oilfield experience, had found significant capital and operating cost savings compared with a more conventional approach and this appeared to be confirmed by Robinson's (1986) survey which suggested that after three years these meters are more economical than positive displacement meters.

To me this seems to be all the more likely today.

20.7 Chapter Conclusions

This chapter was initially written prior to 1994 (Baker 1994) and was revised in preparation of the first edition of this handbook. The very significant developments in the technology since then can be gauged by the large amount of references added in this edition, particularly in Appendix 20.A, and dated since 2000. I have attempted to include the key areas of development, and to give a selection of the references to published work. I recognise that I have, very likely, missed some which I should have included. In parallel with this chapter, Dr Wang and I have prepared a more thorough review (Wang and Baker 2014) to which the interested reader should refer.

The meter lends itself to the use of sophisticated computer models which take in the fine detail of the construction as well as allowing for fluid behaviour, interaction of fluid and tube, compressibility, homogeneity and other features. I have placed much of this material in Appendix 20.A. As suggested in Chapter 24 the optimisation of models should also introduce factors relating to the precision of the manufacturing options, as these are likely to have an influence on final instrument performance. Some work has already been published in this area. Combined with new materials this is leading to production methods which provide an increasingly high-quality product with wide applicability.

These meters are, clearly, most at home with single-phase liquid flows, and we should expect to see uncertainty of the ±0.05% level claimed increasingly widely for 30:1 turndown or greater.

For gases the meter appears to be finding an important niche. In some areas the meter is likely to be in competition with ultrasonic and thermal techniques. While neither of these offers direct mass flow, both approach it and may be less costly.

Discussion of the accuracy of a gas Coriolis meter is implicit in some of the research reported. Effects due to compressibility and other features of the gas could mean that, for the highest accuracy, the meter becomes sensitive to type of gas.

For two or more components in the flow, there are clear problems in the use of these meters, although recent research is addressing this problem with some apparent success. The fundamental problem is the possibility that the components can move relative to each other. Relative motion should be essentially eliminated if the phases are of similar density and the liquid viscosity is high so as to inhibit bubble relative movement. However, the new approach of using multiphase models to interpret the meter signals and, in addition, neural network methods, may prove to be very fruitful.

The meter has developed from the U-tube designs which have been so successful, through various configurations to the very significant development of the single straight tube meter Figure 20.13 (see Hussain and Farrant 1994; Rieder and Drahm 1996; Yamashita 1996). See also Stansfeld et al. (1988) and Harrie (1991) for details of an early design, which appears to have been withdrawn. This trend has resulted in Endress+Hauser's viscosity measurement facility within the meter in addition to the other more common measurements of temperature and density. Interestingly, the problems with straight single-tube designs have led to recent designs with bent tubes as an apparent compromise between U-tube and straight tube.

A silicon resonant sensor structure for Coriolis mass-flow measurements of very small size has been described (Enoksson, Stemme and Stemme 1997), and this may be an important development for the future. Clark et al. (2006a, 2006b) described work on a micro-machined Coriolis meter and the evaluation of the meter using the simulation at Brunel University. Its range appeared to be about 0.1 to 10 mg/s. Recent work has been reported on the development of a sensor-chip Coriolis meter (Wiegerink et al. 2012; cf. Sparks et al. 2003; Sparreboom et al. 2013; Lötters et al. 2013) and a commercial design may be available now or in the near future.

What will be the next major development? Although the single straight tube has been produced by several manufacturers and its development is discussed in various papers (cf. Drahm 1998 and Van Cleve et al. 2000), tube shape continues to be an area of apparent development. Finite element (FE) analysis appears to have been important in enabling the designers to overcome operational problems. Some of the recent work suggests that the single tube vibration can be in pipe cross-section rather than in longitudinal bending; see Hussain (2000a). Could the next stage be shorter pipes, or multiple sensors; see Hussain (2000b)? Is there a need to return to the vibrating elements within tubes (Hemp 1994b) or travelling wave devices which have been suggested in the past? Could tubes be set into torsional motion with

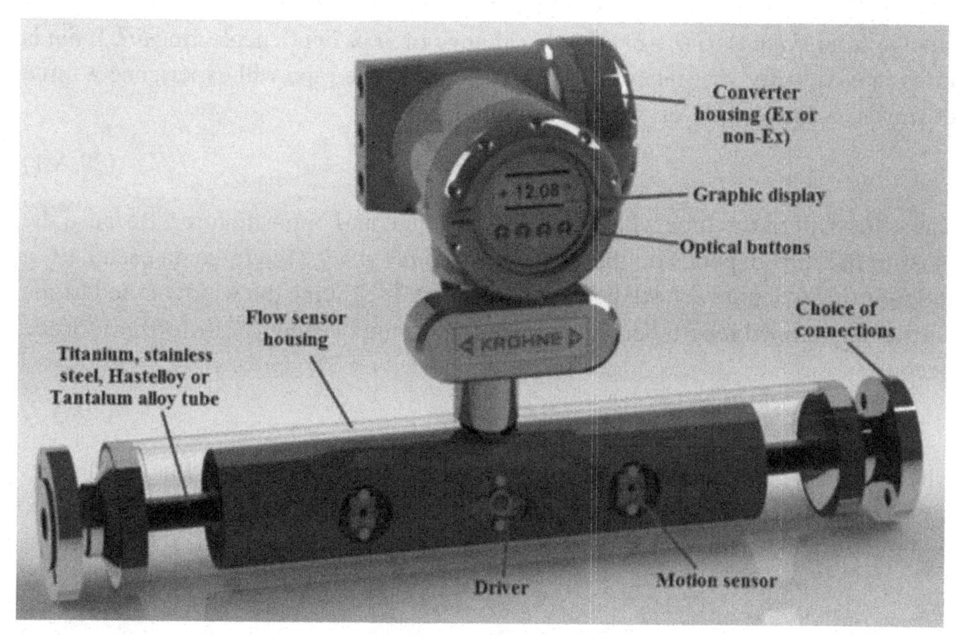

Figure 20.13. Single straight-tube Coriolis meter (reproduced with permission of KROHNE Ltd).

internal segmental arrangements – a cross between the angular momentum devices and the Coriolis?

Matthews and Ayling (1992) described a meter which had tines, rather like an elongated tuning fork, which stretched the length of the meter. The Coriolis principle is still valid, and Hemp's work on the oscillating element meter is presumably relevant. The meter has, as a result, an essentially straight spool piece with the tines projecting into it. The initial work was on a 4 in (100 mm) meter and the plans were that larger sizes should be developed capable of use on refined single phase liquids and gases.

Can the manufacturers reduce the price of these instruments yet further (Blickley 1995)?

Paton (1998) reviewed the capability and methods being used in European standards laboratories to calibrate and test these meters and to determine their potential as a transfer standard. Their potential since then has been amply demonstrated.

Could a clamp-on meter be successful? These are briefly mentioned in Chapter 22 (cf. Bartstra 2006).

Appendix 20.A Notes on the Theory of Coriolis Meters

20.A.1 Simple Theory

If the flow velocity is V, Figure 20.4, and the angular velocity caused by the vibration is Ω upwards, then a mass of δm at a radius r will experience an upwards velocity, $r\Omega$, which increases at $r+\delta r$ to $(r+\delta r)\Omega$. The angular momentum of the mass, δm, will change from $r^2\,\delta m\,\Omega$ to $(r+\delta r)^2\delta m\,\Omega$, a change of $2r\,\delta r\,\delta m\,\Omega$, neglecting δr^2. If δm is moving at velocity V, it takes $\delta r/V$ seconds, and so the pipe will experience a force downwards due to the Coriolis acceleration of:

$$F = 2\Omega V\,\delta m \tag{20.A.1}$$

The other half of the tube will experience an equal and opposite force. Because the mass of the tube depends on the fluid density, $\delta m = \rho A\delta r'$, the force on each half of the tube due to length $\delta r'$ will be equal to $2\Omega\rho AV\delta r'$. Taking the width of the U-tube (or the unwrapped length between sensors) as d, the twisting (or distorting) torque becomes

$$T = 2K\Omega\rho AVdl \tag{20.A.2}$$

where l is the length of each half of the tube and K allows for the fact that the distorting Coriolis forces will not form a straight integration. The mass flow is $q_{\mathrm{m}} = \rho AV$, and the torque can be related to the mass flow rate by

$$q_{\mathrm{m}} = T/(2K\Omega dl) \tag{20.A.3}$$

We can now introduce the oscillating motion by putting $\Omega = \Omega_0\cos\omega t$, where ω is the driving frequency and t is time, and hence

$$T = 2K\Omega_0\rho AVdl\cos\omega t \tag{20.A.4}$$

If, for the special case of the U-tube, we relate the twisting torque to the twist in the tube, θ, (ignoring damping effects etc.) by

$$I_{\mathrm{s}}\frac{d^2\theta}{dt^2} + K_{\mathrm{s}}\theta = T \tag{20.A.5}$$

where I_{s} is the inertia and K_{s} is the spring constant of the U-tube in twisting oscillation, and assume a solution of the form $\theta = \theta_0\cos\omega t$ we obtain an expression for θ_0 by equating T in Equations (20.A.4) and (20.A.5)

$$\theta_0 = \frac{2K\Omega_0\rho AVdl}{K_{\mathrm{s}} - I_{\mathrm{s}}\omega^2} \tag{20.A.6}$$

where θ will be in phase with Ω. It is then apparent that while the twisting torque is at its maximum at the mid point, the twist is zero at the extremities of the oscillation. This is illustrated by Blumenthal (1984) and shown in Figure 20.5. The angle of twist is of order 1/100 degrees.

The free vibration frequency in twisting occurs when $K_{\mathrm{s}} = I_{\mathrm{s}}\omega^2$ or

$$\omega_{\mathrm{s}} = \sqrt{K_{\mathrm{s}}/I_{\mathrm{s}}} \tag{20.A.7}$$

We can now obtain a relationship between q_{m} and θ_0

$$q_{\mathrm{m}} = \frac{K_{\mathrm{s}}(1-\omega^2/\omega_{\mathrm{s}}^2)\theta_0}{2K\Omega_0 ld} \tag{20.A.8}$$

(Damping would prevent the number in the numerator from becoming zero when $\omega = \omega_s$ We can go a step further and relate the twist, θ_0, to the time difference for transit of the two sides of the U-tube. Because the velocity of the U-tube at the sensor is approximately $\Omega_0 l$, and the displacement due to the twist is $\theta_0 d$, the time of transit of the two sides will differ by

$$\tau = \theta_0 d/(\Omega_0 l) \tag{20.A.9}$$

Hence the mass flow is given in terms of the transit time difference by

$$q_m = \frac{K_s \tau(1 - \omega^2/\omega_s^2)}{2Kd^2} \tag{20.A.10}$$

A similar equation will give the relationship for a straight tube. Note that the actual value of angular velocity, Ω, has been eliminated from this expression. The amplitude of the vibration is typically between 60 micrometers and 1 mm and the frequency in the range 80 to 1,100 Hz. The higher frequency tends to be well above most common mechanical vibrations. In most meters two tubes oscillate in anti-phase, acting like a tuning fork. When flow occurs the forces created result in phase shifts in the motion of the two halves of the oscillating tube and the passage of the halves of the tube past fixed points will therefore be displaced in time and the time difference will be measurable.

A similar equation to (20.A.6) can also be used for the oscillation without flow, and for this the natural frequency of the tube will be given by

$$\omega_u = \sqrt{K_u/I_u} \tag{20.A.11}$$

where K_u is the spring constant of the tube in normal oscillation and I_u is the inertia in that plane. Since I_u is proportional to the mass of the tube and therefore related to the density of the fluid, the natural frequency can be used to obtain density of the fluid. In particular the frequency is little affected by flow so that the density can be obtained with flow (Raszillier and Raszillier 1991).

20.A.2 Note on Hemp's Weight Vector Theory

Hemp (1994b) has developed a weight vector theory for the Coriolis meter (cf. appendices in electromagnetic, ultrasonic and thermal chapters). The flowmeter signal is given by

$$\Delta\phi = \int \mathbf{V}.\mathbf{W}\, dv \tag{20.A.12}$$

where $\Delta\phi$ is the phase difference between the total velocities at the two sensing points. The weight vector is then given by

$$\mathbf{W} = -\rho\left[\left(\mathbf{v}^{(2)} \cdot \nabla\right)\mathbf{v}^{(1)} - \left(\mathbf{v}^{(1)} \cdot \nabla\right)\mathbf{v}^{(2)}\right] \tag{20.A.13}$$

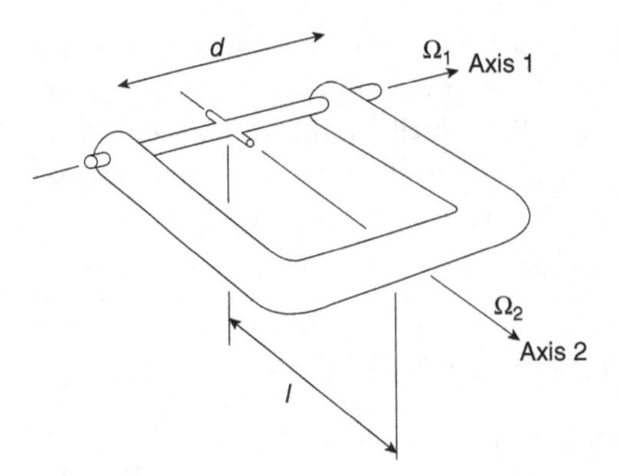

Figure 20.A.1. Geometry of Hemp's U-tube meter (after Hemp 1994).

where $\mathbf{v}^{(1)}$ and $\mathbf{v}^{(2)}$ are oscillatory velocity fields set up in the stationary fluid by the driving transducer and by equal and opposite unit alternating forces applied at the sensing points. (see the last paragraph of the fourth section of Hemp 1994b).

Hemp commented that the theory had been applied at that date (1994) only to two relatively simple flowmeter configurations. The first was for a U-tube meter, Figure 20.A.1, where the tube was rigid and rotation took place about two shafts: one at the "top" of the U, axis 1; the other "vertically" through the centre of the U and parallel with the sides of the U, axis 2. The oscillatory velocity fields, $\mathbf{v}^{(1)}$ and $\mathbf{v}^{(2)}$, were respectively those that accompany a rotation of angular velocity Ω_1 about axis 1 due to the driving force, and a rotation of angular velocity Ω_2 about axis 2 due to forces of unit size applied to the sensing points and in the direction of the Coriolis forces.

Hemp obtained the value of $\mathbf{W}$ as $\pm\frac{1}{2}\rho d\Omega_1\Omega_2$ for the "sides" of the U-tube and parallel with the sides, and as $-\rho l\Omega_1\Omega_2$ along the straight "bottom" of the U and in the same direction as the rotational vector. Hemp commented that since $\mathbf{W}$ is constant and parallel to the tube axis in each straight section and in the direction of flow, perfect averaging of the flow is achieved. The effect of flow in the bends had not been calculated.

The second configuration was for a rigid elliptical cylinder which might be mounted across the diameter of a pipe. Hemp and Hendry confirmed that the weight vector for a vibrating element flow sensor of this sort would mainly be sensitive to flow near the element, and, in a pipe, would be prone to velocity profile effects.

Hemp also stated that the theory was limited to small amplitude vibrations. He suggested that, for short vibrating tube Coriolis meters and in vibrating element sensors, the assumption was probably adequate, but that for traditional meters with long tubes amplitudes could be of order 1/10th of a tube diameter, and might introduce errors into the theory. The theory may also need to be extended to allow for compressibility and turbulence.

As with all Hemp's theories, this is very elegant. It appears to confirm, for the U-tube meter, that installation effects resulting from profile distortion (as opposed to vibration, very high turbulence or pulsation) are likely to be, in general, of minor concern in this type of flowmeter (cf. Hemp and Hendry 1995), but of major concern in vibrating element sensors.

20.A.3 Theoretical Developments

The fundamental equations relating to the flow in a pipe have been set out by Laithier and Paidoussis (1981) (cf. Stack et al 1993). Raszillier and Durst (1991) derived the equation of motion for a straight-tube meter. Sultan and Hemp (1989) developed a theory for a U-tube meter, and obtained impressive agreement with experimental data although indicating various limitations of the theory and further areas of work still needed. [Hemp's (1988) reciprocity theorem to reduce zero drift errors may also be applied to Coriolis meters. See also Hemp and Yeung (2003) on modification of earlier theories to allow for viscosity.]

Durst and Raszillier (1990) have computed the perturbation of the flow in a pipe which is rotating and considered the relevance of flow field and forces which result to Coriolis meters. Raszillier and Raszillier (1991) used dimensional arguments to show the possible interdependence of velocity and density. Hemp (1996) used dimensional analysis to explore design requirements for use with low mass flow rates such as for gas near ambient conditions.

Raszillier and Durst (1991) undertook further analysis of the motion of the Coriolis flow tube. They used a perturbation solution. They derived the equation

$$\left(\rho_f A_f + \rho_p A_p\right)\frac{\partial^2 u}{\partial t^2} + 2\rho_f A_f v_0 \frac{\partial^2 u}{\partial x \partial t} + \left(\rho_f A_f v_0^2 - T\right)\frac{\partial^2 u}{\partial x^2} + EI\frac{\partial^4 u}{\partial x^4} = 0 \qquad (20.A.14)$$

where we have replaced the fluid mass per unit length by $\rho_f A_f$ and the tube mass per unit length by $\rho_p A_p$. u is the transverse movement of the pipe, v_0 is the velocity of the fluid "string"; E is Young's modulus, I is pipe moment of inertia in the direction of vibration and T is the combined tension in the pipe and in the fluid. They took the boundary conditions for u and $\partial u/\partial x$ as zero at the ends of the pipe. They suggested that, as the tension plays no essential role, they should neglect it. In the equation, the first term is a linear acceleration term, the second is the Coriolis term since $\rho_f A_f v_0$ is the mass flow rate of the fluid, and the third term $\left(\rho_f A_f v_0^2\right)\frac{\partial^2 u}{\partial x^2}$ is a centrifugal term due to curvature of the pipe. The authors introduced the question of when a meter obeying this equation is, indeed, a Coriolis effect meter.

They solved the vibration of the pipe with fluid stationary ($v_0 = 0$). The amplitude of the first mode is shown in Figure 20.A.2.

Raszillier and Durst (1991) then used a perturbation method to obtain the Coriolis deflection, making the assumption that the centrifugal term might be neglected for sufficiently low velocities.

The shape of the Coriolis perturbation to the pipe primary bending mode of vibration is shown in Figure 20.A.3. The perturbation, which is smaller in amplitude than the mode shown in Figure 20.A.2, is superimposed on the primary mode, but the amplitude of the primary mode has a time dependence of $\sin(\omega t)$, while the

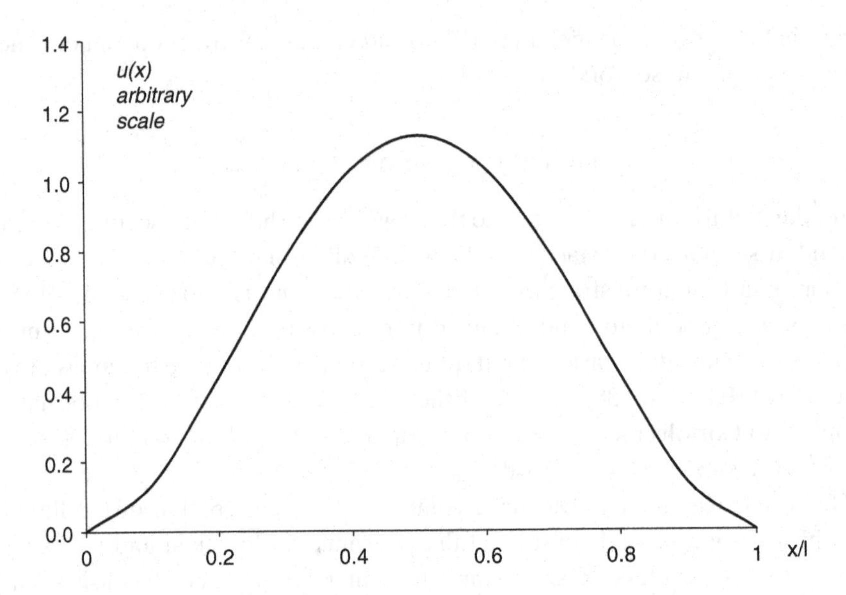

Figure 20.A.2. Normalised amplitude of the first mode of vibration of the pipe with $v_0 = 0$ (after Raszillier and Durst 1991).

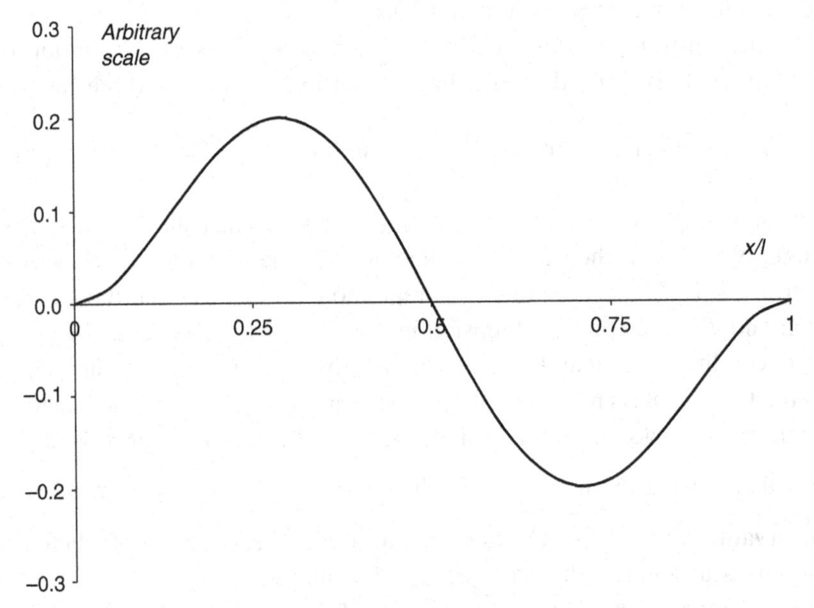

Figure 20.A.3. Shape of the mode caused by the Coriolis acceleration (after Raszillier and Durst 1991).

perturbation amplitude has a time dependence of $\cos(\omega t)$. The perturbation is, therefore, at a maximum when the pipe is at its mean position and zero when the pipe is at its maximum deflection. The paper implies that the ratio of the amplitude of the perturbation mode to the amplitude of the primary mode is of order $\frac{v}{l\omega}\frac{\rho_f A_f}{\rho_p A_p + \rho_f A_f}$, where v is the flow velocity, l is the pipe length, ω is the frequency, and the second fraction is the ratio of fluid mass to total mass of pipe and fluid. Figure 20.5 provides a diagrammatic equivalent motion for the U-tube design.

Raszillier, Allenborn and Durst (1993) have suggested that by influencing the vibration spectrum the sensitivity of the instrument may be increased. Raszillier, Allenborn and Durst (1994) investigated the effect of a concentrated mass at the middle of the Coriolis pipe segment as required for the purpose of symmetric excitation of the vibration. The flowmeter factor K was found to be almost independent of the mass up to fairly high values compared with the mass of the fluid-filled pipe segment, although the frequency of the fundamental mode is strongly influenced. Lange et al. (1994) examined the effect of detector masses on the calibration of Coriolis flowmeters. They showed that the position of the detectors must be chosen with care since it may have implications for the accuracy to which the calibration of the meter remains independent of the fluid density. The calibration may become dependent on fluid density if the detector masses are not negligible compared with the pipe mass. The authors suggested the most appropriate positions for the first symmetric and the first antisymmetric vibration modes (see also Sultan 1992).

Keita (1994) gave a useful discussion of the equations used to predict behaviour. He also concluded that the main source of calibration shift in the gas meter was due to pressure, and that compressibility effects were small.

The details of the meter design and their importance in understanding the behaviour of the meter have led to the use of finite element (FE) methods to model the meter (Watt 1991; Stack and Cunningham 1993; Hulbert, Darnell and Brereton 1995; Belhadj, Cheesewright, and Clark 2000; cf. Cheesewright and Clark 1998; Wang, Baker and Hussain 2006b).

An important example of the outcome of the FE analysis of the Coriolis flowmeter was the development of the single straight tube meter (cf. patent by Hussain and Rolph 1994). Two particular approaches are possible to the FE method of solution. One of these uses the solution of the flow field, and also the solution of the structural field, and iteration allows one to be fed into the other either by forces applied by the fluid, or mesh distortions on the flow caused by the structure. The other is to use a coupled method where the coupling is applied in the derivation of the equations for solution by the FE method. See Pawlas and Pankratz (1994) for an iterative approach linking CFD analysis to structural analysis of the rotation of the flow tube, and hence the pressure distribution on the tube wall. See also Kumar et al. (2010), who used a fluid-structure interaction simulation to investigate a deviation at low Reynolds numbers.

Wang and Baker (2003/2004) used a direct coupled approach for the single tube development. The transverse deflection, u, and the rotational deflection, θ, are given by the equations of motion

$$\left(\rho_f A_f+\rho_p A_p\right)\frac{\partial^2 u}{\partial t^2}+2\rho_f A_f v_0\frac{\partial^2 u}{\partial x\partial t}+\left(\rho_f A_f v_0^2-\sigma_0 A_p\right)\frac{\partial^2 u}{\partial x^2}+kGA_p\left(\frac{\partial\theta}{\partial x}-\frac{\partial^2 u}{\partial x^2}\right)=0$$

$$\left(\rho_f I_f+\rho_p I_p\right)\frac{\partial^2\theta}{\partial t^2}-kGA_p\left(\frac{\partial u}{\partial x}-\theta\right)-\left(E+\sigma_0\right)I_p\frac{\partial^2\theta}{\partial x^2}=0 \tag{20.A.15}$$

where the fluid is assumed to be incompressible, to have uniform and constant velocity throughout the flow tube and to move with the tube of cross-section A_f, to have density ρ_f and a rotational inertia I_f. The initial stress in the pipe is σ_0, the

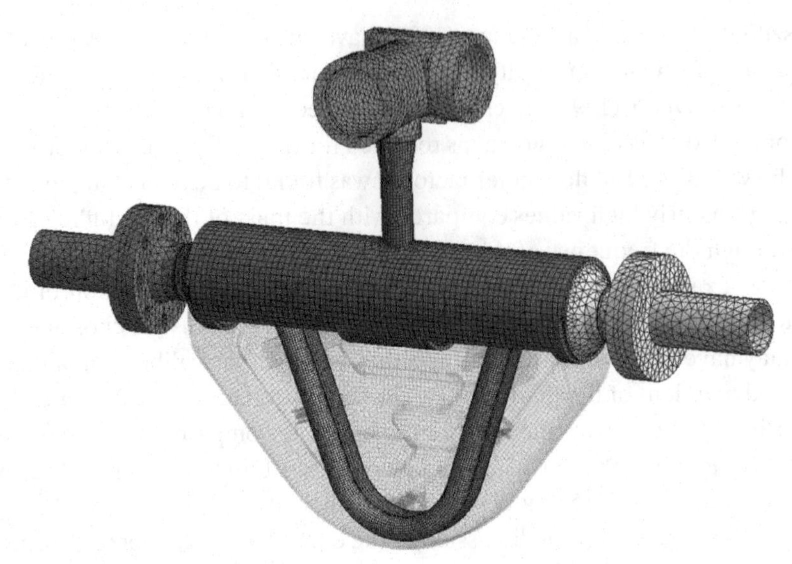

Figure 20.A.4. A typical finite element (FE) model for a complete Coriolis flowmeter (KROHNE's latest generation of twin bent-tube flowmeters) together with adjacent process connection and pipes (Reproduced with the kind permission of KROHNE).

cross-section of the pipe is A_p, the density of the pipe is ρ_p, the rotational inertia is I_p, Young's modulus is E, shear modulus G, and shear correction factor is k. A general purpose finite element program was used which allowed the insertion of the mass, damping and stiffness matrices to be applied. Further details of this approach are given by Wang and Baker (2003/2004), who reference the sources of the approach they used. The FE package used provided a matrix-based element in which the properties could be defined by stiffness, damping and mass matrices. The analytical solution of Raszillier and Durst's (1991) could be used to check the FE results. An example of an FE mesh is shown in Fig 20.A.4.

As is apparent from these developments, the Coriolis meter lends itself to the use of sophisticated computer models which take in the fine detail of the construction as well as allowing for fluid behaviour, compressibility, homogeneity and other features. In their work Hulbert et al. (1995) noted that axial tension terms were important. Keita (2000) used fluid structure interaction for a straight pipe and got good agreement with previous data. See also Cunningham and Hensley (2001). Kutin and Bajsić (2001) discussed stability boundary effect near to what they appear to have termed the critical flow rate.

The optimisation of these models, including new materials, should also introduce factors relating to the precision of the manufacture, which in turn influences instrument performance. Wang and Baker (2003/2004) examined the effect of manufacturing variation on the performance of a single tube flowmeter. Enz (2010; cf. 2011) looked further at the effect of manufacturing variation in the positioning of driver and detectors on the phase shift, and suggested that this could also affect zero stability and the sensitivity of the meter.

(See Drahm 1998 on the Endress + Hauser single straight tube meter.)

In addition to mass flow rate the meter may provide: temperature, density, volumetric flow rate and viscosity. Drahm and Bjønnes (2003) used the torsional vibration of a single-tube Coriolis meter, which causes a consequent shear stress on the fluid at the wall of the tube leading to a rotational motion of the fluid, the consequent effect on the torque providing a means to obtain the viscosity of the fluid. See also Drahm and Staudt (2004), Hussain (2003) and Standiford and Lee (2010).

Hemp and Sultan (1989) explained the concept of effective mass of bubbles in a liquid and calculated the expected error. Compared with Grumski and Bajura's (1984) measurements the theory overestimated the error possibly due to interaction between bubbles. Liu et al. (2001) described a neural network to correct mass flow errors caused by two-phase flow in a digital Coriolis mass flowmeter. They based this on internally observed parameters. They suggested that for a limited range of flow, temperature and void fractions, the mass flow errors were reduced from 20% to within 2%. They also suggested that while the basic trend followed the bubble model, there were sufficient differences to require a more detailed model. More recently Rieder et al. (2005) discussed measurement errors in two-phase flows and developed a theory using a moving resonator model to supplement the "bubble effect" model of Hemp. The model appeared to provide an explanation of errors in various situations and suggested errors of 2.4% for mass flow and –0.4% for density even with gas concentration of 1.5% by volume and a pressure of 5 bar. Lower pressure or increased gas fraction increased these figures significantly. Basse (2014) reviewed the theory for errors due to particle/bubble effects. The force due to a particle/bubble in an oscillating fluid is derived which can be used to obtain an estimate of the meter error.

Wang et al. (2006a) described a simulation which enabled Coriolis designs to be tested in a virtual mode. It allowed dynamic response, signal processing and new designs to be checked and developed. The FE code also allowed pseudo data to be generated for specific points on the Coriolis flow tubes. The model had the potential for simulating the effect of both corrosion and deposition within the flow tubes.

Wang et al. (2006b) reported on an advanced numerical model for single straight-tube Coriolis flowmeters. The predictions of the model agreed well with the data from the manufactured meters. Cheesewright and Shaw (2006) discussed the accuracy of such methods and suggested an approach to achieve high-quality results.

Wang and Hussain (2007) discussed recent numerical research and development of twin straight-tube meters with prototype comparisons.

Ruoff, Hodapp and Kück (2014) developed an FE beam model for arbitrary-shaped pipe geometry allowing for unsteady flow conditions. They claimed that the results indicated the high accuracy which their model could achieve.

20.A.4 Coriolis Flowmeter Reviews

In a special issue of *IEE Computing and Control Engineering* a series of summaries of papers given at an Oxford seminar were introduced by Tombs (2003). Some of the papers are referenced in this chapter. Also a special issue of the *Journal of Flow*

Measurement and Instrumentation (Cheesewright and Tombs 2006) was dedicated to Coriolis meters. The editors of it noted that an earlier issue (Volume 5, Number 4, 1994 edited by Raszillier and Durst) had identified four subjects in urgent need of improved understanding:

- the behaviour of the instrument in non-stationary flows,
- the influence of fluid compressibility,
- corrections in the measuring principle for high flow rates and
- the behaviour of meters in two-phase flow.

They noted that there had been some work published on the first two topics with a reasonable gain in understanding, but they considered that the matter of high flow rates was not of great concern at that time, and that the problem of two-phase flow would take rather longer to solve.

Anklin, Drahm and Rieder (2006) gave an updated review (following that by Baker 1994) which had some useful explanations and descriptions of the meters available. Other contributions not included earlier in the chapter include:

- Investigation of non-proportional damping on zero shifts by perturbation of an FE model of the meter to simulate the damping (Timothy and Cunningham 1997);
- the use of three accelerometers, and the application of wave decomposition theory to the vibration signals. The method appeared to depend essentially on the same theoretical basis as the Coriolis meter, and could indicate an extension of the Coriolis method (Kim and Kim 1996);
- investigation of complete body rotation of the Coriolis meter which led to signal errors and should be avoided (Ma and Eidenschink 2001);
- investigation of the effect of speed of sound on Coriolis mass flowmeters, resulting in a correction for high speed gas flows which may be of the order of a percent (Anklin et al. 2000);
- suggestion that the dynamic stiffness matrix method was capable of modelling a Coriolis mass flowmeter with a general plane-shaped pipe: not allowing for damping which could be added (Samer and Fan 2010);
- use of fibre-optic curvature sensors in a meter based on the Coriolis principle (Dakić et al. 2013);
- use of a vibrating tube to explore other parameters, as well as mass flow, which might be deduced (Zhang 2012).

A very recent and thorough review is by Wang and Baker (2014).

21

Probes for Local Velocity Measurement in Liquids and Gases

21.1 Introduction

We resort to probes for flow measurement for three main reasons:

i. To provide a low-cost method of flow monitoring;
ii. To provide an in situ calibration;
iii. To obtain fine detail of the flow in the pipe (velocity profile, swirl and turbulence).

For our present purposes, we shall not consider (iii). It is relevant to fluid mechanics research, mainly uses hot-wire and laser Doppler anemometers, and is, therefore, outside the scope of this book. We will concentrate on work relating to bulk flow measurement. In situ calibration techniques are covered in Chapter 4. In this chapter, we shall consider the various devices that have been used and are, or have been, commercially available either as a local flow monitor or as the probe for in situ calibration. However, before considering these, we need to understand what is being measured.

i. Ideally, the probe measures the local velocity and obtains a representative value of the mean velocity from the shape of the velocity profile. Alternatively, it may measure the complete profile.
ii. In practice, the probe may be in error because:
   - it was calibrated in an approximately uniform profile or on the pipe axis but will need to measure velocity profiles that result in a significant variation of velocity across the sensing head;
   - it will alter the flow by its presence, may shed vortices and may oscillate as a result; and
   - the flow pattern will continue to change as the probe head is inserted further and further into the pipe and as it approaches the opposite side of the pipe.

In some cases, these flows may be analysed theoretically, but in most cases one would expect to calibrate the probe, and this calibration will be different at the pipe axis and at the pipe wall.

iii. The probe will reduce the flow area of the duct by its presence, and the area will continually decrease as the probe is inserted further. This blockage will need to be allowed for. The blockage will also differ between incompressible and compressible fluids.

Our object in using a probe is to obtain the best measure of the local flow rate that can be obtained. We shall need to remember both the limitations of the probe and the uncertainty of the relationship between the reading and the mean velocity. We also need to remember that, to deduce the mean flow rate in the duct, there is uncertainty in the measurement of the cross-section of the pipe and the blockage caused by the probe.

The variation on one probe, according to tests by the National Engineering Laboratory in Scotland, suggested that blockage at the centre for pipe sizes from 250 to 750 mm ranged between about ±5% and the change as the probe moves from the centre to the wall can be of order 10%.

For any new probe design, there is clearly a need for some careful work as to

a. what velocity the probe is measuring in an ideal uniform flow;
b. how the signal will change in a real, non-uniform pipe flow, and when close to the wall; and
c. how the flow in the duct will be altered by the probe and how the local probe measurement would relate to the flow in the duct if no probe were present.

In Section 2.3 and, specifically, in Table 2.2, we noted that the velocity in a turbulent flow at about the 0.76 radius point (0.24 of a radius in from the wall) is practically equal to the mean flow in the pipe. So this suggests that the probe should be set at this point. However, this may result in reduced precision due to the shape of the flow profile in this region (b), the level of turbulence and the proximity of the wall. (cf. Woo and O'Neal 2006 on effect of an elbow upstream)

21.2 Differential Pressure Probes – Pitot Probes

Using Equation (2.10), but introducing a factor k for non-ideal conditions, we can obtain the pressure rise when flow comes to rest in the entrance of a pitot tube (Figure 21.1) as

$$\Delta p = \frac{1}{2} k \rho V^2 \tag{21.1}$$

For ideal conditions, $k = 1$. Δp is the difference between the stagnation pressure at the mouth of the pitot tube and the static pressure; in this case, it was obtained with a wall tapping in the same cross-sectional plane as the mouth of the pitot tube. The static pressure at the wall tapping will be the same as the value at the pitot tube, assuming that there is no cross-flow in the tube. The measurement assumes that any pressure effects caused by gravity between the pitot tube entry and the wall tapping will be equal for both the total and the static measurements and will, therefore, cancel out.

Figure 21.2 shows two forms of pitot-static tube used to obtain the dynamic pressure or velocity head. Figure 21.2(a) is a diagram of a tube recommended by Bean

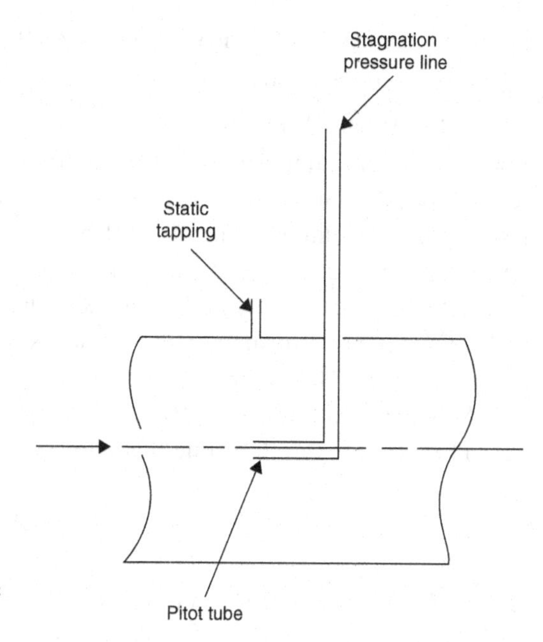

Figure 21.1. Separate pitot and static tubes.

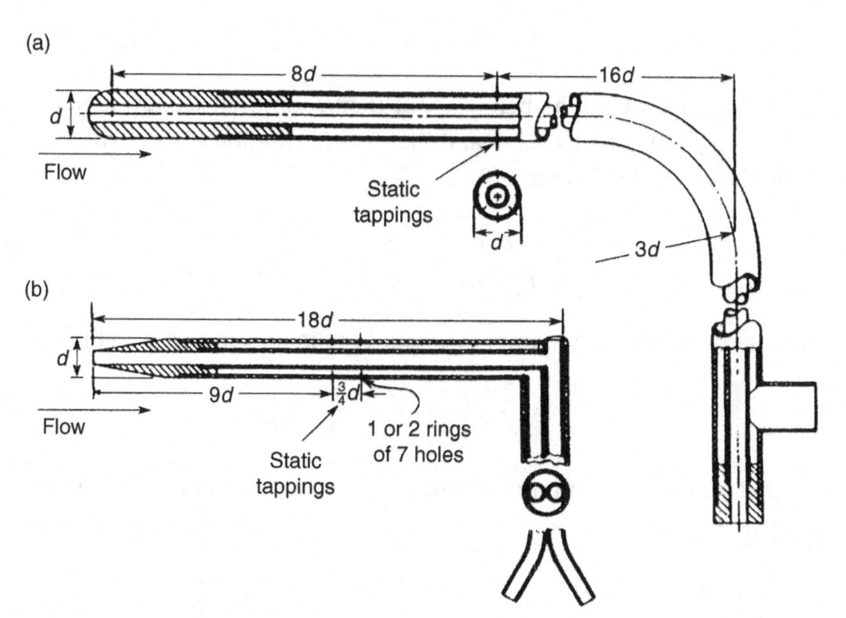

Figure 21.2. Two designs of pitot-static tube recommended by Bean, both similar to NPL designs (reproduced from Bean 1971 with permission of ASME): **(a)** Round-nosed pitot-static tube; **(b)** conical-nosed pitot-static tube.

(1971). A National Physical Laboratory (UK) design appears to differ in that it has an ellipsoidal nose, which is purportedly less sensitive to lack of smoothness between the tip and the cylindrical section. Figure 21.2(b) shows an alternative design. Note in Figure 21.2 the coaxial arrangement of the tubes so that the total pressure is carried in the centre of the tube and the static pressure is carried in the outer annulus. Note also the sensing holes for the static pressure. Bean (1971) recommends that d should be between about 5 and 8 mm $\left(\frac{3}{16}\text{ to }\frac{5}{16}\text{ in.}\right)$.

At low Reynolds number based on the tube diameter (<100) when the effect of viscosity becomes important, k diverges from unity. Goldstein (1965) gave k values for conical entry and round-nosed designs in the range 1.020–1.055 at about 0.6 m/s (Re ≈ 300), a minimum value of $k = 0.991$ at about 3–3.5 m/s (Re 1,600–1,900), and a value within about 0.1% of unity for 6–27 m/s (Re 3,000–10,000). Bean (1971) also reckoned that, for correct alignment (less than 12°), the reading should be within 0.3%. For further information, the reader is referred to ISO 3966 (2008) or Ower and Pankhurst (1966).

At high velocities in a gas, there are compressibility effects. The simple Equation (21.1) has to be replaced by one that allows for these effects. Using Equation (2.16)

$$\frac{p_0}{p} = \left(1 + \frac{\gamma - 1}{2}M^2\right)^{\gamma/(\gamma-1)} \tag{2.16}$$

where $M^2 = \rho V^2/(\gamma p)$ for a perfect gas, we obtain a modified equation for V

$$V = \sqrt{\frac{2\gamma}{\gamma-1}\frac{p}{\rho}\left\{\left(\frac{p_0}{p}\right)^{(\gamma-1)/\gamma} - 1\right\}} \tag{21.2}$$

Bean (1971) suggested that, below about 15 m/s for air at sea level, there is negligible difference between this value and that from Equation (21.1) (with $k = 1$). We can see this if we rewrite Equation (21.2) as

$$V^2 = \frac{2\gamma}{\gamma-1}\frac{p}{\rho}\left\{\left(1 + \frac{\Delta p}{p}\right)^{(\gamma-1)/\gamma} - 1\right\} \tag{21.3}$$

If we then use the binomial theorem to expand the inner bracket, we obtain

$$V^2 = \frac{2\gamma}{\gamma-1}\frac{p}{\rho}\left\{1 + \frac{\gamma-1}{\gamma}\frac{\Delta p}{p} - 1\right\} \tag{21.4}$$

assuming that we can neglect terms of order $(\Delta p/p)^2$ and greater. This reduces to

$$V^2 = \frac{2\Delta p}{\rho} \tag{21.5}$$

In neglecting terms of order $(\Delta p/p)^2$ or greater powers, we have neglected terms of order $\left(\frac{1}{2}\rho V^2/p\right)^2$ or greater. If $\rho = 1.3$, $V = 15$ m/s, and $p = 10^5$ Pa (1 bar), then the error resulting from ignoring $\frac{1}{2}\rho V^2/p$ is of order 0.15%, confirming Bean's value.

One major problem with the pitot tube is the effect of unsteady flow on its reading, and because turbulence is very common in normal industrial flows, the pitot tube will tend to read high (cf. Goldstein 1936). MacMillan (1954) discussed the effects of low Reynolds number and noted (1957) that in a shear flow there was an apparent

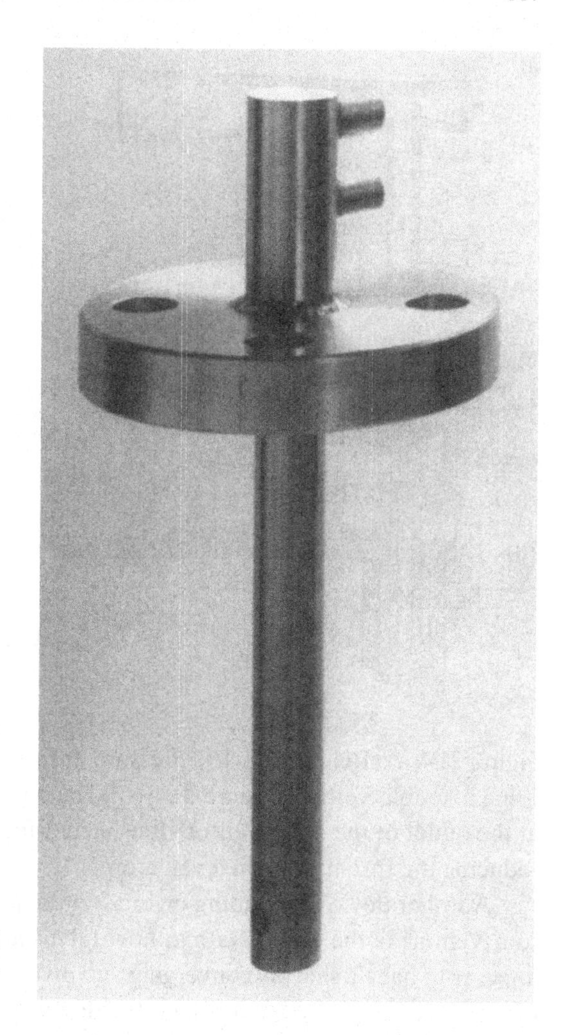

Figure 21.3. Probe-type differential pressure sensor [reproduced with permission of Laaser (UK) Ltd.].

shift in the effective centre of the pitot tube of $0.15d$ in the direction of high velocity, where d is the outside diameter of the pitot tube.

One manufacturer offered a device that is not strictly a pitot tube (Figure 21.3) but consists of a cylindrical probe with sensing at the end. The differential pressure resulting from the flow acts on a spring-loaded diaphragm, the movement of which is sensed by an inductive displacement transducer. Bean (1971) referred to designs with upstream and downstream pointing tubes (cf. Cutler 1982).

21.3 Differential Pressure Probes – Pitot-Venturi Probes

In order to increase the pressure difference in the pitot tube in low flow rates, the pitot-Venturi has been used. In this device, the total pressure is compared with a depressed pressure created by the flow at the throat of the Venturi. This is shown in

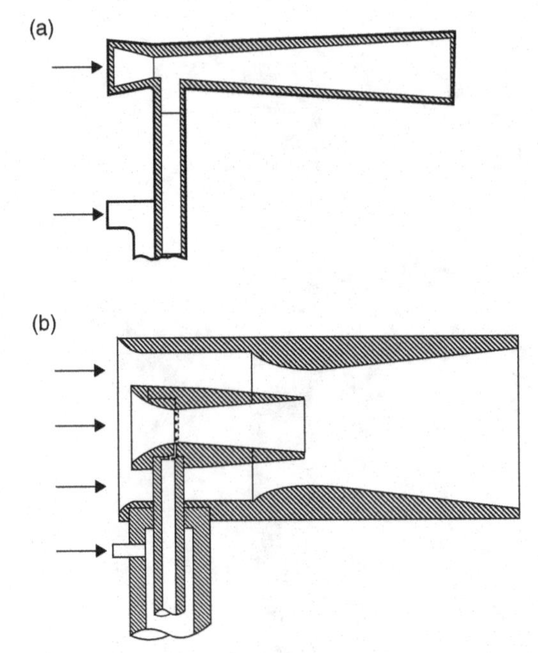

Figure 21.4. Pitot-Venturis (Bean 1971; reproduced with permission of ASME): **(a)** Single Venturi. **(b)** Double Venturi.

Figure 21.4(a) (Bean 1971). The pressure difference can be increased even further by using a double Venturi [Figure 21.4(b)]. The outer Venturi causes a reduced pressure at the outlet of the inner Venturi, thus increasing the flow through the inner one and reducing the throat pressure even more.

Another device depending on differential pressure and having some similarities to a Venturi is the Elliot-Nathan flow tube illustrated by Hayward (1977c), which appears to have a sharply converging and diverging Venturi, in the diverging portion of which is a hollow cone in which the downstream pressure is measured. Hayward commented that it is claimed to have an outstanding performance. However, strictly this is a full-bore meter because its installation would have to be in a spool piece.

Cairney (1991) described the use of a bidirectional pitot-Venturi inserted into the bled-steam line of a generating plant in which there was a possibility of a flow reversal due to a major transient in the system. The device consisted of two very small Venturi meters with axes parallel but facing opposite directions in which the differential pressure was measured between the throats. The device was sensitive to Reynolds number and turbulence level.

21.4 Insertion Target Meter

The flow acted on the disc (Figure 21.5), which was supported by a lever. The small movement of the lever was sensed by an inductive displacement, counterweight or magnetic coupling to a pneumatic sensor. Liquid flows could be measured in a range of about 0.05–4 m/s and gas flows from about 0.6 m/s up to about 5 m/s. It may have been applicable to temperatures up to 500°C. Measurement uncertainty for gas flows was claimed as ±1.5% measured value.

21.5 Insertion Turbine Meter

21.5.1 General Description of Industrial Design

The insertion turbine flowmeter is a small diameter turbine and pickup on the end of a relatively small diameter probe, usually inserted into the pipe via an isolation valve. The propeller may be able to cope with velocities for liquids of 0.1–12 m/s or possibly higher, and for gases of 1 m/s or less to 50 m/s or possibly more, with a range for one probe of about 10:1. An operating range of as much as 100:1 may be quoted in some designs. The turbine probe, shown in Figure 21.6 retracted into the threaded pipe fitting, may have ball bearings, jewelled pivot, tungsten carbide journal or other. Ideally, this will result in a rotational speed for a limited range proportional to the flow rate past the head. There is likely to be a minimum velocity for which the propeller will turn and another for which it can be expected to give results within the normal precision. For very low drag, the pickup may be radio frequency or other low drag type. A typical manufacturer's specification might be:

Diameter of rotor	15–30 mm
Linearity	±0.5–1% of full scale
Repeatability	0.05–0.25%
Turndown	10:1
Construction	stainless steel
Temperature range	–200 to 150°C

The turbine is inserted through an isolation valve of through-bore type with a clear concentric opening of diameter depending on the meter. The complete device consists, typically, of a small stainless steel turbine on the end of the stainless steel insertion tube (Figure 21.6). External handles may indicate the alignment of the probe, and this should be axial to avoid errors.

The insertion tube passes through an outer tube, which seals the line pressure. When this pressure tube is inserted into the valve, the valve can be opened, and the turbine can be inserted through the valve into the line. Fluid seal mechanisms need to avoid leakage but allow the probe to be inserted. In high-pressure applications, inappropriate selection can result in the probe being expelled from the process line, or at least moved from its set position.

These probes are suitable for water and hydrocarbons, particularly the latter because of their lubricating properties. They are not likely to be suitable for high velocities or hot water.

If such a probe is to be inserted and used as a single reading device for pipe flow monitoring, then the position of the probe in the pipe will need to be carefully considered, as discussed earlier.

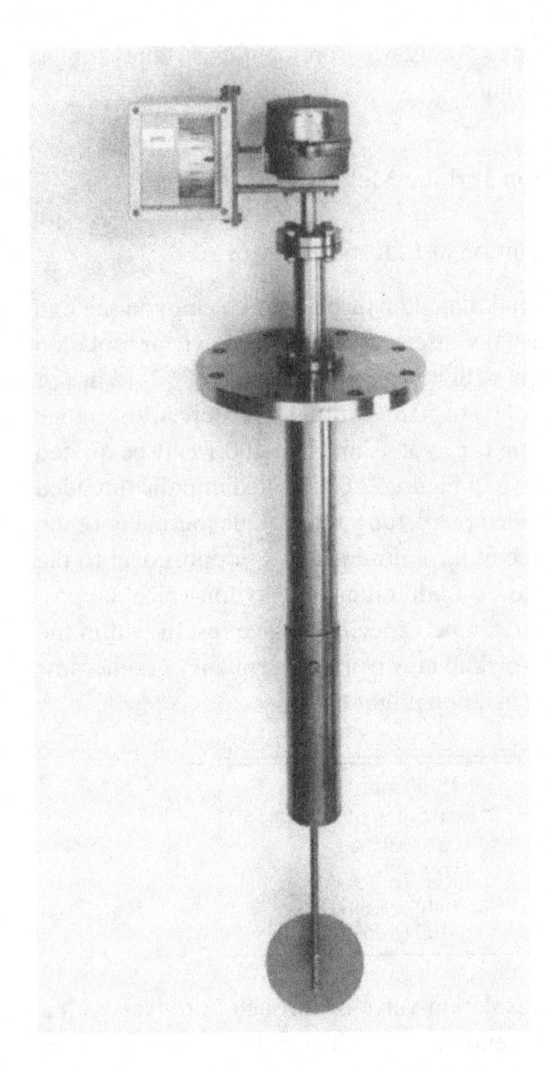

Figure 21.5. Insertion target disc flow probe [reproduced with permission of Laaser (UK) Ltd.].

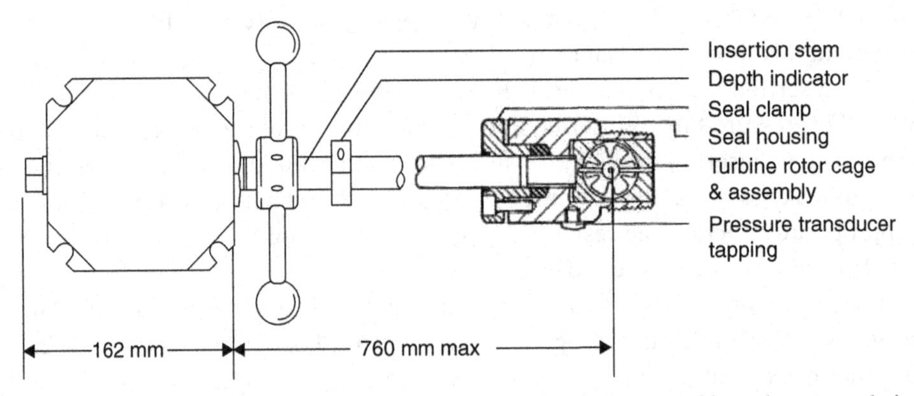

Figure 21.6. Insertion turbine probe showing the turbine cage retracted into the screwed pipe fitting (reproduced with permission from ONIX Measurement). Note: Figure is drawn for clarity, not correct alignment.

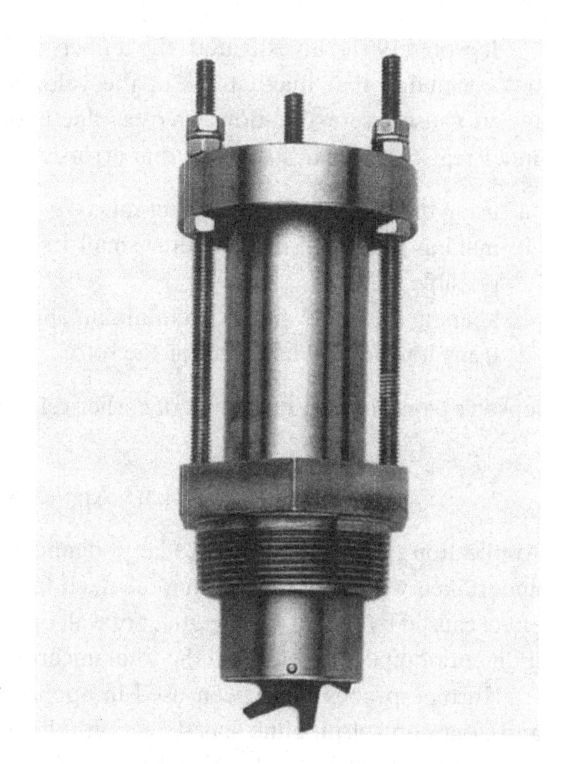

Figure 21.7. Paddle-wheel-type flow sensor (reproduced with permission of Peek Measurement).

To obtain a full flow profile, a series of measurements can be made, and the results can be integrated to obtain the flow rate. There is usually a limit to the proximity to the pipe wall at which measurements can be made. However, the nonslip condition at the wall should allow a curve to be fitted for this region. The profile shape thus determined allows the computation of volumetric flow. In this mode, they are also used for in situ calibration, although the overall uncertainty is likely to be in the range ±2–5%.

Various other designs of wall-mounted probe may be available for hot-tapping or fitting into a T-piece in the wall. An example is shown in Figure 21.7. These paddle-wheel-type probes essentially measure flow near the wall of the pipe, and the deduction of mean flow in the pipe will be obtained by reference to the manufacturer's literature.

21.5.2 Flow-Induced Oscillation and Pulsating Flow

Small-diameter probes at the centre of a pipeline of large diameter and with high velocity flow, are prone to oscillate. This may appear as a pulsation but can lead to failure of the support tube in extreme cases.

Ower (1937) developed a theory for a vane anemometer, showed that it overestimated average air speed, and obtained an expression for the error. Experiments confirmed the theoretical conclusions.

Jepson (1967) investigated the effect of pulsating flow on current meters and concluded that fluctuations of the velocity vector parallel to the mean flow always cause overestimation, whereas fluctuations perpendicular generally cause under-registration. He suggested that errors could be reduced by

a. using blades with a large aspect ratio,
b. making the tip-to-tip diameter small to keep the blade thickness as small as possible,
c. keeping the blade angle to a minimum, and
d. using low density material for the rotor.

Jepson's paper is a useful source of earlier references on current meters.

21.5.3 Applications

Application of turbine probes to large-diameter pipe flow measurement should be undertaken with care, recognising the need for an integration routine, the blockage effect caused by the probe, the effect of wall on probe calibration, the errors in measurement of pipe diameter and the total uncertainty of final value of flow rate.

Turbine probes have been used in applications such as compressor efficiency and surge control, pipeline leak detection, odorisers, samplers and checking throughput. Some have claimed that this device can be used for custody transfer to measure with high accuracy and reduce cost by avoiding the purchase, installation and maintenance of a large-bore flowmeter. In my opinion, this is rarely likely to be achievable. However, if calibrated in situ, the probe may provide a long-term check on performance, although short-term repeatability may not be so good.

Another device was developed for crude oil flow measurement [Figure 21.8(a)] and consisted of a rotor of up to 140 mm diameter that is capable of coping with wax and fibrous materials in the flow. For a flow range of 0.15–12 m/s, the uncertainty was claimed to be about ±0.3% at the top of the range and ±5% at the bottom with a repeatability of ±0.15%. The insertion flowmeter in Figure 21.8(b) was designed for hot tapping for gas, steam or liquid.

Raustein and Fosse (1991) used insertion turbine meters for fuel gas flow measurements and estimated measurement uncertainty as ±2–5%. They have also been used for flare gas metering where it appeared that uncertainty was likely to be in the range ±5–10%.

21.6 Insertion Vortex Probes

A commercial vortex-shedding insertion probe, which has been available in the past, had a rectangular slot in the end of the insertion bar (about 32 mm diameter), and the bluff body spanned this perpendicular to the axis of the bar. The insertion bar was retractable and of length up to about 900 mm. A turndown of 20:1 was claimed with repeatability of 1–1.5% of reading.

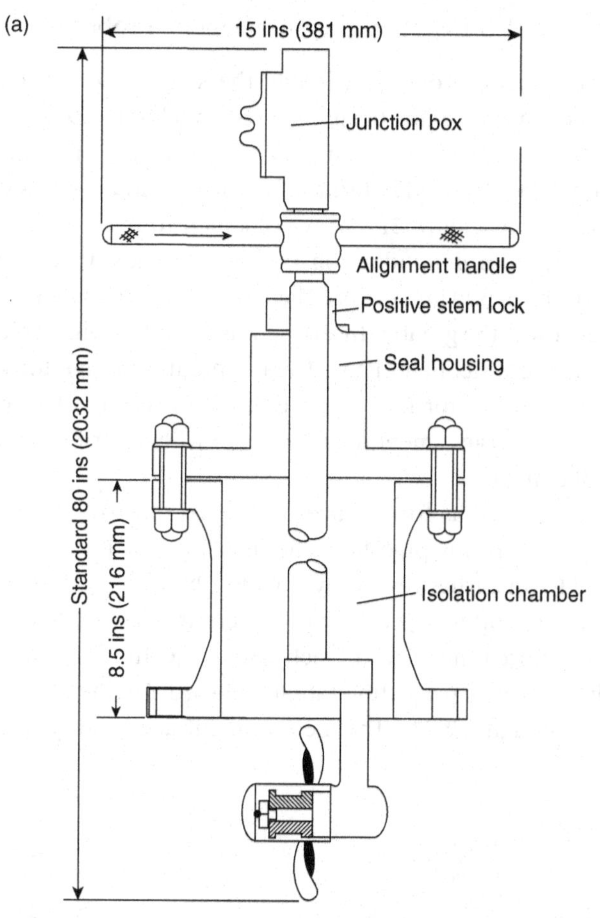

Figure 21.8. **(a)** Mauflo crude oil insertion flowmeter (reproduced with permission from F.M.A. Ltd.).

Figure 21.8. **(b)** Rotor insertion flowmeter for hot tapping with gas, steam or liquid (reproduced with permission from Spirax-Sarco).

21.7 Insertion Electromagnetic Probes

Insertion electromagnetic probes have been the subject of considerable development, and their performance has been analysed by Hemp (1995b) and Zhang and Hemp (1994, 1995) to obtain the sensitivity of its signal to proximity to the wall. Industrially, it has also advanced in terms of accuracy and of low power usage.

The probe shown in Figure 21.9(a) was claimed to have a range up to about 5 m/s, the actual maximum depending on the insertion length with maximum flow allowed for insertions of 300 mm or less. The uncertainty was given as ±2% of rate or ±2 mm/s, whichever was the greater. Insertion was up to 1 m with typical materials for wetted parts of stainless steel and PVC. The temperature range for water was given as 0–60°C. It was designed for local measurement of water velocity. Figure 21.9(b) shows the installation arrangement (cf. Baird 1993 who described an insertion probe for pipes of 25–300 mm).

These devices in various forms have been tried out for current flow measurement, for ships' logs and for pipe flows for many years. For more references, the reader is referred to an earlier review by Baker (1983).

One manufacturer offered probes with different sizes of discus-shaped head, in one case with a spherical head with electrodes projecting radially, and in another with a ring with electrodes projecting radially inward. The heads were directionally sensitive and operated at 128 Hz. The measurement uncertainty was claimed as 1%

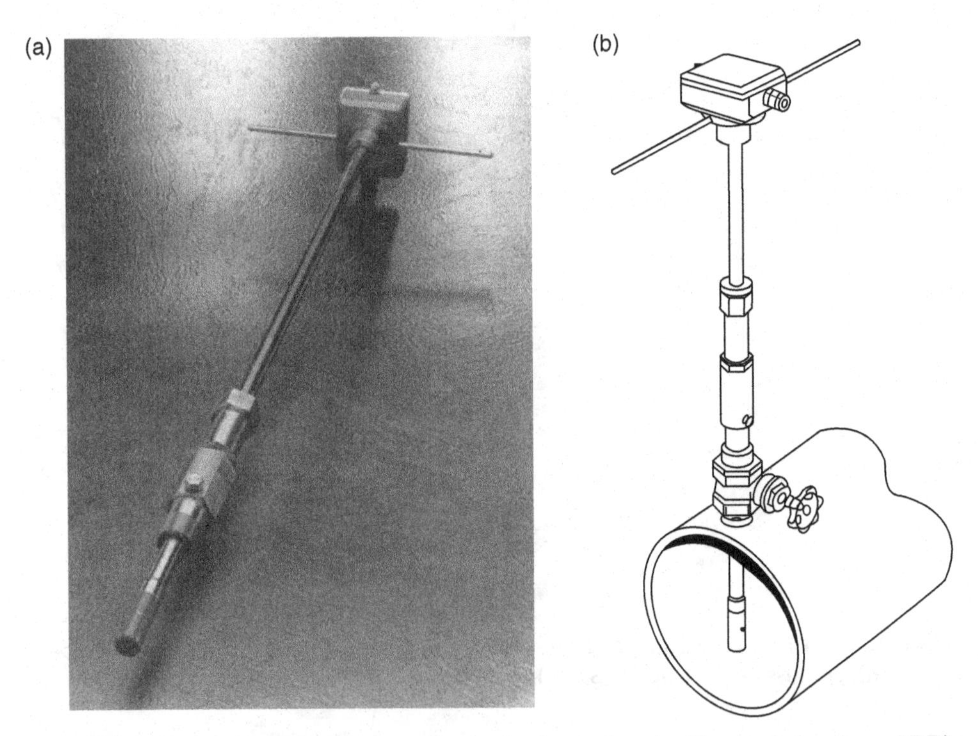

Figure 21.9. Electromagnetic-type insertion probe (reproduced with permission from ABB): **(a)** Pipe insertion probe. **(b)** Insertion arrangement.

of reading + 5 mm/s, and the direction was claimed to be better than 5° (cf. Bowden and Fairbairn 1956).

Another insertion probe, but for axial insertion, was developed for blood flow in arteries (Mills 1966, cf. Baker 1968b).

The application of a probe, such as that in Figure 21.9, to in situ calibration of electromagnetic flowmeters is discussed at length by Thomas et al. (2004a) and Thomas, Kobryn and Franklin (2004b). The probe provides the means to obtain the flow profile in the pipe. In theory, for well-developed profiles, a single measurement might suffice [Section 2.3 and also Geropp and Odenthal (2001)]. In practice, this is unlikely to be adequate and the profile will need to be measured with the probe. Thomas et al. (2004a) set out an approach with tailored software which allowed the profile measurement to be done with greater control and precision. They discussed (Thomas et al. 2004b) the likely precision of the probe which, probably realistically, ranged from ±2% to ±5%, and even to ±8%. They also listed the factors affecting the overall precision of the mean flow measurements: probe accuracy, pipe diameter, pipe area, profile distortion, position and blockage errors. Fletcher, Nicholson and Smith (2000) confirmed the importance of manufacturers' blockage factors. Cascetta, Palombo and Scalabrini (2003) found that the probe was within ±2 % when on the pipe centreline.

Ohnuki and Akimoto (1995) used an electromagnetic probe to obtain the radial distribution of liquid velocity and direction in a vertical air-water flow and claimed errors were within ±10%.

Hu et al. (2010c) analysed and tested for the effect of duct wall proximity.

21.8 Insertion Ultrasonic Probes

The Doppler ultrasonic principle lends itself to applications as a flow probe, provided there are sufficient scatterers in the flow; most of the reservations have already been set out in Chapter 14. See also references in Section 14.4.2 on pulsed Doppler.

A very much more precise instrument, depending on the transit-time principle, was developed by Rawes and Sanderson (1997). The device consisted of a bar, with ultrasonic transducers on each end, inserted through a fitting in the pipe wall. Once in, the bar was turned into the flow so that it was parallel with the pipe axis. It was then possible to send ultrasonic pulses from one transducer via reflections on the pipe wall to the other, both up- and downstream. This might lead to quite wide flow coverage and flow averaging. Its performance, even downstream of disturbing fittings, appeared to be better than ±2.5% for higher flow rates. Figure 21.10 shows the essentials of this meter. Further work (Rawes and Sanderson (1998) made use of a triangular path to interrogate the flow at the half-radius position, and reduce the effect of flow profile.

Olsen (1991) reported on the use of transit-time techniques to measure air velocity. The frequencies used were 3.1 Hz (sonic) and 40 kHz (ultrasonic). He claimed that the precision was better than a pitot-static tube and that the small size, simplicity and low price compared well with other methods.

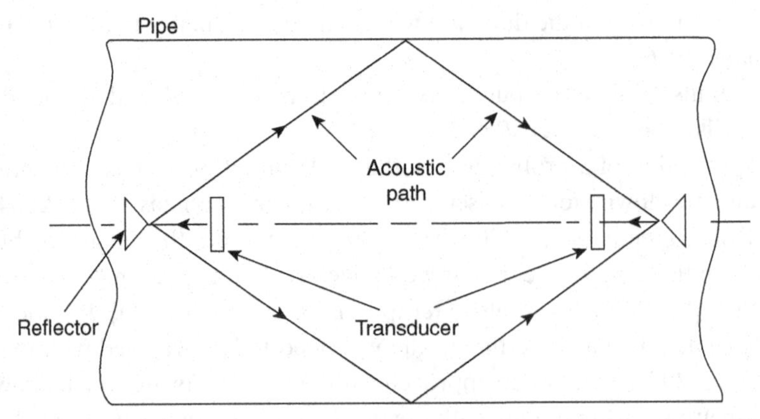

Figure 21.10. Ultrasonic-type insertion probe (after Rawes and Sanderson 1997).

21.9 Thermal Probes

Thermal probes are used so frequently as part of a full-bore flowmeter spool piece that they have been included in Chapter 18.

21.10 Chapter Conclusions

The turbine insertion probe has been an essential workhorse of the industry and is likely to continue to be, particularly where non-conducting liquids and gases are involved.

The latest developments in electromagnetic probes are very promising, and further designs and developments may well appear.

The novel ultrasonic meter of Rawes and Sanderson (1997) may well set a new standard for precision in the measurement of bulk flows using insertion methods, if further developed.

No doubt a further area of development would be the secondary equipment and its ability to obtain the best averaging of flow traverse data or of the ultrasonic response.

One of the weakest parts of in situ calibration using insertion probes tends to be the measurement of internal pipe dimensions. The ultrasonic techniques should provide a precise means of achieving the cross-sectional area.

The problems of blockage and the effect of wall proximity on the flow around the probe may lend themselves to computational modelling. However, the ultrasonic techniques may, again, offer a means to obtain the profile modifications due to the probe.

22

Verification and In Situ Methods for Checking Calibration

22.1 Introduction

Checking calibration is increasingly important and possibly mandatory. To achieve this there are, at least, four possible approaches:

1. remove meter and recalibrate on a certified test stand;
2. in situ calibration as discussed in Chapter 4. To obtain the required uncertainty, this may mean comparing the meter with a transfer standard in series with it in the line. This would require that installations of such a meter would be possible so as to check calibration (Chapter 4). But there may be major problems in so doing and limitations to the safety and accuracy of the process;
3. use a clamp-on meter as transfer standard. This is an attractive option but, as indicated in Chapters 14 and 15, may not allow the level of uncertainty needed for many modern flowmeters;
4. implement verification procedures.

We shall review verification and then look at other possibilities.

22.2 Verification

This is, essentially, condition monitoring applied in a very precise manner to flowmeters. The procedure is, increasingly, a continuous one, which can trigger well-focussed maintenance.

It should be remembered that differential pressure devices, such as orifice meters with removable plates (Figure 5.16), have a form of in situ verification.

Probably the first modern meter to introduce an early version, simulation, was the electromagnetic flowmeter (EMFM). A signal simulating the flow signal was introduced at the electrodes and the ability of the amplifier to measure it correctly was checked. The idea of simulation possibly stemmed from the simplicity of Equation (12.2) and the sense that, if the performance of the amplifier which measures the induced voltage could be checked and the magnetic field measured, the meter's operation would be confirmed.

The pressure to achieve in situ calibration has caused flowmeter companies to develop simulation techniques and to extend them to all the modern types of flowmeter. The effect of this has been to explore the extent to which the new devices can be used as an interim verification between calibrations, allowing the period between calibration to be extended by testing and verifying all the major functions of flowmeters on site. This leads to three matters (Baker, Moore and Thomas 2006a):

- the predictive uncertainty achievable;
- the validity of the technical approach;
- the management and administration of the process.

The initial generation of verification devices required a separate computer-based device to be linked to the flowmeter in the field. This approach appears to be changing so as to incorporate the verification systems within the meter itself. Thus the systems may assess the condition of the flowmeter, may have an internet or similar link back to the user and the manufacturer and may highlight maintenance needs. Delsing (2004) reviewed self-diagnostic flowmeter possibilities and the use of sensor networks.

We consider, first, the EMFM. The main items to be verified are:

- electrode connections and lack of contamination on the electrode surface. Some manufacturers have, in the past, provided means to clean the electrode, but today there may be a move to design and shape the electrode to reduce the problem of fouling.
- the integrity of electrical leads: earths, shielding;
- simulation of flow signal into electrode/amplifier circuit and checking its constancy of performance and output data;
- the constancy of the magnetic field and driver circuit during operation and since leaving the factory;
- the interpretation of changes in the raw signal components.

The field may be checked by measuring voltage, current, resistance, inductance, rise and decay times, or by using a magnetic field measurement sensor. It will need to be linked back to a certified magnetic field at the manufacturer's plant. The amplifier will require high-quality simulation signals and output signals, including digital, again with certified measuring instruments. The development in signal analysis will allow the output signal to be analysed and deductions to be made from it relating to the meter's systems and the installation arrangements. The integrity of leads etc. will relate to safety and to satisfactory operation of the meter.

Can an uncertainty be traced on this approach? There should be traceability through this system and an uncertainty should be deducible to indicate whether the meter has changed since it was first verified at the time of manufacture and calibration. An estimate of a possible uncertainty budget, which would be reasonable for such a device, was given by Baker et al. (2006a), including indicative uncertainties of injected voltage, of that due to square wave and that of the A/D converter. It was assumed that the relevant uncertainties applied equally to initial setup at manufacture and subsequently in situ, and should be allowed for in each case. This gave a total of ±0.48%. For magnetic field checks, again for manufacture and in situ, the estimated uncertainty was ±0.56%. Allowing for temperature variation, the total uncertainty was ±0.84%, to which a manufacturer would be likely to add a safety factor. One manufacturer appears to suggest that a 4% change in certain key parameters may flag up a warning. Another appeared to suggest that if the meter passed the verification procedure it would be within 2% of the calibration value. These figures should be treated with caution.

The main omission from this procedure is the state of the flow tube and changes which may have occurred to it which are not thrown up by the procedure. Johnson (2002) discussed the use of one of the first systems and possible benefits:

- ISO certification of the complete flow metering system without having to remove the EMFM's primary sensor from its location;
- keeping the meters on line thus reducing down time;
- verifying that the original calibration data was still valid;
- periodic calibration and performance checks providing a diagnostic and condition monitoring tool.

While the system provides a valuable check as indicated by Johnson (2002), there are aspects of the meter condition which do not appear to be checked by some or all of the systems, such as the state of the flow tube itself. I have proposed a possible method of checking this (Baker 2005, 2011) which makes use of the concept of the virtual current (Section 12.A.3), but it has not, to my knowledge, been implemented. (See also Baker, Moore and Wang (2005), where it was suggested that installation positions for piezos could be built into another meter, say a vortex meter, to allow an ultrasonic velocity check to be undertaken.) A further area of development which may increase confidence is the analysis of the raw signal. The information in the raw data has been recognised for a long time (Higham, Fell and Ajaya 1986; Higham and Johnston 1992).

Thus, the traceable instrument measurement and verification systems are likely to include signal simulation, measurement of electrical properties, precise timing and, in this case, a traceable magnetic field. The system used in the manufacture and testing will need to be managed to a high standard.

For thermal flowmeters there will also be systems to check the integrity of temperature sensors: position and operation.

For the vortex and ultrasonic flowmeters there will, in addition, be means of checking integrity of sensors and piezos, pulse counters and high-precision timing devices. Endress+Hauser use, instead of a piezo in the vortex meter, a dual switch capacitance (DSC). The ultrasonic meter will also require (Mahadeva 2009):

- check of zero crossing or other method of identifying elapsed transit time;
- comparison of path-elapsed times;
- cross-talk between transducers.

For the Coriolis flowmeter many of these systems will be required, but in addition verification of the original calibration may include (Wang and Baker 2014):

- checking several (resonant and off-resonant) frequencies;
- checking parameters associated with the frequencies;
- comparison of recorded oscillatory response with previously recorded responses and predetermined threshold values (Baker et al. 2009).

Also

- a verification method based on the correlation between tube stiffness and the flow calibration factor was proposed (Cunningham 2009) using off-resonance frequencies
- corrosion and erosion may be tracked based on stiffness.
- embedded electronics may estimate stiffness change for the user (Cunningham 2009; see also Rensing and Cunningham 2010).

My hesitation on the completeness of the verification system for the EMFM may have equivalent limitations for the other meters. The protocols tend to emphasise electrical systems as being sufficient to check the integrity. They may be necessary, but there is a serious question as to whether they are sufficient. Are there serious, even if rare, problems which could occur with the flow tube and related hardware that might not show up on the sophisticated verification systems available?

On the other hand, the great advances in analysing the signal and deducing from this the condition and installation of the meter should allow increasing confidence in the completeness of new verification techniques.

22.3 Non-Invasive, Non-Intrusive and Clamp-On Flowmeter Alternatives

A useful distinction has been made which defined non-intrusive to imply that the meters had a clear bore, without any components protruding into the bore, while non-invasive implied that the pipe had not been altered, essentially "clamp-on" meters, which do not require any break in the pipe (Sanderson et al. 1988).

Ultrasonic and sonar clamp-on flowmeters are discussed at length in Chapters 14 and 15. They are becoming increasingly well established, and there is a good prospect of their uncertainty reducing.

This section briefly reviews the possibility that other technologies might emerge, or other methods might become available.

22.3.1 Use of Existing Pipe Work

Pipe Bends – Pressure Loss

The use of a bend as a method of obtaining flow rate has been discussed in Chapter 8 and has been used for many years. This has involved use with pressure tappings. If bends are to be used, can pressure tappings be inserted (hot tapped)?

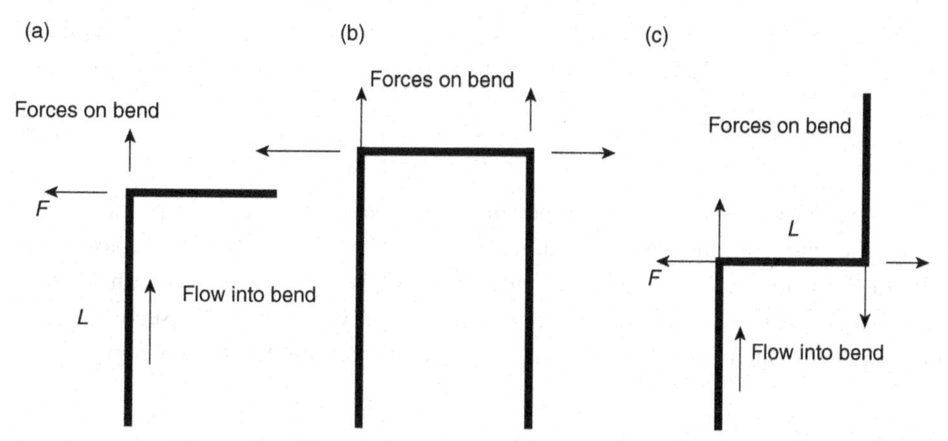

Figure 22.1. Diagram to illustrate the forces on bends due to flow. Further forces will exist where the bends are not in a single plane.

Pipe Bends – Distortion

In Figure 22.1, three bend configurations are illustrated with forces which will result from the change in direction of the internal flows. Depending on the pipe supports in Figure 22.1(a), it is conceivable that the position of the bend could be subject to a very small displacement due to the momentum change of the flow at the bend. However, the force causing the displacement is likely to be very much smaller, for instance, than the fluctuating force due to the internal pressure and is unlikely to be measureable. Figure 22.1(b and c) gives other configurations. In some cases, forces may cancel out, be negligible compared with pressure in the line and be inadequate to create a measurable tensile strain in the pipework.

The single bend is unlikely to exist in an industrial setting. The feature worth considering in Figure 22.1(b and c) is that a length of pipe is put under tension. However, the internal pressure is likely to swamp the momentum change term. The pipe extension may be within the scope of a strain gauge. However, to obtain the momentum term and hence the velocity would require that the static pressure and fluid density in the pipe were known. Could a strain gauge around the pipe provide internal pressure?

These suggestions would appear unrealistic at present. For such measurements, the value of the pipe parameters may not be known with precision. In addition to the unknown pressure in the pipework, there may be pressure fluctuations, vibrations and mechanical creep, and it is likely that the pipework will be very securely fixed to prevent any movement.

Losses Along a Length of Pipe

The pressure loss down a pipe would be small but might be measurable if a sufficient run of pipe with known losses were available. To obtain the loss coefficient for the pipe run, the loss coefficient for each component of the piping system could be obtained (Miller 1990). The loss coefficient is given by:

$$K = \frac{\Delta p}{\rho V^2/2} \tag{22.1}$$

If we obtain the K factor for the pipe run, and if we can measure the pressure drop along the pipe run and know the density, then Equation (22.1) would allow us to obtain the velocity. The accuracy of this approach is not likely to be high as there are likely to be many approximations relating to pipework and fluid. Such measurements provide the basis of the virtual flowmeter briefly mentioned in Chapter 17.

22.3.2 Other Effects: Neural Networks, Tracers, Cross-Correlation

It might be worth investigating the possible application of noise and vibration in flow systems (Chapter 15). Can these be "calibrated" against a temporary master meter, to provide a simple inherent flow sensor? If the calibration conditions differed from the operational, could the changes be modelled and predicted?

There are references, some of which are relevant to this approach. Gribok et al. (2001) discussed what they referred to as an "inferential measurement" system for feedwater flow rate which made use of currently available data from a power plant such as: pump speeds, pump and heater inlet and outlet temperatures, turbine pressures etc. Roverso, Ruan and Fantoni (2002) mentioned artificial neural networks as a means of obtaining the relationship of the signal to the flow rate. Ruan et al. (2003) were interested in computational intelligence schemes (neural networks). Roverso and Ruan (2004) further discussed the enhancement of cross-correlation methods using artificial neural networks.

22.3.3 Other Flowmeter Types in Current use

Vortex

There would appear to be only one approach to such a flowmeter, and that would be the installation of a shedding bar within the bore of the pipe. This could be much thinner than the normal size of a shedding body, making insertion through a tapping point possible, but the shed vortices and stability of shedding might be reduced if the body were too thin.

Electromagnetic (EMFM)

Medical researchers have used "clamp-on" EMFMs in the past – "cuff" type blood flowmeters which could be placed around a blood vessel without breaking it (Figure 22.2). The electrodes made contact with the outside of the vessel, which had a conductivity similar to the blood, and a magnetic field was created across the blood vessel (Kolin 1941; Denison and Spencer 1956; Shercliff 1962). Similar

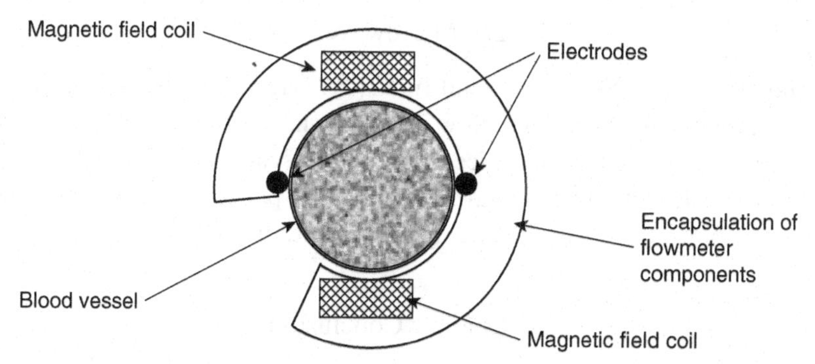

Figure 22.2. Approximate diagram of cuff-type electromagnetic flowmeter.

approaches have been used when an EMFM has been applied to liquid metals (Baker 1977). The preference in such applications is that the conductivity of the pipe wall (assuming that the pipe is not lined on the inside) should be less than or approximately equal to the conductivity of the liquid. If these conditions are not met, but the flow takes place in a non-conducting and non-magnetic tube, then the insertion of electrodes through the tube wall would be an option, if acceptable for other reasons.

Coriolis

Retrofitting to create a clamp-on Coriolis meter has been proposed (Hussain and Rolph 2002; Drahm et al. 2002). Kalotay (1994) and later Bartstra (2006) patented aspects of clamp-on Coriolis mass flowmeters, the latter using in situ calibration (cf. Normen 2003). The inventions consisted of a piece of pipe with an actuator creating the vibration and a series of sensors measuring the response. The system was clamped to the pipe.

An alternative approach is the shell type Coriolis flowmeter (Section 20.1.4), but this may be subject to equal or greater constraints, and greater flow measurement uncertainty.

Thermal Flowmeters

The limitations for a thermal flowmeter are likely to be the thermal conductivity of the pipe wall and consequent rate of response. Approximate calculations suggested that the required temperature would be unrealistically high to achieve the necessary heat flow to the liquid from outside the pipe, and an adequate temperature change to allow the temperature of the heated liquid to be sensed through the pipe wall. But see Zhu and Min (1999).

22.4 Probes and Tracers

Probes have been discussed in Chapter 21 and tracers in Chapter 4.

22.5 Microwaves

Hamid and Stuchly (1975) and Hrin and Tuma (1977) tested, respectively, microwave devices for monitoring particulate flow in pipes which were transparent to microwaves, and a microwave cavity device in which the scattered waves were Doppler shifted by particulate matter and gave a measure of particulate loading. The latter was aimed at applications with high temperature and/or pressure.

22.6 Chapter Conclusions

This chapter has outlined the possibilities of verification. It seems likely that the techniques and systems will move increasingly towards a valid calibration check. Other methods of in situ checking have been reviewed, some of which are very marginal, while others may develop into possible methods.

23

Remote Data Access Systems

23.1 Introduction

This chapter does not attempt more than a cursory review of this topic, as it is outside the author's main area of expertise, and its existence here is a result of very helpful advice and guidance from Edward Jukes of KROHNE.

This book is about flowmeters which depend, in the main, on mechanical, thermal and/or fluid mechanical effects. One example of mainstream meters, where the signal is electrical throughout, is the electromagnetic meter.

The response of the various meters to flow may then be used to generate an electrical output: milliamp, pulse/frequency and/or digital signal output. Once the electrical output signal is available, we shall need to be able to access it using modern communication methods. This chapter gives a summary of this in an attempt to brief the potential user.

The aim of this book has been to bring together information on the performance of these meters based on theoretical, experimental and industrial experience. I have made the assumption that the electrical signal interpretation, which is based in the meter, is capable of doing the job set by the designer of interpreting the meter output into a standard electrical form. The developments in design of electrical systems are assumed to be such that the reader will not benefit from a detailed description or that the circuit will be changing so fast that any statement will be out of date.

It is now possible to send complete information on parameter values by digital signals which do not need to be converted from a mA value to a flow rate, but provide the flow rate in the required units. In addition the sensor can be addressed, and will respond to questions or commands. The communication has thus become two-way.

Communications technology has been transformed in the past 15 years, to the extent that most of us spend much of our time making use of digital communication devices. This chapter attempts very briefly to indicate how this revolution has affected the instrument communications field.

We have moved from the days when instruments were read at their application sites to the situation where most instruments are read from a computer screen. Remote data access has, therefore, become part of normal procedure in industry.

It allows monitoring of: flowmeter and other instrument outputs, totalizers where appropriate, trends, alarm settings and meter condition and operation. It also allows remote monitoring of the complete set of instruments on the plant, analysis of the meter and plant performance, data recording, more rapid response time where necessary, and may allow remote repair to instruments and generally increased efficiency.

Within this system a key component will be a suitable communications architecture. The user will need to ensure that the communication protocol selected is compatible with both the instrument and the plant or interface equipment to which the instrument is to be connected.

23.2 Types of Device – Simple and Intelligent

Twenty-five years ago, we were concerned with the cleanness of the signal which came from a meter, and which provided just one parameter, the flow. The advent of computing power and particularly the microprocessor has allowed more information from the signal to be interrogated, and multiple outputs to be handled. Thus interest is now being shown in the details of the noise in the signal, which previously we smoothed out and discarded.

The microprocessor has allowed a further, and major, step. It is now possible to build into the instrument means for checking the validity of the signal, for adjusting range and for selecting the form of signal. The meter may, therefore, be capable of both transmitting and receiving information.

For such meters, the terms *smart instrument* and *intelligent instrument* were coined, and associated with various degrees of on-board computational power. A definition of these terms can be found in Higham and Johnston (1992), but the continual developments resulting from microprocessor power mean that the distinction is likely to become less relevant or used. They suggested that *smart* be applied to sensors which automatically compensated for nonlinearity, temperature effects etc., and could be configured before installation to provide a predetermined range and span. However, they would normally need to be taken out of service to be reconfigured, although some later devices could be reconfigured via their built-in display.

The term *intelligent* included the smart features, but also implied two-way communication with the control system either by a separate communication port or by a superimposed digital signal on the 4–20 mA signal. This greatly increases the versatility of the sensor, allowing self-diagnostics and other routines and commands to be sent from the control system to the instrument. It is likely that many recent instruments fall, at least, into this category.

An important attribute of the intelligent sensor is addressability, which is a necessary part of the bus communication, but also will put time constraints on the signal transmission each way. Thus the instrument's microprocessor:

- processes signals from one or more sensors and replaces complex analogue electronics;
- is capable of re-ranging, diagnosing problems etc.;

- oversees the communication of the sensor with the wider control system;
- manages display functions if present.

The microprocessor then allows enhanced accuracy by putting in the calibration curve, which provides:

- higher turndown ratios leading to increased rangeability – but beware hidden errors resulting from the higher turndown;
- improved accuracy over a broader range of process conditions e.g. temperature, by including complex compensation algorithms;
- increased manufacturing flexibility;
- increased reliability, a quicker identification of any problems, and an improved maintenance philosophy.

The result of the advance in IT is that the instrument is likely to hold: configuration information, calibration information, condition monitoring procedures, advanced diagnostics.

The information on the meter configuration allows changes to be made from a separate computer screen of internal parameters which would previously have been set at the meter. In Chapter 22, the verification procedures were considered and are likely to be embedded in the meter systems, and these will lead to diagnostics and alarms. Maintenance will be an important part of this.

23.3 Simple Signal Types

The flowmeter will usually have either simple signal types such as:

- an analogue signal of 0–20 or 4–20 mA, or a voltage, or a resistance;
- a frequency or pulse signal – this may be of high or low frequency or both;

or intelligent signal types such as:

- a communication method which allows the sensor to provide more than one piece of information, and the user, through a computer, to instruct the sensor (and other devices) in various ways.

Milliamp Signals

The user's external milliamp circuit is connected to the meter circuit by breaking the internal mA circuit and connecting the user's extension circuit between the meter's internal high-quality resistor, R_1 in Figure 23.1, and a second meter terminal. A twin lead can be taken from these two connections and led to additional external instrumentation. The twin lead can then be connected in series with/across a further resistor of high quality which provides a voltage measurable by a voltmeter.

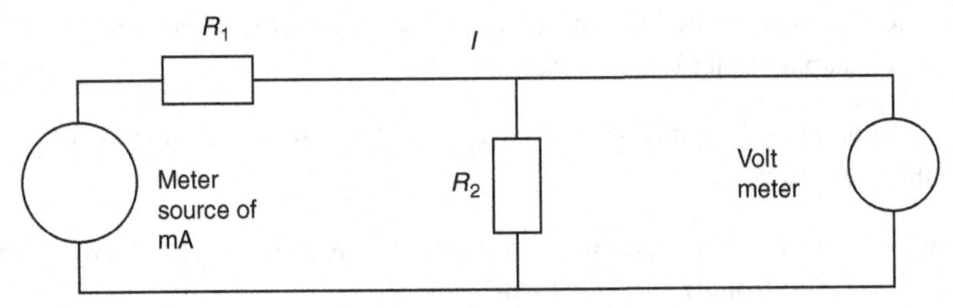

Figure 23.1. Setting up mA.

To set up this arrangement, the meter can usually be set to give, precisely, 4 mA or 20 mA or other values required. Figure 23.1 gives a simple milliamp circuit. If R_2 is set to the correct resistance, then

$$V = IR_2$$

For a constant resistance, V will be linearly related to q and similarly the current I will also relate to q, thus we can write

$$q = C_1 I + C_2$$

where C_1 and C_2 are constants. If I_1, which might be 4 mA, is equivalent to q_1, probably zero, and I_2, which might be 20 mA, is equivalent to q_2, probably the maximum flow rate required, then we can relate them by this equation, which can be set in the output software to give

$$q_1 = C_1 I_1 + C_2$$

and

$$q_2 = C_1 I_2 + C_2$$

so that

$$C_1 = \frac{q_1 - q_2}{I_1 - I_2}$$

and

$$C_2 = q_1 - C_1 I_1$$

and hence we can find the required values of C_1 and C_2. This allows us to set up the mA output without being affected by possible uncertainties in the precise value of the resistor, R_2, used in the external circuit.

Pulse and Frequency

The meter is likely to offer these options and to provide a means of setting the maximum expected flow rate to be that at a predetermined pulse/frequency output. Pulse outputs in particular are used for batching applications. In this case, the user must be careful to set the quanta per pulse value such that ±1 pulse is an insignificant part of the overall batch size thus contributing only a small error.

23.4 Intelligent Signals

An instrument fitted with intelligent signal types can be directly connected into the bus architecture of various possible bus types: HART, FOUNDATION Fieldbus, Profibus, Ethernet/IP, Modbus etc.

It is also possible to fit a converter interface between a simple signal output and a bus network. Such interfaces or interface devices may also include a web server and allow direct wireless communication with the GSM (global systems for mobile communications) worldwide mobile communications network.

The bus operates by reading the output of each instrument in turn, or as called, and each instrument has an address which ensures that the output is correctly assigned and that inward "messages" are delivered to the right instrument. Some of the bus options are given in the following list:

Fieldbus is a communication system specially designed for networking of transducers (Brignell and White 1994). The International Electrotechnical Commission (IEC) is responsible for defining a standard. Fieldbus implies the internationally agreed standard. The protocol model has three layers:

i) Physical layer – with two communication speeds: low speed of 31.25k bits per second for long distance; high speed of 1M bits per second for short distance. Each uses a voltage signal for connection to the bus. An alternative option for intrinsically safe applications may be a current mode;
ii) Data link layer – using a token bus principle;
iii) Applications layer – to make the Fieldbus network appear transparent to the applications programme in each transducer connected to the bus.

Various protocols such as Profibus and FOUNDATION Fieldbus are compatible with the Fieldbus system.

HART (Highway Addressable Remote Transducer) digital communications protocol developed by Rosemount (Brignell and White 1994) uses digital techniques to allow an intelligent instrument to adjust range and span, and undertake flow calculations, to apply self-diagnostics and to communicate (Howarth 1994). HART uses point-to-point or all-digital modes. In the former, it superimposes digital data on a 4–20 mA signal by two defined frequencies of ±0.5 mA amplitude: 1,200 Hz for binary 1 and 2,200 for binary 0. Message rates are typically 3/s. It uses a master/slave protocol (Brignell and White 1994) in which the master is the only device which may initiate a message transaction. Up to two masters can communicate with connected field instruments, typically a handheld communicator and a control system or PC-based workstation. At the applications layer, HART commands are of three types: universal commands to read an identifier, variable or current; common practice commands such as calibrate or perform self-test; transmitter-specific commands such as start/stop totalizer, read density calibration factor.

Wood (1994) described HART and Profibus as examples of proven specifications which provided background and influence for the modern fieldbus proposals. Profibus was set up by a group including AEG, ASEA Brown Boveri, Honeywell, Klochner-Moeller, Lands & Gyr, Robert Bosch and Siemens, with the objective of creating a digital fieldbus standard consistent with the ISO/OSI model in ISO 7498 (Squirrell 1994).

23.5 Selection of Signal Type

The nature of the measurement/data may affect the communication approach (Imrie 1994). Is the variable to be used in a control or time-critical application? What is its likely rate of change? Is continuous trend monitoring required or is it event driven? Can power saving be used?

The location will determine security, protection, safety, interference, availability of services such as power and signal networks and the access requirements of ownership of land.

The consequence of data loss, the rate of data transfer, the response time, one-way or two-way transfer and the size of data package will all affect choice.

The nature of the application will be important: control or monitoring, regulatory and insurance, safety, effect on product or services.

23.6 Communication Systems

In a simplistic way we divide the communication system into a series of levels (Figure 23.2).

Level 1 – the instrument signals discussed in Section 23.3

Level 2 – this may involve some bus protocol which collects the output data from each instrument, passes it along the bus to the identified receiver, and carries returning messages to the instruments

Level 3 – The bus protocol will communicate with web servers and work stations which will process the data into a form which provides information about production, safety or prepares reports for engineers and management

Level 4 – will provide transmission locally or internationally, so that operators anywhere around the globe may be able to interrogate instruments.

It should be noted that these levels are a very simple and minimal set, and if the reader wishes to pursue this he or she should access the standard and protocol documents to obtain the full number of levels of operation which are generally used.

23.7 Remote Access

The communication at level 4 will probably be, at least in part, from antennae sending and receiving radio waves. Thus communication for a conventional plant may well initially be via copper wires, or where there are dangers due to explosive

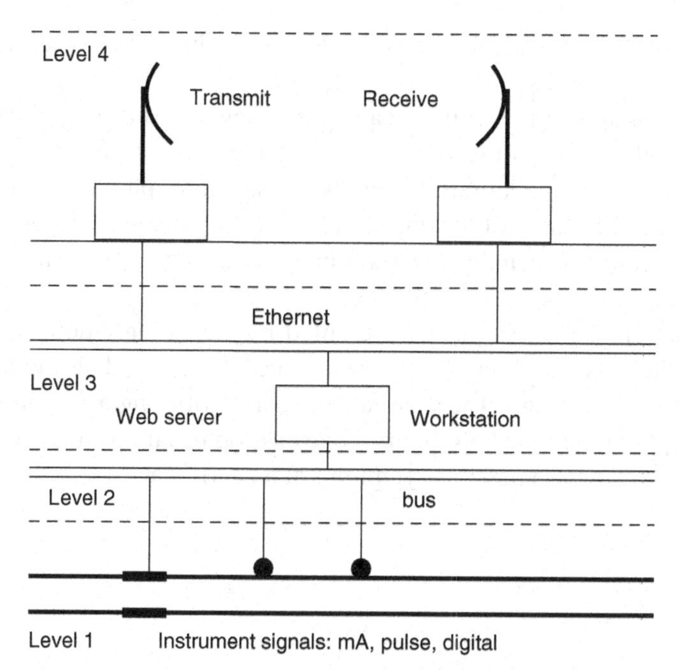

Figure 23.2. Levels for a modern communication system as used in this chapter.

environments, via optical cables. Transmission system may be by the public switch telephone network (PSTN) For oil and gas, transmission may be by microwave which requires line of sight between transmitting and receiving stations, or by radio waves bounced off the troposphere or linked by satellite. A utility may develop means for sending the signals down the power lines which link every house, while an international oil or gas company may be able to afford radio links via their own satellite or rented satellite space.

Combinations of these and other methods will be selected depending on the most appropriate route for the particular application. For added security, dual methods may be used, providing a fall-back in case one method fails. The options are likely to increase.

The bus/highway will increasingly link computers with control and communication capability, and smaller modems with handheld devices for field checking. The development of such data highways is likely, eventually, to lead to the elimination of all but digital information. Standardisation will address the data rates of the bus which will affect the speed of response of the instruments.

The sensor will, therefore, be capable of linking with computer control systems anywhere in the world or beyond and the "wire" connections may become antennae. Even the problems of power supply may be overcome with local power generation.

23.8 Future Implications

From a mechanical engineer's viewpoint, there appear to be a number of fairly obvious ways in which the power of IT will continue to influence the instrumentation and control field. These are likely to be driven by the fast-advancing computer technologies.

The visual graphics, the virtual reality developments and the windows environment of modern computers will demand of the process engineering designer increasingly detailed simulations of hardware, software, physical behaviour and safety scenarios with chaos theory implications. In many cases the expert system will provide the operator with an instant assessment of the most likely causes of failure, or of unusual operations.

While such information will be presented by increasingly detailed computer graphics, behind it will be an ever more detailed understanding of the complex nature of flow phenomena. The computer experts, with engineers and scientists, should attempt to model various theories of catastrophic failure, to enable the computer system to recognise possible symptoms at an early stage.

24

Final Considerations

This chapter contains some personal comments on research, manufacturing variation, quality, existing and new flow measurement challenges, micro-engineering devices and new techniques for existing and new flow metering concepts.

I pose the following questions with brief explanation based on my knowledge of the flow measurement industry.

24.1 Is there an Opportunity to Develop New Designs in Collaboration with the Science Base?

There is a need and an opportunity for the flow metering industry on one hand, and the science base as it relates to instrumentation on the other, to exploit effective means for technology transfer and to bridge the gap between them, which in the process will raise the profile of the instrumentation sector and ensure that government is aware of its importance. Past experience of encouraging collaboration suggests that industry and the science base do not always appreciate the value of working together. Fundamental and unsolved issues remain in the technology which would benefit from collaboration between industry and the science base. Such collaboration should be mutually beneficial in making the other party aware of important developments.

This gap could also be reduced by facilitating exchanges of people between industry and the science base. Industrial collaboration is an essential part of ensuring that academic engineering and applied science research is focused on the real problems of industry, and the cross-fertilisation will be very likely to generate new ideas.

Over the past 15 years or so, I have found close collaboration with several companies in research and teaching extremely rewarding.

24.2 Is Manufacture of High Enough Quality?

The quality of an instrument, and thus its ultimate accuracy, will clearly be affected by the manufacturing process. The manufacturing variation may have predictable consequences for the flowmeter accuracy. Slight variation may require calibration of each instrument. Instruments off the same production process might have different characteristics which may require small adjustments to the product before final calibration. Some changes may cause finished instruments to have different random errors.

The production of instrumentation is a special case in which every finished product may be measured (calibrated) and also the effect of the production process on the final accuracy of the instrument may, in some cases, be predicted and used to specify the production requirements. Although a special case, it may be important for other products; the ability to measure dimensional precision and variation at each stage of the production may lead to higher quality in other products. A special issue of the *Journal of Flow Measurement and Instrumentation* (Vol. 12(2) with an editorial by Baker and Gregory 2001) gave examples of the manufacture of flowmeters (see also Baker and Gregory 2002).

Modern instrumentation frequently is defined by a precise analytical expression or algorithm. The output signal should, therefore, be related to the measured parameter with a high level of confidence. This confidence comes from the design, but also from the quality of manufacture. It is this combination which creates a high quality instrument, in other words, an accurate instrument with a small value of signal uncertainty.

It is, therefore, reasonable to assume that we may be able to deduce from the manufacturing process the effect it has on the parameters which control precision, and we should be able to identify a method of production of the instrument which will optimise the likelihood that the finished product is of specified quality. We should be able to use an error equation for changes in each parameter (Baker 2004b). This could be used forwards or backwards to predict errors or to limit errors.

Examples of the effect of manufacturing variation of construction on flowmeter quality are:

a) Orifice plate: variation in the sharpness of the leading edge will probably be due to manufacturing variation, but may also be due to slight and random variation in the quality of the material.
b) Variable area flowmeter: while the glass tubes are made on a mandrel, and should have a well-controlled tolerance, the positioning of the scale will affect the accuracy, and since the device is essentially nonlinear, it may determine the shape of the characteristic. See Baker and Sorbie (2001) and Baker (2004a).
c) Positive displacement flowmeter: variation of the clearance between the rotors and the chamber may cause a change in performance (cf. Morton 2009; Morton, Baker and Hutchings 2011; Morton, Hutchings and Baker 2014a; Morton, Baker and Hutchings, 2014b).
d) Electromagnetic flowmeter: the construction of the coil and its positioning relative to the flow tube and electrodes will affect the performance, and may be, to some extent, random (cf. Baker 2004b).

e) Ultrasonic flowmeter: the placing of the transducer cavities and the resulting position of the transducers may result in a manufacturing variation between meters off the same production line.
f) Thermal meter: in some designs of flowmeter, it may not be possible fully to test embedded components before completion of the instrument. Such components may be critical to the final accuracy of the finished product (Baker and Gimson 2001).
g) Coriolis flowmeter: during manufacture the positioning of the sensors on the flow tube, together with other masses, tube dimensions etc. will cause consequent variation in the meter response (Wang and Baker 2003/4). Improved robustness was observed due to tensile pre-stressing. Lange et al. (1994) showed that the calibration constant is sensitive to the positioning of the detector masses. They showed that the dependence of the calibration on the fluid density can be rather strong if the positions of the detector masses are not chosen very carefully. Thus to retain the calibration constant within, say, 0.1% of the target value, the position of the detector masses for a pipe of half-length 240 mm may need to be of order 0.1 mm.

In developing a design and production programme for a particular flowmeter the company should be clear about the relationship between the primary construction variables and the resulting performance of the flowmeter. To focus on the quality of the end product and its measurement is essential to the achievement of high quality and this should feed back into the company (Baker 2004b). The final calibration of the flowmeter at the end of the production line gives a unique opportunity to use this measure resulting from the calibration as a check on the overall production quality.

In cases where the meters are for use on liquid, a gravimetric system as described in Chapter 4 may be used. With valuable input from a flowmeter manufacturer we developed a flow calibration rig (now at the University of Warwick, UK) which would be suitable, in various sizes, for the calibration at the end of the manufacturing process (Section 4.A.6). In other cases, it may be more convenient to use transfer standards or pipe provers. A calibration system for thermal mass flowmeters using critical nozzles was described by Caron (1995). A small value of the uncertainty of the calibration facility is an essential part of ensuring the quality and consistency of the product, and should be built into the logging system for the documentation of each unit produced.

Thornton (2000) comments: ‘One way to improve quality is to minimise the impact part and process variation has on final product quality. Although companies know they must reduce variation, they are still struggling with executing coherent variation management strategies. … It was found that many problems with industry implementation are due to a lack of quantitative models that enable a design team to make quick and accurate decisions.’ Thornton (1999a, 1999b, 2000) identified a mathematical framework for the effect of manufacturing variation on a manufacturing process, and considered selection plans for variation reduction. This focus on the quality of the product may also influence the workforce, through job satisfaction and high skill levels and an overall ethos of integrity in the company.

24.3 Does the Company's Business Fall within ISO 9000 and/or ISO 17025?

This section endeavours to reflect my reading of the two standards as a layman and not as an expert. However those reading this should check with an assessment and accreditation organisation as to the appropriateness for their business of one or both of the two standards. In preparing this section, I have benefitted from discussions with co-authors of an unpublished paper[1], and also from very useful comments by Geoff Howe[2], who read this section, and of which I have made use.

The difference and commonality of ISO 9001 and ISO 17025 may be of interest and concern to manufacturing companies which maintain calibration laboratories. There is an expectation that companies will achieve ISO 9001 accreditation, and yet it appears that a company with calibration facilities should also have accreditation for its calibration work, which is essentially covered by ISO 17025. However, it should be noted that accreditation is voluntary and commitment to achieve it should be influenced by their clients' expectations. For both standards there should be indications of continuous improvement.

ISO 17025 (which may have replaced ISO/IEC[3] Guide 25 and EN45001) is a management system which aims to assure competence of testing and calibration laboratories. Companies with calibration facilities (laboratories) which have ISO 9001 or intend to obtain ISO 9001 approval should also consider obtaining ISO 17025 approval if the business case justifies the accreditation costs or supports the business projections. If the company decides to seek approval, then ISO 17025 should be integrated into its Quality Management System (QMS). The company will then have one management system consisting of ISO 9001 and ISO17025. This is easier to maintain and to improve.

ISO 9001 concerns good business practice. All requirements are mandatory for accreditation. Some requirements cover all aspects of the business and may require formally documented procedures. ISO 9001 also mentions such things as verification, monitoring and the validation of equipment. ISO 17025 (Section 4) is generally in line with the requirements of 9001, however, section 5 covers the technical requirements and is more focussed on laboratory controls, including clauses for which an equivalent does not appear to be in ISO 9001. These may address the methods for calibration in terms of validation and selection, uncertainty of measurement and traceability.

1 This section has been based on extracts from Augustin, S.-A., Segler, B. and Baker, R. C. The application of ISO 9001 & ISO 17025 in instrument manufacture with some consideration of calibration as a source of quality measurement. Unpublished draft paper.

2 Geoff Howe, personal communication (2014).

3 IEC = International Electrotechnical Commission.

Both contain similar clauses and recognise the importance of a customer-focussed quality system. The provision of training and resources for personnel are also important.

Is there a need to implement both standards? The answer appears to depend on the nature of the company's business:

- if the company business is manufacture of flowmeters, then ISO 9001 may be sufficient;
- if the company business is solely sales of flowmeters, then ISO 9001 may be sufficient;
- if the company business is manufacture and calibration of precision flowmeters, then ISO 9001 and ISO 17025 may be required;
- if the company is solely concerned with calibration of flowmeters, then ISO 17025 may be sufficient on its own.

Accreditation to ISO 17025 requires an assessment by an accrediting body (Brewer 2005), and may require benchmarking against a similar laboratory nationally or internationally. In some cases, certification of personnel may be a requirement.

Brewer (2005) noted that an accredited laboratory shared many of the quality requirements of ISO 9001 and similar standards, but ISO 17025 also had unique requirements. Both standards address the aspect of internal audits. Aspects covered by ISO 17025 appeared to include: customer confidentiality, customer feedback to be addressed and inter-laboratory comparisons to be available.

Rogers (1995) and King (2004) found, with regard to ISO 17025, that a QMS such as the ISO 9000 family of standards was not sufficient to meet the technical requirements of calibration and testing laboratories.

BOC decided (King 2004) to adopt ISO 17025 for each of its laboratory operations which certified traceable calibration gas mixtures. This recognised that whereas ISO 9001 certification was a recognition of conformance to a quality system, ISO 17025 accreditation was the recognition of laboratory competence. However, it should be recognised, as mentioned earlier, that ISO 9001 certification recognises, not only conformance, but also the demonstration of continuous improvement of the business.

A final question: does the company make full use of the calibration data from the production line to assess its manufacturing quality?

24.4 What are the New Flow Measurement Challenges?

Medlock commented (1986) that interest in fluid flow dates back thousands of years BC, and he notes that in Egypt irrigation was vital. By 2,000 BC it appears that the first primitive water meters used by the Egyptians were crude forms of weirs, presumably to control the flows in a network of irrigation channels.

The history of flow measurement perhaps started with the observations of our ancient ancestors as to the behaviour of flows over weirs. One of the controls of water in Roman times appears to have been by the rate of flow through a pipe of a particular size and possibly with a known head across it.

The measurement of water flow appears to have dominated until the 19th century, when gas and oil started to become important and developments in industry in the 20th and 21st centuries required flow measurement in many areas and of many fluids.

The likelihood of water shortages may bring us back to where we appear to have started, but coupled with the shortage of other liquids and gases, and the need to measure flows in new and urgent applications such as renewable energy, climate change prevention and carbon capture.

The oil and gas industry seems to ask for more accurate meters capable of operating in adverse conditions, in multiphase flow and in subsea installations with possibly less usable reserves, and in addition there is the detailed audit, transfer, monitoring and control at each stage of the production process.

Many other fluids used by industry: steam flow, gases used in the manufacture of solid-state devices, a wide range of chemicals, mining solid/liquid flows, hygienic and medical products etc. will need more meter options and developments in signal analysis.

In addition, the increased precision of flowmeters over the past 15 years or so means that some flowmeters are now virtually on a par with the uncertainty of the best calibration facilities. This presents a challenge for the calibration laboratories and for the philosophy of calibrating flowmeters.

More development of meters capable of compensating for changes in fluid properties (Kinghorn 1996) may be needed.

24.5 What Developments Should We Expect in Micro-Engineering Devices?

An area of growing interest is that of micro-machined flow sensors. The first sensor based on silicon technology appeared in about the 1970s. Nguyen (1997) divided the sensors into two categories: non-thermal and thermal. The non-thermal used forces, pressure drop measured with capacitive or piezo resistive sensors and Coriolis acceleration. Legtenberg, Bouwstra and Fluitman (1991) described a resonating micro-bridge mass flow sensor suspended inside a micro-flow channel. Although over 20 years since publication, it gives an example of the early developments (Figure 24.1). See also Alvarado, Mireles and Soriano (2009) and Kim, Majumdar and Kim (2007a). Clark, Wang and Cheesewright (2006a, 2006b) discussed the evaluation of micro-machined Coriolis meters. Wiegerink et al. (2012) discussed a micro Coriolis mass flowmeter. It is likely that such a device will be on the market in the near future.

Another early device based on heat transfer is shown in Figure 24.2. It was a low-cost silicon sensor (Nguyen and Kiehnscherf 1995) for mass flow measurement of liquids and gases. Recent papers on thermal designs are in Chapter 18.

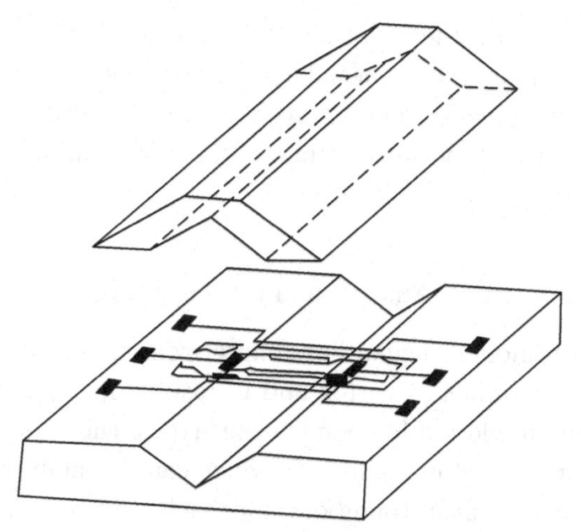

Figure 24.1. Resonating microbridge mass flow sensor. The channel height is 420 μm (reproduced with the agreement of Elsevier from Legtenberg et al. 1991).

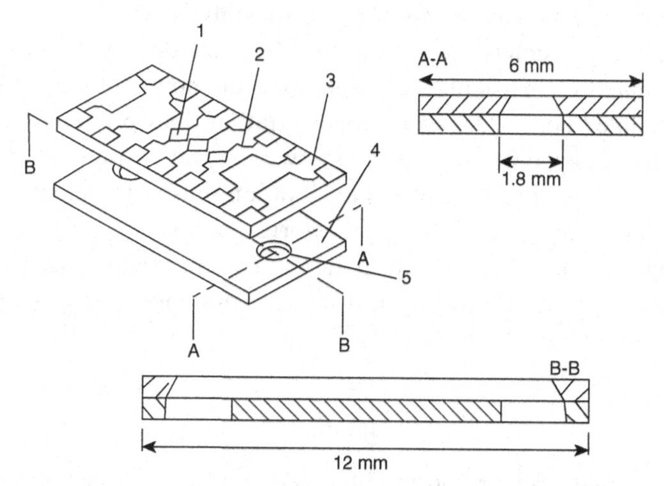

Figure 24.2. Thermal flow sensor: 1. polysilicon resistance; 2. aluminium conductor; 3. Pyrex plate; 5. flow inlet and outlet (reproduced with the agreement of Elsevier from Nguyen and Kiehnscherf 1995).

24.6 Which Techniques for Existing and New Flow Metering Concepts Should Aid Developments?

The dominating themes evident in preparing this edition are:

Computational Design Methods

The application of computational methods to the Coriolis meter has resulted in the development period, which in the 20th century tended to be about 30 years, reduced by half or more. I see little reason why this should not also be a trend in other types of meter, giving them a new lease of life and new applications. There is also the modelling of the flowmeter under normal and exceptional circumstances and during critical flow events, and Kinghorn (1996) foresaw the possibility of reducing capital investment resulting from flowmeter installation by developing meters less susceptible to installation.

Signal Analysis and Data Handling

The possibilities with modern signal handling appear more and more exciting, as shown by some of the trends in vortex and ultrasonic meters. Again there seems little reason why this should not happen in other types. The idea that the meter signal, far from being smoothed, contains data which can reveal information about the meter and the process is transforming our approach. This data may be interpreted by powerful computational methods to indicate change, correct for installation and identify other factors to improve meters and widen their applications.

Such developments should make use of theoretical physical models linking the flowmeter/flow system behaviour with the signal analysis.

An intriguing question is what would have been the design of flowmeters which are now dominated by mechanical engineering if the power of the microprocessor had been available? For instance, the 'finger print' of the vibration of a pipe might be stored in the microprocessor for various flow rates and be used as a look-up table, while the actual sensor might be a simple vibration sensor. Developments like this appear to be happening in, for instance, the vortex meter.

Collecting data on correct plant operation, together with models of the plant flows, and checking current flows against the data and model, may allow any plant changes to be identified (cf. Chapter 15).

Materials

The use of new materials are opening new applications, for instance the ceramic tube for an electromagnetic flowmeter, or the use by Abe et al. (2013) of carbon fibre–reinforced plastic tube as a meter body because of its excellent characteristics, particularly the high attenuation of ultrasound signals

Super Accurate Machines

The achievement of improved uncertainty as in the case of the critical Venturi meter, by a super-accurate lathe, may be useful in other designs.

Verification

I have included a new chapter to cover this area, but I mention it here as a reminder that verification of meters is an important development which needs to be continued to a stage where calibration periods can be yet further extended.

New and Old Concepts Reworked

New ideas are likely to appear, but my reading suggests that some old ideas have come again and been dusted off and reused! There are interesting examples: the Venturi has a new and important role in the measurement of wet-gas and multiphase flows; precision manufacture and miniaturisation of the critical Venturi nozzle reinforces its value in gas calibration; new developments in positive displacement meters, new analysis of the turbine meter, old ideas renewed for the electromagnetic flowmeter.

The new tools are affecting many types of meter, resulting in new capabilities. The new technologies in magnetic resonance and sonar meters are likely to be joined by other new meters for new applications.

The use of pipeline noise, discussed in Chapter 15 and raised in the first edition (Baker 2000, Section 21.5.2), appears to be similar to suggestions of Ibarz et al. (2008) for a sound-based system for water to monitor household usage. It computed the volume flow using pattern recognition.

24.7 Closing Remarks

There are some basic considerations which management should address:

a) For current products
   - Are they pre-eminent technologically?
   - Is the production system appropriate and capable of low variation?
   - Is the uncertainty of the calibration system good enough for the product quality?
   - Has the market been mapped adequately and do the products and the production system respond adequately?
   - How do the company's products compare with those of competitors?

b) For new products
   - Assuming that the questions in (a) have been answered to give an optimum market understanding, the gap in market coverage should be clear.
   - The manufacturer of a proposed new product or design will need to be aware of other new entrants to the market. The existing industry competitors should be identified and the weaknesses of their products, which will be met by strengths in any new designs, should be assessed.

c) For staff
   - The company should review, and if necessary introduce, skill development opportunities for the design and production staff, so that those whose responsibility it is to ensure that products are of the highest quality and that the quality is continually improving, will be well able to do so.

Flow measurement is a large and varied subject. It uses many areas of classical physics in the many and varied sensors. It attempts to measure mean flows in pipes downstream of complex pipework. It attempts to cope with highly adverse conditions both of fluid and of environment. From long experience, I have found that most of the questions posed in industrial training courses are unanswerable, and that the solutions to the questions would merit a major research programme.

There is, therefore, a role for the science base, in terms of the development of new devices, the application of more sophisticated signal processing and data handling techniques, but also to unravel some basic problems.

I hope that this book will provide a significant contribution and benefit for those people whose work is with flowmeters, from invention to application, and will also make a contribution to the quality of the product, and to the financial success of the enterprises, so that the industrial flows of fluids of all sorts will be measured more precisely and reliably.

References

Abe, J. and Yoshinaga, A. (1991) *Flow measurement by using fluidic resistor*. FLUCOME '91 3rd Triennial International Symposium on Fluid Control, Measurement and Visualization, San Francisco, USA: 559–564.

Abe, T., Hayashi, T., Kikawa, N., Tsuruta, D., Kobayashi, Y., Takamoto, M., Shimada, T., Doihara, R. and Terao, Y. (2013) *Development of a new clamp-on ultrasonic flowmeter*. FLOMEKO 2013, 16th International Flow Measurement Conference, Paris, France.

Abernathy, F. H. and Kronauer, R. E. (1962) The formation of vortex streets. *Journal of Fluid Mechanics*, 13: 1–20.

Abro, E., Kleppe, K. and Vikshåland, J. (2009) *Recent field experiences using multiphase meters for fiscal allocation*. 27th International North Sea Flow Measurement Workshop, Tonsberg, Norway, 20–23 October, pp. 67–93.

Adejuyigba, B., Uvwo, I., Liu, J., Ekpecham, O., Scott, S. L., Lansangan, R. and Dutton, R. (2004) *Investigation of three-phase flow measurement capabilities of a Coriolis flow meter. 4th North American Conference on Multiphase Technology*, pp. 161–176.

Adiletta, G., Cascetta, F., della Valle, S., Pagano, S. and Vigo, P. (1993) Twin rigid straight pipe Coriolis mass flowmeter. *Measurement*, 11: 289–308.

Advisory Committee on Aeronautics (1916–1917) Reports and Memoranda, No 295: 675.

AGA-8 (1992) Compressibility and supercompressibility for natural gas and other hydrocarbon gases, Transmission Measurement Committee Report No 8 AGA Catalog No XQ 1285, Arlington, Virginia, USA.

AGA-9 (2007) Measurement of gas by multipath ultrasonic meters, AGA Report no. 9, American Gas Association, Transmission Measurement Committee (April).

Agarwal, A. K. and Turgeon, M. (1984) New MicroMotion mass flowmeters. *Energy Progress*, 4: 221–224.

Agricola, J. B. (1997) *Gas well flowline measurement by ultrasonic flow meter*. North Sea Flow Measurement Workshop Kristiansand, Norway, Paper 16.

Ahmad, K., Baker, R. C. and Goulas, A. (1986) Computation and experimental results of wear in a slurry pump impeller. *Proceedings of the Institution of Mechanical Engineers, Part C*, 200: 439–445.

Ahn, Y.-C., Do Oh, B. and Kim, M. H. (2003) A current-sensing electromagnetic flowmeter for two-phase flow and numerical simulation of the three-dimensional virtual potential distribution: I. *Fundamentals and Annular Flow. Measurement Science and Technology*, 14(3): 239–250.

Akashi, K., Watanabe, H. and Koga, K. (1978) Flow rate measurement in pipe line with many bends. *Technical Review*, June.

Akresh, M., Reindl, L. and Vasic, M. (2010) *Extended measurement range of vortex flow meter in high turbulent range*. 15th Flow Measurement Conference (FLOMEKO), 13–15 October, Taipei, Taiwan, Paper B2-1.

Al-Asmi, K. and Castro, I. P. (1992) Vortex shedding in oscillatory flow: geometrical effects. *Journal of Flow Measurement & Instrumentation*, 3: 187–202.

Al-Bouri, H. A., Samizo, N., Bakhteyar, Z. and Alvi, A. (2005) Application of new multiphase flow measurement systems with satellite-based monitoring in offshore Khafji Field. *International Petroleum Technology Conference Proceedings*, pp. 431–440.

Al-Khamis, M. N., Al-Nojaim, A. A. and Al-Marhoun, M. A. (2002) Performance evaluation of Coriolis mass flowmeters. *Journal of Energy Resources Technology, Transactions of the ASME*, 124(2): 90–94.

Al-Khazraji, Y. A. and Baker, R. C. (1979) Analysis of the performance of three large-electrode electromagnetic flowmeters. *Journal of Physics D*, 12: 1423–1434.

Al-Khazraji, Y. A. and Hemp, J. (1980) Electromagnetic flowmeters and methods of measuring flow. Patent application No. 8011624.

Al-Khazraji, Y. A., Al-Rabeh, R. H., Baker, R. C. and Hemp, J. (1978) *Comparison of the effect of a distorted profile on electromagnetic, ultrasonic and differential pressure flowmeters*. FLOMEKO 1978 – Proceedings of the Conference on Flow Measurement of Fluids, Groningen, The Netherlands, (Amsterdam North-Holland Publishing Company): 215–222.

Al-lababidi, S. and Sanderson, M. L. (2004) Transit time ultrasonic modelling in gas/liquid intermittent flow using slug existence conditions and void fraction analysis. 12th International Conference on Flow Measurement FLOMEKO, Guilin, China.

Al-Mubarak, A. M. (1997) A new method in calculating water cut and oil and water volumes using Coriolis meters. SPE 38785 SPE Conference, 5–8 October, San Antonio, Texas, USA.

Al-Rabeh, R. H. (1981) The design and performance of electromagnetic flowmeters. PhD Thesis, Imperial College, University of London, England.

Al-Rabeh, R. H. and Baker, R. C. (1979) *Optimisation of conventional electromagnetic flowmeters*. Fluid Mechanics Silver Jubilee Conference, National Engineering Laboratory, Glasgow, Scotland, Paper 6.1.

(1986) On the ability to dry calibrate an electromagnetic flowmeter. *Journal of Physics E: Scientific Instruments*, 19: 203–206.

Al-Rabeh, R. H. and Hemp, J. (1981) *A new method for measuring the flow rate of insulating fluids*. International Conference on Advances in Flow Measurement Techniques, BHRA, Paper K3: 267–278.

Al-Rabeh, R. H., Baker, R. C. and Hemp, J. (1978) Induced flow measurement theory for poorly conducting fluids. *Proceedings of the Royal Society A*, 361: 93–107.

Al-Taweel, A. B., Barlow, S. G. and Aggour, M. A. (1997) *Development and testing of multiphase metering unit utilizing Coriolis meters*. Proceedings SPE (38785) Annual Technical Conference and Exhibition, San Antonio, TX, USA, 365–376.

Ali, M., Bohari, M., de Leeuw, H. and Nilsson, C. (2003) Multiphase flow measurement using tracer technology at Dulang Oil Field, Malaysia. *SPE – Asia Pacific Oil and Gas Conference*, 351–359.

Altendorf, M. (1998) Coriolis technique in the pulp and paper industry. *Chemical Engineering World*, 33(5): 87–89.

Altfeld, K. (2002) *Custody transfer metering in the liberalised European market*. 5th International symposium of Fluid Flow Measurement, Arlington, VA, USA.

Alvarado, J., Mireles Jr., J. and Soriano, G. (2009) Development and characterization of a capacitance-based microscale flowmeter. *Flow Measurement and Instrumentation*, 20(2): 81–84.

Amabili, M. and Garziera, R. (2004) Coriolis mass flow meter having a thin-walled measuring tube. US Patent 6,805,013 B2.

Amadi-Echendu, J. E. and Zhu, H. (1992) Signal analysis applied to vortex flowmeters. *IEEE Transactions on Instrumentation and Measurement*, 41.6: 1001–1004.

Amadi-Echendu, J. E., Zhu, H. and Higham, E. H. (1993) Analysis of signals from vortex flowmeters. *Journal of Flow Measurement & Instrumentation*, 4: 225–232.

Amare, T. (1995) Electromagnetic flowmeter for dielectric liquids. PhD thesis, Cranfield University, UK.

(1999) Design of an electromagnetic flowmeter for insulating liquids. *Measurement Science and Technology*, 10: 755.

Amata, Y. (1986) Electromagnetic flowmeter of the residual magnetism type. US Patent 4,601,209.

Amini, A. and Owen, I. (1995) The use of critical flow Venturi nozzles with saturated wet steam. *Journal of Flow Measurement and Instrumentation*, 6: 41–47.

Anabtawi, A. L. and Howlett, R. J. (2000) Detection of blade contamination in turbine flowmeters using neural networks. *Proceedings of the International Conference on Knowledge-based Intelligent Electronic Systems*, KES 2: 635–639.

Andersen, O., Miller, G., Harvey, R. and Stewart, D. (2004) Two-component Coriolis measurement of oil and water at low velocities. In: 22nd North Sea flow metering workshop.

Anklin, M., Drahm, W. and Rieder, A. (2006) Coriolis mass flowmeters: overview of the current state of the art and latest research. *Journal of Flow Measurement and Instrumentation*, 17(6): 317–323.

Anklin, M., Eckert, G., Sorokin, S. and Wenger, A. (2000) Effect of finite medium speed of sound on Coriolis mass flowmeters. FLOMEKO '2000 the 10th International Conference on Flow Measurement, Salvador, Brazil: Paper C11.

Anon. (1966a) Measuring beefeater gin. *Control*, 10(96): 292.

(1966b) Positive displacement flowmeter for low rates. *Control*, 10(98): 437.

(1988) NEL launches offshore multiphase flowmeter project. *Process Engineering*, October: 33.

(1993) Non-intrusive multiphase flow meter. *Offshore Research Focus*, No. 98: 9.

(1994) Multiphase flow meter. *Offshore Research Focus*, No. 103: 2.

(1999) New fuel meter particle filter system from Pierburg. *Diesel Progress – North American Edition*, 65(2): 32.

(2002) A consumers' guide to full-bore magnetic flowmeters. *World Pumps*, (428): 46–49.

(2003) Reincarnation of 2-wire flowmeter. *PACE – Process and Control Engineering*, 56(9): 27–30.

ANSI/API 2530. (1985) Manual of Petroleum Measurement Standards, Chapter 14. Natural Gas Fluids Measurement Section 3 – Orifice Metering of Natural Gas and Other Related Hydrocarbon Fuels.

Ao, S. and Freeke, J. (2003) GE-Panametrics clamp-on flow meter, results in industrial gas applications. 21st North Sea Flow Metering Workshop 2003, Paper 14.

Ao, X. S., Martson, J., Kucmas, P., Khrakovsky, O. and Li, X. S. (2002) *Ultrasonic clamp-on flow measurement of natural gas, steam and compressed air*. 5th International Symposium of Fluid Flow Measurement, Arlington, VA, USA.

Ao, X. S., Caravana, R., Furlong, E. R., Khrakovsky, O. A., McDonald, B. E., Mollo, N. J. and Shen, L. (2011) Method and system for multi-path ultrasonic flow rate measurement. US Patent 7,942,068.

Ao, X. S. Khrakovsky, O. A., Frail, C. A. and Ma, A. Y. (2014) Ultrasonic coupler assembly. US Patent 8635913B2. 28 January 2014.

Aoki, T., Nukui, K., Okamura, S. and Kimura, Y. (1996) *Development of fluidic gas meter (improvement of sensitivity at lower flow rate using U-shaped target sensor)*. FLOMEKO '96 Proceedings of the 8th International Conference on Flow Measurement, Beijing, China: 138–143.

API (2005) Measurement of liquid hydrocarbons by displacement meter systems. American Petroleum Institute (API). *Manual of Petroleum Measurement Standards*, Chapter 5.2.

(2011) Measurement of liquid hydrocarbons by ultrasonic flow meters. American Petroleum Institute (API). *Manual of Petroleum Measurement Standards*, Chapter 5.8.

Araimi, N. M., Jha, N. K. and Al-Zakwani, S. (2005) Exploration well testing with a Venturi/dual energy gamma ray multiphase flow meter – A case study from Oman. SPE Asia Pacific Oil and Gas Conference and Exhibition – Proceedings, pp. 505–512.

Arasi, J. A. (1989) Coriolis mass flowmeter passes NGL line field trials. *Oil & Gas Journal*, January: 59–61.

Arnberg, B. and Ishibashi, M. (2001) Review of flow nozzle and Venturi designs and discharge coefficient. Proc. 47th Intern. Instrumentation Symposium, paper no. 1200.

Arnberg, B. T., Britton, C. L. and Seidl, W. F. (1973) Discharge coefficient correlations for circular-arc Venturi flowmeters at critical (sonic) flow. ASME Paper No 73-WA/FM-8.

Arnold, K. B. and Molz, F. J. (2000) In-well hydraulics of the electromagnetic borehole flowmeter: further studies. *Ground Water Monitoring and Remediation*, 20(1): 52–55.

Arnold, R. M. and Pitts, R. W. (1981) Fluid flow meters for mixed liquid and gas. US Patent 4,272,982, 16 June.

Arvoh, B. K., Asdahl, S., Rabe, K., Ergon, R. and Halstensen, M. (2012) Online estimation of reject gas and liquid flow rates in compact flotation units for produced water treatment. *Flow Measurement and Instrumentation*, 24: 63–70.

Ashauer, M., Glosch, H., Hedrich, F., Hey, N., Sandmaier, H. and Lang, W. (1999) Thermal flow sensor for liquids and gases based on combinations of two principles. *Sensors and Actuators*, 73: 7–13.

Asher, R. C. (1983) Ultrasonic transducers for chemical and process plant. *Physics Technology*, 14: 19–23.

ASME/ANSI MFC-7M (1987) *Measurement of gas flow by means of critical flow Venturi nozzles*. New York: ASME.

Aston, N. A. J. and Evans, G. V. (1975) The integrated pulse velocity method applied to the calibration of flowmeters in high pressure natural gas systems. Conference on Fluid Flow Measurement in the Mid 1970s, Paper D-2: Scotland: National Engineering Laboratory.

Athane, B. (1994) Implications of legal metrology in flow measurement for the utilities. *Journal of Flow Measurement and Instrumentation*, 5(2): 67–69.

Atkinson, D. I., Berard, M. and Segeral, G. (2000) Qualification of a nonintrusive multiphase flow meter in viscous flows. Proceedings 2000 SPE Annual Technical Conference and Exhibition – Production Operations and Engineering General, P1: 533–544.

Atkinson, K. N. (1992) A software tool to calculate the over-registration error of a turbine meter in pulsating flow. *Journal of Flow Measurement and Instrumentation*, 3: 167–172.

Awbery, J. H. and Griffiths, E. (1926–1927) Further experiments with the Ewing ball-and-tube flowmeter. *Proceedings of the Royal Society*, 47: 1.

Baird, J. (1993) Innovations in magnetic flowmetering. *Advances in Instrum. & Control: International Conf. & Exhibition (ISA), Chicago, USA*, 48: 879–884.

Baker, P. D. (1983) *Positive displacement liquid meters*. International School of Hydrocarbon Measurement (ISHM), University of Oklahoma, 12–14 April.

Baker, R. C. (1968a) On the potential distribution resulting from flow across a magnetic field projecting from a plane wall. *Journal of Fluid Mechanics*, 33: 73–86.

(1968b) Solutions of the electromagnetic flowmeter equation for cylindrical geometries. *British Journal of Applied Physics*, 1, Ser. 2: 895–899.

(1969) Flow measurement with motion induced magnetic field at low magnetic Reynolds number. *Magnetohydrodynamics*, No. 3: 69–73 (in Russian).

(1970a) Effects of non-uniform conductivity fluids in electromagnetic flowmeters. *Journal of Physics D*, 3: 637–639.

(1970b) Linearity of motion-induced-magnetic-field flowmeter. *Proc. I.E.E.*, 117: 629–633.

(1973) Numerical analysis of the electromagnetic flowmeter. *Proc. I.E.E.*, 120: 1039–1043.

(1976) Some recent developments in ultrasonic and electromagnetic pipe flow measurement at Imperial College. *ACTA IMEKO*: 339–348.

(1977) Electromagnetic flowmeters for fast reactors. *Progress in Nuclear Energy*, 1: 41–61.

(1982) *Electromagnetic flowmeters. Developments in Flow Measurement – 1*, ed. R. W. W. Scott, Applied Science Publishers Ltd, Chapter 7: 209–251.

(1983) *A review of recent developments in electromagnetic flow measurement.* Third Beer-Sheva Seminar on MHD – Flows and Turbulence, Ben-Gurion University of the

Negev, Beer-Sheva, Israel, 23–27 March 1981. See also in *Prog. in Astro. & Aero.* AIAA Inc, 84: 225–259.

(1985) Principles and practice of electromagnetic flow measurement. Technisches Messen, tm 52 Jahrgang, Heft 1: 4–12 (in German).

(1988a/1989) *An introductory guide to flow measurement.* Mechanical Engineering Publications Limited.

(1988b) Measuring multiphase flow. *Chemical Engineer*, October: 39–45.

(1989) Multi-phase flow moves on. *Control and Instrumentation*, February: 35 and 37.

(1991a) Response of bulk flowmeters to multiphase flow. *Proceedings of the Institution of Mechanical Engineers, Part C: Journal of Mechanical Engineering Science*, 205: 217–229.

(1991b) Turbine and related flowmeters: Part I – Industrial practice. *Journal of Flow Measurement and Instrumentation*, 2: 147–162.

(1993) Turbine flowmeters: Part II – Theoretical and experimental published information. *Journal of Flow Measurement and Instrumentation*, 4: 123–144.

(1994) Coriolis flowmeters: industrial practice and published information. *Flow Measurement and Instrumentation*, 5(4): 229–246.

(1996) *An introductory guide to industrial flow.* London: Mechanical Engineering Publications Limited.

(1998) *Flowmeter. Instruments of science, an historical encyclopedia*, Eds. R. Bud and D. J. Warner. New York and London: Garland Publishing Inc., 245–247.

(2000) *Flow measurement handbook.* 1st Edition. New York: Cambridge University Press.

(2002b) *An introductory guide to flow measurement.* 2nd Edition. Professional Engineering Press (now available from John Wiley & Sons).

(2003) *An introductory guide to flow measurement (ASME Edition).* ASME Press.

(2004a) The impact of component variation in the manufacturing process on variable area (VA) flowmeter performance. *Journal of Flow Measurement and Instrumentation*, 15(4): 207–213.

(2004b) Variation in flowmeter manufacture: some observations and lessons. *Proceedings of the Institution of Mechanical Engineers, Part B: Journal of Engineering Manufacture*, 218: 961–975.

(2005) In-situ calibration verification device and method for electromagnetic flowmeters, International Patent Application number PCT/GB2005/04399 dated 15 November 2005 (taking priority from UK Patent Application number GB0425186.4 dated 15 November 2004).

(2011) On the concept of virtual current as a means to enhance verification of electromagnetic flowmeters. *Measurement Science and Technology*, 22: 105403.

Baker, R. C. and Deacon, J. E. (1983) Tests on turbine, vortex and electromagnetic flowmeters in 2-phase air-water upward flow. International. Conference. on Physical Modelling of Multi-Phase Flow, Coventry, England, BHRA Fluid Engineering: paper H1, 337–352.

Baker, R. C. and Gimson, C. (2001) The effects of manufacturing methods on the precision of insertion and in-line thermal mass flowmeters. *Journal of Flow Measurement and Instrumentation*, 12(2): 113–121.

Baker, R. C. and Gregory, M. J. (2001) Editorial: special issue of the journal focusing on flowmeter manufacture. *Journal of Flow Measurement and Instrumentation*, 12(2): 85–87.

(2002) Flowmeter manufacture: some observations on variation and robustness. 5th International Symposium on Fluid Flow Measurement (ISFFM), Washington, DC, USA.

Baker, R. C. and Hayes, E. R. (1985) Multiphase measurement problems and techniques for crude oil production systems. *Petroleum Review*, November.

Baker, R. C. and Higham, E. H. (1992) Flowmeters. UK Patent No. 2 231 669 B.

Baker, R. C. and Morris, M. V. (1985) Positive-displacement meters for liquids. *Transactions of the Institute of Measurement and Control*, 7: 209–220.

Baker, R. C. and Saunders, P. M. (1977) Tests on flux distortion flowmeters using an analogue rig to simulate flow with a boundary layer. *Annals of Nuclear Energy*, 4: 457–464.

Baker, R. C. and Smith, D. J. M. (1990) *Flowmeter specification guidelines*. Cranfield, England: FLOMIC Report No 14.

Baker, R. C. and Sorbie, I. (2001) A review of the impact of component variation in the manufacturing process on variable area (VA) flowmeter performance. *Journal of Flow Measurement and Instrumentation*, 12(2): 101–112.

Baker, R. C. and Tarabad, M. (1978) The performance of electromagnetic flowmeters with magnetic slurries. *Journal of Physics D Applied Physics*, 11: 167–175.

Baker, R. C. and Thompson, E. J. (1975) A two beam ultrasonic phase-shift flowmeter. Conference on Fluid Flow Measurement in the Mid 1970s, National Engineering Laboratory, East Kilbride Glasgow, Scotland.

(1978) Measurement of fluid flow. US Patent 4,078,428, 14 March.

Baker, R. C., Moore, P. I. and Thomas, A. L. (2004a) Verification for electromagnetic flowmeters – its current state and its potential. Presented at FLOMEKO 2004, Guilin, China.

(2006a) Electronic verification of electromagnetic flowmeters in the water industry. *Water Management* 159 (WM4): 245–251.

Baker, R. C., Moore, P. I. and Wang, T. (2004b) Rethinking dry calibration. Presented at FLOMEKO 2004, Guilin.

(2005) In situ calibration. *Sensor Review*, 25(3): 197–201.

Baker, R. C., Deacon, J. E., Lenn, C. P. and Smart, M. D. (1985) The effect on electromagnetic flowmeters of a change in fluid. *Measurement and Control*, 8: 5–10.

Baker, R. C., Wang, T., Moore, P. I. and Nurse, A. (2006b) Observations on the design and development of a water flow rig related to calibration in the manufacturing process. *Journal of Flow Measurement and Instrumentation*, 17: 171–178.

Baker, R. C., Wang, T., Hussain, Y. A. and Woodhouse, J. (2009) Method for testing a mass flow rate meter. US Patent 7,603,885 B2, 20 October.

Baker, R. C., Gautrey, D. P., Mahadeva, D. V., Sennitt, S. D. and Thorne, A. J. (2013) Case study of the electrical hardware and software for a flowmeter calibration facility. *Journal of Flow Measurement and Instrumentation*, 29: 9–18. http://dx.doi.org/10.1016/j.flowmeasinst.2012.09.001

Baldwin, P. and Rivera, I. (2012) Ultrasonic clamp-on flow metering – 50 years and counting. 8th International Symposium on Fluid Flow Measurement – June.

Ball, J. M. (1977) *Viscosity effects on the turbine flowmeter*. Proceedings of the Symposium on Flow Measurement in Open Channels and Closed Conduits. NBS, Gaithersburg, MD, USA: 847–869.

Balla, J. and Takáras, F. (2003) Long term operational experience of turbine meters. 21st North Sea Flow Metering Workshop 2003, Paper 16.

Balls, B. W. and Brown, K. J. (1959) The magnetic flowmeter. *Symposium on Flow Measurement: Trans. Soc. Instrument Technology*, June: 123–130.

Barbe, J., Dijoux, F., Yardin, C. and Mace, T. (2010) Measurement of helium micro flow-rates with high accuracy for gas chromatography. 15th Flow Measurement Conference (FLOMEKO), 13–15 October 2010, Taipei, Taiwan, Paper A8-5.

Barnes, R. G. (1982) Positive displacement liquid meters. *Advances in Instrumentation*, 37, Part 3, Proc ISA Int Conf and Exhibit (Philadelphia USA, 18–21 October 1982), Research Triangle Park, USA, *Instrum Soc Am*: 1197–1204.

Barry, J. J., Sheikoleslami, M. Z. and Patel, B. R. (1992) Numerical simulation of flow through orifice meters, Gas Research Institute, GRI-92/0060.1.

Barton, N. A. and Boam, D. (2002) In-service performance of ultrasonic flowmeters – Application and validation of CFD modeling methods, Report No. 2002/72, National Engineering Laboratory.

Barton, N. A. and Brown, G. (1999) Velocity distribution effects on ultrasonic flowmeters – Part 2 Determination by computational and experimental methods, Report No. 348/99, National Engineering Laboratory.

Barton, N. and Parry, A. (2012) Using CFD to understand multiphase and wet-gas measurement. North Sea Flow Metering Workshop. Paper 9.2.

Barton, N. and Peebles, B. (2005) Redesign of the ETAP gas export system. FLOMEKO 2005 13th International Flow Measurement Conference, Peebles, Scotland, Paper 6.3.

Barton, N., Reader-Harris, M., Hodges, D. and Coull, C. (2002) The effect of varying Reynolds number on a Zanker flow conditioner plate. 20th North Sea Flow Measurement Workshop 22–25 October 2002, St Andrews, Scotland, Paper 6.2.

Barton, N. A., Gibson, J. G. and Reader-Harris, M. J. (2004) Overview of CFD modelling of orifice-plate flowmeters. 22nd North Sea Flow Metering Workshop, St Andrews, Scotland, 26–28 October 2004, Paper 6.3.

Barton, N., Hodgkinson, E. and Reader-Harris, M. (2005) Estimation of the measurement error of eccentrically installed orifice plates. 23rd North Sea Flow Metering Workshop 2005, Paper 4, pp. 35–44.

Bartstra, R. W. (2006) Clamp-on Coriolis mass flow meter using in-situ calibration. EP 1623192.

Basil, M., Stobie, G. and Letton, W. (2007) *A new approach to MPFM performance assessment in heavy oil.* 25th International North Sea Flow Measurement Workshop, Energy Institute, Oslo, Norway, 311–326.

Basrawi, Y. F. (2003a) Coriolis force mass flow measurement devices. *Proceedings of the International Instrumentation Symposium*, 49: 237–247.

(2003b) Coriolis force mass flow measurement devices. *Annual ISA Analysis Division Symposium – Proceedings*, 444: 161–172.

Basse, N. T. (2014) A review of the theory of Coriolis flowmeter measurement errors due to entrained particles. *Journal of Flow Measurement and Instrumentation*, 37: 107–118.

Batchelor, G. K. (1967) *An introduction to fluid dynamics*. Cambridge University Press.

Bates, C. J. (1981) Laser Doppler anemometry measurements of the flow through an orifice plate. *Flow: Its Measurement and Control in Science and Industry*, 2: 59–68.

(1999) Upstream installation and misalignment effects on the performance of a modified electromagnetic flowmeter. *Journal of Flow Measurement and Instrumentation*, 10: 79–89.

(2000a) The performance of a modified electromagnetic flowmeter when abutted to a smaller misaligned upstream diameter pipe. FLOMEKO '2000 the 10th International Conference on Flow Measurement, Salvador, Brazil: Paper B6.

(2000b) Performance of two electromagnetic flowmeters mounted downstream of a 90° mitre bend/reducer combination. *Measurement: Journal of the International Measurement Confederation*, 27: 197–206.

Bates, C. J. and Turner, R. B. (2003) Fluid flow studies associated with a new electromagnetic flowmeter. *Measurement: Journal of the International Measurement Confederation*, 33(1): 85–94.

Bates, I. P. (1991) Field use of K-lab flow conditioner. Proceedings of the North Sea Flow Measurement Workshop 1991 (two volumes), Norwegian Society of Chartered Engineers, 22–24 October 1991 (9th in the series).

Battye, J. S. (1993) The correlation flowmeter – a detailed investigation of an attempt to improve its performance. FLOMEKO '93 Proceedings of the 6th International Conference on Flow Measurement, Korea: 492–499.

(2001) Acoustic considerations effecting the design of demodulators for the ultrasonic correlation flow meter. *Sensors and Actuators*, A88(1): 29–40.

Baucom, W. E. (1979) Evaluation of a Coriolis mass flow meter for pulverized coal flows. Prepared by The Energy Conservation Division, The University of Tennessee Space Institute, Tullahoma, Tennessee 37388, for the U.S. Dept. of Energy (Contract No DE-AC02 – 79ET10815, December).

Baumann, S. and Vaterlaus, H.-P. (2002) Hot-tapping for the ultrasonic flowmeters at North Umpqua, USA. *International Journal on Hydropower and Dams*, 9(3): 102–104.

Baumgartner, M. (2001) Heat metering by ultrasound (in German). *Euroheart and Power/Femwarme International*, 30(3): 53–55.

Baumoel, J. (1994) *Use of clamp-on transit-time ultrasonic flowmeters in aircraft mass fuel flow, hydraulic fluid leak detection, & ground support applications*. Instrumentation in the Aerospace Industry: Proceedings of the International Symposium ISA, Research Triangle Pk, NC, USA: 243–260.

(1996) *Pipeline management using networked clamp-on transit-time flowmeters*. Proceedings of the International Pipeline Conference, Calgary, Canada, ASME, NY, USA, 2: 1123–1128.

(2002) Performance of wide beam clamp-on ultrasonic flowmeters in API proving tests (clamp-on custody transfer by API method). 5th International symposium of Fluid Flow Measurement, Arlington, VA, USA.

Bean, H. S. (1971) *Fluid meters, their theory and application*. 6th Edition. New York: ASME.

Beaulieu, A., Faucault, E., Braud, P., Micheau, P. and Szeger, P. (2011) A flowmeter for unsteady liquid flow measurements. *Journal of Flow Measurement and Instrumentation*, 22(2): 131–137.

Beck, M. S. and Plaskowski, A. (1987) *Cross correlation flowmeters – their design and application*. Bristol, UK: Adam Hilger.

Beeson, J. (1995) Ultrasonics meters improve NorAm's custody transfer. *Pipeline & Gas Journal*, 222(7): 21–24.

Beg, N. and Toral, H. (1993) Off-site calibration of a two-phase pattern recognition flowmeter. *Int. J. Mult. Flow*, 19: 999–1012.

Belhadj, A., Cheesewright, R. and Clark, C. (2000) The simulation of Coriolis meter response to pulsating flow using a general purpose FE code. *Journal of Fluids and Structures*, 14: 613–634.

Bell, M. J. and MacLeod, M. (2012) Assessment of particle erosion in Coriolis meters. North Sea Flow Metering Workshop, Paper 2.1.

Bellinga, H. and Delhez, F. J. (1993) Experience with a high-capacity piston prover as a primary standard for high-pressure gas flow measurement. *Journal of Flow Measurement and Instrumentation*, 4: 85–90.

Bellinga, H. et al. (1981) Using a piston prover as a primary standard in high-pressure gas metering. Flow, Its Measurement and Control in Science and Industry, Flow 81 Conference, St Louis, USA.

Benabdelkarim, M. and Galiana, C. (1991) Nonradioactive densitometer for continuous monitoring of cement mixing process. Proceedings of the First International Conference on Health, Safety and Environment in Oil and Gas Exploration and Production, Society of Petroleum Engineers of AIME, Texas, USA: 539–545.

Benard, C. J. (1988) *Handbook of fluid flowmetering*. Morden, England: The Trade & Technical Press Limited.

Benes, P. and Zehnula, K. (2000) New design of the two-phase flowmeters. *Sensors and Actuators A: Physical*, 86: 220–225.

Benhadj, R. and Ouazzane, A. K. (2002) Flow conditioners design and their effects in reducing flow metering errors. *Sensor Review*, 22(3): 223–231.

Benkova, M., Makovnik, S., Mikulecky, I. and Zamecnik, V. (2010) Bell prover – calibration and monitoring of time stability. 15th Flow Measurement Conference (FLOMEKO), 13–15 October 2010, Taipei, Taiwan, Paper A1-3.

Bennett, R. (1996) Positive displacement diaphragm gas meter is industry workhorse, *Pipe Line & Gas Industry*, June: 51–53.

Bentley, J. P. and Benson, R. A. (1993) Design conditions for optimal dual bluff body vortex flowmeters. *Journal of Flow Measurement and Instrumentation*, 4: 205–214.

Bentley, J. P. and Nichols, A. R. (1990) The mapping of vortex fields around single and dual bluff bodies. *Journal of Flow Measurement and Instrumentation*, 1: 278–286.

Bentley, J. P., Benson, R. A. and Shanks, A. J. (1996) The development of dual bluff body flowmeters. *Journal of Flow Measurement and Instrumentation*, 7: 85–90.

Bera, S. C. and Kumar Ray, J. (2001) An approach to the design and fabrication of a microprocessor based flow meter using resistance and semiconductor probe. *IETE Technical Review (Institution of Electronics and Telecommunication Engineers, India)*, 18(5): 355–360.

Bera, S. C., Kumar Ray, J. and Chattopadhyay, S. (2004) A modified inductive pick-up type technique of measurement in a vortex flowmeter. *Measurement: Journal of the International Measurement Confederation*, 35(1): 19–24.

Berg, R. F. (2004) Quartz capillary flow meter for gases. *Review of Scientific Instruments*, 75(3): 772–779.

(2005) Simple flow meter and viscometer of high accuracy for gases. *Metrologia*, 42(1): 11–23.

(2008) Capillary flow meter for calibrating spinning rotor gauges. *Journal of Vacuum Science and Technology A: Vacuum, Surfaces and Films*, 26(5): 1161–1165.

Berg, R. F. and Tison, S. A. (2002) *Two primary standards for low flows of gases*. 5th International symposium of Fluid Flow Measurement, Arlington, VA, USA.

Bernier, R. N. and Brennen, C. E. (1983) Use of the electromagnetic flowmeter in a two-phase flow. *Int. J. Multiphase Flow*, 9: 251–257.

Berrebi, J.; Van Deventer, J. and Delsing, J. (2002) Detection of the error generated by a single elbow on an ultrasonic flow meter. *ISA TECH/EXPO Technology Update Conference Proceedings*, 424–425: 1268–1279.

Berrebi, J., Martinsson, P.-E., Willatzen, M. and Delsing, J. (2004a) Ultrasonic flow metering errors due to pulsating flow. *Flow Measurement and Instrumentation*, 15(3): 179–185.

Berrebi, J., van Deventer, J. and Delsing, J. (2004b) Reducing the flow measurement error caused by pulsations in flows. *Flow Measurement and Instrumentation*, 15(5–6): 311–315.

Betts, K. S. (1990) Mass flow sensors. *Measuring up to new applications. Mechanical Engineering*, 112: 72–75.

Bevir, M. K. (1970) The theory of induced voltage electromagnetic flowmeters. *Journal of Fluid Mechanics*, 43: 577.

(1972) The effect of conducting pipe connections and surrounding liquid on the sensitivity of electromagnetic flowmeters. *Journal of Physics D*, 5: 717.

Bevir, M. K., O'Sullivan, V. T. and Wyatt, D. G. (1981) Computation of electromagnetic flowmeter characteristics from magnetic field data. *Journal of Physics D*, 14: 376.

Bezděk, M., Rieder, A., Landes, H., Lerch, R. and Drahm, W. (2003) Numerical analysis of wave propagation in an ultrasonic flowmeter. Paper presented at First Congress, Alps Adria Acoustic Associtrion, 573–80.

Bezděk, M., Rieder, A., Landes, H., Strunz, T. and Lerch, R. (2004) A novel numerical method for simulating wave propagation in moving media. Paper presented at IEEE Ultrasonics Symposium, 2, 934–937.

Bezděk, M., Landes, H., Rieder, A. and Lerch, R. (2007) A coupled finite-element boundary-integral method for simulating ultrasonic flowmeters. *IEEE Transactions on Ultrasonics, Ferroelectrics and Frequency Control*, 54(3): 636–646.

Bignell, N. (1994) A secondary standard ultrasonic gas flowmeter. FLOMEKO '94 Flow Measurement in the Mid-90s, NEL, Glasgow, Scotland: Paper 7.2.

(1996a) Comparison techniques for small sonic nozzles. *Journal of Flow Measurement and Instrumentation*, 7: 109–114.

(2000) Ultrasonic domestic gas meters – a review. FLOMEKO'2000 the 10th International Conference on Flow Measurement, Salvador, Brazil: Paper A1.

Bignell, N. and Takamoto, M. (2000) Editorial of a special issue on Sonic Nozzles. *Journal of Flow Measurement and Instrumentation*, 11(4): 255.

Bignell, N., Collings, A. F., Taylor, K. J. H., Martin, B. J., Braathen, C. W., Peterson, M. and Welsh, C. (1993a) An ultrasonic domestic gas meter. FLOMEKO '93 Proceedings of the 6th International Conference on Flow Measurement, Korea: 403–409.

Bignell, N., Collings, A. F., Taylor, K. J. H., Martin, B. J., Braathen, C. W., Peterson, M. and Welsh, C. (1993b) Calibration of ultrasonic domestic gas meter. FLOMEKO '93 Proceedings of the 6th International Conference on Flow Measurement, Korea: 410–415.

BIPM/IEC/IFCC/ISO/IUPAC/IUPAP/OIML (1993) *Guide to the expression of uncertainty in measurement*, 1st ed., ISO. See updated version at BIPM (JCGM 100: 2008).

BIPM [Internet]. The BIPM key comparison database calibration and measurement capabilities – CMCs (Appendix C). Available from: http://kcdb.bipm.org/AppendixC/default.asp.

Birch, J. R. and Lemon, D. D. (1995) Non-intrusive flow measurement techniques for hydroelectric applications. *ASCE Waterpower – Proceedings of the Int Conf on Hydropower*, 3: 2049–2058.

Birker, B. (1989) *Theory, design and performance of the straight tube mass flowmeter*. Mass Flow Measurement Direct and Indirect, London, England.

Birkhofer, B. H., Jeelani, S. A. K., Windhab, E. J., Ouriev, B., Lisner, K.-J., Braun, P. and Zeng, Y. (2008) Monitoring of fat crystallization process using UVP-PD technique. *Flow Measurement and Instrumentation*, 19(3–4): 163–169.

Blackett, P. M. S. and Henry, P. S. H. (1930) A flow method for comparing specific heats of gases. *Part II – the Theory of the Method. Proc. R. Soc. London Series A*, 126: 333–354.

Blickley, G. J. (1995) Coriolis for the masses. *Control Engineering*, 42(7): 40–41.

Blows, L. G. (1981) Towards a better turbine flowmeter. International Conference on Advances in Flow Measurement Techniques, Warwick, England (Publ by BHRA Fluid Eng, Cranfield, England): 307–318.

Blumenthal, I. (1984) Improving productivity through mass flow measurement and control. Enhancing Productivity, Proceedings of the Pacific Cascade Instrumentation '84 Exhibition and Symposium, ISA: 163–168.

(1985) Direct mass flow rate and density monitoring using a Coriolis/gyroscopic sensor base. *Tappi Journal (USA)*, 68(11): 82–84.

Bø, Ø. L., Nyfors, E., Løland, T. and Couput, J.-P. (2002) New compact wet gas meter based on a microwave water detection technique and differential pressure flow measurement. 20th North Sea Flow Measurement Workshop, 22–25 October 2002, St Andrews, Scotland, Paper 4.1.

Bobovnik, G., Kutin, J. and Bajsić, I. (2005) Estimation of velocity profile effects in the shell-type Coriolis flowmeters using CFD simulations. *Flow Measurement and Instrumentation*, 16(6): 365–373.

Bobovnik, G., Kutin, J., Mole, N., Štok, B. and Bajsić, I. (2013) Numerical analysis of installation effects in Coriolis flowmeters: A case study of a short straight tube full-bore design. *Flow Measurement and Instrumentation*, 34: 142–150.

Boer, A. H. and Volmer, W. (1997) Test results Krohne 8" ultrasonic flowmeter. North Sea Flow Measurement Workshop Kristiansand, Norway: Paper 32.

Bonfig, K. W., Hofman, F., Reinhoold, I. and Feuerstein, M. (1975) A new method of magnetic inductive flow measurement. Conference on Flow Measurement in the Mid-1970s: Paper E-3.

Bonner, J. A. (1977) Pulsating effects in turbine meter. *Pipe Line Industry*, March: 57–62.

(1993) A new international standard, ISO 9951: the measurement of gas flow in closed conduits – turbine meters. *Journal of Flow Measurement and Instrumentation*, 4: 99–100.

Bonner, J. A. and Lee, W. F. Z. (1992) *The history of the gas turbine meter. AGA Distribution and Transmission Conference*, Kansas City, Missouri.

Borkar, K., Venugopal, A. and Prabhu, S. V. (2013) Pressure measurement technique and installation effects on the performance of wafer cone design. *Flow Measurement and Instrumentation*, 30: 52–59.

Bosio, J., Wilcox, P. L., Erdal, A. and Sinding, H. (1990) Gas flowmeters repeatability and accuracy might be impeded by elemental sulphur deposition. North Sea Flow Measurement Workshop, National Engineering Laboratory, Scotland, 23–25.

Botros, K. K., Jungowski, W. M. and Petela, G. (1992) Gauge line effects and DP transmitter response to flow pulsation through orifice plate. *Journal of Flow Measurement and Instrumentation*, 3: 130–144.

Boucher, R. F. (1995) Minimum flow optimization of fluidic flowmeters. *Measurement Science and Technology*, 16: 872–879.

Boucher, R. F. and Mazharoglu, C. (1988) Low Reynolds number fluidic flowmetering. *Journal of Physics E: Scientific Instruments*, 21: 977–989.

Boucher, R. F., Churchill, D., Mazharoglu, C. and Parkinson, G. J. (1991) A fluidic by-pass Venturi meter. FLUCOME '91 3rd Triennial International Symposium on Fluid Control, Measurement and Visualization, San Francisco, USA: 565–569.

Bowden, K. F. and Fairbairn, L. A. (1956) Measurement of turbulent fluctuations and Reynolds stresses in a tidal current. *Proceedings of the Royal Society*, 237: 422.

Boyer, C. and Lemonnier, H. (1996) Design of a flow metering process for two-phase dispersed flows. *International Journal of Multiphase Flow*, 22(4): 713–732.

Boyle, K. C. (2002) Accurate measurement for gypsum. *Global Gypsum,* 2002: 28–31.

Bragg, M. I. and Lynnworth, L. C. (1994) Internally-nonprotruding one-port ultrasonic flow sensors for air and some other gases. Control '94, Conference Publication No. 389, IEE: 1241–1247.

Brain, T. J. S. (1978) The calibration of meters with gases. Short Course notes on The Principles and Practice of Flow Measurement, Lecture No 7, National Engineering Laboratory, East Kilbride.

Brain, T. J. S. and MacDonald, L. M. (1975) Evaluation of the performance of small-scale critical flow Venturis using the NEL gravimetric gas flow standard test facility. Fluid Flow Measurement in the Mid '70s, Edinburgh, Scotland: 103–125.

Brain, T. J. S. and Reid, J. (1978) Primary calibrations of critical flow Venturi nozzles in high-pressure gas. FLOMEKO 1978 Flow Measurement of Fluids, Groningen, The Netherlands: 55–64.

(1980) Primary calibrations of critical flow Venturi nozzles in high-pressure gas. NEL Report No 666.

Brain, T. J. S., Reid, J. and MacDonald, C. (1975) Further development of the NEL pulsed gas-ionization flowmeter. Conference on Fluid Flow Measurement in the Mid 1970's, Paper D-4.

Branch, J. C. (1995) The effect of an upstream short radius elbow and pressure tap location on orifice discharge coefficients. *Journal of Flow Measurement and Instrumentation*, 6: 157–162.

Brassier, P. and Hosten, B. (2000) An ultrasonic gas flowmeter using high frequency transducers and correlation technique. FLOMEKO'2000 the 10th International Conference on Flow Measurement, Salvador, Brazil: Paper B4.

Brassier, P., Hosten, B. and Vulovic, F. (2001) High-frequency transducers and correlation method to enhance ultrasonic gas flow metering. *Journal of Flow Measurement and Instrumentation*, 12(3): 201–211.

Bremser, W., Hässelbarth W., Hirlehei, U., Hotze, H.-J. and Wendt, G. (2002) *Traceability and uncertainty of the German national flow rate measurement standard pigsar.* 5th International symposium of Fluid Flow Measurement, Arlington, VA, USA.

Brennan, J. A., McFaddin, S. E., Sindt, C. F. and Kothari, K. M. (1989) The influence of swirling flow on orifice and turbine flowmeter performance. *Journal of Flow Measurement and Instrumentation*, 1: 5–8.

Brennan, J. A., Sindt, C. F., Lewis, M. A. and Scott, J. L. (1991) Choosing flow conditioners and their location for orifice flow measurement. *Journal of Flow Measurement and Instrumentation*, 2: 40–44.

Brewer, H. (2005) Why your lab should be ISO/IEC 17025-Accredited. Quality Digest, May 2005, 49–52.

Brignell, J. and White, N. (1994) *Intelligent sensor systems.* Institute of Physics Publishing, Bristol and Philadelphia.

Britton, C. L. and Caron, R. W. (1997) Unchoking pressure ratio for critical flow Venturis. ASME Fluids Engineering Division Summer Meeting FEDSM'97, Paper 3004.

Britton, C. and Mesnard, D. (1982) A performance survey of round- and diamond-shaped averaging pitot-type primaries. *Measurement & Control*, 15: 341–350.

Britton, C., Seidl, W. and Kinney, J. (2002) Experimental wet gas data for a Herschel style Venturi. 5th International Symposium of Fluid Flow Measurement, Arlington, VA, USA.

Britton, C. L., Kinney, J. and Savidge, J. L. (2004) Wet gas flow measurements with mixtures of natural gas, hydrocarbon liquids and water. 22nd North Sea Flow Metering Workshop 2004, Paper 8.3

Brobeil, W., Frohlich, R., Schafer, R. and Schulz, K. H. (1993) Flow measurements in slurry applications with switched DC-fields magflowmeters. FLOMEKO'93 Proceedings of the 6th International Conference on Flow Measurement, Korea Research Institute of Standards and Science: 635–641.

Brockhaus, H. (2005) Magnetisch Induktiver Durchflussmesser mit erweiterter Selbstu berwachung. VDI Berichte, No 1883:723–731 (in German).

Brockhaus, H., van der Pol, R., Schoth, U. and Klein, J. W. (1996) Capacitive electro magnetic flowmeter (EMF) using microsystems technology. FLOMEKO'96 Proceedings of the 8th International Conference on Flow Measurement, Beijing, China: 635–641.

Brown, A. F. and Kronberger, H. (1947) A sensitive recording calorimetric mass flowmeter. *Journal of Scientific Instruments*, 24: 151–155.

Brown, G. J. (1997) Factors affecting the performance of ultrasonic flowmeters. North Sea Flow Measurement Workshop Kristiansand, Norway: Paper 33.

Brown, G., Augenstein, D., Estrada, H., Laird, C. and Cousins, T. (2010) Thermal gradient effects on ultrasonic flowmeters in the laminar regime. 28th International North Sea Flow Metering Workshop, Paper 6.1.

Brown, G. J. and Coull, G. J. (2001) Benefits and limitations of ultrasonic meters for upstream oil and gas production. 19th North Sea Flow Measurement Workshop, Kristiansand, Norway, 22–25 October 2001, Paper 19.

Brown, G. J. and Griffith, B. (2012) the effects of flow conditioning on the performance of multipath ultrasonic meters. 8th International Symposium in Fluid Flow Measurement, June 20–22, 2012.

(2013) A new flow conditioner for 4-path ultrasonic flowmeters. FLOMEKO 2013, 16th International Flow Measurement Conference, Paris, France.

Brown, G. J., Reader-Harris, M. J., Gibson, J. and Stobie, G. J. (2000) Correction of the readings of an orifice-plate installed in reverse orientation. 18th North Sea Flow Measurement Workshop. Gleneagles, Scotland, 2000.

Brown, G., Augenstein, D. and Cousins, T. (2006) *Velocity profile effects on multipath ultrasonic meters*. Proc. 6th International Symposium on Fluid Flow Measurement (ISFFM), Querétero, Mexico.

Brown, G., Estrada, H., Augenstein, D. and Cousins, T. (2007) *LNG metering using 8-path ultrasonic meters*. 25th International North Sea Flow Measurement Workshop, Energy Institute, Oslo, Norway, pp. 327–349.

Brown, G. J., Cousins, T., Griffith, B. and Augenstein, D. R. (2009a) Comparison of multipath ultrasonic meter calibration data from two liquid hydrocarbon facilities and one water facility. 27th International North Sea Flow Measurement Workshop, Tonsberg, Norway, 20–23 October 2009, pp. 325–340.

Brown, G. J., Cousins, T., Augenstein, D. A. and Estrada, H. (2009b) A multipath ultrasonic meter with reducing nozzle for improved performance in the laminar/turbulent transition region. 27th International North Sea Flow Measurement Workshop, Tonsberg, Norway, 20–23 October 2009, pp. 361–385.

Bruschi, P., Nizza, N. and Piotto, M. (2006) Measurement and modelling of pulsed mode flow meter for liquids based on a single chip probe. *Sensors and Actuators, A: Physical*, 132(1 SPEC. ISS): 188–194.

Buchanan, R. K. (2003) Non-invasive cryogenic flow measurement. *Proceedings of the International Instrumentation Symposium*, 49: 257–265.

Buckle, U., Durst, F., Howe, B. and Melling, A. (1992) Investigation of a floating element flowmeter. *Flow Measurement and Instrumentation*, 3: 215–225.

Buckle, U., Durst, F., Kochner, H. and Melling, A. (1995) Further investigation of a floating element flowmeter. *Flow Measurement and Instrumentation*, 6: 75–78.

Bucknell, R. L. (1963) Calibration systems and turbine type flow transducers for cryogenic flow measurements. Advances in cryogenic Engineering, Plenum Press, New York, USA, 8: 360–370.

Buhidma, A. and Pal, R. (1996) Flow measurement of two-phase oil-in-water emulsions using wedge meters and segmental orifice meters. *Chemical Engineering Journal*, 63(1): 59–64.

Buonanno, G. (2000) On the field characterization of static domestic gas flowmeters. *Measurement: Journal of the International Measurement Confederation*, 27: 277–285.

Butterworth, D. and Hewitt, G. F. (1977) *Two-phase flow and heat transfer*. Harwell Series, Oxford University Press.

Buttle, R. S. and Kimpton, A. (1989) ESKOM's flow calibration facility. FLOMEKO 5th Internaitonal Conference on Flow Measurement: 1–10.

Caetano, E., Pinheiro, J. A. and Moreira, C. C. (1997) MMS 1200: cooperation on a subsea mutliphase flowmeter application. Offshore Technology Conference, Houston, TX, USA, 4:115–123.

Caetano, E., Pinheiro, J. A., da Costa e Silva, C. B., Kuchpil, C. and Dykesteen, E. (2000) Subsea multiphase flowmetering offshore Brazil. FLOMEKO'2000 the 10th International Conference on Flow Measurement, Salvador, Brazil: Paper D5.

Cai, S., Toral, H., Sinta, D. and Tajak, M. (2004) Experience in field tuning and operation of a multiphase meter based on neural net characterization of flow conditions. 12th International Conference on Flow Measurement FLOMEKO Guilin China.

Cairney, W. D. (1991) Typical flow measurement problems and their solution in the electricity supply industry. *Journal of Flow Measurement and Instrumentation*, 2: 217–224.

Calamante, F., Thomas, D. L., Pell, G. S., Wiersma, J. and Turner, R. (1999) Measuring cerebral blood flow using magnetic resonance imaging techniques. *Journal of Cerebral Blood Flow & Metabolism*, 19: 701–735.

Calcatelli, A., Raiteri, G. and Rumiano, G. (2003) The IMGC-CNR flowmeter for automatic measurements of low-range gas flows. *Measurement: Journal of the International Measurement Confederation*, 34(2): 121–132.

Calogirou, A., Boekhoven, J. and Henkes, R. A. W. M. (2001) Effect of wall roughness changes on ultrasonic gas flowmeters. *Journal of Flow Measurement and Instrumentation*, 12(3): 219–229.

Campion, P. J., Burns, J. E. and Williams, A. (1973) *A code of practice for the statement of accuracy*. London: HMSO.

Carlander, C. and Delsing, J. (2000) Installation effects on an ultrasonic flowmeter with implications for self diagnostics. *Journal of Flow Measurement and Instrumentation*, 11(2): 109–122.

Caron, R. W. (1995) Use of sonic nozzles in a manufacturing environment. Proceedings of the Int Instrumentation Symposium, ISA: 543–558.

(2001) Realistic gas flow measurement traceability. *Proceedings of the International Instrumentation Symposium* 47: 477–491.

Caron, R. W., Britton, C. L., Connolly, T., Hodges, C. and Kegel, T. (2002) A novel primary flow standard for compressible flow calibration: initial testing and calibration. 5th International symposium of Fluid Flow Measurement, Arlington, VA, USA.

Carpenter, B. (1990) Choose the right material for mass flow meters. *Chemical Engineering Progress*, 86, No. 10: 55–60.

Carter, R. M., Yan, Y. and Cameron, S. D. (2005) On-line measurement of particle size distribution and mass flow rate of particles in a pneumatic suspension using combined imaging and electrostatic sensors. *Flow Measurement and Instrumentation*, 16(5): 309–314.

Cascetta, F. (1994) Application of a portable clamp-on ultrasonic flowmeter in the water industry. *Journal of Flow Measurement and Instrumentation*, 5: 191–194.

(1996) Effect of fluid pressure on Coriolis mass flowmeter's performance. *ISA Transactions*, 35: 365–370.

Cascetta, F. and Vigo, P. (1988) *Flowmeters – a comprehensive survey and guide to selection*. ISA Publ, Research Triangle Park, NC, USA.

Cascetta, F., della Valle, S., Guido, A. R. and Vigo, P. (1989a) *A new type of Coriolis acceleration mass flowmeter*. Proc IMEKO XI, Houston: 511–521.

(1989b) A Coriolis mass flowmeter based on a new type of elastic suspension. *Measurement*, 7: 182–191.

Cascetta, F., Cignolo, G., Goria, R., Martini, G., Rivetta, A. and Vigo, P. (1992) Experimental intercomparison of Coriolis mass flowmeters. *Trans Institute of Measurement & Control*,

14: 99–107. (Also International Conference on Flow Measurement of Commercially Important Fluids, IBC Technical Services Ltd, London, Feb/Mar 1990.)

Cascetta, F., Palombo, A. and Scalabrini, G. (2003) Water flow measurement in large bore pipes: An experimental comparison between two different types of insertion flowmeters. *ISA Transactions*, 42(2): 171–179.

Casperson, C. (1975) The vortex flowmeter. Conference on Fluid Flow Measurement in the Mid 1970's, National Engineering Laboratory, Glasgow, Scotland: Paper C-1.

(1993) A new flowmeter for measurement of gas flow at very low rates. FLOMEKO'93 Proceedings of the 6th International Conference on Flow Measurement, Korea: 306–312.

Catherine, L. (2002) *Dynamic measurement of volumes in hydrocarbons pipelines*. 5th International symposium of Fluid Flow Measurement, Arlington, VA, USA.

Cazin, J., Couput, J-P., Dudézert, C., Escande, J., Gajan, P., Lepeau, A. and Strzelecki, A. (2005) Lessons from wet gas flow metering systems using differential measurement devices: testing and flow modelling results. 23rd North Sea Flow Metering Workshop 2005, Paper 12, pp. 185–197.

Cha, J-E., Ahn, Y-C. and Kim, M-H. (2001) Flow measurement with an electromagnetic flowmeter in two-phase bubbly and slug flow regimes. *Flow Measurement and Instrumentation*, 12(5–6): 329–339.

Cha, J-E., Ahn, Y-C., Seo, K-W., Nam, H-Y., Choi, J-H. and Kim, M-H. (2003a) An experimental study on the characteristics of electromagnetic flowmeters in the liquid metal two-phase flow. *Flow Measurement and Instrumentation*, 14(4–5): 201–209.

Cha, J-E., Ahn, Y-C., Seo, K-W., Nam, H. Y., Choi, J. H. and Kim, M. H. (2003b) The performance of electromagnetic flowmeters in a liquid metal two-phase flow (F34). *Journal of Nuclear Science and Technology*, 40(10): 744–753.

Chahine, K. (2005) Comparison of low pressure gas flow standards. Flomeko 2005 13th International Flow Measurement Conference, Peebles, Scotland, Paper 8.3.

Chahine, K. C. and Ballico, M. (2010) Assessment of reproducibility and linearity of the NMIA bell prover using a high flowrate sonic nozzle array. 15th Flow Measurement Conference (FLOMEKO), 13–15 October 2010, Taipei, Taiwan, Paper A1-2.

Chambers, J. (1994) The EMC test house – before and after. *Engineering,* May 1994: 39–42.

Chanaud, R. C. (1965) Observations of oscillatory motion in certain swirling flows. *Journal of Fluid Mechanics*, 21: 111–127.

Chapman, N. R. and Etheridge, D. W. (1993) *A step change in domestic metering technology from leather diaphragms to ultrasonics*. Flow Measurement for the Utilities, Amsterdam, The Netherlands.

Chateau, F. (1991) Mass delivery meters using the Coriolis effect for batching hydrocarbon binders. Bulletin de Liaison des Laboratoires des Ponts et Chaussees, No. 175: 13–19.

Cheesewright, R. and Clark, C. (1998) The effect of flow pulsations on Coriolis mass flow meters. *Journal of Fluids and Structures*, 12: 1025–1039.

(2002) Experimental investigation of the influence of external vibrations on Coriolis mass flow meters. *American Society of Mechanical Engineers, Applied Mechanics Division, AMD*, 253(2): 1109–1117.

Cheesewright, R. and Shaw, S. (2006) Uncertainties associated with finite element modelling of Coriolis mass flow meters. *Flow Measurement and Instrumentation*, 17(6): 335–347.

Cheesewright, R. and Tombs, M. S. (2006) Editorial. *Flow Measurement and Instrumentation*, 17(6): 315–316.

Cheesewright, R., Atkinson, K. N., Clark, C., ter Horst, G. J. P., Mottram, R. C. and Viljeer, J. (1996) Field tests of correction procedures for turbine flowmeters in pulsatile flows. *Journal of Flow Measurement and Instrumentation*, 7: 7–17.

Cheesewright, R., Bisset, D. and Clark, C. (1998) Factors which influence the variability of turbine flowmeter signal characteristics. *Journal of Flow Measurement and Instrumentation*, 9: 83–89.

Cheesewright, R., Clark, C. and Bisset, D. (1999) Understanding the experimental response of Coriolis mass flowmeters to flow pulsations. *Journal of Flow Measurement and Instrumentation*, 10: 207–215.

(2000) The identification of external factors which influence the calibration of Coriolis mass flowmeters. *Journal of Flow Measurement and Instrumentation*, 11: 1–10.

Cheesewright, R., Belhadj, A. and Clark, C. (2003a) Effect of mechanical vibrations on Coriolis mass flow meters. *Journal of Dynamic Systems, Measurement and Control, Transactions of the ASME*, 125(1): 103–113.

Cheesewright, R., Clark, C., Belhadj, A. and Hou, Y. Y. (2003b) The dynamic response of Coriolis mass flow meters. *Journal of Fluids and Structures*, 18(2): 165–178.

Chen, J., Karras, M., Tahkola, E. and Tombery, J. (1993) Flow measurement of medium consistency pulp suspension by cross-correlation flowmeter. FLOMEKO '93 Proceedings of the 6th International Conference on Flow Measurement, Korea: 506–508.

Chen, J.-L., Lin, J.-W., Chen, P., Wei, C.-Y. and Huang, Y.-C. (2010) Numerical simulation on the flow field of a vortex flowmeter with various upstream conditions. 15th Flow Measurement Conference (FLOMEKO), October 13–15, 2010, Taipei, Taiwan, Paper B2-5.

Chesnoy, A. B. (1993) Sonic nozzles meter natural gas at K-Lab. *Journal of Flow Measurement and Instrumentation*, 4: 73–76.

Cheung, W.-S., Kwon, H.-S., Park, K.-A. and Paik, J.-S. (2001) Acoustic flowmeter for the measurement of the mean flow velocity in pipes. *Journal of the Acoustical Society of America*, 110(5 I): 2308–2314.

Chien, S.-F. and Schrodt, J. L. G. (1995) Determination of steam quality and flow rate using pressure data from an orifice meter and a critical flowmeter. *SPE Production & Facilities*, 10(2): 76–81.

Chisholm, D. (1967) Flow of incompressible two-phase mixtures through sharp-edged orifices. *Journal of Mechanical Engineering Science*, 9: 72–78.

(1977) Two-phase flow through sharp-edged orifices. *Journal of Mechanical Engineering Science*, 19: 128–130.

Chisholm, D. and Leishman, J. M. (1969) Metering of wet steam. *Chemical and Process Engineering*, 50: 103–106.

Chisholm, D. and Watson, G. C. (1966) The flow of steam/water mixtures through sharp-edged orifices. NEL Report No 213, East Kilbride, Glasgow.

Choi, H. M., Park, K.-A., Oh, Y. K. and Choi, Y. M. (2009) Improvement and uncertainty evaluation of mercury sealed piston prover using laser interferometer. *Journal of Flow Measurement and Instrumentation*, 20(4–5): 200–205.

(2010) Uncertainty evaluation procedure and intercomparison of bell provers as a calibration system for gas flow meters. *Journal of Flow Measurement and Instrumentation*, 21(4): 488–496.

Choi, Y. M., Park, K. A. and Park, S. O. (1997) Interference effect between sonic nozzles. *Journal of Flow Measurement and Instrumentation*, 8(2): 113–119.

Choi, Y. M., Park, K. A., Park, J. T., Choi, H. M. and Park, S. O. (1999) Interference effects of three sonic nozzles of different throat diameters in the same meter tube. *Journal of Flow Measurement and Instrumentation*, 10: 175–181.

Christensen, T. A. and Willatzen, M. (2010) Investigating the effect of magnetic pipes connected to electromagnetic flowmeters using experimentally validated finite element models. *Journal of Flow Measurement and Instrumentation*, 21(1): 62–69.

Chun, S., Yoon, B.-R., Kang, W. and Kwon, H.-S. (2013) Accuracy enhancement of a combined V/Z clamp-on ultrasonic flow meter. FLOMEKO 2013, 16th International Flow Measurement Conference, Paris, France.

Chung, J., Grigoropoulos, C. P. and Greif, R. (2003) Infrared thermal velocimetry for nonintrusive flow measurement in silicon microfluidic devices. *Review of Scientific Instruments*, 74(5): 2911–2917.

Chunhui Li, C. and Johnson, A. (2010) Bilateral comparison confirms NIM's and NIST's gas flow capabilities. 15th International Flow Measurement Conference FLOMEKO 2010, Taiwan, Paper A1-1.

Cignolo, G., Alasia, F., Capelli, A., Goria, R. and La Piana, G. (2002) A primary standard piston prover for measurement of very small gas flows. 5th International symposium of Fluid Flow Measurement, Arlington, VA, USA.

Clarijs, M. C., Bom, V. R., Van Eijk, C. W. E., Kolar, Z. I. and Scheers, L. M. (2003) X-ray spectrum generation for a multiphase flow meter. *IEEE Transactions on Nuclear Science*, 50 III(4): 713–717.

Clark, C. (1992) The measurement of dynamic differential pressure with reference to the determination of pulsating flows using DP devices. *Journal of Flow Measurement and Instrumentation*, 3: 145–150.

Clark, C. and Cheesewright, R. (2003a) Coriolis flow meters/the potential for outstanding dynamic performance. *Measurement and Control*, 36(9): 275–277+281.

(2003b) The influence upon Coriolis mass flow meters of external vibrations at selected frequencies. *Flow Measurement and Instrumentation*, 14(1–2): 33–42.

(2006) Experimental determination of the dynamic response of Coriolis mass flow meters. *Flow Measurement and Instrumentation*, 17(1): 39–47.

Clark, C., Cheesewright, R. and Hou, Y. Y. (2003) The dynamic response of Coriolis flowmeters. *IEE Computing and Control Engineering*, 14(4): 37.

Clark, C., Wang, S. and Cheesewright, R. (2006a) The dynamic performance of a micro-machined Coriolis flow meter. *2006 NSTI Nanotechnology Conference and Trade Show – NSTI Nanotech 2006 Technical Proceedings*, 3: 336–339.

(2006b) The performance characteristics of a micro-machined Coriolis flow meter: An evaluation by simulation. *Flow Measurement and Instrumentation*, 17(6): 325–333.

Clark, C., Zamora, M., Cheesewright, R. and Henry, M. (2006c) The dynamic performance of a new ultra-fast response Coriolis flow meter. *Flow Measurement and Instrumentation*, 17(6): 391–398.

Clarke, D. W. (1998) Non-linear control of the oscillation amplitude of a Coriolis mass-flowmeter. *European Journal of Control*, 4: 196–207.

Clarke, D. W. and Ghaoud, T. (2002) Validation of vortex flowmeters. *Computing and Control Engineering Journal*, 13(5): 237–241.

(2003) A dual phase-locked loop for vortex flow metering. *Flow Measurement and Instrumentation*, 14(1–2): 1–11.

Clarke, D. W. and Hemp, J. (2009) Eddy-current effects in an electromagnetic flowmeter. *Flow Measurement and Instrumentation*, 20(1): 22–37.

Clayton, C. G., Ball, A. M., Clark, W. E. and Spencer, E. A. (1962a) The accurate measurement of turbulent flow in pipes – using radioactive isotopes – the isotope dilution method. Proc. Symp. on Flow Measurement in Closed Conduits, HMSO, 2: Paper E-3.

Clayton, C. G., Clark, W. E. and Ball, A. M. (1962b) The accurate measurement of turbulent flow in pipes – using radioactive isotopes – using the isotope velocity method and the effect of some restriction on optimum operation. Proc. Symp. on Flow Measurement in Closed Conduits, HMSO, 2: Paper E-4.

Cloy, J. K. (2002) Ultrasonic, transit time flow measurement technology. *Proceedings of the Annual Symposium on Instrumentation for the Process Industries*, 57: 235.

Cobu, T., Berg, R. F., Wright, J. D. and Moldover, M. R. (2010) Modeling Laminar Flow Meters for Process Gases. 15th Flow Measurement Conference (FLOMEKO), 13–15 October 2010, Taipei, Taiwan, Paper A2-1.

Cochran, T. B., Feiveson, H. A., Patterson, W., Pshakin, G., Ramana, M. V., Schneider, M., Suzuki, T. and von Hippel, F. (2010) Fast breeder reactor programs: history and status. A research report of the International Panel on Fissile Materials February 2010.

Cole, J. H. (1985) *Drag turbine mass flowmeter development*. Flow, Its Measurement and Control in Science and Industry, ISA Conf, St Louis, USA, 2: 441–451.

Cole, K. D. (2006) *Flush-mounted steady-periodic heated film with application to fluid-flow measurement*. American Society of Mechanical Engineers, Heat Transfer Division, (Publication) HTD, International Mechanical Engineering Congress and Exposition, IMECE2006 – Heat Transfer, Chicago, IL, United States.

Coleman, H. W. and Steele, W. G. (1999) *Experimentation and uncertainty analysis*. 2nd ed. Wiley Inter Science.

Coleman, M. C. (1956) Variable area flow meters. *Transactions of the Institution of Chemical Engineers*, 34: 339.

Collings, A. F., Bignell, N., Hews Taylor, K. J. and Marting, B. J. (1993) *Ultrasonic metering of gas flows*. Proceedings of the Ultrasonic International Conference, Vienna, Austria, Butterworth Heinemann: 205–208.

Collins, D. B. and Gacesa, M. (1970) Measurement of steam quality in two-phase up flow with Venturi meters. ASME Paper No 70-FE-6.

Collis, D. C. and Williams, M. J. (1959) Two-dimensional convection from heated wires at low Reynolds numbers. *Journal of Fluid Mechanics*, 6: 357.

Conrad, F. and Trostmann, E. (1981) A servo controlled volume rate flowmeter. BHRA Int. Conf. on Advances in Flow Measurement Techniques, Warwick, UK: Paper H3.

Cook, C. E. and Bernfeld, (1967) *Radar signals*. Academic Press, New York.

Cooley, C., Couput, J.-P., Letton, C. and Hall, J. (2003) Wet Gas allocation on the canyon express project. 21st North Sea Flow Metering Workshop 2003, Paper 7.

Corneliussen, S. co-ordinated by Dahl, E. (2005) Handbook of multiphase flow metering 3rd Revision. Norwegian Society for oil and gas measurement & The Norwegian Society of Chartered Technical and Scientific Professionals ISBN 82-91341-89-3.

Coull, C. J. and Barton, N. A. (2002) Investigation of the installation effect on ultrasonic flowmeters and evaluation of computational fluid dynamics prediction methods. 20th North Sea Flow Measurement Workshop, 22–25 October 2002, St Andrews, Scotland, Paper 6.4

Coull, C. J. and Boam, D. (2002) In-Service performance of ultrasonic flowmeters – Investigation of installation effect, Report No. 2002/55, National Engineering Laboratory, 2002.

Coull, C. and Miller, G. (2005) Multiphase measurement using ultrasonic technology – testing a conceptual design. Flomeko 2005 13th International Flow Measurement Conference, Peebles, Scotland, Paper 3.1 .

Coulthard, J. and Yan, Y. (1993a) Vortex wake transit time measurements for flow metering. *Journal of Flow Measurement & Instrumentation*, 4: 269–272.

(1993b) Comparisons of different bluff bodies in vortex wake transit time measurements. *Journal of Flow Measurement & Instrumentation*, 4: 273–276.

Cousins, T. (1971) The performance of long bore orifices at low Reynolds numbers. *Modern Developments in Flow Measurement*, Paper 4.1: 160–179.

(1975) An experimental investigation into the design parameters of the Dall tube. Conference on Fluid Flow Measurement in the Mid-1970s, National Engineering Laboratory, Glasgow, Scotland: Paper J2.

(1977) Vortex meters. Transducer 77 Conference.

et al. (1973) A linear and accurate flowmeter using vortex shedding. Symposium Power Fluidics for Process Control, University of Surrey, Guildford, England: 45–56.

Cousins, T. and Hayward, A. T. J. (1993) Development of the T-ring vortex meter. *Journal of Flow Measurement & Instrumentation*, 4: 197–204.

Cousins, T., Hayward, A. T. J. and Scott, R. (1989) Design and performance of a new vortex shedding flowmeter. FLOMEKO '89 Proceedings of the 5th International Conference on Flow Measurement, Dusseldorf, Germany: 151–167.

Cousins, T., Augenstein, D. and Eagle, S. (2005) The effect of water in oil on the performance of a four path chordal ultrasonic flow meter in horizontal flow lines. 23rd North Sea Flow Metering Workshop 2005, Paper 14, pp. 209–223.

Cowley, M. D. (1965) Flowmetering by a motion-induced magnetic field. *Journal of Scientific Instruments*, 42: 406–409.

Cox, T. J. and Wyatt, D. G. (1984) An electromagnetic flowmeter with insulated electrodes of large surface area. *Journal of Physics E: Scientific Instruments*, 17: 488–503.

Crainic, M. S., Cornel, M. and Ilie, D. (2000) Ferrofluids flow transducer for liquids. *Journal of Flow Measurement & Instrumentation*, 11(2): 101–108.

Cramer, A., Zhang, C. and Eckert, S. (2004) Local flow structures in liquid metals measured by ultrasonic Doppler velocimetry. *Flow Measurement and Instrumentation*, 15: 145–153.

Crawshaw, J. and Chambers, J. (1984) A concise course in A-level statistics. Stanley Thornes (Publishers) Ltd.

Cristancho, D. E., Coy, L. A. Hall, K. R. and Iglesias-Silva, G. A. (2010) An alternative formulation of the standard orifice equation for natural gas. *Journal of Flow Measurement and Instrumentation*, 21(3): 299–301.

Cruz-Maya, J. A., Sánchez-Silva, F. and Quinto-Diez, P. (2006) A new correlation to determine the discharge coefficient of a critical Venturi nozzle with turbulent boundary layer. *Journal of Flow Measurement and Instrumentation*, 17(5): 258–266.

Cui, L., (2010) A kind of New Calibration Method for the Volume of Bell Prover....15th Flow Measurement Conference (FLOMEKO), October 13–15, 2010 Taipei, Taiwan, Paper A1-4.

Cunningham, T. J. (2009) *An in-situ verification technology for Coriolis flowmeters*. 7th ISSFM, Alaska, USA.

Cunningham, T. J. and Hensley, D. P. (2001) *Using IMAT and Matlab for Coriolis flowmeter design*. Proceedings of the International Modal Analysis Conference (IMAC), Kissimmee, USA, 1: 165–71.

Cunningham, W. J. and Astami, K. (1993) The effect of ultrasonic frequency on the accuracy of gas flowmeters. *Measurement Science and Technology*, 4: 1476–1478.

Cushing, V. (1958) Induction flowmeter (for use with dielectrics). *Review of Scientific Instruments*, 29: 692.

Cushing, V.(1965) Electromagnetic flowmeter. *Review of Scientific Instruments*, 36: 1142.

(1971) Electromagnetic flowmeters. Symposium on Flow – Its Measurement and Control in Science and Industry, Pittsburgh, PA, Paper 2-4-38.

(2001) Electromagnetic flowmeter for all fluids. *ISA TECH/EXPO Technology Update Conference Proceedings*, 416: 309–319.

Cushing, V. (2002) Electromagnetic flowmeter for insulating liquids. *Conference Record – IEEE Instrumentation and Measurement Technology Conference*, 1: 103–108.

Cutler, G. D. (September 1982) Averaging Pitot-type primaries. *Measurement & Control*, 15: 436–437.

Dahlström, M. J. (2000) *Effortless oil ultrasonic fiscal meter operation: Krohne Altosonic-V, with master meter approach. Vignis-Snorre Crossover and the Snorre B Export Station.* North Sea Flow Measurement Workshop, National Engineering Laboratory, East Kilbride, Scotland.

(2003) Flare Meter Monitoring Method, and Flare Meter High Velocity Extender Experience from Snorre TLP high pressure flare estimator pilot. 21st North Sea Flow Metering Workshop 2003, Paper 27.

Dakić, B. M., Bajić, J. S., Stupar, D. Z., Slankamenac, M. P. and Živanov, M. B. (2013) A novel fiber-optic mass flow sensor. *Key Engineering Materials*, 543: 231–234.

Dall, H. E. (1962) Flow tubes and non-standard devices for flow measurement with some coefficient considerations. Proceedings of a Symposium on Flow Measurement in Closed Conduits, HMSO, Edinburgh: Paper D-1: 385–394.

Dane, H. J. and Wilsack, R. (1999) Upstream pipe wall roughness influence on ultrasonic flow measurement. 17th North Sea Flow Measurement Workshop, Paper 9, pp. 106–116.

Danen, G. W. A. (ed.) (1985) *Shell flowmeter engineering handbook*. 2nd ed. McGraw Hill.

David, C. (2012) Evaluation of water flow measurement performance of portable ultrasonic flowmeter technology. *International Symposium on Fluid Flow Measurement*.

Davis, R. W. and Mattingly, G. B. (1997) *Numerical modelling of turbulent flow through thin orifice plates*. Symposium on Flow in Open Channels and Closed Conduits, Gaithersburg, USA.

Davis, R. W., Moore, E. F. and Purtell, L. P. (1984) A numerical-experimental study of confined flow around rectangular cylinders. *Physics of Fluids*, 27: 46–59.

Davis, T. C. E. (1990) *Fiscal measurement and proving experience with Coriolis meters*. North Sea Flow Measurement Workshop, National Engineering Laboratory, Scotland.

de Boer, G. and Lansing, J. (1997) Dry calibration of ultrasonic gas flow meters. North Sea Flow Measurement Workshop, Kristiansand, Norway: Paper 18.

De Boom, R. J. (1996) *Flowmeter lifecycle costs, Advances in Instrumentation and Control*, Chicago, 621–630.

de Carvalho, J. G. and Antunes, B. de C. (2000) Multi-phase flowmetering using twin helicoidal rotors – the 'multi-stream' meter. FLOMEKO'2000 the 10th International Conference on Flow Measurement, Salvador, Brazil: Paper B4.

de Jager, P. A., Hemminga, M. A. and Sonneveld A. (1978) Novel method for determination of flow velocities with pulsed nuclear magnetic resonance. *Review of Scientific Instruments*, 49: 1217.

de Jong, J. (1978) Comparison of some 500 mm diameter flowmeters. FLOMEKO 1978 – Proceedings of the Conference on Flow Measurement of Fluids, Groningen, The Netherlands: 565.

de Jong, S. and van der Kam, P. M. A. (1993) High pressure recalibration of turbine meters. FLOMEKO '93 Proceedings of the 6th International Conference on Flow Measurement: 121–128.

de Kraker, F. (1989) *The performance of a Coriolis meter when used to measure density*. Mass Flow Measurement Direct and Indirect, London, England.

de Leeuw, H. (1994) *Wet gas flow measurement by means of a Venturi meter and a tracer technique*. North Sea Flow Measurement Workshop, Peebles, Scotland.

de Leeuw, H. and Dybdahl, B. (2002) Field installation of smartvent wet gas flow meters at Bintang, Malaysia. 20th North Sea Flow Measurement Workshop, 22–25 October 2002, St Andrews, Scotland, Paper 4.3.

de Leeuw, H.R. (1997) Liquid correction of Venturi meter readings in wet gas flow. North Sea Flow Measurement Workshop Kristiansand, Norway: Paper 21.

de Leeuw, R., Kamal, M. and Dybdahl, B. (2005) Operational Experience of Smartvent Wet Gas Metering in SE-Asia. 4th South East Asia Hydrocarbon Flow Measurement Workshop, 7–11 March 2005. http://www.gbv.de/dms/tib-ub-hannover/493689834.pdf

de Leeuw, R., Steven, R. and van Maanen, H. (2011) Venturi Meters and Wet Gas Flow. North Sea Flow Measurement Workshop, 25–28 October 2011, Page 1 of 23.

de Vries, H., Loogmann, L. L., van Dellen, K. and Broekgaarden, G. J. (1989) Ultrasonic gas flow measurements with reflection mode in underground pipelines. FLOMEKO '89 Proceedings of the 5th International Conference on Flow Measurement, Dusseldorf, Germany: 325–332.

Deacon, J. E. (1983) Electromagnetic flowmeter installation tests. IMEKO Budapest, Paper E3: 85–91.

Dean, R. W. (1988a) *Field experience using Coriolis mass meters II*. North Sea Metering Workshop, National Engineering Laboratory, East Kilbride, Scotland.

(1988b) Advantages of mass measurement for petroleum liquids. *Petroleum Review*, 42: 32–33.

Dean, T. L., Dowty, E. L. and Jiskoot, M. A. (1990a) The design manufacture and testing of the Texaco subsea three phase metering system. North Sea Flow Measurement Workshop, National Engineering Laboratory, 23–25 October.

Dean, T. L., Dowty, E. L. and Jiskoot, R. J. J. (1990b) The development of a subsea three phase metering system. European Oil and Gas Conference, Palermo, October.

Decker, M. M. (1960) The gyroscopic mass flowmeter. *Engineers Digest*, 21(7).

Decker, H., Leenhoven, T. and Danen, H. (2001) Multibeam ultrasonic flowmeter-for custody transfer at a tank farm: a long term trial project. *ISA TECH/EXPO Technology Update Conference Proceedings*, 413 I:815–25.

Del Grosso, V. A. and Spurlock, E. M. (1957) The feasibility of using wholly external ultrasonics to measure fluid flow within thick walled metal pipes. NRL Report 4967.

Delajoud, P., Girard, M. and Blair, M. (2005) The implementation of toroidal throat Venturi nozzles to maximize precision in gas flow transfer standards. FLOMEKO 2005 13th International Flow Measurement Conference, Peebles, Scotland, Paper 5.2.

Delenne, B., Mouton, G., Pritchard, M., Huppertz, M., Ciok, K., van den Heuvel, A., Folkestad, T., Vieth, D., Lezuan, F. and Marini, G. (2004) Evaluation of flow conditioners – ultrasonic meters combinations. 22nd North Sea Flow Metering Workshop 2004, Paper 3.2.

Deloughry, R., Young, M., Pickup, E. and Barratt, L. (2001) Variable density flowmeter for loading road tankers using process tomography. *Proceedings of SPIE – The International Society for Optical Engineering*, 4188: 273–283.

Delsing, J. (1991) The zero-flow performance of a sing-around ultrasonic flowmeter. *Journal of Flow Measurement and Instrumentation*, 2: 205–208.

(2004) The prospect of self-diagnosing flow meters. 12th International Conference on Flow Measurement FLOMEKO Guilin China.

(2006) Flow measurement facilities. *Flow Measurement and Instrumentation*, 17(3): 139.

Démolis, J., Escande, J., Gajan, P. and Strzelecki, A. (1998) The use of ultrasonic tomography to characterise internal flows in pipes. FLOMEKO '98 Proceedings of the 9th International Conference on Flow Measurement, Lund, Sweden, 369–373.

Deng, W., Jiang, S., Liu, R., Huang L., Zhao, P., Lei, L. and Jin, K. (2010) Calibration and verification of MEMS mass flow meters for custody transfer. 15th Flow Measurement Conference (FLOMEKO), 13–15 October 2010, Taipei, Taiwan, Paper A8-1.

Denison, A. B. and Spencer, M. P. (1956) Factors involved in intact vessel electromagnetic flow recording. *Federation Proceedings*, 15: 46.

Dijstelbergen, H. H. (1964) Rotameter dynamics. *Chemical Engineering Science*, 19: 853.

(1970) The performance of a swirl meter. *Journal of Physics E: Scientific Instruments*, 3: 886–888.

(1982) Gas meters. Developments in Flow Measurement – 1, Ed. R. W. W. Scott, Applied Science Publishers: Chapter 5.

Dijstelbergen, H. and Richards, R. (2012) A novel concept of an old idea: electronic twin turbine meters for custody transfer. 8th ISFFM.

Dijstelbergen, H. H. and van der Beek, M. P. (1998) A new reference meter for gasmeter calibrations. FLOMEKO '98 Proceedings of the 9th International Conference on Flow Measurement, Lund, Sweden, 37–42.

Dilthey, U., Scheller, W. and Brandenburg, A. (1996) Flow measuring systems for automated adhesive application. *Welding & Cutting*, 48(7): 137–138.

Dimaczek, G., Fassbinder, H-G., Emmel, A. and Kupfer, R. (1994) High-precision Coriolis mass flowmeter for bulk material two-phase flows. *Journal of Flow Measurement and Instrumentation*, 5: 295–202.

Dixey, M. (1993) Putting reliability at the centre of maintenance. *Professional Engineering* June 1993: 23–25.

Dobrowolski, B., Kabaciński, M. and Pospolita, J. (2005) A mathematical model of the self-averaging Pitot tube. *Flow Measurement and Instrumentation*, 16(4): 251–265.

Doihara, R., Shimada, T., Terao, Y. and Takamoto, M. (2004) Development of diverting system employing a rotating double wing method. 12th International Conference on Flow Measurement FLOMEKO, Guilin, China.

(2006) Development of weighing tank system employing rotating double wing diverter. *Flow Measurement and Instrumentation*, 17(3): 141–152.

Dominick, J., Durst, F., Raszillier, H. and Zeisel, H. (1987) A method to measure mass and volume flow rates of two-phase flows. *International Journal ofMultiphase Flow*, 13: 685–698.

Doney, B. (1999a) EMF flow measurement in partially filled pipes. *Sensors (Peterborough, NH)*, 16(10): 65–66, 68.

(1999b) Electromagnetic flow measurement in partially filled pipes. *Water/Engineering and Management*, 146(11): 32–34.

Dong, F., Xu, Y. B., Xu, L. J., Hua, L. and Qiao, X. T. (2005) Application of dual-plane ERT system and cross-correlation technique to measure gas-liquid flows in vertical upward pipe. *Flow Measurement and Instrumentation*, 16(2–3): 191–197.

Dong, W. and Zong Hu, L. (2002) Gas-liquid two-phase flow measurement using ESM. *Experimental Thermal and Fluid Science*, 26(6–7): 827–832.

Dongwei, W. (2004) Technical reformation of gas flow primary standard with pVTt method. 12th International Conference on Flow Measurement FLOMEKO Guilin, China.

dos Reis, E. and Goldstein Jr., L. (2008) On the measurement of mass flow rate of horizontal two-phase flows in the proximity of the transition lines which separates two different flow patterns. *Flow Measurement and Instrumentation*, 19(5): 269–282.

Dou, J., Guo, J. and Gokulnath R. (2005) Is it a MUST to add upstream devices for high GVF multiphase? 23rd North Sea Flow Metering Workshop 2005, Paper 19, pp. 289–303.

Drahm, W. (1998) New single straight tube Coriolis mass flowmeter without installation restrictions. FLOMEKO '98 Proceedings of the 9th International Conference on Flow Measurement, Lund, Sweden, 243–248.

Drahm, W. and Bjønnes, H. (2003) A Coriolis mass flowmeter with direct viscosity measurement. *IEEE Computing and Control Engineering*, 14(4): 42–43.

Drahm, W. and Staudt, W. (2004) Coriolis mass flowmeters: state of the art review and innovations. 12th International Conference on Flow Measurement FLOMEKO, Guilin, China.

Drahm, W., Reider, A., Wenger, A. and Koudal, O. (2002)Method and corresponding sensors for measuring mass flow rate. US Patent 6360614, 26 March 2002.

Drenthen, J. G. (1989) Device for determining the flow Velocity of a medium in a cylindrical conduit. US Patent 4,831,884.

Drenthen, J. (1996) Method and device for determining characteristics of the flow of a medium. US Patent 5,546,812.

Drenthen, J. G. and de Boer, G. (2001) The manufacturing of ultrasonic gas flow meters. *Journal of Flow Measurement and Instrumentation*, 12(2): 89–99.

Drenthen, J. G. and Huijsmans, F. J. J. (1993) Gassonic-400 & P Sonic & Q Sonic ultrasonic gas flow meters. FLOMEKO '93 Proceedings of the 6th International Conference on Flow Measurement, Korea: 285–298.

Drenthen, J. G., Kurth, M. van Kooster, J. and Vermeulen, M. (2009) Reducing installation effects on ultrasonic flow meters. 27th International North Sea Flow Measurement Workshop, Tonsberg, Norway, 20–23 October 2009, pp. 187–204.

Drenthen, J. G., Kurth, M. and Vermeulen, M. (2009) A novel design of a 12-chord ultrasonic gas flow meter. Canadian School of Hydrocarbon Measurement CSHM Calgary 4/2009 Paper 9030.

Drenthen, J. G., Vermeulen, M., Kurth, M. and den Hollander, H. (2010) Ultrasonic flow meter diagnostics and the impact of fouling. 15th Flow Measurement Conference (FLOMEKO), 13–15 October 2010, Taipei, Taiwan, Paper A4-3.

(2011) The detection of corrosion and fouling and the operational influence on ultrasonic flow meters using reflecting paths. North Sea Flow Metering Workshop.

Drysdale, A., Frederiksen, J. and Rasmussen, M. (2005) New on-site calibration technique for large flow meters using laser Doppler velocimetry. Flomeko 2005 13th International Flow Measurement Conference, Peebles, Scotland, Paper 8.1.

DTI (1993) Product standards – electromagnetic compatibility. Department of Trade and Industry UK Regulations.

Dunn, Donald G., Hoge, K., Liolios, G. and Klein, M. (2003) Alkylation unit optimization using Coriolis mass flow meters. Proceedings of the Annual Symposium on Instrumentation for the Process Industries, pp. 15–21.

Durcan, L. P. (1998) *Development of Baseline Stability in an Electromagnetic Flowmeter for Dielectric Liquids*. Cranfield University Press.

Durst, F. and Raszillier, H. (1990) Flow in a rotating straight pipe, with a view on Coriolis mass flow meters. *Journal of Fluids Engineering, Transactions of ASME*, 112: 149–154.

Dyakowski, T. (1996) Process tomography applied to multi-phase flow measurement. *Measurement Science and Technology*, 7: 343–353.

Dykesteen, E. (1992) Multiphase metering. *Chemical Engineering Research & Design*, 70.1: 32–7.

(2000) Multiphase meters offshore Malaysia. *Journal of Offshore Technology*, 8.

(2001) Status and trends in technology and applications. Keynote paper 19th North Sea Flow Measurement Workshop, Kristiansand, Norway, 22–25 October 2001, Keynote paper. ISBN: 978-1-61567-868-6.

Dykesteen, E., Hallanger, A., Hammer, E., Samnoy, E. and Thorn, R. (1985) Non-intrusive three-component ratio measurement using an impedance sensor. *Journal of Physics E: Scientific Instruments*, 18: 1985.

Eccles, A., Green, N. and Porkess, R. (1993a) *MEI structured mathematics – statistics 2.* Hodder & Stoughton.

(1993b) *MEI structured mathematics – statistics 3.* Hodder & Stoughton.

Eide, J. M. (1991) Operational experience, compact prover as a portable calibration unit. Proceedings of the North Sea Flow Measurement Workshop, Norwegian Society of Chartered Engineers.

Eide, J. M. and Gwaspari, S. C. (1996) *Comparison test and calibration of Coriolis meters.* North Sea Flow Measurement Workshop, Peebles, Scotland.

Einhellig, R.F., Schmitt, C. and Fitzwater, J. (2002) Flow measurement opportunities using irrigation pipe elbows. *Hydraulic Measurements and Experimental Methods*, 1111–1118.

Elkins, C. J. and Alley, M. T. (2007) Magnetic resonance velocimetry: applications of magnetic resonance imaging in the measurement of fluid motion. *Experiments in Fluids* (2007) 43: 823–858.

Elliott, K. (2004) API's microprocessor based flowmeter testing programme. 22nd North Sea Flow Metering Workshop 2004, Paper 5.4.

Elperin, T., Fominykh, A. and Klochko, M. (2002) Performance of a Venturi meter in gas-liquid flow in the presence of dissolved gases. *Flow Measurement and Instrumentation*, 13(1–2): 13–16.

El-Wahed, A. K. and Sproston, J. L. (1991) The influence of shedder shape on the performance of the electrostatic vortex flowmeter. *Journal of Flow Measurement & Instrumentation*, 2: 169–179.

El-Wahed, A. K., Johnson, M. W. and Sproston, J. L. (1993) Numerical study of vortex shedding from different shaped bluff bodies. *Journal of Flow Measurement & Instrumentation*, 4: 233–240.

Endress, U., *et al.* (1989) *Flow handbook. Reinach*, Switzerland: Flowtec AG (English edition).

Endress+Hauser (2006) *Flow handbook 3rd Edition*. Endress+Hauser Flowtec AG.

Engel, R. (2002) Dynamic weighing – improvements in gravimetric liquid flowmeter calibration. 5th International symposium of Fluid Flow Measurement, Arlington, VA, USA.

(2010) Water density determination in high-accuracy flowmeter calibration – measurement uncertainties and practical aspects. 15th Flow Measurement Conference (FLOMEKO), October 13–15, 2010 Taipei, Taiwan, Paper B3-1.

Engel, R. and Baade, H.-J. (2012) Water density determination in high-accuracy flowmeter calibration – measurement uncertainties and practical aspects. *Flow Measurement and Instrumentation*, 25: 40–53.

Engelbert, C., Scheulen, R. and Incontri, J. (2007) Successful flow measurement in varnish production. *World Pumps*, (488): 24–25.

Enoksson, P., Stemme, G. and Stemme, E. (1997) A silicon resonant sensor structure for Coriolis mass-flow measurements. *Journal of Microelectromechanical Systems*, 6(2): 119–125.

Enz, S. (2010) Effect of asymmetric actuator and detector position on Coriolis flowmeter and measured phase shift. *Journal of Flow Measurement and Instrumentation*, 21(4):497–503.

Enz, S., Thomsen, J. J. and Neumeyer, S. (2011) Experimental investigation of zero phase shift effects for Coriolis flowmeter due to pipe imperfections. *Journal of Flow Measurement and Instrumentation*, 22(1): 1–9.

Erdal, A. (1997) A numerical investigation of different parameters that affect the performance of a flow conditioner. *Journal of Flow Measurement and Instrumentation*, 8(2): 93–102.

Erdal, A. and Cabrol, J. F. (1991) Comparison of repeatability, reproducibility and linearity for turbine, Coriolis and ultrasonic meters tested at 100 bars on natural gas. *Proceedings of the North Sea Flow Measurement Workshop, Norwegian Society of Chartered Engineers*, October 22–24.

Erdal, A. and Andersson, H. I. (1997) Numerical aspects of flow computation through orifices. *Journal of Flow Measurement and Instrumentation*, 8: 27–37.

Erdal, A., Lindholm, D. and Thomassen, D. (1994) *Development of a flow conditioner*. North Sea Flow Measurement Workshop, Peebles, Scotland.

Eren, H. (1995) Particle concentration characteristics and density measurements of slurries using electromagnetic flowmeters. *IEEE Transactions on Instrumentation and Measurement*, 44, No. 3: 783–786.

Eren, H., Lowe, A. M. and Basharan, B. (2002) Processing ultrasonic signals to identify fluid contents in transit-time flowmeters. *Conference Record – IEEE Instrumentation and Measurement Technology Conference*, 2:1491–5.

Erickson, G. P. and Graber, J. C. (1983) Ultrasonic flowmeters for hydroelectric plants. *Mechanical Engineering*, November: 84–88.

Esbensen, K. H., Halstensen, M., Lied, T. T., Saudland, A., Svalestuen, J., de Silva, S. and Hope, B. (1998) Acoustic chemometrics — from noise to information. *Chemometrics and Intelligent Laboratory Systems*, 44: 61–76.

Esbensen, K. H., Hope, B., Lied, T. T., Halstensen, M., Gravermoen, T. and Sundberg, K. (1999) Acoustic chemometrics for fluid flow quantifications-II: a small constriction will go a long way. *Journal of Chemometrics*, 13: 209–236.

Espina, P. (2005) Results of the North American natural gas flow calibration laboratory comparison: CEESI – SwRI – TCC. Flomeko 2005 13th International Flow Measurement Conference, Peebles, Scotland, Paper 2.3.

Espina, P. G. and Baumoel, D. (2003) Latest advances in ultrasonic flow measurement of natural gas using externally mounted, non-intrusive sensors. 21st North Sea Flow Metering Workshop 2003, Paper 13.

Estrada, H. (2002) *The effect of in-service velocity profiles on flow measurement systems of several types*. 5th International symposium of Fluid Flow Measurement, Arlington, VA, USA.

Estrada, H., Cousins, T. and Augenstein, D. (2004) Installation effects and diagnostic interpretation using the Caldon ultrasonic meter. 22nd North Sea Flow Metering Workshop 2004, Paper 4.1.

Evans, R. P. and Blotter, J. (2002) Mass Flow measurement using flow induced pipe vibration. *Proceedings of the International Instrumentation Symposium*, 48: 231–240.

Evans, R. P., Blotter, J. D. and Stephens, A. G. (2004) Flow rate measurements using flow-induced pipe vibration. *Journal of Fluids Engineering*, 126: 280–285.

Evans, R., Ifft, S. and Hodges, D. (2007) Wet gas performance of differential pressure flowmeters. 25th International North Sea Flow Measurement Workshop, Energy Institute, Oslo, Norway, 138–151.

Ewing, J. A. (1924–25) A ball-and-tube flowmeter. *Proceedings of the Royal Society*, 45: 308.

Fage, A. and Johansen, F. C. (1928) The structure of vortex sheets. *Philosophical Magazine*, 5.

Fahlenbock, T. D. (2005) Coriolis mass flow meter: high accuracy for high flow rates. *Powder and Bulk Engineering*, 19(9): 29–33.

Fairclough, A. M. (2000) Wet gas well testing. FLOMEKO'2000 the 10th International Conference on Flow Measurement, Salvador, Brazil: Paper D9.

Fakouhi, A. (1977) The influence of viscosity on turbine flow meter calibration curves. PhD thesis, University of Southampton.

Falcone, G., Hewitt, G. F., Alimonti, C. and Harrison, B. (2001) multiphase flow metering: current trends and future developments. SPE Annual Technical Conference and Exhibition, 30 September-3 October 2001, New Orleans, Louisiana.

(2002) Multiphase flow metering: current trends and future developments. *JPT, Journal of Petroleum Technology*, 54(4): 77–84.

(2005) Multiphase flow metering: 4 years on. 23rd North Sea Flow Metering Workshop 2005, Paper 18, pp. 277–288.

Falcone, G., Hewitt, G. F. and Alimonti, C. (2009) Multiphase flow metering: principles and applications. Developments in Petroleum Science, Volume 54, Elsevier Science ISBN: 0-444-52991-8. http://www.elsevier.com/wps/find/bookdescription.cws_home/714238/description#description

Falvey, H. T. (1983) Effect of gradients on acoustic velocity meter. *Journal of Hydraulic Engineering*, 109: 1441–1453.

Fan, S. and Song, M. (2003) Analysis on response of straight tube Coriolis mass-flow meter under pulsating flow. *Beijing Hangkong Hangtian Daxue Xuebao/Journal of Beijing University of Aeronautics and Astronautics*, 29: 67–71 (in Chinese)

Fang, T., Xie, D. and Liang, G. (2008) Studies on the flow characteristic of fluidic flow meter. *Yi Qi Yi Biao Xue Bao/Chinese Journal of Scientific Instrument*, 29(SUPPL. 2): 617–620. (in Chinese)

Faraday, M. (1832) Experimental researches in electricity. *Philosophical Transactions of the Royal Society*, 15: 175–177.

Fathimani, A., Marko, P. E., T., Tominaga, K., Hauser, E., Yee, F. and Malcolm, S. (2007) Power uprates through ultrasonic feed water flow measurement – CANDU-specific opportunities and challenges. Canadian Nuclear Society – 28th Annual Conference of the Canadian Nuclear Society and 31st CNS/CNA Student Conference 2007: "Embracing the Future: Canada's Nuclear Renewal and Growth", Saint John, N.B., Canada, 995–1006.

Feng, C.-C., Lin, W.-T. and Yang, C.-T. (2010) Laminar flow element type flow meter with straight glass capillary. 15th Flow Measurement Conference (FLOMEKO), October 13–15, 2010 Taipei, Taiwan, Paper A2-5.

Fenwick, J. S. and Jepson, P. (1975) The problems and needs in large volume gas measurement. Transducer '75 Conference.

Ferreira, V. C. S. (1988) Flow patterns inside a turbine type flowmeter. PhD thesis, Cranfield Institute of Technology, England.

Ferron, A.G. (1962) Velocity profile effects on the discharge coefficient of pressure differential meters. *Journal of Basic Engineering*,85(3): 338–342.

Finlayson, A. J. (1992) Industrial review: selection of flowmeters for nuclear fuel processing. *Journal of Flow Measurement and Instrumentation*, 3(1): 3–8.

Finnof, C., Stainton, D., Saenz, C. P. and Smith, J. E. (1976) Apparatus and method for measuring fluid mass flow. British Patent 1,535,817.

Fischbacker, R. E. (1959) The ultrasonic flowmeter. *Trans. Soc. Inst. Tech.*, 11: 114.

Fischer, C. (1994) Development of a metering system for total mass flow and compositional measurements of multiphase/multicomponent flows such as oil/water/air mixtures. *Journal of Flow Measurement and Instrumentation*, 5: 31–42.

Fischer, R. (1995) Calculation of the discharge characteristic of an orifice for gas-liquid annular-mist flow. *International Journal of Multiphase Flow*, 21.5: 817–835.

Fischer, S., Schmitt, P., Ensminger, D., Abda, F. and Pallares, A. (2008) A new velocity estimation method using spectral identification of noise. *Flow Measurement and Instrumentation*, 19(3–4): 197–203.

Fischer, S., Rebattet, C. and Dufour, D. (2013) Applicability of ultrasonic pulsed doppler for fast flow-metering. FLOMEKO 2013, 16th International Flow Measurement Conference, Paris.

Fish, D. J. (2007) The importance of discerning the impact of new measurement technology. 25th International North Sea Flow Measurement Workshop, Energy Institute, Oslo, Norway, 1–8.

Fletcher, S. I., Nicholson, I. G. and Smith, D. J. M. (2000) An investigation into the effects of installation on the performance of insertion flowmeters. *Journal of Flow Measurement and Instrumentation*, 11: 19–39.

Fling, W. A. and Whetstone, J. R. (1985) Development of basic orifice discharge coefficients. 64th Annual GPA Convention, Houston, Texas.

Folkestad, T. (2001) Testing a 12″ Krohne 5-path Altosonic V ultrasonic liquid flow meter on Osenberg crude oil and on heavy crude oil. 19th North Sea Flow Measurement Workshop, Kristiansand, Norway, 22–25 October 2001, Paper 15.

Fosså, Ø., Stobie, G. and Wee, A. (2009) Successful implementation and use of multiphase meters. 27th International North Sea Flow Measurement Workshop, Tonsberg, Norway, 20–23 October 2009, pp. 107–130.

Fosse, S., Ullebust, B. and Ekerhovd, H. (2008) The importance of proper internal surface and alignment of upstream metering tubes, when metering light hydrocarbons with turbine meter. 26th International North Sea Flow Measurement Workshop, 21–24 October 2008, St Andrews, Scotland, Paper 1.1.

Frail, C. (2005) Clamp-on gas flow measurement using ultrasonic flow pattern recognition. *Technical Papers of ISA*, 459: 851–859.

Frank, R., Mazars, J. and Rique, R. (1977) Determination of mass flowrate and quality using a Venturi and turbine meter. *Proceedings of the Institution of Mechanical Engineers, Part C*, 200.

Franke, S., Büttner, L., Czaskre, J., Räbiger, D. and Eckert, S. (2010) Ultrasound Doppler system for two-dimensional flow mapping in liquid metals. *Journal of Flow Measurement and Instrumentation*, 21(3): 402–409.

Frankvoort, W. (1989) *Results of the evaluation of the performance of mass flow meters using a prover loop*. Mass Flow Measurement Direct and Indirect, Proc Int Conf Mass Flow Measurement, IBC Pbl, London, England.

Frantzen, K. H. and Dykesteen, E. (1990) Field experience with CMI multiphase fraction meter. North Sea Flow Measurement Workshop, National Engineering Laboratory, Scotland.

Fridjonsson, E. O., Stanwix, P. L. and Johns, M. L. (2014) Earth's field NMR flow meter: preliminary quantitative measurements. *Journal of Magnetic Resonance*, 245: 110–115.

Frøysa, K.-E. and Lunde, P. (2004) Mass and energy measurement of natural gas using ultrasonic flow meters. Recent results. Proc. of the 27th Scandinavian Symposium on Physical Acoustics, Norway, 25–28 January 2004.

(2005) Density and calorific value measurement in natural gas using ultrasonic flow meters. 23rd North Sea Flow Metering Workshop 2005, Paper 5, pp. 45–67.

Frøysa, K.-E., Lunde, P. and Vestrheim, M. (2001) A ray theory approach to investigate the influence of flow velocity profiles on transit times in ultrasonic flow meters for gas and liquid. Proc. 19th International North Sea Flow Measurement Workshop, Kristiansand, Norway, 22–25 October 2001, Paper 21.

Frøysa, K.-E., Lunde, P. Paulsen, A. and Jacobsen, E. (2006a) Density and calorific value measurement in natural gas using ultrasonic flow meters. Results from testing on various North Sea gas field data. 24th International North Sea Flow Measurement Workshop, 24–27 October 2006.

Frøysa, K.-E., Lunde, P., Hallanger, A. and Sand, I. Ø. (2006b) Mass and energy measurement of natural gas using ultrasonic flow meters. Results from testing on various North Sea gas field data. Proc. of the 29th Scandinavian Symposium on Physical Acoustics, Norway, 29 January–1st February 2006.

Frøysa, K.-E., Vågenes, A. L. H., Sørli, B. and Jørgenvik, H. (2007) Uncertainty analysis of emissions from the Statoil Mongstad oil refinery. 25th International North Sea Flow Measurement Workshop, Energy Institute, Oslo, Norway, pp. 386–400.

Frøysa, K.-E., Hallanger, A. and Paulsen, A. (2008) Installation effects on the Easington ultrasonic fiscal metering station. 26th International North Sea Flow Measurement Workshop, 21–24 October 2008, St Andrews, Scotland, Paper 6.3.

Frøystein, T., Kvandal, H. and Aakre, H. (2005) Dual energy gamma tomography system for high pressure multiphase flow. *Flow Measurement and Instrumentation*, 16(2–3): 99–112.

Fu, X. and Yang, H. (2001) Study on hydrodynamic vibration in dual bluff body vortex flowmeter. *Chinese Journal of Chemical Engineering*, 9(2): 123–128.

Fueki, M., Tanaka, Y., Nishi, T. and Yamazaki, D. (1998) Development of a multiphase flowmeter without radioactive source. 30th Offshore Technology Conference, Houston, TX, USA: 463–470.

Fulton, J., Hammer, E. A. and Haugs, A. (1987) *Deflection of orifice plates at high differential pressure*. North Sea Flow Metering Workshop, Stavanger.
Funck, B. and Mitzkus, A. (1996). Acoustic transfer function of the clamp-on flowmeter. *IEEE Transactions on Ultrasonics, Ferroelectrics and Frequency Control*, 43(4): 569–575.
Furness, R. A. (1982) *Turbine flowmeters. Developments in Flow Measurement – 1*, Ed. R. W. W. Scott, Applied Science Publishers: 171–207.
(1989) *The application, standardisation and future use of Coriolis type mass meters in the oil and process industries*. Mass Flow Measurement Direct and Indirect, Proc Int Conf Mass Flow Measurement, IBC Pbl, London, England.
(1991) BS 7405: The principles of flowmeter selection. *Journal of Flow Measurement and Instrumentation*, 2(4): 233–242.
(2001) Review of "Flow measurement handbook" by R. C. Baker. *Journal of Flow Measurement and Instrumentation*, 12(3): 233–234.
(2003) Assessing leakage in water supply networks using flowmeters. *Water Engineering and Management*, 150(3): 26-29-36.
Furness, R. A. and Jelffs, P. A. M. (1991) Flowmeters – their role in loss reduction in refining. *Petroleum Review*, 45: 544–549.
Furuichi, N., Sato, H., Terao, Y. and Takamoto, M. (2009) A new calibration facility for water flowrate at high Reynolds number. *Flow Measurement and Instrumentation*, 20(1): 38–47.
Fyrippi, I., Owen, I. and Escudier, M. P. (2004) Flowmetering of non-Newtonian liquids. *Flow Measurement and Instrumentation*, 15(3): 131–138.
Gadshiev, E. M., Grigor'yants, S. E., Gusein-zade, K. P. and Smirnov, V. P. (1988) Metrological support to hot-water meters in use and during production. *Measurement Techniques*, 151–154.
Gainsford, S. (1990) *Tested performance of the Hitec/Multi-Fluid water fraction meter*. North Sea Flow Measurement Workshop, National Engineering Laboratory, Scotland.
Gajan, P., Mottram, R. C., Hebrard, P., Andriamihafy, H. and Platet, B. (1992) The influence of pulsating flows on orifice plate flowmeters. *Journal of Flow Measurement and Instrumentation*, 3: 118–129.
Gallagher, J. E. (1990a) The A.G.A. Report No.3 Orifice Plate Discharge Coefficient Equation. Second International Symposium on Fluid Flow Measurement, Calgary, 6–8 June 1990.
(1990b) The A.G.A. Report No.3 Orifice Plate Discharge Coefficient Equation. North Sea Flow Metering Workshop, 23-25 October 1990. National Engineering Laboratory, East Kilbride, Scotland.
Gallagher, J. E. and Saunders, M. P. (2002) Intelligent ultrasonic flowmeters. 5th International symposium of Fluid Flow Measurement, Arlington, VA, USA.
Gallagher, J. E., LaNasa, P. J. and Beaty, R. E. (1994) *The Gallagher flow conditioner*. North Sea Flow Measurement Workshop, Peebles, Scotland.
Gallagher, J. E., Saunders, P.E. and Saunders, M. P. (2002) High performance flow conditioners bring unparalleled accuracy to metering stations. 5th International symposium of Fluid Flow Measurement, Arlington, VA, USA.
Geach, D. and Jamieson, A. W. (2005) Wet gas measurement in the Southern North Sea. 23rd North Sea Flow Metering Workshop 2005, Paper 11, pp. 163–183.
Geng, Y., Zheng, J. and Shi, T. (2006) Study on the metering characteristics of a slotted nozzle for wet gas flow. *Flow Measurement and Instrumentation*, 17(2): 123–128.
Genthe, W. K. (1974) The nuclear magnetic resonance flowmeter process flow measurement experiences. *Flow Measurement and Control in Science and Industry*, ISA pp849–856.
George, D. L. (2002) Turbine meter research in support of the revision of AGA Report No. 7. 5th International symposium of Fluid Flow Measurement, Arlington, VA, USA.
Geropp D. (1971) *Laminare Grenzschichten in ebenen und rotationssymmetrischen Lavalduesen*. Deutsche Luft- und Raumfahrt Forschungsbericht, 71–90.
Geropp, D. and Odenthal, H.-J. (2001) Flow rate measurements in turbulent pipe flows with minimal loss of pressure using a defect-law. *Journal of Flow Measurement and Instrumentation*, 12(1): 1–7.

Gerrard, D. (1979) Measure viscous flows over 150:1 turndown by PD meter techniques. *Control and Instrumentation*, 11(4): 39–41.

Gerrard, J. H. (1966) The mechanics of the formation region of vortices behind bluff bodies. *Journal of Fluid Mechanics*, 25: 401–413.

Ghaoud, T. and Clarke, D. W. (2002) Modelling and tracking a vortex flow-meter signal. *Flow Measurement and Instrumentation*, 13(3): 103–117.

Gibson, J. (2009) Validation of the CFD method for determining the measurement error in flare gas ultrasonic meter installations. 27th International North Sea Flow Measurement Workshop, Tonsberg, Norway 20–23 October 2009, pp. 292–308.

Ginesi, D. (1990) Flow measurement solved with Venturi-cone meter. *Intech*, February: 30–32.

(1991) Choosing the best flowmeter. *Chemical Engineering, NY*, 98.4: 88–100.

Ginesi, D. and Annarummo, C. (1994) Application and installation guidelines for volumetric and mass flowmeters. *ISA Transactions*, 33.1: 61–72.

Gold, R. C., Miller, J. S. S. and Priddy, W. J. (1991) *Measurement of multiphase well fluids by positive displacement meter*. Offshore Europe Conference, Aberdeen, Scotland: SPE Paper 23065.

Gol'dgammer, K. A., Terent'ev, A. T. and Zalaliev, M. I. (1990) Effect of magnetic properties of a medium on the metrological characteristics of an NMR meter for measurement of gas-liquid flows. *Measurement Techniques*, 33.7: 676–9.

Goldstein, S. (1936) A note on the measurement of total head and static pressure. *Proceedings of the Royal Society, Series A*, 155: 570–575.

(1965) *Modern developments in fluid dynamics*. Dover Publications Inc.

Goodrich, L. D. (1979) *Design and performance of the drag disc turbine transducer*. International Colloquium, Idaho Falls, USA, June.

Gopal, M. and Jepson, P. (1996) Development of a novel non-intrusive, ultrasonic flowmeter for wet gas pipelines. *Proceedings of the ASME Fluids Engineering Division Summer Meeting, San Diego, CA, USA*, 236(1): 647–652.

Gorny, L. J., Gillis, K. A. and Moldover, M. R. (2012) Testing long-wavelength acoustic flowmeter concepts for flue gas flows. 8th International Symposium on Fluid Flow Measurement.

Goujon-Durand, S. (1995) Linearity of the vortex meter as a function of fluid viscosity. *Journal of Flow Measurement and Instrumentation*, 6: 235–238.

Gralenski, N. (2004) Creating a better mass flow meter. *Solid State Technology*, 47(5): 26–28.

Grattan, E., Rooney, D. H. and Simpson, H. C. (1981) Two-phase flow through gate valves and orifice plates. NEL, East Kilbride, Scotland, Report No 678.

Gray, D. E., Benjamin, N. M. P. and Chapman, B. N. (1991) Effects of environmental and installation specific factors on process gas delivery via mass flow controller with an emphasis on real time behaviour. *Proceedings of SPIE, Int Soc Optical Eng*, 1392: 402–410.

Gray, J. O. and Sanderson, M. L. (1970) Electromagnetic differential flowmeter. *Electronic Letters*, 6(7): 194.

Grego, G. and Muciaccia, F. (2008) Choice of the best position of 4 path acoustic systems in circular sections downstream curves or with inadequate straight pipes and disturbed speed profiles. Proceeding of 7th International Conference on Hydraulic Efficiency Measurements, IGHEM (International Group for Hydraulic Efficiency Measure), Milan, Italy.

Gregor, J., Norman, R. S., Bass, R. L. and Spark, C. R. (1993) Establishment of a new natural gas metering research facility for improving flow measurement accuracy under field conditions. FLOMEKO '93 Proceedings of the 6th International Conference on Flow Measurement: 27–41.

Gregory, D., West, M., Paton, R., Casimiro, R., Boo, S., Low, Y. K., Henry, M., Tombs, M., Duta, M., Zhou, F., Zamora, M., Mercado, R. and Machacek, M. (2008) Two-phase flow metering using a large Coriolis mass flow meter applied to ship fuel bunkering. *Measurement and Control*, 41(7): 208–212.

Grendstad, J., Eide, J. and Salvesen, P. (1991) Testing of Coriolis meters for metering of oil, condensate and gas. *Proceedings of the North Sea Flow Measurement Workshop*.
Grenier, P. (1991) Effects of unsteady phenomena on flow metering. *Journal of Flow Measurement and Instrumentation*, 2: 74–80.
Grey, J. (1956) Transient response of the turbine flowmeter. *Jet Propulsion*, February: 98–100.
Gribok, A. V., Attieh, I. K., Hines, J. W. and Uhrig, R. E. (2001) Regularization of feedwater flow rate evaluation for Venturi meter fouling problem in nuclear power plants. *Nuclear Technology*, 134(1): 3–14.
Griffin, D. (2009) New challenges in oil & gas measurement. 27th International North Sea Flow Measurement Workshop, Tonsberg, Norway, 20–23 October 2009, pp. 1–12.
Griffith, B., Cousins, T. and Augenstein, C. (2005) The effect of flow conditioners on the performance of multi-path ultrasonic flowmeters. Flomeko 2005 13th International Flow Measurement Conference, Peebles, Scotland, Paper 7.2.
Griffiths, A. and Newcombe, J. (1970) Large-volume gas measurement. 36th Autumn Research Meeting of the Institution of Gas Engineers.
Griffiths, C. and Silverwood, P. A. (1986) Selection and application of flow measurement instrumentation. *HYDRIL Production Technology Division*, Bulletin 5126-A.
Grimley, T. A. (1996) *Multipath ultrasonic flow meter performance*. North Sea Flow Measurement Workshop, Peebles, Scotland.
(1997) Performance testing of ultrasonic flow meters. North Sea Flow Measurement Workshop, Kristiansand, Norway: Paper 19.
(2002) Installation effects testing of Coriolis flow meters with natural gas. 5th International symposium of Fluid Flow Measurement, Arlington, VA, USA.
Grini, P. G., Maehlum, H. S. and Brendeng, E. (1994) In situ calibration of Coriolis flowmeters for high-pressure gas flow calorimetry. *Journal of Flow Measurement and Instrumentation*, 5: 285–288.
Grumski, J. T. and Bajura, R. A. (1984) Performance of a Coriolis-type mass flowmeter in the measurement of two-phase (air-liquid) mixtures. Mass Flow Measurements ASME Winter Annual Meeting, New Orleans, USA.
Guan, J., Zhang, H. and Hu, C. (2002) Multi-electrode electromagnetic flowmeter. *Proceedings of the Second International Symposium on Instrumentation Science and Technology*, 1:1/325-1/330.
Guilbert, A. R. and Sanderson, M. L. (1996a) Novel ultrasonic mass flowmeter for liquids. *IEE Colloq.* (Dig.), No. 092:8/1–8/4.
(1996b) The development of curved reflective surfaces for ultrasonic beam redirection in high speed gas flow measurement. *Ultrasonics*, 34: 441–445.
Guilbert, A. R., Law, M. and Sanderson, M. L. (1996) A novel ultrasonic/thermal clamp-on flowmeter for low liquid flowrates in small diameter pipes. *Ultrasonics*, 34: 435–439.
Guizot, J. L. (2003) 18 Ultrasonic paths for liquid flow measurement. 21st North Sea Flow Metering Workshop 2003, Paper 28.
Guo, J. and Heslop, M. J. (2004) Diffusion problems of soap-film flowmeter when measuring very low-rate gas flow. *Flow Measurement and Instrumentation*, 15(5–6): 331–334.
Gurevich, Y, Lopez, A., Askari, V., Safavi-Ardibili, V. and Zobin, D. (2002) Performance evaluation and field application of clamp-on ultrasonic cross-correlation flow meter, CROSSFLOWTM. 5th International symposium of Fluid Flow Measurement, Arlington, VA, USA.
Gwaspari, S. C. (1990) *Multiple regression footprinting of meter factors*. North Sea Flow Measurement Workshop, National Engineering Laboratory, Scotland.
Gysling, D. L. (2007) An aeroelastic model of Coriolis mass and density meters operating on aerated mixtures. *Flow Measurement and Instrumentation*, 18(2): 69–77.
Gysling, D. L. and Loose, D. H. (2003) Sonar-based, clamp-on flow meter for gas and liquid applications. BI0036 Rev. B – ISA EXPO 2003 http://www.cidra.com/sites/default/files/document_library/BI0036_Final_ISA_Houston_June172003_October_event.pdf

Gysling, D. and Mueller, E. (2004) Application of sonar-based, clamp-on flow meter in oil sand processing, ISA 2004 Exhibit and Conference.

Gysling, D. L., Loose, D. H. and van der Spek, A. (2005) Clamp-on sonar-based volumetric flow rate and gas volume fraction measurement for industrial applications. Flomeko 2005 13th International Flow Measurement Conference, Peebles, Scotland, Poster Session.

Gysling, D. L., Loose, D. H., Morlino, N. and van der Spek, A. (2007) Wet gas metering using sonar-based flow meters and piping pressure loss gradients. 25th International North Sea Flow Measurement Workshop, Energy Institute, Oslo, Norway, 79–96.

H. M. Customs and Excise (1995) Mineral (hydrocarbon) oils: Duty and VAT: Warehousing and related procedures. Notice 179.

Hafner, P. (1985) New developments in magflowmeters. International Conference on Flow Measurement in the Mid-1980s, National Engineering Laboratory, Glasgow, Scotland: Paper 9.1.

Haga, J., Sotono, Y. and Hanai, J. (1995) Development of a fuel vapor flow meter. Japan Society of Automotive Engineers, *JSAE Review*, 16.2: 185–187.

Hagenmeyer, H., Schulz, K-H., Wenger, A. and Keita, M. (1994) Design of an advanced Coriolis mass flowmeter using hoop mode. FLOMEKO'94 Conference on Flowmeasurement in the Mid 90s, NEL, Scotland.

Hahn, B. V. (1968) Theory of the sliding-vane meter, *Siemens Review* XXXV (9): 362–366.

Hakansson, E. and Delsing, J. (1992) Effects of flow disturbance on an ultrasonic gas flowmeter. *Journal of Flow Measurement and Instrumentation*, 3: 227–234.

(1994) Effects of pulsating flow on an ultrasonic gas flowmeter. *Journal of Flow Measurement and Instrumentation*, 5(2): 93–101.

Hall, A.R.W. (2001) Performance of Venturi meters in multiphase flow. *Proceedings of the Engineering Technology Conference on Energy*, B: 975–994.

Hall, A. and Shaw, C. (1988) Field experience of two phase flow measurement. North Sea Flow Metering Workshop, paper 2.3.

Hall, A. R. W., Whitaker, T. S. and Millington, B. C. (1997) Multiphase flowmetering: current status and future developments. 29th Offshore Technology Conference, Houston, TX, USA, 4:545–552.

Hall, A., Griffin, D. and Steven, R. (2007) A discussion on wet gas flow parameter definitions. 25th International North Sea Flow Measurement Workshop, Energy Institute, Oslo, Norway, 113–137.

Hall, J., Letton, C. and Webb, R. A. (2008) Deepwater measurement verification – a deepstar-RPSEA mandate. 26th International North Sea Flow Measurement Workshop 21–24 October 2008, St Andrews, Scotland, Paper 3.3.

Hall, J., Zanker, K. and Kelner, E. (2010) When should a gas ultrasonic flow meter be recalibrated? 28th International North Sea Flow Metering Workshop, Paper 5.1.

Hall, R. (1990) Measuring mass flow and density with Coriolis meters. *InTech*, 37(4): 45–46.

Hallanger, A. (2002) CFD Analyses of the influence of flow conditioners on liquid ultrasonic flow metering. Oseberg Sør – A Case Study. 20th North Sea Flow Measurement Workshop 22–25 October 2002, St Andrews, Scotland, Paper 6.3.

Hallanger, A., Frøysa, K.-E. and Lunde, P. (2002) CFD simulation and installation effects for ultrasonic flow meters in pipes with bends. *International Journal of Applied Mechanics and Engineering*, 7(1): 33–64.

Halttunen, J. (1990) Installation effects on ultrasonic and electromagnetic flowmeters: a model-based approach. *Journal of Flow Measurement and Instrumentation*, 1: 287–292.

Hamblett, L. S. J. (1970) Flowmeters. Displacement and inferential types. Glenfield Gazette, No. 231: 24–27.

Hamid, A. and Stuchly, S. S. (1975) Microwave Doppler effect flow monitor. *IEE Transactions in Industrial Electronics and Control Instrumentation*, IECI-22(2): 224–228.

Hamidullin, V., Malakhanov, R. and Khamidoullina, E. (2001) Statics and dynamics of ultrasonic flowmeters as sensing elements for power control systems. *IEEE Conference on Control Applications – Proceedings*, 680–685.

Hampel, U., Speck, M., Koch, D., Menz, H.-J., Mayer, H.-G., Fietz, J., Hoppe, D., Schleicher, E., Zippe, C. and, H.-M. (2005) Experimental ultra fast X-ray computed tomography with linearly scanned electron beam source. *Flow Measurement and Instrumentation*, 16(2–3):65–72.

Han, I. Y., Kim, D-K. and Kim, S. J. (2005) Study on the transient characteristics of the sensor tube of a thermal mass flow meter. *International Journal of Heat and Mass Transfer*, 48(13): 2583–2592.

Haneef, I., Ali, S. Z., Udrea, F., Coull, J. D. and Hodson, H. P. (2007) High performance SOI-CMOS wall shear stress sensors. IEEE Sensors Conference, 28–31 October 2007, Atlanta, USA.

Hannisdal, N-E. (1991) Metering study to reduce topsides weight. Proceedings of the North Sea Flow Measurement Workshop, Norwegian Society of Chartered Engineers: 22–24.

Hans, V. (2002a) State and research results of ultrasonic gas flow measurement. *Proceedings of the Second International Symposium on Instrumentation Science and Technology*, 1:1/054-1/066.

Hans, V. H. (2002b) New aspects of the arrangement and geometry of bluff bodies in ultrasonic vortex flow meters. *Conference Record – IEEE Instrumentation and Measurement Technology Conference*, 2: 1661–1664.

Hans, V. (2003a) Ultrasonic gas flow measurement. *Proceedings of the ASME/JSME Joint Fluids Engineering Conference*, 1 A: 31–35.

(2003b) Digital processing of complex modulated ultrasonic signals in flow measurement. *Proceedings of SPIE – The International Society for Optical Engineering*, 5253: 334–338.

Hans, V. and Lin, Y. (2005) Self – Monitoring ultrasonic vortex and correlation gas flow meter. *Conference Record – IEEE Instrumentation and Measurement Technology Conference*, 3: 2276–2280.

Hans, V. and Windorfer, H. (2003) Comparison of pressure and ultrasound measurements in vortex flow meters. *Measurement: Journal of the International Measurement Confederation*, 33(2): 121–133.

Hans, V., Poppen, G., von Lavante, E. and Perpéet, S. (1998a) Vortex flowmeters and ultrasound detection: signal processing and influence of bluff body geometry. *Journal of Flow Measurement and Instrumentation*, 9(2): 79–82.

Hans, V., Windorfer, H., von Lavante, E. and Perpéet, S. (1998b) Experimental and numerical optimization of acoustic signals associated with ultrasound measurement of vortex frequencies: signal processing and influence of bluff body geometry. FLOMEKO '98 Proceedings of the 9th International Conference on Flow Measurement, Lund, Sweden, 363–367.

Harmuth, B. and Erbe, V. (2002) Time of flight flow meters as a reliable and cost effective monitoring alternative in sewer systems. *Water Science and Technology*, 46(6–7): 397–402.

Harper, K., Lansing, J. and Dietz, T. (2009) Field experience of ultrasonic flow meter use in CO2-rich applications. 27th International North Sea Flow Measurement Workshop, Tonsberg, Norway, 20–23 October 2009, pp. 220–234.

(2012) Field experience of ultrasonic flow meter use in CO2-rich applications. 8th International Symposium on Fluid Flow Measurement.

Harrie, P. M. (1991) Mass flow and density. *Measurement & Control*, April.

Harriger, J. W. (1966) High-pressure measurement. 32nd Autumn Research Meeting of the Institution of Gas Engineers.

Harrison, P. (1978a) The calibration of flowmeters with liquids. Short Course notes on The Principles and Practice of Flow Measurement, Lecture No 6, National Engineering Laboratory, East Kilbride.

(1978b) National standards, transfer standards, traceability and the BCS. Short Course notes on The Principles and Practice of Flow Measurement, Lecture No 8, National Engineering Laboratory, East Kilbride.

Harrison, P. and Williamson, J. (1985) Accuracy of flowmeters used in a survey of domestic water consumption in Scotland in October 1982. Proc Flow Measurement for Water Supply, London.

Harrold, D. (2001) Coriolis flowmeters. *New and improved: What's that mean? Control Engineering*, 48(11): 24–33.
Hartley, P. E., Roach, G. J., Stewart, D., Watt, J. S., Zastawny, H. W. and Ellis, W. K. (1995) Trial of a gamma-ray multiphase flowmeter on the West Kingfish oil platform. *Nuclear Geophysics*, 9(6): 533–552.
Hastings, C. R. (1968) LE flowmeter – a new device for measuring liquid flow rates. *Westinghouse Engineer*, 28(6): 183.
(1970) The LE acoustic flowmeter: an application to discharge measurement. *New England Water Works Association*, 84: 127.
Hauser, E., Estrada, H. and Regan, Jennifer (2004) Impact of flow velocity profile on nuclear plant feed water flow measurement accuracy; results of recent laboratory testing. *Instrumentation, Control, and Automation in the Power Industry, Proceedings*, 45(ISA 421): 109–118.
Hayes, E. R. (1988) The prediction of droplet motion and breakup using a vortex model for turbulent flows. PhD thesis, Cranfield Institute of Technology.
Hayward, A. T. J. (1977a) *Flowmeters: a basic guide and source-book for users*. The Macmillan Press Ltd, London, England.
Hayward, A. T. J. (1977b) Measuring the repeatability of flowmeters. NEL Report No. 636, Dept. of Industry, UK.
(1977c) *Repeatability and accuracy*. Mechanical Engineering Publications Limited, London.
(1979) *Flowmeters: A basic guide and source-book for users*. The Macmillan Press, London.
Hayward, A. T. J. and Furness, R. A. (1989) A portable gravimeter prover for the in-line proving of direct mass flowmeters. Mass Flow Measurement Direct and Indirect, Proc Int Conf Mass Flow Measurement, IBC Pbl, London, England.
Haywood, R. W. (1968) *Thermodynamic tables in SI units*. Cambridge University Press.
He, D. and Bai, B. (2012) Numerical investigation of wet gas flow in Venturi meter. *Flow Measurement and Instrumentation*, 28: 2–6.
Head, V. P. (1946–1947) An extension of rotameter theory and its application in new practical fields. *Instrument Practice*. Dec 1946: 64–71, and Feb 1947: 135–141.
(1956) A practical pulsation threshold for flowmeters. *Transactions on ASME*, 78: 1471–1479.
Hebrard, P., Malard, L. and Strzelecki, A. (1992) Experimental study of a vortex flowmeter in pulsatile flow conditions. *Journal of Flow Measurement & Instrumentation*, 3: 173–186.
Heinrichs, K. (1991) Flow measurement by a new push-pull swirlmeter. *Sensors & Actuators, A: Physics*, 27: 809–813.
Hemp, J. (1975) Improved magnetic field for an electromagnetic flowmeter with point electrodes. *Journal of Physics D Applied Physics*, 8: 983–1002.
(1979) British Patent Specification No 2017914A.
(1982) Theory of transit time ultrasonic flowmeters. *Journal of Sound Vibration*, 84(1): 1133–1147.
(1988) Flowmeters and reciprocity. *QJMAM*, 41(4): 503–520.
(1991) Theory of eddy currents in electromagnetic flowmeters. *Journal of Physics D Applied Physics*, 24: 244–251.
(1994a) Weight vector for thermal diffusion flowmeters, Part 1: General theory. *Journal of Flow Measurement and Instrumentation*, 5(3): 217–222.
(1994b) The weight vector theory of Coriolis mass flowmeters. *Journal of Flow Measurement and Instrumentation*, 5(4): 247–253.
(1994c) Error in potential measurements due to nonuniform contact impedence of electrodes. *Quarterly Journal of Mechanics and Applied Mathematics*, 47(1): 175–182.
(1995a) Weight vector for thermal diffusion flowmeters, Part 2: Application to a particular configuration. *Journal of Flow Measurement and Instrumentation*, 6: 149–156.
(1995b) Theory of a simple electromagnetic velocity probe with prediction of the effect on sensitivity of a nearby wall. *Measurement Science and Technology*, 6: 376–382.
(1996) A theoretical investigation into the feasibility of Coriolis mass flowmeters for low density fluids. FLOMEKO'96 Proceedings of the 8th International Conference on Flow Measurement, Beijing, China: 265–270.

(1998) A review of the weight vector theory of transit time ultrasonic flowmeters. FLOMEKO'98 9th International Conference on Flow Measurement, Lund, Sweden.

(2001a) A technique for low cost calibration of large electromagnetic flowmeters. *Journal of Flow Measurement and Instrumentation*, 12(2): 123–134.

(2001b) Calculation of the sensitivity of a straight tube Coriolis mass flowmeter with free ends. *Flow Measurement and Instrumentation*, 12(5–6): 411–420.

Hemp, J. and Hendry, L. A. (1995) The weight vector theory of Coriolis mass flowmeters – Part 2. *Boundary source of secondary vibration. Journal of Flow Measurement and Instrumentation*, 6: 259–264.

Hemp, J. and Kutin, J. (2006) Theory of errors in Coriolis flowmeter readings due to compressibility of the fluid being metered. *Flow Measurement and Instrumentation*, 17(6): 359–369.

Hemp, J. and Sanderson, M. L. (1981) Electromagnetic flowmeters – a state of the art review. BHRA International Conference on Advances in Flow Measurement Techniques, Coventry, England, Paper E1: 319–340.

Hemp, J. and Sultan, G. (1989) On the theory and performance of Coriolis mass flowmeters. Mass Flow Measurement Direct and Indirect, Proc Int Conf Mass Flow Measurement, IBC Pbl, London, England.

Hemp, J. and Versteeg, H. K. (1986) Prediction of electromagnetic flowmeter characteristics. *Journal of Physics D: Applied Physics* 19: 1459–1476.

Hemp, J. and Wyatt, D. G. (1981) A basis for comparing the sensitivity of different electromagnetic flowmeters to velocity distribution. *Journal of Fluid Mechanics*, 112: 189–201.

Hemp, J. and Yeung, H. (2003) Coriolis meters in two phase conditions. *IEE Computing and Control Engineering*, 14(4): 36.

Hemp, J. and Youngs, I. (2003) Problems in the theory and design of electromagnetic flowmeters for dielectric liquids. Part 3a. Modelling of zero drift due to flux linkage between coil and electrode cables. *Flow Measurement and Instrumentation*, 14(3): 65–78.

Hemp, J., Sanderson, M. L., Koptioug, A. V., Liang, B., Sweetland, D. J. and Al-Rabeh, R. H. (2002) Problems in the theory and design of electromagnetic flowmeters for dielectric liquids. Part 1: Experimental assessment of static charge noise levels and signal-to-noise ratios. *Flow Measurement and Instrumentation*, 13(4): 143–153.

Hemp, J., Yeung, H. and Kassi, L. (2003) Coriolis meters in two phase conditions. IEE One-day Seminar on Advanced Coriolis Mass Flow Metering, 8 July 2003.

Hendrix, A. R. (1982) Positive displacement flowmeters: high performance -with a little care. *In Tech*: 47–49.

Henke, R. W. (1955) Positive displacement meters. *Control Engineering*, 2(5): 56–64.

Henry, M. P. (1995) Self-validation improves Coriolis flowmeter. *Control Engineering*, 42(6 May).

Henry, M. (2000) Self-validating digital Coriolis mass flow meter. *Computing and Control Engineering Journal*, 11(5): 219–227.

(2001) On-line compensation in a digital Coriolis mass flow meter. *Flow Measurement and Instrumentation*, 12(2): 147–161.

(2003a) Coriolis meter digital transmitter technology. *IEE Computing and Control Engineering*, 14(4): 34–35.

(2003b) Coriolis flow transmitter and dynamic response performance – digital transmitter. *Measurement & Control*, 36(9): 278–281.

Henry, M. P., Clarke, D. W., Archer, N., Bowles, J., Leahy, M. J., Liu, R. P., Vignos, J. and Zhou, F. B. (2000) A self-validating digital Coriolis mass-flowmeter: an overview. *Control Engineering Practice*, 8: 487–506.

Henry, M., Clark, C. and Cheesewright, B. (2003a) Pushing Coriolis mass flowmeters to the limit. *IEE Computing and Control Engineering*, 14(3): 24–28.

Henry, M. P., Clark, C., Duta, M., Cheesewright, R. and Tombs, M. (2003b) Response of a Coriolis mass flow meter to step changes in flow rate. *Flow Measurement and Instrumentation*, 14(3): 109–118.

Henry, M., Duta, M., Tombs, M., Yeung, H. and Mattar, W. (2004) How a Coriolis mass flow meter can operate in two-phase (gas/liquid) flow. *Technical Papers of ISA*, 454: 17–30.

Henry, M., Tombs, M., Duta, M., Zhou, F., Mercado, R., Kenyery, F., Shen, J., Morles, M., Garcia, C. and Lansangan, R. M. (2006) Two-phase flow metering of heavy oil using a Coriolis mass flow meter: a case study. *Flow Measurement and Instrumentation*, 17(6): 399–413.

Henry, M., Tombs, M., Zamora, M. and Zhou, F. (2013) Coriolis mass flow metering for three-phase flow: A case study. *Flow Measurement and Instrumentation*, 30: 112–122 http://dx.doi.org/10.1016/j.flowmeasinst.2013.01.003.

Heritage, J. E. (1989) The performance of transit time ultrasonic flowmeters under good and disturbed flow conditions. *Journal of Flow Measurement and Instrumentation*, 1: 24–30.

Hermant, C. (1962) Application of flow measurement by the comparative salt-dilution to the determination of turbine efficiency. Proc. Symp. on Flow Measurement in Closed Conduits, HMSO, 2 Paper E-2.

Herremans, P., Hoogendijk, C. J., Boer, A. H. and van Bekkum, A. J. (1989) Ultrasonic flowmeter. US Patent 4838127A.

Herschy, R. W. (1995) *Streamflow measurement*. Second edition, E & F N Spon.

Herzl, P. J. (1982) A vortex flowmeter with 2 active vortex generators. *Advances in Instrumentation*, 37: 1205–1216.

Herzog, J. P. (1992) An optical fibre vortex sensor for flow rate measurements. *Sensors & Actuators, A: Phys.*, 32: 696–700.

Herzog, M. W., Brobeil, W., Schafer, R. and Meyre, S. (1993) *Breakthroughs in the design of the next generation of electromagnetic water meters*. Flow Measurement for the utilities, Amsterdam, The Netherlands.

Hetsroni, G. (1981) *Handbook of multiphase systems*. McGraw Hill.

Higham, E. H. and Johnston, J. S. (1992) A review of 'smart' and 'intelligent' flowmetering systems. FLOMIC Report No 18.

Higham, E. H., Fell, R. and Ajaya, A. (1986) Signal analysis and intelligent flowmeters. *Measurement and Control*: 47–50.

Higson, D. J. (1964) The transient performance of turbine flowmeters in water. *Journal of Scientific Instruments*, 42: 337–342.

Hilgenstock, A. and Ernst, R. (1996) Analysis of installation effects by means of computational fluid dynamics – CFD vs experiments? *Journal of Flow Measurement and Instrumentation*, 7(3/4): 161–171.

Himpe, U., Gotte, B. and Schatz, M. (1994) Influence of upstream bends on the discharge coefficients of classical Venturi tubes and orifice plates. *Journal of Flow Measurement and Instrumentation*, 5: 209–216.

Hirnschrodt, M., von Jena, A., Vontz, T., Fischer, B., Lerch, R. and Meixner, H. (2000) Time domain evaluation of resonance antireflection (RAR) signals for ultrasonic density measurement. *IEEE Transactions on Ultrasonics, Ferroelectrics, and Frequency Control*, 47(6): 1530–1539.

Ho, Y. L., Chen, J. Y., Shaw, J. H. and Yang, C. T. (2005) Design, CFD investigation and implementation of a novel diverter mechanism for water flow measurement. *Key Engineering Materials*, 295–296: 521–526.

Hobbs, J. M. and Humphreys, J. S. (1990) The effect of orifice plate geometry upon discharge coefficient. *Journal of Flow Measurement and Instrumentation*, 1: 133–140.

Hochreiter, H. M. (1958) Dimensionless correlation of coefficients of turbine type flowmeters. *Trans. ASME*: 1363–1368.

Hodges, C., Britton, C., Johansen, W. and Steven, R. (2010) Cone DP meter calibration issues. 15th Flow Measurement Conference (FLOMEKO), 13–15 October Taipei, Taiwan, Paper B4-2.

Hofman, F. (1993) Magnetic flowmeter with flowshaping flowtube. FLOMEKO'93 Proceedings of the 6th International Conference on Flow Measurement, Korea Research Institute of Standards and Science: 445–451.

Hofstede, H., Hogendoorn, J., Danen, H. and Tetzner, R. (2004) ALTOSONIC III –dedicated three beam liquid ultrasonic flowmeter for custody transfer and pipeline leak detection. 12th International Conference on Flow Measurement FLOMEKO Guilin China.

Hogendoorn, J., Laan, D., Hofstede, H. and Danen, H. (2004) ALTOSONIC III – a dedicated three-beam ultrasonic flowmeter for custody transfer of liquid hydrocarbons. 22nd North Sea Flow Metering Workshop 2004, Paper 4.2.

Hogendoorn, J., Boer, A., Hofstede, D. L. H. and Danen, H. (2005) Flow disturbances and flow conditioners: the effect on multi-beam ultrasonic flowmeters. 23rd North Sea Flow Metering Workshop 2005, Paper 16, pages 241–252.

Hogendoorn, J., Boer, A. and Danen, H. (2007) An ultrasonic flowmeter for custody transfer measurement of LNG: a challenge for design and calibration. 25th International North Sea Flow Measurement Workshop, Energy Institute, Oslo, Norway, 350–365.

Hogendoorn, J. Tawackolian, K., van Brakel, P., van Klooster, J. and Drenthen, J. (2009) High viscosity hydrocarbon flow measurement, a challenge for ultrasonic flow meters? 27th International North Sea Flow Measurement Workshop, Tonsberg, Norway, 20–23 October 2009, pp. 341–360.

Hogendoorn, J., Boer, A., Appel, M., de Jong, H. and de Leeuw, R. (2013) Magnetic resonance technology: a new concept for multiphase flow measurement. 31st International North Sea Flow Measurement Workshop, 22-25 October 2013, Tønsberg, Norway

Hogendoorn, J., Boer, A., Zoeteweij, M., Bousché, O., Tromp, R., de Leeuw, R., Moeleker, P., Appel, M. and de Jong, H. (2015) Magnetic resonance multiphase flowmeter: measuring principle and broad range test results. South East Asia Flow Measurement Conference 3–4 March 2015.

Hogrefe, W., Kirchhof, U., Mannherz, E., Marchewka, W., Mecke, U., Otto, F., Rakebrandt, K.-H., Thone, A. and Wegener, H.-J. (1995) *Guide to flowmeasurements*. Bailey-Fischer & Porter GmbH Gottingen.

Holden, J. L. and Peters, R. J. W. (1990) *Practical experiences using ultrasonic flowmeters on high pressure gas*. North Sea Flow Measurement Workshop, National Engineering Laboratory, Scotland.

Holden, J. L. and Peters, R. J. W. (1991) Practical experiences using ultrasonic flowmeters on high pressure gas. *Journal of Flow Measurement and Instrumentation*, 2: 69–73.

Holm, M., Stang, J. and Delsing, J. (1995) Simulation of flowmeter calibration factors for various installation effects. *Measurement:* 15(4): 235–244.

Hooper, L. J. (1962) Discharge measurements by the Allen salt-velocity method. Proc. Symp. on Flow Measurement in Closed Conduits, HMSO, 2: Paper E-1.

Hopkins, D., Savage, P. F. and Fox, E. (1995) Problems encountered during research into flow rate, pattern of water consumption and unaccounted-for water losses in urban areas. *Journal of Flow Measurement and Instrumentation* 6: 173–179.

Hornak, J. P. (1997–1999) The basics of NMR. http://www.cis.rit.edu/htbooks/nmr/inside.htm

Horner, B., Mesch, F. and Trachtler, A. (1996) A multi-sensor induction flowmeter reducing errors due to non-axisymmetric flow profiles. *Measurement Science and Technology*, 7: 354–360.

Hoshikawa, S., Ishikawa, H., Nakao, S. and Takamoto, M. (2005) Development of an ultrasonic flowmeter for hydrogen gas. Flomeko 2005 13th International Flow Measurement Conference, Peebles, Scotland, Paper 7.1.

House, R. K. and Johnson, R. T. (1986) Practical application of hydrostatic fluid bearing design principles to a turbine flow meter. Instrumentation in the Aerospace Industry, Proceedings of the 32nd International Instrumentation Symposium, Seattle, Washington, USA. (Also in ISA Transactions 26, No 3, pp. 59–63, 1987.)

Howarth, M. (1994) HART – Standard for 4–20mA digital communications. *Measurement & Control*, 27(1): 5–8.

Hrin, G. P. and Tuma, D. T. (1977) Doppler microwave cavity resonator for particulate loading. *IEEE Trans Instrument Measurement*, IM26 13–17.

Hu, C.-C. and Lin, W.-T. (2009) Performance test of KOH-etched silicon sonic nozzles. *Flow Measurement and Instrumentation*, 20(3): 122–126.

Hu, C.-C., Lin, W.-T., Su, C.-M. and Liu, W.-J. (2010a) Discharge characteristics of small sonic nozzles in the shape of pyramidal convergent and conical divergent. 15th Flow Measurement Conference (FLOMEKO), 13–15 October 2010, Taipei, Taiwan, Paper A6-5.

Hu, C.-C., Lin, W.-T. and Su, C.-M. (2011) Flow characteristics of pyramidal shaped small sonic nozzles. *Journal of Flow Measurement and Instrumentation*, 22(1): 64–70.

Hu, C.-C., Lin, W.-T., Su, C.-M. and Liu, W.-J. (2012) Discharge characteristics of small sonic nozzles in the shape of pyramidal convergent and conical divergent. *Journal of Flow Measurement and Instrumentation*, 25: 26–31.

Hu, H.-M., Wang, C. and Meng, T. (2010b) Numerical approach to estimate the accuracy of ultrasonic flowmeter under disturbed flow condition. 15th Flow Measurement Conference (FLOMEKO), 13–15 October 2010, Taipei, Taiwan, Paper A10-1.

Hu, H., Zhang, L., Meng, T. and Wang, C. (2015) Mechanism analysis and estimation tool of installation effect on multipath ultrasonic flowmeter. *9th International symposium of Fluid Flow Measurement*, Arlington, VA, USA.

Hu, L., Zou, J., Zhu, Z. C., Fu, X., Ruyan, X. D. and Wang, C. Y. (2010c) Analytical calculation of the measuring error of an electromagnetic velocity probe caused by a side channel wall. *Flow Measurement and Instrumentation*, 21(4): 435–442.

Huang, L., Chen, C. C., Yao, Y., Wang, G., Feng, Y., Wei, K., Deng, W., Jiang, C., Ruan, J. and Jiang, S. (2010) All Electronic MEMS flow meters for city gas applications. 15th Flow Measurement Conference (FLOMEKO), October 13–15, 2010 Taipei, Taiwan, Paper A8-2.

Huang, Z., Xie, D., Zhang, H. and Li, H. (2005) Gas-oil two-phase flow measurement using an electrical capacitance tomography system and a Venturi meter. *Flow Measurement and Instrumentation*, 16(2–3): 177–182.

Huijsing, J. H., van Dorp, A. L. C. and Loos, P. J. G. (1988) Thermal mass-flow meter. *Journal of Physics E: Scientific Instruments*, 21: 994–997.

Hulbert, G. M., Darnell, I. and Brereton, G. J. (1995) Numerical and experimental analysis of Coriolis mass flowmeters. AIAA/ASME/ASCE/AHS Structures, Structural Dynamics and Materials Conference – Collection of Technical Papers, 3: (AIAA-95-1384-CP), pp. 1889–1893.

Hulin, J-P. and Foussat, A. J. M. (1983) Vortex flowmeter behaviour in liquid-liquid two-phase flow. International Conference on Physical Modelling of Multi-Phase Flow, Coventry, England, BHRA Fluid Engineering, Paper H3: 377–390.

Hulin, J-P., Fierfort, C. and Condol, R. (1982) Experimental study of vortex emission behind bluff obstacles in a gas liquid vertical two-phase flow. *International Journal of Multiphase Flow*, 8: 475–490.

Hunter, J. J. and Green, W. L. (1975) Blockage and its effect on a drag plate flowmeter. Conference on Fluid Flow Measurement in the Mid 1970's, National Engineering Laboratory, East Kilbride, Scotland: Paper C-2.

Hussain, Y. A. (2000a) Mass flowmeter which operates according to the Coriolis principle. US Patent 6,041,665.

(2000b) Mass flow rate measuring instrument. US Patent 6,082,202.

(2003) Single straight tube mass flowmeter using 'adaptive sensor technology. *IEE Computing and Control Engineering*, 14(4): 40–41.

Hussain, Y. A. and Baker, R. C. (1985) Optimised non-contact electromagnetic flowmeter. *Journal of Physics E: Scientific Instruments*, 18: 210–213.

Hussain, Y. A. and Farrant, D. (1994) *Coriolis mass flow measurement using single straight tube*. Proceedings of a Study Day on Massflow at RAI (Studiedag 'Massflow'), MRBT, Amsterdam.

Hussain, Y. A. and Rolph, C. N. (1994) Mass flowmeter. US Patent 5, 365, 794.

(2002) Mass flow meter. US Patent 2002020227, 21 February 2002.

Hussain, Y., Rolph, C. and Wang, T. (2010) Mass flowmeter.US2010050783.

Hussein, I. B. and Owen, I. (1991) Calibration of flowmeters in superheated and wet steam. *Journal of Flow Measurement and Instrumentation*, 2: 209–216.

Hussein, I. B., Owen, I. and Amin, A. M. (1992) Energy metering system for high quality saturated steam. *Journal of Flow Measurement and Instrumentation*, 3: 235–240.

Hutchings, I. M. (1992) Tribology: friction and wear of engineering materials. Arnold 1992, reprinted by Butterworth-Heineman 2001.

Hutton, S. P. (1974) The effect of inlet flow conditions on the accuracy of flowmeters. *Int. Mech. E Conf. Publ.* 4: 1–8.

(1986) The effects of fluid viscosity on turbine meter calibration. Flow Measurement in the Mid 80s, NEL, Scotland: Paper 1.1.

Hwang, S.-Y., Lee, H.-J. and Kim, H.-D. (2004) Application of multi-path ultarsonic oil flowmeter using a new weighting factor method. 12th International Conference on Flow Measurement FLOMEKO Guilin China.

Ibarz, A., Bauer, G., Casas, R., Marco, A. and Lukowicz, P. (2008) Design and evaluation of a sound based water flow measurement system. 3rd European Conference on Smart Sensing and Context, EuroSSC: 41–54.

Ibrahim, S., Green, R. G. and Dutton, K. (2000) Optical fibre sensors for flow measurement. *Proceedings of SPIE – The International Society for Optical Engineering*, 4074: 372–376.

Ifft, S. A. (1996) Partially closed valve effects on the V-cone flowmeter. FLOMEKO '96 Proceedings of the 8th International Conference on Flow Measurement, Beijing, China: 49–54.

Ifft, S. A. and Mikklesen, E. D. (1993) Pipe elbow effects on V-cone flow meter FED. *Fluid Meas Instrum ASME*, 161.

Igarashi, T. (1986) Fluid flow around a bluff body used for a Karman vortex flowmeter. *Fluid Control and Measurement*, 2: 1003–1008.

Ilha, A., Doria, M. M. and Aibe, V. Y. (2010) Treatment of the time dependent residual layer and its effects on the calibration procedures of liquids and gases inside a volume prover. 15th Flow Measurement Conference (FLOMEKO), 13–15 October 2010, Taipei, Taiwan, Paper B3-3.

Imrie, A. (1994) Communication options in the water industry. 27(7): 221–224.

Incontri, J. (2005) Recent advances in electromagnetic flow measurement technology. *Technical Papers of ISA*, 459: 163–168.

Inkley, F. A., Walden, D. C. and Scott, D. J. (1980) Flow characteristics of vortex shedding meters. *Measurement & Control*, 13: 166–170.

Iooss, B., Lhuillier, C. and Jeanneau, H. (2002) Numerical simulation of transit-time ultrasonic flowmeters: Uncertainties due to flow profile and fluid turbulence. *Ultrasonics*, 40(9): 1009–1015.

ISA (1959) Terminology, dimensions and safety practices for indicating variable area meters (Rotameters): RP16.1 Glass tube, RP16.2 Metal tube, RP16.3 Extension type glass tube. Instrument Society of America, Report (recommended practice) RP 16.1.2.3.

(1960) Nomenclature and terminology for extension type variable area meters (Rotameters). Instrument Society of America, Report (tentative recommended practice) RP 16.4.

(1961a) Installation, operation, maintenance instructions for glass tube variable area meters (Rotameters). Instrument Society of America, Report (recommended practice) RP 16.5.

(1961b) Methods and equipment for calibration of variable area meters (Rotameters). Instrument Society of America, Report (tentative recommended practice) RP 16.6.

Ishibashi, M. (2002) Proposal of fluid dynamical standard (FDS) for gas flow-rate. 5th International symposium of Fluid Flow Measurement, Arlington, VA, USA.

Ishibashi, M. and Funaki, T. (2013) Boundary layer transition in high precision critical nozzles of various shapes. Proceedings of FLOMEKO2013 Small nozzles

Ishibashi, M. and Morioka, T. (2006) The renewed airflow standard system in Japan for 5–1000 m3/h. *Flow Measurement and Instrumentation*, 17(3): 153–161.

(2010) Dependence of the flow velocity field in critical nozzles on the pressure ratio. 15th Flow Measurement Conference (FLOMEKO), 13–15 October 2010, Taipei, Taiwan, Paper A6-4.

(2012) Velocity field measurements in critical nozzles using Recovery Temperature Anemometry (RTA). *Flow Measurement and Instrumentation*, 25: 15–25.

Ishibashi, M. and Takamoto, M. (1997) Very accurate analytical calculation of the discharge coefficients of critical Venturi nozzles with laminar boundary layer. Proceedings of FLUCOME

(2000a) Methods to calibrate a critical nozzle and flowmeter using reference critical nozzles. *Flow Measurement and Instrumentation*, 11(4): 293–303.

Ishibashi, M. and Takamoto, M. (2000b) Theoretical discharge coefficient of a critical circular-arc nozzle with laminar boundary layer and its verification by measurements using super-accurate nozzles. *Flow Measurement and Instrumentation*, 11(4): 305–313.

(2001)Discharge coefficient of super-accurate critical nozzles accompanied with the boundary layer transition measured by reference super-accurate critical nozzles connected in series. Proc. ASME FEDSM01 paper No. 18036.

Ishikawa, H., Takamoto, M., Shimizu, K., Monji, H. and Matsui, G. (2000) Development of a new ultrasonic liquid flowmeter for very low flow rate applicable to a thin pipe. *IEEE International Symposium on Semiconductor Manufacturing Conference, Proceedings*, (1): 383–386.

Islam, M., Seshadri, V., Singh, S. N. and Hasan, M. M. (2003) Skewed velocity profile effect on turbine flowmeter performance. *Proceedings of the Institution of Mechanical Engineers, Part E: Journal of Process Mechanical Engineering*, 217(1): 25–32.

ISO Standards – the reader is referred to the ISO website on the internet.

Ito, D., Kikura, H. and Antomi, M. (2011) Micro wire-mesh sensor for two-phase flow measurement in a rectangular narrow channel. *Flow Measurement and Instrumentation*, 22(5): 377–382.

Ito, H., Watanabe, Y. and Shoji, Y. (1985) A long-radius inlet nozzle for flow measurement. *Journal of Physics E: Scientific Instruments*, 18: 88–91.

Itoh, I. and Ohki, S. (1993) Mass flowmeter detecting fluctuations in lift generated by vortex shedding. *Journal of Flow Measurement & Instrumentation*, 4: 215–224.

Jackson, G. A., Gibson, J. R. and Holmes, R. (1989) Three-path ultrasonic flowmeter for small-diameter pipelines. *Journal of Physics E: Scientific Instruments*, 22.8: 645–650.

(1991) Three-path ultrasonic flow meter with fluid velocity profile identification. *Measurement Science & Technology*, 2.7: 635–642.

Jacobsen, E., Denstad, H., Downing, A., Daniel, P. and Tudge, M. (2004) Validation and operational experience of a dualstream II wet gas meter in a subsea application on the Statoil Mikkel Field. 22nd North Sea Flow Metering Workshop 2004, Paper 8.2.

Jacobson, S. A., Lynnworth, L. C. and Korba, J. M. (1988) Differential correlation analyzer. US Patent 4,787,252 (Nov. 29, 1988).

Jae, H. H., Woo, H. J., Jung, Y. L., Ho, C. J. and Hang, B. K. (2006) Feed water flow measurement and experience using clamp-on transit-time ultrasonic flow meter in nuclear power plants. 5th International Topical Meeting on Nuclear Plant Instrumentation Controls, and Human Machine Interface Technology (NPIC and HMIT 2006), 2006: 284–291.

Jalbert, P. A. (1999) Evaluation of a dual rotor turbine fuel flow meter (G176). *Instrumentation in the Aerospace Industry: Proceedings of the International Symposium*, 427–437.

James, R. (1965/1966) Metering of steam/water two-phase flow by sharp-edged orifices. *Proc I Mech E*, 180: 549–572.

Jamieson, A. W. (2001) Wet gas metering – the unexpected challenge: Status and trends on technology and applications. Paper 5, 19th North Sea Flow Measurement Workshop, Kristiansand, Norway, 22–25 October 2001.

Jamieson, A. W., Johnson, P. A., Spearman, E. P. and Sattary, J. A. (1996) Unpredicted behaviour of Venturi flowmeter in gas at high Reynolds numbers. Proc. of 14th North Sea Flow Measurement Workshop, Peebles, Scotland, paper 1.5.

Jeanneau, H. and Piguet, M. (2000) Pipe flow modelling for ultrasonic flow measurement. FLOMEKO '2000 the 10th International Conference on Flow Measurement, Salvador, Brazil: Paper B9.

Jenkins, D. M., Lysak, P. D., Capone, D. E., Brown, W. L. and Askari, V. (2006) Ultrasonic cross-correlation flow measurement: Theory, noise contamination mechanisms, and a noise mitigation technique. *International Conference on Nuclear Engineering, Proceedings, ICONE*, 2006(14): 7p.

Jepson, P. (1964) Transient response of a helical flowmeter. *Journal of Mechanical Engineering Science*, 6: 317–320.

(1967) Currentmeter errors under pulsating flow conditions. *Journal of Mechanical Engineering Science*, 9: 45–54.

Jepson, P. and Bean, P. G. (1969) Effect of upstream velocity profiles on turbine flowmeter registration. *Journal of Mechanical Engineering Science*, 11: 503–510.

Jepson, P. and Chamberlain, D. (1977) Operating high pressure orifice metering installations. FLO-CON 77 Proceedings of the Symposium – the Application of Flow Measuring Techniques, Brighton, UK: 285–319.

Jepson, P. and Chipchase, R. (1975) Effect of plate buckling on orifice meter accuracy. *Journal of Mechanical Engineering Science*, 17.

Jeronymo, C. E. and Aibe, V. Y. (2010) Implementation of quadruple-timing pulse interpolation applied to compact piston provers. 15th Flow Measurement Conference (FLOMEKO), 13–15 October 2010, Taipei, Taiwan, Paper A5-1.

Jian, D. and Karcher, C. (2012) Electromagnetic flow measurements in liquid metals using time-of-flight Lorentz force velocimetry. *Measurement Science & Technology*, 23(7) 074021(14pp).

Jian, W. (2004) The middle range gas flow standard of SPRING Singapore. 12th International Conference on Flow Measurement FLOMEKO Guilin, China.

Jin, N., Miao, L. and Li, W. (2006) Symbolic sequence analysis method of gas/liquid two-phase flow measurement fluctuation signals. Proceedings of IEEE ICIA 2006–2006 IEEE International Conference on Information Acquisition, Weihai, Shandong, China, 1321–1326.

Jitschin, W. (2004) Gas flow measurement by the thin orifice and the classical Venturi tube. *Vacuum*, 76(1): 89–100.

Jitschin, W., Ronzheimer, M. and Khodabakhshi, S. (1999) Gas flow measurement by means of orifices and Venturi tubes. *Vacuum* 53(1–2): 181–185.

Johannessen, A. A. (1993) *Evaluation of ultrasonic liquid flowmeters*. North Sea Flow Measurement Workshop, Bergen, Norway.

(2001) Flare gas metering – measurement challenges at hand. 19th North Sea Flow Measurement Workshop, Kristiansand, Norway, 22–25 October 2001, Paper 14.

Johansen, E. S., Hall, A. R. W., Ünalmis, Ö. H., Rodriguez, D. J., Vera, A. and Ramakrishnan, V. (2007) A prototype wet-gas and multiphase flowmeter. 25th International North Sea Flow Measurement Workshop, Energy Institute, Oslo, Norway, 97–112.

Johansen, G. A. and Tjugum, S.-A. (2007) Fluid composition analysis by multiple gamma-ray beam and modality measurements. 25th International North Sea Flow Measurement Workshop, Energy Institute, Oslo, Norway, 69–78.

Johansen, W. R. (2010) The effect of using real gas absolute viscosity and isentropic exponent on orifice flow measurement: proposed adoption of REFPROP 8.0 as a standard for the natural gas industry. 15th Flow Measurement Conference (FLOMEKO), 13–15 October 2010, Taipei, Taiwan, Paper B8-1.

John, H., Hain, K., Bruderie, F., Reimann, J. and Vollmer, T. (1982) Tests of an advanced true mass flow meter (TMFM) in gas-liquid flow. Measurement in Polyphase Flows-1982, AIAA/ASME Joint Fluids, Plasma, Thermophysics and Heat Transfer Conference, St Louis, Missouri, USA: 55–60.

Johnson, A. and Wright, L. (2006) Evaluation of theoretical CFV flow models in the laminar, turbulent, and transition flow regimes. Proc. 6th International Symposium On Fluid Flow Measurement (ISFFM), Querétero, Mexico.

Johnson, A. and Wright, J. (2008) Comparison between theoretical CFV flow models and NIST's primary flow data in the laminar, turbulent, and transition flow regimes. *ASME Journal of. Fluids Engineering,* 130: 271202-1-11.

Johnson, A. N., Espina , P. I., Mattingly, G. E., Wright , J. D. and , Merkle , C. L. (1998) Numerical characterization of the discharge coefficient in critical nozzles. *Proc. NCSL Workshop & Symposium*, Session 4E: 407–442.

Johnson, A. N., Crowley, C. J. and Yeh, T. T. (2010) Uncertainty analysis of NIST's 20 liter hydrocarbon liquid flow standard. 15th Flow Measurement Conference (FLOMEKO), 13–15 October 2010, Taipei, Taiwan, Paper B5-4.

Johnson, A., Harman, E. and Boyd, J. (2013) Blow-down calibration of a large 8 path ultrasonic flow meter under quasi-steady flow conditions. FLOMEKO 2013, 16th International Flow Measurement Conference, Paris

Johnson, M. W. and Farroll, S. (1995) Development of a turbine meter for two-phase flow measurement in vertical pipes. *Journal of Flow Measurement and Instrumentation*, 6: 279–282.

Johnson, R. (2002) Keeping meters on line: accurately measuring drinking water and sewage. *Water Engineering and Management*, 149(9): 14–19.

Jones, F. E. (1992) Application of the equation to the treatment of laminar flowmeter calibration data. *Industrial Metrology*, 2: 91–96.

Jongerius, P. F. M., van der Beek, M. P. and van der Grinten, J. G. M. (1993) Calibration facilities for industrial gas flow meters in The Netherlands. *Flow Measurement and Instrumentation*, 4: 77–84.

Joshi, S. G. (1991) Surface-acoustic-wave (SAW) flow sensor. *IEEE Trans on Ultrasonics, Ferroelectrics & Frequency Control*, 38.2: 148–154.

Joslin, G. (1879) British Patent, No 2428.

Jousten, K., Menzer, H. and Niepraschk, R. (2002) A new fully automated gas flowmeter at the PTB for flow rates between 10^{-13} mol/s and 10^{-6} mol/s. *Metrologia*, 39(6): 519–529.

Jung, J. C. and Seong, P. H. (2005) Estimation of the flow profile correction factor of a transit-time ultrasonic flow meter for the feed water flow measurement in a nuclear power plant. *IEEE Transactions on Nuclear Science*, 52(3 II): 714–718.

Jungowski, W. M. and Weiss, M. H. (1996) Effects of flow pulsation on a single-rotor turbine meter. *Journal of Fluids Engineering, Transactions ASME*, 118: 198–201.

Kabaciński, M. and Pospolita, J. (2008) Numerical and experimental research on new cross-sections of averaging Pitot tubes. *Flow Measurement and Instrumentation*, 19(1): 17–27.

(2011) Experimental research into a new design of flow-averaging tube. *Flow Measurement and Instrumentation*, 22(5): 421–427.

Kalivoda, R. J. and Lunde, P. (2005) Liquid ultrasonic flow meters for crude oil measurement. 23rd North Sea Flow Metering Workshop 2005, Paper 17, pages 253–275.

Kalotay, P. (1994) On-line viscosity measurement using Coriolis mass flowmeters. *Journal of Flow Measurement and Instrumentation*, 5: 303–308.

Kaltsas, G. and Nassiopoulou, A. G. (2004) Gas flow meter for application in medical equipment for respiratory control: Study of the housing. *Sensors and Actuators, A: Physical*, 110(1–3): 413–422.

Kaltsas, G., Katsikogiannis, P., Asimakopoulos, P. and Nassiopoulou, A.G. (2007) A smart flow measurement system for flow evaluation with multiple signals in different operation modes. *Measurement Science and Technology*, 18(11): 3617–3624.

Karlsson, K. and Delsing, J. (2013) The gap discharge transducer as a sound pulse emitter in an ultrasonic gas flow meter. FLOMEKO 2013, 16th International Flow Measurement Conference, Paris, France.

Karnik, U. (1995) A compact orifice meter/flow conditioner package, Proceedings of the American Gas Association Operating Section, Las Vegas, 95-OP-009:564–585.

(2000) Centaur round robin test traceability of Transcanada calibrations facility. FLOMEKO'2000 the 10th International Conference on Flow Measurement, Salvador, Brazil: Paper A5.

Karnik, U. and Geerligs, J. (2002) Effect of steps and roughness on multi-path ultrasonic meters. 5th International symposium of Fluid Flow Measurement, Arlington, VA, USA.

Karnik, U., Jungowski, W. M. and Botros, K. (1991) Effects of flow characteristics downstream of elbow/flow conditioner on orifice meter accuracy. Proceedings of the 9th North Sea Flow Measurement Workshop, Norwegian Society of Chartered Engineers.

Karnik, U., Jungowski, W. M. and Botros, K. (1994) Effect of turbulence on orifice meter performance. *Journal of Offshore Mechanics and Arctic Engineering*, 116.2: 77–85.

Katys, G. P. (1964) *Continuous measurement of unsteady flow*. Chapter 2: 37-?. Pergamon Press Ltd, London, England.

Katz, L. (1971) Improved flowmeter accuracy with electromechanical feedback. Proc. 1st Symp. on Flow – Its Measurement and Control in Science and Industry, Pittsburgh, Penn, USA. (Published by ISA 1, Pt. 2, 1974: 669–678.)

Kawaguchi, T., Aiba, T., Tsukada, K., Tsuzuki, N., Kikura, H., Sugita, K. and Umezawa, S. (2015) Non-intrusive measurement of steam flow rate in a steel pipe by means of a clamp-on ultrasound flowmeter. 9th international symposium of Fluid Flow Measurement, Arlington, VA, USA.

Kawano, T., Miyata, T., Shikuya, N., Takahashi, S., Handoh, M., Itoh, I. and Biles, B. (1992) Intelligent flowmeter. Proceedings of the Conference on Advances in Instrumentation and Control, (ISA) Houston, USA: 997–1009.

Kay, J. M. and Nedderman, R. M. (1974) *An introduction to fluid mechanics and heat transfer*. Cambridge University Press.

Kaye, G. W. C. and Laby, T. H. (1966) *Tables of physical and chemical constants*. Longmans.

Keech, R. P. (1982) The KPC multichannel correlation signal processor for velocity measurement. *Transactions of the Institute of Measurement and Control*, 4: 43-52

Kegel, T. (2002a) Uncertainty analysis of a volumetric primary standard. 5th International symposium of Fluid Flow Measurement, Arlington, VA, USA.

(2002b) Large scale calibration facility. 5th International symposium of Fluid Flow Measurement, Arlington, VA, USA.

Kegel, T. and Cousins, T. (2012) *Different requirements and methods for calibrating gas and liquid ultrasonic flow custody transfer meters*. North Sea Flow Measurement Workshop, Scotland 2012.

Kegel, T. and English, S. (2011) A proposed ultrasonic meter recalibration interval. 29th North Sea Flow Metering Workshop, Paper 23.

Keita, N. M. (1989a) The zero drift effect in Coriolis mass flow meter. Mass Flow Measurement Direct and Indirect, Proc Int Conf Mass Flow Measurement, IBC Pbl, London, England.

(1989b) Contribution to the understanding of the zero shift effects in Coriolis mass flowmeters. *Flow Measurement and Instrumentation*, 1: 39–43.

(1990) *Performance of Coriolis mass flowmeters in the metering of light fluids*. Int Conf Flow Measurement of Commercially Important Fluids, London, England.

(1994) Behaviour of straight pipe Coriolis mass flowmeters in the metering of gas: theoretical predictions with experimental verification. *Journal of Flow Measurement and Instrumentation*, 5: 289–294.

(2000) Ab initio simulation of Coriolis mass flowmeter. FLOMEKO'2000 the 10th International Conference on Flow Measurement, Salvador, Brazil: Paper C9.

Kettle, R. J., Ross, D. and Deznan, D. (2002) The multiphase flowmeter, a tool for well performance diagnostics and production optimization. SPE – Asia Pacific Oil and Gas Conference, 591–598.

Keyser, D. R. (1973) The calibration correlation function for positive displacement liquid meters. *Trans. ASME*, 95, Series I, No. 2: 180–188.

Kiehl, W. (1991) Difference measurement using Coriolis mass flowmeters. *Journal of Flow Measurement and Instrumentation*, 2: 135–138.

Kiehl, W. and Gartner, U. (1989) Two Coriolis meters in one line. Mass Flow Measurement Direct and Indirect, Proc Int Conf Mass Flow Measurement, IBC Pbl, London, England.

Kikura, H., Yamanaka, G. and Aritomi, M. (2004) Effect of measurement volume size on turbulent flow measurement using ultrasonic Doppler method. *Experiments in Fluids*, 36(1): 187–196.

Kim, B.-C., Pak, B.-C., Cho, N.-H., Chi, D.-S., Choi, H.-M., Choi, Y.-M. and Park, K.-A. (1997) Effects of cavitation and plate thickness on small diameter ratio orifice meters. *Journal of Flow Measurement and Instrumentation*, 8(2): 85–92.

Kim, C. H., Lee, D. K. and Paik, J. S. (1993a) Mean velocity measurement of pipe flow by ultrasonic correlation. FLOMEKO '93 Proceedings of the 6th International Conference on Flow Measurement, Korea: 550–558.

Kim, D. K., Han, I. Y. and Kim, S. J. (2003) Study on the transient characteristics of the sensor tube of a thermal mass flow meter. American Society of Mechanical Engineers, Heat Transfer Division, (Publication) HTD, 374(3): 79–86.

(2007b) Study on the steady-state characteristics of the sensor tube of a thermal mass flow meter. *International Journal of Heat and Mass Transfer*, 50(5–6): 1206–1211.

Kim, D.-K., Majumdar, A. and Kim, S. J. (2007a) Electrokinetic flow meter. *Sensors and Actuators, A: Physical*, 136(1): 80–89.

Kim, H-M., Kim, K-Y., Her, J-Y. and Ha, Y-C. (2002) Three-dimensional flow analysis for estimation of measuring error of orifice flowmeter due to upstream flow distortion. American Society of Mechanical Engineers, Fluids Engineering Division (Publication) FED, 257(1 A): 121–126.

Kim, J., Ahn, Y.-C. and Kim, M. H. (2009) Measurement of void fraction and bubble speed on slug flow with three-ring conductance probes. *Journal of Flow Measurement and Instrumentation*, 20(3): 103–109.

Kim, J.-H., Kim, H.-D. and Park, K.-A. (2006) Computational/experimental study of a variable critical nozzle flow. 12th International Conference on Flow Measurement FLOMEKO Guilin China. *Flow Measurement and Instrumentation* 17(2): 81–86.

Kim, R. K., Swain, J. C., Kramer, G. S., Cooper, D. L., Schuluer, L. E. and Haubert, T. D. (1993b) Progress on development of a compact gas meter. FLOMEKO '93 Proceedings of the 6th International Conference on Flow Measurement, Korea Research Institute of Standards and Science: 393–402.

Kim, Y. and O'Neal, D. L. (1995) Comparison of critical flow models for estimating two-phase flow of HCFC22 and HFC134a through short tube orifices. *International Journal of Refrigeration*, 18.7: 447–455.

Kim, Y.-K. and Kim, Y.-H. (1996) A three accelerometer method for the measurement of flow rate in pipe. *J. Acoust. Soc. Am.*, 100(2 Pt 1): 717–726. Advances in Instrumentation and Control: International Conference and Exhibition, New Orleans, LA, USA, 50(2): 643–651.

Kimpton, S. and Niazi, A. (2008) Thermal lagging – the impact on temperature measurement. 26th International North Sea Flow Measurement Workshop, 21–24 October 2008, St Andrews, Scotland, Paper 8.3.

King, D. C. (2004) Benefits of ISO 17025 accreditation for providers and users of calibration gases used in environmental testing. Air Waste Management Association, 97th Annual Conference, 2004, 127(2004), pp. 3651–62.

King, J. O. and Rollwitz, W. L. (1983) Magnetic resonance measurement of flowing coal. *Trans ISA* 22: 69–76.

King, L. V. (1914) On the convection of heat from small cylinders in a stream of fluid. *Philosophical Transactions of the Royal Society*, A214: 373-432.

King, N. W. (1988) Multi-phase flow measurement at NEL. *Measurement & Control*, 21.8.

(1990) Subsea multi-phase flow metering a challenge for the offshore industry? Subsea 90 International Conference, London.

King, N. W., Sidney, J. K. and Coulthard, J. (1988) *Cross-correlation flow measurements in oil-air mixtures*. 2nd Int Conf Flow Measurement, BHRA, London, UK.

Kinghorn, F. C. (1982) *The analysis and assessment of data. Developments in Flow Measurement-1*, ed R. W. W. Scott, London: Applied Science Publishers: 307–326.

(1986) The expansibility correction for orifice plates: EEC data. International Conference on Flow Measurement in the Mid 80's Paper 5.2.

(1988) Challenging areas in flow measurement. *Meas. Control Measurement and Control*, 21(8): 229–235.

(1996) Industrial needs for cost-effective flow measurement. FLOMEKO'96 Proc. 8th Int. Conf. on Flow Measurement, Beijing, China: 741–50.

Kinghorn, F. C., McHugh, A. and Dyet, W. D. (1991) The use of etoile flow straighteners with orifice plates in swirling flow. *Journal of Flow Measurement and Instrumentation*, 2: 162–168.

Kivilis, S. S. and Reshetnikov, V. A. (1965) Effect of a stabilised flow profile on the error of ultrasonic flowmeters. *Measurement Techniques*, No. 3: 276.

Kjolberg, S. A. and Berentsen, H. (1997) *The Porsgrunn 2 test programme of multiphase meters: general results and examples of different meter performance*. North Sea Flow Measurement Workshop Kristiansand, Norway: Paper 2.

Kleppe, K. and Danielsen, H. B. (1993) *Scaling problems in the oil metering system at the Veslefrikk Field*. North Sea Flow Measurement Workshop, Bergen, Norway.

Kneisley, G., Lansing, J. and Dietz, T. (2009) Ultrasonic meter condition based monitoring – a fully automated solution. 27th International North Sea Flow Measurement Workshop, Tonsberg, Norway, 20–23 October 2009, pp. 253–274.

Kocbach, J. (2000) Finite element modeling of ultrasonic piezoelectric transducers: Influence of geometry and material parameters on vibration, response functions and radiated field. Doctoral dissertation, University of Bergen, Department of Physics, September 2000.

Koechner, H. and Melling, A. (2000) Numerical simulation of ultrasonic flowmeters. *Acta Acustica (Stuttgart)*, 86(1): 39–48.

Koechner, H., Melling, A. and Baumgartner, M. (1996) Optical flow field investigation for design improvements of an ultrasonic gas meter. *Journal of Flow Measurement and Instrumentation*, 7: 133–140.

Koizumi, H. and Serizawa, M. (2008) A micro flowmeter based on the velocity measurement of a locally accelerated thermal flow in an upwardly directed Hagen-Poiseuille flow. *Flow Measurement and Instrumentation*, 19(6): 370–376.

Kolahi, K., Gast, Th. and Rock, H. (1994) Coriolis mass flowmeasurement of gas under normal conditions. *Journal of Flow Measurement and Instrumentation*, 5: 275–283.

Kolin, A. (1936) An electromagnetic flowmeter. The principle of the method and its application to blood flow measurement. *Proc. Soc. Exp. Biol., NY*, 35: 53.

(1941) An AC flowmeter for measurement of blood flow in intact blood vessels. *Proc. Soc. Exp. Biol., NY*, 46: 235.

Komiya, K., Higuchi, F. and Ohtani, K. (1988) Characteristic of a thermal gas flowmeter. *Review of Scientific Instruments*, 59: 477–479.

Kondo, M. and Takahashi, M. (2005) Metallurgical study on electro-magnetic flow meter and pump for liquid lead-bismuth flow. *Progress in Nuclear Energy*, 47(1–4): 639–647.

Koning, H., Van Essen, G. J. and Smid, J. (1989) Time behaviour of turbine meters – statistical analysis of (re)calibration results of turbine meters. FLOMEKO'89 Proceedings of the 5th International Conference on Flow Measurement: 333–340.

Koschmieder, F. and Röck, H. (2010) Compensation method applied to Coriolis mass flow metering. 15th Flow Measurement Conference (FLOMEKO), 13–15 October 2010, Taipei, Taiwan, Paper A9-5.

Koudal, O., Bitto, E. and Wenger, A. (1996) A solution to the problem of installation sensitivity of small Coriolis mass flowmeters. FLOMEKO'96 Proceedings of the 8th International Conference on Flow Measurement, Beijing, China: 256–259.

Krafft, R., Hemp, J. and Sanderson, M. L. (1996) Investigation into the use of the electromagnetic flowmeter for two-phase flow measurements. *Advances in Sensors for Fluid Flow Measurement, IEE Colloquium (Digest)* No. 092: 5/1–5/4.

Kragas, T. K., Mayeu, C., Gysling, D. L., van der Spek, A. and Bostick III, F. X. (2002) Downhole fibre-optic multiphase flowmeter. *Journal of Petroleum Technology*, 54(12): 59–61.

Kragas, T. K., Johansen, E. S., Hassanali, H. and Da Costa, S. L. (2003a) Installation and data analysis of a downhole, fibre optic flowmeter at Mahogany Field, offshore Trinidad. Proceedings of the SPE Latin American and Caribbean Petroleum Engineering Conference, pp. 163–176.

Kragas, T. K., Mayeu, C. W., Gysling, D. L., van der Spek, A. M. and Bostick III, F. X. (2003b) Downhole fibre-optic flowmeter: design, operating principle, testing, and field installations. *SPE Production and Facilities*, 18(4): 257–268.

Krajcin, I., Uhrig, M., Wrath, A., Dietz, T. and Herrmann, V. (2007) Impact of regulator noise on ultrasonic flow meters in natural gas. 25th International North Sea Flow Measurement Workshop, Energy Institute, Oslo, Norway, 206–224.

Kramer, R. and Mickan, B. (2012) The Application of small sonic nozzles in test rigs with natural gas. Proc. 8th International Symposium on Fluid Flow Measurement (ISFFM), Colorado, USA, 20–22 June.

Kramer, R., Mickan, B. and Schmidt, R. (2010) The application of critical nozzles in series for the determination of small flow rates and the generation of gas mixtures. 15th Flow Measurement Conference (FLOMEKO), 13–15 October 2010, Taipei, Taiwan, Paper A6-3.

Krassow, H., Campabadal, F. and Lora-Tamayo, E. (1999) Smart-orifice meter: a mini head meter for volume flow measurement. *Flow Measurement and Instrumentation,* 10(2): 109–115.

Kratirov, V., Jamieson, A., Blaney, S. and Yeung, H. (2006) Neftemer – a versatile and cost effective multiphase meter, 24th International North Sea Flow Measurement Workshop, 24–27 October 2006.

Kratzer, W. and Kefer, V. (1988) Two phase flow instrumentation: a survey and operational experience with new and easy-to-handle devices. *Cranfield Short Course Lecture.*

Krautkramer, J. and Krautkramer, H. (1990) *Ultrasonic testing of materials.* Springer-Verlag.

Kristensen, B. D., Lofsei, C. and Frøysa, K-E. (1997) Testing of noise suppression system for multipath ultrasonic gas flow meters. North Sea Flow Measurement Workshop Kristiansand, Norway: Paper 17.

Kritz, J. (1955) An ultrasonic flowmeter for liquids. *Proc. ISA*, 10: 1-55-15-3.

Krokstad, A., Strøm, S. and Sørsdal, S. (1968) Calculating the acoustical room response by the use of a ray tracing technique. *Journal of Sound and Vibration*, 8(1): 118–125.

Kruger, G. J., Birke, A. and Weiss, R. (1996) Nuclear magnetic resonance (NMR) two-phase mass flow measurements. *Journal of Flow Measurement and Instrumentation*, 7: 25–37.

Kuchler, Heiko (1999) Special durchflussmessung. Mit volldampf sparen: Ultraschall-messsystem ermoghcht messung ohne druckverlust (Translation: Flow measurement special. Save with full steam: Ultrasound measurement system enables measurement without pressure drop). Chemie-Technik (Heidelberg), 28(8): S20–S21 (in German).

Kumar, P. and Ming Bing, M. W. (2011) A CFD study of low pressure wet gas metering using slotted orifice meters. *Journal of Flow Measurement and Instrumentation*, 22(1): 33–42.

Kumar, V., Anklin, M. and Schwenter, B. (2010) Fluid-structure interaction (fsi) simulations on the sensitivity of Coriolis flow meter under low Reynolds number flows. 15th Flow Measurement Conference (FLOMEKO), 13–15 October 2010, Taipei, Taiwan, Paper A9-2.

Kunze, J. W., Storm, R. and Wang, T. (2014) Coriolis mass flow measurement with entrained gas. Proceedings of Sensors and Measuring Systems 2014 17 ITG/GMA Symposium

Kuo, C.-Y., Ho, Y.-L., Dietz, T., Wang, W.-B. Yang, F.-R., Su, C.-M. and Shaw, J.-H. (2012) Calibration of ultrasonic flow meter using blow-down type high pressure gas flow standard. *International Symposium on Fluid Flow Measurement* (2012).

Kuoppamäki, R. (2003) Guidelines for efficient improvement of accuracy in oil and gas flow measurements. 21st North Sea Flow Metering Workshop 2003, Paper 2.

Kupnik, M., O'Leary, P., Schröder, A. and Rungger, I. (2003) Numerical simulation of ultrasonic transit-time flowmeter performance in high temperature gas flows. *IEEE Ultrasonics Symposium*, 2: 1354–1359.

Kupnik, M., Schröder, A., O'Leary, P., Benes, E. and Gröschl, M. (2004) An Ultrasonic Transit-Time Gas Flowmeter for Automotive Applications *IEEE Sensors 2004 Proceedings* IEEE1: 451–454.

Kupnik, M., Schröder, A. and Gröschl, M. (2006a) PS-16 adaptive asymmetric double-path ultrasonic transit-time gas flowmeter. *IEEE Ultrasonic Symposium*, 2429–2432.

Kupnik, M., Schröder, A., O'Leary, P., Benes, E. and Gröschl, M.(2006b) Adaptive pulse repetition frequency tfor an ultrasonic transit-time gas flowmeter for hot pulsating gases. *IEEE Sensors Journal*, 6(4), August 2006: 906–915.

Kupyna, A., Rukke, E.-O. Schüller, R. B. and Isaksson, T. (2008) The effect of flow rate, accelerometer location and temperature in acoustic chemometrics on liquid flow: spectral changes and robustness of the prediction models. *Chemometrics and Intelligent Laboratory Systems*, 2008; 93: 87–97.

Kurnadi, D. and Trisnobudi, A. (2006) A multi-path ultrasonic transit time flow meter using a tomography method for gas flow velocity profile measurement. *Particle and Particle Systems Characterization*, 23(3–4): 330–338.

Kuromori, K. et al. (1988) ADMAG series magnetic flowmeters using dual frequency excitation. *Yokogawa Technical Report*, 32(3): 129–134 (In Japanese) cf 5th FLOMEKO 1989, pp 135–42.

Kuromori, K., Goro, Sh. and Matsunaga, Y. (1989) Advanced magnetic flowmeters with dual excitation. Proc. 5th International MEKQ Conference on Flow Measurement, Dusseldorf, VDI Verlag, pp. 135–142.

Kurz, J. L. (1992) Characteristics and applications of industrial thermal mass flow transmitters. Proceedings – Annual Symposium on Instrumentation for the Process Industries (TX, USA): 107–113.

Kutin, J. and Bajsić, I. (1999) Characteristics of the shell-type Coriolis flowmeter. *Journal of Sound and Vibration*, 228(2): 227–242.

Kutin, J. and Bajsić, I. (2001) Stability boundary effect in Coriolis meters. *Journal of Flow Measurement and Instrumentation*, 12 (1): 65–73.

Kutin, J., Hemp, J., Bobovnik, G. and Bajsić, I. (2005a) Weight vector study of velocity profile effects in straight-tube Coriolis flowmeters employing different circumferential modes. *Flow Measurement and Instrumentation*, 16(6): 375–385.

Kutin, J., Bajsić, I., Bobovnik, G. and Hemp, J. (2005b) Modelling and evaluation of velocity profile effects on Coriolis flowmeters. Flomeko 2005 13th International Flow Measurement Conference, Peebles, Scotland, Paper 6.4.

Kutin, J., Bobovnik, G., Hemp, J. and Bajsić, I. (2006) Velocity profile effects in Coriolis mass flowmeters: recent findings and open questions. *Flow Measurement and Instrumentation*, 17(6): 349–358.

Kwietniewski, M. and Miszta-Kruk, K. (2005) Selected methods of flow measurement for the purposes of wastewater networks monitoring. *Proceedings of SPIE – The International Society for Optical Engineering*, 5775: 511–519.

Laan, D. (2012) Ultrasonic Flowmeters, latest developments now used in praxis. International Symposium on 8th International Symposium on Fluid Flow Measurement.

Lai, M-C., Lee, T., Xu, J. S. and Kwak, S. (1991) Inlet flow characterization of thermal mass air flow meters. *SAE Transactions*, 100: 813–819.

Laithier, B. E. and Paidoussis, M. P. (1981) The equations of motion of initially stressed Timoshenko tubular beams conveying fluid. *Journal of Sound and Vibration*, 79, (2): 175–195.

Lamb, H. (1925) *The dynamical theory of sound*. Edward Arnold Ltd (see also Dover Publications Inc, New York 1960).

Laneville, A., Strzelecki, A., Gajan, P. and Hebrard, P. (1993) Signal quality of a vortex flowmeter exposed to swirling flows. *Journal of Flow Measurement & Instrumentation*, 4: 151–154.

Lange, U., Levien, A., Pankratz, T. and Raszillier, H. (1994) Effect of detector masses on calibration of Coriolis flowmeters. *Journal of Flow Measurement and Instrumentation*, 5: 255–262.

Langsholt, M. and Thomassen, D. (1991) The computation of turbulent flow through pipe fittings and the decay of the disturbed flow in a downstream straight pipe. *Journal of Flow Measurement and Instrumentation*, 2: 45–55.

Lansing, J. (2000) Smart monitoring and diagnostics for ultrasonic meters, NSFMW 2000, Gleneagles, Scotland.

(2002a) Dirty vs. clean ultrasonic gas flow meter performance. Proceedings of the Annual ACM-SIAM Symposium on Discrete Algorithms, 989–999.

(2002b) Dirty vs. clean ultrasonic gas flow meters results. 5th International symposium of Fluid Flow Measurement, Arlington, VA, USA.

(2012) How today's gas ultrasonic meter handles compressor pulsations. 8th International Symposium On Fluid Flow Measurement.

Lansing, J. and Mooney, T. (2004) Dirty vs. clean ultrasonic gas flow meter performance. 22nd North Sea Flow Metering Workshop 2004, Paper 2.3.

Lansing, J., Herrmann, V. and Dietz, T. (2007) The relevance of two different path layouts for diagnostic purposes in one ultrasonic meter. 25th International North Sea Flow Measurement Workshop, Energy Institute, Oslo, Norway, 418–434.

Lansing, J., Ehrlich, A. and Dietz, T. (2009) Examination of ultrasonic flow meter in CO2-rich applications. Proc. 7th International Symposium on Fluid Flow Measurement (ISFFM), Anchorage, Alaska, USA.

Lapszewicz, J. A. (1991) Device for measurement of volumetric flow rates of gas mixtures. *Measurement Science & Technology*, 2.8: 815–817.

Laribi, B., Wauters, P. and Aichouni, M. (2003) Experimental study of the decay of swirling turbulent pipe flow and its effect on orifice meter performance. *Proceedings of the ASME Fluids Engineering Division Summer Meeting*, 1: 93–96.

Lashkari, S. and Kruczek, B. (2008) Development of a fully automated soap flowmeter for micro flow measurements. *Flow Measurement and Instrumentation*, 19(6): 397–403.

Laws, E. M. (1990) Flow conditioning – a new development. *Journal of Flow Measurement and Instrumentation*, 1: 165–170.

(1991) A further study of flow through tube bundles. FLUCOME '91, 3rd Triennial International Symposium on Fluid Control, ASME, Measurement and Visualization, San Francisco, Cal USA: 635–641.

Laws, E. M. and Harris, R. (1993) Evaluation of a swirl-vor-tab flow conditioner. *Journal of Flow Measurement and Instrumentation*, 4: 101–108.

Laws, E. M. and Ouazzane, A. (1992) Effect of plate depth on the performance of a Zanker flow straightener. *Journal of Flow Measurement and Instrumentation*, 3: 257–269.

(1995a) A further investigation into flow conditioner design yielding compact installations for orifice plate flow metering. *Journal of Flow Measurement and Instrumentation*, 6: 187–199.

Laws, E. M. and Ouazzane, A. K. (1995b) A further study into the effect of length on the Zanker flow conditioner. *Journal of Flow Measurement and Instrumentation*, 6: 217–224.

(1995c) A preliminary study into the effect of length on the performance of the Etoile flow straightener. *Journal of Flow Measurement and Instrumentation*, 6: 225–233.

Lawson, B. (1988) *Field experience using Coriolis mass meters I*. North Sea Metering Workshop, National Engineering Laboratory, East Kilbride, Scotland.

Le Brusquet, L. and Oksman, J. (1999) Improving the accuracy of tracer flow-measurement techniques by using an inverse-problem approach. *Measurement Science and Technology*, 10: 559–563.

Leder, A. (1996) LDA-measurements in the near wake flow of floats for variable-area flowmeters. FLOMEKO '96 Proceedings of the 8th International Conference on Flow Measurement, Beijing, China: 468–473.

Lee, B., Cheesewright, R. and Clark, C. (2004) The dynamic response of small turbine flowmeters in liquid flows. *Flow Measurement and Instrumentation*, 15(5–6): 239–248.

Lee, W. F. Z. and Evans, H. J. (1965) Density effect and Reynolds number effect on gas turbine flowmeters. *Journal of Basic Engineering, Trans. ASME*: 1043–1057.

Lee, W. F. Z. and Evans, H. J. (1970) A field method of determining gas turbine meter performance. *Journal of Basic Engineering, Trans. ASME*: 724–731.

Lee, W. F. Z. and Karlby, H. (1960) A study of viscosity effect and its compensation on turbinetype flowmeters. *Journal of Basic Engineering*, 1960: 717–727.

Lee, W. F. Z., Kirik, M. J. and Bonner, J. A. (1975) Gas turbine flowmeter measurement of pulsating flow. *Journal of Engineering for Power, Trans ASME*: 531–539.

Lee, W. F. Z., Blakeslee, D. C. and White, R. V. (1982) A self-correcting and self-checking gas turbine meter. *Journal of Fluids Engineering*, 104: 143–148.

Leeungculsatien, T. and Lucas, G.P. (2013) Measurement of velocity profiles in multiphase flow using a multi-electrode electromagnetic flowmeter. *Flow Measurement and Instrumentation*, 31: 86–95.

Leggett, R. B., Borling, D. C., Powers, B. S., Shehata, K., Halvorsen, M. and AboElenain, A. (1996) Multiphase flowmeter successfully measures three-phase flow at extremely high gas column fractions – Gulf of Suez, Egypt, European Petroleum Conference, Milan, Society of Petroleum Engineers, SPE 36837: 215–226.

Legtenberg, R., Bouwstra, S. and Fluitman, J. H. J. (1991) Resonating microbridge mass flow sensor with low-temperature glass-bonded cap wafer. *Sensors & Actuators, A: Physical*, 27.1–3, 723–727.

Lehde, H. and Lang, W T. (1948) AC electromagnetic induction flow meter. US Patent 2,435,043.

Lemon, D. D. (1995) Measuring intake flows in hydroelectric plants with an acoustic scintillation flowmeter. *ASCE Waterpower – Proceedings of the International Conference on Hydropower*, 3: 2039–2048.

Lenn, C. P. and Oddie, G. M. (1990) *The use of ultrasonic methods for monitoring secondary components (solid, liquid and gas) entrained in bulk liquid flows*. International Conference on Basic Principles and industrial Applications of Multiphase Flow, IBC Technical Services Ltd, London.

Letton, W., Svaeren, J. A. and Conort, G. (1997) Topside and subsea experiences with the multiphase flowmeter, Proceedings SPE Annual Technical Conference, 345–357.

Levien, A. and Dudiak, A. (1995) How Coriolis meter design affects field performance. ISA TECH/EXPO Technology Update Conference Proceedings, 1995.

Lewis, D. C. G. (1975) Further development of a low-loss flowmetering device (Epiflo) based on the pressure difference principle. Conference on Fluid Flow Measurement in the Mid-1970s, National Engineering Laboratory, Glasgow, Scotland: Paper J3.

Li, B., Yao, J. and Li, X. (2003) The analysis and application of the rectangular electromagnetic flowmeter. *Conference Record – IEEE Instrumentation and Measurement Technology Conference*, 1: 490–499.

Li, C. and Mickan, B. (2012a) The investigation on the flow characteristic of small MEMS nozzle. Proc. 8th International Symposium on Fluid Flow Measurement (ISFFM).

(2012b) The humidity effect on the calibration of discharge coefficient of sonic nozzle by means of pVTt facility. Proc. 8th International Symposium on Fluid Flow Measurement (ISFFM).

Li, C. H., Peng, X. F. and Wang, C. (2010) Influence of diffuser angle on discharge coefficient of sonic nozzles for flow-rate measurements. *Flow Measurement and Instrumentation*, 21(4): 531–537.

Li, X., Huang, Z.-Y., Wang, B.-L. and Li, H.-Q. (2008a) Using single-phase flowmeters in oil-water two-phase flow measurement. *Kung Cheng Je Wu Li Hsueh Pao/Journal of Engineering Thermophysics*, 29(11): 1872–1874. (in Chinese)

Li, Y., Meng, L., Wang, J. and Geng, Y. (2008b) A new type of wet gas online flow meter based on dual slotted orifice plate. Proceedings of SPIE – The International Society for Optical Engineering, 7127. 7th International Symposium on Instrumentation and Control Technology: Sensors and Instruments, Computer Simulation, and Artificial Intelligence, Beijing, China.

Li, Y., Wang, J. and Geng, Y. (2009) Study on wet gas online flow rate measurement based on dual slotted orifice plate. *Flow Measurement and Instrumentation*, 20(4–5): 168–173.

Li, Y. T. and Lee, S. Y. (1953) A fast-response true-mass-rate flowmeter. *Transactions on ASME*, 75: 835–841.

Li, Z. and Wang, C. (2004) On the standardization of elbow flow meters. 12th International Conference on Flow Measurement FLOMEKO Guilin China.

Lim, J. M., Yoon, B.H., Oh, Y. K. and Park, K-A. (2009) The humidity effect on air flow-rates in a critical flow Venturi nozzle. Proc. 7th International Symposium On Fluid Flow Measurement (ISFFM), Anchorage, Alaska, USA.

Lim, J.-M., Yoon, B.-H., Jpmg, S., Choi, H.-M. and Park, K.-A. (2010) Step-down procedure of sonic nozzle calibration at low Reynolds numbers. *Journal of Flow Measurement and Instrumentation*, 21(3): 340–346.

Lim, J. M., Yoon, B. H., Oh, Y. K. and Park, K.-A. (2011) The humidity effect on air flow rates in a critical flow Venturi nozzle. *Flow Measurement and Instrumentation*, 22(5): 402–405.

Lim, K. W. (2004) An experimental study on the characteristics of oval gear flowmeters. 12th International Conference on Flow Measurement FLOMEKO Guilin China.

(2005) Equipment stability effects on the uncertainty analysis of oil flow standard system. FLOMEKO 2005 13th International Flow Measurement Conference, Peebles, Scotland, Poster Session.

Lin, Y. and Hans, V. (2004) Influence of inclination of bluff body in flowmeters. 12th International Conference on Flow Measurement FLOMEKO Guilin China.

(2006) Improvement of ultrasonic cross-correlation measurement of gas flow by bluff body generated vortices. Metrology for a sustainable development: XVIII IMEKO World Congress, Rio de Janeiro, Brazil.

(2007) Self-monitoring ultrasonic gas flow meter based on vortex and correlation method. *IEEE Transactions on Instrumentation and Measurement*, 56(6): 2420–2424.

Lin, Z. H. (1982) Two-phase flow measurements with sharp-edged orifices. *International Journal of Multiphase Flow*, 8: 683–693.

Lisi, E.L. (1974) Mass flow meter. US Patent 3,785,204, Filed April 6th, 1972, Awarded January 15th, 1974.

Liu, C. Y., Lua, A. C., Chan, W. K. and Wong, Y. W. (1995) Theoretical and experimental investigations of capacitance variable area flowmeter. *Transactions of the Institute of Measurement and Control*, 17.2: 84–89.

Liu, J., Olsson, G. and Mattiasson, B. (2004) A volumetric meter for monitoring of low gas flow rate from laboratory-scale biogas reactors. *Sensors and Actuators, B: Chemical*, 97(2–3): 369–372.

Liu, K. T. and Revus, D. E. (1988) Net-oil computer improves water-cut determination. *Oil and Gas Journal*, Dec.

Liu, K. T., Canfield, D. R. and Conley, J. T. (1986) Application of a mass flow meter for allocation measurement of crude oil production. *SPE Production Engineering*, 3(4): 633–636.

Liu, R. P., Fuent, M. J., Henry, M. P. and Duta, M. D. (2001) A neural network to correct mass flow errors caused by two-phase flow in a digital Coriolis mass flowmeter. *Journal of Flow Measurement and Instrumentation*, 12(1): 53–63.

Liu, Y., Lynnworth, L. C. and Zimmerman, M. A. (1998) Buffer waveguides for flow measurement in hot fluids. *Ultrasonics*, 36: 305–315.

Livelli, G. (2002) DP flow measurement best practices for better plant safety, availability and efficiency. *ISA Monterrey*, 31–40.

Lloyd, K. E., Guthrie, B. D. and Peters, R. J. W. (2002) A flowmeter calibration facility developed at the University of Iowa to evaluate custody transfer steam flowmeters with a cone differential pressure meter used as the metering standard. Joint Power Generator Conference, Phoenix, Arizona, USA, No: IJPGC2002-26029. (See also ISA TECH/EXPO Technology Update Conference Proceedings, 422:383–95, and ISA TECH/EXPO Technology Update Conference Proceedings, 424–425:733–45.)

Løland, T., Sætran, L. R., Olsen, R., Gran, I. R. and Sakariassen, R. (1998) Cavity flow correction for the ultrasonic flowmeter. FLOMEKO '98 Proceedings of the 9th International Conference on Flow Measurement, Lund, Sweden, 127–131.

Lötters, J. C., Lammerink, T. S. J., Pap, M. G., Sanders, R. G. P., de Boer, M. J., Mouris, A. J. and Wiegerink, R. J. (2013) Integrated micro Wobbe index meter towards on-chip energy content measurement. IEEE 26th International Conference on Micro Electro Mechanical Systems, MEMS 2013, 20–24 January 2013, Taipei, Taiwan.

Lovelock, B. G. (2001) Steam flow measurement using alcohol tracers. *Geothermics*, 30(6): 641–654.

Lowell, F. C. and Walsh, J. J. (1991) Performance analysis of multipath acoustic flowmeters under various hydraulic conditions. *Proceedings of the International Conference on Hydropower*, Part 3: 2041–2050.

Lua, A. C. and Zheng, Z. (2003) Numerical simulations and experimental studies on a target fluidic flowmeter. *Flow Measurement and Instrumentation*, 14(1–2): 43–49.

Lucas, G. P. and Turner, J. T. (1985) Influence of cylinder geometry on the quality of its vortex shedding signal. FLOMEKO '85, Paper C4: 81–83.

Lucas, G. P. and Jin, N. D. (2001) Measurement of the homogeneous velocity of inclined oil-in-water flows using a resistance cross correlation flow meter. *Measurement Science and Technology*, 12(9): 1529–1537.

Lunde, P. and Frøysa, K.-E. (2002) Mass and energy measurement of gas using ultrasonic flow meters. Proc. of the 25th Scandinavian Symposium on Physical Acoustics, Norway, 27–30 January 2002.

Lunde, P., Frøysa, K.-E. and Vestrheim, M. (2000) Challenges for improved accuracy and traceability in ultrasonic fiscal flow metering. Proc. of 18th Intern. North Sea Flow Measurem. Workshop, Gleneagles, Scotland.

Lunde, P., Frøysa, K. E. Kippersrud, R. A. and Vestrheim, M. (2003) Transient diffraction effects in ultrasonic meters for volumetric, mass and energy flow measurement of natural gas. 21st North Sea Flow Metering Workshop 2003, Paper 3.

Lunde, P., Vestrheim, M., Bø, R., Smørgrav, S. and Abrahamsen, A. (2005) Reciprocity and its utilization in ultrasonic flowmeter. 23rd North Sea Flow Metering Workshop 2005, Paper 7, pp. 85–112.

Lunde, P., Frøysa, K.-E. and Folkestad, T. (2007) Pressure and temperature effects for Ormen Lange ultrasonic gas flow meters. 25th International North Sea Flow Measurement Workshop, Energy Institute, Oslo, Norway, 97–112.

Lunde, P., Frøysa, K-E., Martinez, V. and Torvanger, Ø. (2008) Pressure and temperature effects for Ormen Lange ultrasonic gas flow meters – results from a follow-up study. 26th International North Sea Flow Measurement Workshop 21–24 October 2008, St Andrews, Scotland, Paper 6.1.

Luntta, E. and Halttunen, J. (1989) Effect of velocity profile on electromagnetic flow measurement. *Sensors & Actuators*, 16.4: 335–344.

(1999) Neural network approach to ultrasonic flow measurements. *Journal of Flow Measurement and Instrumentation*, 10: 33–43.

Luo, S., Liu, Y. and Feng, G. (2002) Application of digital signal processing method in ultrasonic flowmeters for real-time monitoring. *Proceedings of the Second International Symposium on Instrumentation Science and Technology*, 3: 3/673-3/677.

(2003) Digital signal processing implementation of a novel ultrasonic Doppler flowmeter. *Proceedings of the International Symposium on Test and Measurement*, 1: 282–286.

Lupeau, A., Platet, B., Gajan, P., Strzelecki, A., Escande, J. and Couput, J.P. (2007) Influence of the presence of an upstream annular liquid film on the wet gas flow measured by a Venturi in a downward vertical configuration. *Journal of Flow Measurement and Instrumentation*, 18(1): 1–11.

Lüscher, B., Staubli, T., Tresch, T. and Gruber, B. (2007). *Accuracy analysis of the acoustic discharge measurement using analytical, spatial velocity profiles*. Hydro07, Granada.

Lygre, A., Vestrheim, M., Lunde, P. and Berge, V. (1987) Numerical simulation of ultrasonic flowmeters. Proc. of Ultrasonics International 1987, Butterworth Scientific Ltd., Guildford, UK (1987), pp. 196–201.

Lygre, A., Folkestad, T., Sakariassen, R. and Aldal, D. (1992) *A new multi-path ultrasonic flow meter for gas*. North Sea Flow Measurement Workshop, East Kilbride, Scotland.

Lynch, F. and Horciza, E. (1995) Flow measurement using low cost portable clamp-on ultrasonic flowmeters. *ASCE Waterpower – Proceedings of the Int Conf on Hydropower*, 1: 766–773.

Lynnworth, L. (1978) Ultrasonic measuring system for differing flow conditions. US Patent 4,103,551

Lynnworth, L. C. (1988) Buffer rod designs for ultrasonic flowmeters at cryogenic and high temperature, plus and minus 2000C. Proceedings of the 34th International Instrumentation Symposium, Albuquerque, USA, (ISA): 697–702.

(1989) *Ultrasonic Measurements for Process Control: Theory, Techniques, Applications*. Academic Press ISBN 0-12-460585-0.

(1990) Flexural wave externally-attached mass flowmeter for two-phase fluids in small-diameter tubing, 1-mm ID to 16-mm ID. Proceedings IEEE Ultrasonics Symposium, Honolulu: 1557–1562.

(1994) Clamp-on transducers for measuring swirl, cross flow and axial flow. *Proceedings IEEE Ultrasonics Symposium, Cannes, France*: 1317–1321.

Lynnworth, L. (1999) High-temperature flow measurement with wetted and clamp-on ultrasonic sensors. *Sensors* (Peterborough, NH), 16(10): 36, 38, 40–42, 44–46, 48, 50–52.

(2000) Ultrasonic buffer/waveguide. US Patent 6,047,602, issued April 11, 2000.

(2001) Clamp-on flowmeters for fluids. *Sensors* (Peterborough, NH), 18(8): 50–59.

Lynnworth, L. C., Hallewell, G. D. and Bragg, M. I. (1994) One-port profiler. FLOMEKO '94 Flow Measurement in the Mid-90s, NEL, Glasgow, Scotland: Paper 7.3.

Lynnworth, L. C., Nguyen, T. H., Smart, C. D. and Khrakovsky, O. A. (1997) Acoustically isolated paired air transducers for 50-, 100-, 200-, or 500-kHz applications. *IEEE Transactions on Ultrasonics, Ferroelectrics, and Frequency Control*, 44: 1087–1100.

Lynnworth, L. C., Cohen, R., Rose, J. L., Kim, J. O. and Furlong, E. R. (2006) Vortex shedder fluid flow sensor. *IEEE Sensors Journal*, 6(6): 1488–1496.

Lysak, P. D., Jenkins, D. M., Capone, D. E. and Brown, W. L. (2008a) Analytical model of an ultrasonic cross-correlation flow meter, part 1: Stochastic modelling of turbulence. *Flow Measurement and Instrumentation*, 19(1): 1–7.

(2008b) Analytical model of an ultrasonic cross-correlation flow meter, part 2: Application. *Flow Measurement and Instrumentation*, 19(1): 41–46.

Ma, Y. and Eidenschink, T. (2001) Motion induced signals of Coriolis flowmeters. *Journal of Flow Measurement and Instrumentation*, 12(3): 213–217.

Macdonald, G. A. (1983) *A vibrating vane mass flowmeter*. 1st European Conf on Sensors and their Applications, UMIST, Manchester: 58–59.

MacMillan, (1954) Viscosity effects on Pitot tubes at low speeds. *Journal of Royal Aeronautical Society,*, 58: 570–572.

(1957) Experiments on Pitot tubes in shear flow. *Aero. Res. Council, Tech. Rpt., R&M 3028*. 58: 570–572.

Maginnis, T. O. (2002) Dynamical effects in expanding volume flow calibrators. 5th International symposium of Fluid Flow Measurement, Arlington, VA, USA.

Mahadeva, D. V. (2009) Studies of the accuracy of clamp-on ultrasonic flowmeters. PhD Thesis, University of Cambridge.

Mahadeva, D. V., Baker, R.C. and Woodhouse, J. (2008) Studies of the accuracy of clamp-on transit time ultrasonic flowmeters. I²MTC 2008 – IEEE International Instrumentation and Measurement Technology Conference, Victoria, Vancouver Island, Canada, 12–15 May.

(2009) Further studies of the accuracy of clamp-on transit-time ultrasonic flowmeters for liquids. *IEEE Transactions on Instrumentation and Measurement*, 58(5): 1602–1609.

Mahadeva, D. V., Huang, S. M., Oddie, G. and Baker, R. C. (2010) Study of the effect of beam spreading on systemic Doppler flow measurement errors. IEEE International Ultrasonics

Symposium (IUS), Sponsored by the IEEE Ultrasonics, Ferroelectrics, and Frequency Control SocietySan Diego, California, 11–14 October 2010.

Mahulikar, S. P. and Sane, S. K. (2005) Theoretical analysis of experimentally observed perplexing calibration characteristics of ball-in-vortex flow-meter. *Journal of Fluids Engineering, Transactions of the ASME*, 127(5): 1021–1028.

Mainardi, H., Barriol, R. and Panday, P. K. (1977) Pulsating duct flow in the presence of an orifice plate. *International Journal of Mechanical Sciences*, 19: 533–546.

Mair, W. A. (1965) The effect of a rear mounted disc on the drag of a blunt body of revolution. *Aeronautical Quarterly*, 16: 350–360.

Majeed, G. H. A. and Aswad, Z. A. A. (1989) A new approach for estimating the orifice discharge coefficient required in Ashford-Pierce correlations. Multiphase Flow – Proc 4th Int Conf, BHRA, Cranfield: 235–255.

Majumdar, A. S. and Gulek, M. (1981) Vortex shedding from single and compound prisms of various configurations. ASME Paper No. 81-WA/FE-6.

Malard, L., Wisnoe, W., Strzelecki, A., Gajan, P. and Hebrard, P. (1991) Air visualizations and flow measurements applied to the study of a vortex flowmeter: influence of grid turbulence and acoustical effects. FLUCOME '91, 3rd Triennial International Symposium on Fluid Control, Measurement and Visualization, San Francisco, Cal, USA: 689–695.

Malinowski, L. and Rup, K. (2008) Measurement of the fluid flow rate with use of an elbow with oval cross section. *Flow Measurement and Instrumentation*, 19(6): 358–363.

Mandard, E., Kouam, D., Battault, R., Remenieras, J-P. and Patat, F. (2008) Methodology for developing a high-precision ultrasound flow meter and fluid velocity profile reconstruction. *IEEE Transactions on Ultrasonics, Ferroelectrics, and Frequency Control*, 55(1): 161–171.

Mandrup-Jensen, L. (1990) *Testing Coriolis mass flowmeters for pattern approval.* North Sea Flow Measurement Workshop, National Engineering Laboratory, Scotland.

Mankin, P. A. (1955) Measurement of liquid flow by positive displacement meters. *Journal of the Southern California Meter Association, Instruments and Automation*: 453–457.

Manshoor, B., Nicolleau, F. C. G. A. and Beck, S. N. M. (2011) The fractal flow conditioner for orifice plate flow meters. *Flow Measurement and Instrumentation*, 22(3): 208–214.

Marfenko, I., Yeh, T. T. and Wright, J. (2006) Diverter uncertainty less than 0.01% for water flow calibrations. Proc. 6th International Symposium on Fluid Flow Measurement (ISFFM), Querétero, Mexico.

Marić, I. (2005) The Joule-Thomson effect in natural gas flow-rate measurement. *Flow Measurement and Instrumentation*, 16(6): 387–395.

(2007) A procedure for the calculation of the natural gas molar heat capacity, the isentropic exponent, and the Joule-Thomson coefficient. *Journal of Flow Measurement and Instrumentation*, 18(1): 18–26.

Marić, I. and Ivek, I. (2010) Compensation for Joule-Thomson effect in flowrate measurement by GMDH polynomial. *Flow Measurement and Instrumentation*, 21(2): 134–142.

Marić, I., Galović, A. and Šmuc, T. (2005) Calculation of natural gas isentropic exponent. *Flow Measurement and Instrumentation*, 16(1): 13–20.

Mark, P. A., Sproston, J. L. and Johnson, M. W. (1990a) Theoretical and experimental studies of two-phase flows in turbine meters. International Conference on Basic Principles and Industrial Applications of Multiphase Flow, IBC Technical Services Ltd, London.

Mark, P. A., Johnson, M. W., Sproston, J. L. and Millington, B. C. (1990b) The turbine meter applied to void fraction determination in two-phase flow. *Flow Measurement and Instrumentation*, 1: 246–252.

Markoja, B. (2011) CIDRA SONARtrac flow meters: an alternative flow measurement technology. Committee on Operation and Maintenance of Nuclear Power Plant / ISTOG Winter Meeting, 5–9 December 2011, Clearwater, Florida, USA.

Maron, R. and O'Keefe, C. (2007) Application of non-intrusive sonar technology in hydrotransport. FLUIMIN IV Taller de Concentaductos, Mineroductos Y Relaveductos, 18–19 October, Vina del Mar, Chile.

Marshad, A. H. and Irvine-Halliday, D. (1994) Intensity-modulated optical-fibre vortex-shedding flowmeter. *Canadian Journal of Electrical and Computing Engineering*, 119: 75–79.
Marshall, Rebekkah (2004) Accurate flow measurement, pure and simple. *Chemical Engineering*, 111(4): 19–22.
Martin, J. J. (1949) Calibration of rotameters. *Chemical Engineering. Progress*, 45: 338.
Martin, P. (2009) Realistic pipe prover volume uncertainty. 27th International North Sea Flow Measurement Workshop, Tonsberg, Norway, 20–23 October 2009, pp. 412–437.
Martinelli, M. and Viktorov, V. (2011) A mini fluidic oscillating flowmeter. *Flow Measurement and Instrumentation*, 22(6): 537–543.
Martinsson, E. and Delsing, J. (2009) Environmental tests of spark discharge emitter for use in ultrasonic gas flow measurements, Proc. 7th International Symposium on Fluid Flow Measurement (ISFFM), Anchorage, Alaska, USA.
(2010) Electric spark discharge as an ultrasonic generator in flow measurement situations. *Journal of Flow Measurement and Instrumentation*, 21(3): 394–401.
Masri, S., Lin, W.-T. and Su, C.-M. (2010) New primary low-pressure gas flow standard at NIMT. 15th Flow Measurement Conference (FLOMEKO), 13–15 October 2010, Taipei, Taiwan, Paper A5-3.
Matson, J., Marioano, C. F., Khrakovsky, O. and Lynnworth, L. (2002) Ultrasonic mass flowmeters using clamp-on or wetted transducers. 5th International symposium of Fluid Flow Measurement, Arlington, VA, USA.
Matsunaga, Y., Goto, S., Kuromori, K. and Ostling, H. (1988) New intelligent magnetic flowmeter with dual frequency excitation. Proceedings of the ISA/88 International Conference and Exhibition: Advances in Instrumentation, ISA, 43, Part 3: 1259–1267.
Matsunaga, Y., Takahashi, S. and Kuromori, K. (1990) Numerical analysis of a vortex flowmeter and comparison with experiment. *Journal of Flow Measurement & Instrumentation*, 1: 106–112.
Mattar, L., Nicholson, M., Aziz, K. and Gregory, G. A. (1979) Orifice metering of two-phase flow. *Journal of Petroleum Technology, August:*: 955–961.
Mattar, W. (2003) Coriolis metering in difficult industrial applications. *IEE Computing and Control Engineering*, 14(4): 44–45.
Mattar, W. M. (2005) Advances in Coriolis technology resolve tough pipeline flow measurement challenges. *Pipeline and Gas Journal*, 232(7): 35–36.
Matter, D., Kleiner, T., Kramer, B. and Sabbattini, B. (2003) Microsensor-based gas flow meter wins innovation prize. *ABB Review*, (3): 49–50.
Matthews, A. J. and Ayling, C. L. (1992) *Compact large bore direct mass flow meters*. North Sea Flow Measurement Workshop, East Kilbride, Scotland.
Mattingly, G. E. (1982) *Primary calibrators, reference and transfer standards*. Developments in Flow Measurement-1, ed. R. W. W. Scott, London: Applied Science Publishers: 31–71.
(1990/1991) Fluid flowrate metrology: Laboratory uncertainties and traceabilities. *Advanced Techniques for Integrated Circuit Processing, Int Soc Optical Eng, Proceedings of SPIE*, 1392: 386–401.
(2009) Improved meter performance characterizations for liquid and gas turbine meters. Proc. 7th International Symposium on Fluid Flow Measurement (ISFFM), Anchorage, Alaska, USA.
Mattingly, G. E. and Yeh, T. T. (1991) Effects of pipe elbows and tube bundles on selected types of flowmeters. *Journal of Flow Measurement and Instrumentation*, 2: 4–13.
Mattingly, G. E., Pontius, P. E., Allion, H. H. and Moore, E. F. (1977) Laboratory study of turbine meter uncertainty. Proceedings of the Symposium on Flow Measurement in Open Channels and Closed Conduits, NBS, Gaithersburg, Md, USA: 33–54.
Mattingly, G. E., Pursley, W. C., Paton, R. and Spencer, E. A. (1978) Steps towards an ideal transfer standard for flow measurement. FLOMEKO Symposioum on Flow, Groningen, The Netherlands, 543–552.
Mattingly, G. E., Yeh, T. T., Robertson, B. and Kothari, K. M. (1987) *NBS research on 'in situ' flowmeter calibrations*. AGA Distribution Transmission Congress, Las Vegas, USA.

McBrien, R. K. (1997) High pressure pulsation effects on orifice meters. ASME Fluids Engineering Division Summer Meeting FEDSM'97, Paper 3700.

McBrien, R. and Geerlings, J. (2005) The performance of a multipath, 8-inch ultrasonic meter in pulsating flow. FLOMEKO 2005 13th International Flow Measurement Conference, Peebles, Scotland, Paper 1.4.

McCarthy, R. (1996) Five-path ultrasonic flowmeter completes one-year field trial. *Pipe Line & Gas Industry*, 79 (4): pp unknown.

McDonald, B. E. and Sui, L. (2013) Ultrasonic flow measurement with integrated temperature measurement compensation. FLOMEKO 2013, 16th International Flow Measurement Conference, Paris, France.

McFaddin, S. E., Sindt, C. F. and Brennan, J. A. (1989) The effect of the location of an in-line tube bundle on orifice flowmeter performance. *Journal of Flow Measurement and Instrumentation*, 1: 9–14.

McKee, R. J. (1992) Pulsation effects on single- and two-rotor turbine meters. *Journal of Flow Measurement and Instrumentation*, 3: 151–166.

McKenzie, G. (1989) The performance of direct mass flow coriolis meters used for fiscal measurement of high value fluids. *Mass Flow Measurement Direct and Indirect, Proc Int Conf Mass Flow Measurement*, IBC Pbl, London, England.

McQuien, G. E., Poplawski, J., O'Keefe, C., Maron, R. and Rothman, P. (2011) Passive sonar flow monitoring. 26th Annual Phosphate Conference, Lakeland Civic Center, Lakeland, Florida, 12–13 October 2011.

Medlock, R. S. (1976) The vortex flowmeter – its development and characteristics. *Austral. J. Instrum. Control*, 24: 24–32.

(1986) The historical development of flow metering. *Measurement & Control*, 19: 11–22.

Medlock, R. (1989) A review of the techniques of mass flow measurement. Cranfield Short Course Lecture. See also Mass Flow Measurement Direct and Indirect, Proc Int Conf Mass Flow Measurement, IBC Pbl, London, England.

Medlock, R. and Furness, R. A. (1990) Mass flow measurement – a state of the art review. *Measurement and Control*, 23: 100–112.

Melick, T. and Robinson, A. D. (2006) The latest technology in air flow measurement for the cement industry. IEEE Cement Industry Technical Conference, Phoenix, AZ, United States, 255–267.

Mencke, D. (1989) Pattern approval of mass flowmeters. Mass Flow Measurement Direct and Indirect, Proc Int Conf Mass Flow Measurement, IBC Pbl, London, England.

(1996) Use of Coriolis mass flowmeters in custody transfer. FLOMEKO'96 Proceedings of the 8th International Conference on Flow Measurement, Beijing, China: 232–237.

Menendez, A., Biscarri, F. and Gomez, A. (1998) Balance equations estimation with bad measurements detection in a water supply net. *Journal of Flow Measurement and Instrumentation*, 9: 193–198.

Meng, L., Li, Y., Zhang, J. and Dong, S. (2011) The development of a multiphase meter without separation based on sloped open channel dynamics. *Journal of Flow Measurement and Instrumentation*, 22(2): 120–125.

Meng, T. and Wang, C. (2004) Comparison of pVTt methods gas flow prover. 12th International Conference on Flow Measurement FLOMEKO Guilin China.

Meng, X-j., Li, S-f. and Li, Z. (2010a) The CFD simulation and experimental research of the V type elbow flowmeter. 15th Flow Measurement Conference (FLOMEKO), 13–15 October 2010, Taipei, Taiwan, Paper B9-5.

Meng, Z., Huang, Z., Wang, B., Ji, H., Li, H. and Yan, Y. (2010b) Air-water two-phase flow measurement using a Venturi meter and an electrical resistance tomography sensor. *Journal of Flow Measurement and Instrumentation*, 21(3): 268–276.

Meribout, M., Al-Rawahi N., Al-Naamany, A., Al-Bimani, A., Al-Busaidi, K. and Meribout, A. (2010) Integration of impedance measurements with acoustic measurements for accurate two phase flow metering in case of high water-cut. *Journal of Flow Measurement and Instrumentation*, 21(1): 8–19.

Merritt, R. (2001) Turbine flowmeters vs. hydrogen sulphide. *Control (Chicago, Ill)*, 14(7): 55–56.
Merzkirch, W. (1999) Special durchfiussmessung. Sinnvoll oder nicht? Stromungsgleichrichter fur die durchflussmessung (Translation: Flow measurement special. Worthwhile or not? Flow straightener for flow rate measurement). Chemie-Technik (Heidelberg), 28(8): S22–S23 (in German).
(2005) (Ed.) *Fluid mechanics of flow metering*. Springer-Verlag.
Messer, M. and Aidun, C. K. (2009) Main effects on the accuracy of Pulsed-Ultrasound-Doppler-Velocimetry in the presence of rigid impermeable walls. *Flow Measurement and Instrumentation*, 20(2): 85–94.
Mi, Y., Ishii, M. and Tsoukalas, L. H. (2001a) Investigation of vertical slug flow with advanced two-phase flow instrumentation. *Nuclear Engineering and Design*, 204(1–3): 69–85.
(2001b) Flow regime identification methodology with neural networks and two-phase flow models. *Nuclear Engineering and Design*, 204(1–3): 87–100.
Miau, J. J. and Hsu, M. T. (1992) Axisymmetric-type vortex shedders for vortex flowmeters. *Journal of Flow Measurement & Instrumentation*, 3: 73–80.
Miau, J. J., Yang, C. C., Chou, J. H. and Lee, K. R. (1993) A T-shaped vortex shedder for a vortex flowmeter. *Journal of Flow Measurement & Instrumentation*, 4: 259–268.
Miau, J. J., Chen, Y. S., Chou, J. H. and Hsieh, W. D. (1997) Effect of flow swirling on a vortex flowmeter, ASME Fluids Engineering Division Summer Meeting FEDSM'97, Paper 3018.
Miau, J. J., Hu, C. C. and Chou, J. H. (2000) Response of a vortex flowmeter to impulsive vibrations. *Flow Measurement and Instrumentation*, 11(1): 41–49.
Miau, J. J., Wu, C. W., Hu, C. C. and Chou, J. H. (2002) A study on signal quality of a vortex flowmeter downstream of two elbows out-of-plane. *Flow Measurement and Instrumentation*, 13(3): 75–85.
Miau, J. J., Yeh, C. F., Hu, C. C. and Chou, J. H. (2004) On measurement uncertainty of a vortex flowmeter. 12th International Conference on Flow Measurement FLOMEKO Guilin China.
Miau, J. J.; Yeh, C. F., Hu, C. C. and Chou, J. H. (2005) On measurement uncertainty of a vortex flowmeter. *Flow Measurement and Instrumentation*, 16(6): 397–404.
Mickan, B. and Kramer, R. (2009) Evaluation of two new volumetric primary standards for gas volume established by PTB. Proc. 7th International Symposium on Fluid Flow Measurement (ISFFM), Anchorage, Alaska, USA.
Mickan, B., Wendt, G., Kramer, R. and Dopheide, D. (1996a) Systematic investigation of flow profiles in pipes and their effects on gas meter behaviour. *Measurement*, 22: 1–14.
(1996b) Systematic investigation of pipe flows and installation effects using laser Doppler anemometry – Part II The effect of disturbed flow profiles on turbine gas meters – a describing empirical model. *Flow Measurement and Instrumentation*, 7(3/4): 151–160.
Mickan, B., Kramer, R., Hans-Hotze, J. and Dopheide, D. (2002) Pigsar – the extended test facility and new German national primary standard for high pressure natural gas. 5th International symposium of Fluid Flow Measurement, Arlington, VA, USA.
Mickan, B., Kramer, R., Kiesewetter, P. and Dopheide, D. (2004) Determination of discharge coefficients of sonic nozzles obtaining low uncertainty without knowledge of throat diameter. 12th International Conference on Flow Measurement FLOMEKO Guilin China.
Mickan, B., Kramer, R. and Dopheide, D. (2006a) Determination of discharge coefficient of critical nozzles based on their geometry and the theory of laminar and turbulent boundary layers. Proc. 6th International Symposium On Fluid Flow Measurement (ISFFM), Querétero, Mexico.
(2006b) The use of micro-nozzles under sonic and subsonic conditions with various gases. Proc. 6th International Symposium On Fluid Flow Measurement (ISFFM), Querétero, Mexico.
(2006c) Comparisons by PTB, NIST and LNE-LADG in air and natural gas with critical Venturi nozzles agree within 0.05%. Proc. 6th ISFFM
Mickan. B., Kramer, R., Vieth, D. and Hinze, H. M. (2007) The use of sonic nozzles under high pressure conditions for scaling the traceability from high flow rates down to low flow rates

and for the link of the volumetric primary references in low and high pressure at PTB. Proc. FLOMEKO 2007.

Mickan, B., Kramer, R., Müller H., Strunck, V., Vieth, D. and Hinze H.-M. (2009) Highest precision for gas meter calibration worldwide: the high pressure gas calibration facility pigsarTM with optimized uncertainty. Proc. 7th International Symposium On Fluid Flow Measurement (ISFFM), Anchorage, Alaska, USA.

Mickan, B., Kramer, K., Strunck, V. and Dietz, T. (2010) Transient response of turbine flow meters during the application at a high pressure piston prover. 15th Flow Measurement Conference (FLOMEKO), 13–15 October 2010, Taipei, Taiwan, Paper A3-3.

Mickan, B., Kramer R. and Li, C. (2012) The critical back pressure ratio of sonic nozzles – the correlation with diffuser geometry and gas composition. 8th International Symposium on Fluid Flow Measurement (ISFFM), Colorado, USA, 20–22 June 2012.

MID (2004) DIRECTIVE on measuring instruments. Directive 2004/22/EC of the European Parliament and of the Council of 31 March 2004.

Millen, M. J., Sowerby, B. D., Coghill, P. J., Ticker, J. R., Kingsley, R. and Grima, C. (2000) Plant tests of an on-line multiple-pipe pulverised coal mass flow measuring system. *Journal of Flow Measurement and Instrumentation*, 11(3): 153–158.

Miller, D. S. (1990) *Internal flow systems*. 2nd ed., Gulf Publishing Company, London.

Miller, G. and Belshaw, B, (2008) An investigation into the performance of Coriolis and ultrasonic meters at liquid viscosities up to 300 cSt. 26th International North Sea Flow Measurement Workshop, 21–24 October 2008, St Andrews, Scotland, Paper 1.4.

Miller, R. and Hanks, E. (2015) Gas ultrasonic meter installation effects and diagnostic indicators "A history of NAFFMC installation effects testing including current testing". 9th International symposium of Fluid Flow Measurement, Arlington, VA, USA.

Miller, R. W. (1996) *Flow measurement engineering handbook*. 3rd ed, McGraw-Hill: New York, USA.

Miller, R. W., DeCarlo, J. P. and Cullen, J. T. (1977) A vortex flowmeter – calibration results and application experiences. *NBS Special Publication 484*, 2: 549–570.

Millington, B. C. and King, N. W. (1988) Further developments of a jet mixer/turbine meter package for the measurement of gas-liquid mixtures. FLUCOME'88 2nd Int Symp on Fluid-Control Measurement Mechanics and Flow Visualization, Sheffield, UK: 474–478.

Millington, B. C., Adams, C. W. and King, N. W. (1986) The effect of upstream installation conditions on the performance of small liquid turbine meters. *International Symposium on Fluid Flow Measurement, AGA, Washington.*

Millington, B. C., Frantzen, K. and Marshall, M. (1993) *The performance of the Fluenta MPFM 900 Phase Fraction Meter.* North Sea Flow Measurement Workshop, Bergen, Norway.

Mills, C. J. (1966) Electromagnetic catheter-tip probe for blood flow measurement (approximate title). *Phys. Med. Biol Physics in Medicine and Biology.*, 11: 323–324.

Minchenya, V., Karcher, C., Kolesnokov, Y. and Thess, A. (2011) Calibration of the Lorentz force flowmeter. *Flow Measurement and Instrumentation*, 22(3): 242–247.

Minemura, K., Egashira, K., Ihara, M., Furuta, H. and Yamamoto, K. (1996) Simultaneous measurement method for volumetric flow rates of both phases of air-water mixture using a turbine flowmeter. *Transactions of the Japan Society of Mechanical Engineers, Part B*, 62(593): 122–129.

Minkin, H. L., Hobart, H. F. and Warshawsky, I. (1966) Performance of turbine type flowmeter in liquid hydrocarbons, NASA TN D-3770.

Mohamed, P. G. and Al-Saif, K. H. (1998) Field trial of a multiphase flowmeter, Society of Petroleum Engineers Annual Technical Conference and Exhibition, New Orleans, SPE 49161 (also in synopsis in 1998 *Journal of Petroleum Technology*, 50: 74–75).

Mohamed, P. G., Al-Saif, K. H. and Mohamed, H. (1999) Field evaluations of different multiphase flow measurement systems, Society of Petroleum Engineers Annual Technical Conference and Exhibition, Houston, TX, USA, SPE 56643 1(P): 553–561.

Mokhtarzadeh-Dehghan, M. R. and Stephens, D. J. (1998) A numerical study of turbulent flow through a variable area orifice meter. *International Journal of Computer Applications in Technology*, 11: 271–280.

Moore, P. and Brown, G. J. (2000) Modelling of transit time ultrasonic flowmeters in theoretical asymmetric flow. FLOMEKO'2000 the 10th International Conference on Flow Measurement, Salvador, Brazil: Paper B11.

Moore, P. I., Brown, G. J. and Stimpson, B. P. (2000) Ultrasonic transit-time flowmeters modeled with theoretical velocity profiles: methodology. *Measurement Science and Technology*, 11(2000): 1802–1811.

Moore, P. I., Johnson, A. N. and Espina, P. I. (2002) Simulations of ultrasonic transit time in a fully developed turbulent flow using a ray-tracing method. In Proceeding of the 22nd North Sea Flow Measurement Workshop.

Moreau, J. (2000) Multiphase flow measurement. *Journal of Offshore Technology*, 8(1): 4 pp.

Morgan, D. V. and Aliyu, Y. H. (1993) An ionic flowmeter for measuring small rates of gas flow. *Measurement Science and Technology*, 4: 1479–1483.

Mori, M., Tezuka, K. and Takeda, Y. (2006) Effects of inner surface roughness and asymmetric pipe flow on accuracy of profile factor for ultrasonic flow meter. Fourteenth International Conference on Nuclear Engineering 2006, ICONE 14, Miami, FL, United States.

Morita, A. and Yoshimura, H. (1996) Method of measuring phase difference in Coriolis mass flowmeter. *Proceedings of the International Conference on Advances in Instrumentation and Control*, 51(1): 631–640.

Moriyama, T., Sukemura, N. and Morishita, K. (2001) Cross correlation mass flowmeter using pulse heating method. *Proceedings of the SICE Annual Conference*, 345–350.

Morris, S. C., Neal, D. R., Foss, J. F. and Cloud, G. L. (2001) A moment-of-momentum flux mass air flow measurement device. *Measurement Science and Technology*, 12(2): N9–N13.

Morrison, G. L. (1997) Flow field development downstream of two in plane elbows. ASME Fluids Engineering Division Summer Meeting FEDSM'97, Paper 3021.

Morrison, G. L. and Brar P. (2005) Ambient temperature effects upon the flow in gas pipelines at low speeds. Paper no. FEDSM2005-77478 Volume 2 pp. 475–482 (8 pages) ASME 2005 Fluids Engineering Division Summer Meeting (FEDSM2005) ISBN: 0-7918-4199-5, 19–23 June 2005, Houston, Texas, USA.

Morrison, G. L. and Hall, K. R. (2000) Consider slotted orifice flowmeters. *Hydrocarbon Processing*, 79(12): 65–66, 68–72.

Morrison, G. L., DeOtte, R. E., Moen, M., Hall, K. R. and Holste, J. C. (1990a) Beta ratio, swirl and Reynolds number dependence of wall pressure in orifice flowmeters. *Journal of Flow Measurement and Instrumentation*, 1: 269–277.

Morrison, G. L., DeOtte, R. E., Panak, D. L. and Nail, G. H. (1990b) Flow field inside an orifice flow meter. *Chemical Engineering Progress*, 86.7: 75–80.

Morrison, G. L., DeOtte, R. E. and Beam, E. J. (1992) Installation effects upon orifice flowmeters. *Journal of Flow Measurement and Instrumentation*, 3: 89–94.

Morrison, G. L., Hall, K. R., Holste, J. C., DeOtte Jr, R. E., Macek, M. L. and Ihfe, L. M. (Oct 1994a) Slotted orifice flowmeter. *AIChE Journal*. 40.10: 1757–1760.

Morrison, G. L., Hall, K., Holste, J. C., Macek, M., Ihfe, L. and DeOtte, R. E. (1994b) Comparison of orifice and slotted plate flow meters. *Journal of Flow Measurement and Instrumentation*, 5: 71–77.

Morrison, G. L., Hauglie, J. and DeOtte, R. E. (1995) Beta ratio, axisymmetric flow distortion and swirl effects upon orifice flow meters. *Journal of Flow Measurement and Instrumentation*, 6: 207–216.

Morrison, G. L., Hall, K. R., Holste, J. C., Ihfe, L., Gaharan, C. and DeOtte, Jr, R. E. (1997) Flow development downstream of a standard tube bundle and three different porous plate flow conditioners. *Journal of Flow Measurement and Instrumentation*, 8(2): 61–76.

Morrison, G. L., Terracina, D., Brewer, C. and Hall, K. R. (2001) Response of a slotted orifice meter to an air/water mixture. *Journal of Flow Measurement and Instrumentation*, 12(3): 175–180.

Morrison, G. L., Hall, K. R. and Flores, A. E. (2002a) Slotted orifice based two phase flow meter. *Proceedings of the Annual Symposium on Instrumentation for the Process Industries*, 57: 73–99.

Morrison, G. L., Hall, K. R., Brewer, C. and Flores, A. (2002b) Universal slotted orifice flow meter flow coefficient equation for single and two phase flow. 5th International symposium of Fluid Flow Measurement, Arlington, VA, USA.

Morrow, T. B. (1996) Orifice meter installation effects: ten-inch sliding flow conditioner tests. GRI Report No. GRI-96/0391, Gas Research Institute, Chicago, Illinois, USA.

(1997) Effects of flow conditioners on orifice meter installation errors, ASME Fluids Engineering Division Summer Meeting FEDSM'97, Paper 3006.

(2004)Gravimetric calibration of critical flow Venturi nozzles. Proc. HT-FED04 Paper No. 56817.

(2005) Multi-path gas ultrasonic flow meter performance at low velocity. Proceedings of 2005 ASME Fluids Engineering Division Summer Meeting, FEDSM2005, 2005: 2207–2211. (2:447–52).

Morrow, T. B., Park, J. T. and McKee, R. J. (1991) Determination of installation effects for a 100 mm orifice meter using a sliding vane technique. *Journal of Flow Measurement and Instrumentation*, 2: 14–20.

Morrow, T. D. and Behring, K. A. (1999) Energy flow measurement technology, and the promise of reduced operating costs. Proc. of 4th International. Symposium. on Fluid Flow Measurem., Denver, Colorado, 27–30 June 1999.

Morse, P. M. and Ingard, K. U. (1968) *Theoretical acoustics*. McGraw Hill Book Company, New York.

Morton, C. E. (2009) Performance and modelling of the oscillating piston flowmeter. PhD Thesis, Engineering Department, University of Cambridge.

Morton, C. E., Baker, R. C. and Hutchings, I. M. (2011) Measurement of liquid film thickness by optical fluorescence and its application to an oscillating piston positive displacement flowmeter. *Measurement Science and Technology*, 22: 125403 (11pp).

Morton, C. E., Hutchings, I. M. and Baker, R. C. (2014a) Experimental tests of a positive displacement flowmeter: I – piston movement and pressure losses. *Flow Measurement and Instrumentation*, 36(2014): 47–56. http://dx.doi.org/10.1016/j.flowmeasinst.2014.01.006

Morton, C. E., Baker, R. C. and Hutchings, I. M. (2014b) Experimental tests of an oscillating circular piston positive displacement flowmeter: II – leakage flows and wear tests. *Flow Measurement and Instrumentation*, 36(2014): 57–63. http://dx.doi.org/10.1016/j.flowmeasinst.2014.01.007

Motta, R. S. N., Schmedt, R. and Souza, L. E. (2011) Enhanced pulverized coal mass flow measurement. *Flow Measurement and Instrumentation*, 22(4): 303–308.

Mottram, R. C. (1981), Measuring pulsating flow with a differential pressure meter. Proc Conf Flow 81 – Its Measurement and Control in Science and Industry, St Louis, Mo, USA, ISA, 2: 347–361.

(1989) Damping criteria for pulsating gas flow measurement. *Journal of Flow Measurement and Instrumentation*, 1: 15–23.

(1991) Vortex flowmeters – installation effects. *Journal of Flow Measurement & Instrumentation*, 2: 56–60.

(1992) Introduction: an overview of pulsating flow measurement. *Journal of Flow Measurement and Instrumentation*, 3: 114–117.

Mottram, R. C. and Hutton, S. P. (1987) Installation effects turbine and vortex flowmeters. FLOMIC Report No 3, Flow Measurement and Instrumentation Consortium.

Mottram, R. C. and Rawat, M. S. (1986) The swirl damping properties of pipe roughness and the implications for orifice meter installation. International Conference on Flow Measurement in the Mid 80's, 9–12 June. Glasgow, NEL.

(1988) Installation effects on vortex flowmeters. *Journal of Measurement & Control*, 21: 241–246.

Mottram, R. C. and Ting, V. C. (1992) *Presentation at 1992 AIChe Spring National Meeting*, New Orleans, IA.

Mudd, J. and Bentley, J. (2002) The development of a multi-channel vortex flow-meter using a twelve-sensor array. *Measurement and Control*, 35(10): 296–298.

Müller, H., Strunck, V., Kramer, R., Mickan, B., Dopheide, D. and Hotze, H.-J. (2004) Germany's new optical primary national standard for natural gas of high pressure at pigsarTM 12th International Conference on Flow Measurement FLOMEKO Guilin China.

Murakami, M., Maruo, K. and Yoshiki, T. (1990) Development of an electromagnetic flowmeter for studying gas-liquid, two-phase flow. *International Chemical Engineering*, 30.4: 699–702.

Murakawa, H., Kikura, H. and Aritomi, M. (2005) Application of ultrasonic Doppler method for bubbly flow measurement using two ultrasonic frequencies). *Experimental Thermal and Fluid Science*, 29(7 SPEC. ISS): 843–850.

(2008) Application of ultrasonic multi-wave method for two-phase bubbly and slug flows. *Flow Measurement and Instrumentation*, 19(3–4): 205–213.

Murakawa, H., Sugimoto, K. and Takenaka, N. (2014) Effects of the number of pulse repetitions and noise on the velocity data from the ultrasonic pulsed Doppler method with different algorithms. *Flow Measurement and Instrumentation*, 40: 9–18.

Murdock, J. W. (1961) Two-phase flow measurement with orifice. ASME Paper 61-GT-27.

Murugesan, K. (2002) Multiphase flow meter: Trends in well performance testing. *Chemical Engineering World*, 37(12): 151–153.

Mus, E. A., Toskey, E. D., Norris, R. J. and Bascoul, S. J. F. (2002) Added value of a multiphase flowmeter in exploration well testing. *SPE Production and Facilities*, 17(4): 197–203.

Music, M., Ahic-Djokic, M., Music, O. and Djemic, Z. (2004) An approximate mathematical model of ultrasound wave modulated by von Karman vortex street. 12th International Conference on Flow Measurement FLOMEKO Guilin China.

Myhr, S. (1991) Field experience with Coriolis mass meter on hydrocarbon liquid. Proceedings of the North Sea Flow Measurement Workshop.

Mylvaganam, K. S. (1989) High-rangeability ultrasonic gas flowmeter for monitoring flare gas. *IEEE Transactions on. Ultrasonics, Ferroelectrics & Frequency Control*, 36.2: 144–149.

Na, M. G., Shin, S. H. and Jung, D. W. (2005a) Design of a software sensor for feed water flow measurement using a fuzzy inference system. *Nuclear Technology*, 150(3): 293–302.

Na, M. G., Lee, Y. J. and Hwang, I. J. (2005b) A smart software sensor for feed water flow measurement monitoring. *IEEE Transactions on Nuclear Science*, 52(6): 3026–3034.

Nakano, K. and Tanaka, Y. (1990) Electrostatic flowsensor. *Journal of Flow Measurement and Instrumentation*, 1: 191–200.

Nakao, S. (2005) Development of the critical nozzle flow meter for high pressure hydrogen gas dispenser at a hydrogen gas station. Flomeko 2005 13th International Flow Measurement Conference, Peebles, Scotland, Paper 5.1.

(2006) Development of the PVTt system for very low gas flow rates. *Flow Measurement and Instrumentation*, 17(3): 193–200.

Nakao, S-I., Yokoi, Y. and Takamoto, M. (1996) Development of a calibration facility for small mass flow rates of gas and the uncertainty of a sonic Venturi transfer standard. *Journal of Flow Measurement and Instrumentation*, 7: 77–83.

(1997) Development of a calibration facility for small mass flow rates of gas and the uncertainty of a sonic Venturi transfer standard. *Flow Measurement and Instrumentation*,. 7(2): 77–83.

Nakao, S.-I., Terao, Y. and Takamoto, M. (2002) Development of the primary flow standard for very low gas flow rates. 5th International symposium of Fluid Flow Measurement, Arlington, VA, USA.

Nath, B. and Löber, W.(1999) High sample rate ultrasonic gas flow meter for pulsating gas flow. Proceedings of the IMEKO-XV Conference, Osaka, Japan, 1999.

Neda, T., Saito, T. and Nukui, K. (1997) Simple flowmeter for undeveloped turbulent flow using multiple micro hot film flow sensors. FED-211, *Fluid Measurement and Instrumentation, ASME*, 211: 87–91.

Nederlof, A. J. (1994) Product certification of the future. *Journal of Flow Measurement and Instrumentation*, 5(2): 115–120.

NEL (1997b) Ultrasonic meters for oil flow measurement. Flow Measurement Guidance Note, No. 6.
Nesse, Ø. (2007) Experience with ultrasonic meters on high viscosity oil. 25th International North Sea Flow Measurement Workshop, Energy Institute, Oslo, Norway, 295–310.
Nesse, Ø. and Bratten, T. (2013) Qualification of fiscal liquid ultrasonic for operation on extended viscosity range. North Sea Flow Metering Workshop, Paper 3.8.
Nesse, Ø., Folkestad, T., Tunheim, H. and Flølo, D. (2003) Operating experience with two ultrasonic gas meters in series. 21st North Sea Flow Metering Workshop 2003, Paper 17.
Neuhaus, M., Looser, H., Burtscher, H., Schrag, D., Hahn, J. and Schoeb, R. (2007) Flow meter for high-purity and aggressive liquids. *Sensors and Actuators, A: Physical*, 134(2): 303–309.
Newcombe, J. and Griffiths, A. (1973) High throughput flowmeters for gas sales and grid control. 12th World Gas Congress, Nice, France, Paper IGU/D: 12–73.
Newcombe, J., Archbold, T. and Jepson, P. (1972) Errors in measuring gas flows at high pressure – recent developments in correcting methods. 38th Autumn Research Meeting of the Institution of Gas Engineers.
Nguyen, N. T. (1997) Micromachined flow sensors – a review. *Journal of Flow Measurement and Instrumentation*, 8: 7–16.
Nguyen, N. T. and Kiehnscherf, R. (1995) Low-cost silicon sensors for mass flow measurement of liquids and gases. *Sensors & Actuators, A: Physical*, A49.1–2: 17–20.
Niazi, A. and Gaskell, M. (2000) *Building confidence with multi-path ultrasonic meters*. North Sea Flow Measurement Workshop, National Engineering Laboratory, East Kilbride, Scotland.
Nicholson, S. (1994) Coriolis mass flow measurement. FLOMEKO'94 Conference on Flowmeasurement in the Mid 90s, NEL, Scotland.
Nilsson, U. R. C. (1998) A new method for finding inaccurate gas flowmeters using billing data: Finding faulty meters using billing data. *Journal of Flow Measurement and Instrumentation*, 9: 237–242.
Nilsson, U. R. C. and Delsing, D. (1998) In situ detection of inaccurate gas flowmeters using a fingerprint technique. *Journal of Flow Measurement and Instrumentation*, 9: 143–152.
Ningde, J., Hua, Z., Shuying, Z. and Xingbin, L. (2005) Turbine meters for measuring oil-gas-water three phase flow in vertical upward pipes. FLOMEKO 2005 13th International Flow Measurement Conference, Peebles, Scotland, Paper 3.3.
Nishigaki, M., Ippommatsu, M., Ikeda, Y. and Nakajima, T. (1995) Measurement principle of the fluidic gas flowmeter. *Measurement Science & Technology*, 6.6: 833–842.
Nishimura, F. R., Kawashima, S. K. and Kagawa, T. T. (2008) Analysis of laminar flow meter with flute-type cross section laminar elements. 2008 Asia Simulation Conference – 7th International Conference on System Simulation and Scientific Computing, ICSC, Beijing, China: 1110–1114.
Nolan, M. E., Gaskell, M. C. and Cheung, W. S. (1985) Further developments of the British Gas ultrasonic flowmeter. Flowmeasurement in the Mid'80s, Paper 11.2.
Noltingk, B. E. ed. (1988) *Instrumentation reference book*. London: Butterworths.
Norman, D. F. (2003) Calibration of a Coriolis mass flow meter using normal modal analysis. International Publication Number WO 2003021205 A1.
Norman, R., Rawat, M. S. and Jepson, P. (1983) Buckling and eccentricity effects on orifice metering accuracy. International Gas Research Conference, London.
(1984) An experimental investigation into the effects of plate eccentricity effects and elastic deformation on orifice metering accuracy. International Conference on the Metering of Natural Gas and Liquefied Hydrocarbon Gases, London.
Norman, R., Graham, P. and Drew, A. W. (1995) Effects of acoustic noise on orifice meters. Proceedings of the 3rd International Symposium on Fluid Flow Measurement, Electronic Flow Measurement Section, San Antonio, Texas, USA.
Nygaard, G., Mylvaganam, S. and Engan, H. E. (2000) Integration of impedance measurements with transit time measurements for ultrasonic gas mass flow metering – Model

and experiments with transducers in different vibration modes. *Proceedings of the IEEE Ultrasonics Symposium*, 1: 475–482.

O'Keefe, C. V., Maron, R., Rothman, P. and Poplawski, J. (2008a) Application of passive sonar technology to minerals processing flow measurement situations. Society for Mining, Metallurgy and Exploration – SME Annual Meeting and Exhibit 2008: "New Horizons – New Challenges", Salt Lake City, UT, United States: 277–284.

O'Keefe, C., Maron, R., Rothman, P. and Poplawski, J. (2008b) Description of non-intrusive sonar array-based technology and its application to unique and difficult slurry and paste flow measurements presented at Paste 2008, Kaskane, Botswana, May 2008.

O'Keefe, C. V., Poplawski, J. and Maron, R. (2008c) Accuracy of non-intrusive sonar array-based technology to solve unique and difficult measurement situations. CMP 2008.

O'Keefe, C.V., Maron, R. and Rothman, P. (2009a) Improved flow and flotation monitoring for process efficiency improvements through new technology utilizing non-invasive passive arrays. COM2009 (Conference of Metallurgists).

O'Keefe, C. V., Maron, R. J., Fernald, M., Bailey, T. and Van der Spek, A. (2009b) New developments in velocity profile measurement and pipe wall wear monitoring for hydrotransport lines. Canadian Mineral Operators Conference (CMP) 2009.

O'Keefe, C. V., Maron, R., Fernald, M. R., Bailey, T, J., van der Spek, A., M., Davis, M. A. and Viega, J. V. (2011) Flow and pipe management using velocity profile measurement and/or pipe wall thickness and wear monitoring. US Patent Application No. US2011/0056298 March 2011.

O'Keefe, C. V., Felix, J., Peacock, R., Huysamen, T. and Thwaites, P. (2012) *The impact of entrained air and enhanced flow measurements at Eland Platinum concentrator*. The South African Institute of Mining and Metallurgy, Platinum.

O'Sullivan, I. J. and Wright, W. M. D. (2002) Ultrasonic measurement of gas flow using electrostatic transducers. *Ultrasonics*, 40(1–8): 407–411.

O'Sullivan, V. T. and Wyatt, D. G. (1983) Computation of electromagnetic flowmeter characteristics from magnetic field data: III rectilinear weight functions. *Journal of Physics D: Applied Physics,* 16: 461–476.

Obayashi, H., Tasaka, Y., Kon, S. and Takeda, Y. (2008) Velocity vector profile measurement using multiple ultrasonic transducers. *Flow Measurement and Instrumentation*, 19(3–4): 189–195.

Oddie, G. and Pearson, J. A. R. (2004) Flow-rate measurement in two-phase flow. *Annual Reviews in Fluid Mechanics*, 36: 149–172.

Oddie, G., Stephenson, K. E. and Fitzgerald, J. B. (2005) Flow characteristic measuring apparatus and method. US Patent 6,854,341 B2 Feb 15 2005.

Ogawa, Y. (2006) Nagare waza no siruku rodo (means Silk road of flow technology or History of flowmeters). Published by Japan Industry Publishing Co.Ltd.

Oguri, Y. (1988) Wedge flowmeters for measuring bi-directional pipe flows. In Mass Flow Measurement, presented at ASME Winter Annual Meeting (Chicago, Illinois, 27 November-2 December 1988) (Eds G E Mattingly and T R Hendrick), FED 73, pp. 1–5 (ASME, New York).

Ohira, K., Nakamichi, K. and Kihara, Y. (2003) Study on the development of a capacitance-type flowmeter for slush hydrogen. *Cryogenics*, 43(10–11): 607–613.

Ohlmer, E. and Schulze, W. (1985) Experience with CENG full-flow turbinemeters for transient two-phase flow measurements under loss-of-coolant experiment conditions. BHRA 2nd International Conference on Multi-phase Flow, London, England, Paper H1: 381–395.

Ohnuki, A. and Akimoto, H. (1995) Application of electromagnetic velocity meter for measuring liquid velocity distribution in air-water two-phase flow along a large vertical pipe, Proceedings of ASME Heat Transfer and Fluids Engineering Divisions (HYD-321/FED-233, ASME, 473–478.

OIML (2004) The International Organization for Legal Metrology, International Recommendation, Weights of Classes E_1, E_2, F_1, F_2, M_1, $M_{1\text{-}2}$, M_2 and M_3, Part 1: Metrological and technical requirements, OIML R111-1, 1–78.

Okabayashi, M. and Yamasaki, H. (1991) Feasibility study of new fluidic gas meters. FLUCOME '91, 3rd Triennial International Symposium on Fluid Control, Measurement and Visualization, San Francisco, Cal, USA: 313–318.

Okada, T. and Nishimura, J. (2000) The effects of electromagnetic flowmeter on continuous digester. *TAPPI Pulping/Process and Product Quality Conference*, pp. 305–323.

Okada, T., Nishimura, J. and Tanabe, S. (2003) The stable flow measurement for stock flow by capacitance magnetic flowmeter. *Kami Pa Gikyoshi/Japan Tappi Journal*, 57(6): 82–89 (in Japanese).

Okland, O. and Berentsen, H. (1994) *Using the MFI multiphase meter for well testing at Gullfaks B*. North Sea Flow Measurement Workshop, Peebles, Scotland.

Okland, O., Kleppe, K., Berentsen and Klemp, H. (1997) Applications of multiphase meters at the Gullfaks Field in the North Sea. Offshore Technology Conference, Houston, TX, USA, 4: 533–544.

Olivier, P. D. (1997) The effects of line pressure on the performance of thermal mass meters, International Instrumentation Symposium, Instrument Society of America, Aerospace Division, Orlando, Florida, 43: 669–680.

(2002) A turbine flow meter that is insensitive to changes in fluid viscosity. 5th International symposium of Fluid Flow Measurement, Arlington, VA, USA.

Olsen, E. (1991) An investigation of sonic and ultrasonic flowmeters with transducers in free stream. *Journal of Flow Measurement and Instrumentation*, 2: 185–187.

Olsen, L. F. (1974) Introduction to liquid flowmetering and the calibration of liquid flowmeters. National Bureau of Standards TN 831.

Ong, J. T., Oyeneyin, M. B., Coutts, E. J. and MacLean, I. M. (2004) In well nuclear magnetic resonance (NMR) multiphase flowmeter in the oil and gas industry. SPE Annual Technical Conference and Exhibition, 26–29 September 2004, Houston, Texas.

Ong, J. T., Aymond, M., Albarado, T., Majid, J., Daniels, P., Jordy, D. and Lafleur, L. (2007) Inverted Venturi: Optimizing recovery through flow measurement. SPE Annual Technical Conference and Exhibition 2007, ATCE 2007, Anaheim, CA, United States: 2490–2503.

Opara, U. and Bajsiae, I. (2001) Concurrent two-phase downflow measurement with an induced electromagnetic flowmeter. *Journal of Hydraulic Research*, 39(1): 93–98.

Oppenheim, A. K. and Chilton, E. G. (1955) Pulsating flow measurement – a literature survey. *Transactions of ASME*, 77: 231–248.

Orlando, V. A. and Jennings, F. B. (1954) The momentum principle measures true mass flow rate. *Transactions of ASME*, 76: 961–965.

O'Rourke, E. L. (1993) The MTI compact electronic gas meter. FLOMEKO '93 Proceedings of the 6th International Conference on Flow Measurement, Korea Research Institute of Standards and Science: 424–433.

(1996) Results of the MTI compact electronic meter test program. FLOMEKO'96 Proceedings of the 8th International Conference on Flow Measurement, Beijing, China: 313–318.

Ostling, H. and Oki, S. (2001) Spectral Signal Processing (SSP) enhances vibration immunity for vortex flowmeters. *ISA TECH/EXPO Technology Update Conference Proceedings*, 416: 411–417.

Owen, I. and Hussein, I. B. (1991) Wet steam flowmeter correction factors. *Journal of Flow Measurement and Instrumentation*, 2: 139–140.

Owen, I., Hussein, I. B. and Amini, A. M. (1991) The impact of water slugs on wet steam flowmeters. *Journal of Flow Measurement and Instrumentation*, 2: 98–104.

Owen, I., Fyrippi, I. and Escudier, M.P. (2003) Flowmetering of shear-thinning non-Newtonian liquids. *Proceedings of the ASME/JSME Joint Fluids Engineering Conference*, 1 A:3–14.

Ower, E. (1937) On the response of a vane anemometer to an air-stream of pulsating speed. *Philosophical Magazine, Series 7*, 23, No. 157.

Ower, E. and Pankhurst, R. C. (1966) *The measurement of air flow*. Pergamon Press.

Ozgoren, M. (2006) Flow structure in the downstream of square and circular cylinders. *Flow Measurement and Instrumentation*, 17(4): 225–235.

Padden, H. (2002) Uncertainty analysis of a high-speed dry piston flow prover. 5th International symposium of Fluid Flow Measurement, Arlington, VA, USA.

(2004) Uncertainties and inter-laboratory comparisons of dry piston gas flow provers. 12th International Conference on Flow Measurement FLOMEKO Guilin China.

Paik, J. S., Lim, K. W. and Lee, K. B. (1990) Calibration of Coriolis mass flowmeters using a dynamic weighing method. *Journal of Flow Measurement and Instrumentation*, 1: 171–175.

Paik, J. S., Mim, C. H. and Lee, D. K. (1994) Effect of variation of pipe velocity profile on the ultrasonic cross-correlation flowmeters. FLOMEKO '94 Flow Measurement in the Mid-90s, NEL, Glasgow, Scotland: Paper 7.1.

Paik, J. S., Park, K. A. and Park, J. T. (1998) Inter-laboratory comparison of sonic nozzles at KRISS. FLOMEKO '98 Proceedings of the 9th International Conference on Flow Measurement, Lund, Sweden, 95–99.

Paik, J., Lee, K. B. and Mattingly, G. (2005) Uncertainties for an inter-comparison of water flow calibration facilities. FLOMEKO 2005 13th International Flow Measurement Conference, Peebles, Scotland, Paper 2.2.

Pal, R. (1993) Flow of oil-in-water emulsions through orifice and Venturi meters. *Industrial & Engineering Chemistry Research*, 32: 1212–1217.

Pal, R. and Rhodes, E. (1985) Methods for metering oil and water production of wells. BHRA 2nd International Conference on Multi-phase Flow, London, Paper H2, 397–411.

Paladino, E. E. and Maliska, C. R. (2002) The effect of the slip velocity on the differential pressure in multiphase Venturi flow meters. *Proceedings of the International Pipeline Conference, IPC*, A: 965–972.

(2011) Computational modelling of bubbly flows in differential pressure flow meters. *Flow Measurement and Instrumentation*, 22(4): 309–318.

Palau, C. V., Arregui, F. J., Palau, G. and Espert, V. (2004) Velocity profile effects on Woltman water meters performance. 12th International Conference on Flow Measurement FLOMEKO Guilin China.

Panicke, M. and Huebel, C. (2009) Measurement & diagnostic capabilities of clamp-on ultrasonic flow meters. 7th International Symposium on Fluid Flow Measurement (ISFFM), Anchorage, Alaska, USA.

Pannell, C. N., Evans, W. A. B. and Jackson, D. A. (1990) A new integration technique for flowmeters with chordal paths. *Journal of Flow Measurement and Instrumentation*, 1: 216–224.

Panneman, H. J. (2001) On-line comparison of the speed of sound at four Dutch metering stations equipped with ultrasonic gas flow meters. 19th North Sea Flow Measurement Workshop, Kristiansand, Norway, 22–25 October 2001, Paper 10.

Parchen, R. R. and Steenbergen, W. (1998) An experimental and numerical study of turbulent swirling pipe flow. *Transactions of ASME*, 120, March 1998.

Park, K., Kim, J., Kauh, S. K., Ro, S. T. and Lee, J. (1997) Measurement of air flow rate by using an integration type ultrasonic flowmeter applicable for spark ignition engine control. *Proceedings of the Institution of Mechanical Engineers*, 211D: 129–135.

Park, K-A. (1995) Effects of inlet shapes of critical Venturi nozzles on discharge coefficients. *Journal of Flow Measurement and Instrumentation*, 6: 15–19.

Park, K. A., Choi, Y. M., Choi, H. M., Cha, T. S. and Yoon, B. H. (2001) The evaluation of critical pressure ratios of sonic nozzles at low Reynolds numbers. *Journal of Flow Measurement and Instrumentation*, 12(1): 37–41.

Park, K-A., Oh, Y., Choi, H. and Lee, D. (2002) Performance enhancement of wet gas flow meter. *American Society of Mechanical Engineers, Fluids Engineering Division (Publication) FED*, 257(1 A): 149–153.

Parker, M. (September 1990) Improving the performance of a glass tube variable area flowmeter. *Measurement & Control*, 23: 211–215.

Parkinson, G. J. (1991) Fluidic flow sensors for industrial applications. *Measurement & Control*, 24: 4–10.

Patel, B. R. and Sheikholeslami, Z. (1986) Numerical modelling of turbulent flow through orifice meters. International Symposium on Fluid Flow Measurement, Washington, DC.

Paton, R. (1988) Calibration techniques for mass flowmeters. *Petroleum Review*, 42.502, Nov: 40–42.

(1998) Calibration techniques for Coriolis mass flowmeters. FLOMEKO '98 Proceedings of the 9th International Conference on Flow Measurement, Lund, Sweden, pp. 505–508.

Patten, T. and Dunphy, K. (2006) Flow measurement in bitter cold: How to use Coriolis meters in cryogenic service. *Chemical Engineering*, 113(7): 48–49.

Paulsen, F. (1991) Prover ball material problems. Proceedings of the North Sea Flow Measurement Workshop, Norwegian Society of Chartered Engineers.

Pavlovic, V., Dimitrijevic, B., Stojcev, M., Golubovic, L. J., Zivkovic, M. and Stamenkovic, L. J. (1997) Realization of the ultrasonic liquid flowmeter based on the pulse-phase method. *Ultrasonics*, 35: 87–102.

Pawlas, G. E. and Pankratz, T. (1994) Fluid mechanics effects in Coriolis mass flowmeters. FLOMEKO'94 Conference on Flowmeasurement in the Mid 90s, NEL, Scotland.

Pawlas, G. and Patten, T. (1995) Gas measurement using Coriolis mass flowmeters. *ISA Advances in Instrumentation and Control: International Conference and Exhibition*, 50.3: 781–790.

Peng, B. H., Miau, J. J., Bao, F., Weng, L. D., Chao, C. C. and Hsu, C. C. (2010) Performance of Vortex Shedding from a Circular Cylinder with a Slit. 15th Flow Measurement Conference (FLOMEKO), 13–15 October 2010, Taipei, Taiwan, Paper B2-4.

(2012a) Performance of vortex shedding from a circular cylinder with a slit normal to the stream. *Flow Measurement and Instrumentation*, 25: 54–62.

Peng, J., Fu, X. and Chen, Y. (2004) Flow measurement by a new type vortex flowmeter of dual triangulate bluff body. *Sensors and Actuators, A: Physical*, 115(1): 53–59.

(2008a) Response of a swirlmeter to oscillatory flow. *Flow Measurement and Instrumentation*, 19(2): 107–115.

(2008b) Experimental investigation of Strouhal number for flows past dual triangulate bluff bodies. *Flow Measurement and Instrumentation*, 19(6): 350–357.

Peng, J., Wang, W. and Fang, M. (2012b) Hilbert-Huang transform (HHT) based analysis of signal characteristics of vortex flowmeter in oscillatory flow. *Flow Measurement and Instrumentation*, 26: 37–45.

Peng, L., Zhang, Y. and Yan, Y. (2008c) Characterization of electrostatic sensors for flow measurement of particulate solids in square-shaped pneumatic conveying pipelines. *Sensors and Actuators, A Physical*, 141(1): 59–67.

Pereira, M. T. and Nunes, M. (1993) Nozzle chamber to measure flow rates up to 5000m3/h. FLOMEKO Proc 6th Int Conf on Flow Measurement, Korea: 380–386.

Pereira, M. T., de Pimenta, M. and Taira, N. M. (1993) Flow metering with a modified sonic nozzle. FLOMEKO '93 Proceedings of the 6th International Conference on Flow Measurement: 372–379.

Perovic, S. and Higham, E.H. (2002) Electromagnetic flowmeters as a source of diagnostic information. *Flow Measurement and Instrumentation*, 13(3): 87–93.

Perovic, S., Higham, E. H. and Unsworth P. J. (2001) Fault detection and flow regime identification based on analysis of signal noise from electromagnetic flowmeters. *Proceedings of the Institution of Mechanical Engineers, Part E: Journal of Process Mechanical Engineering*, 215(4): 283–293.

Peters, F. and Kuralt, T. (1995) A gas flowmeter of high linearity. *Journal of Flow Measurement and Instrumentation*, 6: 29–32.

Peters, J. and Schook, C. A. (1981) Electromagnetic sensing of slurry concentration. *The Canadian Journal of Chemical Engineering*, 59: 430–437.

Peters, M. C. A. M., Braal, F. M., Limpens, C. H. L. and van Bokhorst, E. (2000) Installation effects on vortex flowmeters – the impact of piping and flow dynamics on the sensor signal. FLOMEKO'2000 the 10th International Conference on Flow Measurement, Salvador, Brazil: Paper E9.

Peters, R. J., Reader-Harris, M. and Stewart, D. (2001) An experimental derivation of an expansibility factor for the V-cone and wafer cone meters. 19th North Sea Flow Measurement Workshop, Kristiansand, Norway, 22–25 October 2001, Paper 23.

Peters, R. J. W., Steven, R., Caldwell, S. and Johansen, B. (2004b) Testing the Wafer V-Cone flowmeters in accordance with API 5.7 "Testing Protocol for Differential Pressure Flow Measurement Devices" in the CEESI Colorado test facility. 12th International Conference on Flow Measurement FLOMEKO Guilin China.

Peters, R. J. W., Steven, R., Caldwell, S. and Johansen, B. (2006) Testing the Wafer V-Cone flowmeters in accordance with API 5.7 "Testing Protocol for Differential Pressure Flow Measurement Devices" in the CEESI Colorado test facility. *Flow Measurement and Instrumentation*, 17(4): 247–254.

Peterson, S., Lightbody, C., Trail, J. and Coughlan, L. (2008) On line condition based monitoring of gas USM's. 26th International North Sea Flow Measurement Workshop, 21–24 October 2008, St Andrews, Scotland, Paper 2.1.

Peyvan, D., Gurevich, Y. and French, C. T. (2002) In-situ calibration for feed water flow measurement. *International Conference on Nuclear Engineering, Proceedings, ICONE*, 1: 137–142.

Pfrehm, R. H. (1981) Improved turbine-meter system measures ethylene accurately. *Oil and Gas Journal*, 79(16): 73–76.

Philip-Chandy, R., Scully, P. J. and Morgan, R. (2000) The design, development and performance characteristics of a fibre optic drag-force flow sensor. *Measurement Science and Technology*, 11: N31–N35.

Place, J. D. and Maurer, R. (1986) Non-invasive fibre optic pick-up for a turbine flowmeter. Conference Fibre Optics 86, London England.

Plache, K. O. (1977) Coriolis/gyroscopic flow meters. ASME 77-WA/FM-4, 1977. (Also Australian Process Engineering, 6, No 9: 47–51, Sept 1978.) (Also Mechanical Engineering March 1979).

Plank, N. (1951) Slippage errors in positive displacement liquid meters. Proc Third World Petroleum Congress, The Hague, Netherlands (Published by E J Brill, Leiden, Netherlands), Section IX: 100–124.

Poiseuille, J. L. M. (1842) *Recueil des savants etrangers*. Academie des Sciences, Paris.

Polo, J., Pallas-Areny, R. and Martin-Vide, J. P. (2001) Analog signal processing in an AC electromagnetic flowmeter. *Conference Record – IEEE Instrumentation and Measurement Technology Conference*, 3: 2136–2139.

(2002) Analog signal processing in an AC electromagnetic flowmeter. *IEEE Transactions on Instrumentation and Measurement*, 51(4): 793–797.

Poornapushpakala, S., Gomathy, C., Sylvia, J. I. and Babu, B. (2014) Design, development and performance testing of fast response electronics for eddy current flowmeter in monitoring sodium flow. *Flow Measurement and Instrumentation*, 38: 98–107.

Pope, J. G. and Wright, J. D. (2014) Performance of Coriolis meters in transient gas flows. *Flow Measurement and Instrumentation*, 37: 42–53.

Pope, J. G., Wright, J. D., Johnson, A. N. and Moldover, M. R. (2012) Extended Lee model for the turbine meter & calibrations with surrogate fluids. *Flow Measurement and Instrumentation*, 24: 71–82.

Pöschel, W. and Engel, R. (1998) The concept of a new primary standard for liquid flow measurement at PTB Braunschweig, FLOMEKO '98 Proceedings of the 9th International Conference on Flow Measurement, Lund, Sweden, 7–12.

Potter, D. M. (1959) Improvements in or relating to mass flow meters. British Patent 860 657, Filed 12th June 1959.
(1961) UK Patent Specification No. 986,831.
Prahu, S. V., Mascomani, R., Balakrishnan, K. and Konnur, M. S. (1996) Effects of upstream pipe fittings on the performance of orifice and conical flowmeters. *Journal of Flow Measurement and Instrumentation*, 7: 49–54.
Prasser, H.-M., Misawa, M. and Tiseanu, I. (2005) Comparison between wire mesh sensor and ultra-fast X-ray tomograph for an air-water flow in a vertical pipe. *Flow Measurement and Instrumentation*, 16(2–3): 73–83.
Priddy, W. J. (1994) Field trials of multiphase metering systems at Prudhoe Bay, Alaska. SPE 69th Annual Technical Conference and Exhibition, New Orleans, USA: 531–543.
Priede, J., Buchenau, D. and Gerbeth, G. (2011a) Contactless electromagnetic phase-shift flowmeter for liquid metals. *Measurement Science and Technology*, 22 (2011): 055402 (11 pp).
(2011b) Single-magnet rotary flowmeter for liquid metals. *Journal of Applied Physics* 110: 034512.
Pritchard, M., Marshall, D. and Wilson, J. (2004) An assessment of the impact of contamination on orifice plate metering accuracy. 22nd North Sea Flow Metering Workshop 2004, Paper 2.2.
Pursley, W. C. (1986) The calibration of flowmeters. *Measurement & Control*, 19(5): 37–45.
Raišutis, R. (2006) Investigation of the flow velocity profile in a metering section of an invasive ultrasonic flowmeter. *Flow Measurement and Instrumentation*, 17(4): 201–206.
Raszillier, H. and Durst, F. (1991) Coriolis-effect in mass flow metering. *Archive of Applied Mechanics*, 61: 192–214.
Raszillier, H. and Raszillier, V. (1991) Dimensional and symmetry analysis of Coriolis mass flowmeters. *Flow Measurement and Instrumentation*, 2: 180–184.
Raszillier, H., Allenborn, N. and Durst, F. (1993) Mode mixing in Coriolis flowmeters. *Archive of Applied Mechanics*, 63(4–5): 219–227.
(1994) Effect of a concentrated mass on Coriolis flowmetering. *Archive of Applied Mechanics*, 64.6: 373–382.
Raustein, O. and Fosse, S. (1991) Measurement of fuel and flare as basis for the CO_2 – tax. Proceedings of the North Sea Flow Measurement Workshop, Norwegian Society of Chartered Engineers.
Rawes, W. and Sanderson, M. L. (1997) *An ultrasonic insertion flowmeter for in-situ calibration, Ultrasonics in Flow Measurement*. Cranfield University, Bedford, England.
(1998) Improvements to an ultrasonic insertion flowmeter for in-situ calibration. FLOMEKO '98 Proceedings of the 9th International Conference on Flow Measurement, Lund, Sweden, pp. 143–147.
Reader-Harris, M. J. (1986) Computation of flow through orifice-plates downstream of rough pipework. Proceedings of International Conference on Flow Measurement in the mid 80's, National Engineering Laboratory, 9–12 June 1986.
(1989) Computation of flow through orifice plates. *Numerical Methods in Laminar and Turbulent Flow*, 6: 1907–1917.
(1994) The decay of swirl in a pipe. *International Journal of Heat and Fluid Flow*, 15(3): 212–217.
(1998) The equation for the expansibility factor for orifice plates. FLOMEKO'98 International Conference on Flow Measurement, Lund, Sweden: 209–214.
Reader-Harris, M. (2012) Wet-gas measurement: ISO/TR 11583. North Sea Flow Metering Workshop
(2015) *Orifice plates and Venturi tubes*. Springer-Verlag GmbH.
Reader-Harris, M. and Addison, D. (2013) Orifice plates with drain holes. FLOMEKO 2013, Paris, or IMEKO- TC9-2013–088.
(2014) Orifice plates with drain holes. North Sea Flow Metering Workshop.

Reader-Harris, M. J. and Brunton, W. C. (2002) The effect of diameter steps in upstream pipework on orifice plate discharge coefficients. 5th International symposium of Fluid Flow Measurement, Arlington, VA, USA.

Reader-Harris, M. and Graham, E. (2009) An improved model for Venturi-tube over-reading in wet gas. 27th International North Sea Flow Measurement Workshop, Tonsberg, Norway, 20–23 October 2009, pp. 131–153. Also at https://www.tekna.no/ikbViewer/Content/783839/Paper%207%20-%20%20E%20Graham.pdf.

Reader-Harris, M. J. and Keegans, W. (1986) Comparison of computation and LDV measurement of flow through orifice and perforated plates, and computation of the effect of rough pipework on orifice plates. Proceedings of the International Symposium on Fluid Flow Measurement, Washington, DC, USA.

Reader-Harris, M. J. and Sattary, J. A. (1990) The orifice plate discharge coefficient equation. *Journal of Flow Measurement and Instrumentation*, 1: 67–76.

(1996) The orifice plate discharge coefficient equation – the equation for ISO5167-1, 1996. Flow Measurement Memo FL/462, September 1996, Equation 11 (National Engineering Laboratory, East Kilbride, Scotland).

Reader-Harris, M. J., Sattary, J. A. and Spearman, E. P. (1995) The orifice plate discharge coefficient equation – further work. *Journal of Flow Measurement and Instrumentation*, 6: 101–114.

Reader-Harris, M. J., Brunton, W. C. and Sattary, J. A. (1997) Installation effects on Venturi tubes. ASME Fluids Engineering Division Summer Meeting FEDSM'97, Paper 3016.

Reader-Harris, M. J., Brunton, W. C., Gibson, J. J., Hodges, D. and Nicholson, I. G. (1999) Venturi tube discharge coefficients. In Proc. 4th Int. Symposium on Fluid Flow Measurement, Denver, Colorado.

Reader-Harris, M. J., Brunton, W. C., Gibson, J. J. and Hodges, D. (2000a) *Discharge coefficients of Venturi tubes with non-standard convergent angles*. FLOMEKO 2000, Salvador, Brazil.

Reader-Harris, M. J., Brunton, W. C., Gibson, J. J., Hodges, D. and Nicholson, I. G. (2000b) Discharge coefficients of Venturi tubes in gas: increasing our understanding. *Proceedings of Flow Metering for Next Millennium, FCRI, Palghat, India.*

Reader-Harris, M. J., Barton, N., Brunton, W. C., Gibson, J. J., Hodges, D., Nicholson, I. G. and Johnson, P. (2000c) The discharge coefficient and through-life performance of Venturi tubes. 18th North Sea Flow Measurement Workshop, Gleneagles, pp. 5.2, October 2000. East Kilbride, Glasgow, National Engineering Laboratory.

Reader-Harris, M. J., Brunton, W. C., Gibson, J. J., Hodges, D. and Nicholson, I. G. (2001) Discharge coefficients of Venturi tubes with standard and non-standard convergent angles. *Journal of Flow Measurement and Instrumentation*, 12(2): 135–145.

Reader-Harris, M. J., Brunton, W. C., Hodges, D. and Nicholson, I. G. (2002) Venturi tubes: improved shape. 20th International. North Sea Flow Measurement Workshop, St Andrews, Scotland, Paper 7.3.

Reader-Harris, M., Brunton, W., Nicholson, I. and Rushworth, R. (2003) Ageing effects on orifice metering. 21st North Sea Flow Metering Workshop 2003, Paper 12.

Reader-Harris, M., Rushworth, R. and Gibson, J. (2004) Installation effects on Venturi tubes of convergent angle 10.5°. 22nd North Sea Flow Metering Workshop 2004, Paper 6.1.

Reader-Harris, M. J., Gibson, J., Hodges, D., Nicholson, I. and Rushworth, R. (2005) Venturi tubes with a 10.5° convergent angle: development of a discharge coefficient equation. FLOMEKO 2005 13th International Flow Measurement Conference, Peebles, Scotland, Paper 5.4.

Reader-Harris, M., Hodges, D. and Rushworth, R. (2008) The effect of drain holes in orifice plates on the discharge coefficient. 26th International North Sea Flow Measurement Workshop, 21–24 October 2008, St Andrews, Scotland, Paper 4.2.

Reader-Harris, M., Barton, N. and Hodges, D. (2010) The effect of contaminated orifice plates on the discharge coefficient. 15th Flow Measurement Conference (FLOMEKO), 13–15 October 2010, Taipei, Taiwan, Paper B4-5.

(2012) The effect of contaminated orifice plates on the discharge coefficient. *Flow Measurement and Instrumentation*, 25: 2–7.

Reeb, B. and Joachim, O. (2002) Development of a diagnostics tool for gas turbine meters: the “Acculert G – II”. 20th North Sea Flow Measurement Workshop, 22–25 October 2002, St Andrews, Scotland, Paper 7.2.

Reid, J. and Pursley, W. C. (1986) An on line prover for the calibration of flowmeters in high pressure gas. International Conference on Flow Measurement in the Mid-80s: Paper 8.3.

Reimann, J., John, H. and Muller, U. (1982) Measurement of two-phase flowrate: a comparison of different techniques. *International Journal of Multiphase Flow*, 8: 33–46.

Reitz, W. C. (1979) Positive displacement meters maintenance. Proc ISA Conf and Exhibit, Advances in Instrumentation. 34, Pt. 2 (Chicago, 22–25 October 1979), Pittsburgh USA, ISA: 259–261.

Reizner, J. R. (2003) Coriolis – the almost perfect flow meter. *IEE Computing and Control Engineering*, 14(4): 28–33.

Rensing, M. and Cunningham, T. J. (2010) *Coriolis flowmeter verification via embedded modal analysis*. IMAC XXVIII, Jacksonville, FL, USA.

Rezende, V. A. and Apple, C. (1997) Coriolis meter for LPG custody transfer at Petrobras. North Sea Flow Measurement Workshop, Kristiansand, Norway, Paper 30.

Rice, J. A. (1988) *Mathematical statistics and data analysis*. Wadsworth & Brooks/Cole Advanced Books & Software, Pacific Grove, California.

Ricken, M. (1989) The swirlmeter – an universal flow measuring instrument. FLOMEKO ’89 Proceedings of the 5th International Conference on Flow Measurement, Dusseldorf, Germany: 295–303.

Ricken, O. and Hans, V. (2004) Influence of the variation of the angle of incidence in vortex-shedding metering. 12th International Conference on Flow Measurement FLOMEKO Guilin China.

Riddle, W. H. (2004) My mass flow meter has gas. *Control (Chicago, Ill)*, 17(2): 14.

Rieder, A. and Drahm, W. (1996) A new type of single straight tube Coriolis mass flowmeter. FLOMEKO’96 Proceedings of the 8th International Conference on Flow Measurement, Beijing, China: 250–254.

Rieder, A., Drahm, W. and Zhu, H. (2005) Coriolis mass flowmeters: on measurement errors in two-phase conditions. FLOMEKO 2005 13th International Flow Measurement Conference, Peebles, Scotland, Paper 3.4.

Riezebos, H. J., Mulder, J. P., Sloet, G. H. and Zwart, R. (2000) Whistling flow straighteners and their influence on US flow meter accuracy. North Sea Flow Measurement Workshop, National Engineering Laboratory, East Kilbride, Scotland.

Rivetti, A., Martini, G., Goria, R., Cignolo, G., Capelli, A. and Alasia, F. (1989) Oil, kerosene and water flowmeter calibration: the integrated IMGC gravimetric/volumetric primary facility. FLOMEKO’89 Proceedings of 5th International Conference on Flow Measurement, Dusseldorf, Germany.

Rivetti, A., Martini, G. and Birello, G. (1994) LHe Venturi flowmeters: practical design criteria and calibration method. *Cryogenics*, 34 Suppl.: 449–452.

Rivkin, I. Y. (1978) Method and apparatus for measuring mass flow rate of individual components of two-phase gas-liquid medium. US Pat. 4 096 745, June.

Robinson, A. C. and Saffman, P. G. (1982) Three-dimensional stability of vortex arrays. *Journal of Fluid Mechanics*, 125: 411–427.

Robinson, C. (1986) Obstructionless flowmeters: smooth sailing for some, rough passage for others. *InTech*, 33(12): 33–36.

Robøle, B., Kvandal, H. K. and Schüller, R. B. (2006) The Norsk hydro multi phase flow loop. A high pressure flow loop for real three-phase hydrocarbon systems *Flow Measurement and Instrumentation*, 17(3): 163–170.

Rogers, J. (1995) Validity of calibration and test data: application of ISO/IEC Guide 25 (EN45001) or the ISO9000 series. *Engineering Science and Education Journal*, 37 Pt. 2: 109–112. See also INSIGHT 37 No 2, February 1995.

Rooney, D. H. (1973) Steam flow through orifices. Report of a meeting at NEL on Two-Phase Flow Through Orifices and Nozzles, Report No 549: 1–17.

Roosnek, N. (2000) Novel digital signal processing techniques for ultrasonic gas flow measurements. *Journal of Flow Measurement and Instrumentation*, 11(2): 89–99.

Rosa, E. S. and Morales, R. E. M. (2004) Experimental and numerical development of a two-phase Venturi flow meter. *Journal of Fluids Engineering, Transactions of the ASME*, 126(3): 457–467.

Rosales, C. and Sanderson, M. L. (2003) Streaming current noise generation in electromagnetic flowmeters measuring conducting fluids. *Flow Measurement and Instrumentation*, 14(3): 97–108.

Rosales, C., Sanderson, M. L. and Hemp, J. (2002a) Problems in the theory and design of electromagnetic flowmeters for dielectric liquids. Part 2a: Theory of noise generation by turbulence modulation of the diffuse ionic charge layer near the pipe wall. *Flow Measurement and Instrumentation*, 13(4): 155–163.

(2002b) Problems in the theory and design of electromagnetic flowmeters for dielectric liquids. Part 2b: Theory of noise generation by charged particles. *Flow Measurement and Instrumentation*, 13(4): 165–171.

Rose, C. and Vass, G. (1995) New developments in flow measurement technology provide solutions to difficult process applications. *ISA Advances in Inst and Control: Int Conf and Exhibition*, 50.3: 791–809.

Roshko, A. (1954) On the development of turbulent wakes from vortex streets. NACA Report 1191.

Rothman, P., O'Keefe, C. and Thomas, A. (2009) Application of unique sonar array based process monitoring measurement equipment for minerals processing applications. BI0407 Rev. A, 10th Mill Operations Conference, Adelaide, Australia, 12–14 October 2009, pp. 365–374.

Roverso, D. and Ruan, D. (2004) Enhancing cross-correlation analysis with artificial neural networks for nuclear power plant feed water flow measurement. *Real-Time Systems*, 27(1): 85–96.

Roverso, D., Ruan, D. and Fantoni, P. F. (2002) Improving feedwater cross-correlation flow measurements in nuclear power plants with artificial neural networks. 5th International FLINS (Fuzzy Logic and Intelligent Techniques in Nuclear Science) Conference on computational intelligent systems for applied research, Ghent, Belgium, 16–18 September 2002, pp. 572–579.

Rowland, J. E. (1989) *Acceleration torque devices*. Cranfield Short Course Lecture, Cranfield Institute of Technology, England.

Ruan, D., Roverso, D., Fantoni, P. F., Sanabrias, J. I., Carrasco, J. A. and Fernandez, L. (2003) Integrating cross-correlation techniques and neural networks for feed water flow measurement. *Progress in Nuclear Energy*, 43(1–4 SPEC): 267–274.

Rubio, E., Solano, J., Torres, F. and Garcia-Nocetti, F. (2006) A proposed warped wigner-ville time frequency distribution applied to Doppler blood flow measurement. Proceedings of the Fourth IASTED International Conference on Biomedical Engineering, 2006: 384–389.

Ruoff, J., Hodapp, M. and Kück, H. (2014) Finite element modelling of Coriolis mass flowmeters with arbitrary pipe geometry and unsteady flow conditions. 37: 119–126.

Rychagov, M. N. and Tereshchenko, S. A. (2002) Ultrasonic flow measurements by multipath measuring spoolpieces: quadrature integration and tomographic reconstruction. 5th International symposium of Fluid Flow Measurement, Arlington, VA, USA.

Rychagov, M. N., Tereshchenko, S., Masloboev, Y., Simon, M. and Lynnworth, L. C. (2002) Mass flowmeters for fluids with density gradient. *Proceedings of the IEEE Ultrasonics Symposium*, 1: 465–470.

Saito, S., Hashimoto, M. and Wada, T. (1993) Development of Karman vortex flowmeter by use of laser diode: measurement of flowrate in a pipe with small cross section. FLOMEKO '93 Proceedings of the 6th International Conference on Flow Measurement, Korea: 335–340.

Sakai, K., Okabayasi, M. and Yasuda, K. (1989) The fluidic flowmeter – a gas flowmeter based on fluidic dynamic oscillation. *Journal of Flow Measurement & Instrumentation*, 1: 44–50.

Sakariassen, R. (1997) On-line quality control of ultrasonic gas flow meters. North Sea Flow Measurement Workshop, Kristiansand, Norway: Paper 15.

Salami, L. A. (1971) Errors in the velocity-area method of measuring asymmetric flows in circular pipes. Proceedings of the International Conference on Modern Developments in Flow Measurement, England: Harwell (Published 1972 by Peter Peregrinus Ltd).

(1984a) Application of a computer to asymmetric flow measurement in circular pipes. *Transactions of the Institute of Measurement and Control*, 6: 197–206.

(1984b) Effect of upstream velocity profile and integral flow straighteners on turbine flowmeters. *International Journal of Heat and Fluid Flow*, 5: 155–165.

(1985) Analysis of swirl, viscosity and temperature effects on turbine flowmeters. *Transactions of the Institute of Measurement and Control*, 7: 183–202.

Salmasi, Z. Z., Jin, W., Gregg, R. D., MacManus, G. and Howarth, C. T. (2001) Electromagnetic flowmeter having low power consumption. US Patent 6,237,424.

Salque, G., Couput, J.-P., Gajan, P., Strzelecki, A. and Fabre, J.-L. (2008) New correction method for wet gas flow metering based on two phase flow modelling: validation on industrial air/oil/water tests at low and high pressure. 26th International North Sea Flow Measurement Workshop, 21–24 October 2008, St Andrews, Scotland, Paper 7.1.

Samer, G. and Fan, S-C. (2010) Modelling of Coriolis mass flow meter of a general plane-shape pipe. *Journal of Flow Measurement and Instrumentation*, 21(1): 8–19.

Sanderson, M. L. (1994) Domestic water metering technology. *Journal of Flow Measurement and Instrumentation*, 5: 107–113.

(1999) Industrial flow measurement by ultrasonics. *Non-Destructive Testing and Condition Monitoring*, 41(1): 16–19.

(2003) Factors affecting the dynamic performance of electromagnetic flowmeters. *Measurement and Control*, 36(9): 270–274.

(2004) Special issue: ultrasonic flowmetering, Editorial. *Flow Measurement and Instrumentation*, 15(3): 128.

Sanderson, M. L. and Al-Rabeh, R. H. (2005) A novel ultrasonic flowmeter for low flowrates in small tubes. Flomeko 2005 13th International Flow Measurement Conference, Peebles, Scotland, Paper 1.1.

Sanderson, M. L. and Hemp, J. (1981) Ultrasonic flowmeters – a review of the state of the art. International Conference on Advances in Flow Measurement Techniques, Coventry, England, Paper G1: 157–178.

Sanderson, M. L. and Sweetland, D. (1991) The effect of four designs of flow conditioner on flowmeter performance. Flow Measurement and Instrumentation Consortium, Category 2A Report No. 1. Cranfield, Bedford: Cranfield Institute of Technology, July 1991.

Sanderson, M. L. and Torley, B. (1985) Error assessment for an intelligent clamp-on transit time ultrasonic flowmeter. International Conference on Flow Measurement in the Mid 80's, NEL Glasgow, Scotland: Paper 11.3.

Sanderson, M. L. and Yeung, H. (2002) Guidelines for the use of ultrasonic non-invasive metering techniques. *Flow Measurement and Instrumentation*, 13: 125–142.

Sanderson M. L., Hemp, J., Coulthard, J. and Henry, R. M. (1988) Non-intrusive flow metering. FLOMIC Report No 4.

Sapra, M. K., Bajaj, M., Kundu, S. M. and Sharma, B. S. V. G. (2011) Experimental and CFD investigation of 100 mm size cone flow elements. *Flow Measurement and Instrumentation*, 22(5): 469–474.

Sato, H. and Watanabe, K. (2000) Experimental study on the use of a vortex whistle as a flowmeter. *IEEE Transactions on Instrumentation and Measurement*, 49: 200–205.

Sato, H., Furuichi, N., Terao, Y. and Takamoto, M. (2005) Basic design of a very large water flow calibration facility for nuclear power application. FLOMEKO 2005 13th International Flow Measurement Conference, Peebles, Scotland, Paper 8.2.

Sato, S., Nukui, K., Ito, S. and Kimura, Y. (1996) Numerical analysis of fluidic oscillation applied to the fluidic gas meter. FLOMEKO'96 Proceedings of the 8th International Conference on Flow Measurement, Beijing, China: 138–143.

Sattary, J. A. (1991) EEC orifice plate programme – installation effects. *Journal of Flow Measurement and Instrumentation*, 2: 21–33.

Sattary, J. A. and Reader-Harris, J. (1997) Computation of flow through Venturi meters. North Sea Flow Measurement Workshop, Kristiansand, Norway: Paper 26.

Sazhin, O. (2013) Novel mass air flow meter for automobile industry base on thermal flow microsensor. *I. Analytical Model and microprocessor. Flow Measurement and Instrumentation 2013,* 30: 60–65.

Scanes, E. P. (1974) A domestic oil flowmeter. *Kent Technical Review*, 11: 31–33.

Scelzo, M. J. A. (2001) Clamp-on ultrasonic flowmeter for gases. *Flow Control*, VII(9): 34–37.

Scheers, A. M. and Wolff, C. J. M. (2002) Production measurement management. 20th North Sea Flow Measurement Workshop, 22–25 October 2002, St Andrews, Scotland, Paper 1.2.

Schieber, W. (1998) The Accutest: a turbine meter with a built-in transfer standard. FLOMEKO '98 Proceedings of the 9th International Conference on Flow Measurement, Lund, Sweden, pp. 509–516.

Schlichting, H. (1979) *Boundary Layer Theory*, McGraw-Hill, New York, ISBN 0-07-055334-3.

Schlumberger (2010) *Fundamentals of multiphase metering*. Schlumberger.

Schluter, Th. and Merzkirch, W. (1996) PIV measurements of the time-averaged flow velocity downstream of flow conditioners in a pipeline. *Journal of Flow Measurement and Instrumentation*, 7(3/4): 173–179.

Schneider, F., Peters, F. and Merzkirch, W. (2003) Quantitative analysis of the cross-correlation ultrasonic flow meter by means of system theory. *Measurement Science and Technology*, 14(5): 573–582.

Schoenborn, E. M. Jr. and Colburn, A. P. (1939) The flow mechanism and performance of the rotameter. *Transactions of the American Institute of Chemical Engineers*, 35: 359–389.

Scott, C. (1984) Sounding out ultrasonic flowmeters. *Control & Instrumentation*, August: 27 and 29.

Scott, R. W. W. (1975a) The use and maintenance of weighing machines in high accuracy liquid flow calibration systems. Conference on Fluid Flow Measurement in the Mid 1970's, Paper B-1: Scotland: National Engineering Laboratory.

(1975b) A practical assessment of the performance of electromagnetic flowmeters. Conference on Fluid Flow Measurement in the mid 1970s, NEL, Scotland: Paper E1.

(1982) *Liquid flow measurement – a general appraisal. Developments in Flow Measurement-1*, ed. R. W. W. Scott, London: Applied Science Publishers: 73–100.

Selvikvag, O. (1997) The Norwegian regulations relating to fiscals measurements of oil and gas – 1997 update. North Sea Flow Measurement Workshop, Kristiansand, Norway: Paper 8.

Shafer, M. R. (1962) Performance characteristics of turbine flowmeters. *Journal of Basic Engineering,*: 471–485.

Shakouchi, T. (1989) New fluidic oscillator, flowmeter, without control port and feedback loop. *Journal of Dynamic Systems, Trans ASME*, 111.3: 535–539.

Sharma, P., Kumar, S. S., Nashine, B. K., Veerasamy, R., Krishnakumar, B., Kalyanasundaram, P. and Vaidyanathan, G. (2010) Development, computer simulation and performance testing in sodium of an eddy current flowmeter. *Annals of Nuclear Energy* 37: 332–338.

Sharma, V., Kumar, G. V., Dash, S. K., Nashine, B. K. and Rajan, K. K. (2012) Modelling of permanent magnet flowmeter for voltage signal estimation and its experimental verification. *Flow Measurement and Instrumentation*, 28: 22–27.

Sheikholeslami, M. Z., Patel, B. R. and Kothari, K. (1988) Numerical modelling of turbulent flow through orifice meters – a parametric study. 2nd International Conference on Flow Measurement, London.

Shen, J. J. S., Vierkandt, S. J. and Ogden, K. A. (2003) Operation and evaluation of a roxar (MFI model) multiphase meter in sour field environment. 21st North Sea Flow Metering Workshop 2003, Paper 23.

Shen, X., Saito, Y., Mishima, K. and Nakamura, H. (2005) Methodological improvement of an intrusive four-sensor probe for the multi-dimensional two-phase flow measurement. *International Journal of Multiphase Flow*, 31(5): 593–617.

Sheppard, T. J. (1994) Solid state gas metering: the future. *Journal of Flow Measurement and Instrumentation*, 5: 103–106.

Shercliff, J. A. (1955) Experiments on the dependence of sensitivity on velocity profile in electromagnetic flowmeters. *Journal of Scientific Instruments*, 32: 441–442.

(1962) *The theory of electromagnetic flow-measurement*. Cambridge University Press (2nd Ed. 1987), Cambridge.

Shields, C. A., Dollard, M., Sridhar, S., Dragnea, G. and Illingworth, M. (2013) Use of SONAR Metering to Optimize Production in Liquid Loading Prone Gas Wells. SPE 166652-MS.

Shim, W. J., Dougherty, T. J. and Cheh, H. Y. (1996) Turbine flowmeter response in two-phase flow. International Conference on Nuclear Engineering Volume 1 – Part B, ASME, 943–953.

Shimada, T., Terao, Y., Takamoto, M., Ono, S. and Gomi, S. (2002) Development of a servo pd oil flowmeter for a transfer standard. 5th International symposium of Fluid Flow Measurement, Arlington, VA, USA.

Shimada, T., Oda, S., Terao, Y. and Takamoto, M. (2003) Development of a new diverter system for liquid flow calibration facilities. *Journal of Flow Measurement and Instrumentation*, 14(3): 89–96.

Shimada, T., Doihara, R., Terao, Y. and Takamoto, M. (2004) Uncertainty analysis of primary standard for hydrocarbon flow at NMIJ FLOMEKO 2004 12th International Conference on Flow Measurement FLOMEKO Guilin China.

(2007) Development of hydrocarbon flow facility as a national standard *Journal of Fluid Science and Technology*, 2(1): 23–34.

(2010a) Establishment of Traceability System for Hydrocarbon Flow in Japan. 15th Flow Measurement Conference (FLOMEKO), 13–15 October 2010, Taipei, Taiwan, Paper B5-3.

Shimada, T., Mahadeva, D. V. and Baker, R.C. (2010b) Further investigation into a water flow rig related to calibration. *Journal of Flow Measurement and Instrumentation*, 21(4): 462–475.

Shimada, T., Doihara, R. and Terao, Y. (2015) Investigation into calibration performance of small volume prover for hydrocarbon flow. *Flow Measurement and Instrumentation*, 41: 174–180.

Shimizu, T. and Takeshima, N. (2001) Numerical study on Faraday-type electromagnetic flowmeter in liquid metal system, (II) analysis of end effect due to saddle-shaped small-sized magnets with FALCON code. *Journal of Nuclear Science and Technology*, 38(1): 19–29.

Shimizu, T., Takeshima, N. and Jimbo, N. (2000) Numerical study on Faraday-type Electromagnetic Flowmeter in liquid metal system, (I) A numerical method based on magnetic field and electric potential field: FALCON code. *Journal of Nuclear Science and Technology*, 37(12): 1038–1048.

Shinder, I. I. and Moldover, M. R. (2009) Dynamic gravitational standard for liquid flow: model and measurements. 7th International Symposium of Fluid Flow Measurement Anchorage, Alaska, USA.

(2010) Feasibility of an accurate dynamic standard for water flow. *Journal of Flow Measurement and Instrumentation*, 21(2): 128–133.

Shufang, H., Yongtao, H. and Lingan, X. (1996) The practice on DN 1400MM Venturi tubes. Proceedings of the 8th International Conference on Flow Measurement, China: Beijing: 37–42.

Shuoping, Z., Zhijie, X. and Baofen, Z. (1996) Flow with differential pressure noise of orifice. FLOMEKO '96 Proceedings of the 8th International Conference on Flow Measurement, Beijing, China: 605–611.

Sidney, J. K., King, N. W. and Coulthard, J. (1988a) Cross-correlation flow measurements in oil-air mixtures. 2nd Int Conf Flow Measurement, BHRA, London.

(1988b) The measurement of individual phase-flowrates using an ultrasonic cross-correlation flowmeter in air-kerosene mixtures. Flucome '88, H S Stephens & Associates, Sheffield, September.

Siegwarth, J. D. (1989) Vortex shedding flowmeters for high velocity liquids. *International Journal of Heat & Fluid Flow*, 10.3: 232–244.

Silva, F. S., Velazquez, M. T. and Ruiz, J. H. (1997) Experimental study for the use of elbows as flowmeters. ASME Fluids Engineering Division Summer Meeting FEDSM'97, Paper 3010.

Simões, E. W., Furlan, R. and Pereira, M. T. (2002) Flow measurement with microfluidic oscillators. 5th International symposium of Fluid Flow Measurement, Arlington, VA, USA.

Simões, E. W., Furlan, R., Brusetti Leminski, R. E., Gongora-Rubio, M. R., Pereira, M. T. and Morimoto, N. I. (2004) Microfluidic oscillator for gas flow control and measurement. 12th International Conference on Flow Measurement FLOMEKO Guilin China.

Simões, E. W., Furlan, R., Brusetti Leminski, R. E., Gongora-Rubio, M. R., Pereira, M. T., Morimoto, N. I. and Santiago-Avilés, J. J. (2005) Microfluidic oscillator for gas flow control and measurement. *Flow Measurement and Instrumentation*, 16(1): 7–12.

Simpson, R. J. (1984) Flexible orifice plates. *Measurement & Control*, 17.

Sims, L. and Rabalais, R. A. (2002) Ultrasonic flow measurement: technology and applications in process and multiple vent stream situations. *Proceedings of the Annual Symposium on Instrumentation for the Process Industries*, 57: 135.

Sims, P. (1992) Mass flowmeter technology benefits blending. *Process & Control Engineering*, 45(5): 32–33.

Sindt, Ch. F., Brennan, J. A., McFaddin, S. E. and Wilson, R. W. (1989) Effect of pipe surface finish on the orifice discharge coefficient. FLOMEKO'89 International Conference on Flow Measurement, Dusseldorf, Germany: 49–56.

Singh, R. K., Singh, S. N. and Seshadri, V. (2009) Study of the effect of vertex angle and upstream swirl on the performance characteristics of cone flowmeter using CFD. *Flow Measurement and Instrumentation*, 20(2): 69–74.

(2010) CFD prediction of the effects of the upstream elbow fittings on the performance of cone flowmeters. *Flow Measurement and Instrumentation*, 21(2): 88–97.

Singh, S. N., Seshadri, V., Singh, R. K. and Gawhade, R. (2006) Effect of upstream flow disturbances on the performance characteristics of a V-cone flowmeter. *Flow Measurement and Instrumentation*, 17(5): 291–297.

Skaardalsmo, K. and Moksnes, P. O. (2003) Phase watcher VX multiphase flowmeter Heidrun experience and analysis. 21st North Sea Flow Metering Workshop 2003, Paper 25.

Skea, A. F. and Hall, A. R. W. (1999a) Effects of gas leaks in oil flow on single-phase flowmeters. *Journal of Flow Measurement and Instrumentation*, 10: 145–150.

(1999b) Effects of water in oil and oil in water on single-phase flowmeters. *Journal of Flow Measurement and Instrumentation*, 10: 151–157.

Skwarek, V. and Hans, V. (2000) The ultrasonic cross-correlation flowmeter – new insights into the physical background. FLOMEKO'2000 the 10th International Conference on Flow Measurement, Salvador, Brazil: Paper B3.

Skwarek, V., Windorfer, H. and Hans, V. (2001) Measuring pulsating flow with ultrasound. *Measurement J. Int. Measurement Confederation*, 29(3): 225–236.

Slater, S. G., Paterson, A. McK. and Marshall, M. F. (1997) *Offshore Technology Conference*, Houston, TX, USA, 4: 523–532.

Sloet, G. and de Nobel, G. (1997) Experiences with ultrasonic meters at the Gasunie export stations. North Sea Flow Measurement Workshop, Kristiansand, Norway: Paper 14.

Sluše, J. and Geršl, J. (2013) Cavitation of lng in ultrasonic flowmeters – CFD modelling. FLOMEKO 2013, 16th International Flow Measurement Conference, Paris, France.

Smalling, J. W., Braswell, L. D., Lynnworth, L. C. and Russell Wallace, D. (1984) Flare gas ultrasonic flow meter. Proc. of 39th Annual Symp. on Instrum. for the Process Industries, 17–20 January 1984, pp. 27–38.

Smalling, J. W., Braswell, L. D. and Lynnworth, L. C. (1986) Apparatus and methods for measuring fluid flow parameters. US Patent 4,596,133, 24 June 1986 (filed 29 July 1983).

Smith, C. R., Greco, J. J. and Hopper, P. B. (1989) Low-loss conditioner for flow distortion/swirl using passive vortex generation devices. FLOMEKO 5th International Flow Measurement Conference, Dusseldorf, Germany: 57–64.

Smith, J. E. (1978) Gyroscopic/Coriolis mass flow meter. *Canadian Controls & Instruments*, 117: 29–31, (follows Willer's (1978) note).

Smith, J. E. and Cage, D. R. (1985) Parallel path Coriolis mass flow rate meter. US Patent 4,491,025, 1st Jan.

Smith, M. and Morfey, C. (1997) The effect of developing flow on the accuracy of an ultrasonic gas meter. Ultrasonics in Flow Measurement, Cranfield University, Bedford, UK.

Smith, R. E. and Matz, R. J. (1962) A theoretical method of determining discharge coefficients for Venturis operating at critical flow conditions. *Journal of Basic Engineering*, 84: 434–446.

Smith, R. V. and Leang, J. T. (1975) Evaluations of correlations for two-phase flowmeters, three current – one new. ASME Paper No 74-WA/FM-5, 1974, and ASME J Eng Power, 1975.

Smith, W. C. and Kowalski, R. R. (1998) Adaptation of commercial Coriolis flowmeters for non-intrusive pressure measurement. *Instrumentation in the Aerospace Industry: Proceedings of the 44th International Symposium*, 44: 87–96.

Smorgrav, A. E. (1990) Multiphase flow meter KO 300 MFM. North Sea Flow Measurement Workshop, National Engineering Laboratory, Scotland.

Smørgrav, S. and Abrahamsen, A. K. (2009) OIML R 137-1, the first ultrasonic meter to be tested to accuracy class 0.5?. 27th International North Sea Flow Measurement Workshop, Tonsberg, Norway, 20–23 October 2009, pp. 235–252.

Soderholm, A. (1999) Bulk solid mass flowmeter using the Coriolis principle. *Powder Handling and Processing*, 11(3): 297–300.

Sofialidis, D. and Prinos, P. (March 1996) Wall suction effects on the structure of fully developed turbulent pipe flow. *Journal of Fluid Engineering*, 118: 33–39.

Sparks, C. R., Durke, R. D. and McKee, R. J. (1989) Pulsation-induced errors in the primary and secondary systems of orifice meters. FLOMEKO'89 International Conference on Flow Measurement, Dusseldorf, Germany: 31–38.

Sparks, D., Smith, R., Cripe, J., Schneider, R. and Najafi, N. (2003) A portable MEMS Coriolis mass flow sensor. IEEEE Sensors Conference, 2003.

Sparreboom, W., van de Geest, J., Katerberg, M., Postma, F., Haneveld, J., Groenesteijn, J., et al. (2013) Compact mass flow meter based on a micro Coriolis flow sensor. *Micromachines*, 4: 22–33.

Spazzini, P. G., Callegaro, L., Pennecchi, F. and Mickan, B. (2010) Comparison of Calibration Curves: an Application Example. 15th Flow Measurement Conference (FLOMEKO), 13–15 October 2010, Taipei, Taiwan, Paper B6-1.

Spearman, E. P., Sattary, J. A. and Reader-Harris, M. J. (1991) A study of flow through a perforated-plate/orifice-meter package in two different pipe configurations using laser Doppler velocimetry. *Journal of Flow Measurement and Instrumentation*, 2: 83–88.

(1996) Comparison of velocity and turbulence profiles downstream of perforated plate flow conditioners. *Journal of Flow Measurement and Instrumentation*, 7(3/4): 181–199.

Spencer, E. A. (1993) Bibliography of the EEC orifice plate project. Report EUR 14885 EN, Commission of the European Communities, Brussels, Belgium.

Spencer, E. A., Heitor, M. V. and Castro, I. P. (1995) Intercomparison of measurements and computations of flow through a contraction and a diffuser. *Journal of Flow Measurement and Instrumentation*, 6: 3–14.

Spink, L. K. (1978) *Principles and practice of flow meter engineering*. 9th ed., Foxboro, MA: The Foxboro Company.

Spitzer, D. W. ed. (1991) *Flow measurement*. Instrument Society of America.

Spragg, W. T. and Seatonberry, B. W. (1975) A radioisotope dilution method for the precise absolute determination of the flowrate of gas under industrial conditions. Conference on

Fluid Flow Measurement in the Mid 1970's, Paper D-1: Scotland: National Engineering Laboratory.

Sproston, J. L., Johnson, M. W. and Pursley, W. C. (1987) Mass flow measurement. FLOMIC Report.

Sproston, J. L., El-Wahed, A. and Johnson, M. W. (1990) An electrostatic vortex-shedding meter. *Journal of Flow Measurement & Instrumentation*, 1: 183–190.

Squirrell, B. (1994) Profibus: a working standard fieldbus. *Measurement & Control*, 27 (1): 9–14.

Stack, C. P. and Cunningham, T. J. (1993) Design and analysis of Coriolis mass flowmeters using MSC/NASTRAN. Presented at Conference on MSC World Users, 1993.

Stack, C. P., Barnett, R. B. and Pawlas, G. E. (1993) A finite-element for the vibration analysis of a fluid-conveying Timoshenko beam AIAA Technical Paper (AIAA-93-1552-CP)', pp. 2120–2129.

Standiford, D. M. and Lee, M. (2010) Inter-laboratory comparison results for coriolis mass flow meter calibration facilities. 15th Flow Measurement Conference (FLOMEKO), 13–15 October 2010, Taipei, Taiwan, Paper A9-3.

Stansfeld, J., Atkinson, I. and Washington, G. (1988) A new mass flow meter and its application to crude oil metering. North Sea Metering Workshop, National Engineering Laboratory, Scotland.

Starling, K. E. and Luongo, J. F. (1997) Electronic implementations of A.G.A. Report No. 8. ASME Fluids Engineering Division Summer Meeting FEDSM'97, Paper 3015.

Staubli, T., Luescher, B., Gruber, P. and Widmer, M. (2008) Optimization of acoustic discharge measurement using CFD. *International Journal on Hydropower & Dams*, 15(2): 109–113.

Steenbergen, W. and Voskamp, W. (1998) The rate of decay of swirl in turbulent pipe flow. *Flow Measurement and Instrumentation*, 9: 67–78.

Stelian, C. (2013) Calibration of a Lorentz force flowmeter by using numerical modeling. *Flow Measurement and Instrumentation.* 33: 36–44.

Steven, R. (2000) Wet gas metering with a horizontally installed Venturi meter. FLOMEKO '2000 the 10th International Conference on Flow Measurement, Salvador, Brazil: Paper D7.

(2007) V-cone wet gas metering. 25th International North Sea Flow Measurement Workshop, Energy Institute, Oslo, Norway, 152–180.

(2008a) A dimensional analysis of two phase flow though a horizontally installed Venturi flow meter. *Flow Measurement and Instrumentation*, 19(6): 342–349.

(2008b) Diagnostic methodologies for generic differential pressure flow meters. 26th International North Sea Flow Measurement Workshop, 21–24 October 2008, St Andrews, Scotland, Paper 4.1.

(2009) Horizontally installed cone differential pressure meter wet gas flow performance. *Flow Measurement and Instrumentation*, 20(4–5): 152–167.

(2009a) Diagnostic capabilities of Δ P cone meter. 7th International Symposium of Fluid Flow Measurement, Anchorage, Alaska, USA.

(2009b) Significantly improved capabilities of DP meter diagnostic methodologies. 27th International North Sea Flow Measurement Workshop, Tonsberg, Norway, 20–23 October 2009, pp. 13–41.

(2010) Diagnostic system for Venturi meters. 15th Flow Measurement Conference (FLOMEKO), 13–15 October 2010 Taipei, Taiwan, Paper B10-4.

Steven, R. and Hall, A. (2009) Orifice plate meter wet gas flow performance. *Flow Measurement and Instrumentation*, 20(4–5): 141–151.

Steven, R. and Hodges, C. (2012) Orifice meter diagnostics – a discussion on theory, laboratory tests, system interface and field results. 8th ISFFM (2012).

Steven, R. and Lawrence, P. A. (2003) Research developments in wet gas metering with V-cone meters. 21st North Sea Flow Metering Workshop 2003, Paper 11.

Steven, R., Kegel, T. and Britton, C. (2005) An update on V-cone meter wet gas flow metering research. FLOMEKO 2005 13th International Flow Measurement Conference, Peebles, Scotland, Paper 3.2 .

Steven, R., Britton, C., Kinney, J. and Pagano, S. (2009) Wedge meters with wet natural gas flows. Proc. 7th International Symposium on Fluid Flow Measurement (ISFFM), Anchorage, Alaska, USA.

Steven, R., Britton, C. and Kinney, J. (2010) 4″, 0.63 beta ratio cone DP meter wet gas performance. 15th Flow Measurement Conference (FLOMEKO), 13–15 October 2010, Taipei, Taiwan, Paper B4-4.

Stewart, D. G. (2002) Performance of Coriolis meters in gas flow. 5th International symposium of Fluid Flow Measurement, Arlington, VA, USA.

Stewart, D. G. and Hodges, D. (2003) Suitability of dry gas metering technology for wet gas metering. 21st North Sea Flow Metering Workshop 2003, Paper 6.

Stewart, D. G., Hodges, D., Steven, R. and Peters, R. J. W. (2002) Wet gas metering with V-cone meters. 20th North Sea Flow Measurement Workshop, 22–25 October 2002, St Andrews, Scotland, Paper 4.2.

Stewart, D. G., Hodges, D. and Brown, G. (2003) Venturi meters in wet gas flow. 21st North Sea Flow Metering Workshop 2003, Paper 4.

Stewart, D. G., Watson, J. T. R. and Vaidya, A. M. (2000) A new correlation for the critical mass flux of natural gas mixtures. *Flow Measurement and Instrumentation*, 11(4): 265–272.

Stidger, Ruth W. (2003) Measurement and instrumentation: flow meter advances push utilities to make changes. *Gas Utility Manager*, 47(7): 12–13.

Stobie, G. J. (1993) *Metering in the real world*. North Sea Flow Measurement Workshop, Bergen, Norway.

Stobie, G. J., Zanker, K. J., Brown, C. and Letton, W. (2001) Flow testing a USM outside its performance envelope. 19th North Sea Flow Measurement Workshop, Kristiansand, Norway, 22–25 October 2001, Paper 8.

Stobie, G., Hart, R., Svedeman, S. and Zanker, K. (2007) Erosion of a Venturi meter with laminar and turbulent flow and low Reynolds number discharge coefficient measurements. 25th International North Sea Flow Measurement Workshop, Energy Institute, Oslo, Norway, 274–294.

Stoll, H. W. (1978) Current trends in flow measurement technology. Proc ISA Pacific Northwest Instrum '78 Symp, Portland Oreg: 89–91.

Stoltenkamp, P. W., Araujo, S. B., Riezebos, H. J., Mulder, J. P. and Hirschberg, A. (2003) Spurious counts in gas volume flow measurements by means of turbine meters. *Journal of Fluids and Structures*, 18(6): 771–781.

Stolz, J. (1978) A universal equation for the calculation of discharge coefficients of orifice plates. Flow Measurement of Fluids (ed. H. H. Dijstelbergen and E. A. Spencer), North Holland, Amsterdam: 519–534.

(1988) The first revision of ISO 5167. North Sea Metering Workshop, NEL, East Kilbride, Scotland: Paper 3.1.

Stone, C. R. and Wright, S. D. (1994) Non-linear and unsteady flow analysis of flow in a viscous flowmeter. *Transactions of the Institute of Measurement and Control*, 16: 128–141.

Storer, J. and Steven, R. (2010) A mass flow meter concept with diagnostic capabilities. 15th Flow Measurement Conference (FLOMEKO), 13–15 October 2010, Taipei, Taiwan, Paper B7-3.

Storer, J., Schroeder, E. and Steven, R. (2009) Advances in vortex shedding flow metering. Proc. 7th International Symposium on Fluid Flow Measurement, Anchorage, Alaska, USA.

Storm, R., Kolahi, K. and Rock, R. (2001) Model based correction of Coriolis mass flowmeters. *Conference Record – IEEE Instrumentation and Measurement Technology Conference*, 2: 1231–1236.

(2002) Model-based correction of Coriolis mass flowmeters. *IEEE Transactions on Instrumentation and Measurement*, 51(4): 605–610.

Stratford, B. S. (1964) The calculation of the discharge coefficient of profiled choked nozzles and the optimum profile for absolute air flow measurement. *Journal of Royal Aeronautical Society*, 68(640): 237–245.

Strawn, C. (1991) Mass meters for liquid measurement. Methods of proving Coriolis mass flowmeters. Proc International School of Hydrocarbon Measurement, University of Oklahoma, Continuing Engineering Education, Norman, USA: 148–150.

Strohrmann, M., Lembke, M., Huftle, G., Konzelmann, U., Lenzing, T., Opitz, B. and Bruckner, J. (2004) Heissfilmluftmassenmesser HFM6 – Prazise Luftmassenmessung fur Kraftfahrzeuganwendungen (Hot film mass air flow meter HFM6 – High precision mass air flow metering for automotive applications). *VDI Berichte*, (1829): 535–542+921 (in German).

Strom, G. R. and Livelli, G. (2001 date uncertain) *Direct mounting allows differential pressure (dp) based flow measurement optimization.* Rosemount, Inc. http://www2.emersonprocess.com/siteadmincenter/PM%20Rosemount%20Documents/ISA-DirectMountingPaper.pdf

Strzelecki, A., Gajan, P., Couput, J. P. and De Laharpe, V. (2000) Behaviour of Venturi meters in two-phase flows. FLOMEKO '2000 the 10th International Conference on Flow Measurement, Salvador, Brazil: Paper D6.

Studzinski, W. and Karnik, U. (1997) Installation effects on orifice meter with no flow conditioner, ASME Fluids Engineering Division Summer Meeting FEDSM '97, Paper 3014.

Sui, L., Nguyen, T. H., Matson, J. E., Espina, P. and Tew, I. (2010) Ultrasonic flowmeter for accurately measuring flare gas over a wide velocity range. 15th Flow Measurement Conference (FLOMEKO), 13–15 October 2010, Taipei, Taiwan, Paper A4-5.

Sui, L., Pfenninger, R. S., Nguyen, T. H., Hobbs, N. A., Sadovnik, I., Su, L. and Matson, J. E. (2013) Ultrasonic flowmeter for coal seam gas application. FLOMEKO 2013, 24-26th September 2013, Paris

Sullivan, J. J., Ewing, J. H. and Jacobs, R. P. (no date) Calibration techniques for thermal-mass flowmeters. (possibly published in Solid State Technology) MKS Instruments Inc, Burlington, MA 01803, USA.

Sultan, G. (1992) Single straight tube Coriolis mass flowmeter. *Journal of Flow Measurement and Instrumentation*, 3: 241–246.

Sultan, G. and Hemp, J. (1989) Modelling of the Coriolis mass flowmeter. *Journal of Sound and Vibration*, 132(3): 473–489.

Summers-Smith, J. D. (1994) *An introductory guide to industrial tribology*. Mechanical Engineering Press, London.

Sun, L. and Zhang, T. (2004) Study on the mathematical model of the turbine flowmeters. 12th International Conference on Flow Measurement FLOMEKO Guilin China.

Sun, L.-J., Qi, L.-X. and Zhang, T. (2010) Numerical simulation and experiment on averaging pitot tube with flow conditioning wing. 15th Flow Measurement Conference (FLOMEKO), 13–15 October 2010, Taipei, Taiwan, Paper B10-2. Possibly cf paper in JFMI).

Sun, M., Liu, S., Lei, J. and Li, Z. (2008) Mass flow measurement of pneumatically conveyed solids using electrical capacitance tomography. *Measurement Science and Technology*, 19(4): 045503.

Sun, Y., Xiong, H., Zhu, R. and Bi, l. (1996) Research and development on wear-resistant orifice plates. Proceedings of the 8th International Conference on Flow Measurement, Beijing, China: 27–31.

Sun, Zhengnai; Li, P., Qiang, X. and Zhao, Y. (2000) Optical fiber interference target flowmeter (II): damper design. *Proceedings of SPIE – The International Society for Optical Engineering*, 4222: 198–201.

Suzuki, N., Nakabori, H. and Yamamoto, M. (1972) Ultrasonic method of flow measurement in large conduits and open channels. Modern Developments in Flow Measurement, Peter Peregrinus Ltd: 115–138.

Suzuki, N., Nakabori, H. and Kitajima, A. (1975) New applications of ultrasonic flowmeters. Flowmeasurement in the Mid '70s, NEL, Glasgow, Scotland: Paper H-3.

Svedin, N., Stemme, E. and Stemme, G. (2001) A static turbine flow meter with a micromachined silicon torque sensor. Proceedings of the 14th IEEE International Conference on Micro Electro Mechanical Systems (MEMS 2001), Interlaken, Switzerland, pp. 208–211.

(2003) A static turbine flow meter with a micro machined silicon torque sensor. *Journal of Microelectromechanical Systems*, 12(6): 937–946.

Svensson, B. and Delsing, J. (1998) Application of ultrasonic clamp-on flowmeters for in situ tests of billing meters in district heating systems. *Journal of Flow Measurement and Instrumentation*, 9(1): 33–41.

Svete, A., Kutin, J. and Bajsić, I. (2009) Static and dynamic characteristics of a hydraulic Wheatstone Bridge. *Flow Measurement and Instrumentation*, 20(6): 264–270.

(2013) Dynamic characteristics of a hydraulic Wheatstone bridge mass flowmeter. 16th International Flow Measurement Conference, FLOMEKO 2013, 24–26 September 2013, Paris.

Svete, A., Kutin, J., Bajsic, I. and Slavic, J. (2012) Development of a liquid flow pulsator. *Flow Measurement and Instrumentation*, 23(1): 1–8.

Swanson, K. (1988) New developments in the measurement of slurries and emulsions using Coriolis effect mass flowmeters. Proc Pacific Cascade Instrum '88 Symposium, ISA, USA.

Szebeszcyk, J. M. (1994) Application of clamp-on ultrasonic flowmeter for industrial flow measurements. *Journal of Flow Measurement and Instrumentation*, 5: 127–131.

Taha, S. M. R. (1994) Digital measurement of the mass-flow rate. *Sensors & Actuators, A: Phys*, 45.2: 139–143.

Tai, S. W., Miau, J. J., Shaw, J. H. and Chen, Z. L. (1993) Signal-quality study of ring-type vortex flowmeters. FLOMEKO '93 Proceedings of the 6th International Conference on Flow Measurement, Korea: 320–326.

Takahashi, S. and Itoh, I. (1993) Intelligent vortex flowmeter. FLOMEKO '93 Proceedings of the 6th International Conference on Flow Measurement, Korea: 313–319.

Takamoto, M. (1996) New flowmeter technology for the next century. FLOMEKO '96 Proceedings of the 8th International Conference on Flow Measurement, China: Beijing: 7–12.

Takamoto, M. and Komiya, K. (1981) Application of a ring to a bluff body of a vortex shedding flowmeter. *Transactions of the Society of Instrument and Control Engineers, Japan*,. 17: 506–510, (in Japanese).

Takamoto, M. and Terao, Y. (1994) Development of a standard vortex shedding flowmeter. FLOMEKO '94 Flow Measurement in the Mid 90's, NEL, Glasgow, Scotland.

Takamoto, M., Ishiashi, M., Watanabe, N., Aschenbrenner, A. and Caldwell, S. (1993a) Intercompariosn tests of gas flowrate standards. FLOMEKO '93 Proceedings of the 6th International Conference on Flow Measurement, Korea: 75–84.

Takamoto, M., Utsumi, H., Watanabe, N. and Terao, Y. (1993b) Installation effects on vortex shedding flowmeters. *Journal of Flow Measurement & Instrumentation*, 4: 277–285.

Takeda, Y. and Shaik, J. (2008) Editorial, Special Issue: ISUD 5: The 5th International Symposium on Ultrasonic Doppler Methods for Fluid Mechanics and Fluid Engineering. *Flow Measurement and Instrumentation*, 19(3–4): 129.

Tan, C. and Dong, F. (2006) Two-phase flow measurement by dual-plane ERT system with drift-flux model and cross correlation technique. Proceedings of the 2006 International Conference on Machine Learning and Cybernetics, ICMLC 2006, 20061443-8.

Tan, P. A. K. (1973) Theoretical and experimental studies of turbine flowmeter. PhD Thesis, University of Southampton.

(1976) Effect of upstream disturbances and velocity profiles on turbine meter performance. I Mech E Conference Paper C77.

Tan, P. A. K. and Hutton, S. P. (1971) Experimental, analytical and tip clearance loss studies in turbine-type flowmeters. Proc. International Conference on Flow Measurements, Harwell PPL Conference Publication 10: 321–346.

Tang, S. P. (1969) Theoretical determination of the discharge coefficients of axisymmetric nozzles under critical flows. Project SQUID Technical report PR-118-PU.

Tarabad, M. and Baker, R. C. (1979) Electromagnetic flowmeters for sodium-cooled nuclear reactors, Paper 6b-6, IMEKO Japan, November 1979.

(1982) Integrating electromagnetic flowmeter for high magnetic Reynolds numbers. *Journal of Physics D Applied Physics*, 15: 739–745.

(1983) Computation of pulsed field electromagnetic flowmeter response to profile change. *Journal of Physics D Applied Physics*, 16: 2103–2111.

Taylor, J. W. and Cassidy, H. P. (1994) Acoustic flowmeter comparison tests at BC Hydro, Canada. FLOMEKO '94 Flow Measurement in the Mid-90s, NEL, Glasgow, Scotland: Paper 7.4.

Taylor, R. D. H. and Nuttall, R. C. H. (1993) On-line well monitoring and its application in a South Oman oil field. *Proc Middle East Oil Show, AIME*, 2: 229–234.

Temperley, N. C., Behnia, M. and Collings, A. F. (2000) Flow patterns in an ultrasonic liquid flowmeter. *Journal of Flow Measurement and Instrumentation*, 11: 11–18.

(2004) Application of computational fluid dynamics and laser Doppler velocimetry to liquid ultrasonic flow meter design. *Flow Measurement and Instrumentation*, 15(3): 155–165.

Terao, Y., Choi, H. M., Edra, R. B. and Chen, Z. L. (1993) An experimental study on flow struucture in vortex flowmeters. FLOMEKO '93 Proceedings of the 6th International Conference on Flow Measurement, Korea: 327–334.

Teufel, M., Trimis, D., Lohmuller, A., Takeda, Y. and Durst, F. (1992) Determination of velocity profiles in oscillating pipe-flows by using laser Doppler velocimetry and ultrasonic measuring devices. *Journal of Flow Measurement and Instrumentation*, 3: 95–102.

Tezuka, K., Mori, M., Suzuki, T. and Kanamine, T. (2008a) Ultrasonic pulse-Doppler flow meter application for hydraulic power plants. *Flow Measurement and Instrumentation*, 19(3–4): 155–162.

Tezuka, K., Mori, M., Suzuki, T. and Takeda, Y. (2008b) Calibration tests of pulse-Doppler flow meter at national standard loops. *Flow Measurement and Instrumentation*, 19(3–4): 181–187.

Thatcher, G., Bentley, P. G. and McGonigal, G. (1970) Sodium flow measurement in PFR. *Nuclear Engineering International*, 15: 822–825.

Theuveny, B. and Walker, J. (2001) Flow meters enhance well test data. *Hart's E and P*, 74(11): 55–56.

Theuveny, B. C., Pithon, J. F., Loicq, O. and Segeral, G. (2002a) Worldwide field experience of mobile well testing services with multiphase flowmeters. *American Society of Mechanical Engineers, Petroleum Division* (Publication) PD, 2: 771–85.

Theuveny, B., Pinguet, B., Pittman, D., Ségéral, G. and Hanssen, B. V. (2002b) Field performance of dual energy spectral gamma ray/Venturi multiphase flowmeters. 5th International symposium of Fluid Flow Measurement, Arlington, VA, USA.

Thinh, N. D. and Evangelisti, J. (1997) Flow modelling and experimental investigation of a vortex shedding flowmeter. ASME Fluids Engineering Division Summer Meeting FEDSM'97, Paper 3011.

Thomas, A., Keech, R., Burt, A. and Yeung, H. (2001) Applying low powered EM metering in the UK water industry—field and lab performance aspects. Flow Measurement 2001, NEL International Conference, Scotland.

Thomas, A., Sheldon, R., Fray, M. K. and Kobryn, P. (2004a) Advances in on-site verification of water flow meters. The intelligent application of electromagnetic insertion probes. 12th International Conference on Flow Measurement FLOMEKO Guilin China.

Thomas, A., Kobryn, P. and Franklin, B. (2004b) Electromagnetic insertion probe calibration. Advances towards a standard. 12th International Conference on Flow Measurement FLOMEKO Guilin China.

Thomas, N. H., Auton, T. R., Sene, K. and Hunt, J. C. R. (1983) Entrapment and transport of bubbles by transient large eddies in multiphase turbulent shear flow. International conference on Physical modelling of multi-phase flow, Coventry, England, BHRA Fluid Engineering, Cranfield, Paper E1: 169–184.

Thomasson, J. A., Pennington, D. A., Pringle, H. C., Columbus, E. P., Thomson, S. J. and Byler, R. K. (1999) Cotton mass flow measurement: experiments with two optical devices. *Applied Engineering in Agriculture*, 15(1):11–7.

Thompson, E. J. (1978) Two beam ultrasonic flow measurement. PhD thesis, University of London.

Thompson, R. E. and Grey, J. (1970) Turbine flowmeter performance model. *Journal of Basic Engineering, Transactions onASME*: 712–723.

Thorn, R., Johansen, G. A. and Hammer, E. A. (1997) Recent developments in three-phase flow measurement. *Measurement Science and Technology*, 8: 691–701.

Thorn, R., Johansen, G. A. and Hjertaker, B. T. (2013) Three-phase flow measurement in the petroleum industry. *Measurement Science and Technology*, 24: 012003 (17pp) http://dx.doi.org/10.1088/0957-0233/24/1/012003.

Thornton, A. C. (1999a) Mathematical framework for the key characteristic process. *Research in Engineering Design*, 11: 145–157.

(1999b) Variation risk management using modelling and simulation. *Transactions on ASME, Journal of Mechanical Design*, 121: 297–304.

(2000) Quantitative selection of variation reduction plans. *Journal of Mechanical Design*, 122(2): 185–193.

Thürlemann, B. (1955) On the electromagnetic speed measurement of fluid. *Helvetica Physica Acta*, 28: 483.

Timothy, J. and Cunningham P. E. (1997) Zero shifts due to non-proportional damping. Proceedings of the 15th International Modal Analysis Conference IMAC, Orlando, FL, USA, 1:237–243.

Ting, V. C. and Ao, X. S, (2002) Evaluation of clamp-on ultrasonic gas transit time flowmeters for natural gas applications. 20th North Sea Flow Measurement Workshop 22–25 October 2002, St Andrews, Scotland, Paper 3.1.

Ting, V. C. and Shen, J. J. S. (1989) Field calibration of orifice meters for natural gas flow. *Journal of Energy Resources Technology, Transactions of ASME*, 111.1: 22–33.

Tison, S. A. and Berndt, L. (1997) High-differential-pressure laminar flowmeter. ASME Fluids Engineering Division Summer Meeting FEDSM'97, Paper 3207.

Tobi, N. V. (1953) British Patent 726 271, 27th May 1953.

Tokarczuk, P. F., Sanderson, M. L. and High, G. (1998) The application of nuclear magnetic resonance (NMR) to multiphase flow metering. FLOMEKO '98 Proceedings of the 9th International Conference on Flow Measurement, Lund, Sweden, 291–296.

Tombs, M. (2003) Coriolis special issue. *IEE Computing and Control Engineering*, 14(4): 27.

Tombs, M., Henry, M., Yeung, H. and Lansangan, R. (2004) Coriolis mass flow meter developments: increasing the range of applications in oil & gas production and processing. 22nd North Sea Flow Metering Workshop 2004, Paper 7.1

Tombs, M., Henry, M., Zhou, F., Lansangan, R. M. and Reese, M. (2006) High precision Coriolis mass flow measurement applied to small volume proving. *Flow Measurement and Instrumentation*, 17(6): 371–82.

Tominaga, K., Yudate, T. and Cormier, M. A. (2005) Reactor power uprate by ultrasonic flow meter. 26th Annual Canadian Nuclear Society Conference and 29th CNS/CNA Student Conference, pp. 1369–1378.

Toral, H., Beg, N. and Archer, J. S. (1990) Multiphase flow metering by software. International Conference on Basic Principles and Industrial Applications of Multiphase Flow, IBC Technical Services Ltd, London, April.

Toral, H., Cai, S., Peters, R. and Steven, R. (2004a) A method for characterization of the turbulence properties of wet gas flow across a V-Cone. 12th International Conference on Flow Measurement FLOMEKO Guilin China.

Toral, H., Cai, S., Steven, R. and Peters, R. (2004b) Characterization of the turbulence properties of wet gas flow in a V-cone meter with neural nets. 22nd North Sea Flow Metering Workshop 2004, Paper 8.4.

Torkildsen, B. H. and Hanssen, B. V. (1996) Practical considerations related to multiphase metering of a well stream. North Sea Flow Measurement Workshop, Peebles, Scotland.

Torkildsen, B. H., Helmers, P. B. and Kanstad, S. K. (1997) Topside and subsea experience with the FRAMO multiphase meters. North Sea Flow Measurement Workshop, Kristiansand, Norway: Paper 3.

Tresch, T., Gruber, P. and Staubli, T. (2006) Comparison of integration methods for multi-path acoustic discharge measurements. 6th International Conference on IGHEM, Portland, USA, 2006.

Trigas, A. and Hope, S. H. (1991) A comparison of the accuracy and process control capability of turbine and Coriolis flowmeters. International Conference on Flow Measurement In Industry and Science, London.

Tritton, D. J. (1988) *Physical fluid dynamics*. 2nd ed., Oxford University Press.

Trostel, B., Clancy, J. and Kegel, T. (2010) Ultrasonic flowmeter calibration intervals. 15th Flow Measurement Conference (FLOMEKO), 13–15 October 2010, Taipei, Taiwan, Paper A4-1.

Trung, M. C., Nishiyama, S. and Anyoji, H. (2007) Application of a 45° bend pipe with a bypass flow meter without changing the flow direction or the system construction. *Transactions of the ASABE*, 50(6): 2051–2057.

Tsuchida, T., Terashima, Y. and Machiyama, T. (1982) The effects of flow velocity profile on the electromagnetic flowmeters. Report of Researches, Nippon Institute of Technology: 101–111.

Tsuchiya, et al. (1970) Karman vortex flow meter. *Bulletin Japanese Society of Mechanical Engineers*, 13: 573–578.

Tsukamoto, H. and Hutton, S. P. (1985) Theoretical prediction of meter factor for a helical turbine flowmeter. Conference on Fluid Control and Measurement, Tokyo, Japan.

Tsutsui, H. and Yamikawa, Y. (1993) Coriolis force mass-flowmeter composed of a straight pipe and an additional resonance vibrator. *Japanese Journal of Applied Physics*. 1, 32(58): 2369–2371.

Tucker, H. G. and Hayes, W. F. (1982) Error analysis of a vibrating pendulum two phase flowmeter for oil well application. Measurement in Polyphase Flows - 1982, AIAA/ASME Joint Fluids, Plasma, Thermophysics and Heat Transfer Conference, St Louis, Missouri: 45–53.

Tullis, P. and Smith, J. (1979) Coriolis flowmeter. NEL Fluid Mechanics Silver Jubilee Conference, East Kilbride, Glasgow: Paper 6.3.

Turkowski, M. (2003) Progress towards the optimisation of a mechanical oscillator flowmeter. *Flow Measurement and Instrumentation*, 14(1–2): 13–21.

(2004) Influence of fluid properties on the characteristics of a mechanical oscillator flowmeter. *Measurement: Journal of the International Measurement Confederation*, 35(1): 11–18.

Turner, D. (1971) A differential pressure flowmeter with linear response. Modern Developments in Flow Measurement, Paper 4.3: 191–199.

Turner, J., Wynne, R. and Hurren, P. (1989) Computation and techniques in flow measurement and their applications to flowmeter diagnostics. FLOMIC, Rept. No. 6.

Turner, J. T., Popiel, C. O. and Robinson, D. I. (1993) Evolution of an improved vortex generator. *Journal of Flow Measurement & Instrumentation*, 4: 249–258.

Tuss, B. (1996) Production evaluation and testing of a high viscosity and high gas volume fraction multiphase meter. North Sea Flow Measurement Workshop, Peebles, Scotland.

(1997) *Wet gas multi-phase measurement*. Offshore Technology Conference, Houston, TX, USA, 4: 517–522.

UKAS(2012) *The expression of uncertainty and confidence in measurement*. 3rd ed., *United Kingdom Accreditation Service*.

Urner, G. (1997) Pressure loss of orifice plates according to ISO 5167-1. *Journal of Flow Measurement and Instrumentation*, 8: 39–41.

van Bloemendaal, K. and van der Kam, P. M. A. (1994) Installation effects on multi-path ultrasonic flow meters: the 'Ultraflow' project. North Sea Flow Measurement Workshop, Peebles, Scotland.

van Bokhorst, E. and Peters, M. C. A. M. (2000) Impact of pulsation sources in pipe systems on multi-path ultrasonic flowmeters. North Sea Flow Measurement Workshop, National Engineering Laboratory, East Kilbride, Scotland.

(2002) A test certificate on the impact of piping and flow dynamic effects on flowmeter accuracy in gas and liquid flows. 5th International symposium of Fluid Flow Measurement, Arlington, VA, USA.

van Cleve, C., Lanham, G., Ollila, C. and Stack, C. (2000) Development and validation of a new single straight tube Coriolis meter. FLOMEKO'2000 the 10th International Conference on Flow Measurement, Salvador, Brazil: Paper D11.

van Dellen, K. (1991) Ultrasonic gas flow meters continue their rise. Proceedings of the North Sea Flow Measurement Workshop, Norwegian Society of Chartered Engineers.

van Deventer, J. (2005) Introduction of a 2 transducer ultrasonic mass flow meter. *Conference Record – IEEE Instrumentation and Measurement Technology Conference*, 2: 1369–1372.

van Deventer, J. and Delsing, J. (2002) Apparent transducer non-reciprocity in an ultrasonic flow meter. *Ultrasonics*, 40(1–8): 403–405.

van Dijk, G. J. A. and Huijsing, J. H. (1995) Bridge-output-to-frequency converter for smart thermal air-flow sensors. *IEEE Transactions on Instrumentation and Measurement*, 44: 881–886.

van Essen, G. J. (2010) Testing the performance of an ultrasonic clamp-on flowmeter. 9th Spith East Asia Hydrocarbon Flow Measurement Workshop, 2nd–4th March.

van Luijk, L. and Riezebos, H. (2014) In situ flow verification by means of US clamp-on technology. US and Coriolis Metering Workshop, Lisbon, Portugal, 27th March 2014. http://www.vsl.nl/sites/default/files/rtf/Lennart_van_Luijk_In_Situ_Flow_Verification_by_Means_of_USM_Clamp_on_Technology%20%281%29.pdf

van Mannen. H. (1999) Cost reduction for wet-gas measurement using the tracer-venturi combination, NEL Natural Gas Metering one day seminar.

van Santen, H., Kolar, Z. I. and Scheers, A. M. (1995) Photon energy selection for dual energy γ - and/or X-ray absorption composition measurements in oil-water-gas mixtures. *Nuclear Geophysics*, 9(3): 193–202.

van Weers, T., van der Beek, M. P. and Landheer, I. J. (1998) Cd – factor of Classical Venturi's: Gaming Technology? 9th Int. Conf. on Flow Measurement, FLOMEKO, Lund, Sweden, June, pp. 203–207.

van Werven, M., Drenthen, J., de Boer, G. and Kurth, M. (2006) Wet gas flow measurement with ultrasonic and differential pressure metering technology. Proc. 6th International Symposium On Fluid Flow Measurement (ISFFM), Querétero, Mexico.

van den Heuvel, A. and Kemmoun, H. F. (2005) Flow measurement errors due to stratified flow conditions. FLOMEKO 2005 13th International Flow Measurement Conference, Peebles, Scotland, Paper 1.3.

van den Heuvel, A., Doorman, F., van den Herik, P., Stehouwer, A. and Kruithof, R. (2009) Calibration errors of ultrasonic meters in the Bernoulli laboratory due to non-isothermal flow conditions. 27th International North Sea Flow Measurement Workshop, Tonsberg, Norway, 20–23 October 2009, pp. 205–219.

van der Grinten, J. G. M. (1990) *Error curves of turbine gas meters*. Netherlands Metrology Institute.

(1994) A comparison of the methods for uncertainty analysis based on ISO 5168 and the Guide prepared by ISO/TAG4/WG3. FLOMEKO'94 Flow Measurement in the mid 90's, East Kilbride, Scotland.

(1997) Recent developments in the uncertainty analysis of flow measurement processes. North Sea Flow Measurement Workshop, Kristiansand, Norway: Paper 11.

van der Grinten, J. (2005) The Reynolds interpolation method for calibrations of turbine gas meters and application to intercomparisons. FLOMEKO 2005 13th International Flow Measurement Conference, Peebles, Scotland, Poster Session.

van der Kam, P. M. A. and Dam, A. M. (1993) Large turbine meters for custody transfer measurement: the renovation of the Gasunie export stations. *Journal of Flow Measurement and Instrumentation*, 4: 91–98.

van der Kam, P. M. A. and De Jong, S. (1994) Gas turbine meters: standardization and operational experiences. *Journal of Flow Measurement and Instrumentation*, 5: 121–126.

van der Kam, P. M. A. and van Dellen, K. (1991) The effect of double bends out of plane on turbine meters. *Journal of Flow Measurement and Instrumentation*, 2: 61–68.

van der Kam, P. M. A., Dam, M. A. and van Dellen, K. (1990) Gasunie selects turbine meters for renovated export metering stations. *Oil and Gas Journal*, 88: 39–44.

Vass, G. E. (1996) Users benefit from newer electronics in today's smarter magnetic flowmeters. *Advances in Instrumentation and Control: International Conference and Exhibition, (ISA)*, 51(1): 641–656.

Vaterlaus, H-P. (1995) A new intelligent ultrasonic flowmeter for closed conduits and open channels. *ASCE Waterpower – Proceedings of the Int Conf on Hydropower*, 2: 999–1008.

VDI/VDE (December 1978) Variable area flowmeters – accuracy. VDI/VDE 3513, Page 2, (In German). Updated edition VDI/VDE 7513-2: 2008.

Vermeulen, M., de Boer, G., van Weelde, A. B., Botte, E. and Dijkmans, R. (2004) Coded Multiple Burst (CMB) signal processing applied to ultrasonic flow meters in applications with high noise levels. 22nd North Sea Flow Metering Workshop 2004, Paper 3.3.

Vermeulen, M. J. M., Drenthen, J. G. and Den Hollander, H. (2012) Expert systems in ultrasonic flow meters. 8th International Symposium on Fluid Flow Measurement.

Vervik, S. (2000) Methods for characterization of gas-coupled ultrasonic sender-receiver measurement systems. Doctoral dissertation, University of Bergen, Department of Physics.

Vestrheim, M. and Vervik, S. (1996) Transit time determination in a measurement system, with effects of transducers. Proc. of 1996 IEEE Intern. Ultrason. Symp.

Vetter, G. and Notzon, S. (1994) Effect of pulsating flow on Coriolis mass flowmeters. *Journal of Flow Measurement and Instrumentation*, 5: 263–273.

Vieth, D., de Boer, G., Buijen van Welden, A. and Huijsman, F. (2001) Test results of a new design ultrasonic gas flow meter. 19th North Sea Flow Measurement Workshop, Kristiansand, Norway, 22–25 October 2001, Paper 7.

Viswanathan, M., Kandaswamy, A., Sreekala, S. K. and Sajna, K. V. (2001) Development, modelling and certain investigations on thermal mass flow meters. *Flow Measurement and Instrumentation*, 12(5–6): 353–360.

Viswanathan, M., Rajesh, R. and Kandaswamy, A. (2002) Design and development of thermal mass flowmeters for high pressure applications. *Flow Measurement and Instrumentation*, 13(3): 95–102.

Vogtlin, B. and Tschabold, P. (undated) Direct measurement of mass flow using the Coriolis force. E&H Flowtec publication.

von Lavante, E. and Mickan, B. (2005) Unsteady transition in critical Venturi nozzles. FLOMEKO 2005 13th International Flow Measurement Conference, Peebles, Scotland, Paper 5.3.

von Lavante, E. and Yao, J. (2010) Numerical investigation of turbulent swirling flows in flow metering configurations. 15th Flow Measurement Conference (FLOMEKO), 13–15 October 2010, Taipei, Taiwan, Paper B9-4.

(2012) Numerical investigation of turbulent swirling flows in axisymmetric internal flow configurations. *Flow Measurement and Instrumentation*, 25: 63–68.

von Lavante, E., Zachcial, A., Nath, B. and Dietrich, H. (2000) Numerical and experimental investigation of unsteady effects in critical Venturi nozzles. *Flow Measurement and Instrumentation*, 11(4): 257–264.

(2001) Unsteady effects in critical nozzles used for flow metering. *Measurement: Journal of the International Measurement Confederation*, 29(1): 1–10.

von Lavante, E., Mickan, B. and Kramer, R. (2004a) Numerical investigation of transition as effects in critical Venturi nozzles. 12th International Conference on Flow Measurement FLOMEKO Guilin China.

von Lavante, E., Banaszak, U. and Lefebvre, M. (2004b) Effect of shape change due to wear on the accuracy of vortex-shedding flow meters. 12th International Conference on Flow Measurement FLOMEKO Guilin China.

von Lavante, E., Banaszak, U., Kettner, T. and Lötz-Dauer, V. (2004c) Numerical simulation of Reynolds number effects in a turbine flow meter. 12th International Conference on Flow Measurement FLOMEKO Guilin China.

von Lavante, E., Banaszak, U., Lötz-Dauer, V., Enste, K., Bergervoet, J. and Dietrich, H. (2006) Theoretical and experimental investigations of rotary piston flow meters. Proc. 6th International Symposium on Fluid Flow Measurement (ISFFM), Querétero, Mexico.

von Lavante, E., Banaszak, U., Yilmaz, M. G. and Ricken O. (2007) Effects of shape change due to wear on the accuracy of vortex-shedding flow meters. 14th International Flow Measurement Conference 2007 (FLOMEKO 2007) Proceedings of a meeting held 18–21 September 2007, Sandton, Gauteng, South Africa.

von Lavante, E., Gedikli, A., Thibaut, A., Tournillon, S. and Krisch, H. (2010) Effects of upstream butterfly valve on the accuracy of a vortex flow meter. 15th Flow Measurement Conference (FLOMEKO), 13–15 October 2010, Taipei, Taiwan, Paper B2-3.

von Lavante, E., Brinkhorst, S., Gedikli, A. and Krisch, H. (2013) Fluid mechanical optimization of a dn25 vortex flow meter with novel vortex detection. FLOMEKO (2013) The 16th International Flow Measurement Conference - 24–26th September 2013 - Paris, France.

Vontz, T. and Magori, V. (1996) Ultrasonic flowmeter for industrial applications using a helical sound path. Proceedings of the IEEE Ultrasonics Symposium, San Antonio, TX, USA, 2: 1047–1050.

Vsesojuzny Nauchno-Issledovatelsky (1976) Method of and apparatus for measuring the mass flowrate of individual components of a gas-liquid medium. British Patent 1,528,232.

Vulovic, F., Vallet, J. P. and Windenberger, C. (2002) The advantages of critical flow Venturi nozzles for the high pressure gas metering. 5th International symposium of Fluid Flow Measurement, Arlington, VA, USA.

Wada, S., Kikura, H. and Aritomi, M. (2006) Pattern recognition and signal processing of ultrasonic echo signal on two-phase flow. *Flow Measurement and Instrumentation*, 17(4): 207–224.

Wagner, J. J. (1988) *Effects of sensor design and application characteristics on Coriolis mass meter performance: an overview*. 2nd Int Conf on Flow Measurement, London, UK.

Walker, J. T. (1992) Advances in Coriolis technology for precision flow and density measurements of industrial fluids. Proc 47th American Symposium on Instrumentation for the Process Industries, Publ. Texas A&M Univ: 69–73.

Walker, J.R. (2001) Diagnostics advance in electromagnetic flow metering. *ISA TECH/EXPO Technology Update Conference Proceedings*, 413 I: 501–510.

Walles, K. F. A. (1975) The long term repeatability of positive displacement liquid flowmeters. Conference on Fluid Flow Measurement in the mid 1970s, National Engineering Laboratory, Scotland: Paper B-4.

Walles, K. F. A. and James, J. H. P. (1985) Performance of a fuel flow calibration system. Int. Conf. on the Metering of Petroleum and its Products, OYEZ, London.

Wallis, R. A. (1961) *Axial flow fans design and practice*. George Newnes Limited, London.

Walsh, J. T. (2004) A report of acoustic transit time accuracy field work performed in North America, 5th International Conference on Hydraulic Efficiency Measurements, Lucerne, 2004.

Walus, S. (2000) Decreasing of volume flow rate measurement error in modified averaging impact tubes. FLOMEKO'2000 the 10th International Conference on Flow Measurement, Salvador, Brazil: Paper B2.

Wang, C., Li, F., Meng, T., Xu, Y., Yang, Y., Jiang, N., Wu, W., Sang, X., Chen, M., Zhou, B., Zang, J., Shen, W. and Xu, L. (2004) Intercomparison tests of gas flow by bell prover. 12th International Conference on Flow Measurement FLOMEKO Guilin China.

Wang, G., Chen, C., Yao, Y. and Huang, L. (2009a) Micromachined mass flow sensor and insertion type flow meters and manufacture methods. US Patent 7536908, 26 May 2009.

Wang, H., Priestman, G. H., Beck, S. B. M. and Boucher, R. F. (1996) Development of fluidic flowmeters for monitoring crude oil production. *Journal of Flow Measurement and Instrumentation*, 7: 91–98.

(1998) A remote measuring flow meter for petroleum and other industrial applications. *Measurement Science and Technology*, 9: 779–789.

Wang, J. (2009) Numerical simulation and verification of weight function of electromagnetic flow meter. *Chinese Journal of Scientific Instrument*, 30(1): 132–137 (in Chinese).

Wang, J. and Gong, C. (2006) Sensing induced voltage of electromagnetic flow meter with multi-electrodes. Proceedings of IEEE ICIA 2006–2006 IEEE International Conference on Information Acquisition, 1031–1036, Weihai, Shandong, China

Wang, J. and Lu, R. (2006) Numerical simulation on weight function of electromagnetic flow meter. *Proceedings of SPIE – The International Society for Optical Engineering*, 6280 II: 628033.

Wang, J. Z., Tian, G. Y. and Lucas, G. P. (2007a) Relationship between velocity profile and distribution of induced potential for an electromagnetic flow meter. *Flow Measurement and Instrumentation*, 18(2): 99–105.

Wang, J. Z., Lucas, G. P. and Tian, G. Y. (2007b) A numerical approach to the determination of electromagnetic flow meter weight functions. *Measurement Science and Technology*, 18(3): 548–554.

Wang, J. Z., Tian, G. Y., Simm, A. and Lucas, G. P. (2009b) Uniform magnetic flux density simulation and verification for a new electromagnetic flow meter. *Nondestructive Testing and Evaluation*, 24(1–2): 143–151.

Wang, L.-J., Hu, L., Fu, X. and Ye, P. (2010) Experimental investigation on zero drift effect in Coriolis mass flowmeters. 15th Flow Measurement Conference (FLOMEKO), 13–15 October 2010, Taipei, Taiwan, Paper A9-1.

Wang, S., Clark, C. and Cheesewright, R. (2006a) Virtual Coriolis flow meter: a tool for simulation and design. *Proceedings of the Institution of Mechanical Engineers, Part C: Journal of Mechanical Engineering Science*, 220(6): 817–835.

Wang, T. and Baker, R. C. (2003/4) Manufacturing variation of the measuring tube in a Coriolis flowmeter. *IEE Computing and Control Engineering*, 14(4): 38–39 and in *IEE Proceedings, Sci. Meas. Technol.* 151(3): 201–204, 2004. (Also in IEE Advanced Coriolis Mass Flow Metering Seminar, Oxford University 2003 in summary.)

(2014) Coriolis flowmeters: a review of developments over the past 20 years, and an assessment of the state of the art and likely future directions. *Flow Measurement and Instrumentation*, 40(2014): 99–123.

Wang, T. and Hussain, Y. A. (2006) Investigation of the batch measurement errors for single-straight tube Coriolis mass flowmeters. *Flow Measurement and Instrumentation*, 17(6): 383–390.

(2007) Latest research and development of twin-straight tube Coriolis mass flowmeters. *Sensor Review*, 27(1): 43–47.

(2009) Coriolis mass flow measurement at cryogenic temperatures. *Journal of Flow Measurement and Instrumentation*, 20(3): 110–115.

(2010a) Extending flow measurement capacity with the straight tube Coriolis technology. 15th Flow Measurement Conference (FLOMEKO), October 13–15, 2010 Taipei, Taiwan, Paper A9-4.

(2010b) Pressure effects on Coriolis mass flowmeters. *Journal of Flow Measurement and Instrumentation*, 21(4): 504–510.

Wang, T., Baker, R. C. and Hussain, Y. A. (2006b) An advanced numerical model for single straight tube Coriolis flowmeters. *Transactions of ASME Journal of Fluids Engineering*, 128: 1346–1350, 2006. (Also presented as A practical numerical model for single straight tube Coriolis meters, at FLOMEKO 2004, Guilin.)

Wangsa, S., Latief, R., Kaura, J., Finley, D. and Ogilvie, A. (2005) Successful field surveillance using portable multi-phase flow meter in a high gas-volume fraction and high water-cut application in east Kalimantan, Indonesia. 67th European Association of Geoscientists and Engineers, EAGE Conference and Exhibition, incorporating SPE EUROPE2005, pp. 2503–2514.

Ward-Smith, A. J. (1980) *Internal fluid flow*. Clarendon Press, Oxford.

Warren, P. B., Al-Dusari, K. H., Zabihi, M. and Al-Abduljabbar, J. M. (2003) field-testing a compact multiphase flow meter – offshore Saudi Arabia. *Proceedings of the Middle East Oil Show*, 13: 993–998.

Washington, G. (1989) *Measuring the flow of wet gas*. North Sea Flow Metering Workshop, Haugesund, Norway.

Watkins, D., Lucchini, F., Weaver, P., Feltresi, E. and Genolini, M. (2014) In line multiphase flow measurement Permian Basin, Texas Field Trial. North Sea Flow Measurement Workshop 2014.

Watson, G. A. and Furness, R. A. (1977) Development and application of the turbine meter. Proc Transducer 77 Conf Flow Measurement Session, Wembley, London.

Watson, G. G., Vaughan, V. E. and McFarlane, M. W. (1967) Two-phase pressure drop with a sharp-edged orifice. NEL Report No 290, East Kilbride, Glasgow.

Watt, J. S. (1993) *Platform trial of a multiphase flow meter*. North Sea Flow Measurement Workshop, Bergen, Norway.

Watt, R. M. (1990) Computational modelling of Coriolis mass flowmeters. North Sea Flow Measurement Workshop, National Engineering Laboratory, Scotland.

(1991) *Modelling of Coriolis mass flowmeters using ANSYS*. ANSYS Users Conf, Pittsburgh, USA.

Way, J. and Wood, I. (2002) Kerr-McGee North Sea (UK) Limited – Gryphon Alpha FPSO Monetary Application for Multiphase Meters. 20th North Sea Flow Measurement Workshop, 22–25 October 2002, St Andrews, Scotland, Discussion session 5.1 (Paper 5.4).

Weager, B. (1993/4) NAMAS-approved flowmetering. *Measurement and Control*, 26(10): 298–301.

Węcel, D., Chmielniak, T. and Kotowicz, J. (2008) Experimental and numerical investigations of the averaging Pitot tube and analysis of installation effects on the flow coefficient. *Flow Measurement and Instrumentation*, 19(5): 301–306.

Wee, A. and Scheers, L. (2009) Measurement of water in a wet gas. 27th International North Sea Flow Measurement Workshop, Tonsberg, Norway, 20–23 October 2009, pp. 154–179.

Wee, A., Berentsen, H., Midttveit, V. R., Moestue, H. and Hide, H. O. (2007) Tomography powered multiphase and wet-gas meter providing measurements used for fiscal metering. 25th International North Sea Flow Measurement Workshop, Energy Institute, Oslo, Norway, 40–63.

Wee, A., Fosså, Ø. and Midttveit, V. R. (2013) Multiphase meter capable of detecting scale on the pipe wall and correcting flow rate measurements. North Sea Flow Metering Workshop, Paper 21.

Weigand, D. E (1972) Magnetometer flow sensor. Argonne National Laboratory, Argonne, Ill., Report ANL-7874.

Weigand, J. (1994) Gas flow measurement using laminar flow elements. *ASHRAE Trans.* 100: 973–979.

Weinig, F. (1932) Stromung durch Profilgitter und einige Anwendungen auf die Stromung in Propellern. *Hydromechanische Probleme des Schiffsantriebs*: 171.

Weiss, M., Studzinski, W. and Attia, J. (2002) Performance evaluation of orifice meter standards for selected T-junction and elbow installations. 5th International symposium of Fluid Flow Measurement, Arlington, VA, USA.

Wemyss, W. A. and Wemyss, A. C. (1975) Development of the Hoverflo: a turbine flowmeter without bearings. NEL Conference on Fluid Flow Measurement in the Mid 1970's, 2, Paper H1.

Wen Dong-xu (1990) Smart fiber optics flowmeter. *Proceedings of SPIE, Int Soc Optical Eng*, 1230: 557–558.

Wendoloski, J. C. (2001) On the theory of acoustic flow measurement. *Journal of the Acoustical Society of America*, 110(2): 724–737.

Wendt, G. and von Lavante, E. (2000)Influence of surface roughness on the flowrate behaviour of small critical Venturi nozzles. Proc. FLOMEKO.

Wendt, G., Mickan, B., Kramer, R. and Dopheide, D. (1996) Systematic investigation of pipe flows and installation effects using laser Doppler anemometry – Part I, *Profile Measurements Downstreaqm of Several Pipe Configurations and Flow Conditioners*. 7(3/4): 141–149.

Wenran, W. and Yunxian, T. (1995) A new method of two-phase flow measurement by orifice plate differential pressure noise. *Journal of Flow Measurement and Instrumentation*, 6: 265–270.

Whitaker, T. S. (1993) A review of multiphase flowmeters and future development potential. FLOMEKO '93 Proceedings of the 6th International Conference on Flow Measurement, Korea: 628–634.

(1996) *Assessment of multiphase flowmeter performance*. North Sea Flow Measurement Workshop, Peebles, Scotland.

Whitaker, T. S. and Millington, B. C. (1993) *Review of multiphase flowmeter projects*. North Sea Flow Measurement Workshop, Bergen, Norway.

Whitaker, T. and Owen, I. (1990) Experience with two designs of differential pressure flowmeters in two-phase flow. International Conference on Basic Principles and Industrial Applications of Multiphase Flow, April 1990 (IBC Technical Services, London).

White, D. F., Rodely, A. E. and McMurtie, C. L. (1974) The vortex shedding flowmeter. Flow, Its Measurement and Control in Science and Industry, Pittsburgh, PA, USA, Instrument Society of America, 1, Pt 2: 967–974.

Whitson, R. J. (2008) A general methodology for geometry related pressure and temperature corrections in ultrasonic time for flight flowmeters. 26th International North Sea Flow Metering Workshop 2008, Paper 6.2.

Widmer, A. E., Fehlmann, R. and Rehwald, W. (1982) A calibration system for calorimetric mass flow devices. *Journal of Physics E: Scientific Instruments*, 15: 213–220.

Wiegerink, R. J. Lammerink, T. S. Groenesteijn, J., Dijkstra, M. and Lotters, J. C. (2012) Micro Coriolis Mass Flow Sensor For Chemical Micropropulsion Systems, in: Micromachines, Enschede, The Netherlands, 2012.

Wiklund, D. and Peluso, M. (2002a) Quantifying and specifying the dynamic response of flowmeter. *ISA TECH/EXPO Technology Update Conference Proceedings*, 422: 463–75. see also ISA TECH/EXPO Technology Update Conference Proceedings, 424–425: 810–22).

(2002b) Flowmeter dynamic response characteristics Part 1: Quantifying dynamic response. 5th International symposium of Fluid Flow Measurement, Arlington, VA, USA.

(2002c) Flowmeter dynamic response characteristics Part 2: Effects in various flow applications. 5th International symposium of Fluid Flow Measurement, Arlington, VA, USA.

Wilcox, P., Barton, N. and Laing, K. (2001) Using computational fluid dynamics to investigate the flow through an offshore gas metering station. 19th North Sea Flow Measurement Workshop, Kristiansand, Norway, 22–25 October 2001, Paper 22.

Willatzen, M. (2001) Temperature gradients and flow-meter performance. *Ultrasonics*, 39(5): 383–389.

(2003) Ultrasonic flowmeters: Temperature gradients and transducer geometry effects. *Ultrasonics*, 41(2): 105–114.

(2004a) Ultrasonic flow measurement and wall acoustic impedance effects. *Ultrasonics*, 41(9): 719–726.

(2004b) Flow acoustics modelling and implications for ultrasonic flow measurement based on the transit-time method. *Ultrasonics*, 41(10): 805–810.

Willatzen, M. and Kamath, H. (2008) Nonlinearities in ultrasonic flow measurement. *Flow Measurement and Instrumentation*, 19(2): 79–84.

Willer, M. D. (1978) Gyroscopic principle key to mass flowmeter. *Canadian Controls & Instruments*, 117(1): 28.

Williams, E. J. (1930) The induction of emfs in a moving fluid by a magnetic field and its application to an investigation of the flow of liquids. *Proceedings of the Royal Society, London, England*, 42: 466–478.

Williams, T. J. (1970) Behaviour of the secondary devices in pulsating flowmeasurement. *Instrumentation Measurement& Control Symposium on the Measurement of Pulsating Flow*: 56–61.

Wilson, H. A. (1901) Phil Mag S5 2 No 7 July pp 144–150.

(1904) Phil Trans A, London, 204 pp 121–137.

Wilson, H. A. and Wilson, M. (1913) Phil Trans A, London, 89 pp 99–106.

Windorfer, H. and Hans, V (2000a) Correlation of ultrasound and pressure in vortex shedding flow-meters. FLOMEKO'2000 the 10th International Conference on Flow Measurement, Salvador, Brazil: Paper E7.

(2000b) Experimental optimisation of bluff bodies in ultrasound vortex shedding flow-meters. FLOMEKO'2000 the 10th International Conference on Flow Measurement, Salvador, Brazil: Paper E8.

Wislicenus, G. F. (1947) *Fluid mechanics of turbomachinery*. McGraw-Hill, New York (subsequent ed. Dover publ, inc. New York).

Withers, V. R., Inkley, F. A. and Chesters, D. A. (1971) Flow characteristics of turbine flowmeters. Proc International Conference on Flow Measurements, Harwell PPL Conference Publication, 10: 305–320.

Withers, V. R., Strang, W. and Allnutt, G. (1996) Practical application of Coriolis meters for offshore tanker loading from the Harding Field. North Sea Flow Measurement Workshop, Peebles, Scotland.

Witlin, W. G. (1979) Theory, design and application of vortex shedding flowmeters. *Proceedings of the Symposium on Measurement Technology for the 80s: Analytical Instrumentation*, 17: 120–125.

Wojtkowiak, J., Kim, W. N. and Hyun, J. M. (1997) Computations of the flow characteristics of a rotating-piston-type flowmeter. *Journal of Flow Measurement and Instrumentation*, 8: 17–25.

Womack, A. (2008) Flow meter selection for improved gas flow measurements. *Heat Treating Progress*, 8(4): 25–29.

Wong, H. A., Rhodes, E. and Scott, D. S. (1981) Flow metering in horizontal, adiabatic, two-phase flow. Proc. 2nd Symp. on Flow: Its Measurement and Control in Science and Industry (ed. W. W. Durgin), ISA, St Louis, 2: 505–516.

Woo, S. and O'Neal, D. L. (2006) The effect of elbows on the accuracy of liquid flow measurement with an insertion flowmeter. ASHRAE Transactions, (Monograph title: ASHRAE Transactions – Technical and Symposium Papers presented at the 2006 Winter Meeting of the American Society of Heating, Refrigerating and Air-Conditioning Engineers) 112(PART 1):195–201.

Wood, G. (1994) Introduction. *Measurement & Control*, 27 (2): 37.

Wood, I. M., Daniel, P. and Downing, A. (2003) Penguin Wet Gas Measurement. 21st North Sea Flow Metering Workshop 2003, Paper 5.

Worch, A. (1998a) A clamp-on ultrasonic cross correlation flowmeter for one-phase flow. *Measurement Science and Technology*, 9: 622–630.

(1998b) A clamp-on ultrasonic cross-correlation flowmeter for two-phase flow, FLOMEKO '98 Proceedings of the 9th International Conference on Flow Measurement, Lund, Sweden, pp. 121–126.

Wright, J. D. (2010) Properties for accurate gas flow measurements. 15th Flow Measurement Conference (FLOMEKO), 13–15 October Taipei, Taiwan, Paper B8-2.

Wright, J. D. and Johnson, A. N. (2000) Uncertainty in primary gas flow standards due to flow work phenomena. FLOMEKO'2000 the 10th International Conference on Flow Measurement, Salvador, Brazil: Paper D11.

Wright, J. D., Kayl, J. P., Johnson, A. N. and Kline, G. M. (2008) *NIST Measurement Services: Gas Flowmeter Calibrations with the Working Gas Flow Standard. NIST Special Publication*, 250–280.

Wright, J. D. Cobu, T., Berg, R. F. and Moldover, M. R. (2012) Calibration of laminar flow meters for process gases. *Flow Measurement and Instrumentation*, 25: 8–14.

Wright, P. H. (1993) The application of sonic (critical flow) nozzles in the gas industry. *Journal of Flow Measurement and Instrumentation*, 4: 67–72.

Wright, W. and Brini, S. (2005) Capacitive ultrasonic transducers for gas flow metering applications. FLOMEKO 2005 13th International Flow Measurement Conference, Peebles, Scotland, Paper 1.2.

Wu, G. B. and Meng, H. (1996) Application and improvement of the Youden analysis in the inter comparison between flowmeter calibration facilities. *Flow Measurement and Instrumentation*, 7(1): 19–24.

Wu, G. and Yan, S. (1996) The calculation of the discharge coefficient of critical Venturi nozzles using the finite element method. FLOMEKO '96 Proceedings of the 8th International Conference on Flow Measurement, China: Beijing: 611–618.

Wyatt, D. G. (1986) Electromagnetic flowmeter sensitivity with two-phase flow. *International Journal of Muultiphase Flow*, 12(6): 109–117.

Wylie, S. R., Shaw, A. and Al-Shamma'a, A. I. (2006) RF sensor for multiphase flow measurement through an oil pipeline. *Measurement Science and Technology*, 17(8): 2141–2149.

Xiaozhang, Z. (1995) New multi-meter system for flow measurement of water-oil-gas mixture. *ISA Advances in Instrumentationand Control: International Conference and Exhibition*, 50.1: 113–120.

Xing, J. and Zhang, T. (2009) Oil-water two-phase flow measurement using vortex flowmeter. *Chinese Journal of Scientific Instrument*, 30(4): 882–886 (in Chinese)

Xu, L. A., Yang, H. L., Zhang, T., Chen, W., Li, J. and Ran, Z. M. (1994) A clamp-on ultrasound cross-correlation flowmeter for liquid/solid two-phase flow measurement. *Journal of Flow Measurement and Instrumentation*, 5: 203–208.

Xu, L. J., Li, X. M., Dong, F., Wang, Y. and Xu, L.A. (2001) Optimum estimation of the mean flow velocity for the multi-electrode inductance flowmeter. *Measurement Science and Technology*, 12(8): 1139–1146.

Xu, L., Wang, Y., Dong, F. and Yan, Y. (2003) On-line monitoring of non-axisymmetric flow profile with a multi-electrode inductance flowmeter. *Conference Record – IEEE Instrumentation and Measurement Technology Conference*, 2: 1541–1546.

Xu, Y. (1992a) Calculation of the flow around turbine flowmeter blades. *Flow Measurement and Instrumentation*, 3: 25–35.

(1992b) A model for the prediction of turbine flowmeter performance. *Flow Measurement and Instrumentation*, 3: 37–43.

Xu, Y., Zhang, T., Wang, H., Liu, Z. and Chen, D. (2004) Computational investigation on the float-type flowmeter in three-dimensional turbulence flow field. 12th International Conference on Flow Measurement FLOMEKO Guilin China.

Xu-bin, Q. (Oct 1993) The simple economic elbow meter for flow measurement. *Measurement & Control*. 26: 245–246.

Xue, G. and Shen, Y. (2008) Study on measurement of oil gas water three phase flow with conductance correlative flow meter. *Proceedings of the IEEE International Conference on Automation and Logistics, ICAL* 2008, 1295–1297.

Yajun, L., Dian, T., Jun, L. and Lumkes, J. (2012) Wear behaviour of piston seals in flow meter of fuel dispenser under different pressure conditions. *Flow Measurement and Instrumentation*, 28: 45–49.

Yamamoto, T., Yao, H. and Kshiro, M. (2004) Advanced hybrid ultrasonic flow meter utilizing pulsed-Doppler method and transit time method. *Technical Papers of ISA*, 454: 163–173.

Yamasaki, H. (1993) Progress in hydrodynamic oscillator type flowmeters. *Journal of Flow Measurement and Instrumentation*, 4: 241–248.

Yamashita, Y. (1996) Development of Coriolis mass flowmeter with a single straight tube as flow tube. FLOMEKO'96 Proceedings of the 8th International Conference on Flow Measurement, Beijing, China: 265–270.

Yan, Y. (1996) Mass flow measurement of bulk solids in pneumatic pipelines. *Measurement Science and Technology*, 7: 1687–1706.

(2000) Flow measurement of particulate solids in pipelines. *Flow Measurement and Instrumentation*, 11(3): 151.

(2005a) Tomographic techniques for multiphase flow measurement. *Flow Measurement and Instrumentation*, 16(2–3): 63.

(2005b) Optical techniques for multiphase flow measurement. *Flow Measurement and Instrumentation*, 16(5): 275.

Yan, Y., Xu, L. and Lee, P. (2006) Mass flow measurement of fine particles in a pneumatic suspension using electrostatic sensing and neural network techniques. *IEEE Transactions on Instrumentation and Measurement*, 55(6): 2330–2334.

Yang, B., Cao, L. and Luo, Y. (2011) Forced oscillation to reduce zero flow error and thermal drift for non-reciprocal operating liquid ultrasonic flow meters. *Flow Measurement and Instrumentation*, 22(4): 257–264.

Yang, C.-T., Chen, J.-Y. and Shaw, J.-H. (2002) CFD simulation of three double-elbow pipe flows. 5th International symposium of Fluid Flow Measurement, Arlington, VA, USA.

Yang, W. Q. and Beck, M. S. (1997) An intelligent cross correlator for pipeline flow velocity measurement. *Journal of Flow Measurement and Instrumentation*, 8: 77–84.

Yao, J., Wang, W. G. and Shi, J. (2011) Study on electromagnetic flowmeter for partially filled flow measurement. 23rd Chinese Control and Decision Conference, 23–25 May 2011, Mianyang, China.

Yao, Y., Chen, C. C., Wu, X. and Huang, L. (2010) MEMS thermal time-of-flight flow meter. 15th Flow Measurement Conference (FLOMEKO), 13–15 October 2010, Taipei, Taiwan, Paper A8-3.

Yeh, T. T. and Mattingly, G. E. (1994) Pipeflow downstream of a reducer and its effects on flowmeters. *Journal of Flow Measurement and Instrumentation*, 5: 181–187.

(1997) Computer simulations of ultrasonic flowmeter performance in ideal and non-ideal pipeflows, ASME Fluids Engineering Division Summer Meeting FEDSM'97, Paper 3012.

(2000) Ultrasonic technology: prospects for improving flow measurements and standards. FLOMEKO'2000 the 10th International Conference on Flow Measurement, Salvador, Brazil: Paper A2.

Yeh, T. T., Espina, P. I. and Osella, S. A. (2001) An intelligent ultrasonic flow meter for improved flow measurement and flow calibration facility. *IEEE Instrumentation and Measurement Technology Conference*, 3: 1741–1746.

Yeung, H., Hemp, J., Henry, M. and Tombs, M. (2004) Coriolis meter in liquid/liquid, liquid/gas and liquid/liquid/gas flows. S.E. Asia Hydrocarbon Flow Measurement Workshop, Singapore, March 2004.

Yinping, J. (2007) Research and application of magnetic suspension technology in wide range turbine flow meter. Proceedings of the 2007 IEEE Sensors Applications Symposium, SAS, San Diego, CA, United States.

Yokota, S., Son, W. C. and Kim, D. T. (1996) Unsteady flow measurement by using a drag-plate-type force flowmeter. FLOMEKO '96 Proceedings of the 8th International Conference on Flow Measurement, Beijing, China: 661–666.

Yoo, S. Y., Lee, S. Y., Yoon, K. Y., Park, K. A. and Paik, J. S. (1993) Experimental study on the factors influencing discharge coefficients of sonic nozzles. FLOMEKO '93 Proceedings of the 6th International Conference on Flow Measurement: 363–371.

Yoshida, Y., Amata, Y. and Frugawa, M. (1993) Development of a partially-filled electromagnetic flowmeter. FLOMEKO'93 Proceedings of the 6th International Conference on Flow Measurement, Korea Research Institute of Standards and Science: 452–459.

Youden, W. J. (1959) Graphical diagnosis of interlaboratory test results. *Journal of Industrial Quality*, 15: 11.

Young, A. (1990) Coriolis flowmeters for accurate measurement of liquid properties. *Advances in Instrumentation, ISA Proceedings*, 45, Pt. 4: 1891–1898.

Yu, H., Gong, J., Li, Y., Liao, Y. and Liao, M. (2000) Optical fibre magneto-optic sensor in the turbine mass flowmeter. *Proceedings of SPIE – The International Society for Optical Engineering*, 4074: 166–170.

Yuan, B.-Z., Nishiyama, S., Fukada, M. and Kanamori, H. (2003) Hydraulic design procedure for bypass flow meters using a pipe bend. *Transactions of the American Society of Agricultural Engineers*, 46(2):279–85.

Yun, W. Y. and Park, M. H. (2008) An approach to reduce measurement uncertainty of fuel channel coolant flow rate for CANDU plants with a CROSSFLOW ultrasonic flow meter. Societe Francaise d'Energie Nucleaire – International Congress on Advances in Nuclear Power Plants – ICAPP 2007, "The Nuclear Renaissance at Work", Report number:7142: 848–852.

Zanker, K. J. (2001) An ultrasonic meter for stratified wet gas service. 19th North Sea Flow Measurement Workshop, Kristiansand, Norway, 22–25 October 2001, Paper 9.

(2006) The calibration, proving and validation of ultrasonic flow meters. Proc. 6th International Symposium on Fluid Flow Measurement (ISFFM), Queretero, Mexico.

Zanker, K. J. and Cousins, T. (1975) The performance and design of vortex meters. Conference on Fluid Flow Measurement in the Mid 1970s, National Engineering Laboratory, East Kilbride, Glasgow, Scotland: Paper C-3.

Zanker, K. J. and Freund, W. R. Jr. (1996) Practical experience with gas ultrasonic flow meters. North Sea Flow Measurement Workshop, Peebles, Scotland.

(2004) A powerful new diagnostic tool for transit time ultrasonic meters. 22nd North Sea Flow Metering Workshop 2004, Paper 3.1.

Zanker, K. and Goodson, D. (2000) Qualification of a flow conditioning device according to the new API 14.3 procedure. *Journal of Flow Measurement and Instrumentation*, 11(2): 79–87.

Zanker, K. J. and Mooney, T. (2003) The transit time difference ultrasonic gas meter – a reassessment. 21st North Sea Flow Metering Workshop 2003, Paper 10.

(2010) Celebrating quarter of a century of gas ultrasonic custody transfer metering. 28th International North Sea Flow Metering Workshop, Paper 9.2

Zedan, M. F. and Teyssandier, R. G. (1990) Effect of errors in pressure tap locations on the discharge coefficient of a flange-tapped orifice plate. *Journal of Flow Measurement and Instrumentation*, 1: 141–148.

Zhang, F., Dong, F. and Tan, C. (2010) High GVF and low pressure gas-liquid two-phase flow measurement based on dual-cone flowmeter. *Journal of Flow Measurement and Instrumentation*, 21(3): 410–417.

Zhang, H., Huang, Y. and Sun, Z. (2006) A study of mass flow rate measurement based on the vortex shedding principle. *Flow Measurement and Instrumentation*, 17(1): 29–38.

Zhang, J., Coulthard, J., Cheng, R. and Keech, R. (2004c) On-line flow measurement and control of pulverised fuel. *Measurement and Control*, 37(9): 273–275.

Zhang, L., Hu, H., Meng, T. and Wang, C. (2013) Effects of flow disturbance on multipath ultrasonic flowmeters. FLOMEKO 2013 Proceedings of 16th International Conference on Flow Measurement, Paris, France.

Zhang, T., Sun, H. and Wu, P. (2004a) Wavelet denoising applied to vortex flowmeters. *Flow Measurement and Instrumentation*, 15(5–6): 325–329.

Zhang, X., Wu, P., Liang, M. and Sun, Y. (2004b) The experimental study and uncertainty analysis on the double turbine mass flowmeter with the corrective property for velocity distribution. 12th International Conference on Flow Measurement FLOMEKO Guilin China.

Zhang, X-Z. (1997) The effect of the phase distribution on the weight function of an electromagnetic flow meter in 2D and in the annular domain. *Measurement Science and Technology*,. 8: 1285–1288.

(1998) Virtual current of an electromagnetic flowmeter in partially filled pipes. *Measurement Science and Technology*, 9: 622–630.

(1999) On the virtual current in an electromagnetic flow meter containing a number of bubbles by two-dimensional analysis. *Measurement Science and Technology*,, 10: 1087–1091.

(2001) Calculation and measurement of the magnetic field in a large diameter electromagnetic flow meter. *ISA TECH/EXPO Technology Update Conference Proceedings*, 416: 287–290.

(2002) Theory to help the use of electromagnetic flow meters. American Society of Mechanical Engineers, Fluids Engineering Division (Publication) FED, 257: 11–14.

(2003) Calculation and measurement of the magnetic field in a large diameter electromagnetic flow meter. *ISA Transactions*, 42(2): 167–170.

(2007) Measurement errors caused by asymmetry in electromagnetic flow meter. FLOMEKO 2007 Proceedings of a meeting held 18–21 September 2007, Sandton, Gauteng, South Africa.

(2010) *Theory and methods for flow measurement by electromagnetic induction*. Tsinghua University Press.

(2012) Preliminary experimental study on multi-parameter measurement of fluid flow by vibrating tube, *Applied Mechanics and Materials*, 241–244: 70–74.

Zhang, X. Z. and Hemp, J. (1994) Measurement of pipe flow by an electromagnetic probe. *ISA Transactions*, 33: 181–184.

(1995) Calculation of the virtual current around an electromagnetic velocity probe using the alternating method of Schwarz. *Flow Measurement and Instrumentation*, 5: 146–149.

Zhang, X.-Z. and Li, Y. (2004) Calculation of the virtual current in an electromagnetic flow meter with one bubble using 3D model. *ISA Transactions*, 43(2): 189–194.

Zhang, X.-Z. and Wang, G.-Q. (2004) Electromagnetic inductive flow pattern reconstruction by means of spectrum expanding. 12th International Conference on Flow Measurement FLOMEKO Guilin China.

Zhang, Y. and Li, Z. (2015) Improving the accuracy of time-difference measurement by reducing the impact of baseline shift. *IEEE Transactions on Instrumentation and Measurement*, 64(11): 3013–3020.

Zhao, W. and Cao, C. (2008) Study of computerized tomography based multiphase flow measurement. Proceedings of the IEEE International Conference on Automation and Logistics, ICAL, Qingdao, China: 2286–2290.

Zhen, W. and Tao, Z. (2008) Computational study of the tangential type turbine flowmeter. *Flow Measurement and Instrumentation*, 19(5): 233–239.

Zheng, D. and Zhang, T. (2008) Research on vortex signal processing based on double-window relaxing notch periodogram. *Flow Measurement and Instrumentation*, 19(2): 85–91.

Zheng, D.-D., Zhang, T., Sun, L.-j., Meng, T., Hu, H.-M. and Wang, C. (2010) Installation effects of ultrasonic flowmeter in single bend pipe. 15th Flow Measurement Conference (FLOMEKO), 13–15 October 2010, Taipei, Taiwan, Paper A10-2.

Zheng, D., Zhang, P. and Xu, T. (2011) Study of acoustic transducer protrusion and recess effects on ultrasonic flowmeter measurement by numerical simulation. *Flow Measurement and Instrumentation*, 22(5): 488–493.

Zheng, D., Zhang, P., Zhang, T. and Zhao, D. (2013) A method based on a novel flow pattern model for the flow adaptability study of ultrasonic flowmeter. *Flow Measurement and Instrumentation*, 29: 25–31.

Zheng, D., Zhao, D. and Mei. J. (2015) Improved numerical integration method for flowrate of ultrasonic flowmeter based on Gauss quadrature for non-ideal flow fields. *Flow Measurement and Instrumentation*, 41: 28–35.

Zheng, G.-B., Jin, N.-D., Jia, X.-H., Lv, P.-J. and Liu, X.-B. (2008a) Gas-liquid two phase flow measurement method based on combination instrument of turbine flowmeter and conductance sensor. *International Journal of Multiphase Flow*, 34(11): 1031–1047.

Zheng, Y., Pugh, J.R., McGlinchey, D. and Ansell, R.O. (2008b) Simulation and experimental study of gas-to-particle heat transfer for non-invasive mass flow measurement. *Measurement: Journal of the International Measurement Confederation*, 41(4): 446–454.

Zhou, S. and Halttunen, J. (2004) Application of electrical impedance tomography in pulp flow measurement. 12th International Conference on Flow Measurement FLOMEKO Guilin China.

Zhu, H-L. and Min, Z. (1999) New simple non-invasive method for flow measurement. *Measurement and Control*, 32(6): 178–80.

Ziani, E. M., Bennouna, M. and Boissier, R. (2004) Ultrasonic flow measurement for irrigation process monitoring. *Proceedings of SPIE – The International Society for Optical Engineering*, 5232: 184–95.

Main Index

accuracy, 4–6, 340, 494, 527, 558, 592, 599. *See also* uncertainty
acoustic impedance. *See* impedance, acoustic
adiabatic compressibility, 426, 506
Allen Salt-Velocity method, 98–99
anemometer
  hot wire, 603
  laser Doppler, 434, 603
applications (examples of flowmeters suitable). *See* Applications Index and Selection Table
audit, 53, 62
  questionnaire, 65–66

bell prover, 85, 106–7
benchmarking, 637
Bernoulli's equation, 35, 166
bias. *See* systematic error
block and bleed valving system, 78–79
BS 1042. *See* ISO 5167
bus protocols, 493, 630

calibration, 1, 107, 489, 494, 498, 527, 536, 558, 585
  accreditation table, 72t4.2
  case study of, 108–13
  dedicated facilities, 72, 101
  dry, 73, 92, 95, 373, 386, 390, 395, 396, 399, 400, 433, 441, 442, 443, 457, 458, 481
  environmental conditions, 70
  expansion techniques, 103
  gas calibration facilities, 105
  gases at very low flows, 107
  gravimetric gas, 105
  in situ, 73, 91–92, 105, 603, 615
  intercomparisons, 104
  large water facility example, 113
  manufacturer's viewpoint, 71
  master meters, 72
  new developments in, 101
  typical laboratory facilities, 70
cavitation, 42, 122, 291–92
CE, 4, 54, 64, 140
CEN/CENELEC, 4
centipoise, 14t1.1.
centistoke, 14t1.1., 299
Central Limit Theorem, 19
clamp-on flowmeters, 95, 437–38
Coanda effect, 352
compact provers. *See* provers, compact
compressibility, 54, 64, 171, 181, 582, 591
compressible flow, 38–40
computational fluid dynamics (CFD), 45, 50–51, 148, 171, 348, 357, 450–52
conditioners. *See* straighteners and conditioners
confidence level, 5–6, 7, 10, 12, 19, 20, 21, 69, 71t4.1, 141, 456, 463, 528
continuity equation, 34
critical flow function, 181, 185, 188
critical nozzles, calibration by, 87

DC systems, 366, 368, 371–72
delay time, 454
differential pressure flowmeter equation, 133, 501, 507
Doppler probes. *See* probes, Doppler
double chronometry. *See* provers, double chronometry method
drain holes, 158
drop test, 93–94
droplets, 42–43
dry calibration, 73, 92, 95, 373, 386, 390, 395, 396, 399, 400, 433, 441, 442, 443, 457, 458, 481

electromagnetic compatibility (EMC), 4, 54
electromagnetic probes. *See* probes, electromagnetic
envelopes, uncertainty, 7, 11–12, 69, 205, 306, 313, 585
error
  types of, 22–23
etoile flow straighteners. *See* straighteners and conditioners
expansibility (expansion) factor, 120–21, 217
expansion techniques, 103

Fanning friction factor, 46
Fieldbus, 629
flow
  compressible, 38–40
  computers, 148
  conditioning, 47
  diverters, 103
  flow profile equations, 46
  laminar, 28, 223, 224, 225, 227, 348
  multiphase (multicomponent), 40–42, 105, 300, 345, 383, 454, 497, 508, 522, 585
  non-Newtonian flow, 28, 47
  oil-in-water, 138
  stratified, 42, 107, 456, 512, 525
  turbulent, 403, 409, 428, 449, 450
  unsteady, 36, 70, 131–36, 157, 228, 299, 308, 326, 454, 601
flowmeter audit, 62
flowmeter envelopes, 11–12
flowmeter selection, 1, 14, 62
  table, 58t3.1.
flowmeter, working definition, 11
flowmeters. *See* Flowmeter Index
FSD. *See* full scale deflection
FSR. *See* full scale reading
full scale deflection (FSD), 8, 11, 12, 13f1.2., 203, 209, 384, 505, 535, 538, 540, 542, 557, 562
full scale reading (FSR), 8

gamma-ray density sensor, 258, 506, 515, 517, 522, 523, 524f17.5.
gas entrapment, 43–45
gas flow standard, 102, 107
gas-to-oil ratio (GOR), 42
Gaussian quadrature, 429. *See also* log linear positioning of measurements
gravimetric gas calibration. *See* calibration, gravimetric gas
Guide, The, 9, 10
gyroscopic device, 101

HART, 629–30
Hodgson number, 37f2.5., 133, 135f5.10.
hot tapping, 466, 612, 620
humidity, 13, 42

impedance, acoustic, 422, 435, 436, 446, 481
impulse lines. *See* pressure impulse lines
in situ calibration. *See* calibration, in situ
inferential mass flow measurement, 173, 315, 465, 501, 622
installation orientation, 295–97, 542–43, 574, 585, 591
instruments, intelligent, 140, 147, 148, 150, 371, 441, 499, 626, 629
instruments, smart, 140, 147, 148, 150, 441, 626
integration methods, 97
integration techniques, 95
intelligent instruments. *See* instruments, intelligent
isentropic exponent, 37, 39, 108, 120, 160, 161, 175, 182, 184
ISO 5167, 119, 124f5.4, 152, 156, 164f6.1, 340
ISO 9000, 636–37
ISO 17025, 636–37

K-factor, 299, 335t11.1., 576

Lamb waves, 437–39, 446
laminar flow. *See* flow, laminar
laminar flow bypass, 223, 531
laser Doppler anemometer. *See* anemometer, laser Doppler
linearity, 7
log linear positioning of measurements, 96. *See also* Gaussian quadrature
loss coefficient, 34, 121, 230, 527, 621, 622
lubricity, 54–55

manufacturing precision, 81
manufacturing variation, 202–3, 633, 634
MAP, 24, 99
mass flow measurement, 3, 15, 474. *See also* inferential mass flow measurement
  direct (true), 501
  indirect, 501
master meter. *See* transfer standard
Meter Factor, 297t10.3.
meter provers. *See* provers, meter
microprocessor-based protection systems (MBPS), 140
microsensors, 530, 537, 545, 638
microwaves, 624
momentum sensing, 507
Monte Carlo method, 10, 479
multicomponent flow. *See* flow, multiphase (multicomponent)
multiphase flows. *See* flow, multiphase (multicomponent)

National Measurement Institute Australia (NMIA), 107
neural network, 173, 214, 316, 452, 465, 510, 514, 525, 526, 528, 592, 601, 622
Newtonian fluid, 27
noise. *See* signal, noise information
non-Newtonian flow, 28, 47
nonslip condition, 28, 29, 391, 611
Normal distribution, 5
nuclear radiation, 520
nuclear sources, 520

oil exploration and processing, 409, 463, 464, 499, 508, 515–28, 585
oil-in-water flows. *See* flow, oil-in-water
opacity, 54, 63
operating envelope, 313, 535
operating range, 7
optical methods, 569

Organisation International de Métrologie Légale (OIML), 3, 4, 256
OSI, 630

paddle-wheel probe. *See* probes, paddle-wheel
particles, 42–43
pattern recognition in two-phase flows, 455
piezometer ring. *See* pressure, piezometer ring
pipe provers. *See* provers, pipe
pipe roughness, 29, 46, 125, 159, 172, 307, 361, 454
pipework features, inlets & bends/elbows, calibration by, 93
piston prover. *See* provers, compact
pitot probe. *See* probes, pitot
pneumatic transmission, 203, 209
precision, 4–5
pressure
  impulse lines, 144, 160
  measurement, 144
  piezometer ring, 143–45
  sealant fluids, 144
  stagnation, 35, 39, 177, 178, 179, 188, 220, 273, 604
  tappings, 116, 117, 124f5.4., 151, 156, 166, 186, 213, 620
pressure transducers, 37, 117, 131, 132, 133, 144, 151, 220
  smart, 60n. d, 140, 150
probes, 95–96
  differential, 604–8
  Doppler, 603
  electromagnetic, 614
  paddle-wheel, 280, 316
  pitot, 604–8
  propeller, 609–11
  target, 609
  thermal, 616
  turbine, 609–11
  ultrasonic, 615, 616f21.10.
  Venturi, 607–8
  vortex, 612
Profibus, 629, 630
profile measurement, 467, 483, 500
protocols
  FOUNDATION Fieldbus, 629
  HART, 629–30
  Profibus, 629, 630
prover spheres, 80
provers
  bell, 85, 106–7
  compact, 85–86, 108
  double chronometry method, 82
  pipe, 80, 81f4.9., 104
  pulse interpolation, 82
proving vessels/tanks, 75–79
pulsation index. *See* flow, pulsation index
pulsation, effects. *See* flow, unsteady
pulse interpolation. *See* provers, pulse interpolation
pvT method of calibration, 71t4.1, 87, 102

quadrature. *See* Gaussian quadrature

random error, 8, 99, 100, 580, 634
range and rangeability, 8, 11, 12, 141, 534
ratio of specific heats. *See* specific heats, ratio of
RCM. *See* Reliability centered Maintenance
reciprocity, 15, 440, 441, 459, 597
reference meters. *See* transfer standard flowmeter
Reliability-centred maintenance, 2
repeatability, 4–6, 578
reproducibility, 4
research, 3, 34, 48, 129, 193, 223, 237, 263, 396–98, 474, 500, 552
  calibration case study, 108–13
  circular oscillating piston positive displacement flowmeter, 273
  Coriolis flowmeter, 578, 585, 592
  Cranfield University, UK, 397, 403
  electromagnetic flowmeters, 396, 398
  orifice plates, 156, 157
  rotary piston positive displacement flowmeter, 271
  Tsinghua University, China, 397
  turbine flowmeters, 307, 317
  ultrasonic flowmeters, 473, 474
  Venturi meters, 172
Reynolds number, 28, 29, 46, 47, 51, 69, 95, 102, 118–20, 126, 128, 327, 394, 429, 450, 451, 453, 477, 494, 606

sampling frequency, 454, 459, 466
SCADA, 523
sealant fluid. *See* pressure, sealant fluids
selection, 56–62
  table, 58t3.1.
sensitivity coefficient, 9, 10–11
sensors. *See* instruments
SI units, 13–14
signals, 301, 338, 384, 439–41, 447, 493, 499, 640
  noise information, 399
  types of, 626–28
smart instruments. *See* instruments, smart
Snell's Law, 433, 445
soap film burette, 88
specific heats
  ratio of, 120, 178, 182, 506, 531
specification
  questionnaire, 63–65
stagnation pressure. *See* pressure, stagnation
stagnation temperature. *See* temperature, stagnation
standard deviation, 8, 9, 17, 19, 21, 99, 172, 308
standard time, 189
standard uncertainty, 9, 21, 102, 105, 106, 112t4.A.1
steam, 40, 136
  effect on orifice plate, 45
Stolz equation. *See* Flowmeter Index, orifice plate
straighteners. *See* straighteners and conditioners

straighteners and conditioners, 31–34, 47–50, 113, 127–28, 455
box (honeycomb), 31f2.2.
etoile, 31f2.2
Gallagher, 34
K-Lab, 32, 33f2.3
Laws, 33f2.3, 34
Mitsubishi, 33f2.3
tube bundle, 31f2.2
Vortab, 33f2.3, 34
Zanker, 33f2.3
strainers, 290
strapping tables, 94
stratified flow. *See* flow, stratified
Strouhal number, 324, 327, 331, 333, 334, 339, 340, 348, 353, 359, 360, 360t11.A.1, 361, 506
Student t value, 9–10, 20
swirl, 31–34
systematic error, 9, 22, 99

temperature
stagnation, 39, 177, 178, 180, 186, 187
thermal mass flow probe. *See* probes, thermal
traceability of calibration, 100, 618
tracers, calibration by, 105
transfer standard flowmeter, 80, 88, 90, 104, 109
true mass flow measurement
See Flowmeter Index
tube bundle flow straightener. *See* straighteners and conditioners
turbulence. *See* flow, turbulent
turndown ratio, 8
two-phase flow. *See* flow, multiphase

ultrasonic probes. *See* probes, ultrasonic
ultrasound, 419–23
uncertainty, 7, 72, 98, 108, 125, 304, 351, 384, 442, 444, 445, 535, 542, 578, 581, 618, *See also* accuracy
pipe provers, 104
units
conversion, 11, 13–14
universal viscosity curve (UVC), 299, 324
unsteady flow. *See* flow, unsteady
upper range value (URV), 8
utilities, 357, 474

variance, 17, 19, 20, 172
velocity in pipes, 360t11.A.1, 485, 487, 493
for various flow rates, 16t1.2.
verification, 398–400, 640
verification, in situ, 73, 92, 341, 474, 587, 617–20, 624
volume sensing, 56, 507
vortab. *See* straighteners and conditioners
vortex probes. *See* probes, vortex
vortex shedding, 223, 286, 328–29, 334, 336, 339, 345, 358–61
vortices
gas entrapment, 43–44

weigh tank, correction factor, 109
weighing method for calibration rigs
conventional value of mass, 111
dynamic, 75
static, 75, 76f4.4.
substitution, 77
weight function/vector for flowmeters, 15, 389, 392–96, 400, 595
Coriolis, 575
electromagnetic, 398
thermal, 548, 551
ultrasonic, 478

Youden analysis, 25f1.A.6, 26, 100

Zanker flow straightener. *See* straighteners and conditioners
zero drift, 12, 388, 389, 403, 580–81, 597

Flowmeter Index

acoustic, 484–500
acoustic chemometrics, 499, 500
acoustic scintillation, 471
angular momentum, 60t3.1., 126, 280, 306, 553–54, 555, 562, 563, 593–94
Annubar. *See* averaging pitot
Arnold and Pitts' patent, 258, 259f9.20.
averaging pitot, 45, 51, 58t3.1, 195, 220–23
axial current, 404

bearingless, 290, 314
Bendix meter, 553
bends, 48–49, 125
Brand and Ginsel, 503f16.2.
bypass meter, 552

charge sensing, 338, 403, 520
clamp-on, 95, 437–38, 491–92, 617, 620, 622–23
classical Venturi. *See* Venturi
cone. *See* V-cone
Coriolis
  cross-talk, 576, 583
  drive mechanism, 569
  flow tube, 569, 585
  meter secondary, 570–72
  pressure sensitivity, 569, 576–77, 578, 580
  secondary containment, 570
  sensor types, 569–70
  temperature sensitivity, 570, 571, 576–77, 578, 580
correlation flowmeter. *See* cross-correlation
critical flow Venturi nozzle, 45, 177–194
  coefficient of discharge, 183–85
  critical flow function, 185
  cylindrical throat, 179
  maximum outlet pressure, 181–83
  toroidal throat, 191–92
Crometer tube. *See* V-cone
cross-correlation, 339
CVM, 258

Dall tubes, 45, 209–12
diaphragm, 204–7
differential pressure, 191, 195, 204–5, 210, 213, 216, 217, 225–26, 227
domestic gas, 85, 204–5, 234
Doppler, 419–20, 434, 437, 464, 466–68, 482–83
  range-gated, 473, 474
drag plate. *See* targer meter

eccentric orifice plates, 51, 128, 168, 379–80
eddy current, 288, 396, 397, 398, 400, 404, 407
elbow. *See* flowmeter, bend
electromagnetic, 362–407
  AC, 366, 369, 370–72, 374, 377, 383, 387, 388
  DC, 366, 370–72, 383, 387
  differential, 377, 385
  dry calibration, 373
  dual frequency, 373, 398
  electrodes, 363–66, 367–70, 373, 374, 375–77, 380–81, 384, 386f12, 391, 396–97
  empty pipe detection, 373, 386
  end shorting, 366
  entrained gases, 377, 385
  flux linking of signal leads, 369, 403
  for liquid metals, 362, 390, 407
  for non-conducting dielectric liquids, 362
  ground/earth links, 367, 370, 378, 387, 401
  liner, 366, 367–68
  magnetic field, 362–67, 369–70, 371, 373, 377–78, 379, 382, 384, 385–86, 391–94
  primary element, 366, 370
  quadrature signal, 369–71, 372, 401
  sensitivity, 385, 389, 394
  sensor (primary) element, 366–70
  Shercliff weight function, 364–66
  transmitter (secondary element), 373
  weight functions, 364–66
  zero drift, 389, 403
Elliot-Nathan, 608
Epiflo, 195, 211f8.12.

flap-type. *See* rotating diaphragm
float-in-tube. *See* variable area
flow computers, 148
flow diverters, 103
flow profile equations, 34, 46
fluidic, 352–55
  as Venturi bypass, 356
force flowmeter, 405–6, 407
fuel flow transmitter, 554–57

gas turbine flowmeter. *See* turbine flowmeter, for gases
Gentile tube, 211, 212f8.14.
Gilflo, 205f8.7.

Hastings, 549–51
helical multiphase, 259
helical rotor, 242–43, 259, 273
high precision gas meter. *See* turbine flowmeter, for gases
high precision liquid meter. *See* turbine flowmeter, for gases
hybrid, 507

inlet, 92, 93, 116, 151, 156, 164, 173, 179, 185–87, 188, 189, 195, 204, 210, 211, 214, 217, 238, 272, 273, 282, 283, 284f10.6., 286, 288, 298, 306, 307, 312, 348, 351, 357, 381, 459, 502, 512, 513, 536, 541, 543, 544, 549, 564, 565, 566, 622
insertion. *See* Main Index, probes
integral, 163, 209, 297, 307, 314
integral orifice, 163, 209
ionic, 390

Katys meter, 553
Kratzer and Kefer meter, 516f17.1.

laminar, 28, 29, 61, 223–27, 530–31
Lehde-Lang, 404, 405f12.A.5., 407
Li and Lee, 560–61
liquid plugs, 261f9.22.
Lorentz force, 407
low loss, 195, 211

magnetic resonance (MR), 59t3.1, 408–18, 519, 520
manometer, 146
mass flow sensing meters, 501–7
Massa Stroon Meter (MSM1), 503
McCrometer tube. *See* V-cone
micro flowmeters, 107, 385, 589, 592, 638
microwave devices, 624
MTI. *See* undulating membrane (MTI)
multiphase, 40–42, 59t3.1, 71, 102, 105, 169, 176, 300–1, 383–84, 409, 416, 418f13.11., 454–55, 471, 472, 474, 497, 501, 508–28, 572, 579f20.11., 585, 592, 641
  helical, 259
  nuclear radiation, 520
multirotor PD, 237

nozzles. *See also* Venturi
  coefficient of discharge, 183–85
  critical flow Venturi. *See* critical flow Venturi nozzle
  ISA1932, 167
  long radius, 166f6.4, 167
  sonic. *See* critical flow Venturi nozzle
nuclear magnetic resonance (NMR). *See* magnetic resonance (MR)
nutating disc, 234, 236–37

orifice plate
  chordal, 128, 129
  coefficient of discharge, 118–20, 156
  conical, 128, 150
  contamination, 157–59, 162
  deflection at high pressure, 129–31
  eccentric, 128
  edge sharpness, 209, 212
  expansibility factor, 120–21
  flange tappings, 123–24
  flatness, 122, 131, 131f5.8
  flow conditioning, 158
  gas conditions, 160
  impulse lines, 144, 160
  lagging pipes, 160
  metering run assembly, 141
  plate carrier assembly, 141
  pressure loss, 121
  pressure measurement, 144–47
  pressure tappings, 123–24, 142–43
  quadrant, 128, 163
  sizing, 142
  slotted, 195, 216–17
  smart orifice meter, 160
  steam quality, 161
  Stolz equation, 118–19, 152
  study of flow through, 148
  thickness, 122–23, 129–31, 160
Orlando and Jennings, 553–54
oscillating circular piston, 57, 237–38, 273–76
oscillating membrane (MTI). *See* undulating membrane
oscillating vane, 195, 216, 356
oval-gear meter, 238–40

paddle wheel, 280, 316, 611
pattern recognition, 455, 511, 525, 528, 641
PD. *See* positive displacement (PD)
Pelton wheel, 195, 216, 280, 314
piston. *See* reciprocating piston
positive displacement (PD), 234–78
  calibration systems, 246–48, 253–54
  pressure effects, 252–53
  slip, 251–52
  temperature effects, 252–53
Potter twin rotor turbine, 553
probe. *See* main index, probes, turbine

quadrant orifice. *See* orifice plate, quadrant
quadrature signal, 369
Quantum Dynamics meter, 314, 418

reciprocating piston PD, 243
Roots. *See* rotary positive displacement gas meter
Rotameter. *See* variable area meter
rotary piston. *See* oscillating circular piston meter
rotary positive displacement gas meter, 257–58
rotating diaphragm, 206, 207f8.8

saddle coil, 404, 405f12.A.4., 407
sensitivity coefficients, 10
sliding vane PD, 240–42, 245–46, 250f9.14., 255–56, 261, 263
slotted orifice, 195, 216–17
sonar, 484–500, 641
  active, 484
  passive, 484–85
sonic nozzle. *See* critical flow Venturi nozzle
spring loaded diaphragm meter, 204–7
spur gear, 244
static charge. *See* charge sensing
surface-acoustic-wave (SAW), 473
swirl meter, 351

target, 208–9
thermal mass flow measurement
  capillary thermal mass flowmeter (CTMF for gas), 530–37
  capillary thermal mass flowmeter (CTMF for liquid), 537–38
  in-line thermal mass flowmeter (ITMF), 541–42
  insertion thermal mass flowmeter (ITMF), 540–41
Torbar. *See* averaging pitot
toroidal throat Venturi nozzles. *See* nozzles, critical flow Venturi
true mass flowmeter, 560–61, *See also* Li and Lee
turbine, 279–326
  dual rotor, 288
  equations, 317–26
  for gases, 303–11
  for liquids, 419–20
twin Venturi, 502–3

ultrasonic, 61, 419–83
  beam sweeping, 472, 472f14.17.
  clamp-on, 437–38
    gases, 419–20, 422, 446–47, 457–63
    liquids, 419–20, 422, 444–46, 456–57
  correlation, 419
  custody transfer, 442, 446, 456, 458, 463, 465, 466
  Doppler, 419–20, 434, 437, 464, 473, 474, 482–83
  mass flow measurement, 426–27
  retrofit, 95, 101, 419, 437
  sing-around, 425
  transducers, 432–36
  transit-time, 419, 423–25
    for pulses, 424–25
undulating membrane (MTI), 262

vane-type, 253, 315
variable area, 196–204
  manufacturing variation, 202
  spring loaded, 204–7
V-cone, 213–15
Venturi, 163–64
  coefficient of discharge, 165, 171
  design details, 167
  two-phase flow, 169–70
  upstream installation spacings, 168
  with fluidic bypass, 356
Venturi nozzle, 163–64, 167, 356f11.16. *See also* critical flow Venturi nozzle
vibrating nozzle, 560–61
vibrating tube, 560–61
viscous. *See* flowmeter, laminar
vortex
  cavitation, 331, 347, 352
  K factor, 329, 331, 339, 341
  measurement of steam flow, 345–48, 351
vortex shedding, 223, 327, 329, 334, 336, 339, 357, 358–61

waste air, 543
water, 311–14
wedge, 212–13, 512
wet gas, 51, 60t3.1, 139, 150, 169, 170, 171, 174–76, 215–16, 217, 234, 334, 346, 464, 493, 509, 511–14, 520, 521t17.2, 641
Wheatstone bridge, hydraulic, 504–5
Woltmann, 279

Flowmeter Application Index

In referring to this index the reader should note that:

a) the applications recorded are those examples which I have encountered in reading the literature on meters **and are obviously not in any sense exhaustive**. In a few general cases I have, therefore, included references to other flowmeters;
b) the references to a particular flowmeter do not imply that other flowmeters are unsuitable, but rather that I was not aware of manufacturers specifically recommending other flowmeters for the application;
c) even if a particular flowmeter is referred to, it does not necessarily mean that it will, in practice, be suitable for the application;
d) the reader should always check suitability of a flowmeter for a particular application with the manufacturer, and the other factors set out in Chapter 3.

abrasive liquids, 57, 387, 388, 496, 574, 579
acids, 350, 362, 388, 468, 588t20.1, 589
additives, 243, 388, 538
adhesives, 240, 243, 245, 263
air, 28, 43, 204, 310, 388, 464, 536, 543, 544, 589
  compressed, 350, 446
  entrained, 487, 492, 496
  preheated and combustion, 540, 543
air conditioning, 540
aircraft, 558
  fluid systems, 466
  fuel flow, 559
alcohol, 240
asphalt, 587
auto industry, 245

bases, 388
batching, 245, 261, 388, 398
beer, 388, 588t20.1
benzene, 350
black liquor, 213, 388, 587
blast furnace flows, 388
bleaching chemicals, 388
blending, 243, 536, 588
boiler applications
  combustion air, 540
  efficiency testing, 505
  preheater, 540
brown stock, 388

calibration. *See* transfer standards
catalysts in petrochemicals, 538
chemical vapour deposition, 536
chemicals, 203, 240, 261, 387, 388
chromatograph flows, 538
clean fluids, 209, 218, 261, 289, 297, 302, 306, 336, 350, 468, 496, 530, 536
coffee extract, 587
combustion gas flows, 540
compressor efficiency and surge control, 612
condensate production, 498
contaminated oil, 468
corrosion inhibitors, 538
corrosive gas, 543
corrosive liquids, 218
cream, 245
creosote, 350
crude oil, 463, 464
cryogenic, 290, 302, 350, 466, 587, 589, 590
custody transfer, 261, 262, 388, 463, 589, 612. *See also* transfer standards

diesel, 245, 350
digester flows, 388
dirty fluids, 57, 212, 213, 261, 290, 306, 352, 454, 461. *See also* gas conditions, dirty; liquids, dirty
dispensing, 261
distillates, 205
domestic gas, 85, 234, 459, 474

drilling mud, 455, 498
dyes, 245, 587

emulsions, 138, 170, 171f6.6, 388, 468
engine
  testing, 243, 350, 436
engine lubricant, 466
enzymes, 538
erosive liquids, 387, 388
ethylene oxide, 587, 589

fats, 261, 587
fertilizers, 261
flavorants, 538
flavouring, 388
flocs, 472
foods and beverages, 203, 302, 388, 496, 543, 588t20.1, 589
  glucose, 240, 468
  ice cream, 388
  juice, 388, 468, 587, 588t20.1
  liquid chocolate, 587
  milk, 388
  peanut butter, 587
  processed egg, 587
  sugar, 388
fuel consumption, 243, 245, 538
fuel oil, 240, 243, 587, 589

gas conditions
  clean, 209, 218, 224, 261, 289, 297, 302, 306, 336, 350, 352, 468, 530, 536
  dirty, 57, 212, 213, 261, 290, 306, 352, 454, 461
gas flow measurement, 190, 191, 499, 536
  blending, 536, 588
  brewery, 543
  burner flow control, 543
  flare, 464, 540
  flue, 543
  high temperature, 436
gases. *See also* air; steam
  acetylene, 204, 310, 536
  ammonia, 204, 536
  argon, 204, 302, 464, 536, 543
  arsine, 536
  butane, 204, 310, 536, 587
  carbon dioxide, 302, 310, 536
  carbon monoxide, 536
  chlorine, 204, 464, 536
  coke-oven, 310
  ethane, 310, 536
  ethylene, 310, 536
  fluorine, 536
  freon, 536
  fuel gas, 310, 612
  HCFC22, 189
  helium, 204, 302, 536, 543
  HFC134a, 189
  high temperature, 436
  hydrogen, 302, 536
  hydrogen coolant flow, 540
  hydrogen sulfide, 536
  krypton, 536
  methane, 350, 446, 536, 540, 543
  natural, 204, 310, 446, 464, 538, 587
  neon, 536
  nitrogen, 302, 306, 310, 464, 536, 543
  nitrous oxide, 536
  oxygen, 204, 302, 310, 536, 543
  propane, 310, 536, 587
  propylene, 536
  refinery, 310
  silane, 536
  xenon, 536
gasoline, 205, 240, 245
glycol, 325, 350
greases, 245
green liquor, 587

heating flows, 540
helium service, liquid, 170
hydraulic oils, 245, 350
hydrocarbons, low viscosity, 350
hygienic, 348, 388, 638

in situ measurement, 94
industrial effluent, 468
industry, 496
  chemical engineering and processing technology, 543
  mechanical engineering, 203, 302
  nuclear, 362, 388, 407, 464, 465, 496, 508, 515, 543
  plant, 204
  power plant, 456, 463, 465, 496
  process, 301, 496, 536
  pulp and paper, 203, 387, 388, 468, 496, 587
  semiconductor, 534, 536, 543
  steelworks nitrogen and oxygen flows, 543
  water, 203, 388, 463, 464, 465, 496, 587
ink, 245
inventory control, 261

kerosene, 245

latex solutions, 388
leak detection, 464, 466, 543, 576, 612
lime slurries, 468
liquids
  clean, 209, 218, 261, 289, 297, 302, 306, 336, 350, 352, 468, 496, 530, 536
  conducting, 387
  dirty, 57, 212, 213, 261, 290, 306, 352, 454, 461
  erosive, 387, 388
  general, 204, 538
  high pressure, 302, 306, 496, 587
  high temperature, 302
  lubricity, 54, 55

metal, 362, 387, 404, 483, 623
organic and inorganic, 240
liquefied gases
CO_2, 350, 587
LPG, 253, 588t20.1
N_2, 350, 588t20.1
O_2, 350, 588t20.1.
low flows, 205, 229, 471, 530, 536, 538
low head, 43
low loss, 195
lubricants, 240, 310, 466
lubrication, 538

medical, 228, 543, 589
mercaptan odorant injection, 243
metals, liquid, 362, 387, 404, 483, 623
microfiltration, 538
milk, 388
mineral oil, 245
molten sulphur, 588t20.1
multiphase, 102, 259, 409, 508, 511, 514, 517, 522, 523, 524, 525, 526, 528, 587

nitrogen purges, 540

odorants into natural gas, 538
odorisers, 612
oil, 238, 302, 464, 515, 526
oil-in-water/water-in-oil, 170, 171f6.6, 455, 514

paints, 240, 245, 261, 468, 589
pigments in, 538
paraffin, 205, 245
paste, 245
petrochemical raw materials, 446
pharmaceuticals, 203, 388, 538
pipeline leak detection. *See* leak detection
pipeline management, 464
plant
coal liquefaction, 587
dosing, 543
pilot, 243
polymers, 240, 243, 388, 587, 588t20.1
polypropylene, 237, 588t20.1
polyurethane, 245
porosity of rocks, 538
potassium hydroxide, 588t20.1
processes involving
chemical vapour deposition, 536
crystal growth, 536
diffusion, 536
ion implantation, 536
plasma etching, 536
sputtering, 536
thermal oxide, 536
pulsating, 103, 201, 228, 326, 348, 454
pulsation, 132, 223
pulverized coal in nitrogen, 584

radioactive flows, 388
reagents
in fermenters, 538
in pharmaceuticals, 538
refinery flows, 310
resin, 245

samplers, 612
sand, 362
sanitary products, 302
sewage, 388, 464, 468
sludge, 213, 218, 388, 468, 543
slurry, 56, 170, 213, 218, 362, 388, 468, 496, 587, 588t20.1
magnetic, 588t20.1
soda, 388
sodium hydroxide, 588t20.1
solid-liquid, 212
solvents, 240, 261, 587
sour crude, 588t20.1
space applications
rocket fuel and oxidiser metering, 466
space vehicle coolant, 466
spray coatings, 388
stack effluent, 543
steam, 204, 446, 463, 464, 540, 543, 562

tar, 205, 350
tar sands, 588t20.1
test rigs, 302
thixotropic, 243
titanium chloride closing, 538
titanium dioxide, 587
toothpaste, 388
transfer standards, 590, 635. *See also* custody transfer

unsteady. *See* pulsating
utilities, 357, 474

velocity, very low, 398, 450
ventilation, 540
viscous fluids, high, 243, 261, 457, 464

waste air, 543
water, 203, 240, 362, 388, 463, 464, 496, 524
distilled, 350
raw, 468
sea, 468, 588t20.1.
sludge, 218, 388
sour, 587
waste, 203, 388, 464, 496
water-in-oil. *See* oil-in-water/water-in-oil
wax, 612
wet-gas, 51, 139, 150, 171, 213, 214, 216, 217, 254, 255, 306, 334, 346, 352, 446, 464, 498, 513, 514, 520, 521t17.2, 641
correlation, 169, 170, 174–76, 215
white liquor, 388

Printed in the United States
By Bookmasters